International Conference on Geological Engineering and Geosciences (ICGoES 2021)

IOP Conference Series: Earth and Environmental Science
Volume 851

Yogyakarta, Indonesia
16 - 18 March 2021

Editors:

Tsutomu Sato
Klaus Reicherter
Yasuo Miyabuchi
Subagyo Pramumijoyo
Teuku Faisal Fathani
Myo Thant
Patrick Lachassagne
Pipat Laowattanabandit
Nguyen Bach Thao
Mohd Ashraf Mohamad Ismail

ISBN: 978-1-7138-4815-8
ISSN: 1755-1307

Printed from e-media with permission by:

Curran Associates, Inc.
57 Morehouse Lane
Red Hook, NY 12571

Some format issues inherent in the e-media version may also appear in this print version.

Institute of Physics
Dirac House, Temple Back
Bristol BS1 6BE UK

Phone: 44 1 17 929 7481
Fax: 44 1 17 920 0979

techtracking@iop.org

Additional copies of this publication are available from:

Curran Associates, Inc.
57 Morehouse Lane
Red Hook, NY 12571 USA
Phone: 845-758-0400
Fax: 845-758-2633
Email: curran@proceedings.com
Web: www.proceedings.com

TABLE OF CONTENTS

IOP Conf. Series: Earth and Environmental Science **851** (2021) 011001 doi:10.1088/1755-1315/851/1/011001

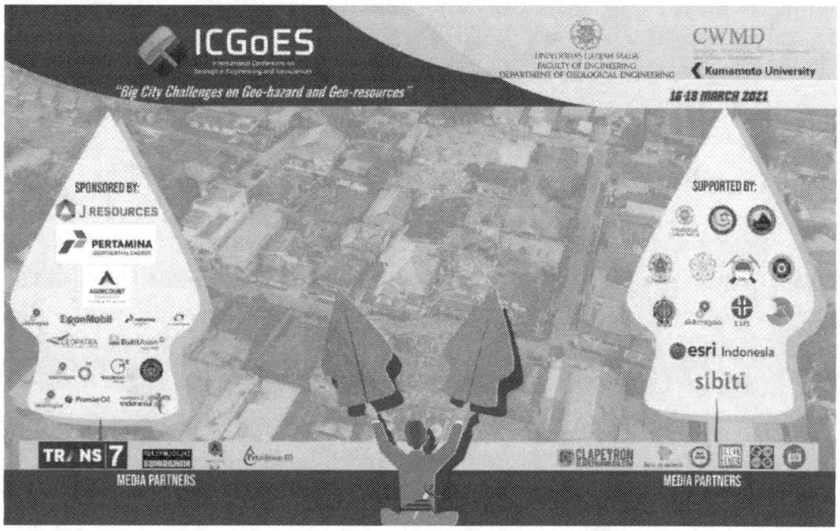

IOP Conf. Series: Earth and Environmental Science **851** (2021) 011001 doi:10.1088/1755-1315/851/1/011001

Preface
International Conference on Geological Engineering and Geosciences
ICGoES 2021

International Conference on Geological Engineering and Geosciences (ICGoES 2021) presents '*Big City Challenges on Geohazard and Georesources*' as the main conference theme. This conference provides a forum for international researchers, academics, practitioners, policymakers and related communities to discuss, share and exchange their latest research and experience progress associated with the field of geological engineering and geosciences. In addition, this conference also aimed to address a common issue of georesources sustainability, renewable energy sources and climate change-related geohazard and disasters in the development of urban areas. Certainly, those issues require an integrated approach and collaborative efforts in particular to build and enhance resilient cities in the respective countries.

ICGoES 2021 provides opportunities to all participants in expanding their network, as an important step to achieve a coordinated approach and interdisciplinary collaboration at the regional and international level in the field of geological engineering and geosciences. We invited research papers on the topic of geohazard and georesources; but not limited to research papers, recent cases, and in-depth reviews of basic geology, applied geology, geophysics, geochemistry, engineering geology, geo-environmental engineering and any related geoscience.

Due to the COVID-19 pandemic situation during the year, it is not easy to hold a regular conference in a specific place. There were travel restrictions and regulations from the World Health Organization and the Indonesian Government to be complied with to cut the spread of the COVID-19 virus in terms of physical distancing. In this uncertain condition, the ICGoES should not be postponed since participants have arranged to address their research findings at this conference. Thus, the committees of ICGoES have decided to switch the conference format into a virtual meeting using ZOOM.

The ICGoES 2021 is held online from 16 to 18 March 2021 and managed by the Department of Geological Engineering, Faculty of Engineering, Universitas Gadjah Mada, Yogyakarta, Indonesia. This virtual international conference invites reputable researchers from all over the world such as Japan, Germany, USA, France, Vietnam, Thailand, Myanmar, Malaysia, Laos, Nepal, UK and the Czech Republic, as our scientific committee and reviewers. As the call of full paper was announced on 15 October 2020, up to 19 February 2021, this conference has attracted more than 70 papers and posters submitted to the EasyChair Submission System. After being reviewed, the scientific committee has selected around 63 papers and posters from diverse countries such as Japan, Germany, Myanmar, France, and Indonesia, to be presented at this conference.

During the plenary session of the conference, we were pleased to welcome five distinguished experts to share their experience and present their insightful keynote speech, that is:

- Prof. Junji Kiyono from Kyoto University, Japan with presentation title '*Toward the construction of a long period earthquake early warning system in Bangkok*',
- Prof. Dwikorita Karnawati, Head of Indonesian Agency for Meteorology, Climatology and Geophysics (BMKG), Indonesia with presentation title '*Earthquake and Tsunami Mitigation in Yogyakarta International Airport, Indonesia*',
- Prof. J. Ramon Arrowsmith from Arizona State University, USA with presentation title '*Sharpening our view of active faults*',

- Prof. Koichiro Watanabe from Kyushu University, Japan with presentation title '*Twenty years research collaboration on georesources between UGM and Kyushu University with expansion to Asia-Africa georesources research network*', and
- Prof. Thomas R. Rüde from RWTH Aachen University, Germany with presentation title '*Water for Big Cities – the challenge between local wells and centralized water supply from rural areas.*'

Following the plenary sessions, we held panel discussions with Geohazard Focus of 'The role of academics in building resilient cities' on the first day, and Georesources Focus of 'Academic collaboration for sustainable development of georesources' on the third/last day of the conference. We divided into 15 parallel sessions for participants to present their research covering the following topics: methods and advances in petroleum and coal geology (2 sessions), anthropogenic geology and environmental degradation (1 session), methods and advances in economic geology (1 session), seismic hazards (1 session), advances in geology and geomorphology (1 session), methods and advances in applied geophysics (1 session), applied geophysics for urban areas (1 session), advance methods in hydrogeology (2 sessions), advances in hazard mitigation (1 session), methods in petrology and petrogenesis (1 session), progress in unconventional georesources (1 session), progress in volcanology and geothermal (1 session), and machine learning in geological engineering and geosciences (1 session). Each parallel session consisted of 4-5 presenters, 1 moderator, 1 timekeeper and 1 technician. An allocation time of 12 minutes maximum for presentation followed by 3 minutes for Q&A session was given to each presenter. Presenters were allowed to submit the recording of the presentation 2 days before the actual schedule if they were not sure about the internet connection. However, during the record is play, the presenter must be still attending live for the following Q&A session.

The ICGoES 2021 virtual conference was technically run very well and smoothly via ZOOM online system by Universitas Gadjah Mada, attended by more than 185 participants in total. No major technical issue from the host side, keynote speakers, as well as participants during the conference. Most of the participants did their presentations live, engaged throughout plenary, panel discussion and parallel sessions. Feedbacks, comments and questions from participants in all sessions were very positive and constructive. The post-conference in-depth review was carried out to refine all papers. Authors were asked to revise their paper according to suggestions and comments from reviewers and considering all constructive inputs and feedbacks during the conference. About 52 papers received final decision to be accepted and submitted for publication into the IOP Conference Series: Earth and Environmental Science.

We would like to acknowledge all of those who supported the ICGoES 2021. We would like to express our sincere gratitude to all scientific committees who made in-depth reviews of full papers, to sponsors who give their support, and to each individual and institutional help that is very important for the success of this conference. Thank you to all the organizing committee and related parties who built this virtual conference feasible and successful.

We are hopeful that the proceedings will serve as important references which will lead not only to scientific research and academic purposes but also in practical, industry and policy aspects associated with the field of geological engineering and geosciences.

Best regards,
Hendy Setiawan, Ph.D.
Chairman of the ICGoES 2021
Universitas Gadjah Mada

IOP Conf. Series: Earth and Environmental Science **851** (2021) 011001 doi:10.1088/1755-1315/851/1/011001

List of Committees

Editorial Board
Prof. Tsutomu Sato (Hokkaido University, Japan)
Univ.-Prof. Dr. Klaus Reicherter (RWTH Aachen University, Germany)
Prof. Yasuo Miyabuchi (Kumamoto University, Japan)
Prof. Dr. Ir. Subagyo Pramumijoyo, DEA. (Universitas Gadjah Mada, Indonesia)
Prof. Ir. Teuku Faisal Fathani, S.T., M.T., Ph.D. IPU. (Universitas Gadjah Mada, Indonesia)
Prof. Dr. Myo Thant, University of Yangon, Myanmar
Dr. Patrick Lachassagne, Université de Montpellier, France
Pipat Laowattanabandit, Ph.D., Chulalongkorn University, Thailand
Dr. Nguyen Bach Thao, Hanoi University of Mining and Geology, Vietnam
Assoc. Prof. Ir. Dr. Mohd Ashraf Mohamad Ismail, Universiti Sains Malaysia, Malaysia

Advisory Boards
Prof. Dr. Ir. Subagyo Pramumijoyo, DEA. (Universitas Gadjah Mada, Indonesia)
Ir. Muhammad Waziz Wildan, M.Sc., Ph.D. (Universitas Gadjah Mada, Indonesia)
Dr.rer.nat. Ir. Heru Hendrayana, IPU. (Universitas Gadjah Mada, Indonesia)
Dr. Ir. Agung Harijoko, S.T., M.Eng., IPM. (Universitas Gadjah Mada, Indonesia)

Scientific Committee
Prof. Tsutomu Sato (Hokkaido University, Japan)
Prof. Toshifumi Igarashi (Hokkaido University, Japan)
Prof. Klaus Reicherter (RWTH Aachen University, Germany)
Prof. Dr. Ir. Subagyo Pramumijoyo, DEA. (Universitas Gadjah Mada, Indonesia)
Prof. Ir. Teuku Faisal Fathani, S.T., M.T., Ph.D., IPU. (Universitas Gadjah Mada, Indonesia)
Prof. Amanda Clarke (Arizona State University, USA)
Prof. Dr. Myo Thant (University of Yangon, Myanmar)
Prof. Dr. Su Su Kyi (University of Yangon, Myanmar)
Prof. Yasuo Miyabuchi (Kumamoto University, Japan)
Prof. Masahide Matsumura (Kumamoto University, Japan)
Dr. Patrick Lachassagne (University of Montpellier, France)
Assoc.Prof. Dr. Ir. Mohd Ashraf Mohamad Ismail (Universiti Sains Malaysia)
Dr. Nguyen Bach Thao (Hanoi University of Mining and Geology, Vietnam)
Dr. Pipat Laowattanabandit (Chulalongkorn University, Thailand)
Dr. Phetnakhone Xaixongdeth (National University of Laos)
Dr. Rishi Parajuli (University of Bristol, UK)
Dr. Afikah Rahim (Universiti Teknologi Malaysia)
Dr. Basanta Raj Adhikari (Tribhuvan University, Nepal)
Dr. Jan Klimeš (The Czech Academy of Sciences, Czech Republic)
Dr. Has Baator (Asia Air Survey, Co. Jp., Japan)
Claude Thielen, M.Sc. (RWTH Aachen University, Germany)
Sara Pena Castellnou, M.Sc. (RWTH Aachen University, Germany)
Dr. Ir. Sugeng Sapto Surjono, S.T., M.T., IPU, ASEAN.Eng. (Universitas Gadjah Mada, Indonesia)
Dr.rer.nat. Ir. Wayan Warmada, IPM. (Universitas Gadjah Mada, Indonesia)
Dr. Ir. Ferian Anggara, S.T., M.Eng. IPM. (Universitas Gadjah Mada, Indonesia)
Dr.rer.nat. Wiwit Suryanto. (Universitas Gadjah Mada, Indonesia)

Dr.Eng. Ir. Agung Setianto, S.T., M.Si., IPM. (Universitas Gadjah Mada, Indonesia)
Dr. Ir. Agung Harijoko, S.T., M.Eng., IPM. (Universitas Gadjah Mada, Indonesia)
Dr.Eng. Ir. Wawan Budianta, S.T., M.Sc., IPM. (Universitas Gadjah Mada, Indonesia)
Dr.rer.nat. Ir. Arifudin Idrus, S.T., M.T., IPU. (Universitas Gadjah Mada, Indonesia)
Dr.Eng. Ir. Wahyu Wilopo, S.T., M.Eng., IPM. (Universitas Gadjah Mada, Indonesia)
Ir. Pri Utami, M.Sc., Ph.D., IPM. (Universitas Gadjah Mada, Indonesia)
Ir. Jarot Setyowiyoto, M.Sc., Ph.D., IPU., ASEAN.Eng. (Universitas Gadjah Mada, Indonesia)
Dr.Eng. Ir. Lucas Donny Setijadji, S.T., M.Sc., IPU. (Universitas Gadjah Mada, Indonesia)
Ir. Salahuddin, S.T., M.Sc., Ph.D., IPM. (Universitas Gadjah Mada, Indonesia)
Ir. I Gde Budi Indrawan, S.T., M.Eng., Ph.D., IPM. (Universitas Gadjah Mada, Indonesia)
Dr.Eng. Ir. Akmaluddin, S.T., M.T. (Universitas Gadjah Mada, Indonesia)
Dr.Eng. Ir. Didit Hadi Barianto. S.T., M.Si., IPM. (Universitas Gadjah Mada, Indonesia)
Dr. Sarju Winardi, S.T., M.T. (Universitas Gadjah Mada, Indonesia)
Dr.rer.nat. Mochamad Nukman, S.T., M.Sc. (Universitas Gadjah Mada, Indonesia)
Dr.rer.nat. Ade Anggraini, S.Si., M.T. (Universitas Gadjah Mada, Indonesia)
Dr.rer.nat. Herlan Darmawan, S.Si., M.Sc. (Universitas Gadjah Mada, Indonesia)
Karlina, S.T., M.Eng., Ph.D. (Universitas Gadjah Mada, Indonesia)
Dr.Eng. Guruh Samodra, S.Si., M.Sc. (Universitas Gadjah Mada, Indonesia)
Dr.Sc. Andung Bayu Sekaranom, S.Si., M.Sc. (Universitas Gadjah Mada, Indonesia)
Dr. Ani Hairani, S.T., M.Eng. (Universitas Muhammadiyah Yogyakarta, Indonesia)
Ir. Moch. Indra Novian, S.T., M.Eng. (Universitas Gadjah Mada, Indonesia)
Hidayat Panuntun, S.T., M.Eng., D.Sc. (Universitas Gadjah Mada, Indonesia)
Rahmadi Hidayat, S.T., M.Eng. (Universitas Gadjah Mada, Indonesia)
Indra Arifianto, S.T., M.Eng. (Universitas Gadjah Mada, Indonesia)
Kartika Palupi Savitri, S.T., M.Sc. (Universitas Gadjah Mada, Indonesia)
Yan Restu Freski, S.T., M.Eng. (Universitas Gadjah Mada, Indonesia)

Organizing Committee

Department of Geological Engineering, Faculty of Engineering, Universitas Gadjah Mada, Indonesia

Chairman	: Hendy Setiawan, S.T., M.Eng., Ph.D.
Co-Chairman	: Esti Handini, S.T., M.Eng., D.Sc.
Secretary	: Wita Dyaswati, A.Md. Maya Kristine Kusumaningtyas, S.Psi. Suci Dandi Pertiwi, S.Fil.
Treasurer	: Monika Anggraeni Pamungkassari, A.Md.
Member	: Dr.rer.nat. Ir. Doni Prakasa Eka Putra, S.T., M.T., IPM. Ir. Anastasia Dewi Titisari, M.T., Ph.D., IPU. Dr.Ing. Ir. Donatus Hendra Amijaya, S.T., M.T., IPM. Ir. Nugroho Imam Setiawan, S.T., M.T., D.Sc., IPM. Ir. Gayatri Indah Marliyani, S.T., M.Sc., Ph.D. IPM. Dr. Haryo Edi Wibowo, S.T., M.Sc. Fahmi Hakim, S.T., M.Sc.RWTH.
Creative Team	: Rocky Tirajean Sembiring, Antonius Chandra Widihermawan, A.Md., Fransisca Noralita, A.Md., Riyantono, Widyatmoko, Eko Hendrawan Arianto, Bagus Wijaya, S.IP., Paulus Winarno, S.Pd., Ghiffary Riza Ramadhan, Bela Islamadina, Dewi Sintia Reka, Fenta Dhia Ayu Prameswari, Zulfa Nurul Azizah Saifudin, and Asep Kuswara.

IOP Conf. Series: Earth and Environmental Science **851** (2021) 011001 doi:10.1088/1755-1315/851/1/011001

Conference Photos

Opening Speech, Dean of Faculty of Engineering, Universitas Gadjah Mada

Master of Ceremony of ICGoES 2021 during Plenary Sesssion

Keynote Speech: Prof. Dwikorita Karnawati, Head of Indonesian Agency for Meteorology, Climatology and Geophysics (BMKG), Indonesia

1st Panel Discussion with Geohazard Focus: The Role of Academics in Building Resilient Cities

Parallel Session Speech: Vanessa Steinritz from RWTH Aachen University, Germany

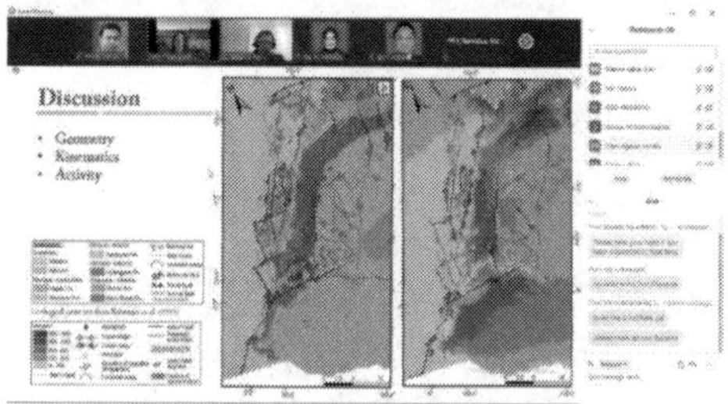

Parallel Session in the topic of Advances in Hazard Mitigation

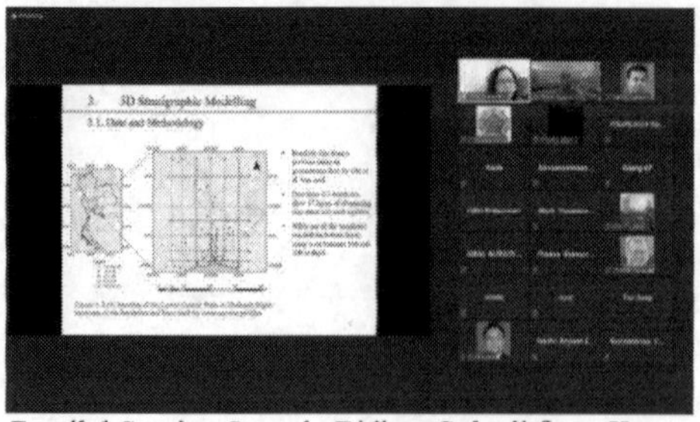

Parallel Session Speech: Bidhya Subedi from Kyoto University, Japan

Luncheon Talk with Topic: Borderless Geology and Geoscience in the face of Changing Earth

Photo Session during the Conference

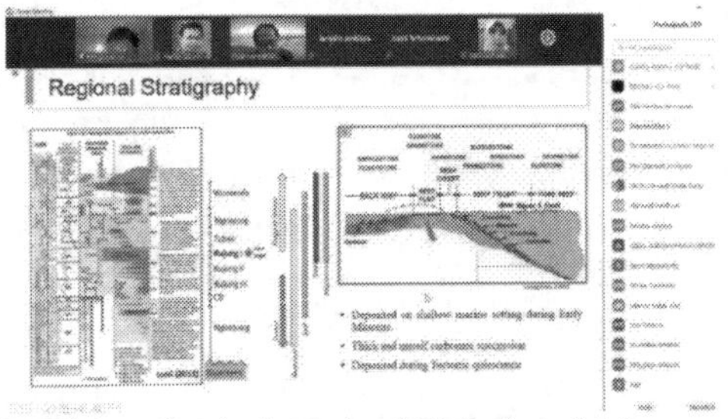

Parallel session in the topic of Methods and Advances in Petroleum and Coal Geology

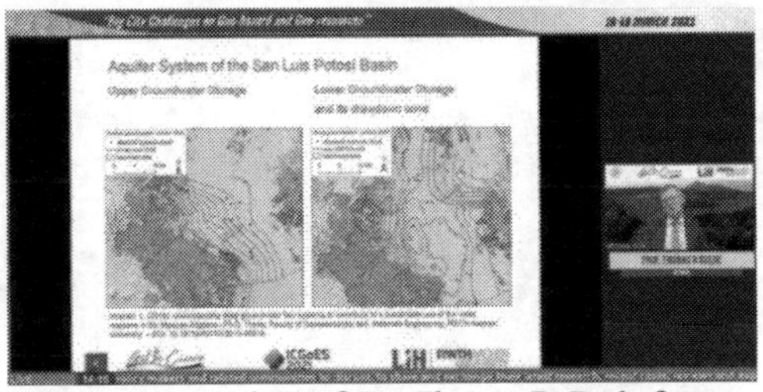

Keynote Speech: Prof. Dr. Thomas R. Rüde from RWTH Aachen University, Germany

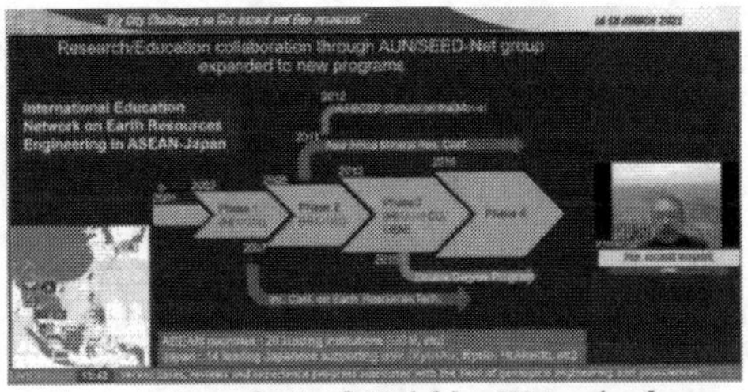

Keynote Speech: Prof. Koichiro Watanabe from Kyushu University, Japan

2nd Panel Discussion with Georesources Focus: Academic Collaboration for Sustainable Development of Georesources

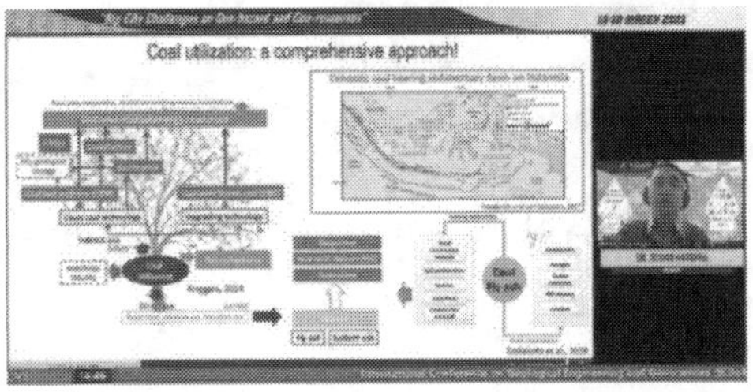

Dr. Ferian Anggara as panelist in the Discussion of Sustainable Development of Georesources

Closing Speech, Head of Departement of Geological Engineering, Universitas Gadjah Mada

Closing and Announcement for Post-Conference Review and Publication Process

IOP Conf. Series: Earth and Environmental Science **851** (2021) 011002 doi:10.1088/1755-1315/851/1/011002

Peer review declaration

All papers published in this volume of IOP Conference Series: Earth and Environmental Science have been peer reviewed through processes administered by the Editors. Reviews were conducted by expert referees to the professional and scientific standards expected of a proceedings journal published by IOP Publishing.

- **Type of peer review:** Double-blind
- **Conference submission management system:**
  Full paper submission was fully managed through EasyChair that can be accessed at https://easychair.org/conferences/?conf=icgoes2021. All required information of the conference can be accessed at https://icgoes.geologi.ugm.ac.id/
- **Number of submissions received:** 73 full papers
- **Number of submissions sent for review:** 73 full papers
- **Number of submissions accepted:** 52 full papers
- **Acceptance Rate (Number of Submissions Accepted / Number of Submissions Received X 100):** 71.23 %
- **Average number of reviews per paper:** 2 reviewers
- **Total number of reviewers involved:** 53 reviewers in total from 10 countries
- **Any additional info on review process:**
  The review process and revision were conducted 2 rounds on average for each paper since the first submission before the conference until the final decision after presentation at the conference. The third-round of review was required only for papers that need major modifications. All reviewers were asked to complete the review within enough time ranged between 10-14 days. While authors were asked to revise according to suggestions by reviewers within time ranged between 10-14 days. After being accepted, English proofread and similarity checking via Turnitin were carried out to ensure the quality prior to submission to IOP EES. About 90.38% (46 papers) of 52 total accepted papers have the similarity of less than or equal to 15 %. Papers with a similarity index exceed 20% were rejected.
- **Contact person for queries:**
  Name: Hendy Setiawan, Ph.D.
  Affiliation: Universitas Gadjah Mada, Faculty of Engineering, Department of Geological Engineering
  Email : hendy.setiawan@ugm.ac.id

IOP Conf. Series: Earth and Environmental Science **851** (2021) 012001 doi:10.1088/1755-1315/851/1/012001

Earthquake risk assessment of the Opak and Merapi-Merbabu active faults to support mitigation program in Yogyakarta province and its vicinity

J Murjaya[1,4*], S Pramumijoyo[2], D Karnawati[1,2], Daryono[1], I Meilano[3], P Supendi[1], S Ahadi[1], G I Marliyani[2], Imananta[1], F Syukur[1], D S Sianipar[1] and A Krisno[1]

[1]Agency for Meteorology, Climatology, and Geophysics, Jakarta 10720, Indonesia
[2]Geological Engineering Department, Faculty of Engineering, Gadjah Mada University, Yogyakarta 55281, Indonesia
[3]Geodesy Research Group, Faculty of Earth Sciences and Technology, Institute Technology of Bandung, Bandung 40132, Indonesia
[4]Geoscience Department, Natural and Science Faculty, University of Indonesia, Depok 16424, Indonesia
*E-mail : murjaya1@gmail.com

Abstract. Based on historical records, Yogyakarta has a high seismic risk related to the earthquake events along active faults, such as the Opak and Merapi-Merbabu Faults. These faults were responsible for several destructive earthquakes in Yogyakarta City and its vicinity and caused fatalities and building damage in the area, e.g., the 2006 (Mw 6.3) Yogyakarta earthquake and earlier in 1943 and 1867. A previous study shows that the Opak Fault has a geodetic slip-rate of 5 mm/y and a potential magnitude Mw 6.6. In addition, the active Merapi-Merbabu Fault has a geodetic slip-rate of 1 mm/y and a potential magnitude Mw 6.6. We used scaling law relations of earthquake parameters and magnitude scenarios to estimate the recurrence time of each fault based on a kinematic model. Our results estimate that the earthquake return period (Tr) for the Opak Fault (Mw 6.6) is ~162 years, the maximum intensity is ~VII-VIII MMI scale, the Peak Ground Acceleration (PGA) is ~36 % g, and Peak Ground Velocity (PGV) is ~ 30 cm/s for a 5 km hypocentral depth. In the meantime, the earthquake return period for the Merapi-Merbabu Fault (Mw 6.6) is estimated to be ~810 years, the maximum intensity is ~ VI-VII MMI, the PGA is ~ 30-36 % g, and the PGV is ~ 21-24 cm/s for a 5 km hypocentre depth. Both faults potentially produce destructive earthquakes (Mw > 6.0) in Yogyakarta City and its vicinity. Therefore, assessments of (paleo) earthquakes are needed of both the Opak Fault and the Merapi-Merbabu Fault to support the long-term earthquake hazard mitigation program.

1. Introduction

Yogyakarta Province in Indonesia had been shaken by many large earthquakes sourced from the Opak Fault which is located in the eastern area of Yogyakarta City. Within several hundred years, several earthquakes destroyed infrastructure in the Yogyakarta and area around it and caused many fatalities. The destructive earthquake in 2006 (Mw 6.3, hypocentral depth 12.5 km) [1] caused more than 5,600 fatalities and around 200,000 different types of buildings collapsed. The earthquake source was located along the Opak Fault which stretches from the Southern Yogyakarta to the Northern Klaten City (Central

Java). The distribution of infrastructural damage had an identical direction to the strike of the Opak Fault. Another destructive earthquake occurred in 1943 with Mw 7.0, the epicenter located in the offshore of Southern Yogyakarta Province (Figure 1). That earthquake caused 213 fatalities, 2,096 injured, and around 28,000 buildings collapsed. Additionally, a destructive earthquake hit Yogyakarta City and its vicinity in 1867 (Table 1). The exact epicenter had not been determined, but based on the historical records about 372 buildings collapsed and 5 people died [2].

There are no earthquake sources data exist in the catalog data for the last 200 years [2] for the Merapi-Merbabu Fault, but geodetic data of the Merapi-Merbabu Fault suggest a geodetic slip rate of 1 mm/y [3]. It means the Merapi-Merbabu Fault is one of the active faults in Central Java that corresponds to a potential magnitude of the earthquake of about Mw 6.6 [3]. Based on the historical seismicity and damage, we pointed out the need of a paleo seismic assessment and monitoring of Opak and Merapi-Merbabu Faults.

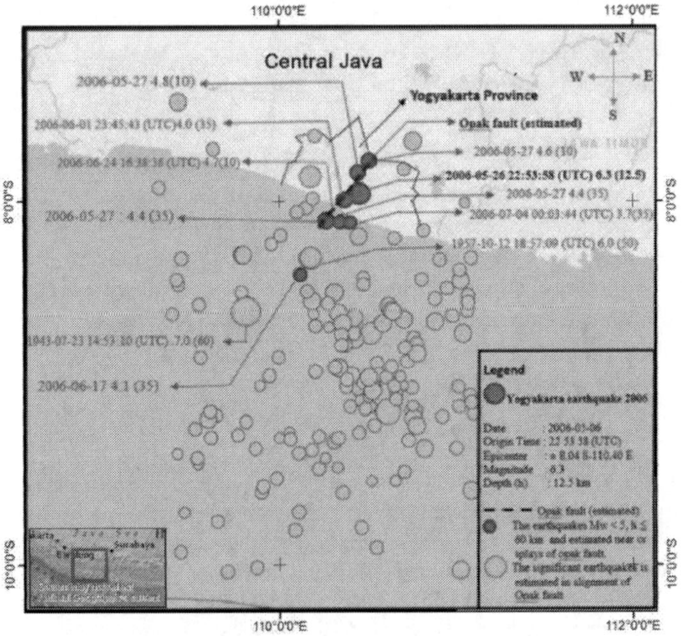

Figure 1. Seismicity of significant earthquakes located near and in Opak Fault in Yogyakarta Province from the time period of 1900 to 2019. The 2006 Yogyakarta destructive earthquake occurred on 22:53:58 UTC, Mw 6.3, hypo central depth 12.5 km.

Table 1. List of significant earthquakes in Yogyakarta Province [2, 4, 5].

No	Event	Magnitude	Depth (km)	Remark
1	Jane 4, 1840	7-8 (estimated)	-	Tsunami in southern coast of Central Java and Yogyakarta
2	October 20, 1859	7-8 (estimated)	-	No remarks
3	June 4, 1867	8	-	De Witte Pal-Golong Gilig and 372 houses collapsed, and 5 people died.
4	July 23, 1943	7	60	213 people died, 2,096 Injured, and more than 2,000 buildings collapse.
5	May 27, 2006	6.3	12.5	More than 5,600 people died, thousand people injured, and more than 200,000 different type building collapsed.

IOP Conf. Series: Earth and Environmental Science **851** (2021) 012001 doi:10.1088/1755-1315/851/1/012001

The tectonics of Java (Yogyakarta Region, part of Southern Java) are dominated by the subduction of the Australian Plate to the north-northeastward beneath the Sunda Plate with a relative velocity of about 5-7 cm/year [6]. The Australian Plate dips to the north-northeastward from the Java Trench, attaining depths of 100-200 km beneath the Java Island and depths of 650-700 km to the north of the island. The earthquake of 26 May 2006 at 22:53:58 UTC occurred at a shallow depth and its epicenter was estimated along the Opak Fault. The Opak Fault is located in the eastern part of Yogyakarta City. It is a strike-slip fault. It has an almost south-north direction from the Southern Coast of Yogyakarta to Prambanan Temple in the north. The neighboring fault is the Merapi-Merbabu Fault (Central Java, it is not showed in fig.1). The fault is located between Merapi and Merbabu mountains and has geodetic slip rate of about 1 mm/y [3].

2. Data and Method

2.1. Data

We used the Indonesian earthquakes catalogue data [1, 2, 4, 5] for all of the earthquakes that occurred along the Opak Fault (Table 1). The geodetic slip-rate data of the Opak Fault which has been published from [3] reveals about 5 mm/y and the potential earthquake magnitude is about Mw 6.6. We used this data (Mw 6.6) and converted to seismic moment to determine the static stress drop and recurrence time estimation of the earthquake in the future

2.2. Fault Length and Rupture Area Estimation

The Yogyakarta earthquakes that occurred in 1867 (Mw 8-estimated) and 2006 (Mw 6.3) indicated that the Opak Fault is active and has the potential to drive large earthquakes in the future. Therefore, the foreshocks and aftershock distribution of the 2006 earthquake (figure 1) were important data to estimate the fault length (L) and rupture area (A) in this paper. Meanwhile, to predict the Tr with targeted magnitude Mw 6.6 and less than Mw 6.6 [3], this paper used the scaling law relation of the earthquake to calculate the L and A. We assumed that the focal mechanism of the earthquake was the strike-slip model, then we estimated the L and A for Mw 6.6 or less using the Wells and Coppersmith [7] formula such as shown in equations (1) and (2).

$$L = 10^{(a+b\,M_w)} \tag{1}$$

and

$$A = 10^{(c+d\,M_w)} \tag{2}$$

Annotation:

a = -3.55; b = 0.74; c = -3.42 and d = 0.9 for strike slip fault

By assuming the average value of L and A, it is possible to calculate the width of fault (W) by divided equations (2) by (1).

2.3. Static Stress Drop Theory

The tectonic energy will be released during a co-seismic event after the critical stage of rock is passed over. This energy is accumulated-along the inter-seismic period since the last large earthquake event has occurred. Figure 2 shows the illustration of built-up stress and released stress. It was begun from the initial stress (σo), the build-up stress along the inter-seismic period until it achieved the critical stress ($\sigma 1$) stage. After the critical stress stage passed over, it would release several stresses called a stress drop.

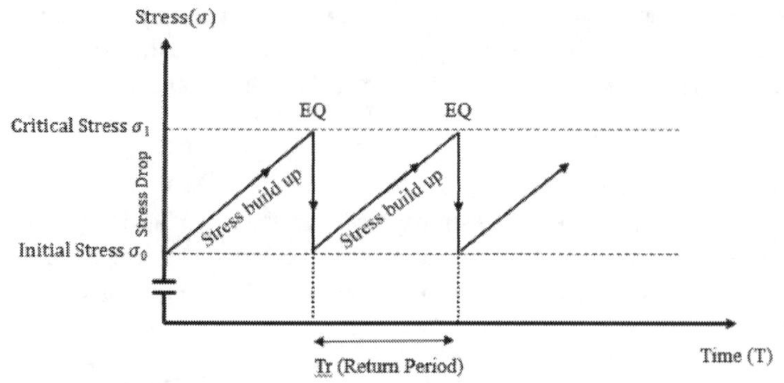

Figure 2. Illustration of the stress drops and return period interval of earthquake relations. The time interval between two events of earthquakes and the difference between the critical stress (σ_1) value and initial stress (σ_0) value is called return period (Tr) and stress drop, respectively. Modified from Kanamori [8].

Furthermore, to describe the stress release effects, we used one type only, i.e. a static stress drop. Then, we used the Kanamori and Anderson [9] formula as follows;

$$\Delta\sigma = c\,\mu\,\frac{\bar{D}}{L} \tag{3}$$

μ is rigidity (SI), $\frac{\bar{D}}{L}$ is strain change, and C is a non-dimensional shape factor. To determine the shape factor (C) and L, we referred to shallow infinite transverse shear (dip-slip) faults. Kanamori and Anderson [9] referred to Starr (1928) and Aki (1966) for C = $\frac{4(\mu+\lambda)}{(2\mu+\lambda)}$ formula, where λ is the Lame constant. We used the definition of a seismic moment as;

$$M_o = \mu S \bar{D} \tag{4}$$

Mo is a static seismic moment (Nm) and S is a rupture area (m^2). Based on the substitute equation (3) and (4), and assumed $\mu = \lambda$, we found the static stress drop formula for dip slip fault as:

$$\Delta\sigma = \frac{2}{\pi}\left(\frac{Mo}{W.W.\ L}\right) \tag{5}$$

Equation (5) is similar to Beeler and Hickman [10] which reveals that static stress drop is calculated using seismic moment measured from seismogram and estimated rupture area, while the recurrence interval is estimated from historic records.

2.4. Seismic Energy and Moment Magnitude Relation

The relation between the surface-wave magnitude (M_s) and the total energy of seismic waves (E_s) had been studied by scientists such as Gutenberg and Richter (1956), Bath (1958), and several revisions Gutenberg and Richter (1956) which is revealed by Kanamori and Anderson [9]. They obtained the relation as:

$$Log\ Es = 1.5\,M_s + 11.8 \tag{6}$$

Then, we used the conversion M_w to M_s with the formula as:

$$M_w = 0.9239\,M_s + 0.5671 \tag{7}$$

The equation (7) was used for magnitude interval $6.2 \leq M_s \leq 8.7$ with $R^2 = 0.8183$.

3. Results and Discussion

As explained above, we used the potential maximum magnitude scenarios of Mw 6.6 for Opak and Merapi-Merbabu Faults by Pusgen (2017) [3] to calculate some fundamental of earthquake risk if the earthquake occurred along Opak and Merapi-Merbabu Faults. The fundamental of earthquake risks is the impact of the earthquake/shakemap, PGA, PGV, and estimation of Tr.

3.1. Fault Length and Static Stress Drop

Table 2 shows the relations between earthquake parameters, such as Mw scenario, Mo, L, and A. These parameters were used to estimate the static stress drop (Δσ) value for focal scenarios in the strike-slip model. The slip model scenario was needed to select a formula to calculate L and A based on Wells and Copper Smith [7].

Table 2. The relations between Mw scenario, Mo, L, A, and Δσ by the scaling law relations of earthquakes.

Mw (scenarios)	Mo (dyne-cm)	Type of focal mechanism	L (km)	A (km^2)	Δ σ (x 10^6 ($dyne/cm^2$)
6.6	7.953×10^{25}	Strike Slip (SS)	21.577	331.131	9.969
6.4	3.981×10^{25}	SS	15.346	218.776	8.130
6.2	1.995×10^{25}	SS	10.914	144.544	6.639
6.0	1.0×10^{25}	SS	7.762	95.499	5.421
5.8	5.012×10^{24}	SS	5.521	63.096	4.427
5.6	2.512×10^{24}	SS	3.926	41.687	3.615
5.4	1.259×10^{24}	SS	2.793	27.542	2.952
5.2	6.231×10^{23}	SS	1.986	18.197	2.381
5.0	3.162×10^{23}	SS	1.412	12.000	1.998

Relations between A and Mo on table 2 the Δσ estimation the strike-slip model for Mw ≤ 6.6 of the scenarios, the result is still applicable [11].

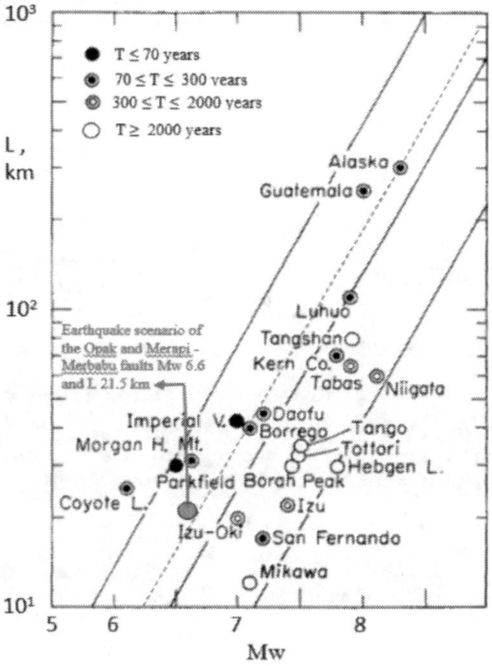

Figure 3. The relations between the moment magnitude, Mw, and the fault length, L. The solid lines indicate the trend for a constant stress drop [11]. We plot maximum credible earthquake 6.6 for Opak and Merapi-Merbabu Faults on this graph, and it is still applicable.

3.2. Recurrence Time Estimation

We used slip rate geodetic data to estimate the earthquake recurrence time with the potential of magnitude scenarios of Mw 6.6. Davidson and Scholz [12] and Kanamori and Anderson [9] calculated the recurrence time of the earthquake by dividing the moment of an earthquake by source time function or moment rate accumulation. By using slip rate geodetic data of the Opak Fault (5 mm/y) and Merapi-Merbabu Fault (1 mm/y), the Tr could be estimated. Table 3 presents the estimation result of the return period of the earthquake that occurred in active Opak and Merapi-Merbabu Faults for several magnitude scenarios.

Table 3. List of the recurrence time estimation for Mw scenarios uses scaling law relation of the earthquake in Opak and Merapi-Merbabu active faults.

Model	Mw (scenarios)	Mo (dyne-cm)	Type of Focal mechanism	Tr OF in years* (slip 5 mm/yr)	Tr MMF in years** (slip 1 mm/yr)
Kinematic	6.6	7.953×10^{25}	Strike Slip (SS) [3]	162	810
	6.4	3.981×10^{25}	SS	121	607
	6.2	1.995×10^{25}	SS	92	460
	6.0	1.0×10^{25}	SS	70	349
	5.8	5.012×10^{24}	SS	53	265
	5.6	2.512×10^{24}	SS	40	200
	5.4	1.259×10^{24}	SS	31	152
	5.2	6.231×10^{23}	SS	24	120
	5.0	3.162×10^{23}	SS	18	90

* Opak Fault, Yogyakarta Indonesia
** Merapi-Merbabu Fault, Central Java Indonesia.

Then diagram (Figure 4) describes the recurrence time of the active Opak Fault with slip rate based on slip rate geodetic data of 5 mm/yr [3] and other parameters based on Table 2.

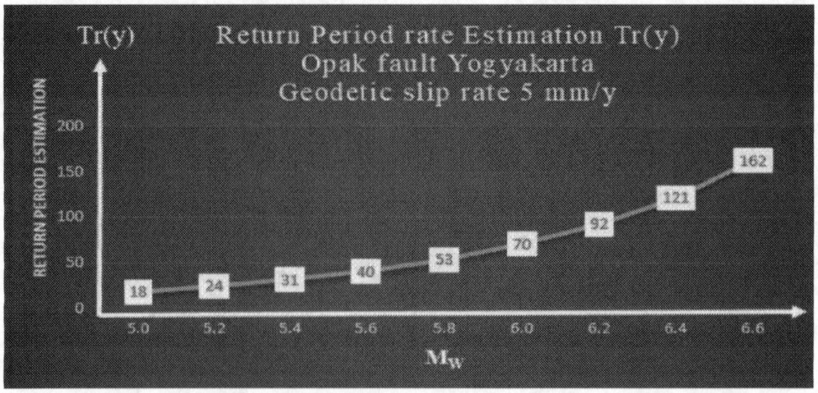

Figure 4. Earthquake recurrence time estimation of the Opak Fault-Yogyakarta. The value of recurrence time was calculated by the kinematic model using a slip rate geodetic of 5 mm/yr [3] and maximum magnitude scenario of Mw 6.6.

Figure 5 explains the recurrence time estimation for Merapi-Merbabu active fault with using a slip rate geodetic data of 1 mm/y [3] and the other parameter.

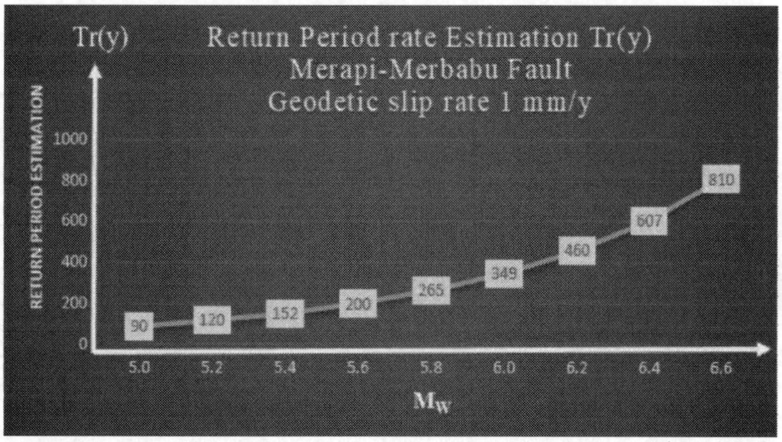

Figure 5. Earthquake recurrence time estimation of the Merapi-Merbabu Fault, central Java. The value of recurrence time was calculated by kinematic model with using a slip rate geodetic of 1 mm/y [3] and maximum magnitude scenario Mw 6.6.

3.3. Potential of Shaking Impact of Earthquake

One of the secondary effects of the earthquake is a shaking impact on infrastructures and the surrounding area. Figure 6 shows our models simulate the shake map, Peak Ground Acceleration (PGA), and Peak Ground Velocity (PGV) caused by strike-slip earthquake scenario (Mw 6.6) of the Opak Fault with the focus depth (h) of about 15 km. The shaking impact, PGA, and PGV were revealed in the MMI scale, in percentage of g (in % g), and in cm/s respectively. This model assumed that the hypocenter was a point source located along the Opak Fault, and it did not consider a focal mechanism. If the earthquake scenario is Mw 6.6 and depth 15 km, it will have the shaking impact of about VII MMI, PGA about 28 % g, and PGV of about 21 cm/s, respectively.

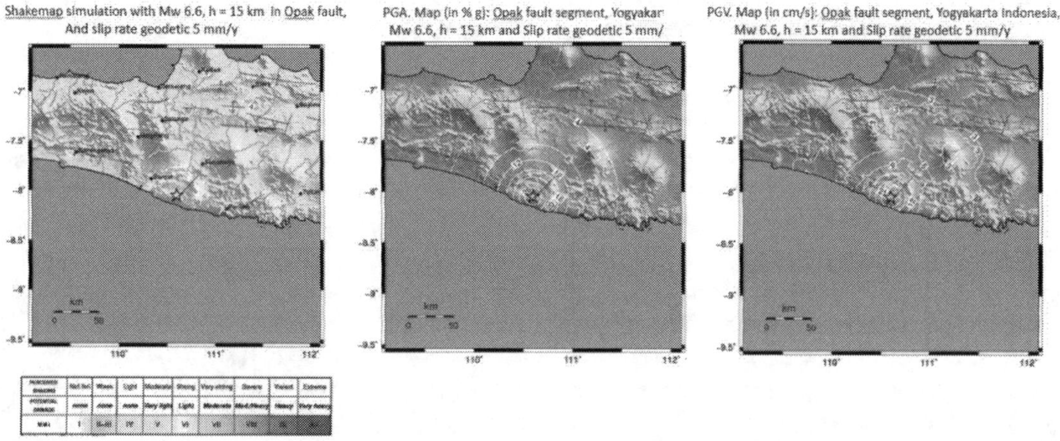

Figure 6. The Shake map, PGA and PGV impacts caused by a strike slip earthquake (Mw 6.6) of the Opak Fault with depth 15 km.

Figure 7 shows our modeled shake map, PGA, and PGV caused by strike-slip earthquake scenario (Mw 6.6) of the Opak Fault and with very shallow depth scenario of about 5 km as the worst scenario. If the earthquake scenario is Mw 6.6 and depth 5 km, it will have the shaking impact of about VII-VIII MMI, PGA of about 36 % g, and PGV of about 30 cm/s, respectively.

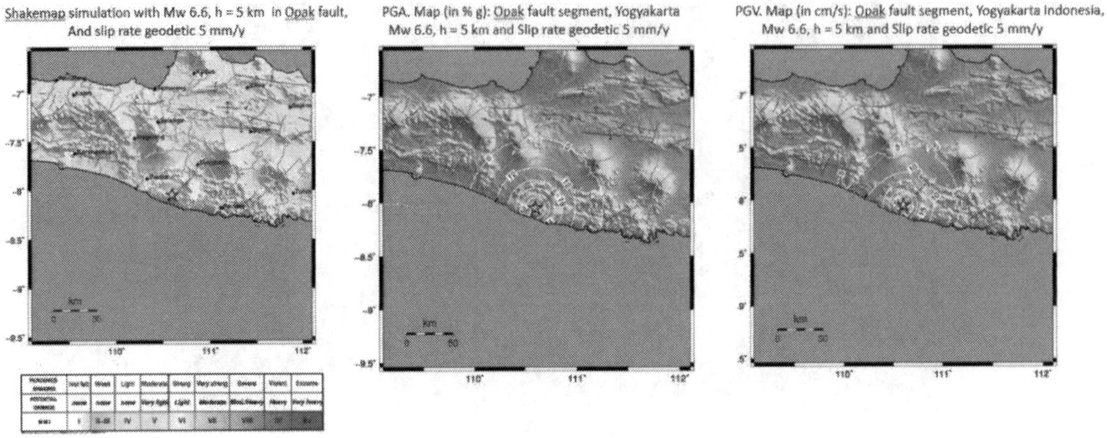

Figure 7. Simulation of the Shake map, PGA, and PGV impacts caused by a strike-slip earthquake (Mw 6.6) of the Opak Fault with hypocenter depth 5 km as the worst scenario.

Figure 8 shows our modeled shake map, PGA, and PGV maps caused by strike-slip earthquake scenario (Mw 6.6) of the Merapi-Merbabu Fault with 15 km depth. If the earthquake scenario is Mw 6.6 and depth 15 km, it will have the shaking impact of about VII MMI, PGA about 24 % g, and PGV of about 18 cm/s, respectively.

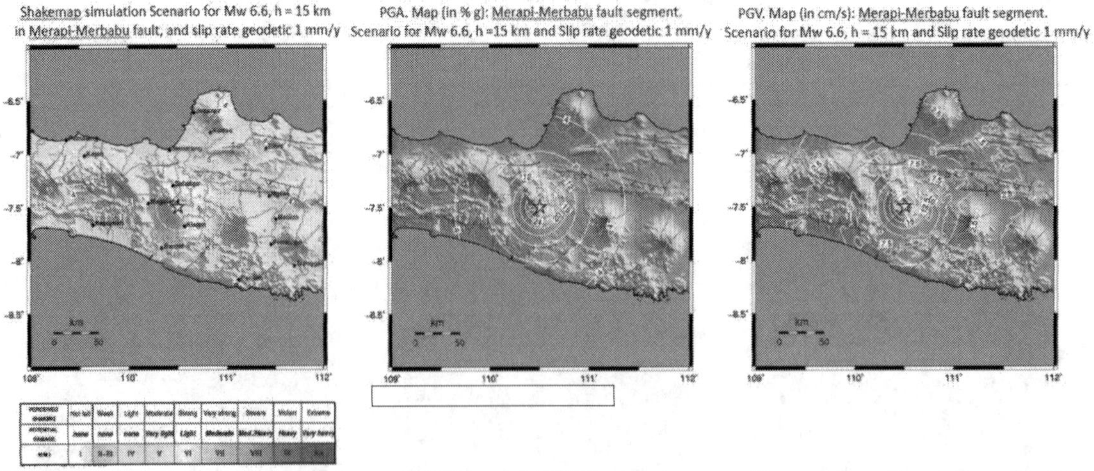

Figure 8. Simulation of the Shake map, PGA, and PGV impacts caused by a strike slip earthquake (Mw 6.6) of the Merapi-Merbabu Fault with hypocentral depth 15 km.

Figure 9 shows our modeled shake map, PGA, and PGV caused by strike-slip earthquake scenario (Mw 6.6) of the Merapi-Merbabu Fault with the depth about 5 km to describe as the worst scenario. If the earthquake scenario is Mw 6.6 and depth 5 km, it will have the shaking impact of about VII-VIII MMI, PGA about 36 % g, and PGV of about 24 cm/s, respectively.

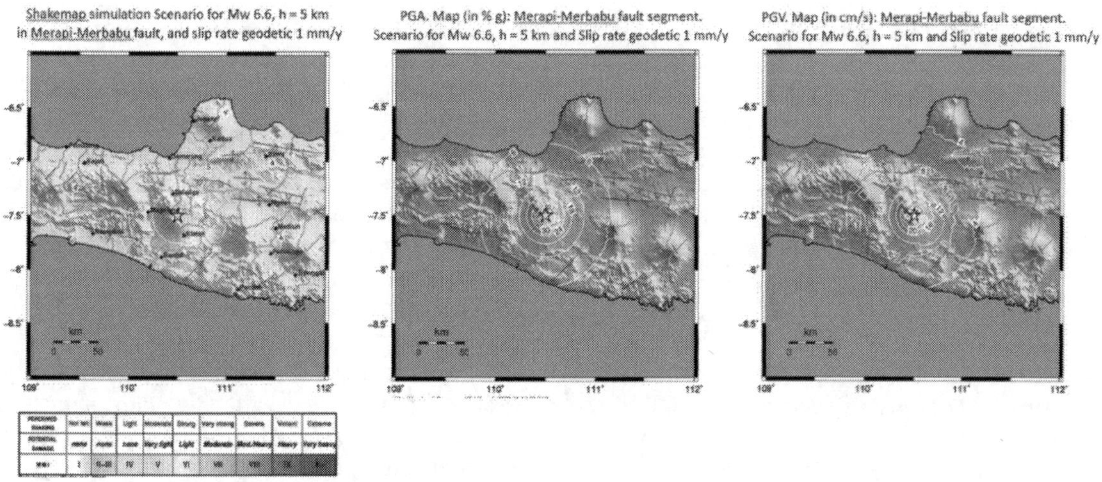

Figure 9. Simulation of the Shake map, PGA, and PGV impacts caused by a strike-slip earthquake (Mw 6.6) of the Merapi-Merbabu Fault with a hypocentral depth 5 km.

The Yogyakarta earthquake on 26 May 2006 with Mw 6.3, depth 12,5 km, and maixmum intensity of VIII MMI by USGS [1], the constructed shakemaps in figure 6, 7, 8, and 9 offered an appropriated estimation. We compared our modeled shaking impact with the field survey results conducted after the 2006 Yogyakarta (Mb 5.9 or Mw 6.3) by Murakami and Pramitasari [13]. They reported the infrastructure damage from south to northeast follows the strike orientation of the Opak Fault. The heavily damage area are spread out starting from Pundong, Imogiri, Bantul, Plered, Piyungan until the Prambanan Temple area. The damage distribution is also consistent with UNOSAT estimation published on their web-site soon after the earthquake [13]. Moreover, the damage distribution matched with the microzonation survey result conducted by Karnawati et al. [14]. They revealed that the very high susceptible zone for future earthquakes is situated in the western part of Pleret, central part of Jetis, Imogiri, and Pundong. All sites were located in Yogyakarta province. Based on this information, we suggest that our modeling result of earthquake impact shaking of the Mw 6.6, depth 5 km, and 15 km, with the epicenter was assumed as a point source at the Opak Fault agreed with the field damage impact of the 2006 Yogyakarta earthquake.

The modeled PGA with Mw 6.6 scenario yield a maximum ~36 % g(depth 5 km) and 28 % g (depth 15 km). We compared this results with field survey results from the Mw6.3 2006 Yogyakarta earthquake (Mb 5.9) conducted by Fathani (2007) [15]. He used the BMKG data (Mb 5.9) and USGS data (Mw 6.3) for 2006 the Yogyakarta earthquake to estimate PGA from several empirical formulas by Donovan (1973), Esteva (1974), Matuschka (1980), Campbell (1981), and Fukushima-Tanaka (1990). In one of his summaries from BMKG data, he determined PGA using the Esteva formula and he found the g (min) value was about 0.233 g (in BPKP site) and g (max) value was 0.322 g (Watu site). Another references were Matuschka (1980) and Fukushima-Tanaka. Matuschka (1980) found that the g(min) value was about 0.241 g and g (max) value was about 0.311 g, while Fukushima-Tanaka (1990) found that the g(min) value was 0.194 g and g(max) value was 0.287 g. These values corresponded with the PGA modeling as explained above. If compared with the earthquake in 1867, which caused the damage in Yogyakarta City, we underwent a difficulty for more interpretation because of incomplete data. The water castle at the Sultan's Palace in Yogyakarta city was heavily damaged, and Gedung Agung as Dutch Governor's residence collapsed. Visser (1922) estimated that the location of the 1967 Yogyakarta earthquake epicenter was following the Opak River [5]. Based on the damaged data, it could be estimated that the magnitude of 1867 earthquake was probably larger than Mw 6. The relations of shaking impact on MMI scale and g value were interpreted by Housner (1970), Gere and Shah (1984), and Yeats et al (1997) in Fathani (2007)[15]. Referring to their result for Ml 6.0, the earthquake impact had a scale of VII-VIII MMI and the ground acceleration was about 0.22 g. Therefore, we interpreted that all shakemap results with Mw 6.6 (Figure 6-9), which show the intensity, was about VII-VIII MMI and the g minimum was about 24 % g.

The high value of estimated MMI, PGV and PGA value suggest that the future earthquake along this fault may cause significant damageto buildings. The possibility of earthquake-induced landslides also needs to be considered to utilize early warning system, monitoring and mitigation program in the future. The Mw6.3 2006 Yogyakarta earthquake caused landslides in Prambanan District and Sengir [16]. With a comparable event in the future scenario of earthquake along the Opak Fault with more than 6.0, the hazards associated with earthquake induced landslides should be considered in some areas, primarily in the high slope area.

4. Conclusion

Based on historical earthquakes, Yogyakarta Region is prone of earthquake hazard region. At least two destructive earthquakes occurred and caused serious damaged in the Yogyakarta City and some areas around it, namely the 2006 earthquake (Mw 6.3) and the 1867 earthquake (unknown magnitude) which likely originated from the Opak Fault. If the active Opak Fault had a slip-rate geodetic of about 5 mm/y and estimated maximum credible earthquake about Mw 6.6, the return period (Tr) of earthquake will be around 162 years (figure 4). Given these parameters, the estimation of the maximum intensity is around VII-VIII MMI, while the PGA is around 36 % g, and PGV is around 30 cm/s (figure 7). Meanwhile, for the Merapi-Merbabu Fault which has a geodetic slip-rate of 1 mm/y and maximum probable earthquake of Mw 6.6, the estimation of earthquake return period (Tr) is around 810 years (Figure 5). These provide an estimation of the maximum intensity value at VI-VII MMI, PGA ~ 30-36 % g and PGV ~ 21-24 cm/s (Figure 9). Results from this study can be utilize to support the landscape planning and the long-term earthquake hazard mitigation program in Yogyakarta region and its vicinity.

Acknowledgments

This study was partially supported by Indonesian Agency for Meteorology, Climatology, and Geophysics (BMKG).

References

[1] United State Geological Survey (USGS) 2020 http//www.usgs.gov earthquake.

[2] BMKG -The Indonesia Agency for Meteorology, Climatology, and Geophysics 2020 Indonesian Earthquakes Catalogue. Jakarta.

[3] PUSGEN (Indonesian Earthquake Study Center) 2017 Peta Sumber dan Bahaya Gempa Indonesia tahun 2017. Consortium of some relevance of Indonesia earthquake institutions.

[4] Soloviev S L and Go C H N 1984 Catalogue of Tsunamis on the Western shore on the Pacific Ocean. Canada Institute for Scientific and Technical Information National Research Council Ottawa, Ontario, Camada KIA OS2.

[5] Newcomb K R and McCann W R 1987 Seismic History and Seismotectonics of Sunda Arc. *Journal of Geophysical Research*, vol. **92**, no. B1, pp 421-439, American Geophysical Union.

[6] Bock Y et al. 2000 Crustal motion in Indonesia from Global Positioning System measurements. *Journal Geophysical Research*, vol.**108**, No. B8, 2367.

[7] Wells D L and Coppersmith K 1994 New Empirical Relationships among Magnitude, Rupture Length, Rupture Width, Rupture Area, and Surface Displacement. *Bulletin of the Seismological Society of America*, Vol. **84**, No. 4, pp. 974-1002, August 1994.

[8] Kanamori H 1994 Mechanics of Earthquakes. *Annu.Rev.Earth Planet.Sci*.**22**:207-37.

[9] Kanamori H and Anderson L D 1975 Theoretical Basis of Some Empirical Relations in Seismology. *Bulletin of the Seismological Society of America*. Vol.**65**, No.5, pp. 1073-1095. October 1975.

[10] Beeler N L and S H Hickman 2001 Earthquake Stress Drop and Laboratory-inferred interseismic strength recovery. *Journal Geophysical Research*, Vol.**106**, No. B12, Pages 30, 701-30,713. December 10, 2001.

[11] Kanamori H and Allen C R 1986 Earthquake Repeat Time and Average Stress Drop, in *Earth Source Mechanics. Monogr. Ser.*, Vol **37**, edited by S. Das et al., pp 227-236, AGU, Washington, D.C.

[12] Davidson F C and Scholz C H 1985 Frequency Moment Distribution of earthquake arc: A Test the Characteristic Earthquake Model. *Bulletin of the Seismological Society of America*. Vol.**75**. No.5. pp. 1349-1361. October 1985.

[13] Murakami H and Pramitasari D 2007 Human Casualty and Damage Distribution in the 2006 Central Java Earthquake. Proceeding AUN/SEED-Net. The Yogyakarta Earthquake 2006. Geological Engineering Department, Faculty of Engineering, Universitas Gadjah Mada. Indonesia.

[14] Karnawati D, Husein S, Pramumijoyo S, Ratdomopurbo A, Matanabe K, and Anderson R 2007 Earthquake Microzonation and Hazard Maps on Bantul Area, Yogyakarta, Indonesia. Proceeding AUN/SEED-Net. The Yogyakarta Earthquake 2006. Geological Engineering Department, Faculty of Engineering, Universitas Gadjah Mada. Indonesia.

[15] Fathani T F 2007 The Determination of Peak Ground Accelaration at Opak-Oyo Fault, Bantul Regency, Yogyakarta. Proceeding AUN/SEED-Net. The Yogyakarta Earthquake 2006. Geological Engineering Department, Faculty of Engineering, Universitas Gadjah Mada. Indonesia.

[16] Karnawati D 2007 Earthquake Induces Landslides at Sengit and Prambanan District, Yogyakarta. Proceeding AUN/SEED-Net. The Yogyakarta Earthquake 2006. Geological Engineering Department, Faculty of Engineering, Universitas Gadjah Mada. Indonesia.

IOP Conf. Series: Earth and Environmental Science **851** (2021) 012002 doi:10.1088/1755-1315/851/1/012002

Reviewing the characteristics of slip behaviour for megathrust earthquake at Sumatera using vertical derivative of GOCE satellite gravity field

Aristo[1*], **Iskandarsyah**[1]

[1]Geophysics Study Program, Department of Geoscience, Faculty of Mathematics and Natural Sciences, Universitas Indonesia, Depok, Jawa Barat, Indonesia

*E-mail: aristo61@office.ui.ac.id, iskandarsyah@sci.ui.ac.id

Abstract. The review of the seismogenic zone characteristics associated with the earthquake rupture process in the Sumatra Subduction Zone has been carried out by various methods. This zone has experienced several major earthquakes; Aceh 2004 Mw=9,1, Nias-Simeulue 2005 Mw=8,6, Bengkulu 2007 Mw=8,5, and Enggano 2000 Mw=7,9. This study focuses on the relationship between density contrast analysis based on gravity data from the GOCE satellite and the slip distribution in four major earthquake rupture zones. Satellite gravity data processing is carried out to obtain data for Gravity disturbance (Gd) and vertical gravity derivatives (Tzz), corrected by topography and sediment effects with different spectrum decomposition to get gravity maps with different depths. Based on the Tzz analysis, the maximum slip of the earthquake rupture is correlated with the minimum Tzz pattern and low-density contrast. In contrast, the rupture ends at the maximum Tzz pattern and high-density contrast. Tzz pattern and Gravity disturbance can describe the barrier and asperity of the Sumatra subduction zone. The schematic maps portray the seismic segmentation of Sumatra Subduction, which have asperities zone along the subduction strike associated with the minimum Tzz and associated with the forearc zone, as well as the barrier related to the maximum Tzz, which is a manifestation of structures (fracture zone and seamount) that are subducted to the oceanic plate.

1. Introduction

The characteristic review of the seismogenic zone associated with the rupture process during an earthquake has been carried out from various studies. Bilek and Lay (1999) previously studied the attributes in the depth-to-depth rupture process and the spatial heterogeneity, and Lay et al. (2012) introduced four-character domains with different seismic radiation characteristics by focusing on variations in the energy drop in a short time along with the dip [2][10]. This model is the main framework in determining the risk for large megathrust earthquakes [14]. Mapping this heterogeneity (along strike and dip) with the density structure in the seismogenic zone will help as a proxy to study the differences in asperity and barrier behavior during the rupture propagation process.

Further research on the probable area and the estimated spread of large earthquake slip in subduction zones will be very useful for risk level analysis. Several authors (e.g., Chlieh et al., 2008, Moreno et al., 2010) estimated that the variation in coupling in subduction zones can provide information about areas that are likely to experience large slip [3][13]. High-coupled sections are fault areas that have the potential to release most of the seismic strain during large earthquakes. Chlieh et al. (2008) found that the Nias-Simeulue earthquake in 2005 originated from a strongly coupled rupture in the megathrust and found a relationship between this coupling and seismic asperity [3]. Métois et al. (2016) found that 5 out of 6 high coupling zones correlated with coseismic asperities in the Chile-Peru subduction zone [11]. Alvarez et al. (2019) shows correlation between relatively low/high Tzz and megathrust slip behavior in Peru-Chilean margin [1]. Low Tzz are related to regions with a high slip, higher degree of seismic coupling and to marginal basins, while high Tzz otherwise.

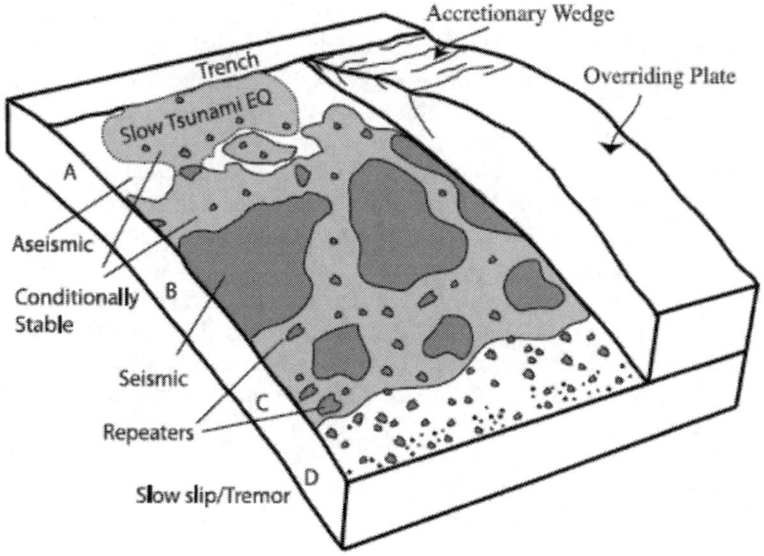

Figure 1. Cross-section schematic explaining the domain to the megathrust rupture characteristics [10]. References: ellipses with grey shading in domain A represent sediment and pore fluids at shallow depths. Dark shaded bodies labelled as seismic areas is the location of unstable shear plane friction. Domain A located close to the trench where tsunami earthquakes or anelastic deformation and stable sliding happen. Domain B is the megathrust center where high slip occurs with short-term energy and is a relatively unstable shear friction (asperity) property. Domain C is the downdip location where moderate slip occurs in small-scale areas of stable shear areas and surrounded by relatively stable areas in earthquake conditions. The transition domain D is the location of slow slip events, low-frequency earthquakes, and seismic tremors.

Research on the structure of the interplate zone has significantly increased using potential methods. Satellite gravity data has the advantage of being able to map the entire megathrust zone without the limitations of terrestrial counting systems (such as GPS, seismological stations) and other satellite systems that only provide earth surface data. In the last 2 decades, Sumatra has experienced several megathrust earthquakes with Mw> 7.5. The use of GOCE satellite data is expected to be an effective method in studying the geodynamics in Sumatra arc system. The advantages by the Gravity Satellite data have made research focused on analyzing and modeling gravity data to reveal the density structure along the megathrust to be compared with the slip behavior of the Sumatra megathrust earthquakes in the last two decades.

2. Methodology

2.1. GOCE Satellite Gravity Data

The most recent GOCE satellite model used in this study is GO_CONS_GCF_2_DIR_R6 (Bruinsma et al., 2013), a combination of GOCE-SGG (Satellite Gravity Gradiometer), GOCE-SST (Satellite-to-Satellite Tracking), GRACE (Gravity Recovery and Climatic Experiment) and LAGEOS (Laser

GEOdynamics Satellite). Anomaly Potential (T) was obtained by reducing the normal potential field with reference ellipsoid (WGS84) with observed potential [7], then reduced by topographic effects to reduce density differences within the crust (Hofmann-Wellenhof and Moritz, 2006, Molodensky et al., 1962):

$$\mathbf{Gd} = \boldsymbol{\delta}\mathbf{g}\,(\mathbf{h},\boldsymbol{\lambda},\boldsymbol{\phi}) = \mathbf{g_{obs}}\,(\mathbf{h},\boldsymbol{\lambda},\boldsymbol{\phi}) - \boldsymbol{\gamma}(\mathbf{h},\boldsymbol{\phi}) - \mathbf{g_t}\,(\mathbf{h},\boldsymbol{\lambda},\boldsymbol{\phi})[\mathbf{mgal}] \qquad (1)$$

where Gd is the Gravity Disturbance, g_{obs} is the observed gravity, g_t is the mass effect of the topography over the ellipsoid, γ is the reference gravity potential. By direct modeling of satellite-only GOCE data, from spherical harmonic coefficients (Janak and Sprlak, 2006) of a grid cell size of 0.05°, Gravity Disturbance (Gd) and Vertical Derivative of Gravity (Tzz) are obtained. Gravity Disturbance data are downloaded directly from the ICGEM (International Center for Global Earth Model) website.

2.2. Topography and Sediment Correction of GOCE Satellite Gravity Data

To eliminate the relation between gravitational signals and topography, topographic effects must be removed from satellite observations (Forsberg and Tscherning, 1997). Topography removal calculations were carried out using a model of the Earth's surface, which also included bathymetry (ETOPO1 data). The calculation also considers the shape of the Earth's curvature to avoid misinterpretation due to the wide area of the study. A topographic density of 2.67 g / cm^3 is used for the mass above sea level, while 1.03 g / cm^3 is used to correct the effects of seawater. Sediment effect reduction was carried out, with a density of 2.4 g / cm^3. Sediment thicknesses were obtained from the National Geophysical Data Center - NOAA data (Whittaker et al., 2013).

2.3. Derivative of GOCE Satellite Gravity Data

Vertical Derivative of Gravity (Tzz) is obtained as the second radial derivative of Gravity Disturbance (Rummel et al., 2011):

$$Tzz = \frac{\partial^2 T}{\partial^2 r}\left(unit : 1\ Eötvös = 10^{-4}\frac{mgal}{m}\right) \qquad (2)$$

Tzz is expressed in Eötvös, in theory has a better resolution than the gravitational vector itself for detecting density variations in the Earth's crust (Li, 2001), especially shallow structures with high-density contrast. It is possible to determine the edges of the anomaly of a mass with detailed and accurate accuracy. Gradiometers are better than gravimeters at detecting anomalies of shorter wavelengths (higher frequency) and the gradient anomaly more localised to the source than the gravity anomaly.

2.4. Topography and Sediment Correction of GOCE Satellite Gravity Data

Spectral analysis is performed by cutting the degree of harmonic expansion to obtain a clearer picture of the body at a certain depth. This spectrum analysis results will lead to response of Tzz's to the mass of the anomaly source at various depths in this study. Featherstone (1997) conducted a spectral analysis of geoid and gravity anomalies and found that cutting the degree/order of the harmonic expansion would decompose the straight-line gravimetric signal with increasing mass causing the gravity signal [4]. Álvarez et al. (2017) derive the formula relating the depth of the signal-causing mass to the various degrees of harmonic expansion for Tzz and find the various depths according to the degree of harmonic expansion cut.

$$Z = \frac{(R_E + H_c)(N-1)}{(N+2)(N+1)} \qquad (3)$$

where R_E = 6371 km is the radius of the Earth, Hc is the height calculation for Tzz and N is the degree/order of the harmonic expansion. Higher orders are associated with shallower sources, while lower orders are associated with deeper sources. The processing result of this harmonic decomposition will be a tool to analyse the increase in-depth with the mass causing the gravity signal in various cases in this study. Table 1 explains the difference in degree/order with the resulting Z depth and its spatial resolution (Alvarez et al., 2017).

Table 1 Depth of the causative mass in the spherical harmonic expansion formula on Tzz.

Degree/ Order N	Spatial resolution $\frac{\lambda}{2} = \pi R/N_{max}$	$Z_l[km]$ for Δ_g (Featherstone, 1997)	$Z_l[km]$ for Tzz (H_c = 7 km)
300	66.72	21.31	20.98
275	72.78	23.251	22.86
250	80.06	25.581	25.11
225	88.95	28.441	27.85
200	100.07	32.011	31.26
175	114.37	36.611	35.62
150	133.43	42.76	41.4

3. Results and analysis

3.1. Aceh Andaman Earthquake 2004 (Mw = 9.1)

The Aceh-Andaman earthquake occurred on 26 December 2004 with Mw = 9.1, and the epicenter was located at 3,295 ° N 95,982 ° E with an estimated hypocenter depth of 30 km [6]. This earthquake occurred along a thrust fault between the Eurasian Plate and the Australian Plate. Within 500 seconds, this fault emitted elastic strains accumulated over centuries due to the subduction of the Indian and Burmese microplates. The 2004 Aceh earthquake had a rupture of more than 1300-1500 km length towards the northwest and a width of c. 150 km [3][6]. Early models of mainshock fault distribution vary in detail. However, they consistently imply that the rupture moves northwestward from the epicenter and substantial fractures occur hundreds of km northwest of the epicenter. The strike of this rupture is N 40° W with a dip of 11° towards the northeast [6].

3.1.1 *Rupture behavior based on Tzz*

In general, there is a geometrical similarity between the maximum slip along the strike rupture and the minimum Tzz value (<-20 Eötvös) in marine forearc. The negative anomaly values continue as the depth increases, as shown by the Tzz map with various depths. The main earthquake rupture points to the northwest with the end of the downdip rupture correlated with positive Tzz anomaly values in the northwest and south of the rupture area. The Fault Zone investigator intersecting the fore bulge is seen at a gravity gradient that divides the higher value Tzz (> +20 Eötvös) with, the lower value Tzz (<+15 Eötvös). According to Robinson et al. (2006), a Fracture Zone Investigator intercepted the trench in a position where there was a halt from the rupture of the 2004 Aceh (south) and 2005 Nias (north) earthquakes, and no foreshocks were recorded in that position. This allows the fracture zone, which subducts the Sumatra trench and affects the earthquake rupture area, and acts as a barrier.

3.1.2 *The direction of movement of the rupture and asperity according to Tzz*

A study conducted by Hayes et al. (2016) and Chlieh et al. (2008) showed a coseismic slip rupture started from the hypocenter (asperity) in the southeastern part of the fault segment and spread northwest to hundreds of kilometers from the epicenter. The minimum Tzz value appears to be related to the geometry of the maximum slip and is the main asperity area of the 2004 Aceh earthquake rupture area. The epicenter of the earthquake is nucleated close to the high Tzz anomaly, and the increased slip to the northwest of the epicenter is limited by the high Tzz anomaly to the northeast and southwest of the rupture area. Analysis of Tzz with different depths up to N = 225 shows a continuity of low Tzz to the northwest of the epicenter, which differs from degrees N = 200 to N = 150. The continuity of the low Tzz shows the path of propagation (direction) and amplification of the 2004 Aceh-Andaman earthquake. According to the Hayes et al. (2016), the hypocenter depth is 30 km close to the downdip rupture, with the updip being northwest of the strike rupture [6].

3.1.3 *Seismic barrier at the SE area of the rupture*

Observations show that the fault rupture did not propagate significantly to the southeast. However, the slip magnitude is quite large in the southeastern part of the rupture, thus suggesting a seismic barrier at this location. In the 2005 Nias earthquake, the maximum slip was about 100 km from this barrier between the 2004 and 2005 earthquake ruptures, pointing to important physical properties of this barrier. The existence of a fracture zone that is predicted to affect rupture has been proposed by Chlieh et al. (2008), Lange et al. (2010) and studied by the British Oceanographic Data Center (2008). In this area, also very little seismic activity indicates a possible seismic gap. Based on the analysis of the Tzz map, there is a relatively high Tzz value at the tip of the rupture spread in the southeast and a narrow minimum

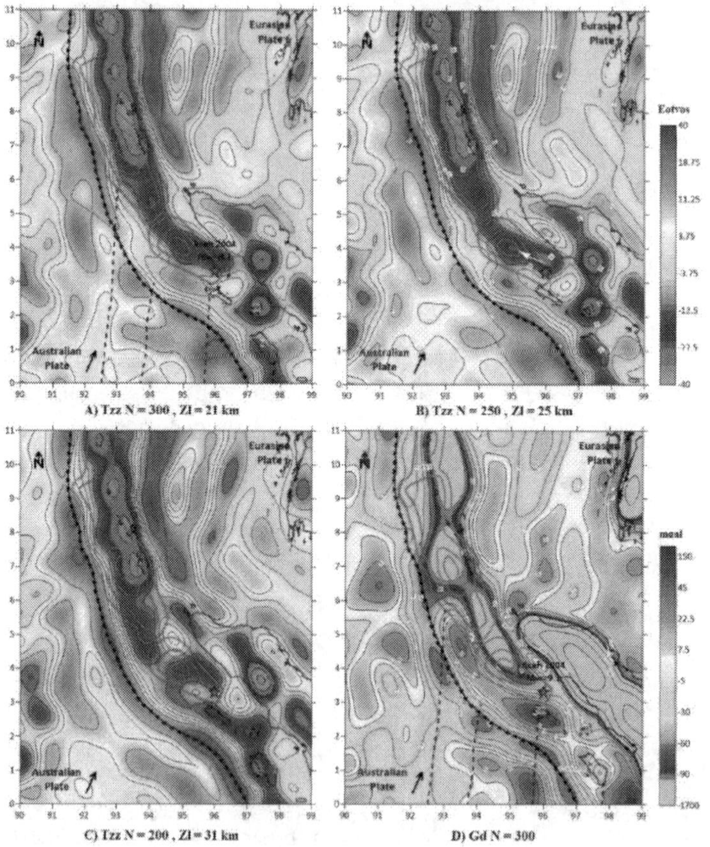

Figure 2. Vertical Derivative of Gravity (Tzz) and Gravity disturbance (Gd) that have been corrected by topography and sediment thickness around Aceh (north Sumatra) to Nicobar Island from the GO_CONS_GCF_2_DIR_6 model (Bruinsma et al., 2013). Figure A) shows Tzz with N = 300, Figure B) shows Tzz with N = 250, Figure C) shows Tzz with N = 200, and Figure D) shows Gd with N = 300-10. Black line is the trench location from Hayes et al. (2016), red star is the epicenter, slip distribution pattern map (5-meter contour) from Chlieh, 2006 from inversion of coseismic slip far-field displacement, Fracture Zone with a black dotted line from Kopp, 2008. SIM: Simeulue, N: Nias. The black arrow line on A) indicates the minimum depreciation of the Tzz value at the end of the rupture. The white arrow line on B) shows the direction of propagation of the 2004 Aceh-Andaman earthquake. The black arrow near Australian Plate represents the subduct direction of Australian Plate.

Tzz pattern in this area. The relatively maximum Tzz in the southeast of the 2004 Aceh earthquake rupture area is predicted to be a seismic barrier. Relatively maximum Tzz is usually associated with various types of oceanic plate roughness (such as seamounts, aseismic ridges) and control seismic segmentation along the margins. Lange et al. (2010) confirmed the existence of a fracture zone that was subducted at a longitude of 96° E and can be regarded as a seismic barrier.

3.1.4 *Gravity disturbance*
Gravity disturbance that has been corrected for topography and sediment thickness from GOCE data up to N = 300 shows that the minimum anomaly correlates with the maximum slip area. The pattern produced by the gravity anomaly is relatively minimum along the trench, similar geometrically to the pattern of the maximum slip distribution. Low-density contrast is at the maximum slip region, whereas high contrast is at the end of the rupture. Gravity disturbance that is relatively minimum and correlates with maximum slip is associated with low density and can reflect the forearc basin.

3.2. Nias-Simeulue Earthquake 2005 (Mw = 8.6)
The Nias-Simeulue earthquake occurred on March 28, 2005, with Mw = 8.6, and the epicenter was located at 2,085° N 97,108° E with an estimated depth of 30 km. The earthquake relates to the activity of a northeast dipping thrust fault at the junction of the Australian and Sunda plates. The rupture area of this earthquake is approximately 340 x 125 km (length x width). This earthquake was most likely triggered by changes in stress due to the 2004 Aceh earthquake. This earthquake occurred about 160 km southeast of the 2004 Aceh earthquake rupture zone. The penultimate earthquake with an Mw = 8.6 occurred in 1861 in this region, which caused a tsunami with a magnitude similar to that of 2005 Nias-Simeulue earthquake. Rupture initially points to the north, then rupture points to the south based on seismogram analysis from Walker et al. (2005).

3.2.1 *Rupture behavior based on Tzz*
The rupture area of the Nias-Simeulue earthquake resulting from GPS inversion (Briggs et al., 2006) shows two patterns of maximum slip to the northeast and southeast. In detail, this pattern is different from the slip distribution produced by Hayes et al. (2016). In general, there are similarities between the maximum slip geometry and the minimum Tzz value in the northeast and southeast of the rupture area. The two ruptures to the northeast and southeast are bounded by the relative maximum Tzz at the minimum Tzz constriction ends in both directions. The negative anomaly values continue as the depth increases, as shown by the Tzz map with various depths. The FZ investigator that cuts the fore bulge is seen from the gravity gradient, which divides the higher Tzz value (> 15 Eötvös) by, the lower Tzz (<15 Eötvös). The map of Lange et al. (2010) shows the existence of a fracture zone at a longitude of 98° E, which extrapolated divided the rupture area of the Nias-Simeulue earthquake.

3.2.2 *Asperity identification*
The Nias-Simeulue earthquake was the result of 2 different simultaneous earthquake rupturing event: Mw = 8.2 to the north of the epicenter and Mw = 8.5 to the south. Most of the coseismic slip occurred about 150 km from the trench and was in a locked fault zone (Hsu, 2006). The maximum slip of this earthquake occurred in the northeastern sector of Nias and the southeast part of Simeulue (Qiang Qiu, 2019). The slip data correlates with the minimum Tzz value found in the rupture zone following the coastline geometry along the continental forearc. The best geometric correlation between minimum Tzz and maximum slip is found at a Z = 31 km depth using the N = 200 model. This is consistent with Hayes et al. (2016) observations, which state high slip (maximum slip = 9 meters) at a depth of 28-40 km. The minimum Tzz continuity at this depth suggests the movement of the Nias-Simeulue earthquake rupture. The positive Tzz pattern and high slip distribution correlate with the high coupling pattern, which is the asperity of this earthquake.

3.2.3 *Barrier identification*

Based on the coupling data, it seems that the seismic barrier zone correlates with the low coupling segment, which functions as a barrier and effectively captures/becomes the final stop of the rupture propagation; this is also found in Metois et al. (2016), Loveless and Meade (2011). The isolated seamount in the southern part may control the slip of earthquakes and may prevent major earthquake events characterized by relatively high Tzz values in the southeastern rupture area.

3.2.4 *Gravity disturbance*

The minimum gravity pattern has a geometrical similarity to the maximum slip pattern from the slip distribution data. The density contrast in the area with low maximum slip, and vice versa at the boundary of the rupture, occurred a significant density change based on the gravity disturbance data. The minimum gravity associated with the northeast maximum slip is dominated by low-density material filling the basin.

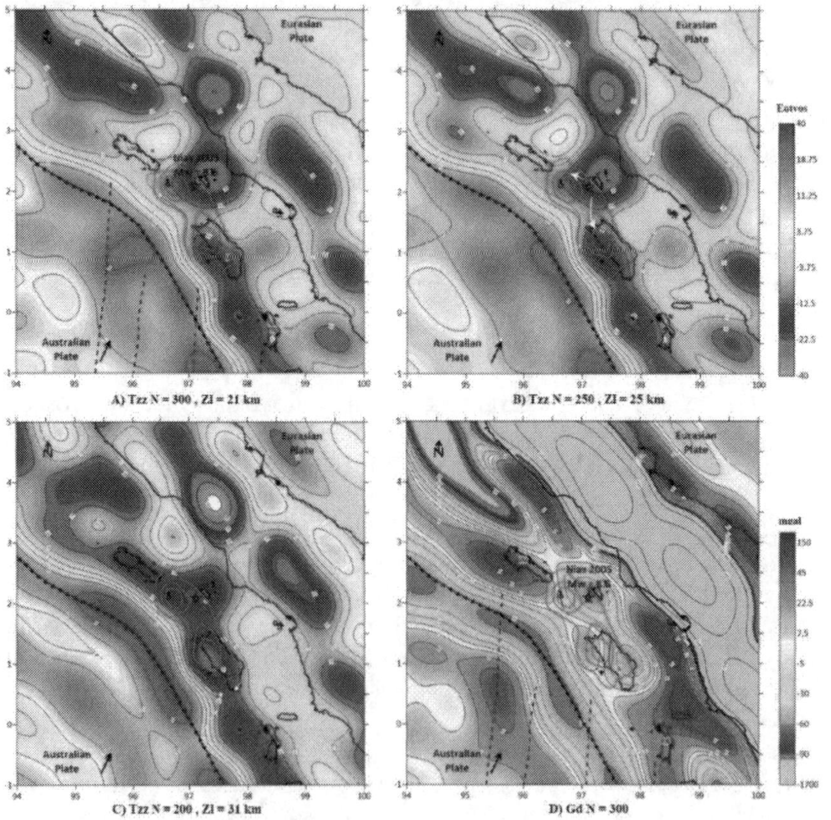

Figure 3. Vertical Derivative of Gravity (Tzz) and Gravity disturbance (Gd) that have been corrected by topography and sediment thickness around Aceh (north Sumatra) to Nias from the GO_CONS_GCF_2_DIR_6 model (Bruinsma et al., 2013). Figure A) shows Tzz with N = 300, Figure B) shows Tzz with N = 250, Figure C) shows Tzz with N = 200, and Figure D) shows Gd with N = 300-10. Reference: black trench location line triangular pattern from Hayes et al. (2016), red star is the epicenter, Slip distribution pattern map (5-meter contour) from Chlieh, 2006 from inversion of coseismic slip far-field displacement, Fracture Zone with black dotted line from Kopp, 2008. SIM: Simeulue, N: Nias, PB = Pulau Batu. The white arrow line at B) represents the direction of propagation of the rupture, the white arrow line at C) represents the minimum shrinkage of Tzz at the end of the rupture. The black arrow near Australian Plate represents the subduct direction of Australian Plate.

3.3. Bengkulu Earthquake 2007 (Mw = 8.5)

On 12 September 2007 the Bengkulu earthquake occured with Mw = 8.5 and the epicenter at 4,438° S 101,367° E. A thrust fault at the boundary between the Australian Plate and the Sunda plate was activated. At the earthquake location, the Australian plate is moving northeast towards the Sunda plate converging about 60 mm / year. The thrust fault was continuous with a rupture area of about 300 x 110 km (length x width), the hypocenter was located at a depth of about 20 km with a slip depth at a depth of 16-36 km [5]. The 2007 earthquake did not result in the rupture that occured in the 1833 earthquake [6].

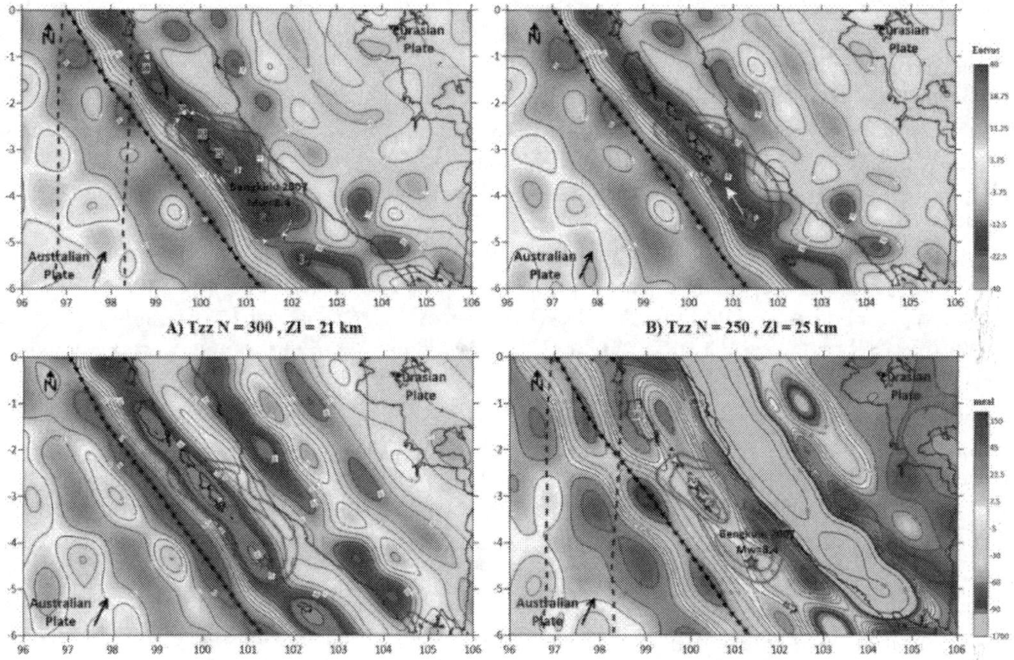

Figure 4. Vertical Derivative of Gravity (Tzz) and Gravity disturbance (Gd) that have been corrected by topography and sediments around Bengkulu (central Sumatra) from the GO_CONS_GCF_2_DIR_6 model (Bruinsma et al., 2013). Figure A) shows Tzz with N = 300, Figure B) shows Tzz with N = 250, Figure C) shows Tzz with N = 200, and Figure D) shows Gd with N = 300-10. Black stripped line = trench location line from Hayes et al. (2016), red star is the epicenter, slip distribution pattern map (5-meter contour) from Chlieh (2006) from inversion of coseismic slip far-field displacement, fracture zone with black dotted line from Kopp (2008). SB: Siberut, SP: Sipura, PU: North Pagai, PS: South Pagai, E: Enggano. The black arrow near Australian Plate represents the subduct direction of Australian Plate.

3.3.1 *Asperity identification*

The 2007 Bengkulu earthquake had a rupture with a length of about 300 km, with a maximum slip between the epicenter and Pagai Island, with large coseismic displacement. The upper boundary of the rupture is located on the west coast of the island belt and caused a large uplift on Pagai Island. The lower boundary of the earthquake rupture points westward away the coastline of Sumatra Island. The geometry of the maximum slip based on the slip distribution data correlates with the minimum Tzz value (<-20 Eötvös) and is parallel to the trench. The geometric correlation between minimum Tzz and maximum continuous slip is up to N = 200. This correlates quite well with slip distribution modeling, which explains large slip at a depth of 16-36 km [5]. The geometry of the maximum slip and minimum Tzz also correlates with the large interplate coupling.

3.3.2 *Barrier identification*
The Asperity of the Bengkulu earthquake correlates with the minimum Tzz, which laterally spreads to the northwest and is bounded by the relative maximum Tzz at the end of the minimum Tzz. In the northwestern direction of the rupture, the narrowing of the minimum Tzz correlates roughly with the end of the earthquake rupture area, although the values are barely different. The existence of a seismic barrier is thought to be due to the fracture zone, which occurs at longitudes of 99° E and 102° E [9] and is geometrically at the lateral ends of both sides of the Bengkulu 2007 fault zone.

3.3.3 *Gravity disturbance*
The maximum slip generated by the slip distribution map is similar in geometry to the high coupling and minimum gravity anomaly value patterns. The density contrast is low in the maximum slip region due to the uniform gravity anomaly pattern in this pattern. In contrast, the high contrast is at the end/tip of the rupture. Gravity disturbance that is relatively minimum and correlates with maximum slip is associated with low density and may reflect forearc basin.

3.4. Enggano Earthquake 2000 (Mw = 7.9)
The 2000 Enggano earthquake occurred on 4 June 2000 with Mw = 7.9 close to Enggano Island. This earthquake happened in the Sumatra subduction zone in the final southeastern position of the rupture area of the M> 8 subduction earthquake in 1833 (Hayes et al., 2016) and was fully coupled (Zachariasen et al., 2000). Based on the broadband modeling, the rupture moved to the southeast of the epicenter up to 100 km from the hypocenter.

3.4.1 *Asperity and barrier identification*
Based on the distribution of Tzz and aftershock values from the earthquake, it is hypothesized that the asperity zone is southeast from the epicenter to about 150 km to the southeast, where the narrowing of the Tzz value is minimum. This adapts to minimum Tzz spread and interseismic coupling, which is moderate and has a similar geometry to minimum Tzz. The Enggano 2000 earthquake barrier is located at the tip of the lateral distribution in the northwest and southeast. The barrier is predicted to be due to the subduction of the fracture zone mapped at a longitude of 102° E by Lange et al. (2010). The southeastern part of the rupture end may also be caused by the fracture zone and become a seismic barrier. The in-depth Tzz map shows that minimum expansion of Tzz to a smaller (deeper) degree.

4. Conclusions
Vertical gravity derivative (Tzz) and gravity disturbance (Gd) can help explain the slip behavior of megathrust earthquakes that occur along the Sumatra subduction zone by correlating slip distribution data, coupling and geological structures. Previous studies have successfully linked high or low coupling properties with the high and low slip rupture distribution patterns of earthquakes. This research mapped the slip rupture zone of the earthquake and the slip magnitude using Tzz data. The minimum Tzz value is closely related to high slip in the rupture zone and correlates well with high coupling. The maximum Tzz value is closely related to decreased earthquake slip at the end of the rupture zone and correlates with low coupling. Rupture propagation can also be explained by looking at the minimum Tzz dispersion pattern, which is confirmed by the direction of the earthquake rupture analyzed. Tzz can provide information on seismic zones, the presence of asperities and barriers along the Sumatra subduction zone. Seismic barrier is clearly visible on the Tzz map, the barrier located on Simeulue Island and Batu Island. Meanwhile, there are two more barriers that are being questioned, those located in the Mentawai Islands and on Enggano Island. Tzz's data shows that in the Mentawai Islands, there is a minimum narrowing of Tzz in the center of the archipelago, but there is no clear boundary. Meanwhile, the barrier on Enggano Island shows a minimum narrowing of Tzz and an increase in the value of Tzz, however, due to the absence of a detailed slip distribution record and this zone having experienced a major earthquake in 1861, it cannot be ascertained that this area is a persistent barrier. This methodology is able to provide information about the physical properties along the roundup of Australian and Eurasian Plates, so as to

map the risk level of the study area. The minimum Tzz distribution along the marine forearc correlates with the maximum slip and can be used to consider the seismicity hazard along the coastline to avoid disaster on high population zones

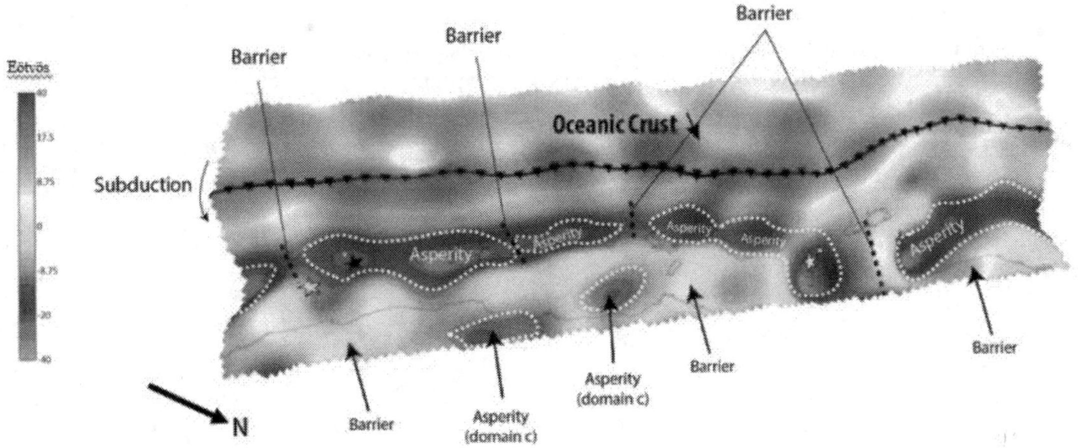

Figure 5. Interpretation of Tzz in Sumatra's subduction with the asperity sites associated with minimum Tzz and the barrier associated with Tzz is relatively maximum and its relationship to the rupture zone. The red star is the epicenter of the 2004 Aceh-Andaman earthquake. The white star is the 2005 Nias-Simeulue earthquake epicenter. The black star is the 2007 Bengkulu, and the light blue color is the Enggano 2000 earthquake epicenter. Minimum Tzz near the oceanic plate is a zone with high slip and high coupling and interpreted as asperity and located in the forearc basin. This zone, according to the subduction domain by [10], is zone B. The minimum Tzz value closer to the Sumatra coastline are in the zone C in Lay et al. (2012).

Acknowledgments
The authors would like to thank Universitas Indonesia for funding this research through PUTI Grant with contract number NKB-5/UN2.RST/HKP.05.00/2020.

References

[1] Alvarez et al. 2019 Reviewing megathrust slip behavior for recent Mw > 8.0 earthquakes along the Peru-Chilean margin from satellite GOCE gravity field derivatives. *Tectonophysics* 769
[2] Bilek S L and Lay T 1999 Rigidity variations with depth along interplate megathrust faults in subduction zones. *Nature*. **400**, 443–446
[3] Chlieh M 2008 Heterogeneous coupling of the Sumatran megathrust constrained by geodetic and paleogeodetic measurements. *Journal Of Geophysical Research*. **113**, B05305
[4] Featherstone W 1997 On the use of the geoid in geophysics: a case study over the northwest shelf of Australia. *Explor. Geophys*. **28**, 52–57.
[5] Gusman et al. 2007 Slip distribution of the 2007 Bengkulu earthquake inferred from tsunami waveforms and InSAR data. *J. Geophys. Res*. **115**, B12316
[6] Hayes et al. 2016 Tectonic summaries of magnitude 7 and greater earthquakes from 2000 to 2015, USGS Open-File Report 2016-1192.
[7] Janak J and Sprlak M 2006 New software for gravity field modelling using spherical harmonic. *Geod. Cartog. Hor*. (Geodetický a kartografický obzo) **52**, 1–8
[8] Kopp H 2008 Lower slope morphology of the Sumatra trench system. *Basin Research* (2008)
[9] Lange D 2010 The Fine Structure of the Subducted Investigator Fracture Zonein Western Sumatra as Seen by Local Seismicity, *Earth Planet.Sc. Lett.*, **298**, 47–56.

[10] Lay T. et al. 2012 Depth varying rupture properties of subduction zone megathrust faults. *J. Geoph. Res., Solid Earth* **117**, B04311.
[11] Métois, 2016 Interseismic coupling, megathrust earthquakes and seismic swarms along the Chilean subduction zone (38°–18°S). *Pure Appl. Geophys.* **173**, 1431.
[12] Molodensky et al. 1962 Methods for Study of the External Gravity Field and Figure of the Earth. Israel Program of Scientific Translations, Jerusalem (Russian original 1960).
[13] Moreno M 2010 Maule earthquake slip correlates with pre-seismic locking of Andean subduction zone, *Nature*, **467**.
[14] Ye et al. 2016 Rupture characteristics of major and great ($Mw \geq 7.0$) megathrust earthquakes from 1990 to 2015: 1. Source parameter scaling relationships. *J. Geophys. Res. Solid Earth* **121**.

IOP Conf. Series: Earth and Environmental Science **851** (2021) 012003 doi:10.1088/1755-1315/851/1/012003

Analysis of the effect of magnetotelluric data quality improvement using rho variance and edit XPR parameters in densely populated areas

A Widodo[1], **W Lestari**[1*], **D D Warnana**[1], **F Syaifuddin**[1], **R S Rivensky**[1] **and R Z Ilmawan**[1]

[1]Geophysical Engineering Department, Institut Teknologi Sepuluh Nopember, Surabaya, East Java, Indonesia 60111

*E-mail: wien@geofisika.its.ac.id

Abstract. Magnetotelluric (MT) research was conducted to map the fault structure in the western part of East Java. The advantage of the magnetotelluric method is its deep penetration because it propagates at a low frequency so that it can detect resistivity anomalies that are thought to be faults. The measuring path of the magnetotelluric method covered the Surabaya and Gresik areas. Research conducted in densely populated areas thus it requires a noise reduction process to improve the quality of the data. Performing the robust processing with Rho Variance parameters and editing XPR is an effort to reduce noise and to improve the quality of magnetotelluric data. The robust processing result of 5 (five) sounding MT data using the Rho Variance parameter can improve the quality of magnetotelluric data by increasing the coherence value of 2.128% (MT01), 2.804% (MT02), 2.378% (MT03), 3.508% (MT04), and 3.129% (MT12). Meanwhile, the editing XPR process can increase the coherence value of 0.184% (MT01), 7.155% (MT02), 9,364% (MT03), 0.784% (MT04), and 0.628% (MT12). Escalation of coherence value by implemented the Rho Variance and Editing XPR means that two methods are sufficient effectively to raise the confidence level of MT data processing and interpretation result.

1. Introduction

Indonesia occupied a very active tectonic zone because the world's three major plates and nine other smaller plates meet each other in Indonesian territory and form complex plate meeting paths [1]. The existence of interactions between those major plates made an impact onto Indonesia as an area that is very prone to earthquakes [2]. Loss of material and fatalities as a result of the earthquake disaster were mostly caused by ignorance of a fault in an area, so there was no anticipatory measure against the movement of the fault. Therefore, active fault mapping is very necessary for the development of spatial planning in Indonesia because from now on, building construction must consider aspects of disasters, one of which is earthquakes.

Geodetic studies using the Global Positioning System (GPS) have been carried out on surface deformation measurements in Java that was identified an active extension of Flores back-arc thrust westward along 300 km throughout East Java as the Kendeng Fault [3]. The results of this study indicate a new major earthquake threat for East Java and a tsunami threat to Bali, Lombok, Nusa Tenggara, and other beaches along the Flores Sea. Information regarding near-surface faults forms the basis for studies of deformation in deeper areas. The western part of East Java, including Surabaya and its surroundings, is traversed by several geological structures including the active Lidah and Kedung Waru faults which are associated with the Tongue Anticline, the Gayungan Anticline and the Kedung Waru Anticline, causing the western part of East Java and its surroundings to have the potential for earthquakes [4]. The National Earthquake Center (PusGeN) updated the 2017 national earthquake map with the discovery of new active faults including in the Kendeng Zone of East Java, including the Surabaya Fault, the Waru Fault, the Blumbang Fault, and the Cepu Fault. The fault moves with a displacement of about 0.05 mm with a magnitude <7 [5]. Earthquake potentials are crucial in determining earthquake hazard, therefore geometric information and location of potential sources are very important [6]. Geodetic studies have been carried out, however, data or research on fault dimensions and depths in the western part of East Java and its surroundings are still incomplete, so further research is needed.

The magnetotelluric method is a passive electromagnetic method for determining the conductivity of the earth's surface from tens of meters to thousands of meters deep [7]. The mapping of subsurface geological structures has been successfully carried out using the magnetotelluric method [8]. The existence of a geological structure is identified by the presence of a resistivity contrast that has a certain directional trend. The contrast of this resistivity value is highly dependent on the fluid or fault which are filled by the material. Based on research related to the priority of the earthquake potential on the Java Island using multi-criteria analysis of decision making [9], The Kendeng Surabaya Fault is one of the main priorities for delineating therefore this research will also depict the subsurface structures in the Surabaya, Gresik and Lamongan areas [9].

Measurement of data carried out in densely populated areas is a challenge due to noise so that it requires a noise reduction process to improve the quality of the data. Coherence analysis by performing robust processing with Rho Variance parameters and edit XPR is an effort to reduce noise and to improve the quality of magnetotelluric data. A study by comparing weighting parameters was conducting into magnetotelluric data on the island of Muna and its surroundings based on coherence analysis research which is the Rho Variance weighting parameter shows more dominating results than the other weighting parameters i.e. Robust No Weigh and Ordinary Coherency thus the Rho Variance was implemented in this study [10]. Due to the importance of further understanding about the coherence analysis, this study will examine the effects of robust processing and XPR editing as an effort to improve the quality of magnetotelluric data on geological structure investigations in East Java.

2. Methodology

2.1. Basic Principal of Magnetotelluric Method

The magnetotelluric method is one of geophysical method that utilizes passive electromagnetic waves by involving measurements of perpendicular fluctuations in the electric and natural magnetic fields on the earth's surface which can be used to determine the conductivity value of rocks beneath the earth's surface from a depth of several meters to hundreds of kilometers. Magnetotelluric measurements record two components of the electric field (E) and three orthogonal components of the magnetic field (H). The magnetotelluric method is carried out using a frequency in the range of 10-5 Hz - 103 Hz [7]. The magnetotelluric method is considered capable of mapping subsurface geological structures [11].

Figure 1 described the electromagnetic wave propagation from transmitter to receiver where the natural electromagnetic field (primary electromagnetic field) as the source of the magnetotelluric method reaches the earth with variations with time. This electromagnetic field induces ore bodies below the earth's surface, causing eddy currents (telluric currents) to arise which produce a secondary electromagnetic field. The receiver at the surface captures the total electromagnetic field as the sum of the primary electromagnetic field and the secondary electromagnetic field [12].

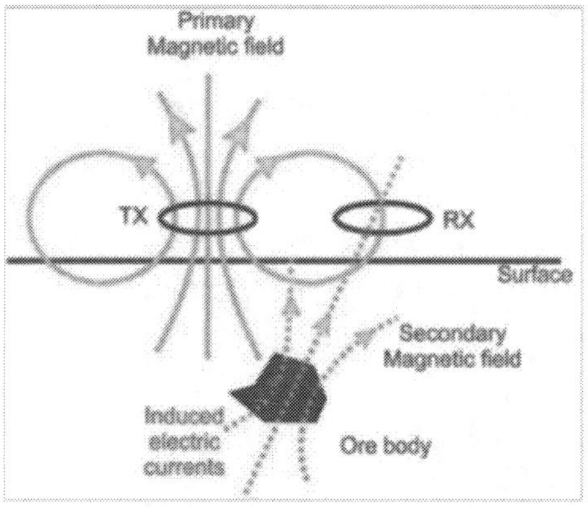

Figure 1. The interaction of EM waves with the subsurface medium.

2.2. Fourier transformation

This process was used to convert data in the time domain into the frequency domain [13] [14]. Data from field measurements are data in the time domain. There are several parameters that cannot be analyzed in the time domain so that it requires a transformation from time domain to frequency domain to further analysis can be carried out. Fourier transform plays a role in converting time series data into frequency domain data as it explained below in equation 1:

$$x(t) = \frac{1}{2\pi}\int_{-\infty}^{\infty} X(\omega)e^{i\omega t}d\omega \quad (1)$$

Where ω is the angular frequency (rad/sec), t represents time (second), $x(t)$ is the function in time domain, $X(\omega)$ is the function in frequency domain; $e^{i\omega t}$ is the sinusioidal signal component.

2.3. Robust processing

Robust processing is a process used to minimize the influence of outliers (noise) by looking for the more dominant data trend direction. Data that is dominated by noise will have a small coherence value, so it can reduce the quality of the data both qualitatively and quantitatively [16]. Robust processing functions is to detect the presence of outliers in data, where outliers are data whose value is far from the overall value so that the data is given a smaller iterative weighting [17]. There are 3 parameters that can be used in the robust processing, namely [18]:

1. No Weight, data that is not weighted in consequence this data has similar type with raw data.
2. Ordinary Coherence, weighted data based on good coherence in the E field and the H field.
3. Rho Variance, is a data weighting process based on variations/changes in the rho value.

Literature study indicated that the Rho Variance weighting parameter shows more dominating results than the other weighting parameters [10] [17] [18]. Further, this research was conducted using the Rho Variance weighting parameter as well to reach the high quality of 2D magnetotelluric data result.

2.4. Crosspower selection (XPR)

Crosspower is a partial data set in the form of pairs of resistivity and phase values with the same frequency. The crosspower selection process aims to make a smooth apparent resistivity curve by selecting and sorting the most suitable data and eliminating data affected by noise [19].

2.5. Magnetoteluric Data Coherence

The coherence of magnetotelluric data is the relationship between the electric field (E) and the magnetic field (H) which is perpendicular to each other. Ideally, the coherence of magnetotelluric data will be worth one, but this is very rare, so if the data is closer to the value 1, the data will be more coherent [18] [19].

2.6. Research Flow Diagram

The general research flow diagram from acquisition data of magnetotelluric survey until its analysis is able to be seen in Figure 2 below.

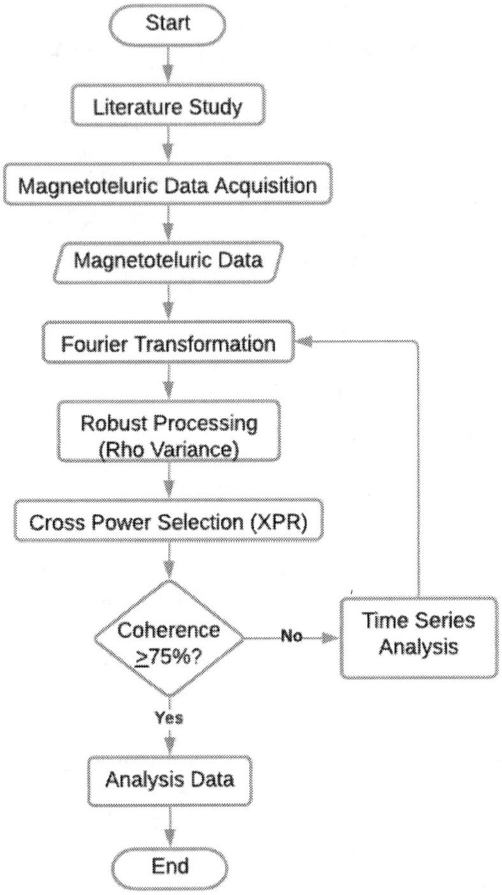

Figure 2. Research Flow Diagram.

3. Study Area

This research was conducted using 5 points of magnetotelluric data covering the Surabaya and Gresik areas as displayed by Figure 3. Those points are in one straight line across perpendicular with the two major Kendeng Faults on the study area consequently those faults are captured by the magnetotelluric measurement besides other geological structures that are possible connected with the Kendeng Fault. The coordinates of the magnetotelluric data are available in Table 1. The flow of this research starts with the measurement of the magnetotelluric data which is then carried out by processing and analyzing the coherence of the data to produce quality data for subsequent purposes.

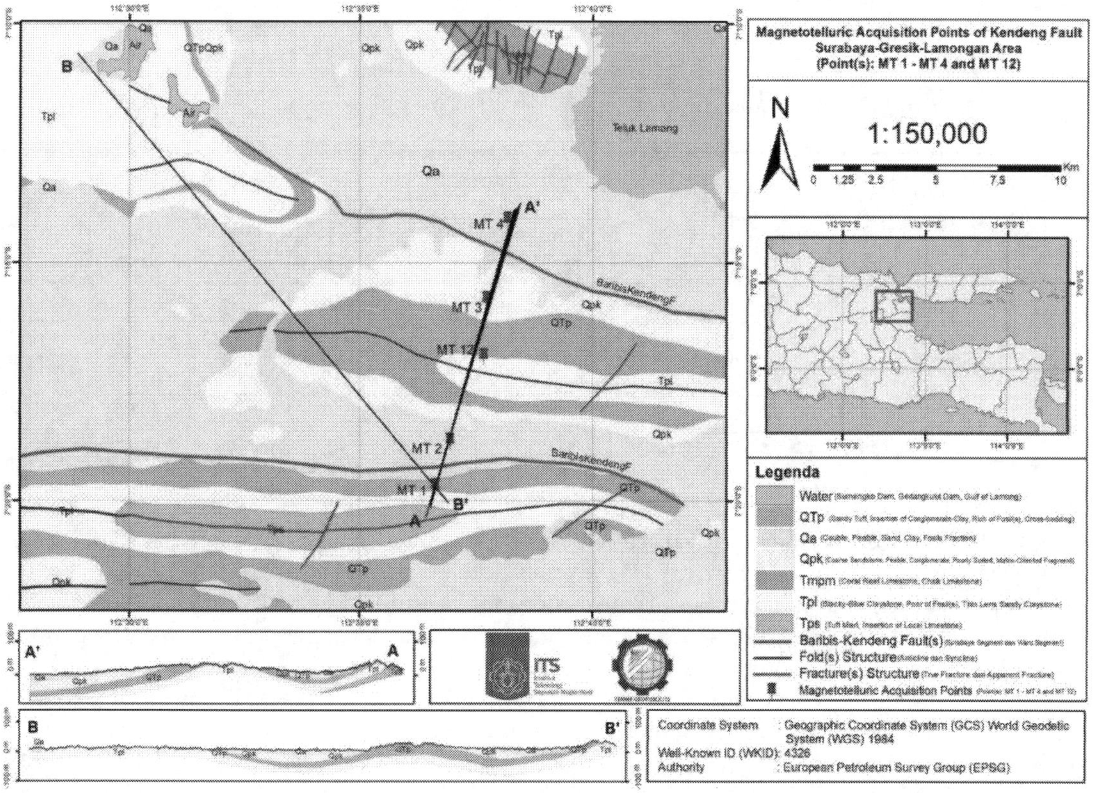

Figure 3. Magnetotelluric Method Measurement Survey in Research Fields (Surabaya-Gresik Areas).

Table 1. Measurement point coordinates.

Name	Northing	Easting	Elevation (m)
MT01	9189473	677870	30
MT02	9191356	678418	14
MT03	9197437	678377	33
MT04	9200255	678600	8
MT12	9194073	678502	39

4. Results and Discussion

According to the percentage value which is resulted from the coherence process, it is the quantitative parameter to inform the robust quality of magnetotelluric field data. If the coherent data value is closer to 1, it is meant the survey parameters of observed data are fit properly thus we captured the subsurface anomaly with a high signal to noise ratio. When the electromagnetic wave propagated beneath the subsurface, the signal might have occurred the interference that causing noise in the measurement data. Therefore, a noise reduction process is needed to obtain high quality data. This study aims to improve the quality of magnetotelluric data by conducting coherence analysis. The coherence analysis carried out included the robust processing using the Rho Variance parameter further it is followed by XPR editing on the entire MT data. In the robust processing, there are three weighting parameters; however, in this study only the Rho Variance parameter was used. As for what is obtained as follows (Table 2).

Table 2. The results of the coherence analysis from 5 points in the study area.

Station	Preliminary data (%)		Average (%)	Rho Variance (%)		Average (%)	XPR Edits (%)		Average (%)
	Rho XY	Rho YX		Rho XY	Rho YX		Rho XY	Rho YX	
MT01	80.917	83.682	82.3	79.902	88.954	84.428	80.955	88.269	84.612
MT02	91.751	68.407	80.079	92.615	73.151	82.883	91.807	88.269	90.038
MT03	77.664	54.387	66.025	78.208	58.598	68.403	79.009	76.526	77.767
MT04	76.741	69.401	73.071	80.086	73.072	76.579	78.513	76.213	77.363
MT12	89.219	78.709	83.964	90.4	83.787	87.093	88.627	86.868	87.747

The results of the coherence analysis carried out by the robust processing using the Rho Variance parameter in this area showed an increase in the average coherence value of the initial data, an increase of 2.79%. The lowest increase in the coherence value of data occurred in the MT01 data from 82.3% to 84.428% or an increase of 2.128%. The highest increase in the coherence value of data occurred in MT04 data from 73,071% to 76,579% or an increase of 3,508%.

The results of the coherence analysis carried out by the XPR editing process showed an increase in the average coherence value of the initial data, an increase of 6.418%. The lowest increase in the coherence value of data occurred in the MT01 data from 82.3% to 84,614% or an increase of 2,313%. The highest increase in data coherence value occurred in MT03 data from 66,025% to 77,767% or an increase of 11,742%. The increase in the highest coherence value can also be seen significantly in the coherence to frequency curve in the following MT03 data (Figures 4 – 6).

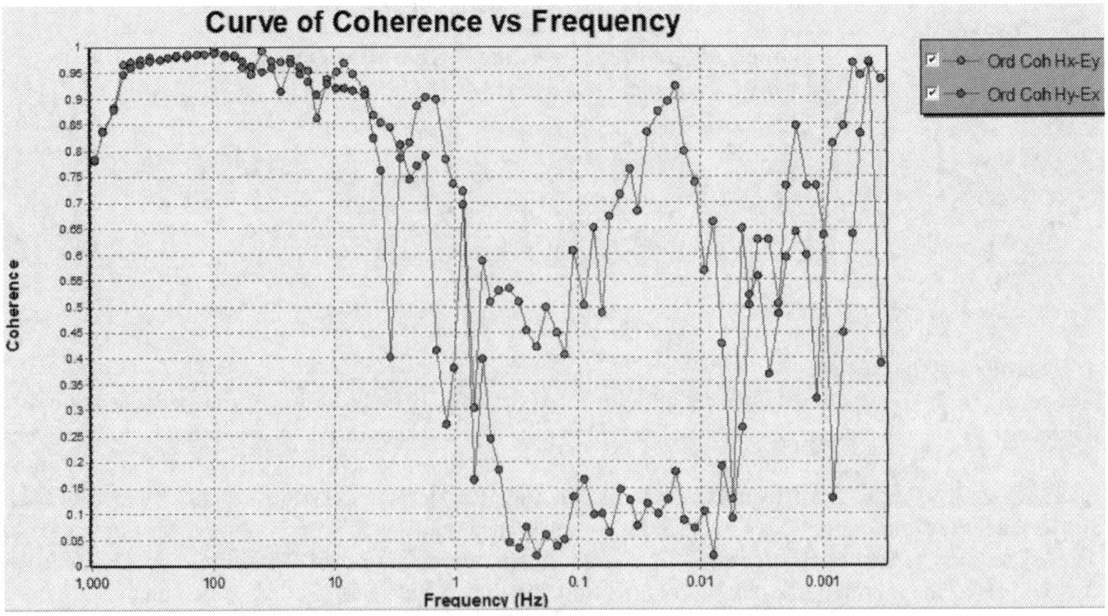

Figure 4. The curve of the coherence values for the Rho-XY (red) and Rho-YX (green) frequencies of the MT3 points.

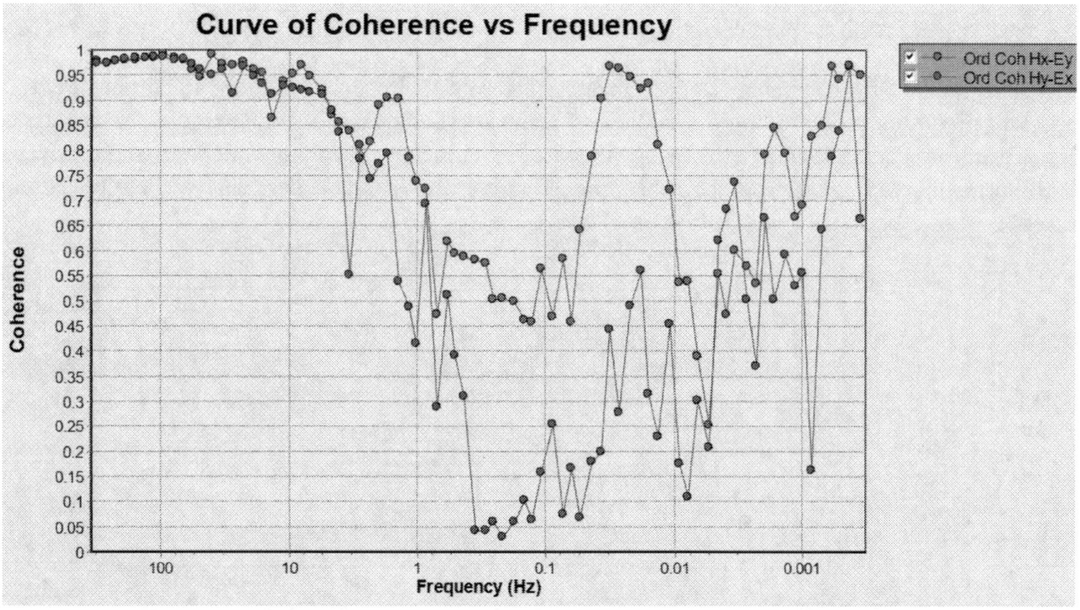

Figure 5. The curve of the coherence values for the Rho-XY (red) and Rho-YX (green) points of MT3 after Robust Processing is carried out using the Rho Variance parameter.

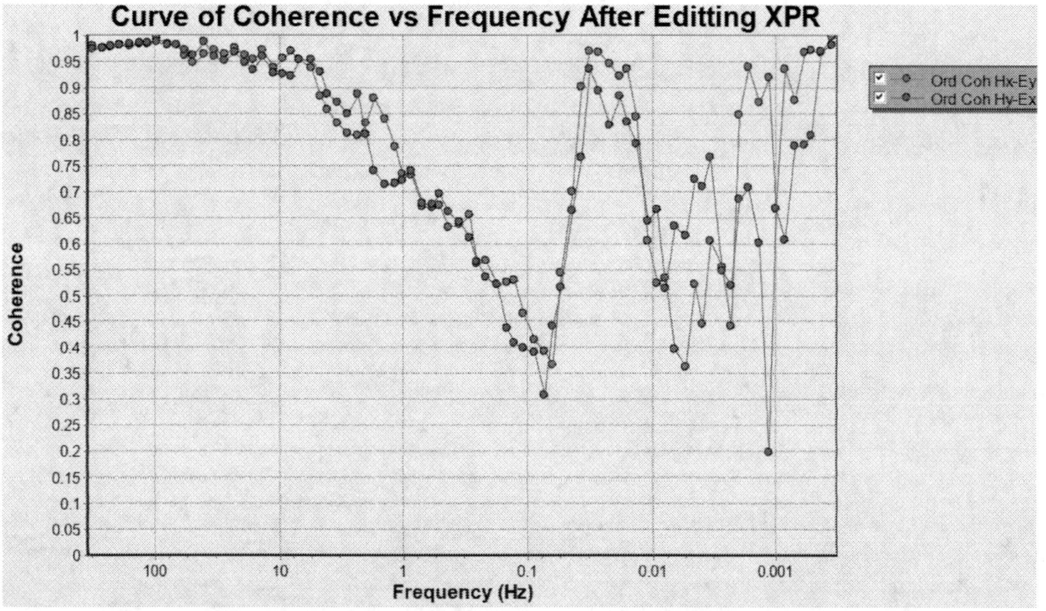

Figure 6. The curve of the coherence values for the Rho-XY (red) and Rho-YX (green) points of MT3 after the process of XPR Editing.

The coherence curve with respect to frequency shows a significant change from 3.4 to 0.00034 Hz. In the initial data curve (Figure 5), the trend of Rho XY and Rho YX curves is very random, even at frequencies <1 - 0.004 Hz the Rho YX curve (green) has very low data coherence (<25%). In the data from Robust Processing (Figure 6), the trend of the Rho XY and Rho YX curves is better but there is still very low data coherence (<25%) at frequencies <0.215 - 0.0055 Hz. The XPR edited data shows the Rho XY and Rho YX curves have almost the same direction as a whole (Figure 7). The XPR edited data shows a very good data coherence value compared to the initial data and the robust processing results.

The increase in the coherence value indicates that robust processing and XPR editing are very important processes in improving the quality of magnetotelluric data. The increase in the coherence value greatly affects the results of the apparent resistivity curve and the phase which tends to follow the trend. Furthermore, the data from the robust processing results will be carried out by the XPR editing process as shown below.

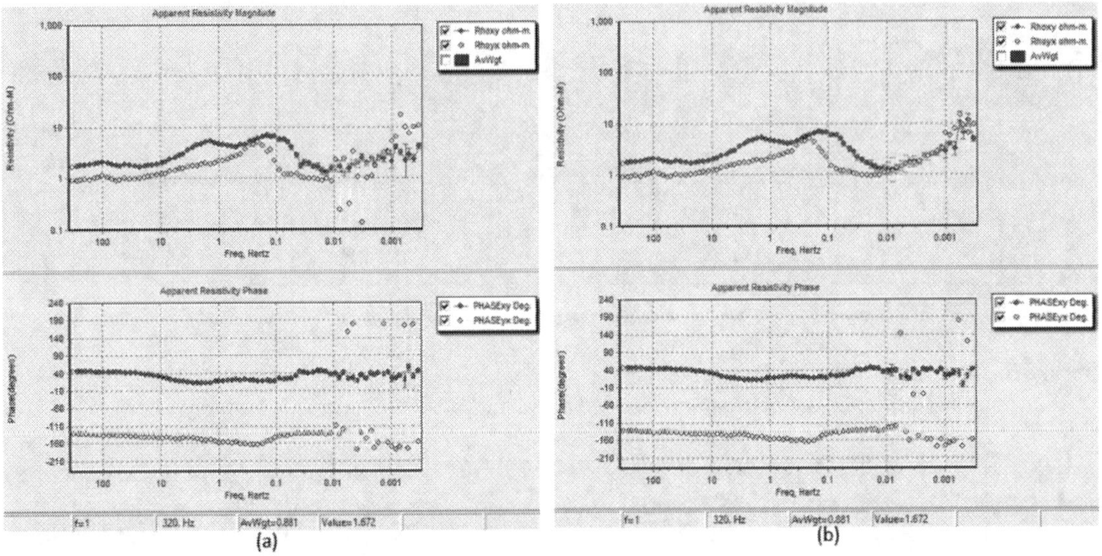

Figure 7. Apparent resistivity and phase curves of the Rho-XY and Rho-YX points MT01: (a) before XPR editing, and (b) after XPR editing.

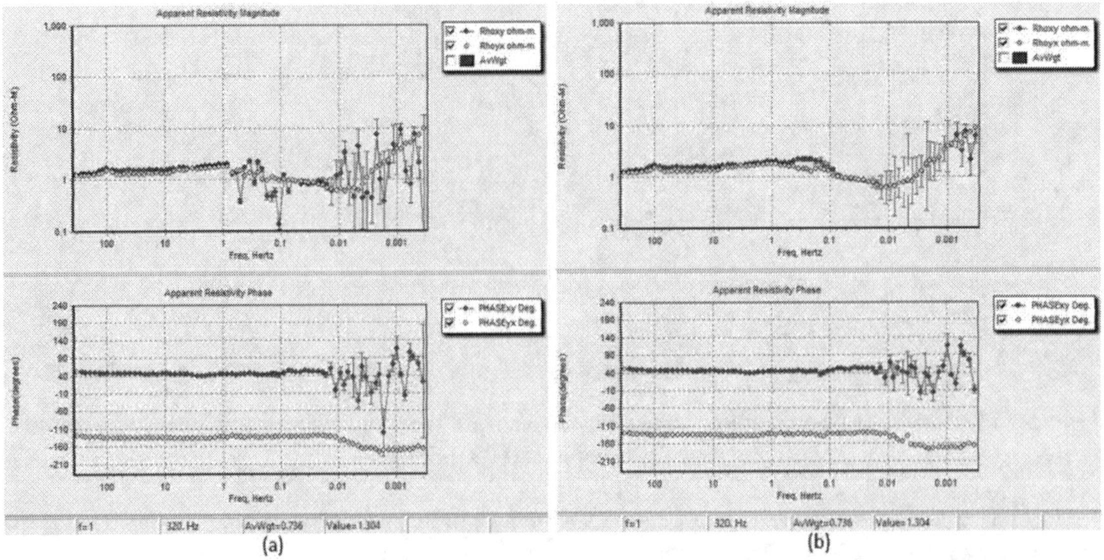

Figure 8. Apparent resistivity and phase curves of the Rho-XY and Rho-YX points MT02: (a) before XPR editing, and (b) after XPR editing.

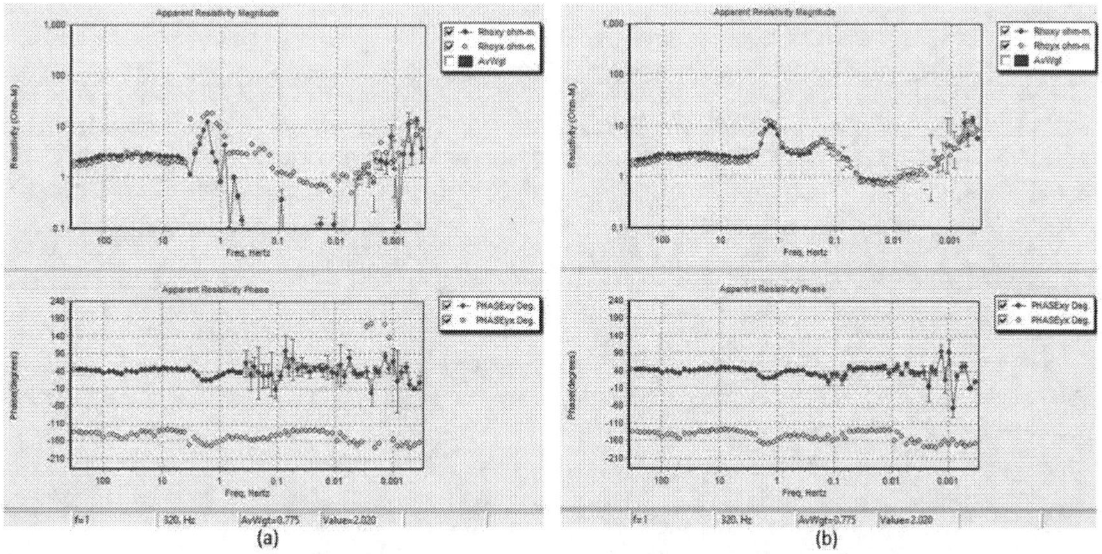

Figure 9. Apparent resistivity and phase curves of the Rho-XY and Rho-YX points MT03: (a) before XPR editing, and (b) after XPR editing.

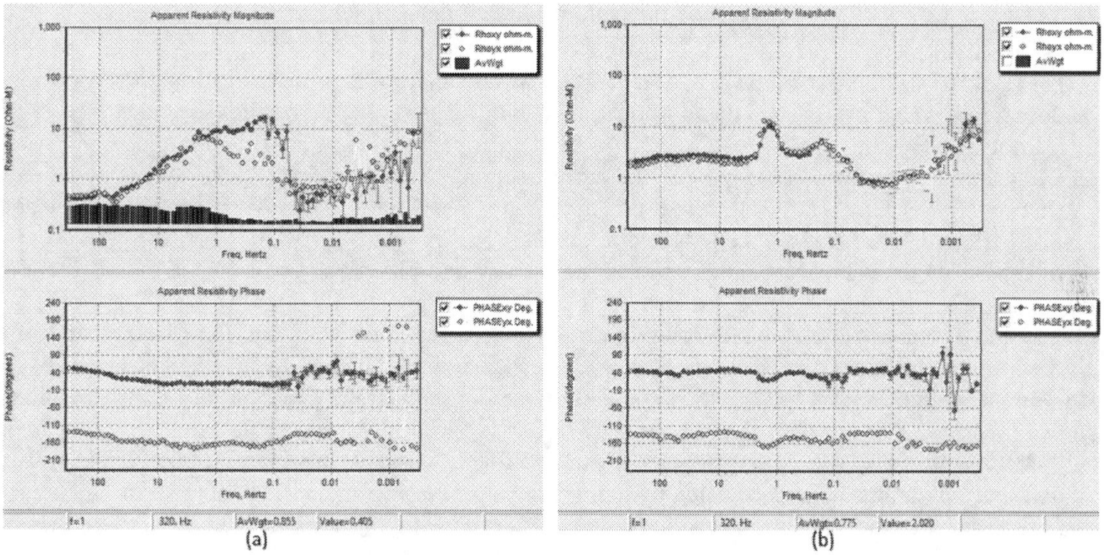

Figure 10. Apparent resistivity and phase curves of the Rho-XY and Rho-YX points MT04: (a) before XPR editing, and (b) after XPR editing.

In the MT01 data (Figure 7), the robust processing data using the Rho Variance parameter shows the apparent resistivity and phase curves of Rho-XY and Rho-YX have a very good trend thus this data does not carry out a significant XPR editing process. This is indicated by the data coherence value after robust processing at the MT01 point of 84.428% and after XPR editing of 84.612% which only increased the coherence value of 0.184%. In the MT02 data (Figure 8), robust processing data shows the apparent resistivity and phase curves of Rho-YX (marked in red color) have a very good trend, however the apparent resistivity and phase curves of Rho-XY (represented in green) have a random trend therefore the XPR editing is required. In prior of XPR editing, the data had the coherence value of 82,883%.

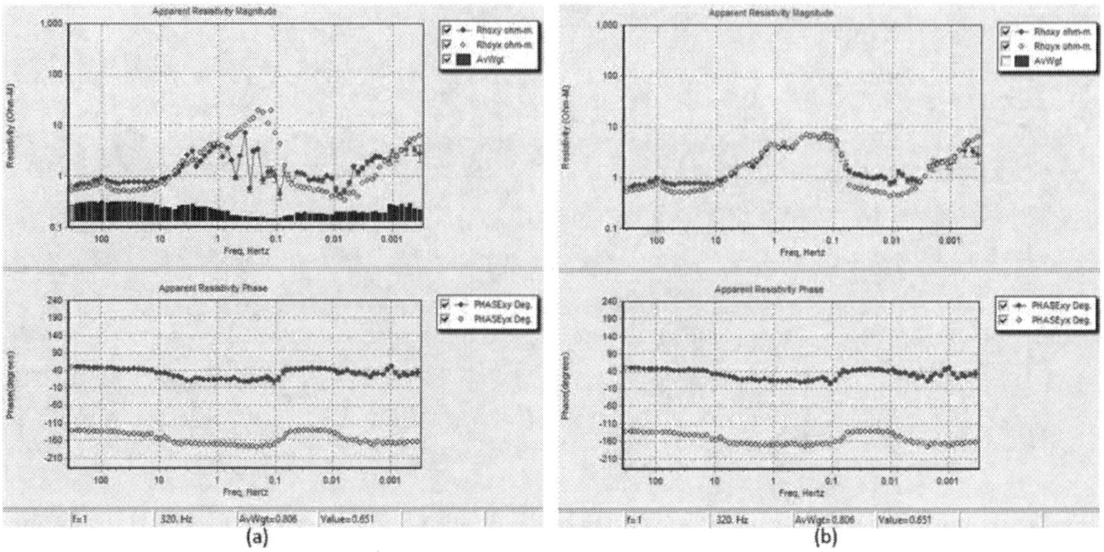

Figure 11. Apparent resistivity and phase curves of the Rho-XY and Rho-YX points MT12: (a) before XPR editing, and (b) after XPR editing

In the MT03 data, the apparent resistivity and phase curves from Rho-XY and Rho-YX have a very random trend, which is predicted contain high noise, while it is supported by the data coherence value after Robust Processing of 68.403%, we can conclude that it should have a XPR editing process. After the XPR editing process was carried out, the data coherence value increased by 9,364% thus the data coherence value became 77,767%. In addition, the apparent resistivity and phase curves of Rho-XY and Rho-YX (Figure 9) are having a better smooth trend.

In the MT04 data (Figure 10), the apparent resistivity and phase curves of Rho-XY and Rho-YX have a feasible trend however the distribution is slightly random, and then the XPR editing process is carried out. The data before the XPR editing process had a coherence value of 76,579% and after the XPR editing process was carried out it became 77,363%, there was an increase in coherence of 0.784%. It is similar with the result of MT12 (Figure 11) which is showed the apparent resistivity and phase curves for Rho-XY and Rho-YX were having a very proper trend where the data coherence value of 87,093% and after the XPR editing process became 87,747%, it means that increased by 0.654 %.

The increase in this coherence value by doing robust processing and XPR editing shows that these two processes are very important to be implemented for reducing noise in the data, in consequence, it can improve the quality of the magnetotelluric data.

5. Conclusion

Based on the data coherence analysis that has been carried out, so the conclusion is:

1. The robust processing using the Rho Variance parameter can improve the quality of magnetotelluric data by increasing the coherence value of 2.128% (MT01), 2.804% (MT02), 2.378% (MT03), 3.508% (MT04), and 3.129% (MT12).
2. The XPR editing process can improve the quality of magnetotelluric data by increasing the coherence value of 0.184% (MT01), 7.155% (MT02), 9,364% (MT03), 0.784% (MT04), and 0.628% (MT12).

References

[1] P. Bird 2003 *An Updated Digital Model of Plate Boundaries, Geochemistry, Geophysics, Geosystems, **4**(3).*

[2] W. Hamilton 1973 "Tectonics of the Indonesian Region," pp. 3–10,.

[3] A. Koulali, S. Susilo, S. McClusky, I. Meilano, P. Cummins, P. Tregoning, G. Lister, J. Efendi, M. A. Syafi'i 2016 Crustal strain partitioning and the associated earthquake hazard in the eastern Sunda-Banda Arc *Geophys Res Lett* **43** pp. 1943–1949

[4] J. B. Supandjono, K. Hasan, H. Panggabean, D. Satria, and Sukardi, "Peta Geologi Lembar Surabaya & Sapulu, Jawa," 1992.

[5] National Earthquake Centre, *Indonesian Earthquake Hazard and Source Map of 2017* (Peta Sumber dan Bahaya Gempa Indonesia Tahun 2017)

[6] A. Widodo, F. Syaifuddin, W. Lestari, and D. D. Warnana 2020 *Earthquake potential source identification using magnetotelluric data of Kendeng thrust Surabaya area E3S Web Conf.*, vol. 156 p. 01002 doi: 10.1051/e3sconf/202015601002.

[7] F. Simpson and K. Bahr 2005 *Practical Magnetotellurics* (Cambridge: Cambridge University Press)

[8] W. Lestari, A. Widodo, D. D. Warnana, F. Syaifuddin, J. P. G. N. Rochman, A. Zarkasyi, and N. S. Setiawan 2020 *Active Fault Delineation Using Magnetotelluric Data in The Western Region of East Java IOP Conf. Ser. Earth Environ. Sci.*, vol. 506, p. 012054, Jun. 2020, doi: 10.1088/1755-1315/506/1/012054.

[9] W. Lestari, A. Widodo, F. Syaifuddin, and D. D. Warnana 2020 *Research priority of the potential earthquake on the java island using decision making analysis E3S Web Conf.*, vol. 156, p. 03003 doi: 10.1051/e3sconf/202015603003.

[10] G. M. L. Junursyah, D. H. Salsabil, and E. Mirnanda 2019 *The Effect of Magnetotelluric Data Quality Improvement in Muna Island Based on Coherence Analysis of 2D Modeling Nat. B Vol 5 No 2 Oktober 2019* (In Indonesian language)

[11] W. Lestari, A. Widodo, D. D. Warnana, F. Syaifuddin, W. Utama, and J. P. G. N. Rochman 2018 *Mapping of Kendeng Thrust Active Fault in East Java Using Magnetotelluric Method* EAGE-HAGI 1st Asia Pasific Meeting on Near Surface Geoscience and Engineering doi: https://doi.org/10.3997/2214-4609.201800424.

[12] M. Unsworth 2005 *New Developments in Conventional Hydrocarbon Exploration with Electromagnetic Methods Focus Article Cont'd.* Focus Article (Coordinated by Helen Isaac)

[13] N. S. Setiawan, A. Widodo, W. Lestari, F. Syaifuddin, A. Zarkasyi, D. D. Warnana and J. P. G. N. Rochman 2020 *Application of empirical mode decomposition (EMD) filtering at magnetotelluric time-series data* Malang, Indonesia p. 040039, doi: 10.1063/5.0015767.

[14] M. I. M. Wachisbu, B. J. Santosa, and J. A. R. Hakim 2015 *Magnetotelluric Data Modeling with Remote Reference for Exploration of the Oil and Gas Basin Case Study: Em-4* vol. 4 p. 4 (In Indonesian language)

[15] J. F. Hermance 1973 *Processing of Magnetotelluric Data Phys. Earth Planet. Inter.*

[16] L. Maryani, G. M. L. Junursyah, and A. Harja 2016 *Time Series Analysis of the Magnetotelluric Method in the Buton Basin, Southeast Sulawesi (Analisis Deret Waktu (Time Series) Metode Magnetotellurik Pada Cekungan Buton, Sulawesi Tenggara)* p. 8

[17] G. M. L. Junursyah and A. Prabowo 2020 *Analysis of Magnetotelluric Data Quality Based on Coherence Parameters Case Study: Magnetotelluric Data in Bandung, West Java* vol. 4, p. 7, (In Indonesian language)

[18] A. R. Hidayat, G. M. L. Junursyah, and A. Harja 2016 *Time Series Analysis for Quality Improvement of Magnetotelluric Data-Geothermal Field Case Study (Analisis Deret Waktu Untuk Peningkatan Kualitas Data Magnetotelurik-Studi Kasus Lapangan Geothermal)* p. 11 (In Indonesian language)

[19] R. Mwakirani 2012 *Magneto-Telluric (MT) Data Processing* p. 7

IOP Conf. Series: Earth and Environmental Science **851** (2021) 012004 doi:10.1088/1755-1315/851/1/012004

Characteristic of 27th September – 7th October 2017 earthquake swarms in Jailolo Volcano, West Halmahera, Indonesia, based on hypocenter and b-value

A P Astuti[1*], N S Arifuddin[1], M I Tahir[2], E M Elsera[2] and M F I Massinai[1]

[1]Department of Geophysics, Hasanuddin University, Indonesia
[2]Meteorological, Climatological, and Geophysical Agency (BMKG) of Indonesia, Makassar, Indonesia
*E-mail: auliapuji24@gmail.com

Abstract. Halmahera is an area with active tectonics, so it has a high level of seismicity. Swarm earthquakes occurred in Jailolo from November to December 2015, and then in 2017, another earthquake swarm occurred from September to October. This earthquake is characterized by an increase in the number of earthquakes in a certain period with a relatively small magnitude, without mainshocks, and occurs in volcanic areas. This research used arrival time from P and S waves recorded at Taide Digital Seismograph (TDS) which was positioned at Ternate Geophysical Station (TNTI). We used cross-section on hypocenter to see the depth distribution using GMT and determination of b-value using ZMAP code. From the results of this study, the variation in the magnitude of the earthquake swarm obtained ranged from 0.7 to 5.0 with a depth of 7.7-12 km. Our results show a b-value of aproximately 1.0 in the area near Jailolo Volcano, 1.0-1.5 in the northwest of Jailolo and 1.0-2.0 in the southeastern part of Jailolo. Based on b-value we obtained, the characteristics of the Jailolo swarm earthquake tend to be influenced by magmatic activity.

1. Introduction

Research on the earthquake swarm in Jailolo has been carried out by several previous researchers, such as Passarelli et al. [1] that study the spatio-temporal pattern of the seismic swarm around Jailolo Volcano on November 2015-February 2016, then Nugraha et al. [2] study the earthquake swarm in Jailolo that occured on September-October 2017, by relocating the hypocenter to obtain more precise location. This study also analyzed data from the swarm earthquake in Jailolo in September-October 2017, but Nugraha et al. [2] analyzed the data by relocating the hypocenter, while we used the b-value to analyze the characteristics of the earthquake.

North Molucca and Molucca have complex tectonic regions caused by the interaction of three large plates, which consist of the Eurasian plate in the west, Indo-Australian plate in the south, Pacific plate in the east, and the micro-Mindanao plate in the north [3]. Molucca Sea was located at the confluence of three major plates in the world, known as Eurasian Plate, Philippine Sea Plate, and Australian Plate [4]. The subduction zone between the Molucca Sea plate in the east and Halmahera in the west caused a series of active volcanoes at Sangihe Arc [5]. As a result of the subduction activity between Sangihe and Halmahera, this region has a high seismicity index.

Based on the time sequence, there are three different types of earthquakes [6]. Type (1) is an earthquake with a mainshock followed by several aftershocks (decreasing magnitude and frequency of occurrence), type (2) is characterized by a slow build-up of seismicity (foreshock) leading to the type 1, and type (3) is characterized by a gradual increase and decay of the seismicity in time without a distinct mainshock [7]. The earthquakes type graph can be seen in figure 1.

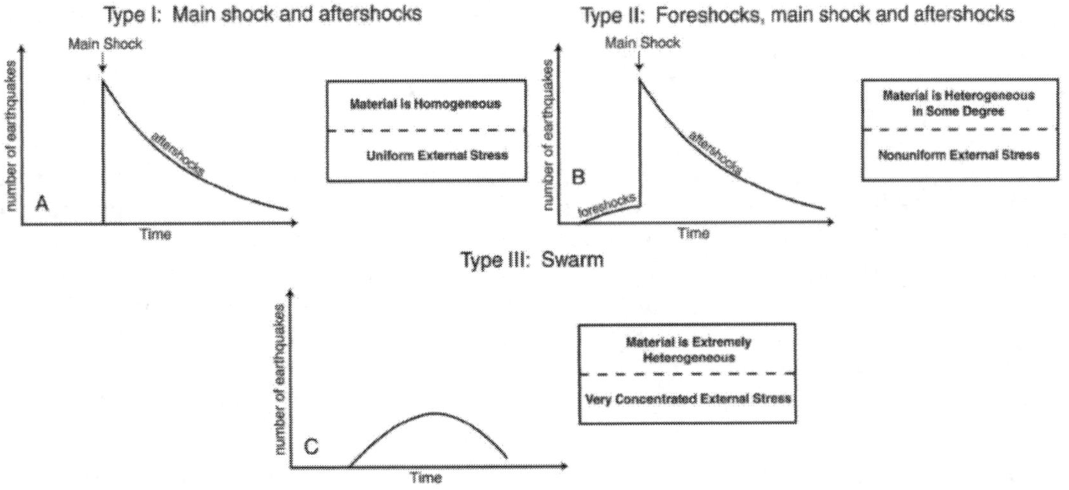

Figure 1 Schematic graph showing the differences in earthquake types based on the time sequence [6,7].

Type (3) is also known as earthquake swarm [6,7]. This is indicated by the increase in the number of earthquakes in a certain period with relatively small magnitudes without mainshocks which generally occur in volcanic areas [6]. Type 3, or swarm, occurs in very heterogeneous materials, has a high fracture density, and highly concentrated external stresses [6,7]. Swarm earthquakes occurred in Jailolo from November to December 2015 and then in 2017, another earthquake swarm occurred from September to October. Another cased of earthquake swarms also occurred in Mamasa, West Sulawesi in November 2018 [8].

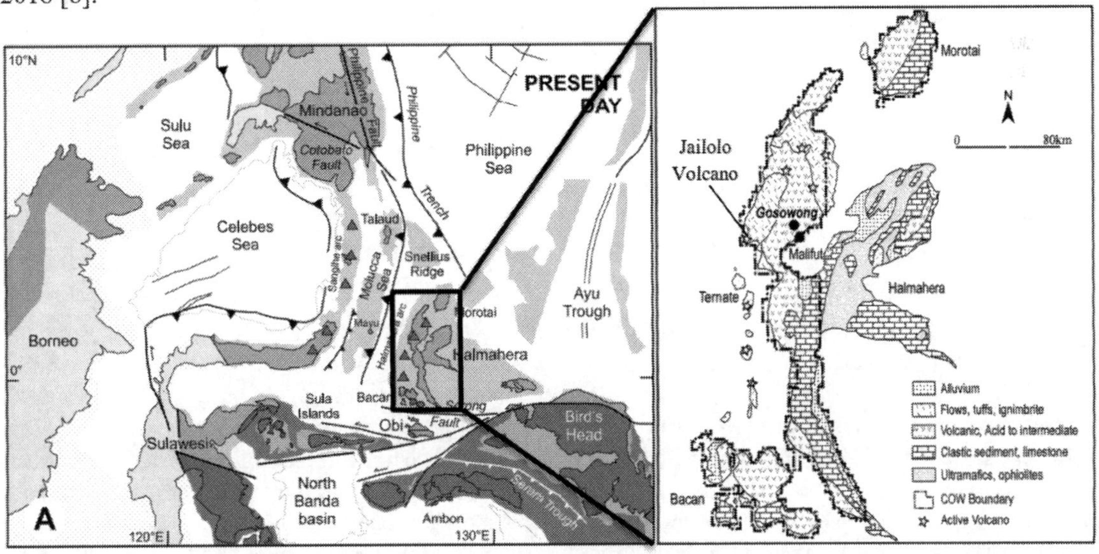

Figure 2 Tectonic and geological map of Halmahera Island. Bold barbed line are subduction zones and small barbed lines are thrusts. Green areas are ophiolite, emergent arc, and accreted crust. Yellow is Eurasian crust and pale yellow is submarine parts of the Eurasian continental margin. Red is Australian origin crust; pale and deep pink areas represent submarine parts of the Australian continental margin. Fine black lines are marine magnetic anomalies, and cyan areas are submarine [9,10].

Halmahera is an area with active tectonics that has a high level of seismicity. This is evidenced by the occurrence of swarm earthquakes with events reaching hundreds of earthquakes in a day. This study aims to determine the distribution of hypocenter and b-value to find the characteristics of the Jailolo swarm earthquake on September 27 - October 7, 2017.

2. Data and Method

This study used the arrival time of the P and S waves recorded on Short Periode Seismograph TDS-303 with sensor type DS-4A, that was placed at the Ternate Geophysical Station. This measurement was carried out by BMKG. This device is a single station seismograph that works as a seismic vibration processing system for monitoring earthquakes in local region. The amount of data used in this study was 941 earthquakes that occurred from 27 September to 7 October 2017, with the most earthquakes occurring on 28 September 2017, which were 275 events. The amount of data in the graph below depends on the number of signals that have been successfully analyzed. As previously explained, the earthquake swarms in Jailolo can reach hundreds of earthquakes a day, so the number of earthquakes that occur can be more than is shown in the graph below. This is also because we only select the waveforms of the P and S waves which can be clearly analyzed to determine the earthquake parameters.

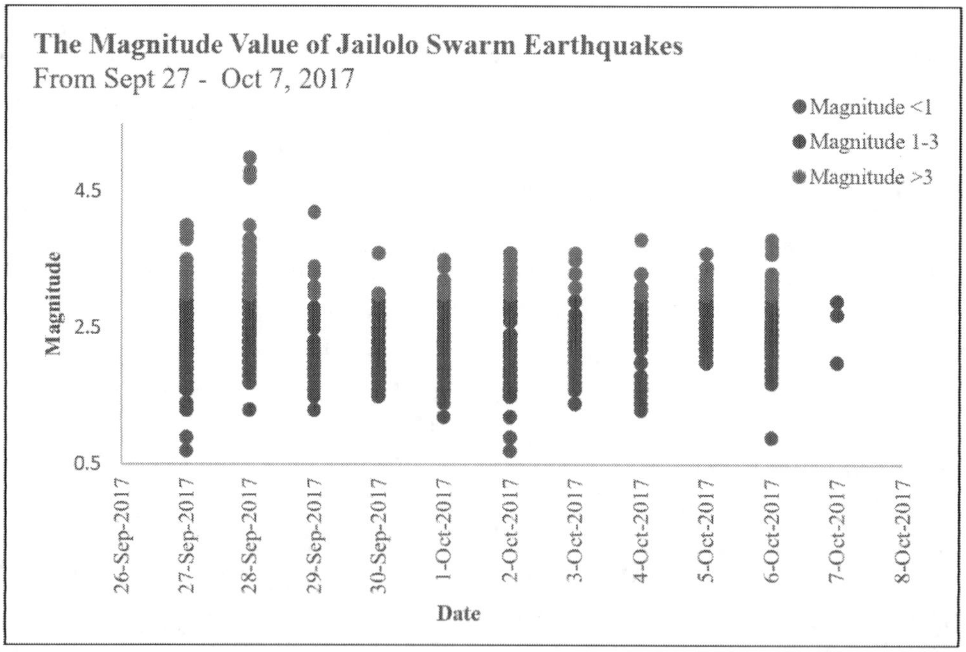

Figure 3 Jailolo earthquakes activity graph that shows variations in magnitude recorded on September 27 to October 7, 2017.

Determination of arrival time of P and S waves was done by picking the amplitude of waveforms using software datapro.exe, which is part of the TDS 5.2 [11]. After the picking process, we obtained the earthquake parameters: origin time, magnitude, hypocenter, and depth. Determination of b-values is based on the relationship between magnitude and earthquake frequency [12-14]. In its development, the determination of these parameters was improved by using the maximum likelihood method [13-18]. This method produces a new parameter called the magnitude of completeness (Mc). Mc is a function of time for all earthquake periods used in the calculation [13-16]. After Mc is determined, the b-value can be determined. Empirically, b-value provides information of seismotectonic implications. This value can be related to regional tectonic, structural heterogeneity in the crust and stress distribution. Thermal gradients also have a relationship that is directly proportional to the b-value. Low b-value indicates an increase of stress which describes a high thermal gradient from thermal rise above the magma chamber or geothermal effect that still exists after volcanic activity in the past[13,17,18]. Scholz [19] states that the b-value decreases with increasing stress level. Farrel et al. [7] states that high b-value found in volcanic areas, the presence of dikes, and and high thermal gradients. In this research, the determination of b-value was done using ZMAP code [20].

3. Results and Discussion

Based on the picking result of P and S wave arrival time, the earthquakes hypocenter are located at $127^0 10'$-$127^0 40'$ E and 1^0-$1^0 10'$ N with the smallest magnitude is 0.7 and the largest magnitude is 5. We use AB and CD cross-sections on the hypocenter to see the earthquake depth distribution, which is illustrated by figure 4.

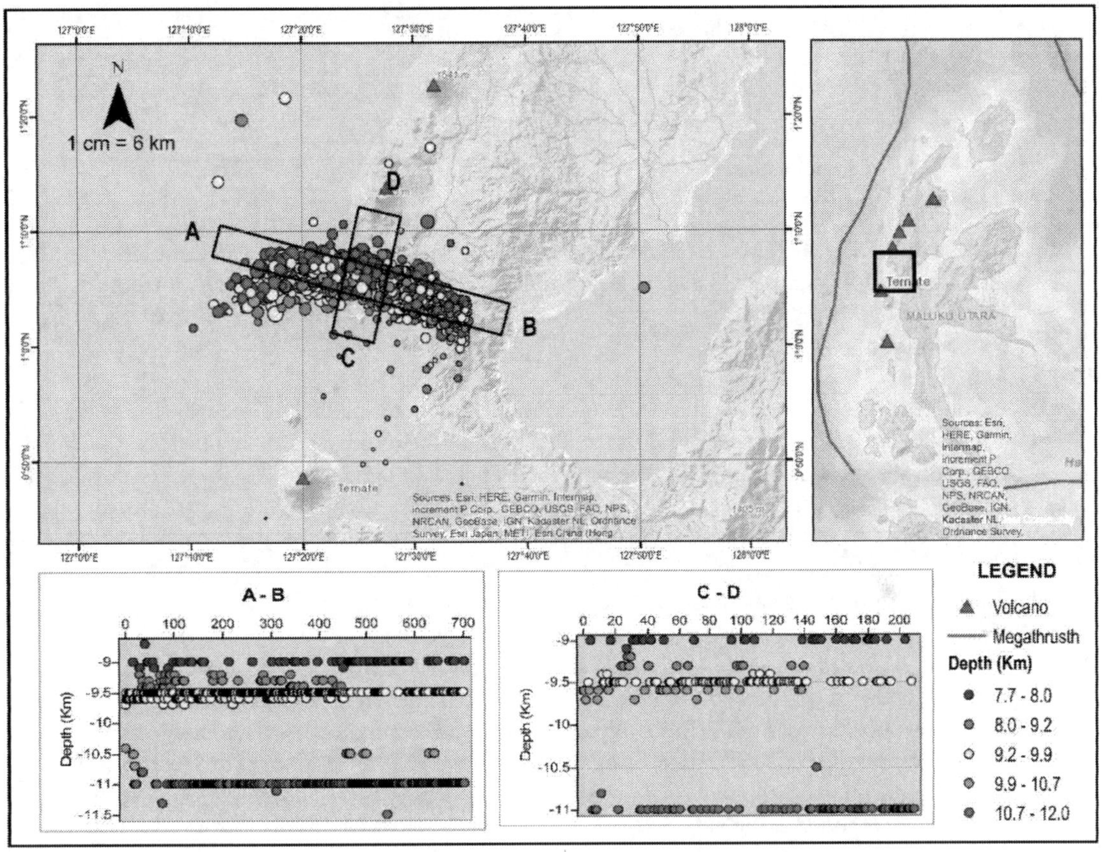

Figure 4 Cross section A-B and C-D to see the earthquake depth distribution. The dots are hypocenter. Varying sizes of the hypocenters represent differences in magnitude (between 0.7-5).

Both results of the cross section show the distribution of earthquakes depth have variation between 7.7-12 km which are included in the shallow earthquake category. Based on the distribution, the earthquakes hypocenter are located around Jailolo volcano. According to [7], earthquake swarms are common in volcanic areas. In this study, there were a fixed depth in the cross-section result, so that the analysis of the causes might be better if the hypocenter is relocated. However, research by Nugraha et al. [2] has conducted the relocation of the hypocenter, but this study did not do so because the data used in this study only consisted of waveforms recorded at one station. A map of the b-value distribution is shown in the figure below to study the characteristics.

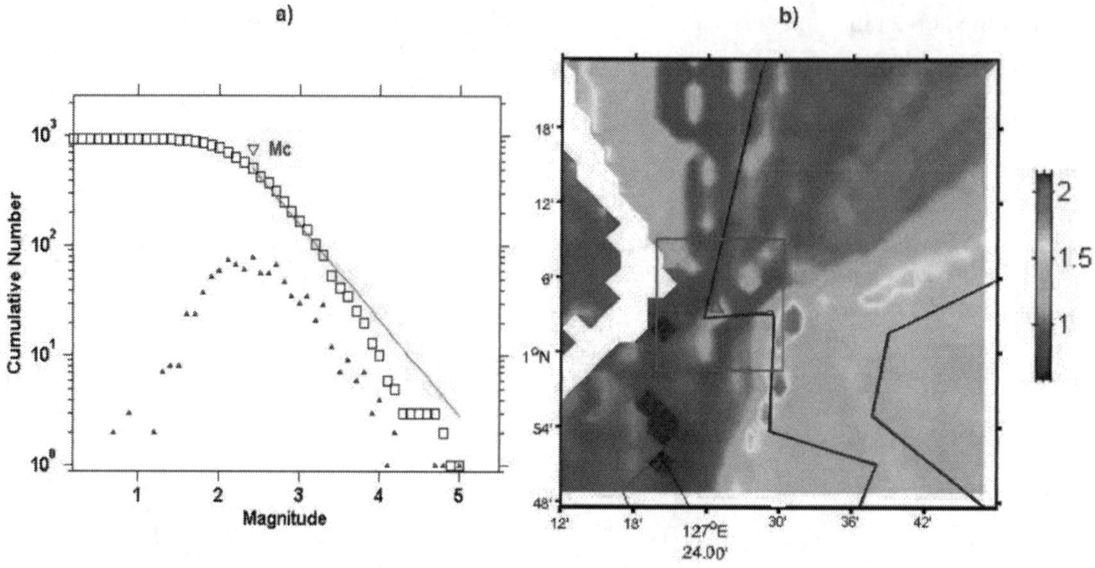

Figure 5 Graph (a) shows the magnitude of completeness (Mc) while (b) shows the b-value distribution in Halmahera. The black line in figure (b) is the boundary of Halmahera Island, the red triangle shows the Jailolo volcano and the red square shows the region of earthquake concentration.

Based on the result of calculation using maximum likelihood method (figure 5), the value of Mc is 2.4 with a b-value of approximately 1.0 in the area near Jailolo Volcano. In the northwest of Jailolo, the b-value ranges between 1.0-1.5 and in the southeast of Jailolo, between 1.0-2.0. According to Farrel et al. [7], a volcanic area has typically high b-value (between the range of 1.0 to 3.0), this b-value has a depth of 7-10 km where earthquakes are generated due to magma activity, whereas Frolic and Davis [21] state that in a tectonic region, the b-value is around 1.0, and Ruhl et al. [22] state that b-values in volcanic and earthquake swarm regions are often much larger than 1.

Research by Yanuarsih et al. [23] states that the activity of the Jailolo swarm earthquake in November to December 2015 was caused by magmatic activity, this is because field observations show that there are several fractures in areas with high b-values, whereas research by Passarelli et al. [1] states that the 2015-2016 seismic swarm at the Jailolo volcano most likely originated from magmatic activity which is affected by the presence of dike. Nugraha et al. [2] interpret that the September to October 2017 swarm earthquake in Jailolo was probably related to changes in stress due to a combination of tectonic activity and deep magma activity in the area.

Based on the b-value we obtained, the characteristics of the Jailolo swarm earthquake tend to be influenced by magmatic activity, because the b-value tends to be high, which is more than 1.0. However, further research should use the relocated hypocenter data to obtain a more accurate b-value and add further deformation analysis on the crust such as analyzing the possibility of fracture caused by dike intrusion, or analyzing the possibility of magma-filled cracks push to the surface (migration of magmatic fluids).

4. Conclusions

Based on this research, the parameters of the earthquake swarms in Jailolo from September 27 to October 7, 2017, have a magnitude variation ranged from 0.7 to 5.0 with a depth of 7.7-12 km. The b-value results were obtained approximately 1 in the area near Jailolo volcano, 1.0-1.5 in the northwest Jailolo and 1.0-2.0 in the southeast Jailolo. Based on our result, the characteristics of the Jailolo swarm earthquake tend to be influenced by magmatic activity. Further analysis of deformation in the crust is needed, such as analyzing the possibility of fracture caused by dike intrusion, or analyzing the possibility of magma-filled cracks push to the surface.

Acknowledgment
We would like to thank the Indonesian Meteorology, Climatology and Geophysics Agency (BMKG Region IV, Makassar) for providing the Jailolo earthquake swarms data that was used in this research. Department of Geophysics Hasanuddin University which has provided support to the authors is also gratefully acknowledged. We highly appreciate all parties who have helped and provided criticisms and suggestions to the authors in this research.

References

[1] Passarelli L, Heryandoko N, Cesca S, Rivalta E, Rasmid, Rohadi S, Dahm T and Milkereit C 2018 Magmatic or Not Magmatic? The 2015–2016 Seismic Swarm at the Long-Dormant Jailolo Volcano, West Halmahera, Indonesia *Frontiers in Earth Science* **6**

[2] Nugraha A D, Supendi P, Widiyantoro S, Daryono, and Wiyono S 2018 Hypocenter Relocation of Earthquake Swarm around Jailolo Volcano, North Molucca, Indonesia using the BMKG Network Data: Time Periods of September 27- October 10, 2017 *AIP Conference Proceedings* **1987**

[3] Masinu A L, Yustesia A and Suwardi 2018 Sistem Tektonik dan Implikasinya Terhadap Gempa Bumi di Pulau Halmahera *Jurnal Pendidikan Geografi* **23** 20-29

[4] PuSGen 2017 *Peta Sumber dan Bahaya Gempa Indonesia Tahun 2017*

[5] McCaffrey R and Silver E A 1980 Crustal Structure of the Molluca Sea Collision Zone, Indonesia in Hayes (eds). The Tectonic and Geologic Evolution of South-East Asian seas and Islands *American Geophysical Union* **23** 161-177

[6] Mogi K 1963 The fracture of a semi-infinite body caused by a inner stress origin and its relation to the earthquake phenomena (second paper) *Bulletin of the Earthquake Research Institute* **41** 615-658

[7] Farrell J, Husen S and Smith R B 2009 Earthquake swarm and b-value characterization of the Yellowstone volcano-tectonic system *Jurnal of Volcanologi and Geothermal Research* **188** 260-276.

[8] Supendi P, Nugraha A D, Widiyantoro S, Abdullah C I, Puspito N T, Palgunadi K H, Daryono D and Wiyono S H 2019 Hypocenter relocation of the aftershocks of the Mw7.5 Palu earthquake September 28, 2018 and earthquake swarms of Mamasa, Sulawesi, Indonesia, using the BMKG Network Data *Geoscience Letters* **6** 1–11

[9] Hall R 2000 Neogene history of collision in the Halmahera region, Indonesia *Proceedings of the Indonesian Petroleum Association*, 487-493.

[10] Carlile J C, Davey G R, Kadir I, Langmead R P, and Rafferty W J 1998 Discovery and exploration of the Gosowong epithermal gold deposit, Halmahera, Indonesia *Journal of Geochemical Exploration* **60** 207–227

[11] Dibyo S 2012 Pengukuran Parameter Gempabumi di Daerah Yogyakarta dengan Menggunakan Seismograph TDS-303S *Thesis (Universitas Indonesia)*

[12] Gutenberg B, Richter C F 1949 *Seismicity of the Earth and Associated Phenomena*, Princeton University Press, Princeton, New Jersey.

[13] Massinai M A, Harimei B, Agustiawati A, Massinai M F I 2019 Seismicity analysis Sulawesi North Arm based on B-Values *Journal of Physics: Conference Series* **1341** 082032

[14] Putri A C, Harimei B, Massinai M F I 2019 Analysis of Seismicity Level in Central Sulawesi Period 1997 - 2017 *IOP Conference Series: Earth and Environmental Science* **279** 012022

[15] Utsu T 1965 A Method for Determining the Value of b in a Formula Log N =a-bM Showing the Magnitude-Frequency Relation for Earthquakes *Geophys. Bull Hokkaido University* **13**, 99-103.

[16] Kalafat D and Görgün E 2019 Source characteristics and b-values of the Tuz Gölü Fault Zone in Central Anatolia, Turkey *Journal of Asian Earth Sci* **179** 337–349

[17] Batte A G and Rümpker G 2019 Spatial mapping of b-value heterogenety beneath the Rwenzori region, Albertine rift: Evidence of magmatic intrusions *J. Volcanol Geoth. Res.* **381** 238 – 245

[18] Gui Z, Bai Y, Wang Z, and Li T 2019 Seismic b-value anomalies in the Sumatra region: Seismotectonic implications *Journal of Asian Earth Sciences* **173** 29-41

IOP Conf. Series: Earth and Environmental Science **851** (2021) 012004 doi:10.1088/1755-1315/851/1/012004

[19] Scholz C H 2002 *The Mechanics of Earthquakes and Faulting, 2nd ed*, New York, Cambridge University Press.

[20] Wiemer S 2001 A Software Package to Analyze Seismicity: ZMAP *Seismological Research Letters* **72** 374-383.

[21] Frolich C and Davis S 1993 Teleseismic b-values: Or, much ado about 1.0 *Journal of Geophysical Research* **98** 631–644.

[22] Ruhl C, Bilek S L, and Stankova-Pursley J 2010 Relocation and characterization of the August 2009 microearthquake swarm above the Socorro magma body in the central Rio Grande Rift *Geophysical Research Letters* **37**

[23] Yanuarsih T P, Nova H, Wandono, Rivai M, Suwardi, Zulfikar, Barkah Y, Lutfi P, and Basri K 2016 Relokasi dan Distribusi b-value Gempabumi Swarm Jailolo, Halmahera Barat *Jurnal Meteorologi dan Geofisika* **17** 199-206

IOP Conf. Series: Earth and Environmental Science **851** (2021) 012005 doi:10.1088/1755-1315/851/1/012005

Active tectonics of the Yogyakarta area (Central Java, Indonesia): preliminary findings obtained from a tectonic-geomorphic evaluation

S Pena-Castellnou[1*], V Steinritz[1], G I Marliyani[2] and K Reicherter[1]

[1]Neotectonics and Natural Hazards, RWTH Aachen University, Aachen, Germany
[2]Geological Engineering Department, Universitas Gadjah Mada, Yogyakarta, Indonesia
*E-mail: s.pena-castellnou@nug.rwth-aachen.de

Abstract. Throughout the island of Java, several shallow faults located in the vicinity of densely populated areas accommodate the tectonic strain generated by the subduction of the Indo-Australian Plate beneath the Sunda Plate. These faults are characterized by a poor surface geomorphological expression due to the tropical climate that masks paleoseismological evidence making the identification of active faults in this environment challenging. We present preliminary data of a tectonic-geomorphic study that aims to identify and characterize active faults in the province of Yogyakarta (Central Java) as a basis to improve seismic hazard assessment. Here we focus on the Opak fault by describing its geomorphology and structure, using field data and remote sensing-based observations to contribute to the understanding of its geometry, kinematics, and tectonic activity. We show preliminary data supporting the Opak Fault as an active SW-NE transtensional left-lateral strike-slip reactivated normal fault consisting of several parallel fault strands. Although geomorphological expression of active tectonics is poor and diffuse, we found numerous field evidence of active tectonics ranging from tilted Quaternary fluvial terrace risers, triangular facets, and linear valleys to peculiar drainage patterns that allow us to provide evidence of the Holocene activity of the Opak Fault.

1. Introduction

Identification of active tectonics geomorphological features represents a challenge in slowly broad deforming areas with tropical climates like Java. Heavy rainfall and high humidity foster intense weathering, the development of thick tropical soils, high erosion rates, shallow water tables, and thick vegetation covers (including anthropogenic fertile rice fields) that erode and bury faults and their associative geomorphology. Hence, tectonic geomorphology in tropical areas differs from the conceptual models that we are used to, typical from drier climatic contexts where geomorphic markers are well exposed and preserved. Moreover, the tectonic strain on-shore of Java is accommodated by slow slip rate faults with long earthquake recurrence intervals [1], resulting in inherent poor surface expression.

Java forms part of the volcanic arc in the Sunda Plate adjacent to the Sunda trench, where the Indo-Australian Plate subdues at a rate of 7 cm/yr [2] (Figure 1a). Subduction controls Java's current stress field, leading to a north-south maximum stress direction, which has generated E-W trending thrust belts and re-activated SW-NE structures as slow strike-slip faults [3] that accommodate the tectonic strain generated by the deeper subduction mechanism within the volcanic arc. Inherited N-S structural features in the west and central Java were considered inactive because of a lack of recorded seismicity [4]. However, events such as the well-known M_w 6.3 shallow crustal earthquake in May 2006 in the province of Yogyakarta, which caused approximately 6000 fatalities, and up to 3 billion dollars of economic losses (e.g. [5]) are evidence of the activity of these faults.

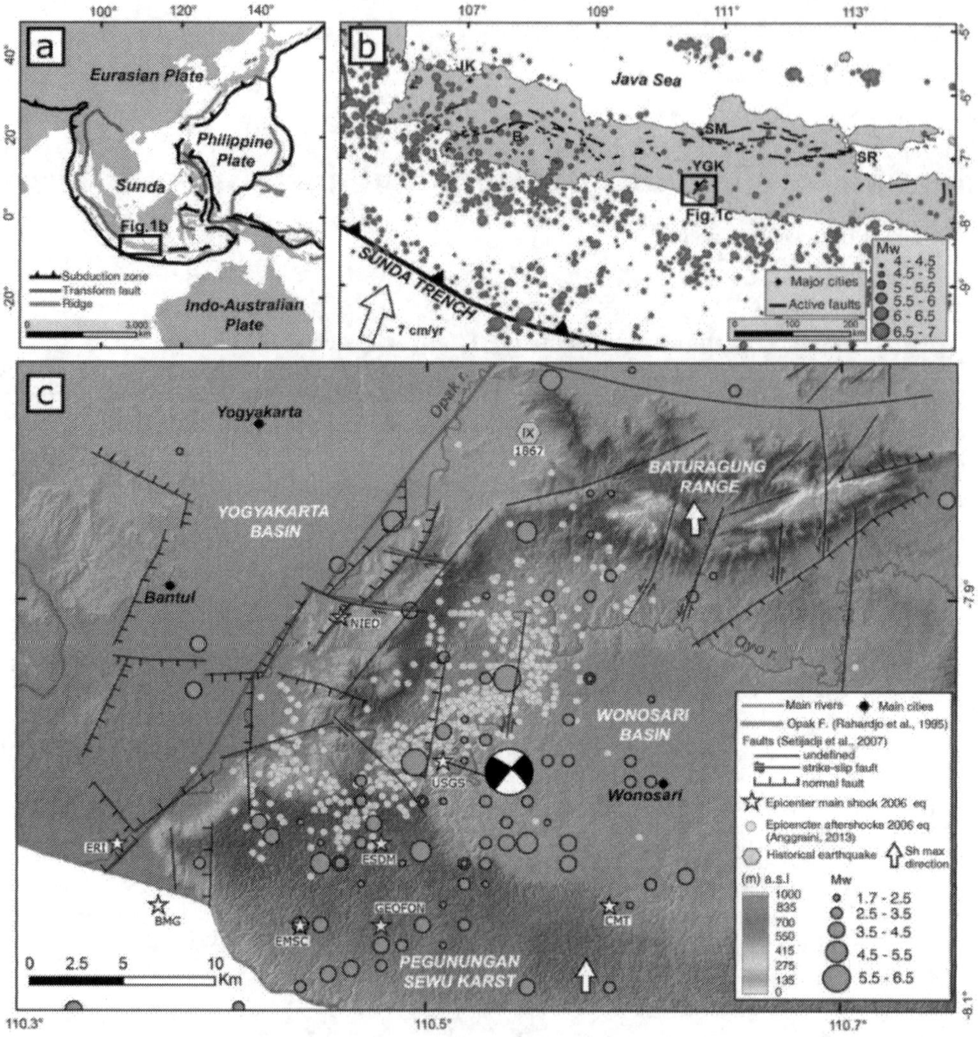

Figure 1. a) Simplified tectonic setting of the Indonesian archipelago. b) Instrumental seismicity of the Java Island of the period 2010-2020 (USGS seismic catalog, https://earthquake.usgs.gov). The arrow and number show the convergence velocity of the Indo-Australian plate relative to the Eurasian plate fixed frame [2]. Black lines [4] indicate active faults. Abbreviations for major cities: JK: Jakarta, B: Bandung, YGK: Yogyakarta, SR: Surabaya, SM: Semarang. c) Overview of the study area, the province of Yogyakarta. Instrumental seismicity data from the online USGS seismic catalog (https://earthquake.usgs.gov) from 2006 to 2010, and from the online BMGK earthquake repository (https://repogempa.bmkg.go.id/) from 2010 to 2020, historical earthquake from [6]. Focal mechanism of the 2006 earthquake from Global CMT Harvard Catalog (https://www.globalcmt.org). In red, the Opak Fault from the regional geological map [7] and in black, faults from Setijadji et al. [8]. The epicenters of the aftershocks that followed the 2006 earthquake are from Anggraini [9]. The maximum horizontal stress direction, derived from earthquake focal mechanisms, from the World Stress Map [10].

The province of Yogyakarta is, with more than four million inhabitants, Indonesia's fourth-most densely populated province and one of the fastest-growing provinces in terms of social and economic aspects [11]. It is exposed to several natural hazards, especially to volcanism and seismicity. The province's northern area is situated at Mount Merapi's foot, one of Java's most active volcanoes. To the east, the boundary between the Southern Mountains and the Yogyakarta Basin (Figure 1c) is represented by the Opak Fault System. This area is characterized by shallow earthquakes of moderate to low magnitude, scant historical records, and absent paleoseismic data. The Opak Fault was thought to be the source of the devastating 2006 earthquake (e.g. [8]; [12]; [13]). However, the causative fault did not rupture the surface, and the major disaster area with a localized concentration of secondary effects (e.g., liquefaction, slope movements, and ground fractures) did not concur with the epicenter locations proposed by different seismological agencies and the aftershock sequence distribution (Figure 1), where no previous fault had been identified (e.g. [8]).

We aim to improve and complete the current tectonic map of the province of Yogyakarta and identify active faults that have the potential to generate seismic events with social impact in the future, as the first step to improve the assessment of seismic hazard in the area. In the present day, the location, geometry, kinematics, and total length of the Opak fault are not well understood; and off-shore continuation remains unclear. Furthermore, in the area with the 2006 aftershock sequence distribution, no previous fault had been identified. We use field-based geomorphological and structural data together with GIS analysis of digital terrain models to recognize and examine geomorphological markers associated with the superficial expression of fault activity to characterize the mapped faults, providing insight into the challenges of the study of active faults in tropical environments.

2. Background - The Opak Fault

The Opak Fault is an SW-NE trending hypothesized normal fault located at the boundary between the Yogyakarta Basin and the Baturagung Range (part of the Southern Mountains) ([7]; [8]) (Figure 1), suspected to be responsible for the formation of a half-graben, the Yogyakarta Basin, where the city of Yogyakarta is based. Geologically, the Baturagung Range consists of volcaniclastic Oligocene and Miocene rocks, limited to the west by the Wonosari Basin, a limestone plateau from the Miocene, bedded clastic to the north, and massive reefal to the south constituting the Pegunungan Sewu Karst. To the east, the Yogyakarta basin is filled with Quaternary sediments of the Merapi volcano, and fluvial and alluvial Quaternary sediments. The seismicity of the area is moderate to low. However, two damaging earthquakes were recorded: the historical MMI IX 1867 earthquake, which resulted in 327 reported casualties and significant damages over the entire areas of Yogyakarta, Klaten, and Surakarta [6]; and the Mw 6.3 2006 earthquake that left more than 6000 casualties, and enormous damages throughout all the Yogyakarta Basin (e.g. [5]). The Opak Fault is considered to be the source of the 1867 earthquake.

The first mapping of the Opak Fault and other structures in its vicinity was carried out by Rahardjo et al. [7] and Surono et al. [14] and presented in the Regional Geology map issued by the Indonesian Center for Geological Research and Development, at a scale of 1: 100,000. The fault is interpreted to be normal and located following a NE-SW direction by the Opak River (hence its name) three kilometers to the west of the Baturagung Range foot (Figure 1c). After the 2006 Yogyakarta earthquake, the Opak Fault became the focus of several research efforts, with the aim of understanding this event and especially determining its source, investigating: hypocenter and aftershocks distribution (e.g. [9]; [13]; [15]), InSAR data to reveal surface deformation [16], GPS data of post-seismic deformation [17], macroseismic data [12], tectonic geomorphology [8], secondary effects such as landslides (e.g. [18]), local site conditions (e.g. [19]), and arqueoseismology [20]. The outcome led to the hypothesis that a fault located approximately ten kilometers to the east of Opak, in the Wonosari Basin where no previous fault was mapped, was likely to be the source of the earthquake.

The most relevant study in terms of characterizing active faults in Yogyakarta is the work of Setijadji et al. [8] that locates the trace of the Opak Fault at the foot of the Baturagung Range as a segmented fault zone consisting of several faults with different styles and orientations. More recent studies have led by Supartoyo et al. [21], presenting a mapping of lineaments and interpretation of ERT profiles across the Opak Fault, where they show that the fault dips to the north-west without providing more details about kinematics or level of activity.

3. Methods

Large and great (Mw > 6.5) recurrent earthquakes produce characteristic geological evidence along faults (e.g. [22]; [23]). Evidence of paleoearthquakes can range from localized deformation along a crustal fault (e.g. fault scarp, laterally offset stream valleys) to indicators of the sudden uplift or subsidence of large regions above a plate boundary fault (e.g. uplifted or subsided shorelines), to stratigraphic and geomorphic effects of strong ground shaking, the so-called environmental earthquake effects, far from the seismogenic fault (e.g. landslides, liquefaction features, tsunami deposits) [22]. We perform detailed geomorphological and structural mapping of Quaternary deposits and different-scale paleoseismic features, combining field data with remote sensing observations based on the 8 m resolution DEMNAS DEM [24] and ~10 cm resolution DEM from unmanned aerial vehicle photogrammetry to identify and characterize active faults in the province of Yogyakarta. The geological mapping was conducted throughout the area comprising the Baturagung Range to collect structural data (bedding dip direction, micro-faults, and fracture patterns) and identify different lithological formations from more than 250 outcrops. At the same time, we corroborated and carefully mapped the geomorphological features previously recognized with remote sensing. Additional imagery was obtained from UAV surveys with the DJI Mavic Pro (12 MP sensor) and high-resolution DEMs that were created with the photogrammetry software AgiSoft PhotoScan (AgiSoft PhotoScan Professional, Version 1.2.6, 2016). We took several samples of charcoal and shells to date different levels of terraces exposed on the shores of Opak and Oyo rivers at one key study site. The dating was based on the radiocarbon method (C_{14}). A total of nine samples were pre-processed in the laboratory at RWTH Aachen University and then sent to analysis at the University of California Irvine, Keck Carbon Cycle AMS Program.

4. Results - Geomorphological signature of ongoing tectonic activity

The province of Yogyakarta has distinct geomorphological landscapes, shaped by the interaction of sedimentation and erosion, volcanic activity, tectonics and, climate. On the center, the Yogyakarta Basin is characterized by gentle and uniform slopes where the Opak River develops in an approximate SW-NE mean direction. To the east rises the Baturagung Range, part of the Southern Mountains, with a linear hillslope of SW-NE direction that bends at its northern part to W-E. South, the Wonosari Basin with a circular shape is characterized generally by a low topography where the Oyo River incises extensively and junctions the Opak River crossing through the Baturagung Range. By the coastline, the reefal plateau forms the Pegunungan Sewu Karst with distinguishing geomorphology of dome-shaped hills result of the intense karstification. Within this landscape, the geomorphological expression of tectonic activity is weak and may have been masked by the aforementioned processes. However, we have identified several faults (Figure 2) based on various paleoseismic features, typical from the classic landform landscape of active normal and strike-slip faults (e.g. [22]; [23]).

The most prominent feature is the SW-NE well-defined western mountain front of the Baturagung Range, with steep slopes (60-75°) that contrast with the gentle eastern front (10-25°). There are secondary parallel lower ridges on the central part of the western front that also form SW-NE lineaments. Other lineaments have been mapped throughout the area corresponding to wind gaps and, linear ridges and valleys (Figure 2). The mapped lineaments exhibit three trends: a main SW-NE orientation, W-E, and a minor NNE-SSW; and vary in length from a hundred meters to dozen of kilometers. Besides, despite the dense tropical vegetation, triangular facets have been inferred (Figure 3c).

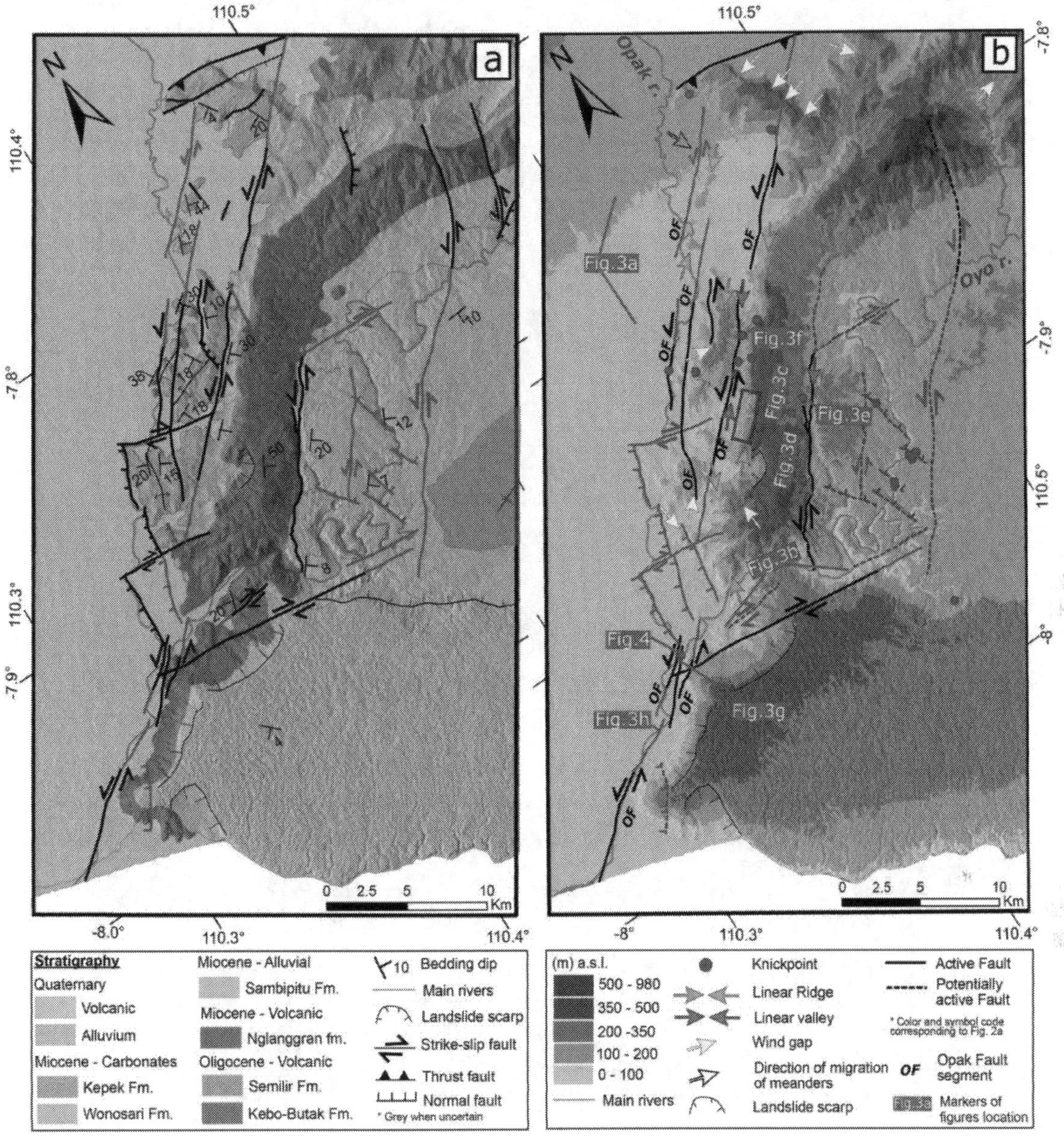

Figure 2. a) Geology of the SW-NE striking front of the Baturagung Range, including structural and stratigraphic data from field observations. Geological units are from Rahardjo et al. [7]. b) Active faults and identified paleoseismic features. Note the drainage patterns of the Oyo and Opak Rivers.

On the Opak River, the river sinuosity is low along its entire course and changes its direction on several locations where tectonic lineations have been mapped (Figure 2). Moreover, we have identified tilt-induced avulsion and migration of meanders to the east, in direction to the major fault trace (Figure 2) and slightly tilted quaternary terraces on several locations (e.g., Figure 3g-h). The Oyo River presents a peculiar drainage pattern often changing its sinuosity, from meandering to straight and back, and its course direction from north-south to east-west following lineations, at various locations along a restricted area (Figure 2); despite the fact that the lithology of calcareous sandstones and limestones does not change along the river's course and there is no dip variation of the layers.

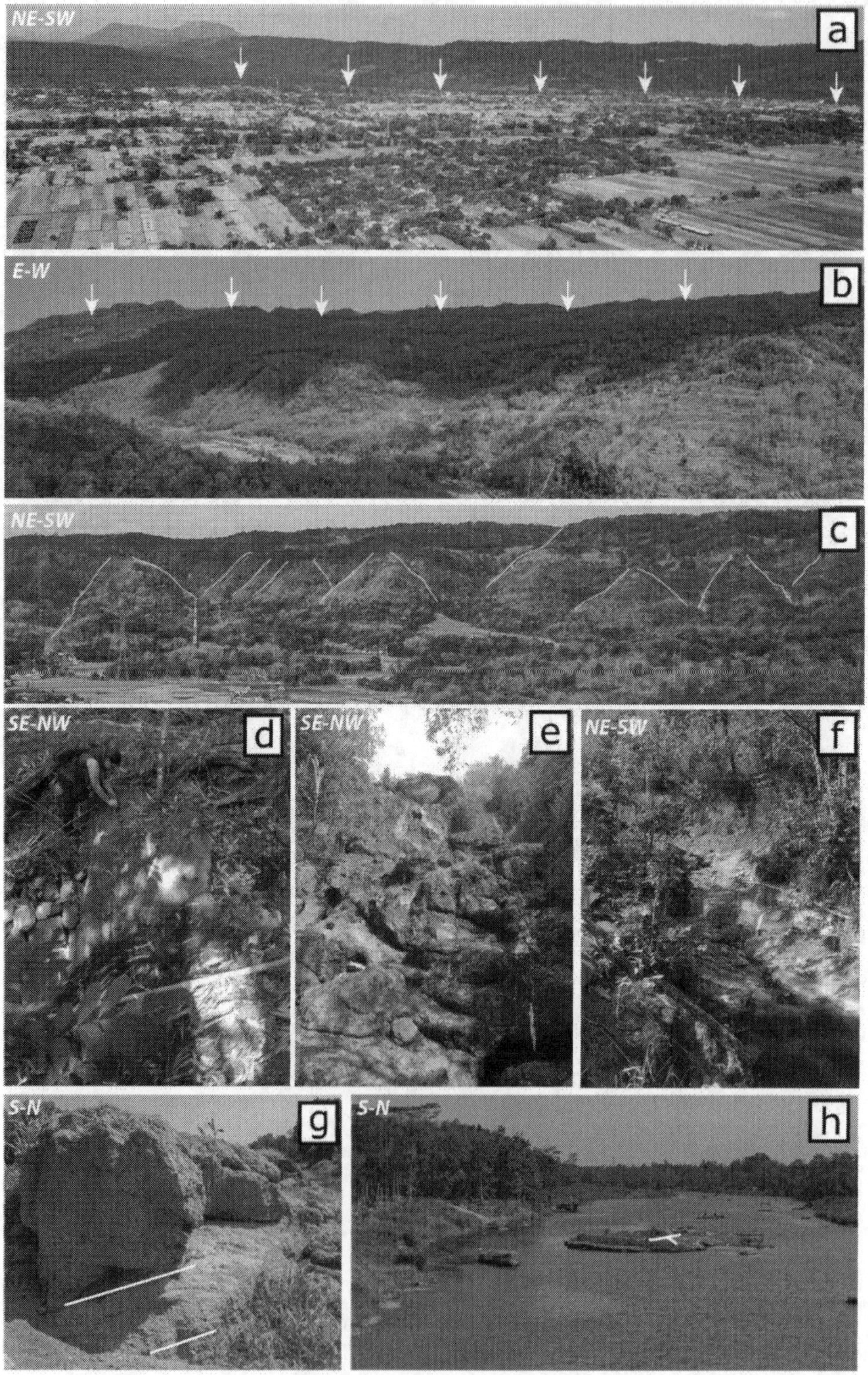

Figure 3. Examples of active tectonic features. a) Geomorphological expression of the Opak Fault; northwestern mountain front of the Baturagung Range. b) E-W fault in the Sewu Karst. c) Triangular facets on the northwestern hillslope of the Baturagung Range. d), e), f) Knickpoints. g), f) Tilted Quaternary terrace of Opak River. For locations see figure 2.

Throughout the streams that form the drainage system, we have identified several knickpoints of different magnitudes ranging from 0.5 meters to 30 meters (e.g. waterfalls), which we interpret to be caused by tectonic. Where the Oyo River meets the Opak River, a clear knickpoint can be identified on Oyo's channel (Figure 4a-b). In this area, we have mapped, described, and dated several degradational fluvial terrace levels that give information on tectonic deformation. On the Opak River's right shore, the terraces appear uplifted with respect to the left shore by about two meters, indicating the existence of a fault by the river. Furthermore, we observe local deformation, such as a folded terrace (of age 6260 ± 30 years BP) that sediments on top of bedrock while tectonic activity was ongoing (Figure 4c).

Figure 4. Landscape view of the Opak-Oyo river junction and inferred fault trace. Shaded colours represent rock units of various ages, the relative ages were determined based on rock characteristic and stratigraphic relationship. Shaded in green: younger Quaternary terrace; in red: older Quaternary terrace; in purple: Miocene bedrock. b) One of observable knickpoint along the Opak River. c) Deformed Quaternary terrace accommodating across the fault.

5. Discussion

Geomorphological observations have helped to identify and map several fault traces in the study area. However, the characterization of these faults in terms of geometry, kinematics, and activity is challenging. Short-term (Quaternary to Holocene) tectonic evidence in the landforms is diffuse or lacking, partly because Quaternary deposits are scarce and intensely modified by anthropogenic activity (mainly agriculture – large extension of fertile rice paddies), resulting in poor preservation of recent faulting activity.

We interpret the Opak Fault as a normal fault composed of several fault strands dipping to the northwest at about 70 to 80°, reactivated during the Quaternary as transtensional strike-slip faults. The Opak Fault is located at the foot of the Baturagung Range and appears to be segmented in different SW-NE segments of different lengths (~1-10 km) that spread within a width of three kilometers. Its geometry has been inferred from the north-western hillslope of the Baturagung Range, which has slopes of about 60-75° and where triangular facets have been identified. Considering the dissection and degradation of the landscape, we interpret that the fault may dip at about 70 to 80° towards the northwest. The same applies to the fault strands located further from the mountain front, where the Opak River incises following the fault traces; the north-western hillslopes of the smaller isolated hills located north of the study area (Figure 2) have the same slope gradient as the main front of the Baturagung Range.

We have mapped volcanic breccia from the Nglanggran formation along the parallel hills on the central part of the Baturagung Range close to the Opak River (Figure 2a) that was not previously documented on the regional geological map [7]. These outcrops are located approximately 300 meters below the Baturagung Range hilltop, where the Nglanggran formation outcrops and extends along the hillslope. This has direct implications on the interpretation of the kinematics history of the Opak Fault, suggesting normal tertiary faulting. The triangular facets on the mountain front of Baturagung Range are also evidence of normal faulting. Besides, we have identified SW-NE linear valleys, more typical of strike-slip landscapes. We have not found any other evidence of strike-slip tectonics, such as laterally displaced streams. However, we believe that the Opak fault has reactivated during the Quaternary as a left-lateral strike-slip fault with a significant normal vertical component in agreement with the present maximum stress direction.

Regarding tectonic activity, we can affirm that several fault segments of the Opak Fault have been active during the Quaternary, evidenced by: 1) tilted Quaternary terraces (Figure 3g-h and 4); 2) the peculiar drainage pattern of Opak River that is very linear and changes its course direction following major lineations - its drainage evolution has been clearly driven and controlled by tectonics; 3) tilt-induced avulsion and migration of meanders in Opak River; and 4) knickpoints – abrupt changes in channel gradient; that are transient as the channel adjusts to its equilibrium profile within time periods of 10^4 to 10^6 years depending on the erosional efficiency of the system and the nature of the perturbation ([25]; [26]). We could not find clear displaced tectonic markers to infer slip rates. However, we believe that the fault slip is slow and has long recurrence intervals since the present-day low seismicity does not agree with the tectonic geomorphological imprint in the landscape. Considering the length of the segments of the Opak Fault and using the empirical equations of Wells and Coppersmith [27], we can expect earthquakes of M_w 6-7.

Concerning the other faults located west of the Baturagung Range, two minor structural trends stand out. North of the Opak and Oyo rivers' junction, lineations with N-S to NNE-SSW directions are interpreted as normal faults that form topographic steps. We have not found any marker to assess its level of activity. Other faults follow an E-W trend (e.g. Figure 3b) and seem to block the Opak River that has to find its way drifting its course towards the west; therefore, we believe that these faults might have been active during the Quaternary.

To the east of the Baturagung Range, a well-defined SW-NE fault delineates the boundary between the Baturagung Range and the Wonosari Basin. Paleoseismic features related to this fault are knickpoints and linear valleys. Structurally, it separates two blocks with different bedding dips suggesting normal faulting during the Tertiary. Activity during the Quaternary cannot be discarded due to the presence of knickpoints and significant topographic morphological expression. Parallel to this fault and three kilometers to the east, along the area where the aftershocks sequence of the 2006 Yogyakarta earthquake concentrated (Figure 1), a notable change in topography may indicate the presence of a fault. On the center of the Wonosari basin, an SW-NE fault that may play a role in the Oyo River's evolution is insinuated by a very subtle lineation. In addition, we identified several NNW-SSW normal faults that control the drainage evolution of the Oyo River, forcing the change of its course direction on several locations.

6. Conclusion

The Opak Fault is an active SW-NE transtensional left-lateral strike-slip reactivated normal fault consisting of several parallel fault strands. Detailed geomorphological and structural mapping of Quaternary deposits and paleoseismic features, combining field data with remote sensing observations, allowed to the preliminary characterization (in terms of geometry, kinematics, and relative activity) of the Opak Fault and other faults in the province of Yogyakarta. Quaternary geomorphological expression of tectonic activity is diffuse as a result of the combination of several factors: Quaternary deposits are scarce and intensely modified by anthropogenic activity; the tropical environment hastens landscape degradation (due to, e.g. intense weathering, high erosion, and sedimentation rates, dense vegetation cover); faults have slow slip rates [28] and long earthquake recurrence intervals (although the period covering the historical record is short, the only documented event is the MMI IX 1867 earthquake) resulting in inherent poor geomorphological manifestation and poor preservation of faulting markers. Despite these factors, we were able to identify several paleoseismic features such as lineaments, tilted Quaternary fluvial terrace risers, triangular facets, linear valleys and ridges, migration of meanders, wind gaps, peculiar drainage patterns, and knickpoints to prove the existence of active faults within the area of Yogyakarta.

Additional studies are needed to provide a better understanding of the hazard associated with these faults. Geophysical studies (e.g. electrical resistivity tomography profiles) are essential to corroborate the locations of the different segments of the faults, as well as to select suitable locations for paleoseismological trenching that will contribute to extending the seismic record and to the determination of slip rates, maximum expected earthquake magnitude and recurrence intervals. We aim to highlight the importance of improving seismic hazard assessment with regional studies characterizing active faults, especially in densely populated areas like Yogyakarta and where faults can be underestimated because of poor geomorphological expression and low seismicity due to long earthquake recurrence intervals.

Acknowledgments

We appreciate and thank the constructive comments from the reviewer. We want to thank our colleagues from the University Gadjah Mada, who helped organize and carry out the field campaign. The authors also would like to thank the Keck Carbon Cycle AMS laboratory for the radiocarbon analysis. This work is part of the project GetIn-CICERO (German Indonesian Geo-Campus in Indonesia for Competence in Education and Research for Organizations) within the cluster of Geohazards and Coastal Risk, funded by the Federal Ministry of Education and Research of Germany (FKZ: 01DP17033), which aims to build up German-Indonesian cooperation for research, industry, and politics.

References

[1] Marliyani G I, Arrowsmith J R, Whipple K X 2016 Characterization of slow slip rate faults in humid areas: Cimandiri fault zone, Indonesia, *J. Geophys. Res. Earth Surf.*, **121**

[2] Argus D F and Gordon R G 1996 Tests of the rigid-plate hypothesis and bounds on intraplate deformation using geodetic data from very long baseline interferometry, *J. Geophys. Res. Solid Earth*, **101**, B6, 13555-13572

[3] Pulunggono A and Martodjojo S 1994 Perubahan tektonik Paleogen – Neogen merupakan peristiwa terpenting di Jawa *Proceedings Geologi dan Geotektonik Pulau Jawa*: 37-50

[4] Marliyani G I 2016 Crustal Deformation in the Overriding Plate of an Orthogonal Subduction System, Thesis, Arizona State University

[5] The Yogyakarta Earthquake of May 27, 2006 edited by Karnawati et al, 2008 Star Publishing company

[6] Utsu T 2002 A list of deadly earthquakes in the World: 1500-2000 in *International Handbook of Earthquake and Engineering Seismology, Part A*, edited by Lee WK, et al., pp. 691-717, Academic Press, San Diego

[7] Rahardjo W and Sukandarrumidi RH, 1995, Peta Geologi Lembar Yogyakarta, Jawa (Yogyakarta Quadrangle Geological Map)

[8] Setijadji L D et al. 2007 Searching for the Active Fault of the Yogyakarta Earthquake of 2006 Using Data Integration on Aftershocks, Cenozoic GeoHistory, and Tectonic Geomorphology, in: Karnawati D, et al., May 27, 2006 The Yogyakarta Earthquake Of. Star Publishing Company, pp. 1–23

[9] Anggraini A 2013 The 26 May 2006 Yogyakarta Earthquake, Aftershocks and Interactions, Thesis, University of Potsdam, Germany.
[10] Heidbach O et al. 2018. The World Stress Map database release 2016: Crustal stress pattern across scales, Tectonophysics **744**, 484–498
[11] Badan Pusat Statistik, 2019 City Population: The population of the provinces, regencies and cities of Indonesia https://www.bps.go.id/ <Last viewed 27 January 2019>
[12] Supartoyo 2006 Gempabumi Yogyakarta tanggal 27 mei 2006 Buletin Berkala Merapi : Vol. 3, No. 2, Edisi Agustus 2006, ISSN 1693-9212, BPPTK PVMBG, Hal 36 - 55
[13] Fukuoka K et al. 2007 Interpretation of the 27 May 2006 Yogyakarta Earthquake Hypocenter and Subsurface Structure Deduced from the Aftershock and Gravity Data in: Karnawati D, et al., May 27, 2006 The Yogyakarta Earthquake Of. Star Publishing Company, pp. 1–23
[14] Surono T B et al. 1992 Geological Map of the Surakarta-Giritonrtro Quadrangle, Java. Geological Research and Development Centre, Bandung.
[15] Diambama A D et al. 2019 Velocity structure of the earthquake zone of the M6.3 Yogyakarta earthquake 2006 from a seismic tomography study, *Geophys. J. Int.* **216**, 439–452
[16] Tsuji T et al. 2009 Earthquake fault of the 26 May 2006 Yogyakarta earthquake observed by SAR interferometry, *Earth Planet Sp* **61**, e29–e32
[17] Abidin H Z et al. 2009 Deformasi Koseismik dan Pascaseismik Gempa Yogyakarta 2006 dari Hasil Survei GPS, *Jurnal Geology Indonesia*, Vol. **4** No. 4, 275-284
[18] Karnawati D and Fathani T F 2007 Mechanism of Earthquake Induced Landslides in Yogyakarta Province, Indonesia in: Karnawati D, et al., May 27, 2006 The Yogyakarta Earthquake Of. Star Publishing Company, pp. 1–23
[19] Walter T R et al. 2008 The 26 May 2006 magnitude 6.4 Yogyakarta earthquake south of Mt. Merapi volcano: Did lahar deposits amplify ground shaking and thus lead to the disaster?, *Geochemistry Geophysics Geosystems Research Letter* Vol. **9**, No. 5
[20] Miklòs and Agus 2014 Damages to the 9th century Prambanan temple caused by the 2006 Yogyakarta earthquake (Java, Indonesia, 5th international INQUA Meeting, Busan, Korea
[21] Supartoyo et al. 2016 Morphotectonic analysis to identify Opak Fault (Preliminary result) Proceedings Jogya Earthquake in Reflection (International Conference for Commemorating 10 years of Yogyakarta Earthquake), Published by UPN Veteran Yogyakarta, pp. 28-35
[22] McCalpin J P 2009 *Paleoseismology,* 2nd ed. Academic Press, Elsevier, Amsterdam (613 pp.)
[23] Burbank D W and Anderson R S 2001 *Tectonic Geomorphology*, Blackwell Science, Malden (USA) (274 pp.)
[24] Badan Informasi Geospasial 2014 DEMNAS. http://tides.big.go.id/DEMNAS/ <last viewed 13 December 2019>
[25] Whipple K X 2001 Fluvial landscape response time: How plausible is steady-state denudation?, American Journal of Science, **301**, 313-325
[26] Whipple KX, et al. 2013 Bedrock Rivers, in: Shroder JF, et al., *Treatise on Geomorphology*, Elsevier Inc. Vol. **9**, pp. 550-573
[27] Wells D L and Coppersmith K J 1994 New empirical relationships among magnitude, rupture length, rupture width, rupture area, and surface displacement, *Bulletin of the Seismological Society of America* **84**, 974–1002
[28] Koulali A et al. 2016 The kinematics of crustal deformation in Java from GPS observations: Implications for fault slip partitioning, *Earth Planet. Sci. Lett*. Vol. **458**, 69-79

IOP Conf. Series: Earth and Environmental Science **851** (2021) 012006 doi:10.1088/1755-1315/851/1/012006

Spatial modeling of tsunami for evacuation planning in Ciletuh Bay, Sukabumi, West Java

Dinda Z Arifah[1], **Supriatna**[1*] **and Ratna Saraswati**[1]

[1]Department of Geography, Faculty of Mathematics and Natural Sciences, Universitas Indonesia, Kampus UI Depok 16424, Indonesia

*E-mail: ysupri@sci.ui.ac.id, dinda.zannuba@ui.ac.id

Abstract. Indonesia is a country located at the meeting of three tectonic plates. This condition makes Indonesia a disaster-prone region, especially caused by the movement of tectonic plates, one of which is the tsunami. Several tsunami occurs were recorded in Indonesia. Tsunami occur on a large scale and causes losses, both environmental damage and fatalities. Therefore, evacuation planning is needed to minimize the number of fatalities when a tsunami occurs. This study aims to find out the potential threat of tsunami hazard in Ciletuh Bay and analyze evacuation planning in the region. The method used in this research is the spatial modeling method which is divided into tsunami inundation modeling, population distribution, and estimated evacuation travel time. The result of this study is that potential tsunami hazard in Ciwaru Village id found in two hamlets with an area of 104.63 Ha. A total of 590 exposed residents can evacuate within 0 – 15 minutes. The selection of evacuation routes with the fastest travel time is needed so that residents can reach the evacuation collection point before the tsunami arrives. The availability of road access in the hamlet makes it easier for residents in tsunami-exposed areas to reach evacuation gathering points.

1. Introduction

Indonesia is a country located at the meeting of three tectonic plates. The three plates are the Eurasian Plate, the Indo-Australian Plate, and the Pacific Plate. The Pacific Plate is located in the northern part of Maluku and Papua, while the Eurasian And Indo-Australian Plates collide in the regions off the coast of Java, Sumatra, and Nusa Tenggara. This condition makes Indonesia a disaster-prone region, especially caused by the movement of tectonic plates that can cause collisions between plates. Collisions between these plates can cause tectonic earthquakes. Earthquakes located on the seafloor trigger a tsunami if the earthquake has considerable strength.

Tsunami becomes one of the disasters with low frequency but great impact on exposed areas [1]. Tsunami events have occurred in various parts of the world for a long time but the impact of this disaster is still difficult to avoid. Therefore, research on tsunamis in various regions of the world is carried out to minimize the impact of the disaster. One of the tsunami-prone areas is the southern waters of Java Island. This is due to the condition of the southern waters of Java Island directly to the Indian Ocean [2]. One of the southern waters of Java Island is the coast in Sukabumi Regency. The potential for tsunami on the coast of Sukabumi is triggered by the active fault Cimandiri. Cimandiri active fault has a rising fault activity that has the potential to be the source route of the earthquake [3]. Several areas are included in the tsunami-prone area, one of which is the Ciletuh Bay. Beaches that do not have protection such as mangroves, coconuts, and other coastal forests have the potential for a larger tsunami due to no wave barriers coming [4].

The process of evacuating people from tsunami-exposed areas to safe evacuation sites is a top priority during tsunami warnings. Between earthquakes and tsunamis that occur there is usually a time lag that can be used to evacuate themselves. Education related to the routes and evacuation places outside the tsunami exposed areas to the community must be done properly to minimize the death toll in the event. The determination of evacuation routes and places can refer to several things such as the estimated time of the tsunami coming, the speed at which people save themselves, and the capacity of the evacuation site. To get planning evacuation routes, a spatial system can be used to support decision making. Geographic Information Systems (GIS) can help make decisions regarding some of the things that were injured during the evacuation process, such as the extent to which tsunami waves reached land, which routes to go, and evacuation places that could be intended to save themselves. This research aims to analyze the potential tsunami hazard threat that occurred in Ciletuh Bay and determine evacuation planning based on modeling to be made.

2. Method

The method used in this research is spatial modeling based on Geographic Information System. Spatial modeling in GIS can help provide the simulation of emergency conditions with various scenarios so that data can be obtained for overall disaster mitigation. The modeling used is tsunami inundation modeling, population distribution, and estimated evacuation time. To find out the best evacuation route, network analysis is carried out. The combination of modeling in this study can be used to find out the evacuation planning in Ciletuh Bay.

Data processing of tsunami inundation area modeling refers to modeling techniques used by BNPB [5]. The distribution of affected areas is made from data processing with mathematical calculations developed by Berryman [6]. Berryman's method is based on the calculation of tsunami height loss every 1-meter distance of puddle height based on the distance to slope and surface roughness. This method helps for spatial analysis of tsunami exposure.

$$H_{loss} = \left(\frac{167\, n^2}{H_0^{\;1/3}}\right) + 5\, Sin\, S$$

Description

H_{loss}	: loss of tsunami height per 1 meter delay distance
n	: coefficient of surface roughness
H_0	: tsunami wave height on the shoreline (m)
S	: the size of the surface slope (degrees)

The population is used for modeling the distribution of the population by processing the number of inhabitants per village and land use. Population distribution modeling is a combination of population data in the village and the distribution of settlements. Population data are obtained from statistical and census data. The right method to use in population distribution modeling in this study is to use the concept of dasymetric mapping [7].

$$X_d = \sum_{i=1}^{n} P_i$$

$$P_i = \sum_{i=1}^{n} P_{ij}$$

$$P_{ij} = \frac{S_{ij}}{\sum_{i,j=1}^{n} S_{ij}} .\ X_d$$

Description

Xd	: Number of populations in administration unit
Pi	: The number of populations in all settlements in each village
Pij	: Number of populations in polygon j a polygon settlement
Sij	: Area of polygon j a polygon settlement

Cost Distance Weighted (CWD) will be used in the calculation of evacuation travel time. In this method, Speed Conservation Value (SCV) will be used in calculating evacuation accessibility. The SCV value itself is the percentage value of the maximum speed obtained from the classification of land cover in each cell grid [8]. This method also required slope data classified in the form of percent units and land use data in the form of the raster. In the calculation of the travel time of this evacuation, the speed of movement used is the slowest speed of movement of refugees. According to the Asian Disaster Preparedness Center [9], the average speed of 1.2 m/s was used and considered more realistic for refugees. With the CWD method, there will be a period of evacuation distance needed by residents in the research area.

3. Result and Discussion

- *Modeling the Potential Hazard of Tsunami in Ciletuh Bay*

In modeling tsunami inundation in the Ciletuh Bay area in Ciwaru Village, it uses two scenarios of maximum tsunami wave potential on the coast. In the first scenario, the maximum wave potential in Ciletuh Bay is as high as 10 meters. This maximum wave potential refers to Regulation of the head of the Disaster Management Agency No. 2 in 2012 [10] about the maximum wave height on the Sukabumi coast is 10 meters. While the second scenario has the potential for maximum tsunami waves in Ciletuh Bay as high as 20 meters. In the second scenario, this is the worst-case scenario of tsunami altitude referring to the maximum potential tsunami altitude in the southern waters of Java Island. The height of the tsunami wave of 20 meters is based on research in 2020 related to tsunami modeling using earthquake records in the south of Java Island [11]. The existence of seismic gap, megathrust and active fault in the south of Java Island allows for tsunami with high waves.

In the tsunami wave scenario of 10 meters, the total tsunami-affected area reaches 57.74 ha. Tsunami-affected areas are further classified into three hazard classes, namely low hazard class, medium hazard class, and high hazard class. In the low hazard class, the affected area reaches 6.54 Ha. Tsunami-affected areas are included in the medium hazard class covering an area of 10.82 ha. Meanwhile, tsunami-affected areas are included in the high hazard class of 30.38 ha. Determination of classification of this hazard class refers to Perka BNPB 2/2012 . The low hazard class has a ≤ 1 inundation grade, a medium hazard class with a grade of 1< an inundation of ≤ 3, and a high hazard class with a > 3. In the second scenario with the potential for maximum tsunami waves as high as 20 meters, the area affected by the tsunami reaches 104.63 ha. In the low hazard class, the affected area reaches 5.17 ha. Tsunami-affected areas are included in the medium hazard class covering an area of 6.89 ha. Meanwhile, tsunami-affected areas are included in a high hazard class of 92.57 ha. For further analysis, the potential tsunami hazard used is a wave height scenario of 20 meters considering this scenario is a worst-case scenario with the potential for maximum tsunami waves.

From the modeling results, we can find out the extent to which the tsunami will damage the area in Ciwaru Village. The absence of coastal forests such as mangroves or coconuts along the coast of Ciletuh Bay caused the absence of barriers that can break the waves when a tsunami occurs. This resulted in the coverage of tsunami inundation areas in the region. As previously known, the shape of the coastline in the form of bays has a greater potential for tsunami height due to the accumulated waves of seawater to the coast. In addition to being influenced by land use in Ciwaru Village, morphological conditions in this village also affect the range of tsunami inundation. Ciwaru village is a lowland area with an altitude of 0-334 meters above sea level. The area is dominated by lowland, except the eastern part of the area which is a hilly area that belongs to the Darma Hills area. Radius 1 km from the shoreline has an altitude of 0-100 meters above sea level. In addition to altitude, the slope of slopes in this region is quite sloping between 0° - 61.4°. The physical condition of this area plus the absence of coastal forests or embankments on the coast makes the reach of tsunami puddles widespread due to the absence of barriers or tsunami breakwaters that come to hit the mainland.

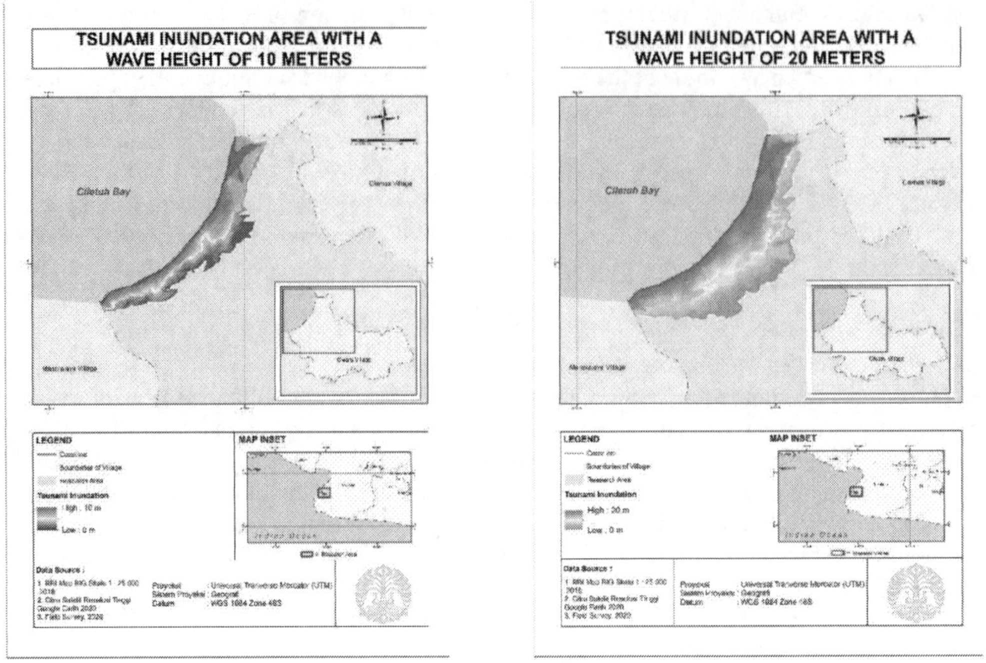

(a) (b)

Figure 1. Map of tsunami inundation areas of 10-meter (a) and 20-meter (b).

- *Population in Tsunami-Exposed Area*

Data processing of potential tsunami hazard threats and population distribution can provide information on the number of people exposed to tsunamis. The people who were exposed to the tsunami were residents who were in the area of tsunami inundation. From the data processing, 590 people were found in tsunami-exposed areas. The number is indigenous in Ciwaru Village. Most of the people in the area are fishermen or traders who sell along Palangpang Beach. In addition to the locals, data on the number of tourists is also needed considering Ciwaru Village is included in the Ciletuh Geopark area. Several tourist attractions attract tourists, such as Palangpang Beach, Cimarinjung Waterfall, Cikanteh Waterfall, and Sodong Waterfall. Based on data from the Pakidulan Sukabumi Association (PAPSI), the number of tourists can reach 819 people [12]. The tourist data is used in the evacuation plan considering that most of the inns are in exposed areas. Also, Palangpang Beach is one of the tourist attractions that attract visitors. The area of tourist attractions included in the tsunami-exposed areas makes the number of tourists also a priority for evacuation to minimize fatalities when the tsunami occurs.

The area exposed to tsunami inundation reaches 600 meters from the shoreline. Meanwhile, residents who are in areas 1 km away from the coastline are still included in the population with high vulnerability Therefore, evacuation planning is not only focused on residents in exposed areas but also on residents who are in the area 1 km from the coastline. This condition caused the number of people who had to be evacuated not only 590 people who were in the area exposed. The total number of people who are the priority of evacuation as many as 2,143 people spread across two hamlets, namely Tegalcaringin Hamlet and Gunungbatu Hamlet.

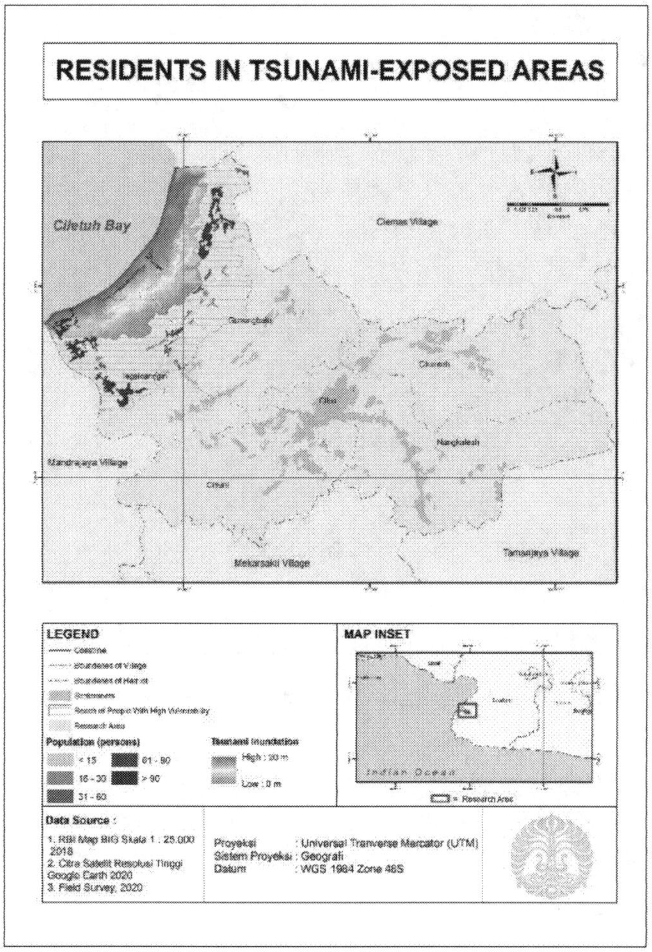

Figure 2. Map of Population in Tsunami-Exposed Area.

- *Estimated Evacuation Travel Time*

Evacuation time in this research is obtained from data processing by the Cost Distance method. The method uses land-use data and slopes that have been weighted by the Speed Conservation Value (SCV). From the results of the data processing, the maximum time needed for evacuation by the arrival time of the tsunami on the coast of Sukabumi Regency is 25 minutes.

After knowing the estimated travel time of tsunami evacuation in Ciletuh Bay can also be known the number of refugees based on the time of evacuation. The analysis was obtained from a combination of estimated evacuation time and distribution of residents in exposed areas. The estimated evacuation time in this area is influenced by the SCV value factor in each land use. Evacuation time is classified into 5 classes, namely minutes to 0-5, minutes to 5-10, minutes to 10-15, minutes to 15-20, and minutes to 20-25. In the first 5 minutes, 269 people could be evacuated from exposed areas to safe zones. The presence of roads with slopes that ramps help accelerate the evacuation of this area. Furthermore, a total of 266 people can be evacuated in the next 5 minutes with an estimated evacuation time of 5-10 minutes. In the estimated evacuation time of 10-15 minutes, there are as many as 55 people who can evacuate. This resident is located in the north of Ciwaru Village. The length of time of evacuation is caused because the slope in the north of the village is steeper than the southern region. All residents of exposed areas were able to evacuate before the tsunami arrived in the Teluk Ciletuh area. The existence of an adequate road network helps facilitate residents in the process of self-evaluation.

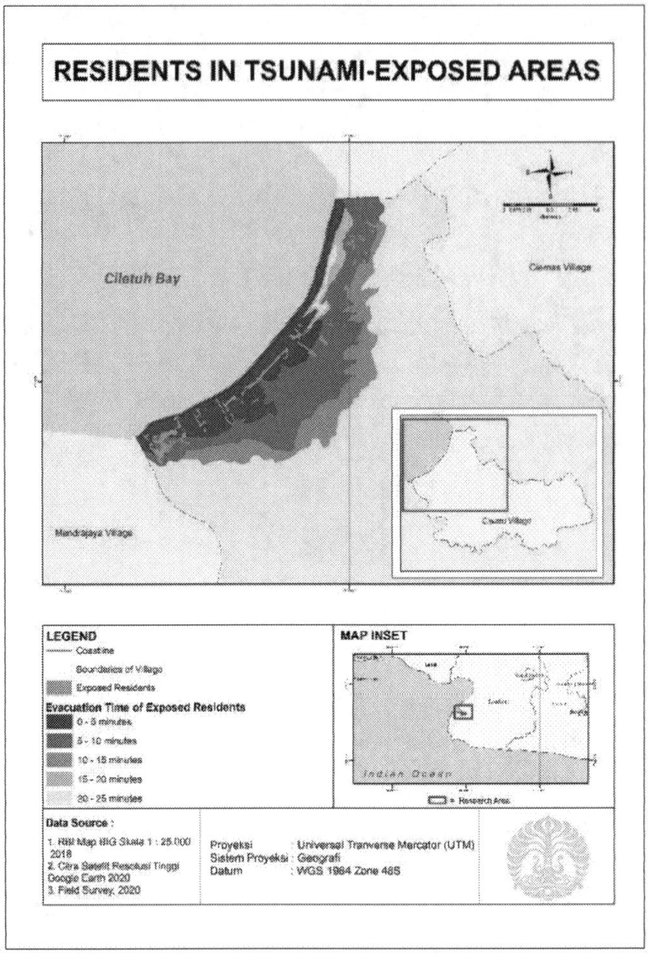

Figure 3. Population Map of Exposed Areas with Estimated Tsunami Evacuation Time.

- *Determination of Evacuation Routes in Tegalcaringin and Gunungbatu Hamlets*

There are five evacuation gathering points, namely Darma Peak, Pasir Haur, Cimarinjung Waterfall area, Ciwaru Village Hall, and MTs Al - Furqon. With five evacuation points, locals have a more diverse choice and can reduce the buildup of refugees in one place. The evacuation route to Cimarinjung Waterfall and Darma Peak area is parallel to the shoreline. This is due to the river crossing the area. In addition, the absence of other roads other than the main road makes the road the only road that can be taken by refugees to the Cimarinjung Waterfall and Darma Peak area.

Tegalcaringin Hamlet is one of the hamlets whose area is partly exposed to the tsunami. Its location directly adjacent to the sea makes this area very vulnerable to tsunami. The physical condition of the area which is a lowland with an altitude of fewer than 100 meters above sea level with sloping slopes is one of the factors causing some of these areas to be exposed to tsunamis. To minimize the death toll when the tsunami occurred, it is necessary to plan the evacuation route of residents in Tegalcaringin Hamlet.Residents in Tegalcaringin Hamlet can evacuate to a safer place, namely cimarinjung waterfall area, Darma Peak, SMAN 1 Ciemas, Ciwaru Village Hall, and MTs Al-Furqon. Each evacuation collection point has different mileage and travel times. To get to the Cimarinjung Waterfall area, the population takes 8 minutes. Residents can evacuate to Darma Peak in 15 minutes. To get to the gathering point of SMAN 1 Ciemas, Ciwaru Village Hall, and MTs Al-Furqon, there are two lines that can be used. The first line through the bridge and the second line is an alternative route if the bridge in the first lane is damaged so that it cannot be passed by residents for evacuation. The distance and time required by residents for evacuation can be seen after table 1.

Gunungbatu Hamlet is one of the areas directly adjacent to the sea. The physical condition of this area in the form of lowland with an altitude of fewer than 100 meters above sea level and a sloping slope caused some areas in this hamlet to be exposed to tsunamis. Planning evacuation routes in this area are needed to minimize fatalities when a tsunami occurs. Residents in Gunungbatu Hamlets can evacuate to the existing gathering point. To get to the gathering point in the Cimarinjung Waterfall area only takes 4 minutes from the settlement area in this hamlet. Residents can also evacuate to Darma Peak with a travel time of 10 minutes. In addition, residents in Gunungbatu Hamlet can also evacuate to the gathering point in the eastern area of this village. Residents can evacuate to Ciwaru village hall gathering point takes 12 minutes. The travel time needed to get to the evacuation point at MTs Al-Furqon is 13 minutes. Meanwhile, to get to the gathering point at SMAN 1 Ciemas requires an evacuation time of 13 minutes as well. The distance and time required by residents for evacuation can be seen after table 2.

Table 1. Distance and Travel Time of The Evacuation of Tegalcaringin Villagers.

Origin	Evacuation Gathering Point	Distance (km)	Travel Time (minutes)
Tegalcaringin Hamlet	Cimarinjung Waterfall	2,7	8
	Darma Peak	4,8	15
	Ciwaru Village Hall	3,7	11
	SMAN 1 Ciemas	3,8	11
	MTS Al-Furqon	4,3	13
	Ciwaru Village Hall *(Alternative routes without crossing the bridge)*	7	20
	SMAN 1 Ciemas *(Alternative routes without crossing the bridge)*	7,2	21
	MTS Al-Furqon *(Alternative routes without crossing the bridge)*	7,5	22

Table 2. Distance and Travel Time of The Evacuation of Gunungbatu Villagers.

Origin	Evacuation Gathering Point	Distance (km)	Travel Time (minutes)
Gunungbatu Hamlet	Cimarinjung Waterfall	1,4	4
	Darma Peak	3,5	10
	Ciwaru Village Hall	4	12
	SMAN 1 Ciemas	4,2	13
	MTS Al-Furqon	4,5	13

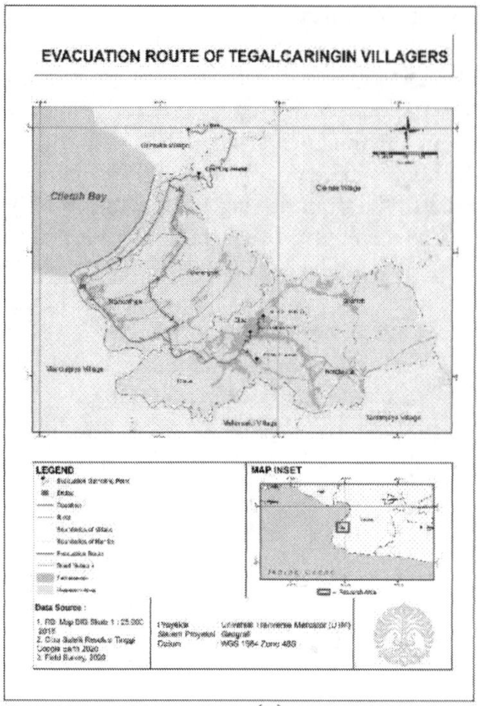

(a)

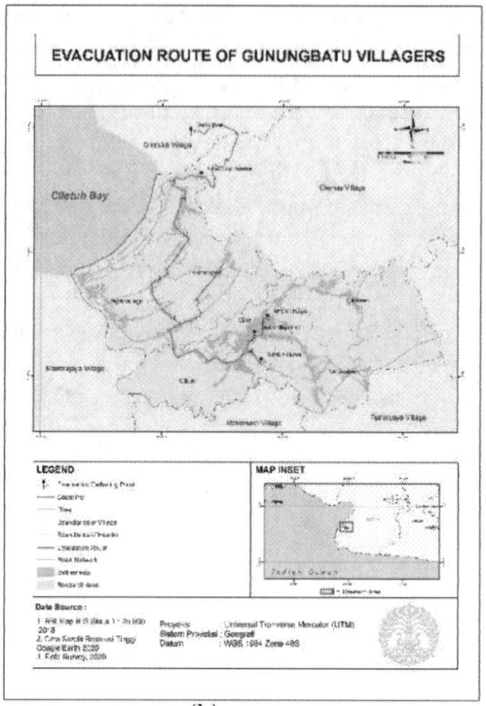

(b)

Figure 4. Map of The Evacuation Route of Tegalcaringin Villagers (a) and Gunungbatu (b).

4. Conclusion

The potential of tsunami hazard in Ciletuh Bay is influenced by the physical condition of the region. Ciwaru village has a slope that is quite sloping with a height of 0-334 meters above sea level affecting the reach of tsunami inundation areas. With a maximum tsunami wave height scenario of 20 meters, the range of tsunami inundation area reaches 104.63 ha. The absence of coastal forests such as mangroves or coconuts along the research area caused the absence of barriers that can break the waves when a tsunami occurs. This condition caused the widespread distribution of inundation areas in Ciwaru Village. There are two areas that the priority of evacuation, namely Tegalcaringin Hamlet and Gunungbatu Hamlet. These two hamlets are areas that have the potential tsunami hazard of Ciletuh Bay. Based on the data processing that has been done, the population exposed in both areas reached 590 people. All residents in exposed areas can evacuate within 0-15 minutes. The selection of evacuation routes with the fastest travel time is needed so that residents can reach the evacuation collection point before the tsunami arrives. The availability of road access in the hamlet makes it easy for residents in tsunami-exposed areas to reach the designated evacuation collection point.

Acknowledgement

Thanks to DRPM of Universitas Indonesia which has supported and funded this research grant PUTI 2020.

References

[1] Imamura F, Boret S P, Suppasri A and Muhari A 2019 Recent occurrences of serious tsunami damage and the future challenges of tsunami disaster risk reduction. *Progress in Disaster Science* **1**, 100009.
[2] Nugraha B and Surbakti H 2009 Simulasi Pola Arus Dua Dimensi di Perairan Teluk Pelabuhan

Ratu pada Bulan September 2004. *Jurnal Kelautan Nasional* Vol. **4**, No. I April 2009.

[3] Sarmili L 2015 The Formation of Accretionary Prisms At Ciletuh Bay in Relation To Cimandiri Fault, West Java. *13*(3), 173–182.

[4] Diposaptono S 2005 Mitigasi Bencana Wilayah Pesisir Berbasis Ekosistem Mangrove. Ditjen Kelautan, Pesisir, dam Pulau-pulau Kecil, DKP.

[5] Badan Nasional Penanggulangan Bencana 2018 Modul Penyusunan Kajian Risiko Bencana Tsunami.

[6] Berryman K *2006 Review of Tsunami hazard and risk in New Zealand.* New Zealand: Institute of Geological and Nuclear Sciences, pp. 64.

[7] Khomarudin M R 2010 Tsunami Risk and Vulnerability. *Dissertation der Fakultät für Geowissenschaften der Ludwig-Maximilians-Universität München.*

[8] Wood N J and Schmidtlein M C 2011 *Anisotropic Path Modeling to Assess Pedestrian-Evacuation Potential from Cascadia-Related Tsunamis in The US Pacific Northwest.* Cascadia: Springer.

[9] Asian Disaster Preparedness Center *2007 Evacuation Routes Tools ArcGIS toolbox - User's manual.* Bangkok: Italian Ministry for the Environment Land and Sea. Department for Environmental Research and Development, 88 pp.

[10] Perka BNPB No. 2 Tahun 2012 tentang Pedoman Umum Pengkajian Risiko Bencana. Badan Nasional Penanggulangan Bencana.

[11] Widiyantoro S, Gunawan E, Muhari A, Rawlinson N, Mori J, Hanifa N R, Susilo S, Supendi P, Shiddiqi H A, Nugraha A D and Putra H E 2020 Implications for megathrust earthquakes and tsunamis from seismic gaps south of Java Indonesia. *Scientific Reports*, **10**(1), 1–11.

[12] Darsiharjo, Supriatna U and Saputra I M 2016 Pengembangan Geopark Ciletuh Berbasis. *Jurnal Manajemen Resort dan Leisure*, **13**(1), 55–60.

IOP Conf. Series: Earth and Environmental Science **851** (2021) 012007 doi:10.1088/1755-1315/851/1/012007

GIS-based study of tsunami risk in the Special Region of Yogyakarta (Central Java, Indonesia)

V Steinritz[1]*, S Pena-Castellnou[1], G I Marliyani[2] and K Reicherter[1]

[1]Neotectonics and Natural Hazards, RWTH Aachen University, Aachen, Germany
[2]Geological Engineering Department, Universitas Gadjah Mada, Yogyakarta, Indonesia
*E-mail: v.steinritz@nug.rwth-aachen.de

Abstract. The island of Java is located above a plate boundary within the Indonesian archipelago as part of the volcanic arc resulting from the subduction of the Indo–Australian plate beneath the Eurasian plate. Due to its emplacement, it is exposed to several geohazards, including active volcanoes and earthquakes, and its secondary effects (e.g., landslides, tsunamis). Tsunamis have repeatedly hit Java; for example, the 17th of July 2006 caused by an Mw 7.8 earthquake off the coast of western Java, or the 22nd of December 2018 triggered by the collapse of the flank of Anak Krakatau during its eruption. In most cases, tsunamis have a destructive coastal impact and can cause a significant loss of lives. Therefore, it is of high importance to determine areas situated in tsunami risk zones and estimate plans for risk reduction and prevention. We present an ArcGIS-based method to calculate tsunami risk zones for the Special Region of Yogyakarta (Central Java), including the calculation of hazard zones, vulnerability zones, and the estimation of highly exposed areas. The final risk map reflects the current risk zones in the occurrence of a tsunami and should be recalculated and re-evaluated if the environmental and socio-economic conditions change. The results can be used as a base for tsunami evacuation plans and site planning.

1. Introduction

Indonesia is an archipelago located in Southeast Asia, in the Pacific ring of fire, the most seismically and volcanically active region in the world. Java Island is located a hundred kilometers north of Java Subduction Zone (figure 1). This area accumulates part of the shortening between the Eurasian and Indo-Australian Plates, which converge at the rate of 5-7 cm/yr ([1]; [2]). This plate-tectonic configuration builds up a high density of active volcanoes along the island and the occurrence of large numbers of earthquakes and its secondary processes (e.g., landslides, liquefaction, tsunamis).

The Special Region of Yogyakarta is located in the central south of the island of Java. With 1207 inhab./km^2 it is the fourth highest densely populated province of Indonesia and one of the fastest-growing provinces regarding social and economic aspects [3]. In the past years, a high urban expansion rate was caused by a rising migration rate and economic growth. Reasons for this include the excellent reputation of Yogyakarta as "The City of Education" and "City of Culture", and the rising number of tourists; Yogyakarta is the second largest tourist destination besides Bali ([4]; [5]). As a consequence, the hazard and exposition of society to natural hazard rises.

The northern region of the province is located on the Merapi volcano's feet, while the southern region is delimited by the coast occupied by settlements, industries, tourism areas, agricultural fields, and the new Yogyakarta International Airport. The province is frequently hit by volcanic eruptions from the active Merapi volcano and earthquakes related to the subduction regime and shallow faults in the vicinity of the area (e.g., Opak fault). Regarding tsunamis, two events affected the coast of Yogyakarta in the historical period. On the 2nd of June of 1994, an Mw 7.9 earthquake occurred in Banyuwangi (East Java) and caused a tsunami with run-up heights over 10 m leaving more than 250 fatalities (e.g. [6]). On the 17th of July 2006, an Mw 7.8 earthquake off the coast of western Java caused a tsunami that affected over 300 km of coastline and caused more than 600 fatalities, with average run-up heights of 5 to 7 m [7].

Subduction zone earthquakes are by far the most typical source of tsunamis. However, there are other sources capable of generating a tsunami, including volcanic eruptions and (submarine) landslides. On the southern coast of Java, no tsunami caused by a submarine landslide has ever been historically documented [8]. Regarding volcanic eruptions, the Krakatau (located to the west of Java, between Java and Sumatra Islands) triggered several tsunamis due to its eruption. In 1883 the run-up height was higher than 20 m in Java, causing more than 35000 fatalities (e.g. [9]). More recently, on the 22nd of December 2018, the flank of Anak Krakatau collapsed during eruption and generated a devastating tsunami on the eastern coast of Java, leaving a death toll of 437 (e.g. [10]).

The previously mentioned local tsunamigenic events and large-scale events such as the 2004 Indian Ocean Tsunami (one of the most disastrous events in recent Indonesian history); together with the recent studies of [11] proposing the likelihood of megathrust events along the Java trench, suggested by seismic gaps (from seismic and geodetic data) in the subduction trench, and the study of [12] that found tsunami deposits evidence at many sites along the southern coastline of Java suggesting a return period of approximately 500 years for this megathrust earthquakes, demonstrate the importance of tsunami risk assessment in this area.

Several worldwide efforts were put into the improvement of tsunami risk assessment in Southern Asia after the 2004 Indian Ocean Tsunami. Rakowsky et al. [13] computed a complete database with tsunami scenarios covering the Sunda Trench as an aid for tsunami inundation studies and the Indonesian tsunami early warning system. Tsunami hazard maps are available for various municipalities of the southern coast of Java including Bantul and Parangtritis in the Special Region of Yogyakarta ([14];[15]). However, tsunami risk maps for the Special Region of Yogyakarta are not freely available. We present a static approach to model risk maps based on hazard, vulnerability, and exposure. The results can be used as a base for tsunami evacuation plans and site planning.

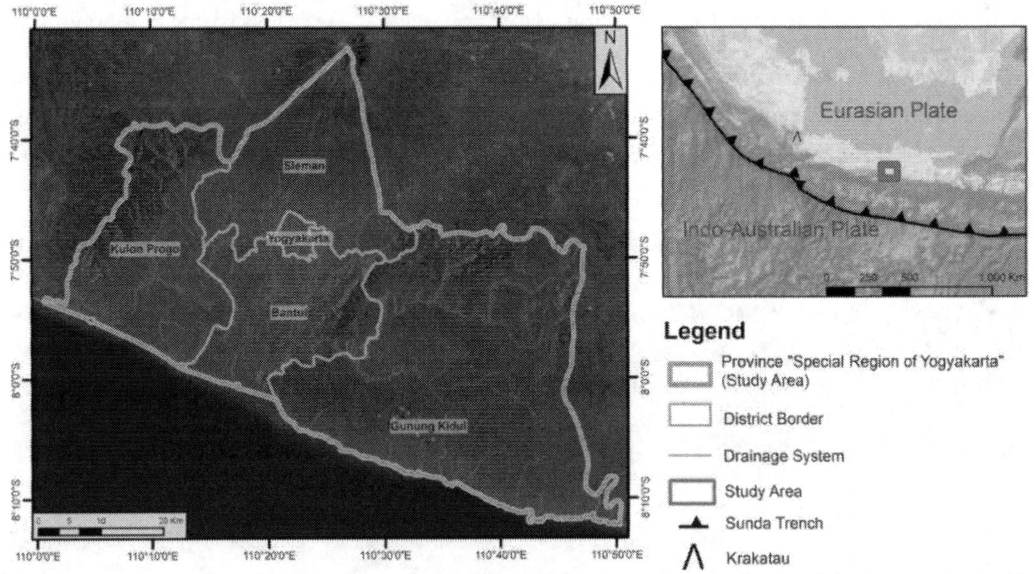

Figure 1. a) Overview of the study area, the Special Region of Yogyakarta with the provinces. b) Simplified tectonic setting. Java Island is located hundred kilometers north of the Sunda Trench.

2. Seismotectonic setting

The Island of Java is located on the Sunda volcanic Arc, due to the subduction of the Indo-Australian Plate beneath the Eurasian plate. The seismicity of Java is characterized mainly by interplate earthquakes that occur along the subduction slab interface in the Wadati-Benioff zone. The Wadati-Benioff zone slides gently, almost horizontal, beneath a melange wedge for the first 150 km from the Java trench, where the dipping angle increases rather sharply to 45° [16]. The slab prolongates to depths of about 650 km dipping approximately 65° northward [17]. This change in the steepness might cause a northward pushing and stress accumulation in the overriding plate [16]. Earthquakes of intermediate-depth (40–250 km) are generated beneath the Java Island, and deep earthquakes (400–650 km) at the back-arc region located north of the island and beneath the Java Sea. Regarding the historical seismicity, the earthquake magnitudes are likely to be in the range of Mw 7.0-7.5[18]. These earthquakes generally have high tsunamigenic potential. In some cases, these events have demonstrated slow moment-release and have been defined as 'tsunami' earthquakes, where rupture is large in the weak crustal layers very close to the seafloor [18].

3. Methods

The term "risk" is described as the probability of harmful consequences or expected losses (deaths, injuries, property, livelihoods, disrupted economic activity, or damaged environment) arising from the interaction between a hazard, vulnerable conditions, and the exposed elements (people, infrastructure, among others [19].

Risk can be expressed in an equation (1) [20] as the sum of 1) hazard, an occurring physical event which can cause damages; 2) vulnerability, the capacity of conditions to resist hazards; and 3) exposure, different features located in hazard-prone areas [19].

$$\text{Risk} = \text{Hazard} * \text{Vulnerability} * \text{Exposure} \tag{1}$$

We use a multi-criteria GIS-based method (figure 2) using the Geographic Information System Software ArcGIS 10.6.1 to determine the tsunami risk in the Special Region of Yogyakarta. We first compute a map for each component of the risk equation (eq 1), considering different criteria and combining them into a final risk map. We followed two approaches: 1) the hazard and vulnerability map were combined to produce an inherent risk map to determine the areas that would be hit by a tsunami; and 2) we include the exposure factor to define the areas that would be more affected in case of a tsunami to assess disaster management. The input data includes Landsat 8 satellite imagery [21], digital elevation model DEMNAS [22] and SRTM [23], land use data ([24]; [25]; [26]), and population density [27]. All data are freely available online.

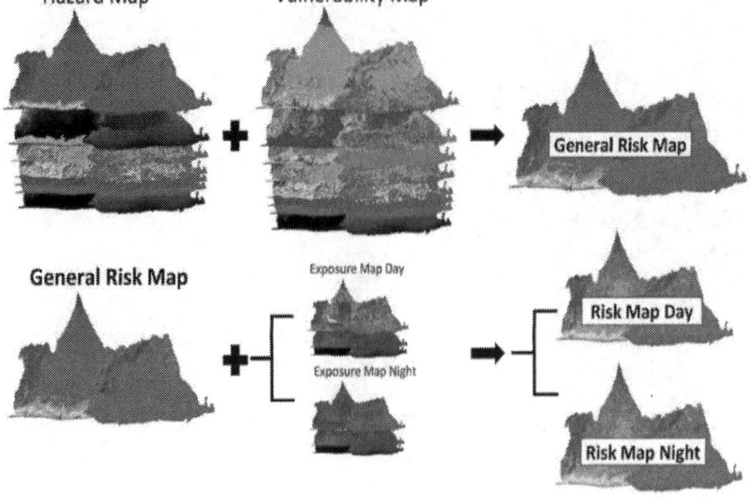

Figure 2. Workflow and input data for each step of the GIS-based tsunami risk calculation.

3.1 *Hazard*

To estimate tsunami hazard zones, we determine how far tsunami waves with different wave heights at coastline will inundate inland, the so-called inundation limit. For calculating the tsunami inundation in ArcGIS, the analysis toolbox "TsunamiInundation_10.1-10.2" from [28] was used. For each raster cell of the digital elevation model of the study area, the included tool in the toolbox, "Tsunami Inundation," calculates the loss of wave height per meter with a usage of equation (2), described by [29]. Next, the maximum inundation width is calculated with equation (3).

$$Y_{loss} = \left(\frac{167n^2}{Y_0^{\frac{1}{3}}}\right) + 5\, sin\, sin\, s_0 \tag{2}$$

$$X_{max} = \frac{Y_0}{Y_{loss}} 10 \tag{3}$$

Where Y_{loss} [m/m] is the loss of the wave height per meter, n [-] is the roughness coefficient, Y_0 [m] is the initial wave height on the shoreline, s_0 is the topographic slope [rad] and, X_{max} [m] is the maximum inundation width.

The Manning roughness coefficient (n) indicates the ground resistance to flood flows in channels and flood plains [30]. This coefficient has a high influence on the tsunami inundation formula (2) [31]. It has been proved that for a lower roughness coefficient the tsunami would travel four times further inland than for higher coefficients [32]. The reason for this is that higher roughness involves higher friction, which directly influences the velocity of the water flowing inland. For calculating the roughness for the tsunami inundation, we use the Manning coefficient formula developed by [32] and modified from [30], which includes several roughnesses to represent the floodplain characteristics:

$$n = (n_b + n_1 + n_2 + n_3 + n_4)\, m \tag{4}$$

Where n is the Manning coefficient, n_b is a base value for the flood plain´s natural bare soil surface (geology), n_1 is a correction factor for surface irregularities (morphology), n_2 is a value for variation in shape and size of the floodplain cross-section (0 for floodplains), n_3 is a value for obstruction, n_4 is a value for vegetation and flow conditions (land use) and, m is a correction factor for sinuosity of the floodplain which takes a value of 1 for floodplains [30].To estimate the parameter n_1 we determine the surface roughness with the Terrain Ruggedness Index (TRI). For parameter n_b and n_4 we consider the geological and land use characteristics of the area and we assign a Manning value proposed by [30] (e.g. coarse gravel – n= 0.03). It is not practicable to determine the parameter n_3 for the study area since it is nearly impossible to distinguish small scale obstructions (debris deposits, exposed roots, logs, isolated boulders, among others) from satellite images. Therefore, we set the value to zero to avoid an overestimation. Then, the single parameters are summed up to a final value n, whereby the parameter n_b is taken into account only with 25% to avoid an overestimation of the geology since the land use overprints the geology.

To compute different hazard tsunami scenarios, the tsunami inundation is calculated using formula (2) for different wave heights on the shoreline. The wave heights are divided into five classes and sorted according to their recurrence period into a scale of hazard classes. By comparing the wave heights of historical tsunamis [33], wave heights of 0 – 5 meters are more probable than 10 – 15 meters. Therefore, the class 0 – 5 meters wave height is attributed to the hazard class "Very High," and so forth.

3.2 Vulnerability

According to the definition of vulnerability we use an environmental multi-criteria approach based on topographic elevation, slope, and roughness derived from land use following the workflow analysis proposed by [34] to estimate the vulnerability of the study area in the event of a tsunami. We produce a map for each criterion in ArcGIS, and then the criteria are weighted regarding their relevance against each to construct a final vulnerability map.

The topographic slope is directly linked to the tsunami inland flow. For gently sloping areas, the vulnerability is higher than for steep sloping areas. A gentle slope represents less resistance for the water to flow; therefore, the tsunami wave's flow velocity and force are not reduced as much [35]. For the slope vulnerability map, we use an older version of SRTM data (3 arc-second) from [23] with a resolution of 90 meters. Newer DEM versions with higher resolutions like DEMNAS [22] are not suitable since vegetation and buildings/structures change the general slope on a very small scale, which leads to a wrong calculation of the topographic slope. We use the "Slope" tool in ArcGIS to compute this parameter. For the vulnerability classification, the classification from [36] were used (also implemented by [34]), where the slope is divided into five groups from low slope gradients associated with a very high vulnerability to high slope gradients associated with a very low vulnerability. Regarding elevation, high topographic regions are less vulnerable to the tsunami inland flow than low topographic areas. The tsunami wave needs a more considerable wave height and energy to reach high topographic areas. The elevation height is based on the DEMNAS DEM. For the elevation vulnerability classification, we use, as proposed in [34], the classification of [37], which classifies the elevation height in five classes, from low elevations associated with a high vulnerability to high elevations associated with a low vulnerability. As an additional criterion to the model of [34], we include the surface roughness and excluded the criteria coastal shape, proximity and direction, since the method is not taking into account a single tsunami point source and the coastal shape is almost straight everywhere. Due to the fact that the roughness influences the vulnerability (areas of low roughness are related to higher vulnerability), the parameter is additionally included here. The roughness coefficient is calculated with the Manning coefficient (as described in section 3.1) and represents its own criterion. The roughness vulnerability classes are set, based on [31], from very high for 0 – 0.01 to very low > 0.1.

The three calculated parameters (elevation, slope, and roughness) are merged using a weighted overlaying analysis to generate a final environmental vulnerability map. To estimate the percentage weighting factor for each criterion, we use a method called "Analytical Hierarchy Process" (AHP) introduced by Thomas Saaty in the 1970s [38]. With this method, the criteria are compared, due to their importance, with each other pairwise [39]. Then, the three parameters are merged in ArcGIS using the tool "Weighted Overlay".

3.3 Exposure
People or infrastructures situated in a tsunami hazard zone are directly exposed to a potential tsunami and, therefore, at risk. We use various criteria to evaluate the socioeconomic exposure to a tsunami in the study area, which includes the categories: buildings and facilities, tourism areas, areas of critical land use, critical infrastructure and, critical transport. All criteria are integrated with ArcGIS into one map. Then each criterion is given an exposure class ("Very High" to "Very Low") for the day-time and one for the night-time. Since the exposure spatially varies from day-time to night-time (e.g. more public places like the beach site are visited during-day time than during night-time), we generated two maps to depict this fact. This evaluation is based on statistical data from [24], [25], [26]; and field observations from the fieldwork carried out in June and July of 2019.

4. Results
The resulting tsunami risk map for the first approach integrates hazard and vulnerability. In general, most of the Special Region of Yogyakarta is categorized as a "Low" and "Very Low" while 1.7% of the total area as "High" and "Very High". High-risk areas are located in the southern coastline of the districts Bantul and Kulon Progo, extending up to five kilometers inland following the main rivers. Contrarily, almost the entire coastline of Gunung Kidul is categorized as "Low" tsunami risk area due to its steep cliffs of a high of 40 meters, in exception of small bays (figure 3).

The tsunami risk map (figure 4 and 5) obtained with the second approach, which incorporates the exposure factor, shows a similar pattern as the inherent risk map. Approximately 134,502 people (3.5%) [27] are located in a "Very High" to "Moderate" risk class area during day-time and approximately 122,974 people (3.2%) [27] during night-time. The areas categorized as "Very High" and "High" are adjacent to the coastline and belong mainly to the socioeconomic categories: "Beach", "Airport" and "Tourism Area". A comparison between the results for day-time and night-time shows that the percentage amounts in the risk class "Very High" decreases from day- to night-time. It is essential to point out that 25.1% of the medical facilities and 17.3% of the schools of Kulon Progo and Bantul districts are located in a categorized "High" and "Moderate" tsunami risk area [27].

Figure 3. General tsunami risk map of Gunung Kidul. The risk classes are ranging from ''Very High'' (red) to ''Very Low'' (dark green).

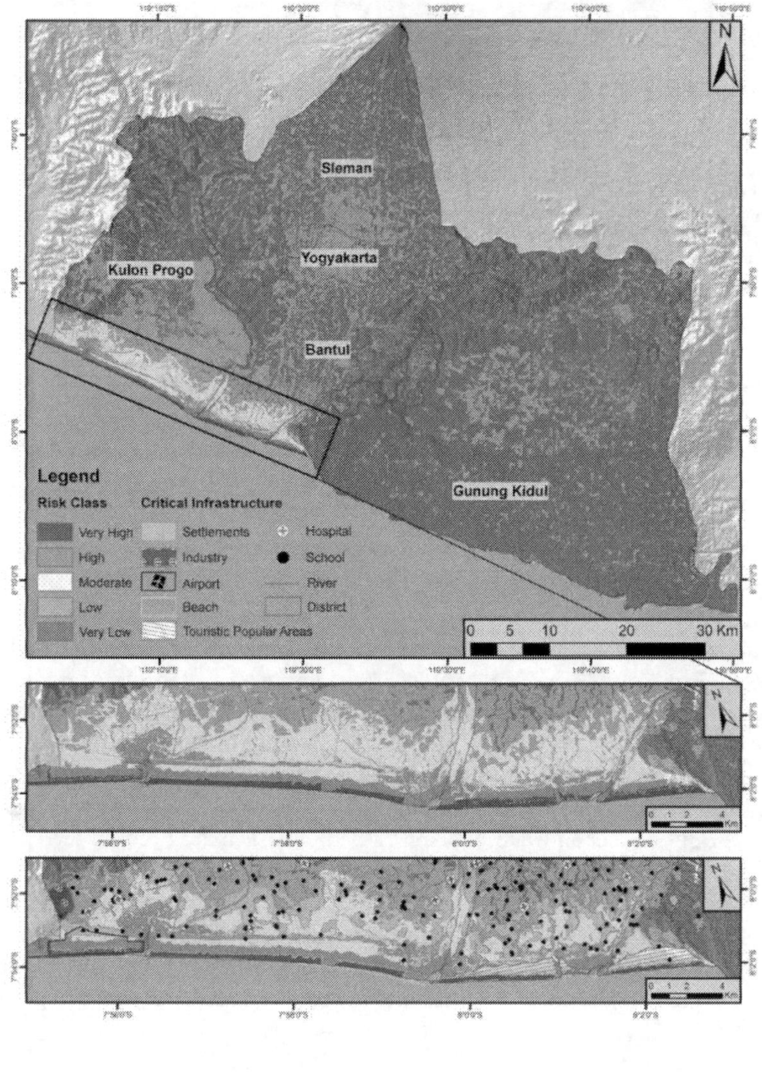

Figure 4. Day-time tsunami risk map of the Special Region of Yogyakarta. The risk classes are ranging from ''Very High'' (red) to ''Very Low'' (dark green).

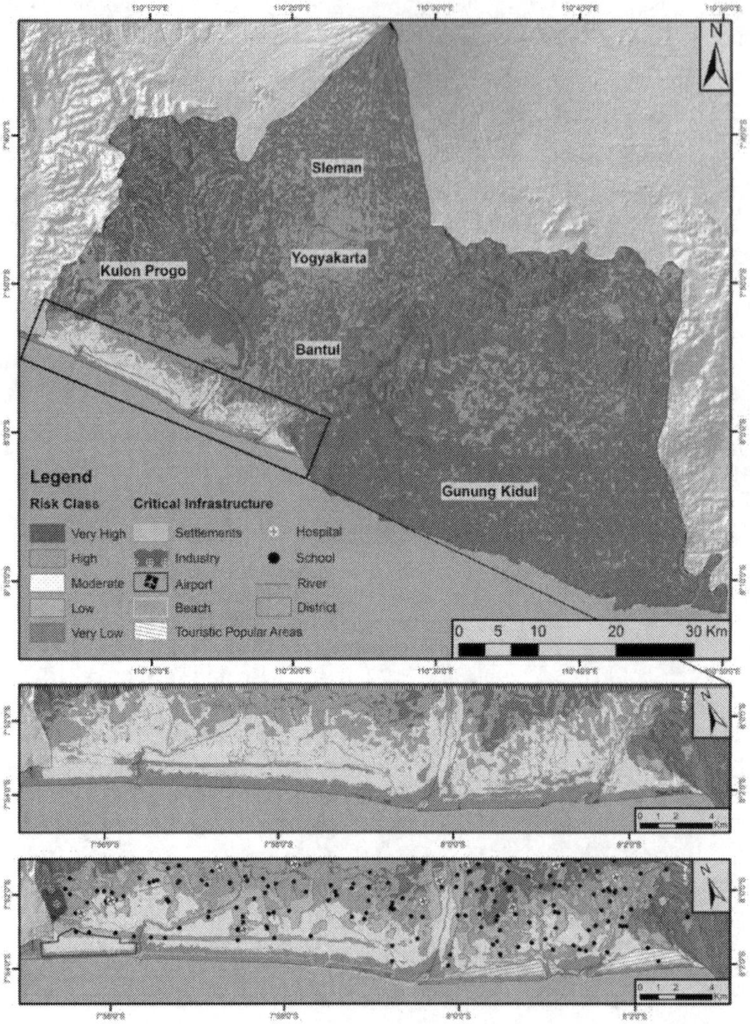

Figure 5. Night-time tsunami risk map of the Special Region of Yogyakarta. The risk classes are ranging from "Very High" (red) to "Very Low" (dark green).

5. Discussion

For the estimation of tsunami risk in the Special Region of Yogyakarta, we made two approaches. The first approach is based on the given natural conditions (slope, elevation, roughness, and proximity from the coastline) where risk areas are calculated considering their inherent terrain condition (what we name here as vulnerability) and location (hazard - inundation). The second approach includes the factor socioeconomic exposure, where risk areas are calculated considering the terrain conditions (vulnerability), the location (hazard – inundation), and where and how many critical infrastructures and people are situated in these areas during day- and night-time. As mentioned above, it is important to differ for day-time and night-time because public places like the beach site are more exposed during the day than during the night whereby at night-time the settlements are more exposed. Both approaches consider different perspectives. If only the second one is taken as a final risk map, it could lead to a misunderstanding.

Risk is complex and depends on a number of parameters. It is different understood depending on the perspective. For example, those construction companies interested in knowing where to build up new constructions in a risk-free tsunami area focus on the inherent tsunami risk map, which is only dependent on the site's natural conditions. Alternatively, local authorities and agencies tasked with managing and responding to actual disasters and developing tsunami evacuation plans are concerned about where and which percentage of the population is emplaced in high tsunami risk zones. Their focus is set on tsunami risk zones categorized, including the exposure.

The same applies to vulnerability. There are several types of vulnerability: physical, economic, social, and environmental. In this study, we only consider the environmental vulnerability and include some of the socioeconomic exposure's factors. We understand our data as a preliminary and fast method to overview the risk area in the Special Region of Yogyakarta. Accurate studies in vulnerability should include other elements such as the construction standards of buildings, socioeconomic status, level of understanding and hazard perception, the warning systems, and preparedness (ability to move away from the flood zone to a safe place), among others. The method we use is strongly dependent on the available input source data. We use open-source data that may not be up-to-date anymore. Therefore, the method is prone to error for example, in calculating the roughness parameter and the exposure factor, which are mainly based on land use data. The results will reflect the potential general tsunami risk zones for the current environmental and socioeconomic situation since we live in a changing world (and more so today).

Many studies focused on numerical modelling on tsunami propagation in the Java arc [13] which is an important parameter for a risk calculation. Since we did not take into account specific tsunami source points in the hazard calculation, the results of our model are not highly accurate. If we would consider specific tsunami source points, the results would increase in accuracy but would also increase the complexity, computing time and loose the general factor. Our general hazard results are still in good agreement with the existent hazard maps for parts of the Special Region of Yogyakarta (Bantul and Parangtritis) ([14];[15]). In advantage of our model is a relatively simplistic approach thus a short computing time is needed, that still delivers adequate results.

6. Conclusion

Tsunami risk zones were calculated for the Special Region of Yogyakarta following a GIS-based workflow. Two approaches were made for risk calculation where the results of the first approach show risk zones based on the environmental terrain conditions, and the results of the second approach include the exposure factor. For the final risk calculation, eight parameters were taken into account: geology, terrain conditions, land use, slope, elevation, tsunami inundation, critical infrastructures, and exposure of people. It is shown for both approaches that most areas of the Special Region of Yogyakarta are located in low or very low tsunami risk areas. Nevertheless, on the southern coastal area, the districts of Kulon Progo and Bantul are situated in high tsunami risk zones susceptible to be directly hit and most severely affected by a tsunami. Since the Special Region of Yogyakarta is prone to tsunami hazards, it is very important to know which areas are situated in risk zones and create risk reduction and preventative measures on this basis. Consideration should also be given to building high tsunami evacuation towers near the coast as well as to expand the early tsunami warning system.

In general, the method is highly dependent on the input source data and their quality. Influencing parameters like the land use must be documented accurately. The results should be taken as approximate and only valid for the present-day due to the changing current environmental conditions including climate change and sea level rise which change the land use and erosion/sedimentation rates. Furthermore, the results show a general risk estimation which is not based on a specific tsunami event. Summarizing it can be used as a fast efficient and easy method.

References

[1] Hall R 2009 Encyclopedia of Islands. University of California Press 454–460

[2] Koulali A et al. 2017 The kinematics of crustal deformation in Java from GPS observations: implications for fault slip partitioning *Earth Planet. Sci. Lett.* **458**, 69–79

[3] Badan Pusat Statistik 2019 City Population: The population of the provinces, regencies and cities of Indonesia https://www.bps.go.id/ <Last viewed 27 January 2019>

[4] McNeill D, Nesheim I and Brouwer F 2012 *Land use policies for sustainable development: Exploring integrated assessment approaches*. Edward Elgar Pub, 292 pp

[5] König H J et al. 2010 Assessing the Impact of Land Use Policy on Urban-Rural Sustainability Using the FoPIA Approach in Yogyakarta, Indonesia. *Sustainability* 2 (7)
[6] Tsuji Y et al. 1995 Field survey of the east Java earthquake and tsunami of June 3, *Pure Appl. Geophys* **144**, 839 – 854
[7] Fritz H M 2007 Extreme runup from the 17 July 2006 Java tsunami. *Geophys. Res. Lett* **34**, Issue 12
[8] Brune S, Ladage S, Babeyko Y, Müller C, Kopp H and Sobolev S.V, 2010 Submarine landslides at the eastern Sunda margin: observations and tsunami impact assessment *Natural hazards*, **54** 547-562.
[9] Self S 1992 Krakatau Revisited: The Course of Events and Interpretation of the 1883 Eruption *Geo-Journal* **28**.2 109-121.
[10] Ye L et al. 2020 The 22 December 2018 tsunami from flank collapse of Anak Krakatau volcano during eruption *Science Advances* 6
[11] Widiyantoro S et al. 2020 Implications for megathrust earthquakes and tsunamis from seismic gaps south of Java Indonesia *Scientific Reports* **10**, 15274
[12] Harris R et al. 2019 Seismic and tsunami risk of the Java trench and implementation of risk reduction strategies. *SEG Tech. Program. Expand. Abstr.*
[13] Rakowsky N et al. 2013 Operational tsunami modelling with TsunAWI – recent developments and applications. *Nat. Hazards Earth Syst. Sci.* **13**, pp. 1629-1642
[14] Permerintah Daerah Kab. Bantul 2008 Tsunami Hazard Map Bantul District D.I.Yogyakarta Province
[15] Subardjo S et al. 2020 Mapping Vulnerability Level Tsunami Coastal Parangtritis Subdistrict, Kretek, Bantul,Yogyakarta. *Journal of Engineering and Applied Sciences*, Vol.**15** pp. 2850 - 2857
[16] Wagner D et al. 2007 Joint inversion of active and passive seismic data in Central Java *Geophys. J. Int.* **170** 923-932
[17] Hamilton W 1979 Tectonics of the Indonesian region *U.S Geological Survey Professional Paper* 1078
[18] Jones E S et al. 2014 Seismicity of the Earth 1900-2012 Java and Vicinity *U.S. Geological Survey Open-File Report* 2010-1083-N, 1 sheet, scale 1:5,000,000
[19] UN/ISDR 2004 Glossary. Basic Terms of Disaster Risk Reduction. Source: http://www.unisdr.org/unisdr/eng/library/lib-terminology-eng%20home.htm <last viewed January 2019>.
[20] Schneiderbauer S and Ehrlich D 2004 *Risk, hazard and people's vulnerability to natural hazards: A review of definitions, concepts and data.*
[21] USGS - U.S. Geological Survey, 2018 Landsat 8 OLI/TIRS C1 Level - 1 scenes LC81190662018278LGN00; LC81200662018269LGN00. *Earth Resources Observation and Science (EROS) Center* https://earthexplorer.usgs.gov/. <last viewed 13 December 2019>
[22] Badan Informasi Geospasial 2014 DEMNAS. http://tides.big.go.id/DEMNAS/ <last viewed 13 December 2019>
[23] USGS - U.S. Geological Survey 2005 Shuttle Radar Topography Mission, 3 Arc- Second scenes SRTM3S08E110V1. *Earth Resources Observation and Science (EROS) Center* https://earthexplorer.usgs.gov/ <last viewed January 2019>
[24] Bappeda DIY 2019 Arahan Pola Ruang Kawasan Budidaya Kabupaten Sleman; Yogyakarta; Bantul; Kulon Progo; Gunung Kidul; Sungai DIY (1 : 25.000); Jaringan Jalan DIY (1 : 50.000) Geoportal DIY http://gis.jogjaprov.go.id/ <last viewed November 2019>
[25] Dinas Kesehatan DIY 2017 Alamat Puskesmas dan Tenaga Medis Puskesmas di DIY Tahun 2017 Geoportal DIY http://gis.jogjaprov.go.id/ <last viewed January 2019>
[26] Badan Informasi Geospasial 2006 Download Rupabumi Indonesia Batas Wilayah 25K data RBI25K.ADMINISTRASI_LN. *Badan Informasi Geospasial* https://portal.ina sdi.or.id/downloadaoi/<last viewed January 2019>
[27] Statistics Indonesia 2019 Population by Regency / City in D.I Yogyakarta (Soul), 2010-2019. Badan Pusat Statistik Provinsi Daerah Istimewa Yogyakarta.

https://yogyakarta.bps.go.id/dynamictable/2017/08/02/32/jumlah-penduduk-menurut-kabupaten-kota-di-d-i-yogyakarta-jiwa-2010-2019.html. <last viewed November 2019>
[28] Anshori 2019 Tsunami-Inundation-Toolbox-ArcGIS-Desktop-10.x. https://github.com/anshori/Tsunami-Inundation-Toolbox-ArcGIS-Desktop-10.x.
[29] McSaveney M and Rattenbury M 2000 Tsunami impact in Hawke's Bay. *Institute of Geological and Nuclear Sience*
[30] Arcement G J and Schneider V R 1989 *Guide for Selecting Manning's Roughness Coefficients for Natural Channels and Flood Plains*.
[31] Hills J G and Mader C L 1997 Tsunami Produced by the Impacts of Small Asteroids *Annals of the New York Academy of Sciences* **822** (1), 381–394.
[32] Chow V T 1959 *Open - Channel Hydraulics* McGraw-Hill New York
[33] National Geophysical Data Center / World Data Service, 1900-2020. NCEI/WDS Global
[34] Sinaga T P T, Nugroho A, Lee Y W and Suh Y 2011 GIS mapping of tsunami vulnerability: Case study of the Jembrana regency in Bali, Indonesia KSCE *Journal of Civil Engineering* 15 (3), 537–543
[35] Kumar T S, Mahendra R S, Nayak S, Radhakrishnan K and Sahu K C 2010 Coastal Vulnerability Assessment for Orissa State, *East Coast of India Journal of Coastal Research* 263 523–534
[36] Zuidam 1986 *Aerial photo-interpretation in terrain analysis and geomorphologic mapping*. Smits Publishers The Hague 442 pp
[37] IIDA K 1963 Magnitude, energy and generation mechanisms of tsunamis and a catalogue of earthquakes associated with tsunamis *Proceedings, Tsunami Meetings Associated with the Tenth Pacific Science Congress*
[38] Saaty T 1994 *Fundamentals of Decision Making and Priority Theory, with the Analytical Hierarchy Process. Pittsburgh*, PA.: *RWS Publications*
[39] Chen C F 2006 Applying the Analytical Hierarchy Process (AHP) Approach to Convention Site *Selection Journal of Travel Research* 45 (2), 167–174

IOP Conf. Series: Earth and Environmental Science **851** (2021) 012008 doi:10.1088/1755-1315/851/1/012008

The physical characteristics of the small volcanic island of Tidore and Hiri to support disaster management in Maluku

M A Marfai[1], B W Mutaqin[1*], D S Hadmoko[1], H Wijayanti[1], F Lavigne[2], and A Faral[2]

[1]Coastal and Watershed Research Group, Faculty of Geography, Universitas Gadjah Mada, Bulaksumur 55281Yogyakarta, Indonesia
[2]Laboratoire de Géographie Physique (LGP), Center National de la Recherche Scientifique (CNRS), Université Paris 1 Panthéon Sorbonne, France
*E-mail: mutaqin@ugm.ac.id

Abstract. This article presents the characteristics of the small islands of Tidore and Hiri, which are vulnerable to the dangers of volcanic eruptions in Indonesia. The physical condition of the area is very important for carrying out a sustainable disaster risk reduction analysis; therefore, climate and topographic data are important to know. The limited availability of data in the study area causes climate data to be obtained from NASA's Earth Science Data Systems (ESDS) and regional topographical presentation of Indonesia's national DEM data. The results show that in the period 2004 to 2014, the temperature in the study area ranged from 292 K to 309 K. The islands of Tidore and Hiri have monthly precipitation between 50 mm to 500 mm.

1. Introduction

As an archipelagic country, Indonesia has the risk of being exposed to complex disasters, both from the sea and land. The various types of hazards that exist cause a high number of vulnerable groups in Indonesia. On the other hand, the level of community preparedness in facing disasters tends to be low [1,2]. Therefore, it requires a spatial arrangement that is in accordance with the characteristics of the local area, both physically, culturally and minimizing the boundaries of an area in order to increase community resilience. Institutionally this has been regulated in Law Number 26 of 2007 concerning Spatial Planning which makes disaster aspects in spatial planning considerations in Indonesia so that disaster-prone areas are the basis for disaster mitigation to improve the safety and comfort of life and community livelihoods. The objective to be achieved from implementing this rule is the creation of disaster management in Indonesia that supports sustainable development and improves the welfare of the Indonesian people.

Every action taken in a disaster analysis must be in accordance with the characteristics of the area, one of which is the characteristics of small islands. Based on the Law of the Republic of Indonesia Number 27 of 2007, a small island is an island with an area of less than or equal to 2,000 km^2, including its ecosystem unity. Volcanic eruption disaster management on a small island (e.g., Ternate, Tidore, and Hiri Island) requires more effort than volcanic eruption disaster management on a larger island (e.g., Java, Sumatra, and Sulawesi). This is due to all the limitations that need to be minimized so that small islands can support their communities to become more resilient to disasters. Some of the limitations they have are as limited of space, limited of natural resources, limited of accessibility, vulnerability to natural disasters, vulnerable to ecosystems damage, difficulties in terms of transportation and communication, limited of internet networks, limited of markets, limited of clean water supply, and high dependence on imports [3]. This is what causes people on small islands to be more vulnerable to disasters, especially people on small volcanic islands to volcanic hazards [4,5].

2. Method

This research begins with the collection of secondary data related to climate data (temperature data, rainfall data), Digital Elevation Model (DEM) data, and analysis of previous research related to the characteristics of the Tidore and Hiri Islands region. The climate data obtained from NASA's Earth Science Data Systems (ESDS), which was downloaded from https://earthdata.nasa.gov/esds, and regional topographical presentation of Indonesia's national DEM data, which was downloaded from http://tides.big.go.id/DEMNAS/.

3. Research Location

Mount Gamalama, Ternate, is one of the active volcanoes in Indonesia (Fig. 1). Based on the potential for volcanic hazards, data from the Indonesian Ministry of Energy and Mineral Resources (ESDM) explained that the Gamalama has a history of eruptions from 1538 to 2018 [6,7]. The first recorded eruption of Gamalama was in 1538, while the most significant eruption occurred in 1775. Gamalama eruptions in 1608, 1771, and 1840 have also triggered tsunamis that affected the surrounding area [8,9,6]. Since then, the activity of Gamalama has resulted in more than 200 death tolls [10]. As a small island with an area of 111 km^2 and 202,262 populations, Ternate is very prone to volcanic hazards from Gamalama, which is currently in a dormant situation [11]. Therefore, when Gamalama erupted in 1980, 40,000 peoples from Ternate were evacuated to Tidore since thick tephra and ashfall covered almost the entire island [12].

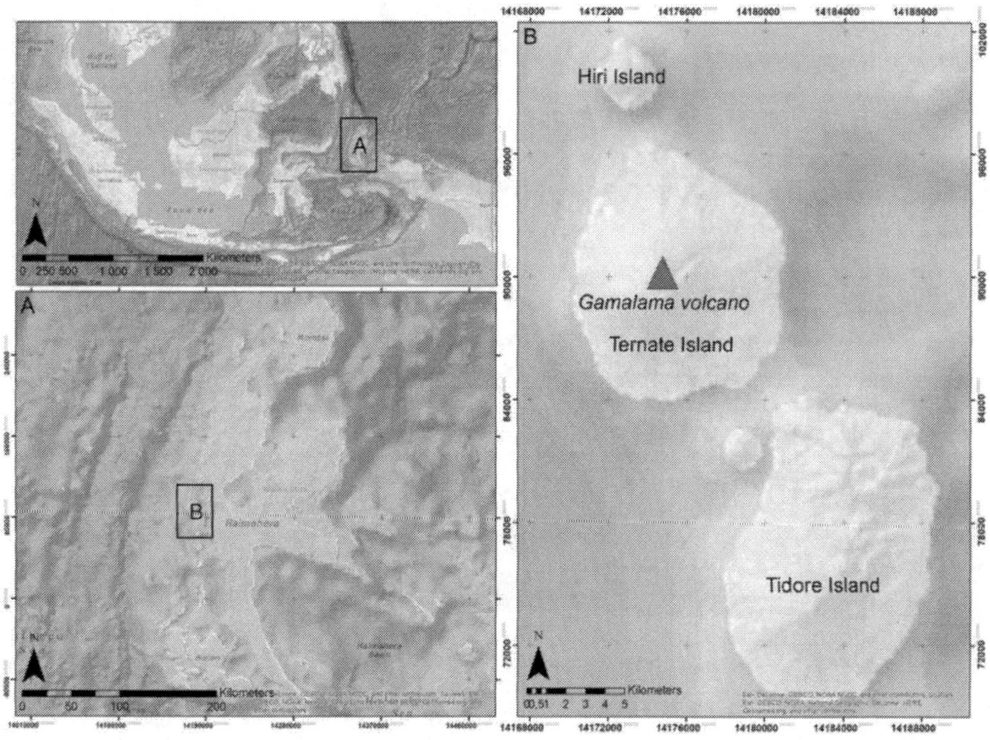

Figure 1. The research location.

IOP Conf. Series: Earth and Environmental Science **851** (2021) 012008 doi:10.1088/1755-1315/851/1/012008

4. Result and Discussion

Research related to volcanic eruptions has been experiencing a fairly good development in recent years in Indonesia, including small volcanic islands such as Ternate Island [13]. Ternate Island is one of small island in Indonesia with an extensive land area of 162.17 km^2 where it is surrounded by 5,547.55 km^2 sea areas. This condition causes sea transportation to and from Ternate Island is the main access. The population growth rate continues to increase causing the population density on Ternate Island in 2021 to reach 1,264.11 per km^2. One of the famous tourist attractions in Ternate is Batu Angus (local name, or charred rock) as a lava flow transferred from the summit down the northern flank to the coastal on March 10-13, 1737 (Fig. 2 and Fig. 3).

Moreover, Gamalama Volcano (1,715 masl) has recorded eruptions that have been fairly routine, occurring almost every year since the first recording was made, namely in the 1538. Between 1990 and 2018, 14 eruptions were recorded, which occurred randomly in uncertain months (Table 1). The Global Volcanism Program notes that these eruption events last from 1 to 15 days. The shortest eruption occurred in 1991 on 5 June, and the eruption occurred for only one day. Then the longest eruption occurred in 2011 from 5 to 23 December. Reporting of the last eruption in October 2018 resulted in gas emissions (mostly water vapor) which caused visitors and residents to stay away from the crater for a radius of 1.5 km and have an alert level of 2 (scale 1-4). Gamalama volcano is monitored by the Center for Volcanology and Geological Hazard Mitigation (PVMBG, also known as the Indonesian Center for Volcanology and Geological Hazard Mitigation, CVGHM).

Table 1. Historical Gamalama eruption's data from 1990 to 2018.

NO.	ERUPTION DATE (START)	ERUPTION DATE (END)	VEI	EVIDENCE
1	2018 Oct 4	2018 Oct 6	1	Historical Observations
2	2016 Aug 3	2016 Aug 4	1	Historical Observations
3	2015 Jul 16	2015 Sep 8	2	Historical Observations
4	2014 Dec 18	2014 Dec 25	2	Historical Observations
5	2012 Sep 15	2012 Sep 17	1	-
6	2011 Dec 5	2011 Dec 23	-	-
7	2008 May 10	[Unknown]	1	-
8	2007 Aug 23	[Unknown]	1	-
9	2003 Jul 31	2003 Oct 2 ± 3 days	2	Historical Observations
10	1996 Jul 2	Unknown	2	Historical Observations
11	1994 Jan 16	1994 Oct 15 (in or after)	2	Historical Observations
12	1993 May 6	1993 May 21	2	Historical Observations
13	1991 Jun 15	[1991 Jun 15]	1	-
14	1990 Apr 25	1990 Apr 26	3	Historical Observations

Sc: volcano.si.edu/global volcanism program

Figure 2. Batu Angus (local name, or charred rock) as a lava flow transferred from the summit down the northern flank to the coastal on March 10-13, 1737
Sc: volcano.si.edu

Figure 3. Batu Angus current conditions as a tourism object

To support disaster risk reduction in coastal areas and small islands due to potential of Mt. Gamalama eruption, especially Hiri Island and Tidore Island, local climate analysis is needed through several parameters, such as temperature and rainfall data [14,15]. The research locations, Tidore and Hiri Islands are part of the climate station on Ternate Island, and there is only one rain station that is actively taking measurements, namely the Sultan Babullah Station, Ternate City, North Maluku. Due to the limited historical data at the research location, rainfall data used in this study comes from data downloaded online via earthdata.nasa.gov with a time span from September 2004 to December 2014. NASA's Earth Science Data Systems (ESDS) facilitates users to access data as a general output product. The average surface temperature at the study site ranged from 292 K to a high of 309 K in the period from September 2004 to December 2014, which is shown in Fig. 4. The lowest temperature that occurs tends to be stable between 292 K to 294 K, while the upper average temperature tends to be more dynamic, from 302 to 309 K.

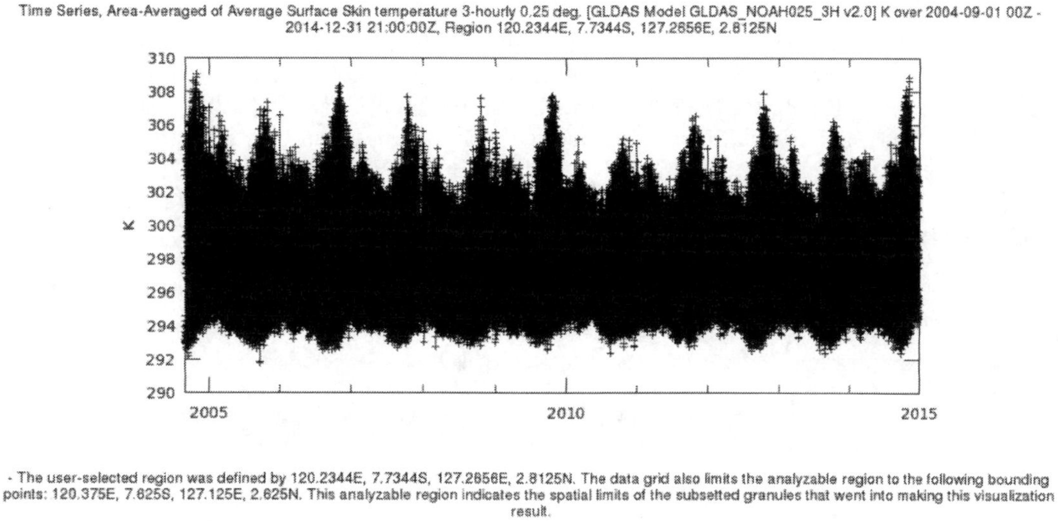

Figure 4. Temperature data of research locations in 2004-2014.

IOP Conf. Series: Earth and Environmental Science **851** (2021) 012008 doi:10.1088/1755-1315/851/1/012008

The precipitation data are presented in Fig 5. The precipitation at the study location fluctuated in ten years, increasing and decreasing in value each month; tropical climatic conditions also influence this in Indonesia. The range of precipitation experienced is from 50 mm per month to 500 mm per month. The lowest precipitation occurred at the end of 2009, which is under 50 mm and the highest occurred in mid-2010, which is 500 mm.

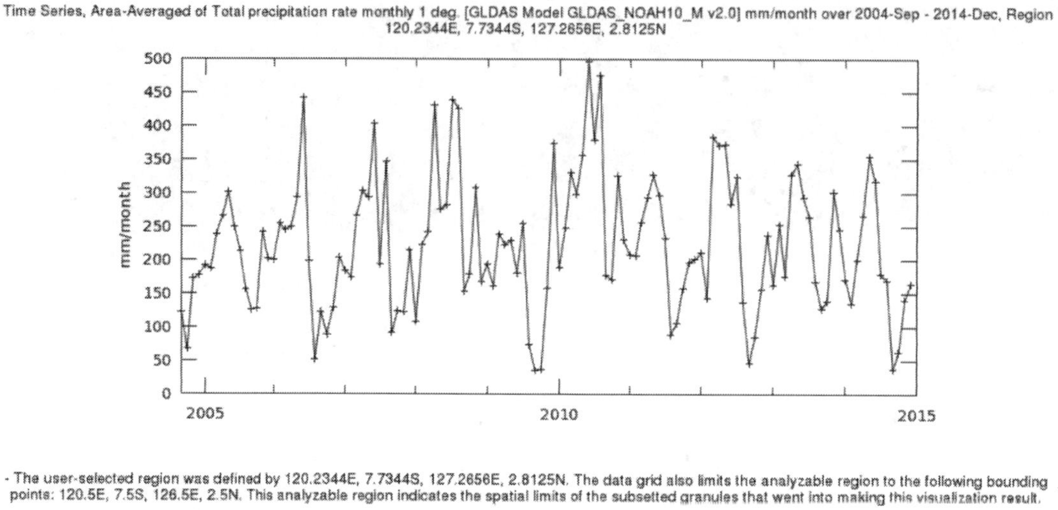

Figure 5. Monthly rainfall data of research locations in 2004-2014.

In answering the first objective, a topographic map is one of the outputs. Topographic maps help in the analysis of potential evacuation sites for the eruption of Mt. Gamalama on Ternate Island. Topographic maps in the study area represent differences in relief aided by the display of contour lines of the area [16]. The use of the distance between contours or contour intervals of 200 meters is presented in Fig 6. Broadly, the dominant topographic expression in the two research locations is stepped topography. This is the intention of occurring on Tidore Island compared to Hiri Island, not only because Tidore Island has a higher altitude than Hiri Island, but also because Tidore Island is bigger than Hiri Island, which causes variations in topographic expressions to be more varied.

Based on geological conditions and the risk of the eruption of Mt. Gamalama, volcano prone to eruption area type is differ into three categories with various range of materials, shown in Fig. 7 [13]. Disaster prone area (DPA) of Mt. Gamalama, which is divided into 3, namely DPA 1, 2 and 3 distributed in several locations. DPA 1 with low risk is located along the basin and the downstream area in the summit, namely in Dufa-Dufa Tabam, Tubo, Kulaba, Bula, Tobololo, Takome, Loto, and Togafo Village. Then DPA 2 with a medium risk level in the form of rock outburst and thick volcano ash in the radius of 3.5 km from the crater. And the last DPA 3 closest to the center of the eruption consists of the eruption column, incandescent rock outburst and lava stream.

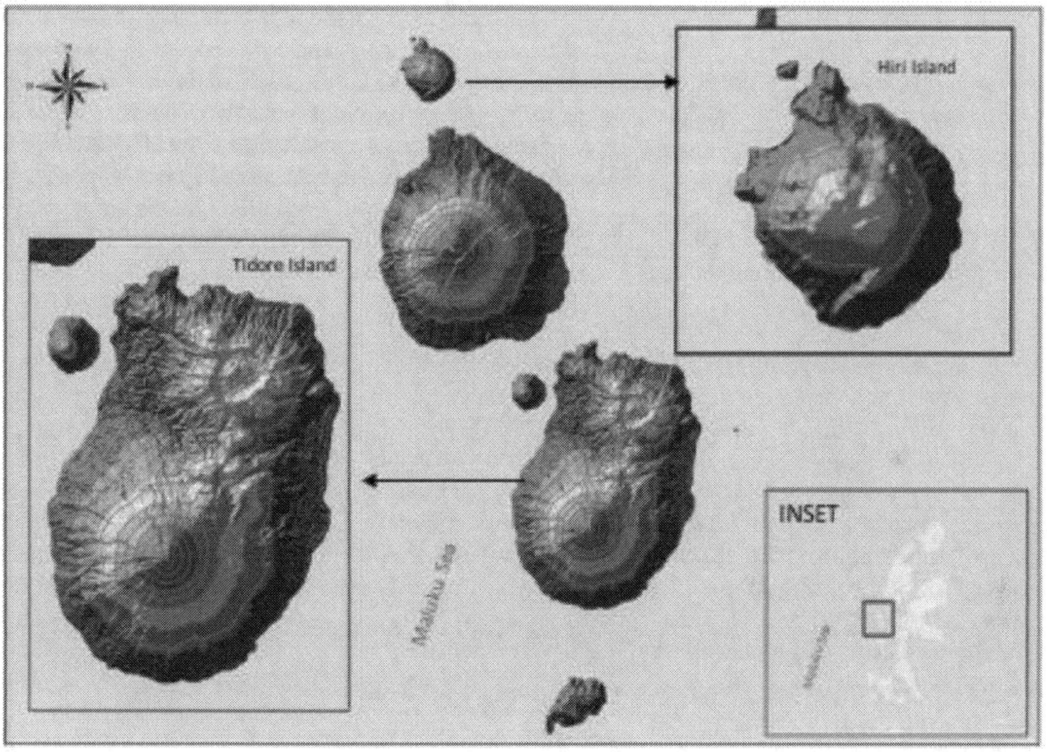

Figure 6. Topographic maps Tidore and Hiri Islands.

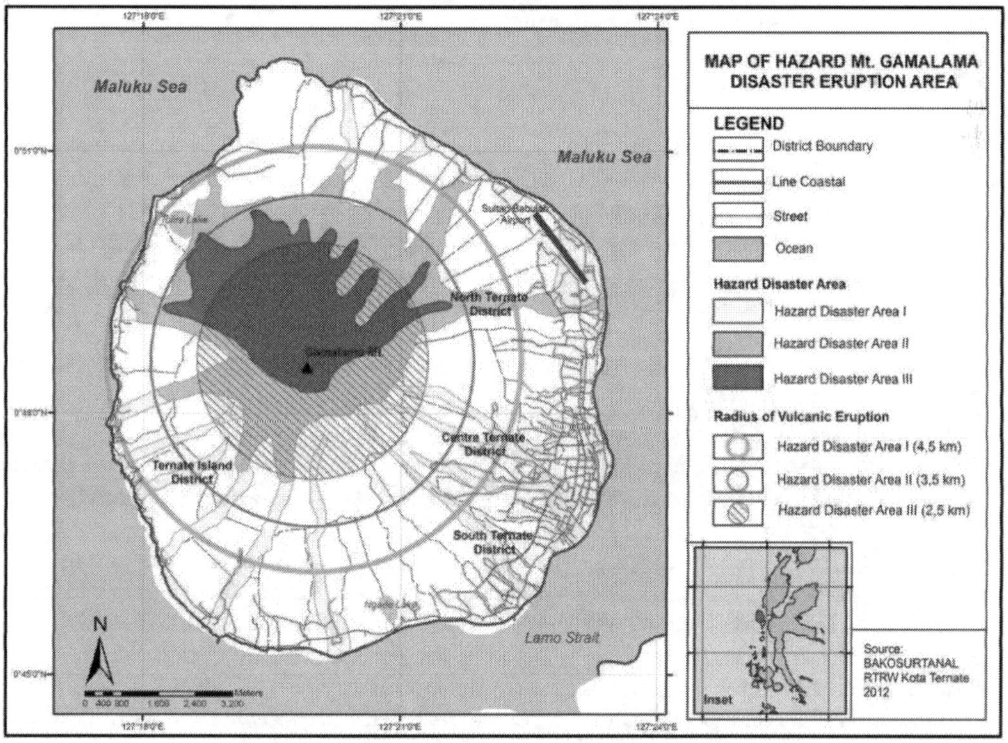

Figure 7. Map of Disaster-prone area Mt. Gamalama [13].

The above eruption disaster risk causes Hiri Island and Tidore Island to be safer locations to be used as evacuation sites for victims of the Mt. Gamalama. These two islands are located in two opposite directions. Hiri Island with an area of 6.69 km^2 is located to the north of the island of Ternate, while Tidore Island with an area of 1,550 km^2 is to the south. The selection of these two islands as alternative sister islands when the Mt. Gamalama volcanic eruption disaster is considered a relatively safer land for refugees, using sea transportation routes in the form of ferries or speedboats that can be provided by the local government of North Maluku Province. Hiri Island is one of the mainland that is equipped with public facilities, for example a source of lighting from PT PLN in North Maluku, this is one of the important factors in determining the location of an eruption evacuation.

5. Conclusion

To sum up, climatic conditions and geological situations greatly influence decision making in the evacuation of Mount Gamalama eruption disaster. Tidore and Hiri Islands were experiencing dynamic climate conditions within ten years (from 2004 to 2014). The results show that in this period, the study area's temperature ranges from 292 K to 309 K. The islands of Tidore and Hiri have monthly precipitation between 50 mm to 500 mm, and the dominant topographic expression in the two research locations is stepped topography. In the context of disaster risk, there is an increase in the elements at risk that can be affected by the activity of Gamalama in the future. Analysis of the physical characteristics of the small volcanic island of Tidore and Hiri is one of the strategies for disaster risk reduction (DRR) in this area. These efforts to reduce the risk related to the Gamalama eruption are helpful for the success of regional and national development and the realization of sustainable mitigation of natural disasters.

Acknowledgments

This research was funded by the grant of National Basic Research from the Indonesian Ministry of Research, Technology and Higher Education - GASTROPODE Project. We thank anonymous reviewers for their constructive comments on the manuscript.

References

[1] Sudibyakto 2000 *Kajian dan Mitigasi Bencana Gempa Bumi : Studi Kasus Gempa Bumi Bengkulu,4 Juni 2000 Jurnal Majalah Geografi Indonesia 2000,XIV,* Fakultas Geografi Universitas Gadjah Mada, Yogyakarta.

[2] Lavigne F 2010 Ulasan Publikasi. dalam: Sunarto.. Marfai. M.A.. dan Mardiatno. D (ed). Penaksiran Multirisiko Bencana di Wilayah Parangtritis: Suatu Analisis Serbacakup untuk Membangun Kepedulian Masyarakat terhadap Berbagai Kejadian Bencana (Yogyakarta: BPFG)

[3] Wilkinson E, Lovell E, Carby B, Barclay J and Robertson R E A 2016 The Dilemmas of Risk-Sensitive Development on a Small Volcanic Island. *Resources* **5** 21. https://doi.org/10.3390/resources5020021.

[4] Schultz J M, Cohen M A, Hermosilla S, Espinel Z and McLean A 2016 Disaster risk reduction and sustainable development for small island developing states. *Disaster Health*, 3:1, 32-44, https://doi.org/10.1080/21665044.2016.1173443.

[5] Mei E T W, Sari I M, Fajarwati A and Safitri D 2017 Assessing the Social Economic and Physical Vulnerabilities to Gamalama Volcano. Advances in Social Science, Education and Humanities Research, *1st International Conference on Geography and Education* (ICGE 2016), Vol. 79, 33-40.

[6] Mutaqin B W, Lavigne F, Hadmoko D S and Ngalawani M N 2019 Volcanic Eruption-Induced Tsunami in Indonesia: A Review, *IOP Conf. Ser.: Earth Environ. Sci.* **256** 012023. https://doi.org/10.1088/1755-1315/256/1/012023.

[7] Hidayat A, Marfai M A and Hadmoko D S 2020 The 2015 Eruption of Gamalama Volcano (Ternate Island-Indonesia): Precursor, Crisis Management, and Community Response, *GeoJournal (2020)*. https://doi.org/10.1007/s10708-020-10237-w.

[8] Soloviev S L, Go Ch N, Kim K S 1992 *Catalog of tsunamis in the Pacific, 1969-1982*. (Academy of Sciences of the USSR, Moscow) 208p.

[9] National Geophysical Data Center / World Data Service (NGDC/WDS). 2018. *Global Historical Tsunami Database*. National Geophysical Data Center, NOAA. https://doi.org/10.7289/V5PN93H7.

[10] Bacharudin R, Martono A and Djuhara A 1996 *Peta kawasan rawan bencana gunungapi Gamalama, Ternate, Maluku Utara*. Bandung: Direktorat Vulkanologi.

[11] Hidayat A, Marfai M A and Hadmoko D S 2020 Eruption on Indonesia's volcanic islands: a Review of Potential Hazards, Fatalities, and Management, *IOP Conf. Series (ICEGE 2019)*, Sci. 485 012061. https://iopscience.iop.org/article/10.1088/1755-1315/485/1/012061/pdf.

[12] Gaudru H 2005 *Potential Impacts of Eruption on Volcanic Islands: Global Approach for Volcanic Risk Mitigation*. SVEUROP, Geneve.

[13] Syiko S F, Ayu T R and Yudono 2013 Evacuation route planning in Mount Gamalama, Ternate Island, Indonesia. The 3rd International Conference on Sustainable Future for Human Security SUSTAIN 2013. *Procedia Environmental Sciences* **17**, 344-353.

[14] Setiawan O 2012 Rainfall and Temperature variability analysis in Bali. *E-journal Analisis Kebijakan Kehutanan* Vol **9** No 1, 66-79.

[15] Bayong T 2004 *Klimatologi*. ITB Press. Bandung.

[16] Rogers J D and Doyle B C 2004 Mapping of Seismically-Induced Landslippage in the Benton Hills and Crowley's Ridge, New Madrid Seismic Zone, Missouri and Arkansas. Department of Geological Sciences & Engineering. The University of Missouri-Rolla.

Geological structures control on the streamflow orientation pattern in the Cijulang area and its surroundings, Garut, West Java, Indonesia

A Setianto[1]*, G Yugamaris[1], I W Warmada[1] and P Bangun[2]

[1]Department of Geological Engineering, Faculty of Engineering, Universitas Gadjah Mada, Yogyakarta, Indonesia.
[2]PT ANTAM Tbk, Unit Geomin, Jakarta Timur, Indonesia.
*E-mail: agung_setianto@ugm.ac.id

Abstract. The Cijulang area is one of the deposits included in the high sulfidation epithermal deposits in Garut Regency, West Java Province. Ore mineralization in high sulfidation epithermal deposits is closely related to the geological structure. The existence of streamflows can be used as a fundamental basis for identifying geological structures. Geological structure identification is done by classifying the orientation angle interval class in stream segments. Each grid unit in the orientation angle interval class stream segment which has the same pattern is controlled by geological factors, including geological structures or the constituent rocks. By utilizing streamflow data, Digital Elevation Model (DEM), and secondary data (e.g., geological features such as structure geology and lithology from previous research data), geological structures can be identified. Faults that can be identified are the Cikahuripan fault, Citando fault, Cisuren fault, and Cibuni fault. These faults affect streamflow by 44.54% in the research area while lithology affects 35.12% of it. The influence of lithology will be greater in lower-order stream segments and the effect of faults will be greater in high order stream segments.

1. Introduction

The geological structure is one of the factors controlling geological processes that occur in an area. A geological structure can be seen from its distinctive morphological appearance, such as presence of fault scarps, the stream position with a certain streamflow pattern, and the continuity of slopes or ridges of morphology. Using remote sensing, geological structure can be observed through streamflow pattern and can facilitate the early stage of exploration. The relationship between geological structures and streamflow patterns can be studied quantitatively to identify the geological structure and determine geological structures effect on the streamflow pattern. In this study, each stream order orientation formed by perennial streams and intermittent streams in a watershed is measured, and its pattern is observed using Digital Elevation Model DEM. The geological structure can be identified from DEM by the appearance on the surface in the form of lineament created by fault scarps, stream deflections, or the shutter ridges. With the geological structure in the rock body, the stream flows easily in the said zone [1]. This is because the fault itself is a weak zone in the rock body. Lineament of morphology can also be controlled by lithology, where in this case, the lineament in the rocks is in the form of streamflow due to slope morphology of an area controlled by rock resistance [2]. Streams can flow in a valley but generally do not have a distinctive orientation and pattern. By comparing some of these aspects, the presence of lineament can determine the control of the geological structure or lithological lineament of a stream segment.

Mineralization in high sulfidation epithermal deposits is controlled by geological structure closely related to lithology. As in the Cijulang Area, mineralization is controlled by lithology and geological structure [3]. In this Area, mineralization is developed in Tertiary (Neogene) sedimentary rocks and altered Quaternary volcanic rocks. The Cijulang Area is located in Talegong and Cidaun Districts, Garut and Cianjur Regencies, West Java Province (Figure 1), which has been explored by PT ANTAM since 1966 [4]. This study aims to identify the geological structure and examine the geological structure's influence on the streamflow patterns developed in the research area. This research can support exploration data in the research area by utilizing the relationship analysis between geological structures and streamflow.

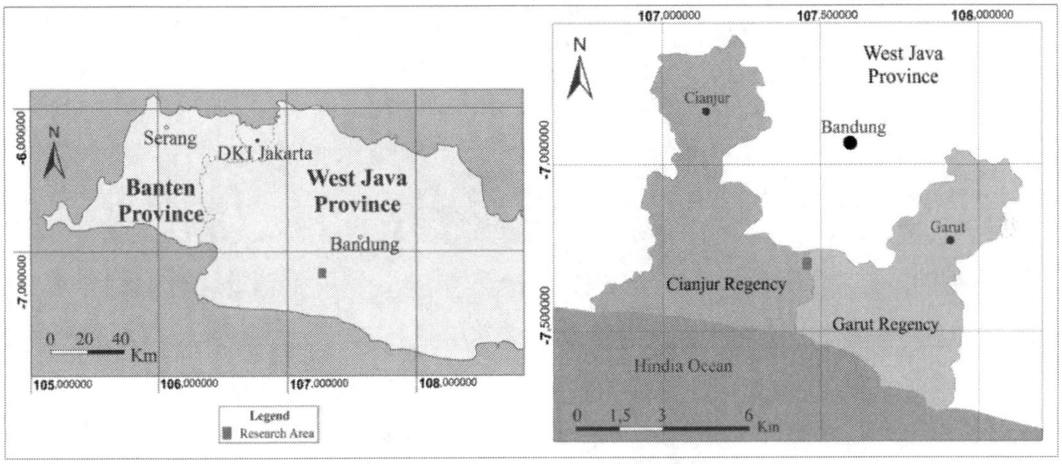

Figure 1. The research area located in Garut and Cianjur Regencies, West Java Province.

2. Regional Geology of The Research Area

Cijulang Area is included in the Southern Mountain Zone of West Java Pangalengan Plateau [5]. The Southern Mountains of West Java is part of the Java Geosyncline's limb, uplifted and tilted towards the south. Pangalengan Plateau then undergoes normal faulting and fracturing towards the Bandung Zone. In this area, the morphology is characterized by strong relief hills with a trellis streamflow pattern. The slope at this location ranges from 15° to 40° and slopes towards the west. The morphology of Cijulang Area is controlled by lithology and geological structure. It can also be seen that there is stream alignment, fault scarps, and cliffs with waterfalls in several locations [5]. The research area constituent rocks are Tertiary sedimentary rocks and Quaternary volcanic rocks, namely andesite, tuff, dacite, polymict breccia, andesite breccia, and diorite (Figure 2) [6]. The geological map made by PT ANTAM to determine the prospect zone of Cijulang Area. Detailed geological mapping had been conducted only on the area around streamflow, mainly on perennial stream (Cikahuripan River) and intermittent stream around it, based on soil and rock sampling. These were intended to prospect the area at an early stage of exploration.

Geological structures that control mineralization are either strike-slip or oblique faults. The peak of structure formed in the research area was formed in Plio-Pleistocene. In this area, mineralization is generally found in tuff, andesite, and dacite, which are generally altered into silicification and advanced argillic alteration. Mineralization in high sulfidation epithermal deposits is generally initiated in the presence of geological structures and permeable rocks, including diatreme breccia pipes [7,8].

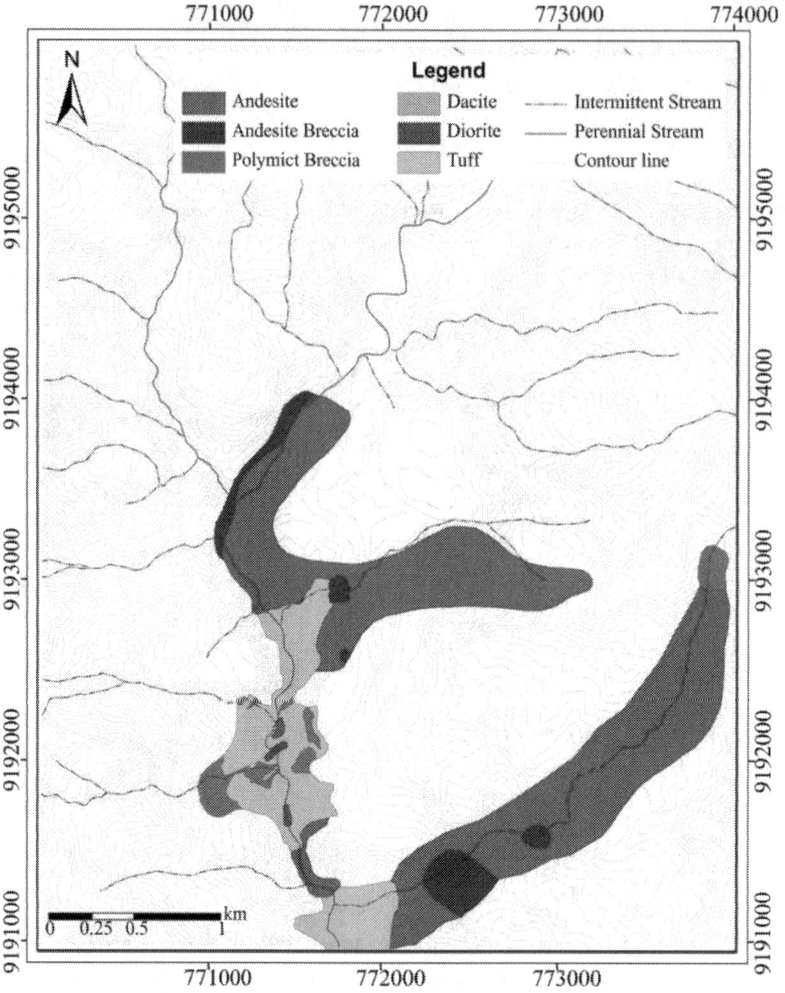

Figure 2. Geological map of the research area [6].

3. Methodology

Identification of geological structure control based on streamflow orientation can be made using a quantitative calculation. This method has previously been applied to the Loucka drainage basin, SE margin of the Bohemian Massif [9]. The degree of correspondence between the direction of drainage network segments and the orientation of geological structure was investigated by measure the length of drainage segment oriented parallel with geological lineation. In this study, the drainage network map used has a scale of 1:25.000. Both perennial and intermittent stream were included. In the research area, streams are divided into segments based on their stream order (Figure 3). Strahler stream ordering method is used in this study. Strahler stream ordering is a method for assessing river size and complexity based on the number and hierarchical relationship of tributaries. The stream without any tributaries is considered a 1st order stream. When two of 1st order stream join, a 2nd order stream is formed. When two of 1st order stream join, a 2nd order stream is formed, and so on. The segment order divided into 4 orders, such as first order, second order, third order, and higher order. Higher order is referred to segment that has stream order higher than third order. The stream segment in the research area is processed by converting the streamflow into a straight line, and the angle of its orientation is measured clockwise the 0° direction, which refers to the north. The measuring angles are divided into 12 classes of intervals, every 15° in an angle range of 0-180°. Interval classes with the same values are grouped into 1 area, where this area results from the conversion from the stream network to a grid size with a pixel size of 50×50 meters. Areas with the same class intervals are then compared with the DEM data resulting from the interpolation of elevation points, streams, and topographic lines. Areas with distinctive pattern in the DEM, such as fault scarps, stream deflections, and shutter ridges, are identified as geological structures. In contrast, those that do not have a distinctive or specific pattern will be grouped into lithological lineation. Each segment is then measured in length. The segment's position parallel to the geological structure or rock lineation is calculated for its length separately. This is done to determine the percentage of stream segment orientation affected by faults or rock lineation. Calculations are then made on each stream order, where the numbering uses ordering by Strahler.

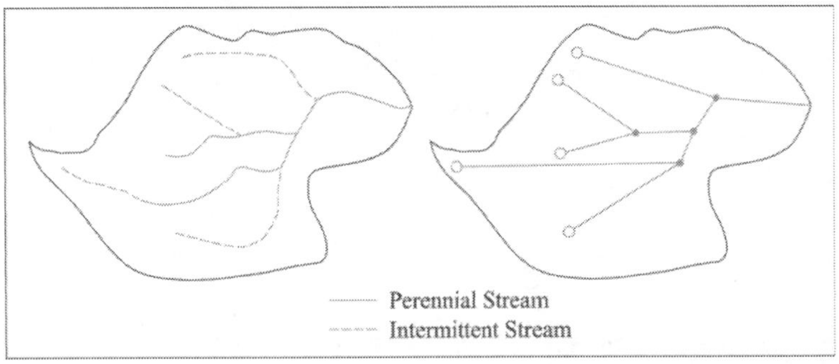

Figure 3. The procedure for converting streamflow pattern, where all parts or segments are converted into straight lines. The orientation and length of each segment are then measured [9].

4. Results & Discussion

The streamflow in the Cijulang area has a trellis pattern (Figure 4a). It has streamflow with the stream direction of the Northwest-Southeast with the direction of the tributaries the West-East and Northeast-Southwest directions. In the research area, the streams were converted to straight lines, and their orientations are then measured (Figure 4b). The stream is then divided into 46 segments. The stream's angular orientation is expressed in interval classes on a grid with a size of 50×50 meters. Grids with the same pattern and value will be grouped to identify geological structures in the research area. Based on this analysis, it is found that several stream segments formed lineament patterns, which are presumed to be the alignments of the geological structures because it has same interval class and distinctive pattern morphology in DEM.

The area that is presumed to be the geological structure's location in the DEM has a valley lineament with a distinctive pattern. Besides, there are also deep valleys in the constituent rocks. This structure was identified as a fault. The fault that can be identified is the Cikahuripan fault with the stream orientation of the Northwest-Southeast and is a dextral oblique fault. Then another identified fault in the research area is the Citando fault with a Southwest-Northeast orientation. This fault is a sinistral strike-slip fault with a reverse fault type. The other structures are dilatational faults, such as the Cibuni and Cisuren faults (Figure 5).

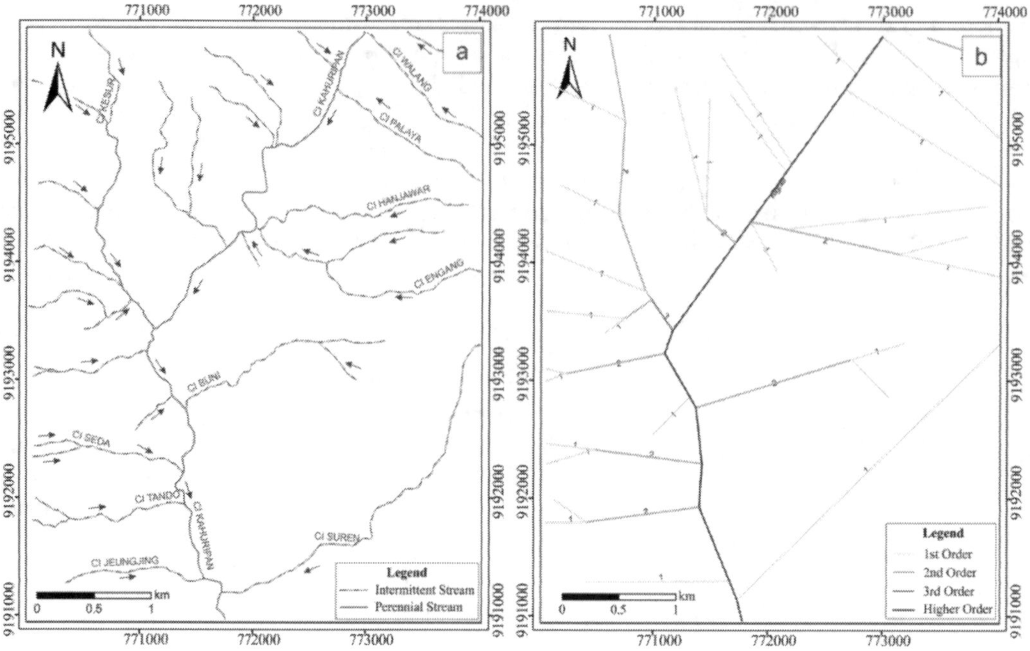

Figure 4. Streamflow in the Cijulang area, (a) the streamflow pattern of the stream in the form of trellis pattern, (b) the results of the stream conversion into straight lines and the stream order numbering.

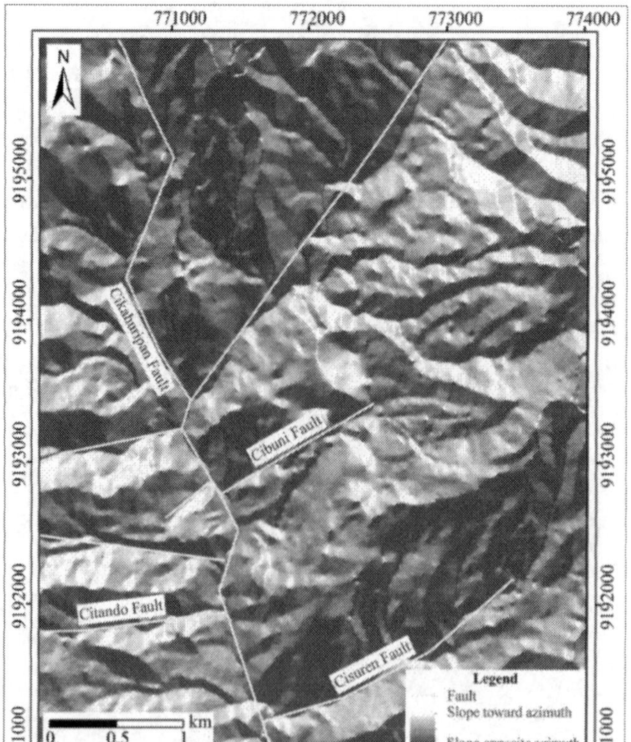

Figure 5. Geological structures in the Cijulang Area identified from the stream flow pattern.

In the research area, several segments are parallel to faults and lithological lineament (Figure 6). In which, 21 segment are considered to be affected by faults and 20 segment are considered to be affected by rock lineament. In contrast, the rest of the segment orientation is not parallel with faults or lithological lineation. The length of the segment parallel to the fault and rock lineation is then measured, and the percentage of segments affected by the presence of the fault or rock lineation is then calculated. The length of the segment parallel to the geological structure and rock lineation is calculated. The percentage calculation is carried out using the total length of the segment in each order. The results of the percentage effect of faults and rock lineation calculation are presented in Table 1.

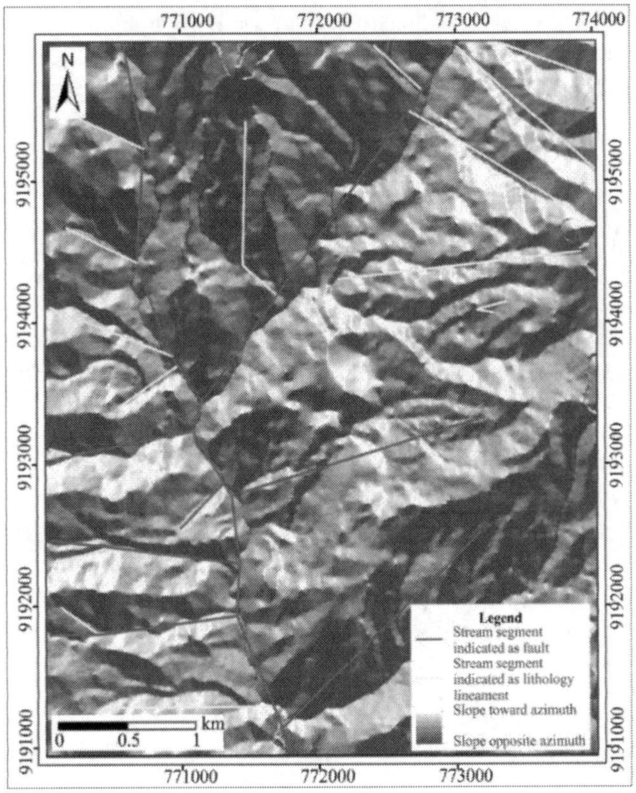

Figure 6. The position of the segments affected by faults and rock lineation.

Table 1. Percentage of the faults and rock lineation effect on streams in the Cijulang area.

Segment	Number of segment	Total length of the segment (km)	Length of segment parallel to the geological feature		Percentage of affected value	
			Fault (km)	Rock lineation (km)	Fault (%)	Rock lineation (%)
1st Order	27	22.396	4.058	11.986	18.123	53.518
2nd Order	12	9.350	6.682	1.317	71.477	14.092
3rd Order	1	0.325	0.325	-	100	-
Higher Order	6	5.805	5.805	-	100	-
Total Segment	46	37.876	16.870	13.303	44.54	35.12

Among the influence of the valley's existence on hills marked by lineation in rocks or geological structures in the form of faults, first order stream in the research area tends to be controlled by constituent rocks (53.52%) when compared to the effect of faults (18.12%) in the research area. This striking comparison can confirm the influence of the dominant constituent rocks in controlling the streamflow. These streamflows occupy small valleys on hills. Second order streams have several differences with first order streams. Second order streams are influenced by the structure with a percentage of 71.48% and are influenced by rocks with a percentage of 14.09%. In this order, streams are controlled by a combination of geological and lithological structures but are dominated by the influence of geological structures. Aside from that, the third order stream and higher orders generally have a position parallel to or coinciding with the faults that have been identified by previous researchers, and there are no segments that correlate singularly with rock lineation without the presence of faults. This is indicated by the presence of a regular stream pattern. In this case, the stream can indicate a geological structure in this research in the form of a fault. Major faults can be seen coinciding with or parallel to the streamflow by this method compared to minor faults.

The developments of morphology and streamflow patterns in the upstream (low order) area are closely related to morphology and lithology development in the research area. Meanwhile, streams that have reached a further order will be more influenced by geological structures. The older major structures generally have a major influence due to their increasingly complex developments. The higher the order of the streams, the higher the fault effects would be. Meanwhile, low order streams have less fault effect, more dominated by the constituent lithological conditions. This is also found in Macka's (2003) research, where faults predominantly affect streams in the high order. In general, the stream segment in the research area is also dominated by faults (44.54%), with an influence from the lithology of the research area's composition (35.12%). These structures will then become weak areas for the migration of hydrothermal fluids that form alteration and mineralization in the research area.

5. Conclusion

Geological structures in the form of faults identified in the Cijulang area are the Cikahuripan fault, the Citando fault, the Cisuren fault, and the Cibuni fault. The streams in the research area were affected by faults with a percentage of 44.54%. The higher the stream order, the higher the geological structure control in the stream's existence, and the lower order stream is influenced dominantly by lithology with a little influence of the geological structure.

Acknowledgments

The author would like to show gratitude to PT Antam Tbk. for their permission and support in the making of this research in the Cijulang area.

References

[1] Azhari A P, Maryanto S and Rachmansyah A 2016 Identifikasi Struktur Geologi dan Pengaruhnya Terhadap Suhu Permukaan Tanah Berdasarkan Data Landsat 8 di Lapangan Panasbumi Blawan *Jurnal Inderaja* **13** 1–12

[2] Noviarni A and Setiawan B 2019 Studi Tingkat Erosi Dengan Analisa Morfometri Studi Khusus Daerah Kemiling, Bandar Lampung *Prosiding Applicable Innovation of Engineering and Science Research 2019* 1181–5

[3] Tun M M, Warmada I W, Idrus A, Harijoko A, Verdiansyah O and Watanabe K 2014 High Sulfidation Epithermal Mineralization and Ore Mineral Assemblages of Cijulang Prospect, West Java, Indonesia *J. SE Appl. Geol.* **6** 29–38

[4] Oo K Y, Warmada W, Titisari A D and Watanabe K 2019 Ore Forming Fluid of Epithermal Quartz Veins at Cisuru Prospect, Papandayan District, West Java, Indonesia *J. Geoscience Eng. Environ. Technol.* **4** 170

[5] Purwanto H S and Verdiansyah O 2013 Karakteristik Endapan Emas Epitermal Sulfidasi Tinggi dan Hubungannya dengan Mineral Lempung Hasil Analisa Spektral, Daerah Cijulang, Kabupaten Garut Provinsi Jawa Barat *MTG* **6** 15

[6] PT ANTAM 2014 *Alterasi Daerah Cijulang, Papandayan, West Java* (Papandayan: PT ANTAM)

[7] Corbett G 2012 Structural Controls to, and Exploration for, Epithermal Au-Ag Deposits *Australian Institute of Geoscientists Bulletin* **56** 43–7

[8] Corbett G J and Leach T M 1998 *Southwest Pacific Rim Gold-Copper Systems: Structure, Alteration, and Mineralization* (Society of Economic Geologists)

[9] Macka Z 2003 Structural control on drainage network orientation an example from the Loucka drainage basin, SE margin of the Bohemian Massif (S Moravia, Czech Rep.) *Landform Analysis* **4** 109–17

IOP Conf. Series: Earth and Environmental Science **851** (2021) 012010 doi:10.1088/1755-1315/851/1/012010

Effect of opening depth of oblique weir with an opening on river bed morphology

H Une[1*], T Ohmoto[2] and K Kuranaga[3]

[1]Graduate School of Science and Technology, Kumamoto University, Japan
[2]Civil and Environmental Engineering
[3]Graduate School of Science and Technology, Kumamoto University, Japan
*E-mail: 181d9201@st.kumamoto-u.ac.jp

Abstract. In this study, we focused on the fact that an oblique weir with an opening activates the fluidity of the sediment compared with a conventional continuous fixed one and examined the removal function of the sediment deposited upstream of the weir. Experimental results showed that the weir with the opening increased the fluidity of the sediment deposited upstream of the weir and lowered the riverbed. The scouring section upstream of the weir was divided into a gradually varied flow section and a local flow section based on the bed topography. In the gradual flow section, the descent of water level and riverbed increases linearly as the opening depth increases. In addition, it was shown that the local scour shapes upstream and downstream of the weir were similar to each other as a result of normalization on the representative scale of vertical and transverse shapes. By creating the opening of the oblique weir, the magnitude of the downward flow generated near the left bank was reduced from 45% to 34% of the non-open weir.

1. Introduction

Cross-river structures such as dams and weirs have been installed for diversion to main rivers and tributaries, irrigation, water intake such as hydroelectric power generation, water depth maintenance, and salt damage prevention.

However, around these structures have been pointed out geo-environment degradation due to a discontinuity of sediment transport. For example, the flow velocity decreases upstream of the river crossing structure due to backwater. Moreover, river-bed levels rise because sediment transport is interrupted by these structures. On the downstream side of the fixed weir, physical and biological environments disappear and tend to suppress natural disturbance. The riverbed on the downstream side is eroded because the supply of sediment from upstream of the fixed weir is interrupted by the weir.

In order to solve these problems, the authors [1], [2] proposed to create an opening in the weir. The following effects can be expected by providing an opening and approaching a natural river.

1) The dredging cost can be reduced by flowing sediment accumulated in the upstream area of the weir for many years at the time of flooding.
2) Reduction of sediment accumulation on the upstream side leads to a decrease in water level at the time of flooding, and improvement of flood control safety can be expected.
3) Due to sediment transport from upstream, improvement of environmental functions can be expected on the downstream side due to recovery of physical and biological environment and increase of natural disturbance.

Figure 1. Ibiyamazeki (Ibiyama-Weir) Source: Google Earth.

On the other hand, it is necessary to study the effect of weirs with openings on riverbed variation and flow. However, few studies have examined the effects of weirs with openings on riverbed changes in previous studies. Zhang et al. [3] conducted a series of experiments using a flume bedded with various materials having different grain sizes and specific gravities and various weirs with an opening and investigated scour characteristics of the channel bed upstream of a submerged weir and local flows over the weir. Sumida et al. [4] performed experiments using a flume having a relatively large width-to-height ratio and investigated the effects of weir opening geometry and size on the channel bed processes upstream of the weir resulting from scouring.

By the way, weirs are generally installed orthogonal to the river channel, but some weirs are installed obliquely to the river channel. Oblique weirs have some advantages as follows [5].

1) In the upstream of the weir, the oblique weir has a lower water level than the orthogonal weir.
2) In the normal stage of water, oblique weir can smoothly withdraw water.
3) If an oblique weir is suitably made, it is possible to prevent side banks during floods.

However, if the oblique weir is not installed properly, it may promote side bank erosion downstream of the weir. As shown in Figure 1, there is also an oblique weir in the Shirakawa River that flows through Kumamoto Prefecture. There are concerns about the rise of the riverbed upstream of the weir and local scouring along the downstream riverbank.

There are several previous studies on oblique weirs. The prediction of the overflow of oblique weirs has also been investigated, Nguyen [6] and Kabiri-Samani [7] have investigated the effects of the geometrical characteristics of the weir and the overflow morphology on the flow coefficient. However, the effect of oblique weirs to riverbed morphology and river flow discharge have not been studied. On the other hand, in the examination of the flow around the oblique weir, Otoshi et al [8] conducted a hydraulic model experiment on the technical grounds related to the linear determination of the "Miyagawa river" in the Matsuda River flowing through Kochi Prefecture. Fukuoka et al. [9] focused on the relationship between the position of the oblique weir and channel shape, examined the effect of the oblique weir on water level and riverbed deformation.

However, no specific solution to the weir problems has been discussed in previous studies. The author [10] investigated the influence of oblique fixed weir with openings on flow and bed fluctuation to improve the flood control and environmental problems of the oblique weir.

As a subsequent study, in this study, this paper is to investigate effect of the opening depth of the oblique weir to on riverbed morphology at the downstream of the weir and sedimentation at the upstream of the weir.

2. Experimental setup

As shown in Figure 2 the experiments were conducted with a 10-m-long, 40-cm-wide, 20-cm-deep variable-slope circulating acrylic flume. The weir shown in Figure 3 is installed at a position 6 m downstream from the upstream end (weir angle $\theta = 30^{\circ}$), and the average particle size d50 = 0.77 mm in the section from the weir to 6 m upstream and 4 m downstream. The initial bed thickness upstream of the weir was 17 cm, which is the same height as the weir height. And the initial bed thickness downstream of the weir was 6 cm lower than the upstream initial bed thickness, 11 cm thick.

The depth of the opening (opening depth) of the weir model was set to D, and experiments were conducted in 7 cases by systematically changing the depth in the range of 0 cm to 6 cm at 1 cm intervals.

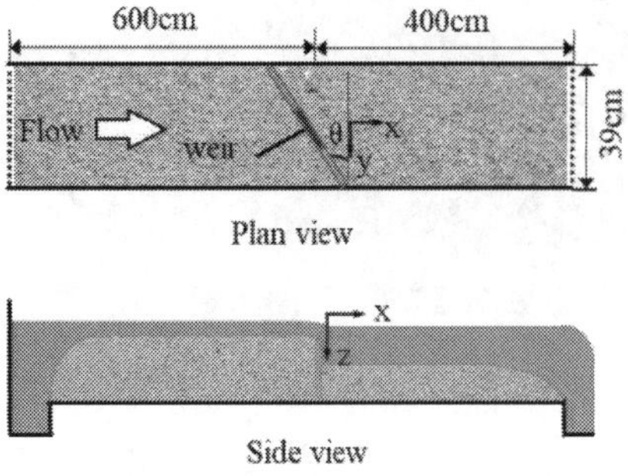

Figure 2. Experimental flume.

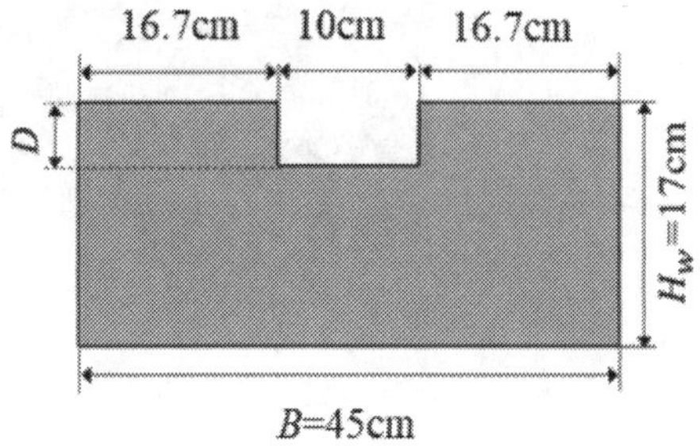

Figure 3. Weir shape and size.

Table 1 shows the initial conditions of the experiment. These hydraulic quantities show the values at the position 3 m upstream and 2 m downstream of the weir, respectively. Due to evaluate how much the sediment deposition upstream of the weir is improved, the flow discharge was set that the nondimensional tractive force τ_* calculated by assuming the law of the wall at 3 m upstream of the weir under the initial conditions was about 3 times the critical nondimensional tractive force τ_{*c} calculated from the mean grain size.

The water level and the bed elevation were measured with a point gauge and an ultrasonic sensor, respectively. The initial water depth downstream of the weir was set to 6.5 cm by manipulating the weir at the downstream end of the experimental channel so that it was below the nondimensional tractive force. The water level and riverbed shape were measured after confirming the equilibrium state of the riverbed. The time to reach equilibrium is 48 hours at an opening depth of D = 0 to 1 cm, 72 hours at D = 2 to 4 cm, and 96 hours at D = 5 to 6 cm, respectively. In order to secure the run-up section, sand was supplied to maintain the continuity of the riverbed surface of the target section in the range of 5 m to 6 m upstream from the weir.

Table 1. Hydraulic Condition.

Discharge: Q(l/s)	5.4	
Bed slope: I	1/300	
Critical depth(cm)	2.69	
Grain size: d_{50} (cm)	0.077	
Friction velocity: u_{*c}(cm/s)	2.06	
	Upstream (x= -300cm)	Downstream (x =200cm)
Flow depth: (cm)	2.8	6.5
Mean velocity (cm)	47.7	21.3
$\tau*/\tau*c$	3.0	0.45
Froude Number	0.9	0.27

In the local flow section, Figure 4 shows the illustration of parameters that were measured to describe the local scour hole in the longitudinal and transversal section. The maximum scouring depths upstream and downstream are dsu and dsd, respectively. The scouring length in the flow axis direction and the scouring length in the transverse direction downstream of the weir were set to lxd and lyd, respectively.

The right-hand coordinate system has its origin at the center of the bottom of the flume, where the opening is located, and the x-axis, y-axis, and z-axis representing the longitudinal direction, the transverse direction, and the vertical direction, respectively. The instantaneous flow velocity components, time average components, and fluctuation components are represented by u, v, and w; U, V, and W; and u', v' and w', respectively. The magnitude of flow velocity |u| was defined by the following equation:

$$|u| = \sqrt{U^2 + V^2} \tag{1}$$

where $|u|$ = magnitude of flow velocity; U = longitudinal velocity component; V = transversal velocity component.

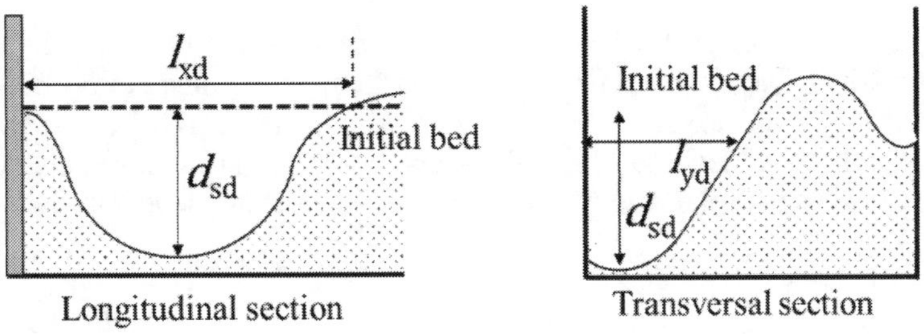

Figure 4. Parameter at local scour hole downstream of the weir.

To investigate the flow structure, point measurement, and multi-point simultaneous measurement of water surface profile and flow velocity were conducted in Case D=0cm and D=6cm shown in Table 1. The surface flow measurement was made by the PIV (particle image velocimetry) method using a video camera system capable of multi-point simultaneous measurement of flow velocity. The point measurement of flow velocity was made with I type and L type two-component electromagnetic current meters. In the flow velocity measurement by the PIV method, the surface flow was shot with the camera positioned directly over the flume. The flow velocity measurement duration was 20 seconds, and nylon particles having a diameter of 100 μm and a specific gravity of 1.02 were used as tracers. Image data were recorded on the hard disk as 59.94-fps (frame per second) 1920×1080-pixel monochrome video image data and processed by the PIV method. Following analogue-to-digital conversion, output signals (100 Hz) from the electromagnetic current meters were statistically processed so that 4,096 data sets were used for each measurement point.

3. Experimental results

Figure 5 shows the longitudinal in riverbed and water level. It can be seen that the longitudinal bed level rapidly changed in the range of x = -30 to 200 cm around the weir, but it was relatively gradual in the range of x = -300 to -30 cm. Therefore, the range of the gradually varied flow section and the local flow section was set to x = -300 to -30 cm and x = -30 to 200 cm, respectively. The average riverbed height upstream of the weir decreases as the opening depth D increases in both the gradual change section and the local flow section.

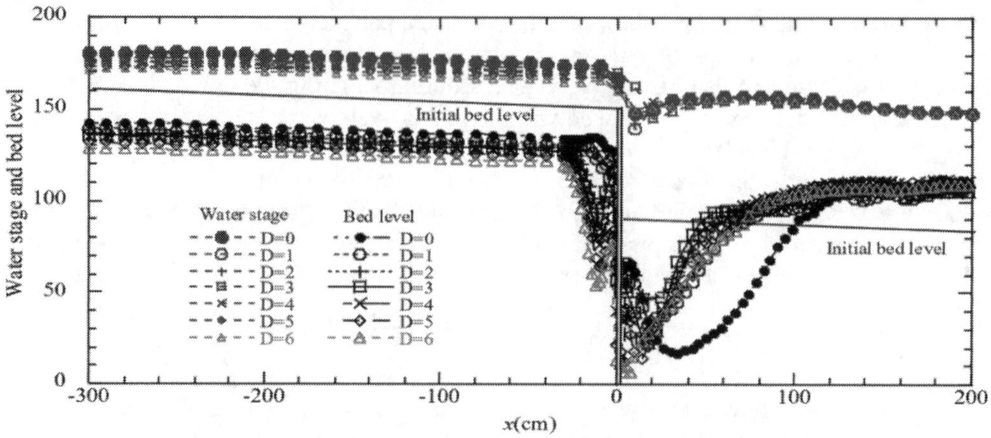

Figure 5. The longitudinal in riverbed and water level.

The average riverbed height of the non-open weir D = 0 m at x = 0 to 120 cm in the local flow section is significantly lower than that of the open weir D = 1 to 6 cm. In the range of opening weir D = 1 to 6 cm, the average bed height decreases as the opening depth increases. In the upstream of the weir, the amount of decrease in the water level increases as the opening depth D increases, but downstream of the weir, the change in the water level due to the difference in the opening depth D is not significantly.

3.1. Gradually varied flow section

Figure 6 shows the relationship between the riverbed degradation and the depth of the opening. Figure 6 (a) shows the effect of the weir opening depth D on the amount of riverbed degradation averaged in the gradual flow section. In the figure, the average riverbed reduction of the orthogonal weir is shown by the black line. As can be seen from the figure, the average riverbed reduction amount Δz_D of the oblique weir increases linearly as the opening depth D increases.

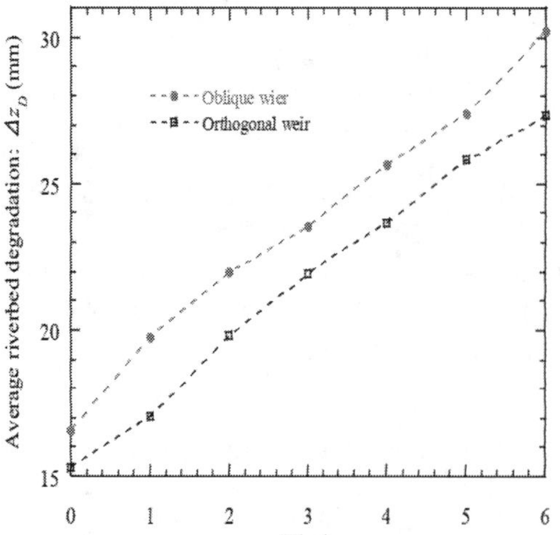

(a) The average riverbed degradation

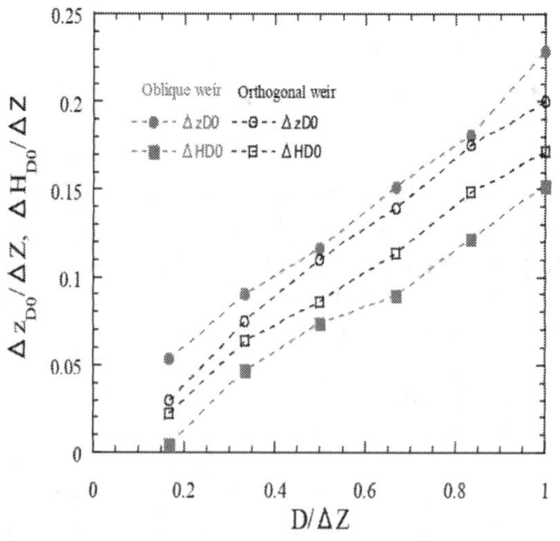

(b) The relative riverbed degradation

Figure 6. The longitudinal in riverbed and water level.

In the case where the weir opening depth D is 60 mm and equal to the riverbed height difference, the amount of decrease from the initial riverbed is 30.3 mm, which corresponds to about 51% of the riverbed height difference. Compared with the orthogonal weir, the average bed height of the oblique weir is about 2 mm, which is about 10% larger than that of the orthogonal weir. It is considered that the critical water depth at the top of the weir is smaller in the oblique weir than in the orthogonal one.

Figure 6 (b) shows the effect of the weir opening depth D on the relative water level degradation and the relative riverbed degradation. The opening depth D is nondimensionalized by the initial riverbed height difference ΔZ = 60 mm. As can be seen that both the relative water level degradation and the relative riverbed degradation at the oblique weir increase linearly as the opening depth $D / \Delta Z$ increases.

The relative riverbed degradation in the oblique weir is larger than the relative water level degradation, and the difference tends to increase slightly as the opening depth $D / \Delta Z$ increases. This shows the same tendency as the result with the orthogonal weir. In the case where the weir opening depth D is 60 mm and equal to the riverbed height difference, the relative riverbed reduction amount corresponds to 23% of the riverbed height difference, which is about 15% larger than the relative water level reduction amount. Compared to the orthogonal weir, in the case where the weir opening depth D is 60 mm and equal to the riverbed height difference, the relative riverbed reduction is about 14% larger at about 1.7 mm and the relative riverbed reduction is about 12% smaller at about 1.2 mm. From these results, the oblique weir has a higher sand removal function than the orthogonal weir.

3.2. Local flow section

Figure 7 shows the contour of the riverbed in the upstream and downstream of the weir. The contour value is the amount of change from the initial riverbed in mm. In all cases, local scour holes are found around the weir. Significant local scour upstream of both weir sleeve parts occurs except for the non-opening weir D = 0 cm.

The depth and area of the scour hole upstream of the weir are larger on the left bank side than on the right bank side, and asymmetry is observed. Also, the depth and area of the scour upstream of the weir increase as the opening depth D increases. It is considered that the sediment upstream is rolled up and transported downstream through the opening.

In addition, as the depth D of the opening increases, the amount of riverbed reduction increases in the gradual flow section, and the height from the riverbed to the weir top increases. As a result, vortices similar to horseshoe-shaped vortices increase the spatial scale and strength. It is guessed that the depth and area of the scour of the scour hole were increased.

On the other hand, a scour hole is observed near the left wall downstream of the oblique weir, and sedimentation is observed near the right wall. The depth and surface spread of the scour hole are the largest at the non-opening weir D = 0 cm. The maximum scouring depth decreases as the opening depth D increases in the range of D = 1 to 3 cm and becomes the smallest when D = 3 cm. In the range of D = 3 to 6 cm, the maximum scouring depth tends to increase as the opening depth D increases.

Downstream of the oblique non-opening weir, as it flows down to the side bank as shown in Figure 7, it converges along the left bank, a concentrated flow occurs, and a vortex with an extraordinarily strong downward flow is formed near the left bank. On the other hand, it is considered that the flow velocity and flow velocity passing through the opening increased in the range of D = 1 to 3 cm by providing the opening, and as a result, the flow velocity concentrated on the side bank decreased.

However, in the range of D = 3 to 6 cm, the depth of the opening is large, and it is relatively close to the riverbed downstream of the weir, suggesting that local scouring may have increased. Sediment deposits downstream of the oblique weir occur near the right bank, and the sediment thickness is almost the same as the riverbed height difference in all experimental cases.

IOP Conf. Series: Earth and Environmental Science **851** (2021) 012010 doi:10.1088/1755-1315/851/1/012010

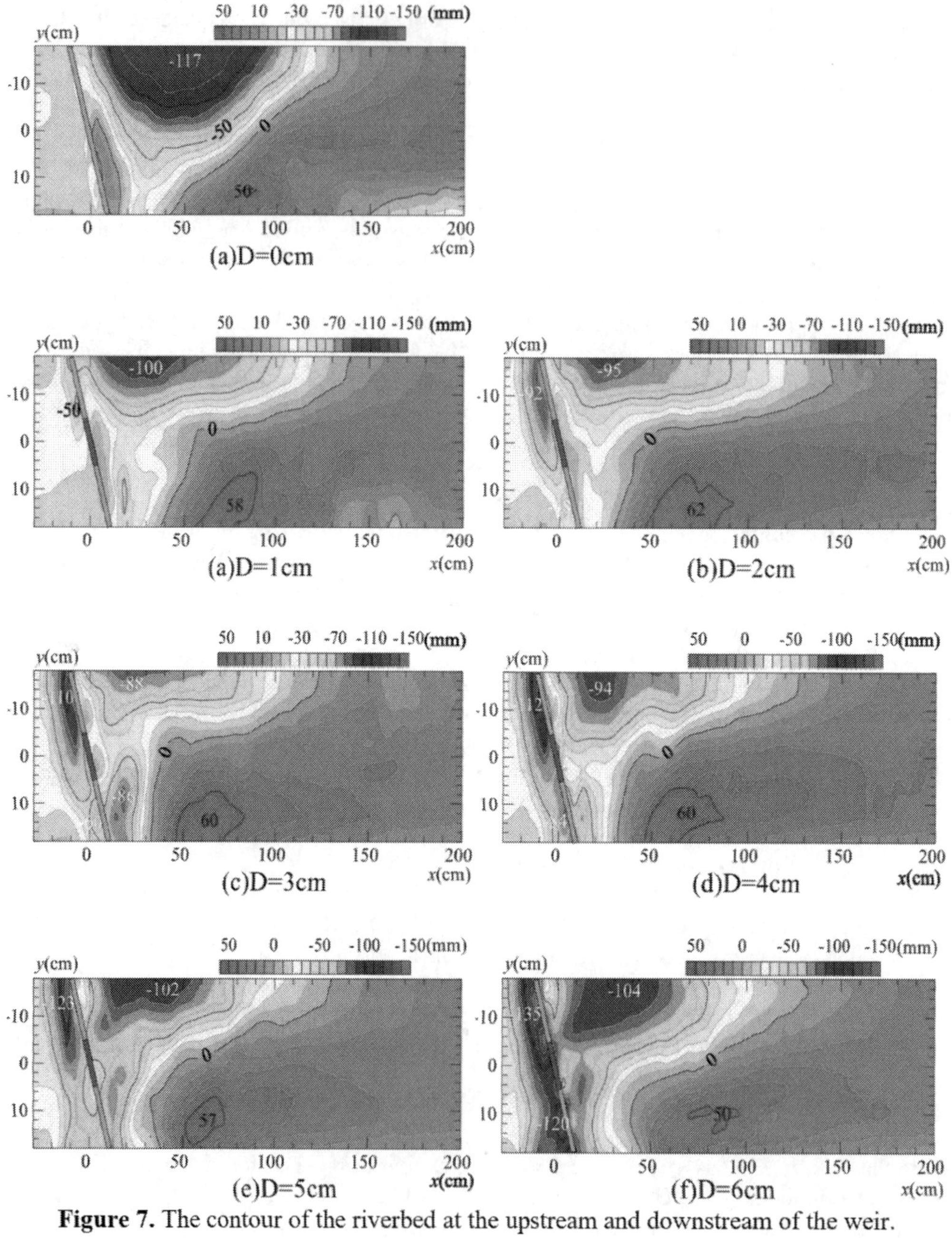

Figure 7. The contour of the riverbed at the upstream and downstream of the weir.

Figure 8 shows the longitudinal and cross-sectional shapes of the riverbed at the maximum scour depth downstream of the weir, respectively. In the vertical cross-sectional shape in Figure 8 (a), the scouring depth and the longitudinal direction x were nondimensionalized by maximum scour depth d_{sd} and scour length l_{xd} downstream of the weir, respectively.

The cross-sectional shape in Figure 8 (b) was nondimensionalized by the parameter shown in Figure 4 The longitudinal shape of the riverbed at the maximum scouring depth downstream of the weir is similar to that of the oblique opening weir D = 1 to 6 cm. The maximum scour depth position occurs near $x / l_{xd} = 0.2$ to 0.35. Scour hole occurs in a wider area in the flow axis direction in the oblique non-opening weir than in the opening weir.

If described with paying attention to the range of $0 < y / l_{yd} < 1.0$, which corresponds to the scouring hole near the left bank, the scouring longitudinal shape of the downstream of the weir is found to be almost similar in all experimental cases.

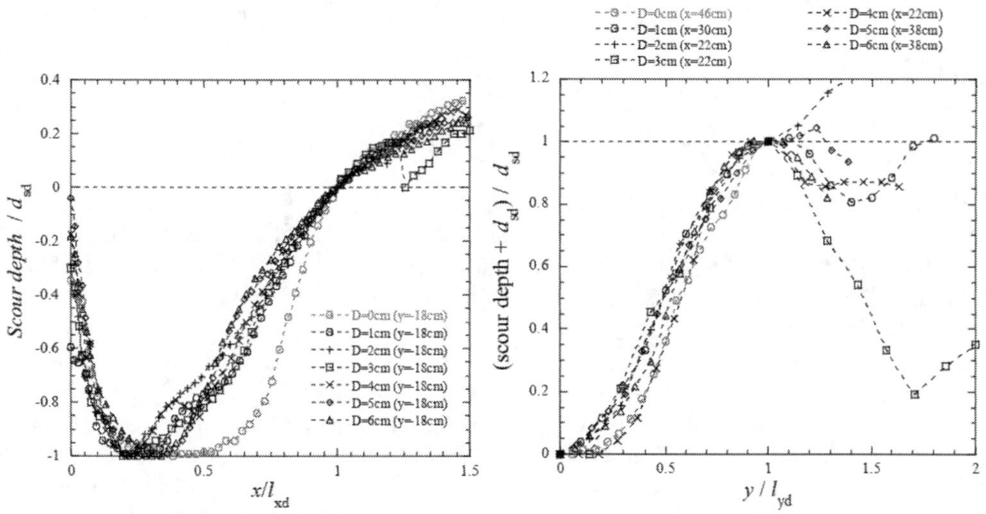

(a) the longitudinal profile (b) The transverse profile

Figure 8. The similarity of local scour hole downstream.

From Figure 9, the maximum scouring depth $d_{su} / \Delta Z$ upstream of the weir shows an increasing tendency as the opening depth *D* increases. The maximum scour depth is the maximum at the opening depth *D* = 6 cm in this experiment, and its size is about 2.3 of the riverbed height difference. The maximum sedimentary thickness downstream of the weir is $d_{sd} / \Delta Z$. It can be seen that sediment deposition downstream of the diagonal weir occurs near the right bank, and the sedimentary thickness is almost the same as the riverbed height difference in all experimental cases.

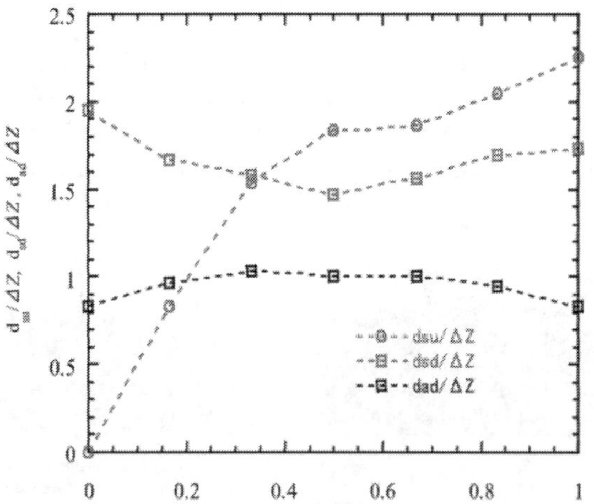

Figure 9. The characteristics of bed deformation and opening depth.

On the other hand, the maximum scouring depth d_{sd} / ΔZ downstream of the weir decreases as the opening depth *D* increases in the range of *D* = 1 to 3 cm and becomes the smallest when *D* = 3 cm. In the range of *D* = 3 to 6 cm, the maximum scouring depth tends to increase as the opening depth *D* increases. The maximum scouring depth d_{sd} / ΔZ is the largest when the non-opening weir *D* = 0 cm, and the maximum scouring depth is about twice as large as the riverbed height difference of 60 mm. On the other hand, in the range of *D* = 1 to 3 cm, the maximum scour depth decreases as the opening depth *D* increases. The case of *D* = 3 cm shows a minimum value, which is 25% compared to the non-opening weir *D* = 0 cm.

3.3. Local flow on the downstream of the weir opening

Figure 10 shows the contour of the horizontal plane of the magnitude of flow velocity |*u*|. The synthetic vector of the main velocity U and the transverse component V of the secondary flow is shown in this figure. By overflowing the weir, it is accelerated at the top of the weir, and a high-velocity region is formed. On the downstream side of the weir, the high-speed region where the average cross-sectional flow velocity in the equilibrium state is greater than |*u*|> 35 cm/s converges along the left bank is observed. On the right bank side downstream of the weir, it can be seen that a backflow area is formed by the vortex separated from the flow concentrated on the left bank side. When x> 100 cm, it becomes an acceleration region again as the riverbed rises. The flow velocity is relatively small on the right bank side directly downstream of the weir and in the range of x = 30 cm to 100 cm and y = -14 cm to 0 cm.

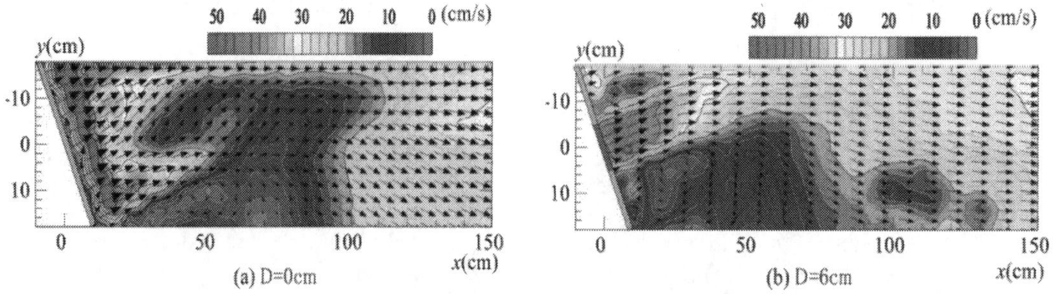

Figure 10. The contour of the horizontal plane of the magnitude of flow velocity |u|.

Figure 11 shows the vertical plane contour of the magnitude of flow velocity |*u*| where maximum scour depth occurred. In the figure, the synthetic vector of the transverse component V and the vertical component W of the secondary flow is shown. A significant vortex is formed in the range of y = -18 cm to 0 cm. The magnitude of the downflow near the left wall is about 45% of the average cross-sectional flow velocity of 35 cm / s, and a strong downflow is generated.

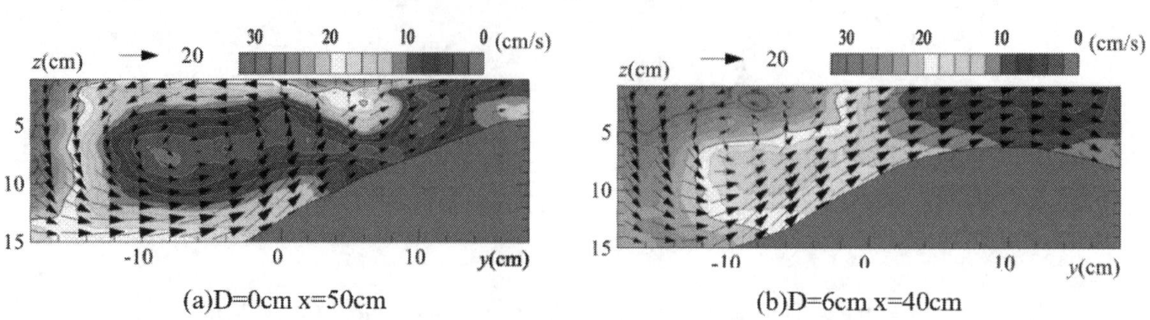

Figure 11. The vertical plane contour of the magnitude of flow velocity |u|.

A remarkable vortex is formed in the range of y = -18 cm to -6 cm near the left bank. The magnitude of the downward flow near the left bank is about 34% of the average cross-sectional flow velocity of 31 cm / s, indicating that the downward flow near the left bank is smaller than that of the non-opening weir. It is considered that as the high-speed region formed in the center of the channel flowed down due to the influence of the opening, the high-speed region was formed over a wide area on the left bank side due to the concentration on the left wall.

4. Conclusions

In this study, the sedimentation function of sediments upstream of the opening depth of the oblique weir and the riverbed fluctuations downstream of the weir were investigated. The main results obtained are as follows.

1) Both the relative water level degradation and the relative bed degradation at the diagonal weir increase linearly as the opening depth increases. Compared with the orthogonal weir, the relative riverbed reduction is about 14% larger in the case where the weir opening depth D is 60 mm and the riverbed height difference is equal. From this, from the viewpoint of the sand removal function upstream of the weir, the diagonal weir has a higher sand removal function than the right-angled weir.
2) The local scour shapes upstream and downstream of the diagonal weir were dimensionless on the representative scales of the longitudinal shape and the transverse shape, and as a result, similar figures were found.
3) The maximum scouring depth is the largest in non-opening weirs, and in particular, the maximum scouring depth is approximately twice as large as the riverbed height difference of 60 mm. On the other hand, it can be seen that the maximum scour depth has a minimum value when D = 3 cm and is reduced by 25% compared to the non-opening weir D = 0 cm.
4) For both the diagonal non-opening weir and the open weir, it was confirmed that there is a strong positional relationship between the maximum value of the vertical component W of the secondary flow and the sandbar.

References

[1] Terunori O, Tatsuhiro Y and Hirotaka U 2016 Effects of weir with an opening on river bed and flow structure. *Journal of Japan Society of Civil Engineers, Ser. B1 (Hydraulic Engineering).* Volume **72** Issue 4 Pages I_811-I_816.

[2] Terunori O and Hirotaka U 2017 Effects of upward weir with an opening on river bed and flow structure. *Journal of Japan Society of Civil Engineers, Ser. B1 (Hydraulic Engineering).* Volume **73** Issue 4 Pages I_703-I_708.

[3] Zhang H, Muto Y, Nakagawa H and Nakanishi S 2012 Weir removal and its influence on hydro-morphological features of upstream channel, *Journal of applied mechanics*, Vol.**15**, No2, pp.591-599.

[4] Sumida H, Muto Y and Tamura T 2015 Channel to development on upstream due to change of partially removed weir, *Journal of Japan Society of Civil Engineers, Ser.B1 (Hydraulic Engineering)*, Vol.**71**, No.4.I925-I930.

[5] Hajime M 1978 The Rationality of Curved Oblique Weirs on the Basis of the Formations of 'Sarekitai' *Transactions of The Japanese Society of Irrigation, Drainage and Reclamation Engineering*, Vol. **1978**, Issue 76, pp. 61-66,a2 (Japanese press)

[6] Nguyen B T 2006 Flow over oblique weirs, TU Delft, Faculty of Civil Engineering and Geosciences, Hydraulic Engineering

[7] Kabiri-Samani A R 2010 Analytical Approach for Flow over an Oblique Weir, *Scientia Iranica*, Vol. **17**, No. 2, pp. 107-117

[8] Ohtoshi K, Matsuda S, Shino K and Shimizu A 1998 Experimental Study on the Shape of a Historical Curvilinear-Oblique Weir, *Historical Studies in Civil Engineering*, Volume **18** Pages 227-234,1998

[9] Fukuoka S, Mishiro S, Aratani S, Nakasuka J, Okada S and Tanaka M 2001 The effect of weir position and its shape on the flow and bed topography. *Journal of Japan Society of Civil Engineers (JSCE), PROCEEDINGS OF HYDRAULIC ENGINEERING*, Vol.**45**, pp.397-402. (Japanese press)

[10] Une H, Ohmoto T, Adachi K and Kondo H 2019 Effects of Oblique Weir with an Opening on Bed Configuration and Flow Structure, *International Congress on Natural Sciences and Engineering 2019*, pp.147-158.

IOP Conf. Series: Earth and Environmental Science **851** (2021) 012011 doi:10.1088/1755-1315/851/1/012011

Thornthwaite and Mather water balance method in Indonesian Tropical Area

H Hendrayana[1*], M Widyastuti[2], I A Riyanto[3], A Nuha[4], M Y Widasmara[3], N Ismayuni[2] and I N Rachmi[2]

[1]Geological Engineering Department, Faculty of Engineering, Universitas Gadjah Mada, Yogyakarta, Indonesia
[2]Environmental Geography Department, Faculty of Geography, Universitas Gadjah Mada, Yogyakarta, Indonesia
[3]Master Program on Planning and Management of Coastal Area and Watershed, Faculty of Geography, Universitas Gadjah Mada, Yogyakarta, Indonesia
[4]Groundwater Working Group (GWWG), Faculty of Engineering, Universitas Gadjah Mada, Yogyakarta, Indonesia
*E-mail: heruha@ugm.ac.id

Abstract. The Thornthwaite and Mather water balance method has been widely applied in the world, one of which is in Indonesia. However, almost all studies using the Thornthwaite and Mather water balance method in Indonesia are not validated. Considering that the Thornthwaite and Mather water balance study makes it easier to study hydrology, especially in areas that do not yet have complete and evenly distributed hydrological and meteorological stations. The research locations are in the Wampu sub-watershed, Serayu sub-watershed, Ayung sub-watershed, and Tondano sub-watershed. The method used is the Thornthwaite and Mather water balance with input data on rainfall, temperature, land use, latitude, and soil texture. The validation test was performed using the RMSE equation from the runoff model compared to the observation runoff (river discharge data). The RMSE results of the Thornthwaite and Mather water balance method in the Ayung sub-watershed (42.42%), Serayu sub-watershed (53.77) Wampu sub-watershed (65.58%) and Tondano sub-watershed (83.68%) were classified as weak correlation categories. The comparison results of the Thornthwaite method in tropical areas had greater rain, potential evapotranspiration, and runoff values than other climatic areas. Based on the result, we need to modify the Thornthwaite and Mather method for tropical areas and add a range of 25-30 years of climate data.

1. Introduction

The Thornthwaite and Mather water balance method has long been used and applied worldwide [1,2]. The Thornthwaite and Mather water balance method has included latitude parameters so that it can be applied in different climatic regions [3,4]. The Thornthwaite and Mather method provides an easy assessment of water balance in any region of the world [5], including regions that have zero water discharge data [6,7]. In addition, the Thornthwaite and Mather method can be used either within natural boundary areas in the form of watersheds [8,9] or within non-natural boundary areas [10,11]. The Thornthwaite and Mather method has been modified several times for more detailed and specific areas [12,13]. The Thornthwaite and Mather method modification was made due to the low accuracy of the original model and measurement, or some mismatches with existing local parameters.

The Thornthwaite and Mather method has been widely applied in Indonesia. There is a tendency that Thornthwaite and Mather water balance studies applied in Indonesia are mostly without any validation. Validation is very important in knowing the accuracy of the Thornthwaite and Mather method so that there are no errors in the analysis and calculation of further studies. Based on this, it is necessary for us to conduct research to validate the applications of the Thornthwaite and Mather method in Indonesia.

In Indonesia, the Thornthwaite and Mather water balance method is widely used by government agencies, academics, and privates for regional and development planning. Many regional and development plans are designed for areas that do not have hydrological measurement stations or in remote areas. Under this condition, the Thornthwaite and Mather method is very important because it is used for preliminary studies before conducting any further and more detailed studies. The preliminary studies are also an initial stage in selecting planning priorities in a region. Therefore, the validity of the Thornthwaite method results in tropical areas, especially in Indonesia, is crucial despite the fact that Indonesia still has a minimal distribution of hydrological stations, both for discharge and rain measurements. The Thornthwaite and Mather method is one of the keys in determining the priorities for initial planning related to hydrology.

Table 1. Study area.

No	Watershed	Regency, Provinces	Mount	Watershed area (km^2)
1	Sub Serayu	Wonosobo, Central Java	Dieng Volcano and Sindoro	136.82
2	Sub Tondano	Minahasa, North Sulawesi	Tondano and Kabat	140.41
3	Sub Wampu	Langkat, North Sumatra	Sinabung Volcano	435.44
4	Sub Ayung	Mambal, Bali	Agung Volcano	275.55

The study locations are in Indonesia, spreading across several islands: Sumatra, Java, Bali, and Sulawesi (Table 1). Climatic conditions on sites. The study locations include tropical climates with annual rainfall ranging from 1000 to 4000 mm / year [14], humidity ranging from 60% to 80% [15], and temperatures ranging from 28 ° to 30 ° C [16]. In addition, the study locations are strongly influenced by the western and eastern monsoonal winds [17]. The study locations are predominantly head-watered on active volcanoes and a small number of them are head-watered on old volcanoes. The dominant land uses in the study locations are in the form of forests and gardens in the upstream part; and settlements and rice fields in the downstream part. Elevational variations in the sub-watersheds in the study locations range from 500 to 2500 MASL (mean average sea level). The shape of the sub-watersheds in the study locations are elongated (Figure 1).

2. Methods

2.1. Water Balance Data

The data used for the water balance study include rainfall, temperature, latitude, land use, and soil texture. Rainfall and temperature data were obtained from the Meteorology, Climatology, and Geophysics Agency (BMKG) with a time span of 10 years (January 2006 - December 2018). Data on land use and latitude were obtained from the Spatial Information Agency (BIG). Soil texture data were obtained from the Ministry of Agriculture's Soil Research Institute. In addition, this research used secondary good discharge data as a validation of the water balance model. Secondary discharge data for the Ayung sub-watershed were obtained from (1993-2001) [18]; the Wampu sub-watershed was obtained from the Public Works Service of North Sumatra Province (2000-2015); the Tondano sub-watershed was obtained from (1989-1993) [19], and Serayu sub-watershed was obtained from (1994-2004) [20].

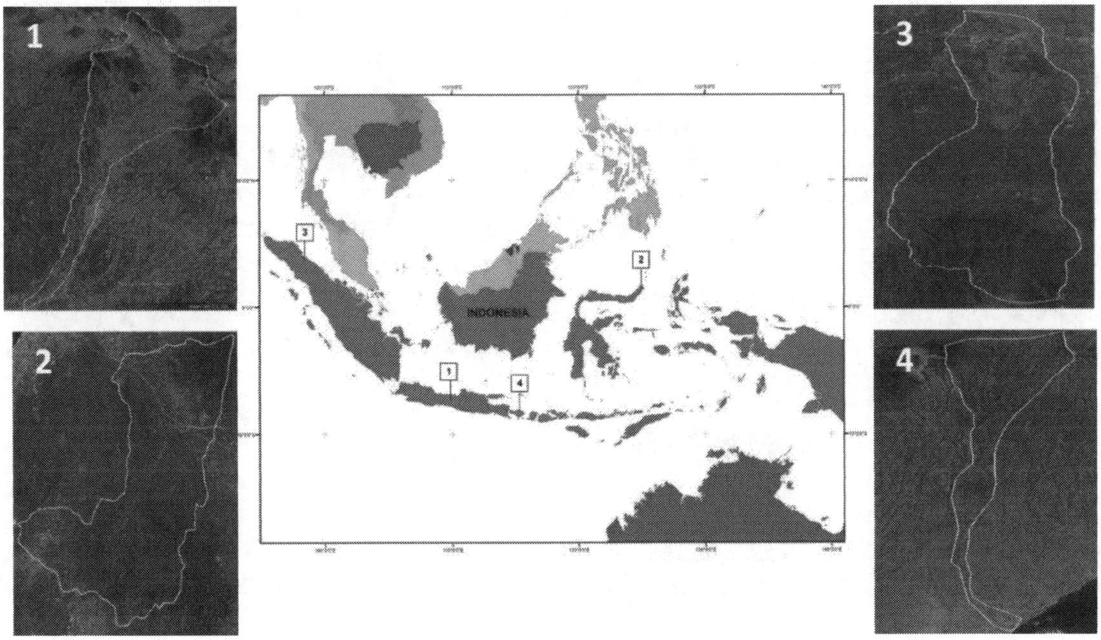

Figure 1. Study Locations.

2.2. Water Balance Calculation

The water balance was calculated using the method formulated by [21]. The first step was to calculate the potential evapotranspiration (Equation 1)

$$PET_0 = 16bi\left(\frac{10T_i}{I}\right)^{\alpha} \qquad \text{[mm/month]} \qquad (1)$$

PET_0 = potential evapotranspiration;
bi = radiation parameter for specific latitude;
T_i = monthly air temperature;
I = annual heat index (Equation 2)
α = complex function of heat index (Equation 3)

$$I = \sum_{i=1}^{12} \left(\frac{T_i}{5}\right)^{1.514} \qquad (2)$$

T_i = monthly temperature

$$\alpha = 6.75 \times 10^{-7}I^3 - 7.71 \times 10^{-5}I^2 + 1.7912 \times 10^{-2}I + 0.49239 \qquad (3)$$

I = annual heat index

The second step was to calculate the difference between P (precipitation) - PET. The result of P-PET reduction was used to calculate the APWL (accumulated potential water loss). If (P-PET) was positive (+), then APWL was 0. If (P-PET) was negative (-) then it was reduced to an absolute price so that it was positive. The third step was to calculate STOR (moisture storage capacity) which is a function of land use, soil texture, and root zone depth. After that, the fourth step was to calculate the

SM formula (the actual storage of soil moisture) which was then divided into two scenarios, Equation 4 and Equation 5.

$$SM = STOR \times e^{-(APWL/STOR)}, \text{ if } APWL \neq 0 \quad (4)$$

$$SM = STOR, \text{ if } APWL = 0 \quad (5)$$

The fifth step was to calculate ΔSM to find out the STOR value, whether negative or positive. If ΔSM was positive (+), then soil moisture storage was supplied from infiltration; if ΔSM was negative (-), then soil moisture storage was reduced due to evapotranspiration. ΔSM was calculated based on Equation 6. The current month's STOR minus the previous month's STOR.

$$\Delta SM = SM_{month} - SM_{previousmonth} \quad (6)$$

The sixth step was to calculate the AET (actual evapotranspiration) value. The AET value was calculated based on the two scenarios described in Equations 7 and 8.

$$\Delta SM > PET, \quad \text{then } AET = PET \quad (7)$$

$$\Delta SM < PET, \quad \text{then } AET = P + \Delta SM \quad (8)$$

The seventh step was to calculate the value of DEF (the water deficit) in Equation 9 and SUR (moisture surplus) in Equation 10. The results of DEF and SUR calculations are more clearly displayed in the graph of the relationship between monthly AET, PET, and P.

$$DEF = PET - AET \quad (9)$$

$$SUR = P - (AET + \Delta SM) \quad (10)$$

Step eight was to perform the R (Runoff) calculation. The assumption used for the calculation of R was that half of the rain falls into the surface, in a form of surface runoff; and half of it enters groundwater (Equation 11). When entering dry seasons, R is still there, R during the dry season is obtained from the groundwater reserve (baseflow). The details of the calculation for equation 11 are described in table 2. The first stage is by dividing ½ the surplus-value of December in the first row of the last column, followed by dividing ½ in the following month (first row of the first column and so on). The surplus in January is divided by ½ in the second row of the first column, then continued to be divided by ½ in the next column, and so on for the other months. The R modelled results are obtained from the total columns for each month (number with bold style). The result of R was then multiplied by the area of the study area.

$$R = \tfrac{1}{2} (SUR_{month} + SUR_{previousmonth}) \quad (11)$$

Table 2. Example calculation R modelled of Ayung Sub-Watershed

	Jan	Feb	Mar	Apr	May	June	July	Aug	Sept	Oct	Nov	Dec
Dec	30.10	15.05	7.53	3.76	1.88	0.94	0.47	0.24	0.12	0.06	0.03	60.21
Jan	84.76	42.38	21.19	10.60	5.30	2.65	1.32	0.66	0.33	0.17	0.08	0.04
Feb	**114.86**	84.90	42.45	21.23	10.61	5.31	2.65	1.33	0.66	0.33	0.17	0.08
Mar		**142.33**	52.15	26.08	13.04	6.52	3.26	1.63	0.81	0.41	0.20	0.10
Apr			**123.32**	17.74	8.87	4.44	2.22	1.11	0.55	0.28	0.14	0.07
May				**79.40**	0.00	0.00	0.00	0.00	0.00	0.00	0.00	0.00
June					**39.70**	0.00	0.00	0.00	0.00	0.00	0.00	0.00
July						**19.85**	0.00	0.00	0.00	0.00	0.00	0.00
Aug							**9.93**	0.00	0.00	0.00	0.00	0.00
Sept								**4.96**	0.00	0.00	0.00	0.00
Oct									**2.48**	0.00	0.00	0.00
Nov										**1.24**	11.08	5.54
											11.70	**66.04**

2.3. Validation of the Water Balance

The Thornthwaite and Mather water balance method validation was carried out by comparing the R results (model) with primary and secondary discharge data (Equation 12). Comparisons were made in the form of calculating the accuracy in the form of the root mean square error (RMSE) percentage. RMSE can be accepted if the score is less than and close to 10% [22,29]. Other results can be found without using RMSE, namely by comparing the model discharge value with observed discharge multiplied by 100%; the closer to 100% the model is, the more accurate the result. The accuracy results can provide an idea of whether the Thornthwaite and Mather water balance methods can be applied somewhere. In addition, the analysis was also carried out by comparing the results of the Thornthwaite and Mather water balance methods in different climatic regions.

$$\%RMSE = \sqrt{\frac{\sum(Vmodelled - Vobs)^2}{n}} \times \frac{100 \times n}{\sum Vobs} \quad (12)$$

3. Results and Discussions

The analysis of the water balance in the Ayung Sub-watershed showed water surpluses in January-April and October-December (Figure 2), and water deficit in May-September. The highest surplus was in February with 169.80 mm or 46,790,587 m^3 / month, while the highest deficit was in August with 131.38 mm or 36,203,668 m^3 / month (Table 2). The runoff model for the Ayung sub-watershed showed that the highest discharge was in February with 142.33 mm or 15.13 m^3 / s and the lowest was in October with 1.24 mm or 0.13 m^3 / s. The score of calculating RMSE (%) of modeled R and observed R in the Ayung sub-watershed was 41.42%. This result indicates a weak correlation between modeled R and observed R. The modeled R values that underestimated R observed were found in June, July, October, November, and December; the modeled R values that were close to the observed R values were observed in January, February, March, August, and September; and the modeled R values that overestimated the observed R were found in April and May (Table 3).

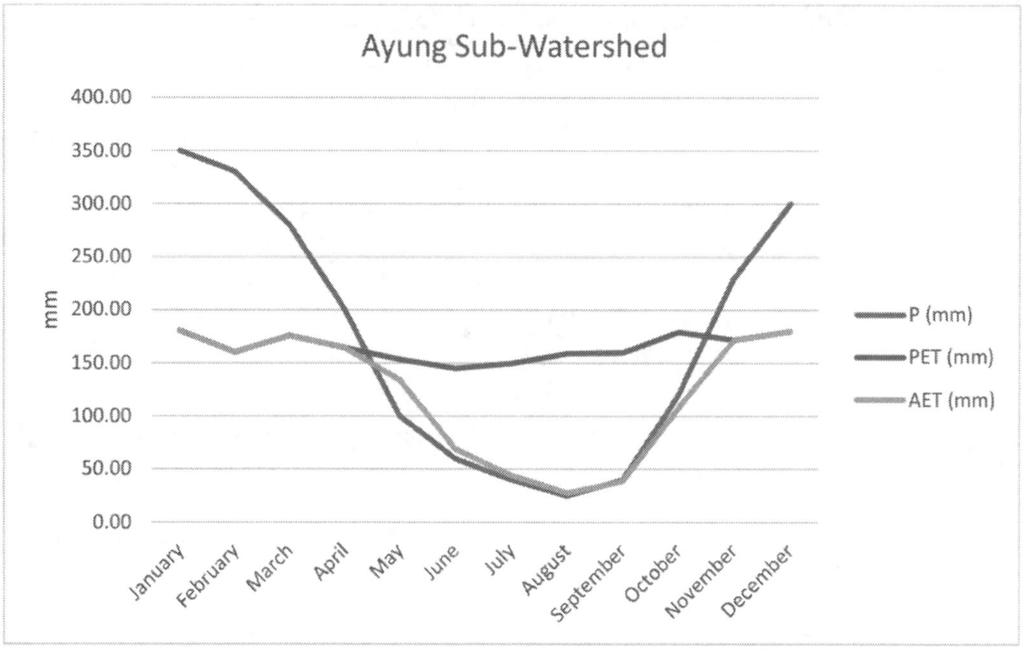

Figure 2. Water Balance Graph of Ayung Sub-Watershed.

The analysis of the water balance in the Wampu Sub-watershed showed water surpluses in April-May and August-December (Figure 3), and water deficit in January-March and June-July. The highest water surplus was in October with 162.50 mm or 70,758,417 m^3 / month, while the highest water deficit was in March with 50.65 mm or 22,055,342 m^3 / month (Table 4). The results of the modeled runoff in Wampu Sub-watershed showed that the highest discharge was in November with 110 mm or 18.84 m^3 / s and the lowest was in April with 2.64 mm or 0.44 m^3 / s. The score of RMSE (%) calculation between modeled R and observed R in the Wampu sub-watershed was 65.58%. This result indicated a weak correlation and low accuracy of the two results. The modeled R values that underestimated observed R was found in January, February, March, April, June, July, and December; the modeled R values that were close to the observed R values were in May, August, September, and November (Table 3); and the overestimated modeled R-value of observed R-value occurred in October.

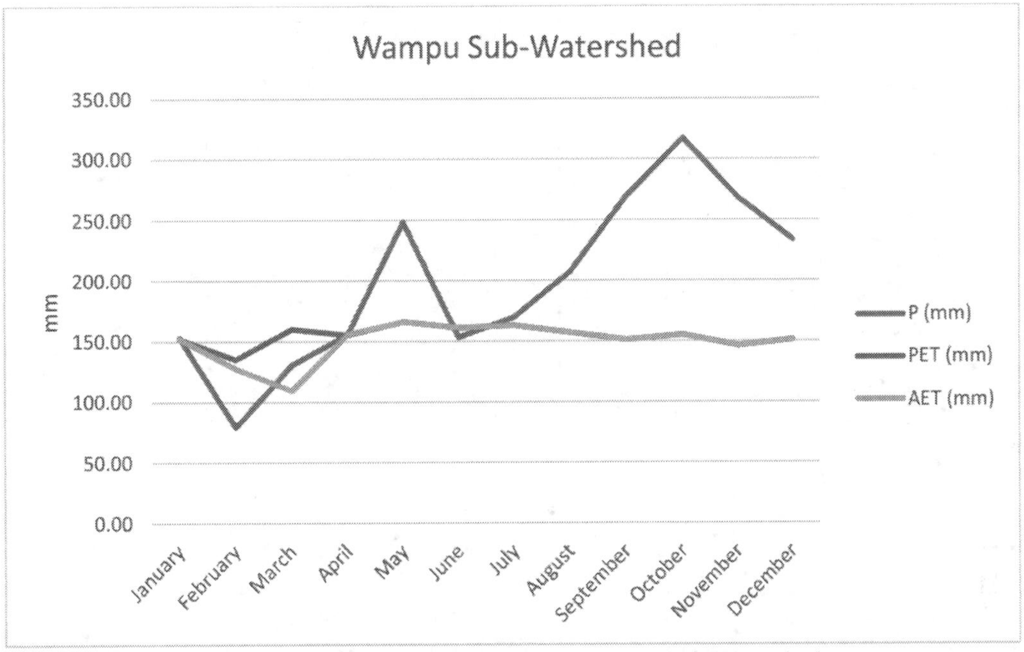

Figure 3. Water Balance Graph of Wampu Sub-Watershed

The analysis of the water balance in the Tondano Sub-watershed showed water surpluses in January-May and October-December (Figure 4), and water deficits in June-September. The highest surplus was in December with 107.85 mm or 15,143,064 m^3 / month and the highest deficit was in September with 45.29 mm or 6,359,560 m^3 / month (Table 5). The modeled runoff for the Tondano sub-watershed indicated that the highest discharge was in April with 124.73 mm or 6.76 m^3 / s and the lowest was in October with 2.30 mm or 0.12 m^3 / s. The score of RMSE (%) calculation between modeled R and observed R in the Tondano sub-watershed was 83.68%. This result indicated a weak and low correlation between the two results. The modeled R values which underestimated observed R were found in May-November and the modeled R that were close to the observed R were in January-April and December (Table 4). There was no overestimated modeled R-value from the observed R.

The analysis of the water balance in the Serayu Sub-watershed showed water surpluses in January-May and October-December (Figure 5), and water deficit in June-September. The highest waster surplus was in December with 399.40 mm or 54,646,666 m^3 / month, and the highest water deficit was in October with 75.30 mm or 10,302,689 m3 / month (Table 5). The analysis of the modeled runoff in the Serayu Sub-watershed revealed that the highest discharge was in March with 324.48 mm or 17.13 m^3 / s, and the lowest was in October with 12.83 mm or 0.677 m^3 / s. The score of RMSE (%) calculation between modeled R and observed R in Serayu sub-watershed was 53.77%. This result indicated a weak correlation. The modeled R values which underestimated observed R were found in June, May-November, while the modeled R values that were close to the observed score were found in January-April and December (Table 6). No modeled R-value was overestimated from the observed R.

Table 3. Water Balance of Ayung Sub-Watershed.

	January	February	March	April	May	June	July	August	September	October	November	December
T (C°)	28.20	28.10	28.30	28.00	26.10	25.80	25.70	26.00	27.10	28.50	27.90	28.00
P (mm)	350.00	330.00	280.00	200.00	100.00	60.00	40.00	25.00	40.00	120.00	230.00	300.00
PET (mm)	180.48	160.20	175.69	164.51	153.35	145.04	149.68	159.04	159.85	179.17	171.68	179.59
P-PET (mm)	169.52	169.80	104.31	35.49	-53.35	-85.04	-109.68	-134.04	-119.85	-59.17	58.32	120.41
APWL (mm)	0.00	0.00	0.00	0.00	53.35	85.04	109.68	134.04	119.85	59.17	0.00	0.00
STOR (mm)	54.75	54.75	54.75	54.75	54.75	54.75	54.75	54.75	54.75	54.75	54.75	54.75
SM (mm)	54.75	54.75	54.75	54.75	20.67	11.58	7.39	4.73	6.13	18.58	54.75	54.75
ΔSM (mm)	0.00	0.00	0.00	0.00	-34.08	-9.08	-4.20	-2.65	1.40	12.45	36.17	0.00
AET (mm)	180.48	160.20	175.69	164.51	134.08	69.08	44.20	27.65	38.60	107.55	171.68	179.59
SUR (mm)	169.52	169.80	104.31	35.49	0.00	0.00	0.00	0.00	0.00	0.00	22.15	120.41
DEF (mm)	0.00	0.00	0.00	0.00	19.27	75.96	105.48	131.38	121.25	71.62	0.00	0.00
R *modelled* (mm)	114.86	142.33	123.32	79.40	39.70	19.85	9.93	4.96	2.48	1.24	11.70	66.04
R *modelled* (m^3/s)	12.21	15.13	13.11	8.44	4.22	2.11	1.06	0.53	0.26	0.13	1.24	7.02
R *observed* (m^3/s)	12.87	20.42	10.01	2.14	0.83	11.37	8.62	1.50	0.62	12.72	7.74	19.80
RMSE						41.42						

Table 4. Water Balance of Wampu Sub-Watershed.

	January	February	March	April	May	June	July	August	September	October	November	December
T (C°)	25.65	25.17	26.46	26.34	26.97	26.96	26.49	26.10	25.83	25.93	25.53	25.59
P (mm)	153.00	79.00	130.50	155.50	248.50	153.00	169.50	207.00	269.00	317.50	268.00	233.50
PET (mm)	152.00	135.00	160.00	155.00	166.00	161.00	163.00	157.00	151.00	155.00	146.00	151.00
P-PET (mm)	1.00	-56.00	-29.50	0.50	82.50	-8.00	6.50	50.00	118.00	162.50	122.00	82.50
APWL (mm)	0.00	56.00	29.50	0.00	0.00	8.00	0.00	0.00	0.00	0.00	0.00	0.00
STOR (mm)	189.00	189.00	189.00	189.00	189.00	189.00	189.00	189.00	189.00	189.00	189.00	189.00
SM (mm)	189.00	140.54	161.69	189.00	189.00	181.17	189.00	189.00	189.00	189.00	189.00	189.00
ΔSM (mm)	0.00	-48.46	21.15	27.31	0.00	-7.83	7.83	0.00	0.00	0.00	0.00	0.00
AET (mm)	152.00	127.46	109.35	155.50	166.00	160.83	163.00	157.00	151.00	155.00	146.00	151.00
SUR (mm)	1.00	0.00	0.00	0.00	82.50	0.00	0.00	50.00	118.00	162.50	122.00	82.50
DEF (mm)	0.00	7.54	50.65	0.00	0.00	0.17	0.00	0.00	0.00	0.00	0.00	0.00
R *modelled* (mm)	21.10	10.56	5.28	2.64	42.57	21.28	18.10	45.70	82.96	105.70	110.00	96.20
R *modelled* (m^3/s)	3.54	1.77	0.89	0.44	7.15	3.57	3.04	7.68	13.94	17.76	18.48	16.16
R *observed* (m^3/s)	24.55	13.72	11.24	12.77	13.01	10.83	12.74	12.27	14.47	15.14	18.28	24.17
RMSE						65.58						

IOP Conf. Series: Earth and Environmental Science **851** (2021) 012011 doi:10.1088/1755-1315/851/1/012011

The water balance graphs, especially the rainfall character, of the Wampu (Sumatra) and Tondano (Sulawesi) sub-watersheds are quite different from the graphs of the Serayu and Ayung sub-watersheds (Java and Bali). This result is also reinforced by research [23] finding that there are differences in rain zoning from 1985-2010 in several regions, including Sumatra (B), Sulawesi (C), and Java and Bali (A). The rain pattern of the water balance of the Wampu Sub-watershed has two peaks in April-May and August December because the pattern in region B is influenced by movements of the southward and northward inter-tropical convergence zone (ITCZ) [24]. Rainfall graphs for Serayu and Ayung sub-watersheds are in accordance with the pattern of region A with lows in the months of May-September and highs in October-April. Meanwhile, the rain pattern in the Tondano sub-watershed is the same as type C, with 1 rainfall peak in June-July, and other highs in January-April and November-December.

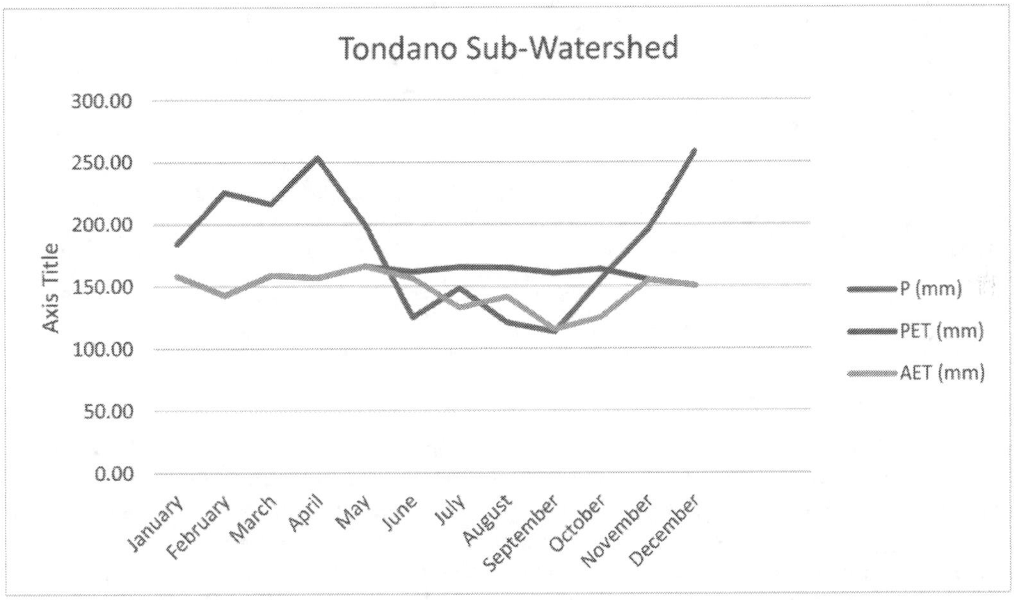

Figure 4. Water Balance Graph of Tondano Sub-Watershed.

Rain in the tropical area showed a higher value compared to other regions. This implicates the runoff value in the four study locations have greater than in other climatic areas. Rainfall in Iran with four monsoons is 296 mm/year and the runoff is 30 mm/year [9] smaller than the study location. Rainfall in the arid US region is 98 mm/year and the mean runoff is 35 mm [10] smaller than the study location. Rainfall in the sub-tropical region of Georgia US is 1,200 mm/year with runoff is 300 mm/years [27], which is still smaller than the study area. Rainfall in the tropical region of Costa Rica is 2697 mm/year and runoff 169 mm/year [26]. These results have a value close to the study location for the rain value, but the runoff is relatively small because the forest becomes the dominant of land use (SM 200). Rainfall in tropical India obtained 1,175 mm/year rain and 255 for the runoff [27]. Rainfall in other parts of India is 1,393 mm / year and Runoff 88 [6] both values are still smaller than the four study locations. These results are still far from the study location for the value of rain and runoff (forest dominance of SM 49). Rainfall in tropical Equador is 843 mm/year and runoff 202.2 mm/years [2] these results are still below the value of the four study locations. Another tropical area is in Kulon Progo Regency that has similarities with the four study locations with 2,600 mm/year of rain and runoff 633 [7].

Other factors causing the different of Water Balance is in the PET, especially in arid and semi-arid areas. The results of the Thornthwaite and Mather method were quite inaccurate in Aswan Dam Egypt because the PET value was underestimated, with RMSE of 80 and 110% [22]. The same method was applied in Portugal and indicated that the results of the water balance method showed underestimated PET and high runoff values [25]. The application of the Thornthwaite and Mather method in the semi-arid region shows that the PET accuracy is by 69% and its modification by adding the Willmot equation resulted in better accuracy, by 77% [12]. PET values in Southwest Asia range from 300-600 mm [13] which are quite larger compared to the tropical regions of this study sites.

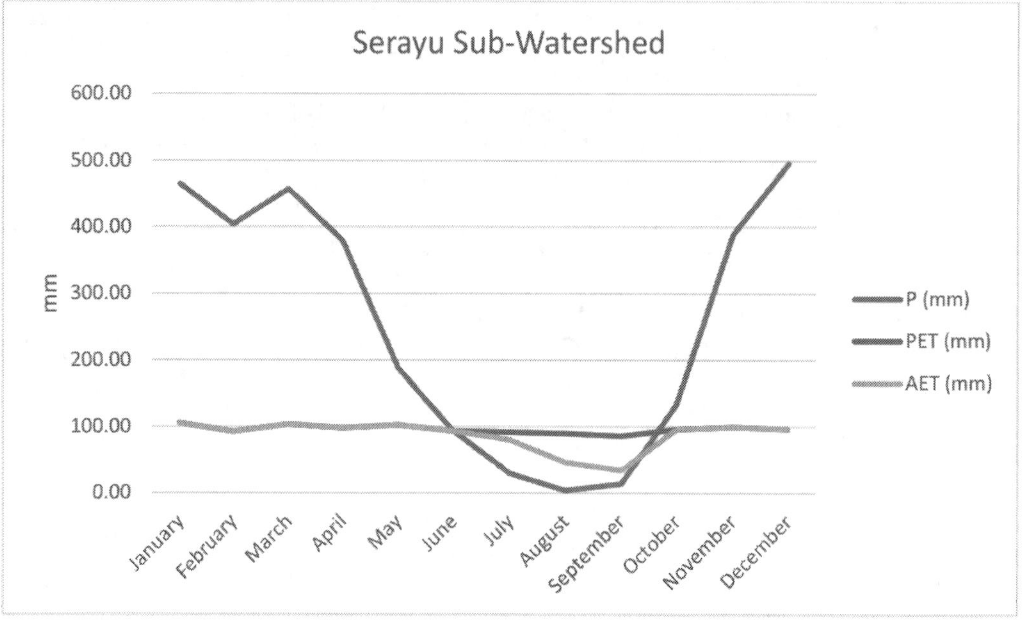

Figure 5. Water Balance Graph of Serayu Sub-Watershed.

However, after being compared with the results of the Thornthwaite and Mather method in other tropical areas, the PET value of this study location is not much different from the values of other areas. With this consideration, the PET value of this study does not need to be corrected. The global yield of the Thornthwaite and Mather method PET in the tropics ranges from 1000-1100 mm / year [1]. In addition, the Thornthwaite and Mather method is also applied in Costa Rica with an accuracy of 40-76% between modeled R and observed R [26]. Other tropical applications are applied in India with underestimated modeled R values compared to the observed R values [27], however, the PET, AET, and Precipitation values are almost close to the values of the study location. The applications of the Thornthwaite and Mather method in subtropical India have an average accuracy of 88.33% [6]. In addition, in Indonesia, its application in Kulon Progo, DIY is characterized by almost the same results in the values of PET, AET, P and R as the 4 sub-watersheds in the study locations [7].

Comparison of the results of other Thornthwaite and Mather method applications shows many variations. The results of Thornthwaite and Mather water balance in the tropical watershed region of Ecuador show PET values ranging from 55-70 mm / month, AET of 24-70 mm / month, and SUR of 7-41 mm / month [2]. These results are close to their values in the tropical region near the ITCZ. Another Thornthwaite and Mather method applied in the arid region of Migan Watershed area (Iran) showed a very small precipitation value compared to our study locations, around 30-50 mm / month; higher AET values than the values our study locations, with values of around 50-400 mm / month; and PET ranged from 30 to 120 mm / month [9]. Other results in the semi-arid area of California, the US showed that PET values range from 30 to 197 mm/month, the surplus is only occurred in January with 291 mm/month, with runoffs occurring for 6 months averagely ranging from 53 to 212 mm / month [10]. These results are also smaller compared to the results in our study locations.

IOP Conf. Series: Earth and Environmental Science **851** (2021) 012011 doi:10.1088/1755-1315/851/1/012011

Table 5. Water Balance of Tondano Sub-Watershed.

	January	February	March	April	May	June	July	August	September	October	November	December
T (C°)	26.28	26.24	26.26	26.67	27.14	27.27	27.03	27.13	27.21	27.00	26.55	25.23
P (mm)	184.02	225.91	216.47	253.71	200.23	125.17	148.53	120.74	113.44	156.04	196.13	258.44
PET (mm)	158.22	142.91	158.88	157.12	166.24	162.00	165.57	165.07	160.64	163.91	155.11	150.59
P-PET (mm)	25.80	83.00	57.59	96.59	33.99	-36.83	-17.04	-44.33	-47.20	-7.87	41.02	107.85
APWL (mm)	0.00	0.00	0.00	0.00	0.00	36.83	17.04	44.33	47.20	7.87	0.00	0.00
STOR (mm)	112.26	112.26	112.26	112.26	112.26	112.26	112.26	112.26	112.26	112.26	112.26	112.26
SM (mm)	112.26	112.26	112.26	112.26	112.26	80.86	96.45	75.64	73.73	104.66	112.26	112.26
ΔSM (mm)	0.00	0.00	0.00	0.00	0.00	-31.40	15.59	-20.81	-1.91	30.93	7.60	0.00
AET (mm)	158.22	142.91	158.88	157.12	166.24	156.57	132.94	141.55	115.35	125.11	155.11	150.59
SUR (mm)	25.80	83.00	57.59	96.59	33.99	0.00	0.00	0.00	0.00	0.00	33.42	107.85
DEF (mm)	0.00	0.00	0.00	0.00	0.00	5.43	32.63	23.52	45.29	38.80	0.00	0.00
R *modelled* (mm)	39.80	96.47	85.07	124.73	72.21	36.10	18.10	9.00	4.51	2.30	17.80	62.80
R *modelled* (m^3/s)	2.16	5.23	4.61	6.76	3.91	1.96	0.98	0.49	0.24	0.12	0.96	3.40
R *observed* (m^3/s)	4.00	10.00	10.00	13.00	12.00	13.00	14.00	13.00	10.00	10.00	8.00	7.00
RMSE							83.68					

Table 6. Water Balance of Serayu Sub-Watershed.

	January	February	March	April	May	June	July	August	September	October	November	December
T (C°)	17.83	17.67	17.85	17.80	18.00	17.66	17.22	16.98	16.80	17.27	17.54	17.38
P (mm)	465.00	404.00	457.00	379.00	189.00	94.00	30.00	4.00	14.00	134.00	389.00	496.00
PET (mm)	105.30	92.60	103.60	98.00	102.60	93.40	91.90	89.90	86.00	96.10	99.60	96.60
P-PET (mm)	359.70	311.40	353.40	281.00	86.40	0.60	-61.90	-85.90	-72.00	37.90	289.40	399.40
APWL (mm)	0.00	0.00	0.00	0.00	0.00	0.00	61.90	85.90	72.00	0.00	0.00	0.00
STOR (mm)	145.00	145.00	145.00	145.00	145.00	145.00	145.00	145.00	145.00	145.00	145.00	145.00
SM (mm)	145.00	145.00	145.00	145.00	145.00	145.00	95.00	52.00	31.85	145.00	145.00	145.00
ΔSM (mm)	0.00	0.00	0.00	0.00	0.00	0.00	-50.00	-43.00	-20.15	113.15	0.00	0.00
AET (mm)	105.30	92.60	103.60	98.00	102.60	93.40	80.38	46.29	34.48	96.10	99.60	96.60
SUR (mm)	359.70	311.40	353.40	281.00	86.40	43.20	21.60	10.80	5.40	2.70	289.40	399.40
DEF (mm)	0.00	0.00	0.00	0.00	0.00	0.00	11.52	43.61	51.52	75.30	0.00	0.00
R *modelled* (mm)	279.70	295.55	324.48	302.74	194.57	118.88	70.24	40.52	22.96	12.83	151.12	275.21
R *modelled* (m^3/s)	14.76	15.60	17.13	15.98	10.27	6.28	3.71	2.139	1.212	0.677	7.98	14.53
R *observed* (m^3/s)	19.00	21.00	19.00	21.00	16.00	12.00	10.00	9.00	8.00	12.00	18.00	18.00
RMSE							53.77					

Another factor that distinguishes the water balance in different climates is the SM value. SM gives different effects on evapotranspiration and annual deficits [30]. Besides that, SM has some impact on the Runoff. The results of the four study areas have an average SM value, around 120-190 mm which is classified as high with the dominance of land use as forests and agriculture. The SM value in the Arid Iran region is 44 mm with the dominant empty land [9] the SM value is still below the study location. In the US semi-arid area, the SM value is 35 mm with the dominant agricultural land use [10] that value is still lower than study location. In the tropics of Ecuador, the value of SM is 30 with the dominant forest land use [2]. The tropical region of Costa Rica has an SM value; 200 [26] which has the same value as the study location.

The topographic factor is not included in calculating the water balance because already in temperature and rain data. The differences in elevation certainly have an impact on variances in temperature and rainfall. River morphometry is not included in the Thornthwaite and Mather water balance calculations because the basic model uses climatic and physical parameters (vegetation and soil texture). Morphometric factors are used to determine the amount of runoff in a watershed, namely the runoff multiplied by the watershed area. The results water balance of this study showing that four watersheds have weak RMSE. The Thornthwaite and Mather method needs to be modified in the tropics so that it is applicable and has an acceptable RMSE. To improve the RMSE scores to go below 50%, it is necessary to add rain and temperature data with a period of 20-30 years [26,6].

4. Conclusions

The RMSE results of the Thornthwaite and Mather method in Indonesian tropical areas showed weak correlation values in the Ayung, Serayu, Bingei and Tondano sub-watersheds. The results of the Thornthwaite and Mather method validation in other tropical areas show that the validation values are inaccurate, and their characteristics are almost the same as the values of our study locations. Based on the result we need to modify the Thornthwaite and Mather method in the tropics.

Acknowledgments

The authors thank the Department of Geological Engineering, Faculty of Engineering, Universitas Gadjah Mada and the Department of Environmental Geography, Faculty of Geography, Universitas Gadjah Mada for the data supporting and the collaborative study.

References

[1] Fisher J B, Whittaker R J and Malhi Y 2011 ET come home: potential evapotranspiration in geographical ecology *Global Ecology dan Biogeography.* **20**(1), 1–18 https://doi.org/10.1111/j.1466-8238.2010.00578.x

[2] Duque-Sarango P, Cajamarca-Rivadeneira R, Wemple B C and Delgado-Fernández M E 2019 Estimation of the water balance of for a small tropical danean catchment *Granja.* **29**(1), 56–69 https://doi.org/10.17163/lgr.n29.2019.05

[3] Willmott C J and Feddema J J 1992 A more rational climatic moisture index *Professional Geographer.* **44**(1), 84–88.

[4] McMahon T A, Peel M C, Lowe L, Srikanthan R and McVicar T R 2013 Estimating actual, potential, reference crop dan pan evaporation using stdanard meteorological data: a pragmatic synthesis *Hydrology dan Earth System Sciences.* **17**(4), 1331–1363 https://doi.org/10.5194/hess-17-1331-2013

[5] Stephenson N L 1990 Climatic control of vegetation distribution: the role of the water balance *American Naturalist.* **135**(5), 649–670 https://doi.org/10.1086/285067

[6] Rao K H V D, Rao V V and Dadhwal V K 2014 Improvement to the Thornthwaite method to study the runoff at a basin scale using temporal remote sensing data *Water Resources Management*, **28**(6), 1567–1578 https://doi.org/10.1007/s11269-014-0564-8

[7] Riyanto I A, Kurniadhini F, Damayanti N L, Arfani R N 2015 Forward modelling manejemen sumberdaya air permukaan DAS Ngrancah In Marfai MA, Hizbaron DR, and Cahyadi (Ed.)

Seminar Nasional Pengelolaan Pesisir & Daerah Aliran Sungai Ke-1,Yogyakarta, 319–325, Badan Penerbit Fakultas Geografi https://www.researchgate.net/publication/323508846_Forward_Modelling_Manejemen_Sumberdaya_Air_Permukaan_DAS_Ngrancah

[8] Widyastuti M, Riyanto I A, Hendrayana H, Muhammad A S 2017 Potensi sumberdaya air sub DAS Serayu In Marfai MA, Hizbaron DR, and Cahyadi (Ed.) *Seminar Nasional ke-3 Pengelolaan Pesisir dan DAS, Yogyakarta,* 60–73, Badan Penerbit Fakultas Geografi https://www.researchgate.net/publication/324525210_POTENSI_SUMBERDAYA_AIR_SUB_DAS_SERAYU

[9] Moghaddam T E and Mohammadkhan S 2017 An estimation of Thornthwaite monthly water *Natural Environment Change* ***3***(1), 71–80 https://doi.org/10.22059/jnec.2017.216461.54

[10] Roy S and Ophori D 2013 Assessment of water balance of the semi-arid region in southern San Joaquin Valley California using Thornthwaite and Mather's model *Journal of Environmental Hydrology*. **20**(15), 1–9.

[11] Rahman A T M S, Ahmed M S, Adnan H M, Kamruzzaman M, Khalek M A, Mazumder Q H, and Jahan C S 2018 Modeling the changes in water balance components of the highly irrigated western part of Bangladesh *Hydrology dan Earth System Sciences*. **22**(8), 4213–4228 https://doi.org/10.5194/hess-22-4213-2018

[12] Sepaskhah A R. and Razzaghi F 2009 Evaluation of the adjusted Thornthwaite dan Hargreaves-Samani methods for estimation of daily evapotranspiration in a semi-arid region of Iran *Archives of Agronomy dan Soil Science*. **55**(1), 51–66 https://doi.org/10.1080/03650340802383148

[13] Rahimi J, Khalili A and Butterbach-Bahl K 2019 Projected changes in modified Thornthwaite climate zones over Southwest Asia using a CMIP5 multi-model ensemble *International Journal of Climatology*. **39**(12), 4575–4594 https://doi.org/10.1002/joc.6088

[14] As-syakur A R, Tanaka T, Osawa T and Mahendra MS 2013 Indonesian rainfall variability observation using TRMM multi-satellite data *International Journal of Remote Sensing*. **34**(21), 7723–7738 https://doi.org/10.1080/01431161.2013.826837

[15] Romps D M 2014 An analytical model for tropical relative humidity *Journal of Climate*. **27**(19), 7432–7449 https://doi.org/10.1175/JCLI-D-14-00255.1

[16] Runtunuwu E and Kondoh A 2016 Assessing global climate variability under coldest dan warmest periods at different latitudinal regions *Indonesian Journal of Agricultural Science*. **9**(1), 7 https://doi.org/10.21082/ijas.v9n1.2008.p7-18

[17] Lee S Y and McBride J L 2016 The progression of the boreal winter monsoon through the western Maritime Continent as differentiated by ENSO phase *Advances in Geosciences*. **42**, 51–60 https://doi.org/10.5194/adgeo-42-51-2016

[18] Triadi I N S, Mudhina M and Handayani K W 2013 Pengelolaan sumberdaya air Tukad Ayung sebagai upaya ketersediaan air *Jurnal Logic*. **13**(2), 114–120

[19] Hikmatullah, Subagyo H, Kurnia U A L 2000 Evaluasi erosi dan siltasi Danau Tondano berdasarkan sifat hidrologi dan tanah dari daerah aliran sungainya *Jurnal Tanah Dan Iklim*. **18**, 13–25

[20] Purnama S 2018 Potensi sumberdaya air Das Serayu *Jurnal Rekayasa Lingkungan*. **6**(3) https://doi.org/10.29122/jrl.v6i3.1942

[21] Thornthwaite CW and Mather JR. 1995 The water balance *Publ. Climato*. **8**, 1–86

[22] El-Shazli A, Hörmann G, Wagner P D and Fohrer N 2018 Comparison of water balance method dan alternative evaporation methods applied to the Aswan High Dam Reservoir *Erde*, **149**(2–3), 117–131 https://doi.org/10.12854/erde-2018-349

[23] Supari, Tangang F, Juneng L and Aldrian E 2017 Observed changes in extreme temperature dan precipitation over Indonesia *International Journal of Climatology*. **37**(4), 1979–1997 https://doi.org/10.1002/joc.4829

[24] Aldrian E and Susanto D R 2003 Identification of three dominant rainfall regions within Indonesia and their relationship to sea surface *Int. J. Climatol.* **23**, 1435–1452 https://doi.org/10.1002/joc.950
[25] Portela M M, Santos J and Studart T M D 2020 Effect of the evapotranspiration of Thornthwaite dan of penman-monteith in the estimation of monthly streamflows based on a monthly water balance model *Current Practice in Fluvial Geomorphology - Dynamics dan Diversity*, 1–14 https://doi.org/10.5772/intechopen.88441
[26] Calvo-Alvarado J C 1986 An evaluation of Thornthwaite's water balance technique in predicting stream runoff in costa rica *Hydrological Sciences Journal.* **31**(1), 51–60 https://doi.org/10.1080/02626668609491027
[27] Singh R K and Prasad V H 2004 Remote sensing dan GIS approach for assessment of the water balance of a watershed *Hydrological Sciences Journal.* **49**(1), 131–141 https://doi.org/10.1623/hysj.49.1.131.53997
[28] Caldwell P V, Kennen J G, Sun G, Kiang J E, Butcher J B, Eddy M C, Hay L E, LaFontaine J H, Hain E F, Nelson S A C and McNulty S G 2015 A Comparison of hydrologic models for ecological flows and water availability *Ecohydrology.* **8**, 1525-1546. https://DOI:10.1002/eco.1602
[29] Moriasi D N, Arnold J G, Van Liew M W, Bingner R L, Harmel R D and Veith T L 2007 Model evaluation guidelines for systematic quantification of accuracy in watershed simulation *American Society of Agriculture and Biological Engineers.* **50**(3), 885-900.
[30] Stephenson N L 1990 Climatic control of vegetation distribution: the role of the water balance *The American Naturalist.* **135**(5), 649-670.

IOP Conf. Series: Earth and Environmental Science **851** (2021) 012012 doi:10.1088/1755-1315/851/1/012012

Comparison of three recurrent neural networks for rainfall-runoff modelling at a snow-dominated watershed

K Yokoo[1], K Ishida[2*], T Nagasato[3], A Ercan[4] and T Tu[5]

[1]Graduated School of Science and Technology, Kumamoto University, Japan
[2]Kumamoto University, Kumamoto, Japan
[3]Department of Civil and Environmental Engineering, Kumamoto University
[4]University of California, Davis, One Shields Avenue Davis, California, United States
[5]School of Civil Engineering, Sun Yat-Sen University, Guangzhou, China
*E-mail: keiishida@kumamoto-u.ac.jp

Abstract. In recent years, rainfall-runoff modelling using LSTM has shown high adaptability. However, LSTM requires far more computational costs than traditional RNN. In addition, a different type of RNN, GRU, has been developed to solve this issue of LSTM. Therefore, this study compares the accuracy of the deep learning methods for rainfall-runoff modelling using three deep learning methods in a snow-dominated area. Besides, the setting of hyperparameters may affect accuracy. The accuracy of these deep learning methods was investigated by trying multiple combinations of hyperparameters. The input data were daily temperature data and precipitation data. The results show that GRU gives the highest accuracy in most combinations.

1. Introduction

Application of deep learning is a hot topic in various research fields. Deep learning is nowadays also applied to many kinds of issues in geoscience as summarised in review papers in high-impact journals [1–3]. These applications showed the high potential of deep learning in geoscience.

Rainfall-runoff modelling including flow discharge forecasting may be an issue that deep learning is frequently applied. There is a type of deep learning that is suitable for time series modelling. It is known as the recurrent neural network (RNN). A method categorized into RNN, which is called Long and Short-Term Memory (LSTM) network, has large potential to model time series that has a long-term dependency. Due to this feature of the LSTM, LSTM has been applied for time series modelling in geoscience, especially, rainfall-runoff modelling. For instance, Kratzert et al. (2019) used meteorological data such as precipitation, air temperature, and radiation as input, and then implemented flow discharge models at multiple watersheds in the United States[4]. Some other research groups applied LSTM for flow forecasting [5–7]. Furthermore, Kao et al. (2020), Li et al. (2020), and Xiang et al. (2020) applied the encoder-decoder version of LSTM for flow forecasting [8–10]. The results of these previous studies show the high applicability of LSTM for rainfall-runoff modelling.

Although LSTM has an advantage of accuracy, it has a disadvantage over the traditional RNN. It is known that LSTM requires much more computational resources than the traditional RNN because of the complex structure of LSTM. Due to this issue of LSTM, another type of RNN with a simpler structure, Gated Recurrent Unit (GRU), was developed by Cho et al. (2014)[11]. Jeong and Park (2019)[12] applied GRU and LSTM for groundwater level modelling. The accuracy of GRU is comparable to LSTM. Thus, GRU may also be another good option for rainfall-runoff modelling.

Zhang et al. (2018)[13] applied GRU for flow discharge forecasting at a sewer, and then they showed similar model accuracy between GRU and LSTM.

On the other hand, comparisons among the deep learning methods in the above studies are generally conducted with a single trained result. For the training process of a deep learning method, the initial states of its learnable parameters such as weights and biases are generally set with random values. The randomness of their initial states may affect the accuracy of the trained results. Meanwhile, a deep learning method including the traditional RNN, LSTM, GRU has model options to be set, which are called hyper-parameters. The setting of hyper-parameters may also affect accuracy. The objective of this study is to compare the model accuracy among the traditional RNN, LSTM, and GRU for rainfall-runoff modelling. In order to consider the effects of the initial states of their learnable parameters and their model configuration, the comparisons are conducted with multiple initial states of their learnable parameters, which are given by random values. Meanwhile, the comparisons are also conducted with multiple combinations of hyper-parameters. As a study watershed, the Ishikari River watershed is selected. The Ishikari River watershed is a snow-dominated watershed, which is located in the Hokkaido region, Japan. There are long-term dependencies between meteorological variables and flow discharge at a snow-dominated watershed. It may be a suitable issue to compare the three methods.

2. Methods

2.1. Long and Short-Term Memory (LSTM) Network[14]

LSTM is described as follows:

$$i_s = \sigma(W_{ii}x_s + \boldsymbol{b}_{ii} + W_{hi}\boldsymbol{h}_{s-1} + \boldsymbol{b}_{hi})$$
$$o_s = \sigma(W_{io}x_s + \boldsymbol{b}_{io} + W_{ho}\boldsymbol{h}_{s-1} + \boldsymbol{b}_{ho})$$
$$f_s = \sigma(W_{if}x_s + \boldsymbol{b}_{if} + W_{hf}\boldsymbol{h}_{s-1} + \boldsymbol{b}_{hf})$$
$$g_s = \tanh(W_{ig}x_s + \boldsymbol{b}_{ig} + w_{\boldsymbol{hg}}\boldsymbol{h}_{s-1} + \boldsymbol{b}_{hg})$$
$$\boldsymbol{c}_s = f_s \cdot \boldsymbol{c}_{s-1} + i_s \cdot g_s$$
$$\boldsymbol{h}_s = o_s \cdot \tanh(\boldsymbol{c}_s)$$

where i_s, f_s, g_s, and o_s are the input, forget, cell, and output gate at time s. $\boldsymbol{h}_s$, $\boldsymbol{c}_s$ are the hidden state and cell state at time s. W_{ii}, W_{io}, W_{if}, W_{ig}, W_{hi}, W_{hf}, W_{hg}, and W_{hg} are weights. $\boldsymbol{b}_{ii}$, $\boldsymbol{b}_{io}$, $\boldsymbol{b}_{if}$, $\boldsymbol{b}_{ig}$, $\boldsymbol{b}_{hi}$, $\boldsymbol{b}_{ho}$, $\boldsymbol{b}_{hf}$, and $\boldsymbol{b}_{hg}$ are biases. When x_s is input, x_s is used $\boldsymbol{h}_{s-1}$ is the hidden state at time s. $\boldsymbol{h}_s$ is a linear transformation of $\boldsymbol{c}_s$. The internal information of $\boldsymbol{c}_{s-1}$ is updated by inputting x_s and $\boldsymbol{h}_{s-1}$. Finally, $\boldsymbol{h}_s$ of each LSTM block is linearly transformed, and the total value becomes the o_s.

2.2. Traditional Recurrent Neural Network (RNN)

The traditional RNN can be written by the following equation:

$$\boldsymbol{h}_s = \tanh(W_{ih}x_s + \boldsymbol{b}_{ih} + W_{hh}\boldsymbol{h}_{s-1} + \boldsymbol{b}_{hh})$$

where $\boldsymbol{h}_s$, x_s are the hidden state and the input at time s. $\boldsymbol{h}_{s-1}$ is the hidden state at the previous time $(s-1)$. $\boldsymbol{W}_{ih}$, $\boldsymbol{W}_{hh}$ are input weights. $\boldsymbol{b}_{ih}$, $\boldsymbol{b}_{hh}$ are biases. The hidden state is updated each time. Finally, the value obtained by linearly transforming the hidden state is given as output.

2.3. Gated Recurrent Unit (GRU)[15]

GRU consists of the following equations:

$$\boldsymbol{\gamma}_s = \boldsymbol{\sigma}(\boldsymbol{W}_{ir}\boldsymbol{x}_s + \boldsymbol{b}_{ir} + \boldsymbol{W}_{hr} + \boldsymbol{h}_{s-1} + \boldsymbol{b}_{hr})$$
$$z_s = \sigma(W_{iz}x_s + \boldsymbol{b}_{iz} + W_{hz}\boldsymbol{h}_{s-1} + \boldsymbol{b}_{hz})$$
$$n_s = \tanh(W_{in}x_s + \boldsymbol{b}_{in} + \gamma_s * (W_{hn}\boldsymbol{h}_{s-1} + \boldsymbol{b}_{hn}))$$

$$\boldsymbol{h}_s = (1 - z_s) * n_s + z_s * \boldsymbol{h}_{s-1}$$

Where γ_s, z_s, and n_s are the reset, update, and new gates at time s. $\boldsymbol{h}_s$ is the hidden state at time s. GRU is a method that has a simpler structure than LSTM.

2.4. Hyperparameters

The above three methods have various hyper-parameters which are their model options to be set. The input data length (IDL) may be an important hyper-parameter. Rainfall-runoff modelling at a snow-dominated watershed is affected by the snow accumulation and melting processes which leads long term dependencies between the meteorological variables and flow discharge. The hidden state length (HSL) may also be an important hyper-parameter. HSL is equivalent to the number of neurons of a traditional neural network. A deep learning method is generally trained by the mini-batch gradient descent method. This study also utilized the mini-batch gradient descent method. In this case, the batch size needs to be set. Meanwhile, the mini-batch gradient descent method requires a loss function to update its learnable parameters, and an optimization algorism to adjust the learning rate. Among these hyper-parameters, this study focused on IDL, HSL, and Optimization algorism. The combinations of hyper-parameters used in this study are tabulated in Table 1-9.

3. Study area and Data

3.1. Study area

The Ishikari River originates from Mt.Ishikari whose altitude is 1,967 meters above sea level. This channel length is 268km, which is the third-largest in Japan. This basin is 14,330km^2, which is the second-largest in Japan. This basin is located at E141°20'45" N43°14'14" alt (Figure 1). This basin which is a snow-accumulated area accumulates a lot of snow in winter. The accumulated-snow period is from November to March, and the snow-melting period is from April to May. The flow discharge in this basin increases significantly compared to other periods.

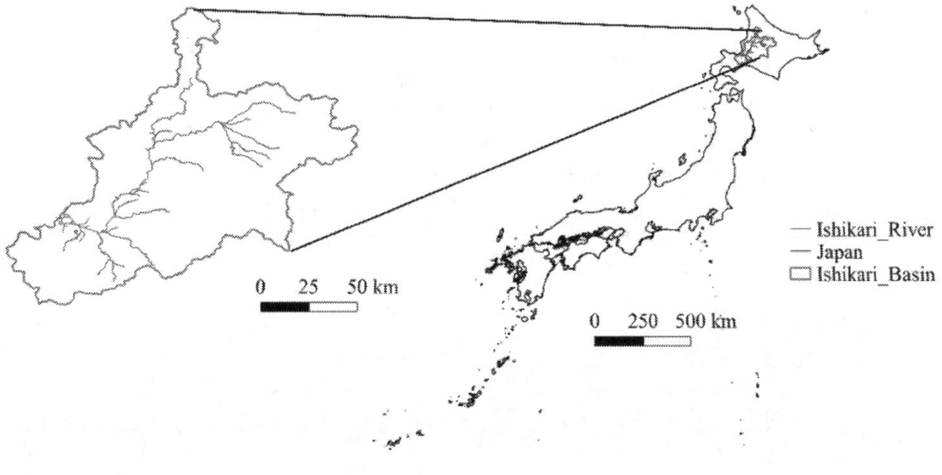

Figure 1. Ishikari basin in Japan.

3.2. Input and target data

This study utilizes precipitation and air temperature data as input. The target data are flow discharge data at the study watershed. These data were obtained from different data sources as follows.

3.2.1. Precipitation data. This study obtained daily precipitation from the Asian Precipitation – Highly Resolved Observational Data Integration Towards Evaluation of Water Resource (APHRODITE) [16]. APHRODITE is a gridded observed precipitation dataset at the daily-scale. Its spatial coverage is entire Japan. Its spatial resolution is approximately 5 km. The temporal coverage is from 1900 to 2015.

3.2.2. Air temperature data. The air temperature data except precipitation data were extracted from an atmospheric reanalysis dataset, the fifth generation of the European Centre for Medium-Range Weather Forecasts Re-analysis (ERA5)[17]. The horizontal resolution of ERA5 is approximately 25 km x 25 km (0.25° x 0.25 °). The temporal resolution of the provided data is hourly. The ERA5 dataset contains many variables. This study obtained air temperature from the single-level data. The basin average value at the study watershed was calculated from the gridded information at each hour. Then the hourly data were converted into the daily scale after adjusting the time difference between the global time and Japan standard time.

3.2.3. Flow Discharge data. The flow discharge data that are used as the target data in this study were obtained at the Ishikari Ohashi station, which is provided by Water Information System (WIS). WIS is operated by the Ministry of Land, Infrastructure, Transport and Tourism, Japan (http://www1.river.go.jp/). The Ishikari Ohashi station is located at E141°32'32" N43°07'20". The river mouth is 26.60 km downstream from the station. The original data were obtained at hourly-scale. The original flow discharge data were also converted to the daily-scale.

4. Comparison procedure

This study utilized a deep learning framework, PyTorch[18], to develop rainfall-runoff models with LSTM, RNN, and GRU. Daily precipitation and daily mean air temperature were given together to the model as input. The target data are the daily flow discharge at the Ishikari Ohashi station. The dataset was divided into three subsets: the training dataset (1998-2009), the validation dataset (2010-2012), and the test dataset (2013-2015). The models by each deep learning method are trained with the dataset with each combination of the hyperparameters (Table 1-9). To consider the randomness contained in the initial states of the learnable parameters, each method with each hyperparameter combination is trained 100 times. The best result with respect to the losses for the validation dataset is picked up and then evaluated with the test dataset. Then, the three methods are compared with respect to statistical values such as root mean square error (RMSE) and Nash-Sutcliffe efficiency (NSE) for the test dataset. The RMSE decreases as the observed and predicted values approach each other. The accuracy of NSE increases as it approaches 1.0. The accuracy is high when NSE = 0.5 or more. The accuracy of R increases as it approaches 1.0. Meanwhile, these statistical values for the training and validation datasets are also investigated to check the fluctuations in the results due to the randomness in the initial states.

5. Results and Discussions

First, IDL and the optimization algorithm were set to be 365 and Adam, respectively. Then the five HSLs were tried with all the methods. Table 1-3 shows the three statistical values (R, NSE, and RMSE) for the test dataset. For each of the three methods, there is not a linear relationship between the model accuracy and the HSL. All three methods show the highest accuracy with HSL=25. For LSTM, R, NSE, and RMSE are 0.899, 0.801, and 173.8 m^3/s, respectively. For RNN, R, NSE, and RMSE are 0.873, 0.759, and 191.1 m^3/s, respectively. For GRU, R, NSE, and RMSE are 0.917, 0.837, and 157.2 m^3/s, respectively. However, the worst results are found with HSL=100 for LSTM, with HSL=100 for RNN, and with HSL=50 for GRU. On the other hand, GRU shows better results than LSTM and RNN with all the HSL. The R-value for GRU is 0.017-0.041 higher than that for LSTM, and 0.043-0.083 higher than that for RNN. The NSE value for GRU is 0.035-0.085 higher than that for LSTM, and 0.078-0.179 higher than that for RNN. The RMSE value for GRU is 15.6-36.5 m^3/s small than that for LSTM, and 33.9-69.8 m^3/s smaller than that for RNN.

Figures 2-4 show R, NSE, and RMSE, respectively, for the training and validation datasets that were obtained by three methods with the five HSLs. LSTM and GRU clearly show better results than RNN with respect to all three statistical values. Furthermore, GRU is slightly better than LSTM with respect to its median values. Meanwhile, LSTM and GRU obtained better accuracy with a longer HSL although there is not such a relation between the model accuracy and HSL for the test dataset as shown above. The simulation for the test dataset was obtained with the trained model that shows the best results with respect to the losses for the valuation dataset among the 100 trained results as explained above. This indicates that the best result for the validation dataset may not be the best for the test dataset although there is some consistency between them.

Next, HSL and the optimization algorithm were set to be 50 and Adam, respectively. Then the five IDLs were tried with all the methods. Table 4-6 shows the three statistical values (R, NSE, and RMSE) for the test dataset. For LSTM, the best results were obtained with IDL=180. R, NSE, and RMSE are 0.894, 0.796, and 176.1 m^3/s, respectively. RNN obtained the best accuracy with IDL=100. R, NSE, and RMSE are 0.877, 0.766, and 188.3 m^3/s, respectively. GRU showed the best with IDL=365. R, NSE, and RMSE are 0.909, 0.813, and 168.5 m^3/s, respectively. LSTM and GRU seem to show better accuracy with a longer IDL. Similar to the above comparisons, GRU shows better results than LSTM and RNN with all the IDLs. The R-value for GRU is 0.004-0.031 higher than that for LSTM, and 0.004-0.059 higher than that for RNN. The NSE value for GRU is 0.005-0.044 higher than that for LSTM, and 0.005-0.100 higher than that for RNN. The RMSE value for GRU is 2.42-18.7 m^3/s small than that for LSTM, and 1.6-39.1 m^3/s smaller than that for RNN. Figures 5-7 show the corresponding R, NSE, and RMSE values, respectively, for the training and validation datasets. The three statistical values are worse with IDL=10 for all three methods. With the other IDLs, the results show similar features to those with different HSLs as shown above.

For the last comparison, HSL and IDL were set to be 50 and 365, respectively, and then four optimization algorithms were tried with all three methods. With respect to the statistical values (R, NSE, and RMSE) for the test period (Table 7-9), LSTM and GRU show the highest accuracy with Adagrad while RNN shows the best with Adam. R, NSE, and RMSE for LSTM with Adagrad are 0.903, 0.808, and 170.8 m^3/s, respectively. These for GRU with Adagrad are 0.912, 0.823, and 164.0 m^3/s, respectively. Unlike the above two comparisons, GRU is not always the best among the three methods. LSTM is the best with RMSprop and Adadelta.

Figures 8-10 show R, NSE, and RMSE, respectively, for the training and validation datasets obtained by the three methods with the four optimization algorithms. The figures show that RNN was stably trained only with Adagrad and Adam. Especially, RNN cannot properly be trained with RMSprop. The training process of GRU was also very unstable with RMSprop. Only LSTM is relatively stable with all the optimization algorithms.

Comparisons between the three methods and each Optimizer have been performed, but none have been compared at the same time. There is not much comparison by HSL or IDL. Therefore, by making this comparison, it was clarified which method is suitable for each hyperparameter.

The results in this study show that GRU has a high potential to obtain high accuracy for rainfall-runoff modelling at a snow-dominated watershed although LSTM has the most complex structure among the three methods. GRU showed the best accuracy with most combinations of hyper-parameters. However, GRU was unstable with RMSprop for the optimization algorithms. These results indicate that the comparisons among the methods should be conducted with some combinations of hyper-parameters.

Table 1. R for each HSL in the test period.

HSL	10	25	50	75	100
LSTM	0.899	0.899	0.878	0.891	0.873
RNN	0.855	0.873	0.850	0.859	0.831
GRU	0.917	0.916	0.909	0.914	0.914

Table 2. NSE for each HSL in the test period.

HSL	10	25	50	75	100
LSTM	0.782	0.801	0.769	0.792	0.747
RNN	0.724	0.759	0.721	0.716	0.653
GRU	0.837	0.837	0.813	0.827	0.832

Table 3. RMSE for each HSL in the test period.

HSL	10	25	50	75	100
LSTM	181.918	173.837	187.159	177.638	195.989
RNN	204.611	191.108	205.596	207.464	229.327
GRU	157.293	157.220	168.482	162.006	159.474

Table 4. R for each IDL in the test period.

IDL	10	50	100	180	365
LSTM	0.721	0.887	0.868	0.894	0.878
RNN	0.722	0.869	0.877	0.840	0.850
GRU	0.726	0.894	0.887	0.898	0.909

Table 5. NSE for each IDL in the test period.

IDL	10	50	100	180	365
LSTM	0.512	0.787	0.747	0.796	0.769
RNN	0.522	0.754	0.766	0.701	0.721
GRU	0.527	0.793	0.779	0.801	0.813

Table 6. RMSE for each IDL in the test period.

IDL	10	50	100	180	365
LSTM	272.157	179.836	197.190	176.133	187.159
RNN	269.401	193.100	188.316	212.840	205.596
GRU	267.827	177.416	183.051	173.710	168.482

Table 7. R for each optimizer in the test period.

OPT	Adadelta	Adagrad	Adam	RMSprop
LSTM	0.845	0.903	0.878	0.894
RNN	0.847	0.841	0.850	0.280
GRU	0.837	0.912	0.909	0.808

Table 8. NSE for each optimizer in the test period.

OPT	Adadelta	Adagrad	Adam	RMSprop
LSTM	0.714	0.808	0.769	0.796
RNN	0.707	0.693	0.721	0.00
GRU	0.701	0.823	0.812	0.631

Table 9. RMSE for each optimizer in the test period.

OPT	Adadelta	Adagrad	Adam	RMSprop
LSTM	208.473	170.787	187.159	175.764
RNN	210.711	215.660	205.596	388.869
GRU	213.057	163.997	168.482	236.647

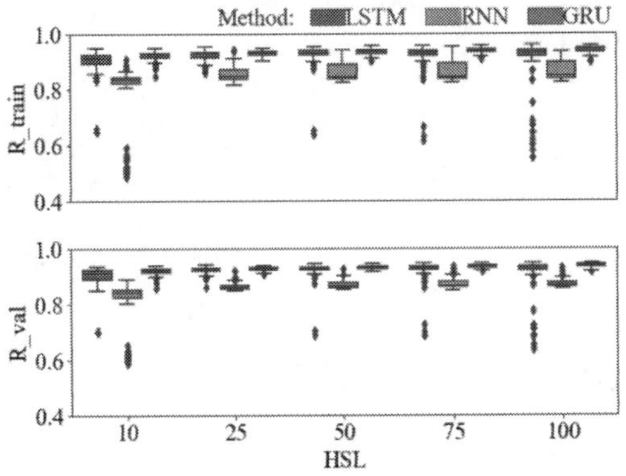

Figure 2. R values for the training and validation period.

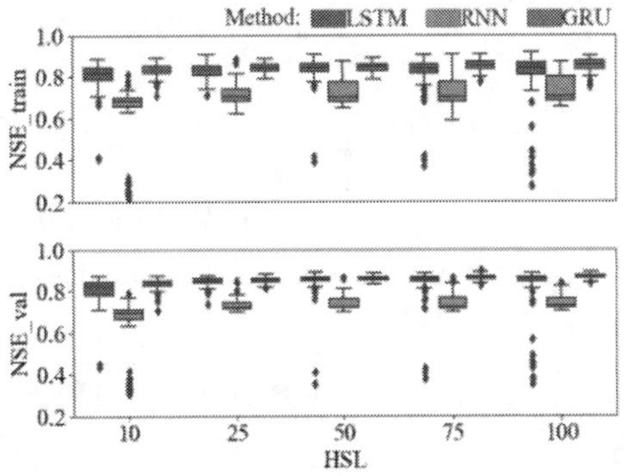

Figure 3. NSE values for the training and validation period.

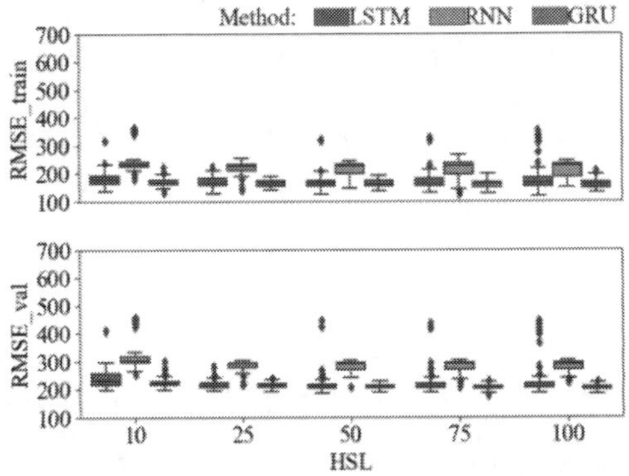

Figure 4. RMSE values for the training and validation period.

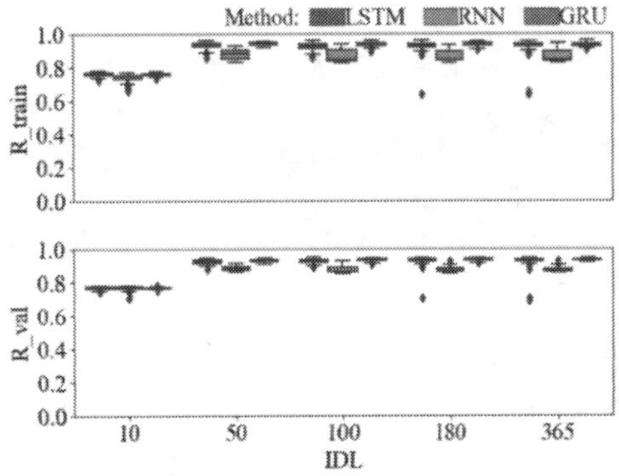

Figure 5. R values for the training and validation period.

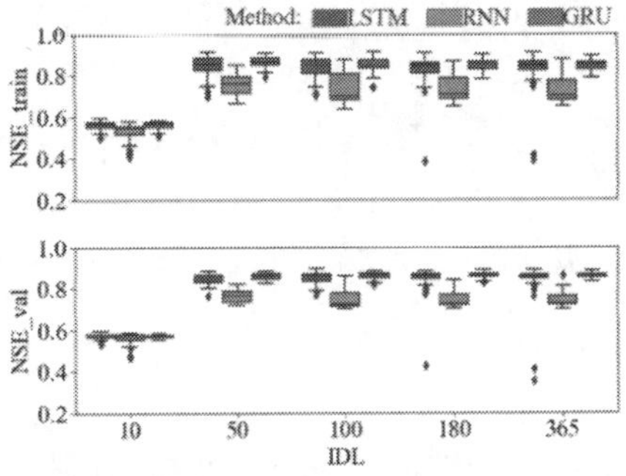

Figure 6. NSE values for the training and validation period.

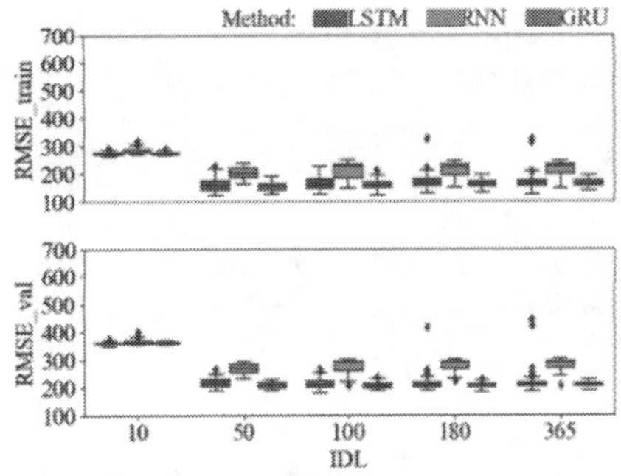

Figure 7. RMSE values for the training and validation period.

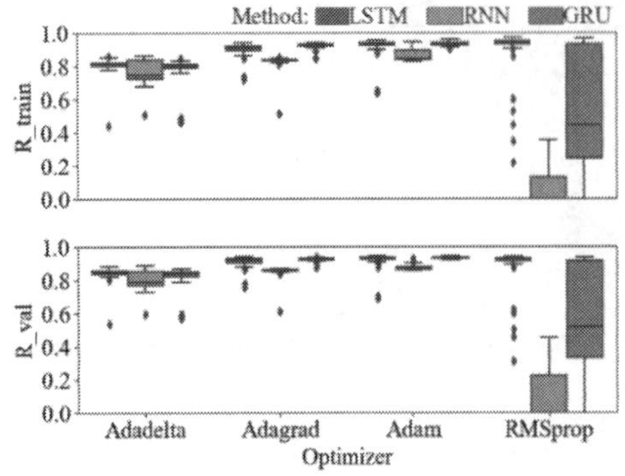

Figure 8. R values for the training and validation period.

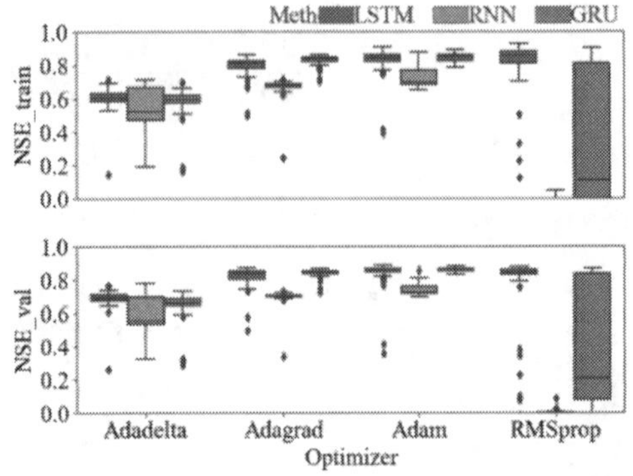

Figure 9. NSE values for the training and validation period.

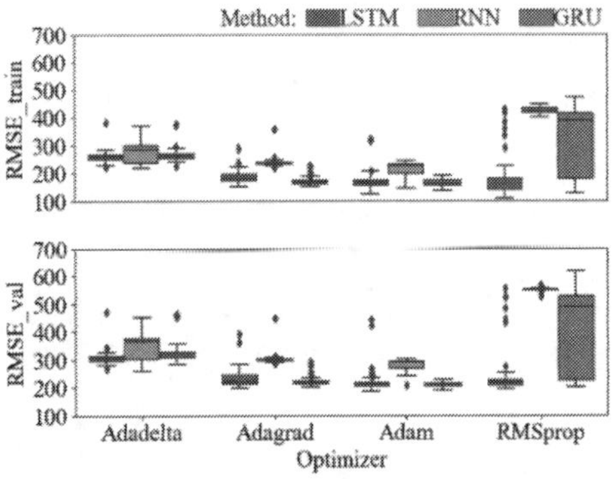

Figure 10. RMSE values for the training and validation period.

6. Conclusions

This study compared three deep learning methods, the traditional RNN, LSTM, and GRU for rainfall-runoff modelling at a snow-dominated watershed, which has long-term dependencies between input data (meteorological data) and the target data (flow discharge). In order to consider the effects of the selections of hyper-parameters, the comparisons were conducted with several combinations of hyper-parameters. The results in this study show that the GRU of the three methods is the best accurate in most cases. Therefore, GRU is the best among the three methods with most of the combinations of the hyper-parameters. On the other hand, GRU is worse than LSTM with a few combinations. This indicates that the comparisons among the methods should be conducted with various combinations of hyper-parameters.

Acknowledgments

This study was partially supported by JSPS KAKENHI Grant-in-Aid for Special Purposes 20K2191600.

References

[1] Reichstein M, Camps-Valls G, Stevens B, Jung M, Denzler J, Carvalhais N and Prabhat 2019 Deep learning and process understanding for data-driven Earth system science *Nature* **566** 195–204

[2] Shen C 2018 A Transdisciplinary Review of Deep Learning Research and Its Relevance for Water Resources Scientists *Water Resour. Res.* **54** 8558–93

[3] Karpatne A, Ebert-Uphoff I, Ravela S, Babaie H A and Kumar V 2019 Machine Learning for the Geosciences: Challenges and Opportunities *IEEE Trans. Knowl. Data Eng.* **31** 1544–54

[4] Kratzert F, Klotz D, Brenner C, Schulz K and Herrnegger M 2018 Rainfall-runoff modelling using Long Short-Term Memory (LSTM) networks *Hydrol. Earth Syst. Sci.* **22** 6005–22

[5] Tian Y, Xu Y-P, Yang Z, Wang G and Zhu Q 2018 Integration of a Parsimonious Hydrological Model with Recurrent Neural Networks for Improved Streamflow Forecasting *Water* **10** 1655

[6] Song T, Ding W, Wu J, Liu H, Zhou H and Chu J 2019 Flash Flood Forecasting Based on Long Short-Term Memory Networks *Water* **12** 109

[7] Zhu S, Luo X, Yuan X and Xu Z 2020 An improved long short-term memory network for streamflow forecasting in the upper Yangtze River *Stoch. Environ. Res. Risk Assess.* 1–17

[8] Li W, Kiaghadi A and Dawson C N 2020 High Temporal Resolution Rainfall Runoff Modelling Using Long-Short-Term-Memory (LSTM) Networks

[9] Kao I-F, Zhou Y, Chang L-C and Chang F-J 2020 Exploring a Long Short-Term Memory based Encoder-Decoder Framework for Multi-Step-Ahead Flood Forecasting *J. Hydrol.* 124631

[10] Xiang Z, Yan J and Demir I 2020 A Rainfall-Runoff Model With LSTM-Based Sequence-to-Sequence Learning *Water Resour. Res.* **56**

[11] Cho K, van Merrienboer B, Gulcehre C, Bahdanau D, Bougares F, Schwenk H and Bengio Y 2014 Learning Phrase Representations using RNN Encoder-Decoder for Statistical Machine Translation

[12] Jeong J and Park E 2019 Comparative applications of data-driven models representing water table fluctuations *J. Hydrol.* **572** 261–73

[13] Zhang D, Lin J, Peng Q, Wang D, Yang T, Sorooshian S, Liu X and Zhuang J 2018 Modeling and simulating of reservoir operation using the artificial neural network, support vector regression, deep learning algorithm *J. Hydrol.* **565** 720–36

[14] Hochreiter S and Schmidhuber J 1997 Long Short-Term Memory *Neural Comput.* **9** 1735–80

[15] Dey R and Salemt F M 2017 Gate-variants of Gated Recurrent Unit (GRU) neural networks *Midwest Symposium on Circuits and Systems* vol 2017-August (Institute of Electrical and Electronics Engineers Inc.) pp 1597–600

[16] Kamiguchi K, Arakawa O, Kitoh A, Yatagai A, Hamada A and Yasutomi N 2010 Development of APHRO_JP, the first Japanese high-resolution daily precipitation product for more than 100 years *Hydrol. Res. Lett.* **4** 60–4

[17] Hersbach H, Bell B, Berrisford P, Hirahara S, Horányi A, Muñoz-Sabater J, Nicolas J, Peubey C, Radu R, Schepers D, Simmons A, Soci C, Abdalla S, Abellan X, Balsamo G, Bechtold P, Biavati G, Bidlot J, Bonavita M, Chiara G, Dahlgren P, Dee D, Diamantakis M, Dragani R, Flemming J, Forbes R, Fuentes M, Geer A, Haimberger L, Healy S, Hogan R J, Hólm E, Janisková M, Keeley S, Laloyaux P, Lopez P, Lupu C, Radnoti G, Rosnay P, Rozum I, Vamborg F, Villaume S and Thépaut J 2020 The ERA5 global reanalysis *Q. J. R. Meteorol. Soc.* **146** 1999–2049

[18] Paszke A, Gross S, Massa F, Lerer A, Bradbury J, Chanan G, Killeen T, Lin Z, Gimelshein N, Antiga L, Desmaison A, Kopf A, Yang E, DeVito Z, Raison M, Tejani A, Chilamkurthy S, Steiner B, Fang L, Bai J and Chintala S 2019 PyTorch: An Imperative Style, High-Performance Deep Learning Library *Advances in Neural Information Processing Systems 32* ed H Wallach, H Larochelle, A Beygelzimer, F d\textquotesingle Alché-Buc, E Fox and R Garnett (Curran Associates, Inc.) pp 8024–35

IOP Conf. Series: Earth and Environmental Science **851** (2021) 012013 doi:10.1088/1755-1315/851/1/012013

Stable isotopes changes in groundwater: a case study in Mudal and Clapar springs, West Progo

T Listyani R A[1*]

[1]Geological Engineering, Yogyakarta National Institute of Technology
[*]E-mail: lis@itny.ac.id

Abstract. Hydroisotope studies were carried out on Mudal and Clapar springs located in the central part of the West Progo Dome. The research was conducted by taking samples of groundwater in each spring for three periods, representing the rainy (2016), dry (2017), and rainy (2018) seasons. Data on stable isotope content of ^{18}O and D were analyzed to see the hydroisotope characteristics of groundwater and their relationship to climate change. The results show that the stable isotope content of groundwater in both springs was relatively stable, with insignificant changes over time and season. Mudal springs tend to show light isotopes, indicating deep aquifer or high elevation recharge, less affected by the season. Clapar spring shows heavy isotopes, which may be sourced from a shallow aquifer with mixing/evaporation processes and are more influenced by the season. Meanwhile, the range value of δD in the two springs shows slightly - totally changes, indicating that the D content also changes due to seasons, although it is small. The δD enrichment shows the medium-large change in both spring springs, but uncertainty in Mudal. However, the D-excess value shows that the dry and rainy season conditions, which may be related to temperature or precipitation, are not much different.

1. Introduction

The study of groundwater has been developed because this natural resource is becoming increasingly important over time, in line with the needs of living things for groundwater. Various groundwater studies have been carried out, both physically and chemically [1-5]. Hydrochemical studies were also developed using various methods, complemented by studies of groundwater isotopes. Isotope analysis is helpful to aid in the interpretation of groundwater flows as well as aid in its genetic interpretation. The results of hydrochemical groundwater analysis can be verified by isotope analysis to produce a better interpretation of the groundwater flow system. In addition, isotope studies have also been developed using stable isotopes ^{18}O and ^{2}H (deuterium / D). One of the hydroisotope studies that can be done is related to the climate aspect in an area.

Stable isotope analysis helps know the origin of groundwater and the interpretation of catchment areas. In addition, stable isotope data can also analyze hydrochemical processes due to seasonal changes [6]. This paper intends to discuss the characteristics of the stable isotopes ^{18}O and D, particularly concerning seasonal changes in the West Progo Hills area. The case study, in this research, was carried out on the Mudal and Clapar springs located in the central part of the West Progo Dome physiography [7].

Mudal spring is at an elevation of 664 meters above sea level (masl), emerging from the limestone aquifer of the Jonggrangan Formation in Banyunganti Hamlet, Jatimulyo Village, Girimulyo Subdistrict. Meanwhile, the Clapar spring is at an elevation of 437 masl, emerging from the andesite breccia aquifer of the Old Andesite Formation in Clapar II Hamlet, Hargowilis Village, Kokap Subdistrict, West Progo Regency (Figure 1).

The study area is included in the physiography of the Dome and Hills Zone in the Central Depression [7]. The center of this dome physiography forms the morphology of the Jonggrangan plateau. The Jonggrangan Formation is quite extensive in this area. Around the Jonggrangan highlands, volcanic rocks from the Old Andesite Formation are exposed (Figure 1) [8].

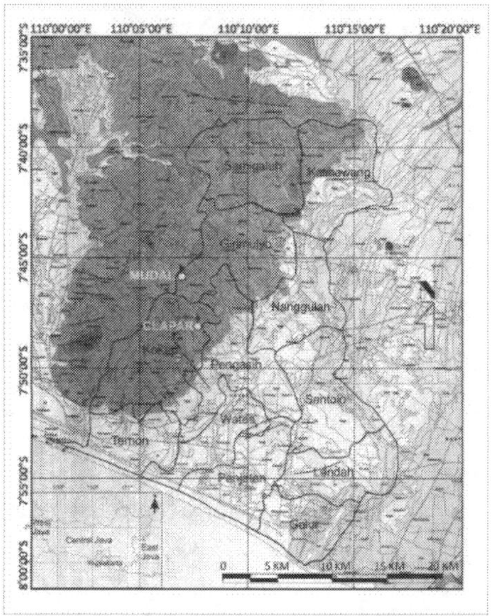

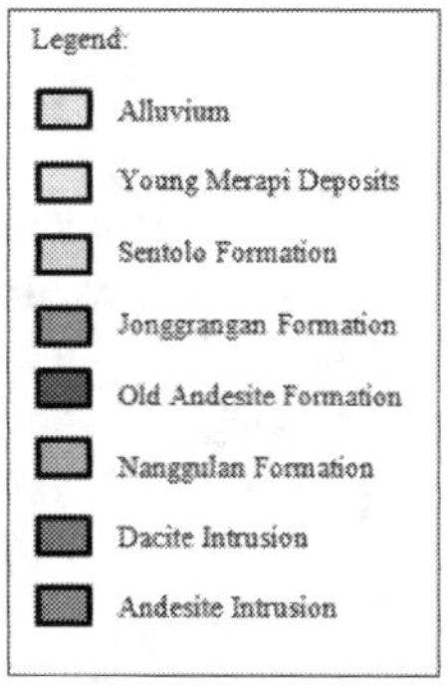

Figure 1. Location of Mudal and Clapar springs in Geological map [8]. Mudal spring is located at Jonggrangan Formation, while Clapar is at Old Andesite Formation.

The regional stratigraphy of the West Progo Mountains from the oldest to the young is composed of the Nanggulan, Old Andesite, Jonggrangan, Sentolo Formations, and Alluvial Deposits [7-9]. The Jonggrangan Formation comprises conglomerates, tuff marl, and limestone sandstones with lignite inserts, layered limestone, and coral limestone. Meanwhile, the Old Andesite Formation comprises andesite breccias, tuffs, lapilli, agglomerates, and intercalation of andesite [8]. Mudal spring appears in the Jonggrangan Formation rocks, while Clapar spring appears in the Old Andesite Formation (Figure 2).

Figure 2. Mudal spring appears in the Jonggrangan Formation (top), while Clapar spring appears in the Old Andesite Formation (bottom).

Although West Progo Hills is classified as a non-groundwater basin [10], many springs can be found even though they generally have small discharges. However, several springs with moderate to large discharge can also be found on these hills. Large discharge can be found in the limestone aquifers of the Jonggrangan Formation. The presence of springs in the West Progo Hills zone is highly controlled by the local topography [11] and lineament factors [12]. Geological lineaments influence the occurrence of springs, especially significantly controlled by the density and distance of lineaments to the location of the springs.

2. Method

The research begins with a hydrogeological survey to determine the geological conditions and springs in the study area. Several springs with small to large discharge are found in the central part of the West Progo area. This area is dominated by limestones of the Jonggrangan Formation and andesite breccias of the Old Andesite Formation. Springs with a large discharge were selected as the sample of this study. Mudal springs have large discharge and represent the aquifer of the Jongrangan Formation, while the Clapar springs are medium/large enough and represent the Old Andesite Formation. This research focuses on isotope studies, but some groundwater hydrochemical data is also taken together with isotope sampling in the field.

Groundwater samples from both springs were taken in three periods: period I in the rainy season in December 2016, period II in the dry season (August 2017), and period III in the rainy season (March 2018). Precipitation of the research area at the time of sampling can be seen in Table 1. The difference in sampling time from each period to the next is around eight months. In each sample, 30 ml of groundwater was put into an airtight bottle (polyethylene) by inserting the bottle into a water source to avoid evaporation.

Table 1. Precipitation data in research area (mm/month) [13-15].

Spring	Dec 2016	Aug 2017	Mar 2018
Mudal	216	2,5	218
Clapar	311	13	152

Isotope testing was carried out at the Hydrology Laboratory, Center for Isotope and Radiation Application (PAIR) - National Nuclear Energy Agency (BATAN), located in Pasar Jumat, South Jakarta. Isotope content in groundwater samples was determined using a Liquid Water Stable Isotope Analyzer (LWIA) type DLT-100 made by LGR (Los Gatos Research) USA. The isotope content analyzed is oxygen-18 (18O) and hydrogen (2H), known as deuterium (D) isotope. A mass spectrometer measured isotope ratios, and the results were referenced against the SMOW standard. The internal standard was calibrated using V-SMOW with an analysis accuracy of ± 0.1 for $\delta^{18}O$ and ± 1‰ for δD [3]. Furthermore, the stable isotope test results were analyzed to determine the changes and interpret the influence of the seasons/climate in the study area.

Isotope data analysis was carried out by looking at the absolute value trend and the relative value of Mudal and Clapar isotope content in three periods. In addition, the $\delta^{18}O$ and δD relationships in groundwater springs compared with meteoric water lines were also analyzed to assist in the genetic evaluation of groundwater in the springs. Analysis of changes in isotope content related to seasonal effects can be done by looking at the δD enrichment and D-excess (d) of the groundwater.

3. Stable isotope review

Isotopes are elements that have the same atomic number but different mass numbers [16]. In nature, isotopes in water can be found as stable or radioactive isotopes. The content of radioactive isotopes in water can determine age, while stable isotopes help determine water genetics [17].

Isotopes contained in water, namely hydrogen atoms (^{1}H, ^{2}H, ^{3}H) and oxygen atoms (^{16}O, ^{17}O, ^{18}O), often be used in hydrogeological studies [18,19]. The abundance of ^{1}H isotope is about 99.985%, ^{2}H is about 0.015%, and ^3H is < 0.001%, while the ^{16}O isotope is about 99.63%, ^{17}O is about 0.0375%, and ^{18}O is around 0.195% [17]. Isotope abundance is measured by the ratio of the deviation from the standard [16]. The stable isotopes ^{18}O and ^2H are present in water in compounds $^1H_2{}^{18}O$ and $^1H^2H^{16}O_2$ [17,20]. Since the abundance of $H_2{}^{18}O$ and $HD^{16}O$ molecules compared to the abundance of $H_2{}^{16}O$ is very small, the measured abundance is usually the relative abundance of an international standard water / SMOW (Standard Mean Ocean Water) [6].

The ^{18}O and D isotopes are often used in the study of chemical processes. This isotope is a stable, non-radioactive isotope and is often used to indicate groundwater sources [6]. The ^{18}O and ^2H isotopes are natural tracers because they are stable [21-23]. That is, they are not affected by the water-rock interaction process at low temperatures [24]. Therefore, isotopes are often used in genetic studies to determine groundwater infiltration zones [5,25-28] and studies of mixing groundwater from different sources [29]. The geological structure control in deep groundwater flow systems can also be determined by groundwater isotope analysis [5].

The ^{18}O and D isotopes are very sensitive to physical processes such as evaporation and condensation. Therefore, the content of these stable isotopes can be used to see the climate effect on springs. The isotopic fractionation process in precipitation is a temperature-dependent process [6]. Thus, if there is a change in seasonal temperature in a place, it will be seen that there is a variation in the stable isotope composition of the precipitation where a light value occurs in a cold month. For the same reason, precipitation will also have a lighter isotope content in the arctic/high latitudes, further away from the sea, and higher elevation places. For every 100 m elevation increase, ^{18}O in the rainwater will decrease by 0.15 - 0.5 ‰, and ^2H will be depleted by 1 - 4‰ [30].

Stable isotope content in rainwater shows a linear relationship in the form of a global meteoric water line. The relationship between $\delta^{18}O$ and δD of the precipitation water follows the equation of the meteoric water line. From the results of the global investigation [31], the equation for the meteoric waterline (GMWL) was known as $\delta D = 8\delta^{18}O + 10‰$. Rainwater tends to contain the stable isotopes $\delta^{18}O$ and δ^2H, which are depleted at higher latitudes. This phenomenon also occurs when the two stable isotopes move deep inland. For this reason, the plot results of the two isotopes yield slightly different slopes known as the local meteoric water line [32].

Based on the research of the recharge area of the underground river water system in Gunungkidul, Yogyakarta [3], it is known that the local meteoric water line (LMWL) equation for the area is $\delta^2H = 7.978\ \delta^{18}O + 8.423$ ‰. This LMWL value is then used for isotope studies in the West Progo area because of its relatively close location and considering that the LMWL value in this area is not yet available.

To see the influence of climate/rainfall, regression line relationships $\delta^{18}O$ and δD groundwater can be plotted together with the global meteoric water line GWML or the local meteoric water line (LMWL). If the groundwater regression line is adjacent to the LMWL, the groundwater is affected by local climate (originating from local precipitation) or topographic effects [1].

4. Result and discussion

4.1. Spring characteristic

Mudal Springs emerge from the Jonggrangan reef limestone aquifer, supported by large porosity permeability. The porosity developed as fracture and channel types. Jonggrangan limestone is dominated by thick to massive layered coral limestones. Around the Mudal springs, these reef limestone outcrops show white to brownish-white color, compact and hard, with some fairly intensive tectonic of joints characteristics. Mudal Springs has a large fluctuation in discharge. The discharges show; moderate magnitude during the dry season but can be very large discharges during the rainy season [11]. When the isotope sampling was carried out, the Mudal spring discharge was measured to be 100 - 236 L/s, but at the end of the dry season (September 2018), it appears that this discharge has decreased drastically to <50 L/s.The spring can be classified as depressions, fractures, or channels type of spring. Mudal spring has a large flow that develops as runoff/rivers. This spring is a parennial spring, although it has a significant change of discharge over the season. Based on its temperature, Mudal is classified as a normal spring. The physicochemical data show groundwater of Mudal spring has a temperature range of 23,1 – 24°C, pH of 6.7 – 8.3, TDS of 225 – 254 ppm, and EC of 380 – 418 µS /cm.

Meanwhile, Clapar springs have smaller dimensions than Mudal springs. Clapar springs emerge from aquifers in andesite breccias and autoclastic / lava breccias of the Old Andesite Formation, supported by fracture and sheeting joints porosity with moderate intensity controlled by low - medium permeability. Clapar springs have fracture-type springs. The discharge of springs is usually small (stagnant) - medium flow rate, with small discharge fluctuation. These springs can be classified as normal springs based on the temperature of the water. The physicochemical data of these springs show a temperature of 23.7 – 24.5°C, with a pH range of 7 – 8.2, TDS of 75 – 97 ppm, and EC of 157 – 185 µS /cm.

4.2. ^{18}O and ^{2}H isotopes contents analysis

Stable isotope content data in Mudal and Clapar spring water can be seen in Table 2 below. Furthermore, the isotope content's absolute value and range value can be analyzed to determine the hydrochemical processes that occur in the groundwater system.

Table 2. Data on stable isotope content of groundwater from the investigated springs.

Spring	I (Dec 2016)		II (Aug 2017)		III (Mar 2018)	
	^{18}O (‰)	D (‰)	^{18}O (‰)	D (‰)	^{18}O (‰)	D (‰)
Mudal	-7.1 ± 0.11	-41.7 ± 0.4	-7.39 ± 0.42	-45.1 ± 3.1	-6.94 ± 0.39	-50.2 ± 1.5
Clapar	-6.25 ± 0.07	-40 ± 1.8	-5.51 ± 0.32	-34.7 ± 1.0	-4.77 ± 0.34	-38.3 ± 3

4.2.1. Absolute value of $\delta^{18}O$ and δD. From period I to III, Mudal springs showed relatively stable O isotope, while D isotope tended to be lighter (Figure 3). Groundwater with light isotope generally flows in deep aquifers or comes from high absorption areas [2], as seen in Mudal springs, which have light D isotope (-50.2‰) in period III (Figure 3; Table 3). It means that groundwater that appears in Mudal springs may flow in deep enough aquifers or originate from precipitation of rainwater that infiltrates at a high enough elevation. The infiltration zone may exist locally because the Mudal springs are indeed at a high enough elevation.

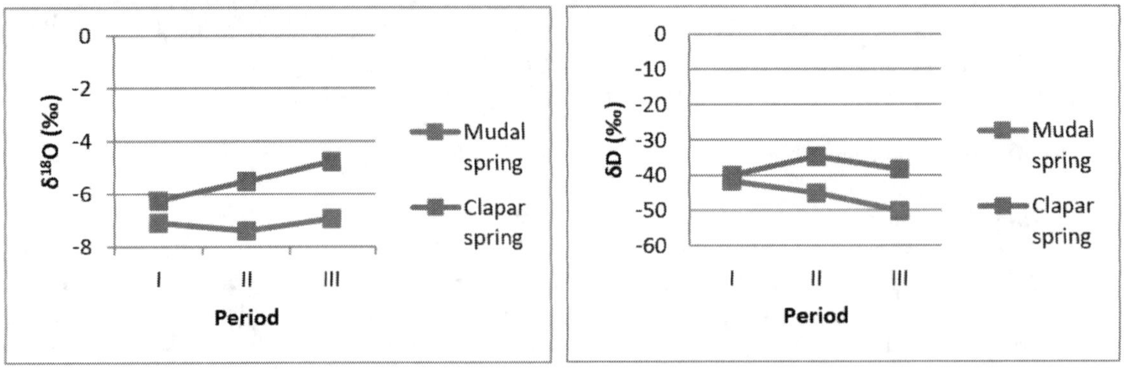

Figure 3. The development of stable isotope content over the three test periods.

Clapar springs have groundwater with ^{18}O heavier from period I to III and a relatively stable D isotope. The heavy isotopes in springs indicate a mixing or evaporation process [1, 2], strongly supported by groundwater isotopes of dug wells in the area [33]. The D isotope indicates a shallow aquifer [2]. Thus, the groundwater in the Clapar springs comes from shallow aquifers that have undergone a mixing or evaporation process.

Table 3. Changes in the stable isotope content of the springs.

Variable	Spring	$\delta^{18}O$ (‰)	δD (‰)
Time	Mudal	down-up, stable relatively difference = 0.45 ‰	get lighter difference = 8.5 ‰
	Clapar	get heavier difference = 1.48 ‰	up - down, stable relatively difference = 5.3 ‰
Season (T-effect)	Mudal	lower when dry	no effect
	Clapar	no effect	higher when dry

Compared to Clapar springs, Mudal springs contain lighter ^{18}O and D isotopes in the three periods studied. It shows that the stable isotopes possessed by the two springs are relatively consistent, whereas the Mudal springs tend to have genetics from deeper aquifers (Table 4).

Table 4. Interpretation of light / heavy isotope content.

Spring	I		II		III		Interpretation
	$\delta^{18}O$ (‰)	δD (‰)	$\delta^{18}O$ (‰)	δD (‰)	$\delta^{18}O$ (‰)	δD (‰)	
Mudal	-7.1	-41.7	-7.39	-45.1	-6.94	-50.2	- Deep aquifer, or - High elevation recharge
Clapar	-6.25	-40	-5.51	-34.7	-4.77	-38.3	- Shallow aquifer - Mixing with runoff or other sources/evaporation

When compared with GMWL and LMWL, it appears that the absolute values of isotopes contained in the Mudal springs at all periods tend to move away from the two meteoric water lines (Figure 4). Clapar springs contain isotopes that tend to be close to the meteoric water line during the rainy (period I) and dry (II) seasons. This considerable deviation in period III for Clapar springs indicates the influence of the water from other sources or runoff.

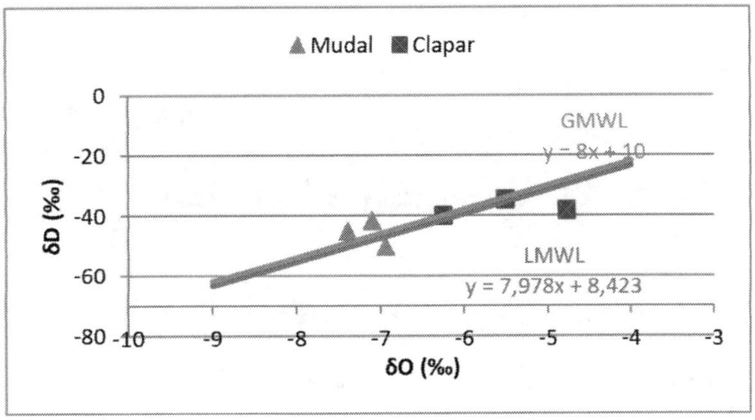

Figure 4. The relation of $\delta^{18}O$ and δD in groundwater of springs.

The interpretation of water sources in the Mudal and Clapar springs is also supported by physicochemical data from the groundwater. Mudal springs release water from deep aquifers characterized by cooler temperatures. According to other researchers, a higher temperature may be sourced from mixing groundwater and surface water [2]. The pH value, which tends to be alkaline, indicates a long interaction with carbonate rocks in the relatively deeper aquifer. This condition is also supported by the greater TDS and EC values of water from the Mudal springs than Clapar.

4.2.2 Range value of $\delta^{18}O$ and δD. The stable isotope content studied showed a short range of values and generally did not have overlapping values (Figure 5). With due regard to the $\delta^{18}O$ range value in all periods, it appears that the groundwater from Mudal springs has isotopes $\delta^{18}O$ is light, while the Clapar springs have value $\delta^{18}O$. The overlapping values in the three periods in Mudal springs indicate that groundwater in these springs is less affected by seasonal changes, while seasonal changes have more effect on Clapar springs.

The widest δD range value occurs in Mudal and Clapar springs at different periods (Figure 5). The δD value, which is relatively stable, light but appears to shift in the Mudal spring, indicates that the groundwater in this spring is less affected by seasonal changes, with relatively deep circulation. As for the springs, Clapar has relatively stable (heavy) δD, which shows significant overlapping in the rainy period, slightly different from the range value in the dry season, indicating that groundwater in these springs is immensely affected by changes in the season. Referring to previous research [2], groundwater with heavy δD as in the Clapar springs can be interpreted as a result of a fairly intensive mixing or evaporation process (Table 5).

Monthly rainfall in the three periods shows that during the rainy season, there is quite a lot of precipitation in both Mudal and Clapar (Table 1). In the dry season (period II), the precipitation is very low. However, the values range $\delta^{18}O$ for Mudal spring did not show any clear changes. The ^{18}O isotope content in Mudal springs is relatively stable and less affected by the amount of precipitation. It also indicates that the Mudal springs are supported by relatively deeper aquifers.

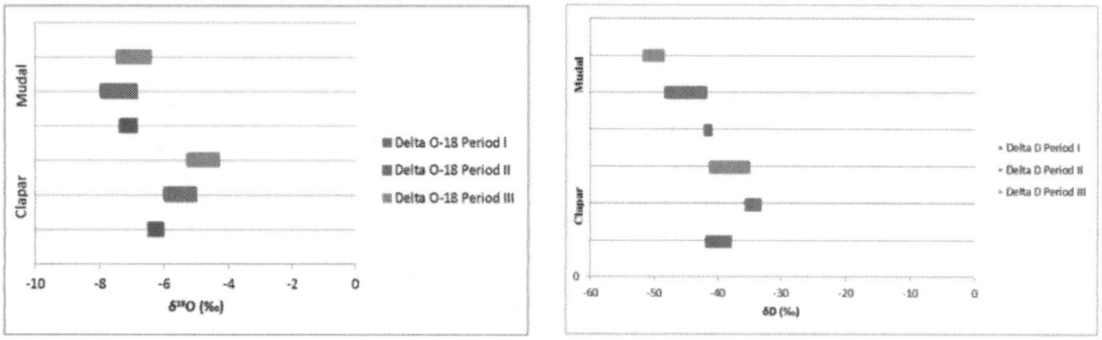

Figure 5. The values range $\delta^{18}O$ and δD for Mudal and Clapar springs. The overlapping values indicate the similarity of δD in different seasons.

Table 5. Value range interpretation $\delta^{18}O$ and δD.

Springs	$\delta^{18}O$	δD	Analysis
Mudal	In short, some overlap	Short and long, shifted	$\delta^{18}O$ and δD are relatively stable/light, less affected by seasonal changes
Clapar	Short - long, shifted, enrichment	Short-a bit long, overlap significantly in the rainy season	$\delta^{18}O$ and δD relatively stable/heavy, affected by season, intensive evaporation/mixing

Change δD of groundwater usually occurs due to isotopic exchange with minerals containing hydrogen, such as gypsum and clay minerals [34,35]. However, data does not support this exchange. The δD in these two materials is unknown, so the cause of the δD change in groundwater is still difficult to determine. Moreover, this variation in value is usually not large, so this exchange is considered insignificant. Furthermore, membrane filtration is associated with increased δD. It is difficult to happen in the study area because this process usually requires high pressure, which is equivalent to a sediment depth of 1.6 km [36]. In sedimentary rock formations, less than 1 km deep membrane filtration is less effective [37].

4.3. The effects of season on $\delta^{18}O$ and δD changes

The process that occurs related to the seasonal effect can be assessed based on the $\delta^{18}O$ against δD of groundwater relationship. In the dry season (period II), the regression line of springs in the study area is very close to the LMWL, indicating that meteoric water isotope content enrichment has not been clearly seen [38]. However, the climatic influence in this dry season can be seen from the presence of d-excess [1]. Further research added that the value of the line gradient is in the range 3-6, indicating an evaporation process [31].

The groundwater line in period III was partly below the LMWL, which indicates that it experienced isotopic enrichment [1,11], for example, due to a fairly intensive evaporation process or mixing with surface water/runoff. The slope of the regression line smaller than the LMWL gradient indicates a variation in evaporation rate. In addition, evaporation may occur in the catchment area along with the infiltration process [2].

4.3.1. The enrichment of ^{18}O and D stable isotopes. Changes in stable isotope content associated with changing seasons can cause an δD or $\delta^{18}O$ enrichment effect. O-18 isotope enrichment during the rainy season relative to the dry season occurs in Mudal springs, while δD isotope enrichment occurs in Clapar springs in the dry season compared to the rainy season (Table 6; Figure 3).

Isotopic enrichment $\delta^{18}O$ in the rainy season relative to the dry season in Mudal springs is related to the isotopic fractionation of carbonate rocks due to water-rock interaction. It was also supported by the greater TDS and EC values of groundwater in the Mudal than Clapar springs, both during the rainy and dry seasons. Enrichment of $\delta^{18}O$ can be caused by carbonate minerals [35,37]. Meanwhile, the δD enrichment of Clapar springs occurs, indicating that seasonality affects the content of these stable isotopes. The season affects the evaporation process, which can enrich the isotopic content of groundwater.

Table 6 shows the degrees of δD enrichment in the springs studied. The degree of δD enrichment is calculated in the dry season (period II) relative to the rainy season, both periods I and III. The magnitude of the changes caused by D isotope enrichment can be seen in Figure 6.

Table 6. The δD enrichment in the dry season relative to δD in the rainy season.

Spring	δD dry (Period II)	δD rainy		Enrichment Degree	Explanation
		Period I	Period III		
Mudal	-45.1	-41.7	-50.2	-3.4 - 5.1	Uncertainty
Clapar	-34.7	-40	-38.3	3.6 – 5.3	Medium-large

*) Negative values indicate enrichment during the rainy season

Figure 6 shows that the Mudal spring has a medium-large δD enrichment (>5‰) but is not related to seasonal changes. The Clapar springs undergo moderate - large changes due to enrichment during the dry season. The δD enrichment in the Clapar springs in the dry season shows a seasonal effect on the D isotope of groundwater. It is confirmed by a shift in δD values that can occur due to seasonal changes [2].

A spring that has δD enrichment >5‰ is classified to have a large change, while moderate change is indicated by D enrichment of >3 - 5‰. Meanwhile, a small change is indicated by δD enrichment of >1 - 3‰ [2]. The enrichment δD <1 indicates no enrichment. If the δD range value is considered, some groundwater samples appear to have shifted (Figure 6). The two springs have shifted slightly - totally change.

4.3.2. The "d" value (δD-excess). Changes in stable isotope content can occur due to the influence of seasons due to temperature differences. Usually, the temperature effect is related to the elevation of an area. However, it is difficult to study the effect of elevation in this study, considering that the two springs studied do not have a contrasting elevation difference. The humidity aspect also cannot be studied considering the absence of data. Isotope data in the two investigated springs showed good $\delta^{18}O$, nor δD varies considerably, both in absolute value and in range. The data that is not much different is generally considered to have no seasonal variation (temperature effect) [1]. However, if we examine one by one, there is a "d" variable which is δD-excess which we can calculate (Table 7). The value of "d" in general can be calculated with the following formula [1].

$$d = \delta D - 8\delta^{18}O \quad (1)$$

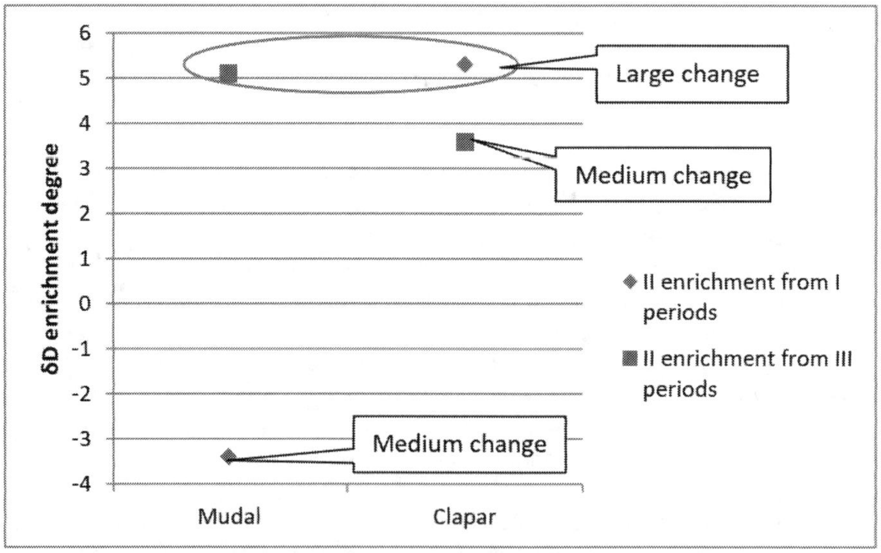

Figure 6. D-enrichment of groundwater in Mudal and Clapar springs.

Table 7. The value of the δD-excess of groundwater.

Spring	δD-excess ("d") (‰)		
	I	II	III
Mudal	15.1	14.02	5.32
Clapar	10	9.38	-0.14

The "d" value or δD excess indicates the presence of D isotope enrichment versus the $\delta^{18}O$ value. The value of "d" is a relatively important parameter concerning the climate of an area. The values of groundwater in the study area in the period I range from 10 and 15.1‰; period II range from 9.38 to 14.02‰; in period III of 5.32‰ in Mudal, indicating that the range of "d" values in Mudal is relatively higher, in all seasons. Clapar springs do not show d excess in period III. In general, the value of d gets lower over time.

In general, δD excess is influenced by air mass which is usually different, where the dry season tends to be dry, while the rainy season has humid air [1]. In rural areas, the isotopic exchange between rainwater and humidity can slightly shift d [39]. However, the d value was not significant for the springs in the study area. However, the d values in the two springs in the two seasons varied, not showing a great difference. This less difference can be interpreted that the humidity in the air during the dry and rainy seasons is not much different. The evapotranspiration conditions can occur quite intensively in the two seasons [33].

In dry conditions, evapotranspiration as a controller for groundwater recharge is usually relatively reduced, while in the rainy season / humid air, evapotranspiration is greater [1]. In addition, many plants are dormant in the dry season in the dry season, while the plants are more developed in the rainy season. Thus, the differences in evapotranspiration and humidity conditions in all seasons were not significant.

In addition, large d values usually occur in high permeability rocks or thin soil, resulting in rapid infiltration [1]. This rapid infiltration causes groundwater to experience no / less evapotranspiration. Mudal spring has a character like this, supported by many fractures, cracks, and dissolving cavities in the limestone that consist of the aquifer of these springs. Mudal aquifers are examples of karst aquifers that usually have conduit characteristics and can have underground rivers due to interconnected conduits. However, a large shifting in d values can occur in both the Jonggrangan and Old Andesite Formations aquifers.

5. Conclusion

This groundwater hydroisotope study was carried out on two selected springs in the West Progo Hills, namely the Mudal springs, which emerged from the limestone of the Jonggrangan Formation and the Clapar springs from the volcanic breccias of the Old Andesite Formation. Both springs have stable isotope content characteristics, relatively stable, with insignificant changes with time and season. Mudal springs have an isotope that tends to be light, indicating deep aquifer or high elevation recharge based on their absolute value. Meanwhile, Clapar spring shows heavier isotopes, which come from shallow aquifers with a mixing/evaporation process and are more influenced by the season. Based on the range value of $\delta^{18}O$ and δD, Mudal springs contain isotopes that are less affected by seasonal changes, while Clapar springs are seasonal. The range value of δD in both springs is slightly - totally change, which means that it changes due to the change of seasons even though it is small. Based on the season, δD enrichment in Mudal shows uncertainty, while Clapar spring has a medium-large change character. Meanwhile, the "d" value varies independently of the season, which can be interpreted that the climate conditions during the dry and rainy seasons in the study area are not much different.

Acknowledgment

This paper is based on the results of the 2016 STTNAS internal research and dissertation data. Therefore, the authors would like to thank STTNAS for the finance of this research.

References

[1] Lee KS, Wenner DB, Lee I 1999 Using H- and O-isotopic data for estimating the relative contributions of rainy and dry season precipitation to groundwater: example from Cheju Island, Korea *J. of Hydrology* **222** pp. 65 - 74, Elsevier Science BV

[2] Alam, BYCSSS, Itoi R, Taguchi S. & Yamashiro R 2014 Spatial variation in groundwater types in Mt. Karang (Java, Indonesia) volcanic aquifer system based on hydrochemical and stable isotope δD and $\delta^{18}O$ Analysis *Modern Applied Sci.* **8** 6 pp 87-102

[3] Satrio, Sidauruk P 2015 Karakteristik air tanah dangkal Kota Semarang pada musim penghujan berdasarkan pendekatan isotop stabil (^{18}O, ^{2}H) dan Kimia Air *J. Ilmiah Aplikasi Isotop dan Radiasi,* **11** 1.

[4] Satrio, Prasetio R, Hadian MSD, Syafri I 2017 Stable isotopes and hydrochemistry approach for determining the salinization pattern of shallow groundwater in alluvium deposit Semarang, Central Java *Indonesian J. on Geosci.* **4** 1 pp 1-10

[5] Setiawan T, Alam BYCSSS, Haryono E, Hendarmawan 2020 Hydrochemical and environmental isotopes analysis for characterizing a complex karst hydrogeological system of Watuputih area, Rembang, Central Java, Indonesia *Hydrogeol. J.* **28** 5, p.1635-1659.

[6] Payne B 1988 *The Basic Principles of Isotope Techniques in Hydrology and Examples of Their Application* (Padova: Centro Internazionale di Idrologia "Dino Tonini", Universita' Degli Studi)

[7] Van Bemmelen RW 1949 *The Geology of Indonesia* Vol. 1A (Netherland: Martinus Nijhoff, The Hague).

[8] Rahardjo W, Sukandarrumidi, Rosidi HMS 1977 *Geological map of Yogyakarta sheet* Scale 1: 100.000 (Bandung: Geological Agency).

[9] Budiadi Ev 2008 *The role of tectonics in controlling the geomorphology of Kulon Progo Mountainous Regions, Yogyakarta* Dissertation (Bandung: Padjadjaran University).

[10] Geological Agency 2011 *Atlas Cekungan Air Tanah Indonesia* (Bandung: Ministry of Energy and Mineral Resources)

[11] Listyani RA T, Sulaksana N, Alam BYCSSS, Sudradjat A 2019 Topographic control on groundwater flow in central of hard water area, West Progo Hills, Indonesia *Int. J. of GEOMATE* **17** 60 pp. 83-89

[12] Listyani RA T, Sulaksana N, Alam BYCSSS, Sudradjat A, Haryanto AD 2018 Lineament Control on Spring Characteristics at Central West Progo Hills, Indonesia, *Int. J. of GEOMATE* **14** 46 pp 177-184

[13] Badan Pusat Statistik Kabupaten Kulon Progo 2017 *Kabupaten Kulon Progo dalam Angka 2016* (Yogyakarta: BPS Kab. Kulon Progo)

[14] Badan Pusat Statistik Kabupaten Kulon Progo 2018 *Kabupaten Kulon Progo dalam Angka 2017* (Yogyakarta: BPS Kab. Kulon Progo)

[15] Badan Pusat Statistik Kabupaten Kulon Progo 2019 *Kabupaten Kulon Progo dalam Angka 2018* (Yogyakarta: BPS Kab. Kulon Progo)

[16] Fritz P & Fontes JC 1980 *Introduction Handbook of Environmental Isotope Geochemistry* Vol. 1 (New York: Elsevier)

[17] IAEA 1981 *Stable Isotope Hydrology* Technical Report Series No. 210 (Vienna: IAEA)

[18] Mazor E 2004 *Chemical and Isotopic Groundwater Hydrology* 3rd Ed. (New York: Marcel Dekker)

[19] Goldscheider N & Drew D 2007 *Methods in Karst Hydrogeology* (London: Taylor & Francis Group)

[20] Hamed Y 2014 Stable isotope ratios in meteoric waters in El Kef Region, Northern Tunisia: implications for changes of moisture sources *Earth Science & Climatic Change* **5** 2-6

[21] Kendall C, McDonnell JJ 1998 *Isotopes Tracers in Catchment Hydrology* (Amsterdam: Elsevier Sci.)

[22] Pu T, He Y, Zhang T, Wu J, Zhu G, Chang L 2013 Isotopic and geochemical evolution of ground and river waters in a karst dominated geological setting: A case study from Lijiang basin, South-Asia monsoon region *Appl. Geochem.* **33** pp 199–212

[23] Murillo RS, Brooks E, Elliot JW, Bolla J 2015 Isotope hydrology and baseflow geochemistry in natural and human altered watersheds in the Inland Pacific North, USA *Isotopes in Environmental and Health Studies* **51** pp 231-254

[24] Marfia AM, Krishnamurthy RV, Atekwana EA, Panton WF 2004 Isotopic and geochemical evolution of groundwater and surface waters in a karst dominated geological setting: a case study from Belize, Central America *Appl. Geochem.* **19** 937946

[25] Listyani RA T 2001 Perkiraan elevasi daerah resapan berdasarkan analisis isotop stabil airtanah (studi kasus : zone akifer III Cekungan Airtanah Jakarta) *Wahana Teknik* **3** 3 Kopertis Wil. V Yogyakarta

[26] Blasch KW, Bryson JR 2007 Distinguishing sources of groundwater recharge by using δ^2H and δ^{18}O *Ground Water* **45** pp 294–308

[27] Mukherjee A, Fryar AE, Rowe HD 2007 Regional-scale stable isotopic signatures of recharge and deep groundwater in the arsenic affected areas of Bengal, India *J. Hydrol.* **334** pp151-161

[28] Singh M, Kumar S, Kumar B, Singh S, Singh IB 2013 Investigation on the hydrodynamics of Ganga Alluvial Plain using environmental isotopes: A case study of the Gomati River Basin, northern India *Hydrogeol. J.* **21** pp 687–700

[29] Coplen T 1993 *Use of Environmental Isotopes in Regional Groundwater Quality* Ed. Alley WM (New York: Van Nostrand Reinhold) pp 227–254

[30] Clark I 2015 *Groundwater Geochemistry and Isotopes* (Florida: Taylor & Francis Group)

[31] Craig H 1961 Isotopic Variations in Meteoric Waters *American Association for The Advancement of Science* **133** 3456

[32] Gaj M, Beyer M, Koeniger P, Wanke H, Hamutoko J, Himmelsbach T 2015 In-situ unsaturated zone stable water isotope (^{2}H and ^{18}O) measurements in semi-arid environments using tunable off-axis integrated cavity output spectroscopy *Hydrol. Earth Syst. Sci. Discuss.* 12 6115-6149

[33] Listyani RA T 2019 *Groundwater flow model based on geology, hydrochemical and stable isotope at central West Progo Dome* Dissertation (Bandung: Faculty of Geological Engineering, Universitas Padjadjaran)

[34] Listyani RA T 2016 Groundwater flow and its isotopic evolution in deep aquifer of Jakarta Groundwater Basin *J. of Geological Sci.* **3**, 1

[35] Clayton RN, Friedman I, Graf DL, Mayeda TK, Meents WF, Shimp NF 1996 The Origin of Saline Formation Waters, Isotopic Composition *J. Geophys. Res.* **71** 16 pp. 3869 - 3882.

[36] Graf DL 1965 Chemical osmosis, reverse osmosis, and the origin of subsurface brines *Geochimica et Cosmochimica Acta* **46** Pergamon Press Ltd., USA.

[37] Drever JI 1988 *The Geochemistry of Natural Waters* 2nd Ed (New Jersey: Prentice Hall Inc.)

[38] Geyh MA 1990 *Isotopic Hydrological Study in the Bandung Basin, Indonesia* Project Report No. 10, Project CTA 108 (Bandung: Environmental Geology for Landuse and Regional Planning)

[39] Liotta M, Farvara R, Valenza M 2006 Isotopic composition of the precipitations in the central Mediterranean: origin marks and orographic precipitation effects atmospheres *J. of Geophys. Res.* **111** D19 AGU Publication

IOP Conf. Series: Earth and Environmental Science **851** (2021) 012014 doi:10.1088/1755-1315/851/1/012014

3D stratigraphic modelling of the Bangkok basin using Kriging on borehole data

B Subedi[1*], T Kitaoka[2] and J Kiyono[1]

[1]Department of Urban Management, Graduate School of Engineering, Kyoto University, Kyoto, Japan
[2]Faculty of Environmental and Urban Engineering, Department of Civil, Environmental and Applied Systems Engineering, Kansai University, Japan
[*]E-mail: subedi.bidhya.24s@st.kyoto-u.ac.jp

Abstract. The Lower Central Plain of Thailand has a deep and highly irregular basement filled with complex layers of alluvial sandy soil and deltaic clay or silt. The Bangkok Metropolitan region with its high level of infrastructure development is located in this plain. With high population density, the problem of land subsidence is critical. This study uses borehole data and the Kriging method to interpolate the data. A detailed 3D stratigraphic model of the basin is presented and several cross sections along two directions. Seventeen layers and the points used for modelling each layer are presented. The model shows eight aquifers in the Bangkok basin, lying between eight layers of clay, and a bottom layer above the basement. The bottom of the lowest aquifer of the Bangkok aquifer system is 610 m deep. The basement extends down to a depth of more than 2000 m in some places. Our 3D model, which extends to the basement, is of interest for mining and site-specific seismic risk analysis. Moreover, our results can be very useful for groundwater and land subsidence studies.

1. Introduction

The central region of the Kingdom of Thailand has seen intense economic development and construction growth. The areas in and around Bangkok are the regional hub of transport and logistics, with high level of development [1]. The Bangkok Metropolitan Region covers an area of 7,762 km^2. The planning region includes Krung Thep Mahanakhon, commonly known as Bangkok and five adjacent provinces: Nakhon Pathom, Nonthaburi, Pathum Thani, Samut Prakan and Samut Sakhon. The region has a growing population – more than 10 million people in 2000, more than 14 million in 2010, and earlier projections estimated a population of 16 million for 2019. The population density in the region is high – 5631 people per km^2 in Krung Thep Mahanakhon [2]. Excessive deep well pumping has led to critical land subsidence in Bangkok [3]. To carry out aquifer studies and implement ground-water resource management, a 3D stratified model and knowledge of the geotechnical properties of the local strata is necessary. For regions with high infrastructure development, such as Bangkok, such information is important in many fields and applications. The mechanical properties of the soil or rock is paramount in geotechnical engineering when selecting the type, depth and size of structural foundations suitable for the bearing capacity of the soil. It is also necessary for landform system studies, for establishing the groundwater table and for soil or water contamination studies. The stratigraphy and geotechnical data are also significant for mining, construction of subsurface structures and material surveys. For seismic risk estimation and risk reduction, it is necessary to study site-specific amplification of seismic waves, which requires knowledge of the local stratigraphic model and the geotechnical properties of the area.

In 2014, 3D geological modelling and a study of the geotechnical characteristics of the subsoils was conducted in Phnom Penh City, Cambodia [4]. In a similar study, a 3D geological model was developed for Bangkok subsoils, to a depth of up to 60 m [5]. Most of the subsurface surveys carried out in the Bangkok basin were for land subsidence studies. Although shallow subsurface profiles of Bangkok have been established, there is lack of information on deeper soil stratification and a detailed layer profile of the basin has not been published. This study uses the available information for deeper strata to construct a stratigraphic model for the deeper layers of Bangkok subsoils.

2. Geological Setting

The Lower Central Plain or Chao Phraya plain (Figure 1) is a large flat low land plain [3]. The approximate width of the Chao Phraya plain in the North–South direction from Chainat to the estuary of the Chao-Phraya river in the Gulf of Thailand is 200 km and the total area of the plain is approximately 36,000 km^2. The plain is mildly sloped with an average elevation of 2 m above sea level [6].

Data from boreholes show that the Chao Phraya basement is large, with a depth of more than 2000 m, and composed of granites and metamorphic rocks. The basement depth varies considerably in the different regions of the basin, including near the estuary of the Chao-Phraya river. In some areas, the depth of basement is unknown. The basement is filled with interbedded alluvial and flood-plain silt of the Chao-Phraya delta. The sequence of aquifers and clays is known only down to 450 m depth. The sequence has eight confined aquifers. The top layer of the sequence is soft marine clay called Bangkok clay, covering an area of 13,800 km^2 of the plain [3].

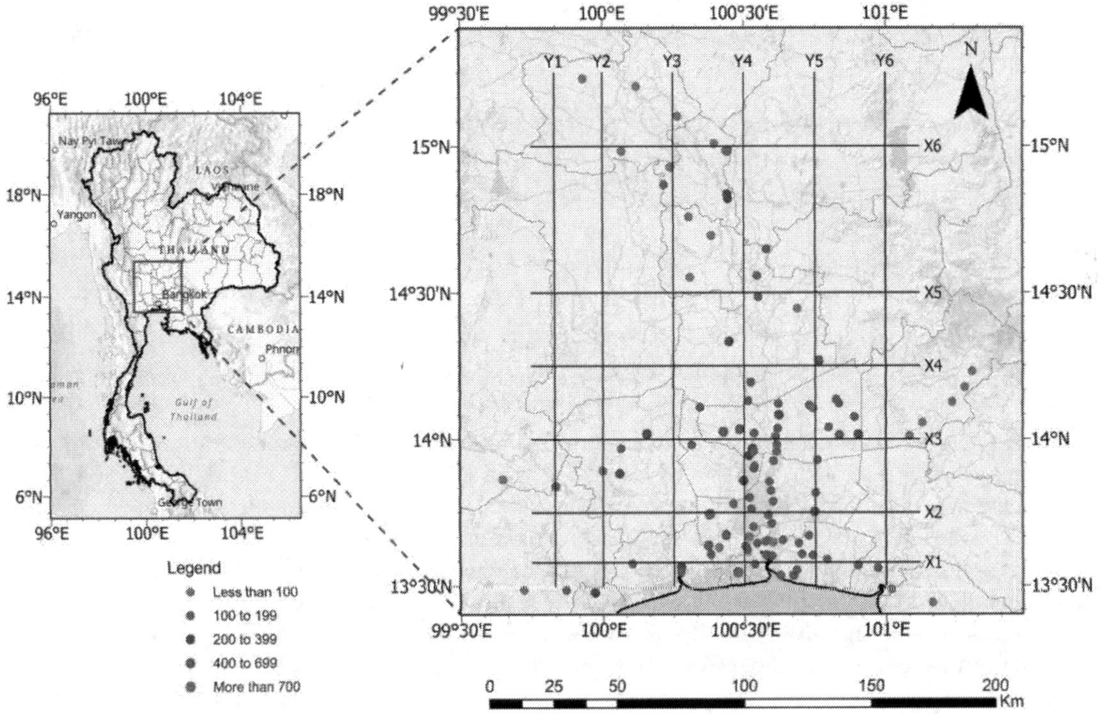

Figure 1. Left: location of the Lower Central Plain in Thailand; Right: locations of the boreholes and lines used for cross-section profiles.

3. 3D Stratigraphic Modelling

3.1. Data and Methodology

Borehole data from a previous study on groundwater flow [7] was used in our study. The data from 113 boreholes (Figure 1) show 17 layers of alternating clay strata and sand aquifers. The 17 layers are listed in descending order in Table 1. While not all the boreholes reached the bottom layer, many were between 100 and 200 m depth. The borehole points for each layer are shown in Figure 2. The properties of layers BC to PN of the Bangkok basin are well established in other literatures. The borehole data used in our study has one more aquifer layer 'T' below layer PN.

To construct the 3D-model, the data for each layer was interpolated between the gridded points in the study area. Adding depths in each layer for every grid point, a 17-layered profile was created.

Table 1. Strata of the Bangkok basin (from top to bottom); hydrogeological classification and layer code given in this study.

Order from top	Hydrogeological Classification	Layer Code
1	Bangkok Clay	BC
2	Bangkok Aquifer	BK
3	Clay 1	C1
4	Phra Pradeng Aquifer	PD
5	Clay 2	C2
6	Nakhon Luang Aquifer	NL
7	Clay 3	C3
8	Nothaburi Aquifer	NB
9	Clay 4	C4

Order from top	Hydrogeological Classification	Layer Code
10	Sam Khok Aquifer	SK
11	Clay 5	C5
12	Phayathai Aquifer	PT
13	Clay 6	C6
14	Thonburi Aquifer	TB
15	Clay 7	C7
16	Pak Nam Aquifer	PN
17	Last layer	T

3.2. Interpolation of data

Interpolation was needed to estimate the stratigraphy where borehole data was unavailable. Among the various interpolation methods available, the Kriging, Inverse Distance Weighting (IDW), Natural Neighbor (NN) and Spline methods were considered in this study. IDW, NN and Spline are deterministic methods, where the estimated values are based on a fixed set of mathematical formulas. IDW and NN use weighted averages of the data for interpolation. NN uses Delauney Triangulation of the input data. The result estimated from the Spline method passes through the input values, but estimates may exceed the range of the input data. Kriging is a geostatistical method, which uses correlation between the data rather than a fixed set of mathematical equations and is an exact interpolator.

For dense and uniformly distributed data locations, all the methods provide good and similar estimated results. For clustered data locations with large gaps, the values cannot be well estimated by any of the methods. However, Kriging gives less weight to the clustered points or treats clusters as single points. This helps to compensate for the effect of clustering and gives better estimates than the other interpolation techniques. Kriging also calculates the estimation error at the estimation points [8]. There are various Kriging methods based on the trend of the data. Simple Kriging (SK) assumes a constant trend with a known mean. Ordinary Kriging (OK) assumes a constant mean in the neighbourhood of the estimation point, rather than the whole domain. Universal Kriging (UK) or Kriging with a trend is a variant of OK where a linear or higher order trend is fitted instead of a local mean in the neighbourhood of the estimation point [8].

In Kriging, the empirical semivariogram calculated is replaced with an acceptable semivariogram model. Choosing the semivariogram model is based on subjective judgement. Spherical and exponential models are appropriate for representing properties with higher levels of short-term variability [8]. For a lag distance 'h', range 'a' and sill 'c', a spherical model can be represented as Equation (1) and an exponential model can be represented by Equation (2):

$$g(h) = \begin{cases} c.\left(1.5\left(\frac{h}{a}\right) - 0.5\left(\frac{h}{a}\right)^3\right) & \text{if } h \leq a \\ c & \text{otherwise} \end{cases} \quad (1)$$

$$g(h) = c.\left(1 - exp\left(\frac{-3h}{a}\right)\right) \quad (2)$$

The borehole data in this study is not uniformly distributed within the study area. There is a large cluster near the estuary of Chao-Phraya basin in the Gulf of Bangkok while in surrounding locations the boreholes are sparsely distributed (Figure 1). The surface of the basin usually does not have good correlation between values separated by large distances. In other words, the variable may have a spatial trend. To analyse such a dataset, the UK method is the most suitable of all the methods mentioned above. We also analysed the data using the IDW, NN and Spline methods. For the clustered dataset, IDW did not give satisfactory estimates. The estimates of the Spline method were far higher than the input values. The estimates of the NN and UK methods were better than those of the IDW and Spline methods. The surfaces produced by the UK method were smoother than those of the NN method. Based on the theoretical suitability as well as visual fitting, UK was used to interpolate all seventeen layers.

In the UK method, a trend surface is fitted and residuals from the trend are calculated. We used the 'gstat' package in R [9]. An empirical variogram is generated from the data and fitted into an existing acceptable semivariogram model. The gstat package has a choice of exponential, spherical, Gaussian or Matern models. In this study, either the exponential or spherical model were used based on the nature of the empirical variogram and the least-weighted sum of the squared errors of the fitted model (SSErr). The range coefficients and partial sill coefficients were fitted manually or automatically by the program, using the default fitting method.

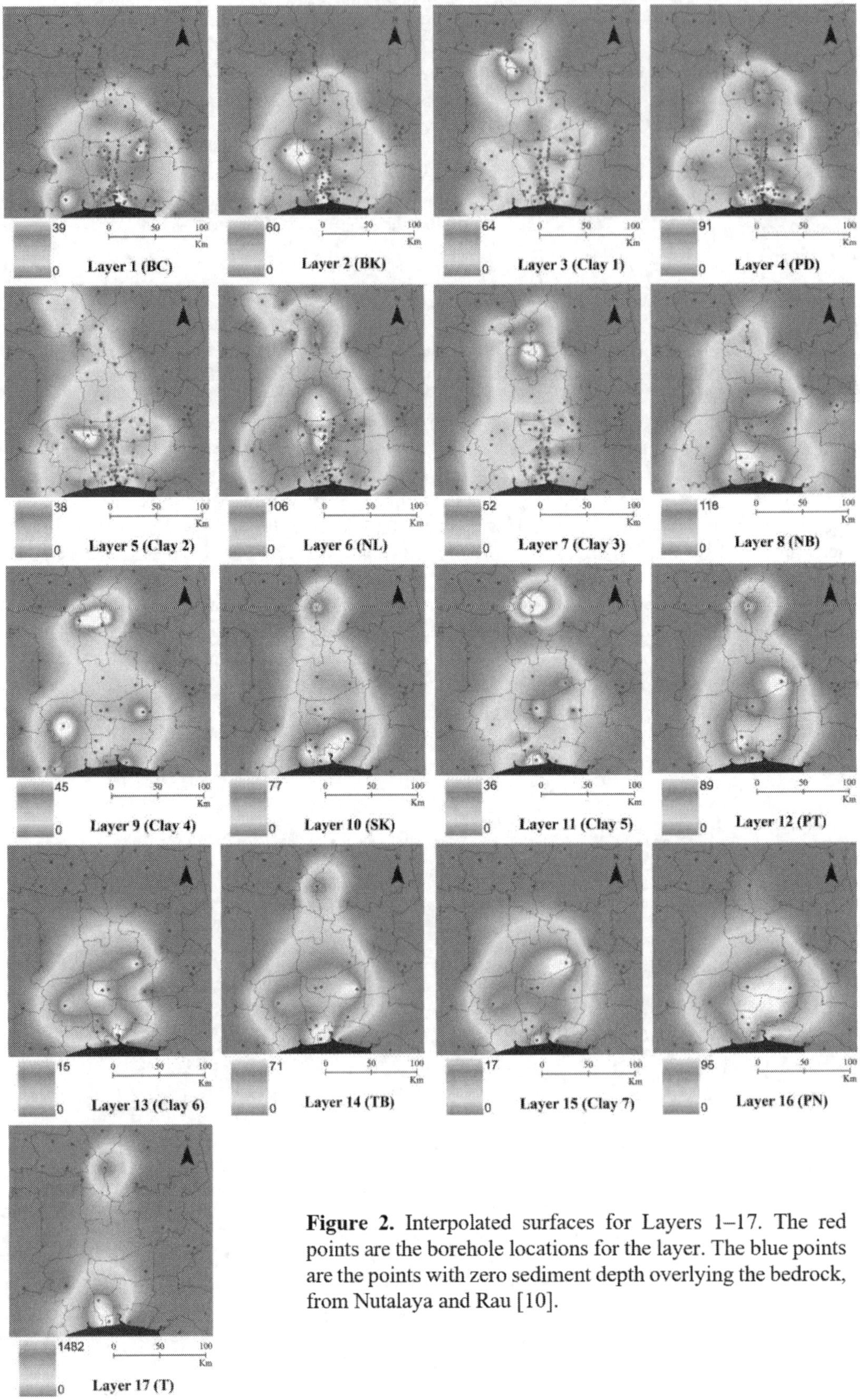

Figure 2. Interpolated surfaces for Layers 1–17. The red points are the borehole locations for the layer. The blue points are the points with zero sediment depth overlying the bedrock, from Nutalaya and Rau [10].

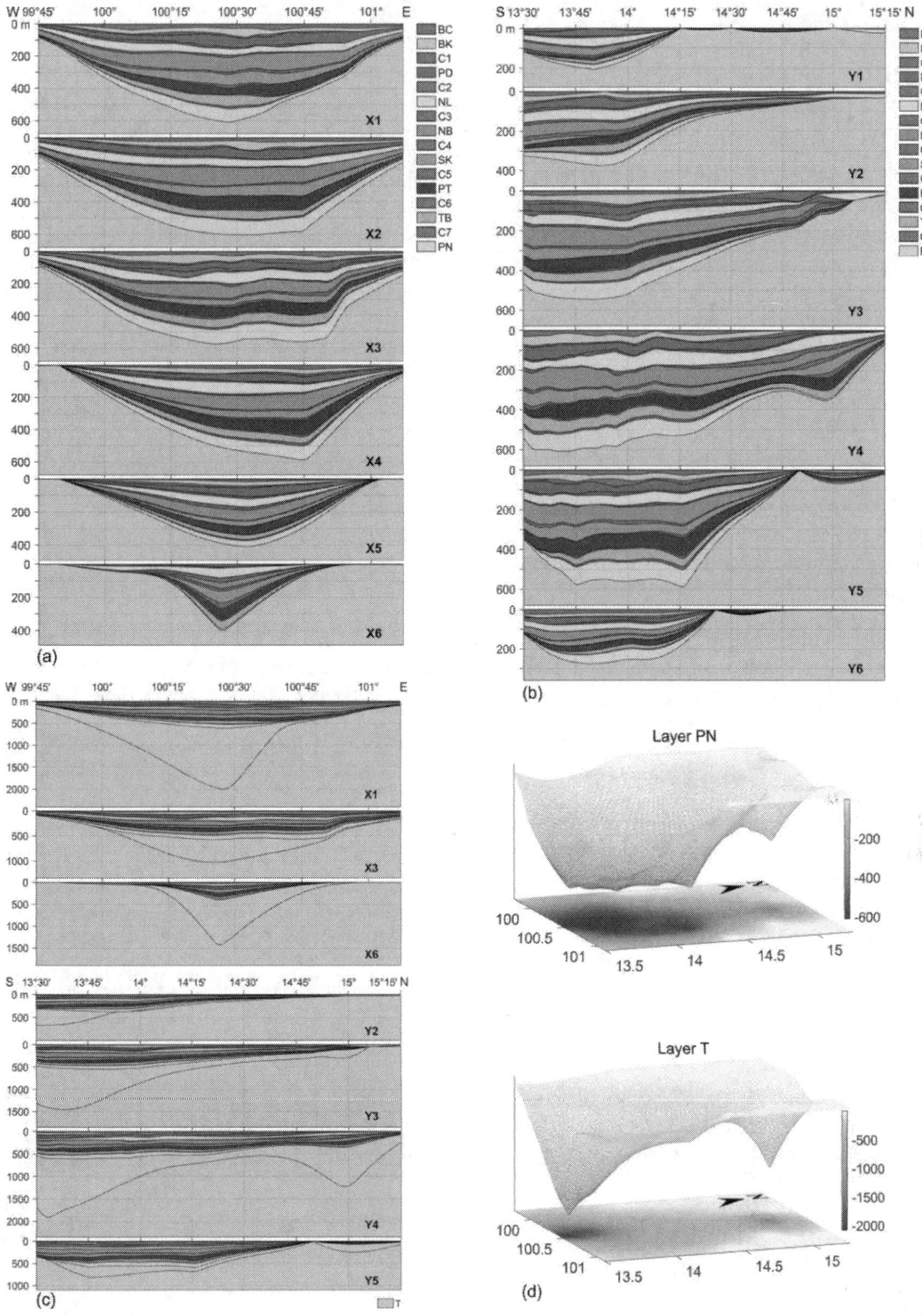

Figure 3. Cross sections from layers BC to PN along six lines in the E–W (a) and N–S (b) directions shown in Figure 1. (c) Cross sections including layer T (d). 3D surface of the bottom of layers PN and T.

4. Results and Discussion

Figure 2 shows the interpolated values of each layer. The red points are the borehole data points for each layer. The blue points represent the points with no sediment overlying the bedrock, based on a previous study [10], which were added to the dataset to confine the region of interpolation. The points are dense in the upper layers of the aquifer system; the point density gradually decreases towards the lower layers, especially in the centre of the study area at the Ang Thong and Phra Nakhon Si Ayutthaya provinces. For the bottom layer, there are no points available in those provinces (Figure 2). However, in the Bangkok metropolitan region data is available for all the layers. Layers NB and NL extend through a large area of the basin whereas layer T is confined to a small area.

The 2D-cross sections along six lines in the East–West direction and six lines along the North–South direction are shown in Figure 3(a) and (b), respectively, from layer BC to PN of the Bangkok aquifer system. The maximum depth of the bottom of the last aquifer layer of the Bangkok aquifer system is 610 m. Research on site-specific seismic amplification requires information of the deeper layers. Figure 3(c) shows the cross-sections including layer T for three lines along the E–W and four lines along the N–S directions. The lines are shown in Figure 1. The depth of the aquifer system decreases towards the north of the basin.

The maximum basement depth is 2109 m at [13.546N, 100.477E]. At one point in the Singburi province, the depth is 1550 m. Sixteen data locations are available for layer T, which is not dense enough in the centre of the study area. The cross section shows a considerable decrease in depth for layer T from 14° to 14.75° in section Y4 of Figure 3(c). The results cannot be verified for layer T in the Singburi region; however, in the area surrounding Krung Thep Mahanakhon, where more data is available, the results are reliable for all the layers.

5. Conclusion

In this study we built a 3D stratigraphic model of the Bangkok basin using borehole data. The model shows 17 layers of varying thickness. The individual layers and the points used for modelling the layers are presented. The Bangkok aquifer system comprises eight aquifers confined by alternating clay layers. The maximum depth of the bottom of the lowest aquifer of the Bangkok aquifer system is 610 m. The Bangkok basin is more than 2000 m deep in places. Cross sections along two directions were produced for clear representation of the results.

The accuracy of such analysis depends on the density of the data. The data is denser in the area surrounding the Krung Thep Mahanakhon, where the model can be used with confidence. The 3D stratigraphic model presented here can have applications in broad fields such as groundwater studies, land subsidence studies, and site-specific seismic risk analysis.

Acknowledgments

We gratefully acknowledge Ota et al. (2016) as the source of the data. We express our sincere thanks to Assoc Prof. A Furukawa (Kyoto University, Japan), Prof. Y Ono (Tottori University, Japan), Assoc Prof. T Ornthammarath (Mahidol University, Thailand), Asst Prof. B Charatpangoon (Chiang Mai University, Thailand), Dr. P Latcharote (Mahidol University, Thailand) and Dr. S Suttisak (Kasetsart University, Thailand) for their helpful comments during our model making. This work is supported by the project "International Collaborative Research on Long-period Earthquake Early Warning for the Bangkok Metropolitan Area". The project is supported by the Supporting Program for Interaction-Based Initiative Team Studies (SPIRITS) funded by Kyoto University.

References

[1] Asian Development Bank 2015 *Thailand Industrialization and Economic Catch-Up: country diagnostic study*
[2] https://www.citypopulation.de/en/thailand/prov/admin/ (accessed on December 25, 2020)
[3] Phien-wej N, Giao P H and Nutalaya P 2006 Land Subsidence in Bangkok, Thailand. *Engineering Geology* **82** p 187–201.

[4] Touch S, Likitlersuang S and Pipatpongsa T 2014 3D Geological Modelling and Geotechnical Characteristics of Phnom Penh Subsoils in Cambodia. *Engineering Geology* **178** p 58-69 10.1016/j.enggeo.2014.06.010.

[5] Ngamcharoen K, Likitlersuang S and Boonyatee T 2016 Development of 3D Geological Modelling for Bangkok Subsoils *the Twenty-Ninth KKHTCNN Symp. on Civil Engineering Hong Kong, China*

[6] Sinsakul S 2000 Late Quaternary geology of the Lower Central Plain, Thailand. *Journal of Asian Earth Sciences* **18**(4) p 415-426.

[7] Ota S, Kitaoka T, Ohtsu H and Soralump C 2016 Establishing a 3D geological model for groundwater flow considering water quality in Bangkok *Proc. of EIT-JSCE Joint Int. Symp. on Geotechnical Engineering for the Establishment of Resilient Society to Natural/Man-Made Disasters 2016*

[8] Bohling G 2007 *Introduction to Geostatistics, Hydrogeophysics: Theory, Methods, and Modeling* (Boise State University, Boise, Idaho)

[9] Gräler B, Pebesma E and Heuvelink G 2016 Spatio-Temporal Interpolation using gstat *The R Journal* 8 p 204-218.

[10] Nutalaya P and Rau J L 1984 Structural Framework of the Chao Phraya Basin, Thailand 1984 *Proc. of the Symp. on Cenozoic Basins Thailand, Geology and Resources, Chiang Mai University*

IOP Conf. Series: Earth and Environmental Science **851** (2021) 012015 doi:10.1088/1755-1315/851/1/012015

Characterization of shallow groundwater chemistry in the Yogyakarta basin, Central Java

L Kämpfner[1], T R Rüde[2], D P Eka Putra[3]

[1, 2]Department of Engineering Geology and Hydrogeology, RWTH Aachen University, Lochnerstr. 4-20, 52064 Aachen, Germany
[3]Department of Geological Engineering, Faculty of Engineering, Universitas Gadjah Mada, Jl.Grafika No.2 Bulaksumur, Sleman, D.I. Yogyakarta, Indonesia
*E-mail: kaempfner@hydro.rwth-aachen.de

Abstract. Sufficient water supply can still be a big challenge in developing countries. A large number of households in Yogyakarta still relay on private dug wells to meet their daily demand. But water from shallow aquifers is prone to pollution from sewage systems, industry and agriculture. To assess the water quality in the basin of Yogyakarta, this study determines the chemical characteristics of the shallow groundwater in anthropogenically influenced and uninfluenced areas in and around the city of Yogyakarta. Therefore, a combination of on-site and laboratory analysis is used to distinguish areas of different water types. The observations give a major ion distribution for the whole research area and show significant anomalies in nitrate and phosphate concentrations especially in the city. Furthermore, high DOC values point out that the aquifer is widely polluted. There is a need for a better groundwater protection and remediation strategy to facilitate future use of those groundwaters.

1. Introduction

The special province of Yogyakarta (Daerah Istimeva Yogyakarta) is located at the south coast in the centre of Java Island, Indonesia. The province and its capital Kota Yogyakarta experienced a high variety of influences from different cultures during its past [1]. Water supply is mainly obtained from an aquifer close to the surface which is therefore prone to pollution. From 1984 to 2013, a groundwater table decline of about three meters was observed [2]. Only 30 % of the population is connected to the city's water supply network which loses about 40 % of its capacity underground [3]. A majority of inhabitants meet their water demand by private dug wells. A small number of households is connected to a sewerage system. High coli bacteria concentrations suggest a high leakage rate from sewage in the city area. The adjusted annual groundwater recharge is about 45.10^6 m^3/a [3]. The groundwater chemistry in these private dug wells is at the focus of this study. In the context of a master thesis within the German-Indonesian GetIn-CICERO cooperation between RWTH Aachen University (Germany) and Universitas Gadjah Mada (Indonesia), a campaign of on-site parameter measurements and laboratory analyses was carried out to assess groundwater quality in the Yogyakarta groundwater basin. We measured on-site parameters from dug wells including the water temperature, electrical conductivity (EC), pH and groundwater level. In each location, water samples were taken and analysed for alkalinity, concentration of the major ions and carbon concentrations. These results are used to outline the types of groundwater present in the aquifer and identify possible sources of pollution. This paper aims to deliver a characterisation of the chemical state of the shallow groundwater in Yogyakarta and to identify possible threats to its groundwater quality.

2. Research Area

The research area is located in the Special Province Yogyakarta. The province is divided into five districts (or "regencies"): Kulon Progo, Bantul, Gunungkidul, Sleman and Kota Yogyakarta. The research area covers Sleman, Kota Yogyakarta and parts of Bantul. It follows the borders of the Yogyakarta groundwater basin. DI Yogyakarta has a population of approximately 3.8 million.

Sleman Regency is located in the northern area of DI Yogyakarta from the city centre (Kota Yogyakarta) up to the slope of Merapi volcano. Sleman Regency has a population density of 2122 inhab. km^{-2}. Kota Yogyakarta is the province capital and has the highest population density in the province by 13290 inhab. km^{-2}. Bantul Regency is located in the south and has a population density of 2009 inhab. km^{-2} [4]. The exact location and outline of the research area is displayed in figure 1. The outline of the groundwater basin is shown in red. Also displayed are the districts, and their subdistricts as well as rivers and observation points. In the northern area of Yogyakarta at the slope of Merapi, farming is a very important land use due to the fertile soil properties of the volcanic products. The precipitation in Yogyakarta averages to 2012 mm/a with 119 days of rain per year. The climate is driven by the alternation of rainy and dry seasons [5].

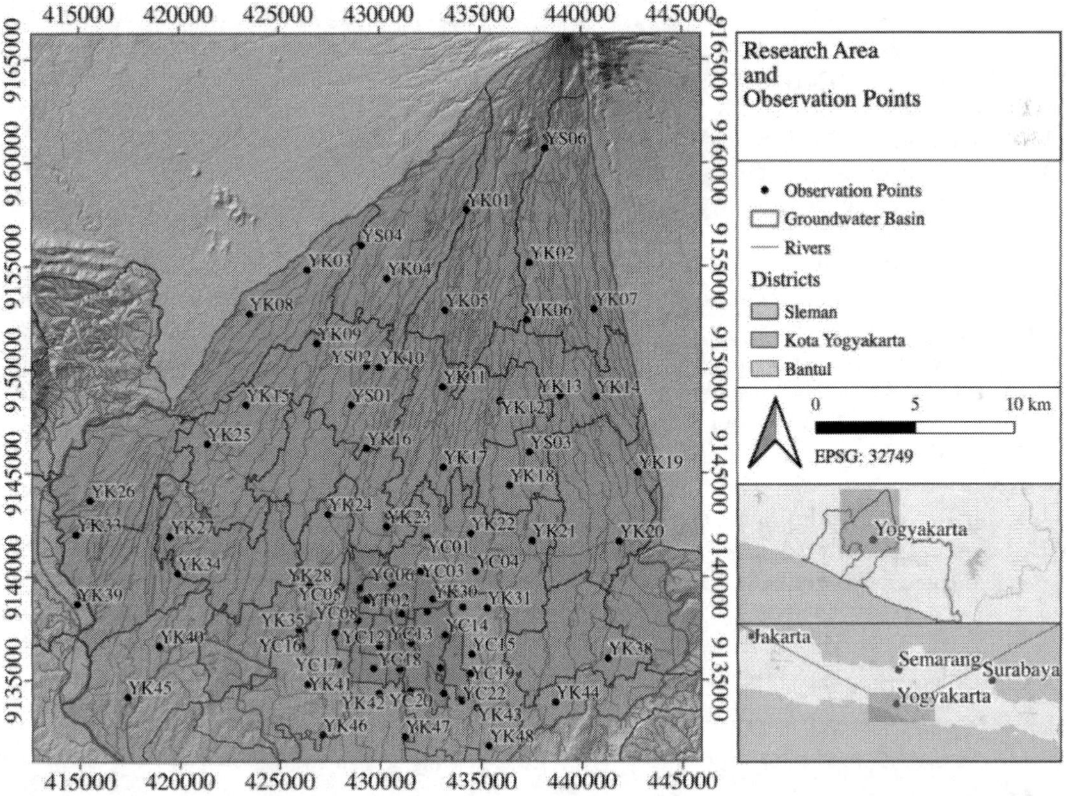

Figure 1. Overview of the research area and location of observation points categorised in the central districts Sleman (north), Kota Yogyakarta (centre) and Bantul (south). EPSG (European Petroleum Service Group Geodesy) is one of the largest geodetic parameter datasets widely used for coordinate reference system identification and is used for the coordinates of the main map and also in figure 5.

3. Hydrogeological Setting

The dominating hydrogeological system in the research area is the Yogyakarta groundwater basin. The basin is constituted by an upper aquifer (Yogyakarta formation) and a lower aquifer (Sleman formation). Their substratum is formed by the Old Merapi volcano in the north and the Sentolo formation in the south. The Yogyakarta Graben is located underneath the central part of Yogyakarta. This Graben system is filled with Young Merapi deposits which form the upper and lower aquifers of the Yogyakarta groundwater basin. Both formations can be subdivided into several Quaternary layers of gravel, sand, and clayey sand. These layers are partially interrupted by low permeable, more clayey layers [6]. This leads to locally changing head differences between the upper and lower aquifers [7].

4. Methodology

On the basis of former studies in this area, a sampling grid was set up including 78 locations. The grid was designed to get an overall impression of the chemistry in the whole research area. A focus was also to gain detailed information about groundwater uninfluenced from anthropogenic activities in the north as well as to get a higher resolution in the densely populated city area of Kota Yogyakarta. Most samples were taken from private dug wells of local households and few from springs.

At each sampling location, a water sample was taken from the dug wells for further laboratory analysis. Furthermore, on-site parameters as water temperature, EC, pH and groundwater level were measured. The water samples were analysed on alkalinity by manual titration. Total organic carbon (TOC), total inorganic carbon (TIC), their dissolved species (DOC, DIC) and total bound nitrogen (TNb) were measured in the GETIN CICERO laboratory at Universitas Gadjah Mada, Yogyakarta, by elemental analyzers. An additional set of samples was sent to RWTH Aachen University in Germany for major ion analysis by ion chromatography.

4.1. Groundwater Sampling

A majority of groundwater samples was taken from private dug wells since there was no sufficient network of groundwater observation wells available to cover the whole research area. Sampling dug wells in private households requires a minimal invasive sampling method. To still obtain water samples which represents the actual groundwater chemistry, without setting up a pumping installation, the sample was taken one meter below groundwater surface. For this purpose, a steel cage for sampling bottles was designed as shown in figure 2. The sampler was attached to a rope and positioned right above the water surface. For sampling, it was then dropped quickly one meter below surface to be filled in that depth and pulled up after it filled completely with water. Samples were not taken deeper, because the depth of most of the wells was unknow and the well sump was to be avoided. Bottles were taken off the cage, sealed and new bottles were placed for each sampling point.

Figure 2. Bottle cage with additional weight designed to sample water from a certain depth in a private dug well. The cage suits a sample bottle with a volume of 1 L.

4.2. On-site Parameters

Parameters like pH, EC and water temperature underlay changes with time after sampling [8]. These parameters were thus measured on each sample right after sampling. On-site measurements were carried out with HI98195 and HI9811-5 probes of the company HANNA instruments which were calibrated daily. Additional to chemical parameters, the groundwater level in each dug well was measured with a water level probe.

4.3. Laboratory Analyses

Alkalinity was obtained in the laboratory by manual titration after German norm DIN 38409-7 using methyl orange and hydrochloric acid. The dissolved species were measured on samples filtered at the lab using a filter size of 0.45 µm. Former studies have shown that DOC concentrations in unpolluted aquifers are within the range of 0.1 – 4 mg/L as C. Where concentrations in shallow aquifers are highly dependent on soil properties which influence organic carbon intake [9]. Inorganic carbon concentrations are compared with the results of the alkalinity measurements. Total bound nitrogen (TNb) represents the sum of nitrite, nitrogen (NO_2-N, NO_3-N), ammoniacal nitrogen (NH_4-N) as well as organic bound nitrogen [10]. Main artificial sources of nitrogen in groundwater are sewage and fertilizers [11]. Therefore, TNb is a good indicator for anthropogenically influenced groundwater chemistry.

5. Results

The observed groundwater temperature ranges between 20.8°C and 28.7°C with a mean of 27.2°C. The groundwater temperature is increasing from north towards the city area in the south as elevation decreases, and thus atmospheric temperature increases. Groundwater elevation is decreasing from north to south, following also the elevation decrease. An overview is given in figure 3 and 4.

Minimum, maximum, mean and standard deviation of pH and electrical conductivity in µS/cm are given in table 1 for each district. The largest range of pH values was observed in Sleman with a significantly higher maximum pH value than in the other districts. The same was observed by the electrical conductivities. Remarkable is that in Sleman the minimum observed electrical conductivity was significantly lower than in any other district.

Table 1. Overview of on-site measurements categorized into district.

	pH [-]			**EC [µS/cm]**		
	Sleman	Yogyakarta	Bantul	Sleman	Yogyakarta	Bantul
Min	5.9	6.1	6.0	93	394	311
Max	7.9	7.1	6.9	881	737	873
Mean	6.6	6.4	6.4	336	516	512
Std. Deviation	0.4	0.3	0.2	143	92	155

Table 2. Overview of TNb and DOC measurements in [mg/L] categorized by district.

	TNb [mg/L]			**DOC [mg/L]**		
	Sleman	Yogyakarta	Bantul	Sleman	Yogyakarta	Bantul
Min	< 0.1	0.6	< 0.9	0.9	1.2	1.5
Max	24.1	40.9	40.9	7.4	3.5	6.5
Mean	4.8	17.9	9.8	2.0	2.0	2.7
Std. Deviation	6.7	9.9	10.7	1.2	0.6	1.3

Out of the measurements of TOC, TIC, DOC, DIC and TNb, table 2 gives an overview of TNb and DOC concentrations in mg/L. These are the more significant parameters measured.

Figure 3 and 4 display samples from the city area and from the main recharge area in the north of the city. Groundwater level is used here but correlates with elevation very significantly ($R^2 = 0.9997$). In the city area data are clustered as the change in groundwater level is low. The trend in temperature is as discussed before. The trend in electrical conductivity is representative to emphasise the anthropogenic influence on the groundwater chemistry that evolves with an increasing population density, and maybe also an increasing trend of water-rock interactions. One outlier is not taken into account, since the observed electrical conductivity is rather high for this elevation.

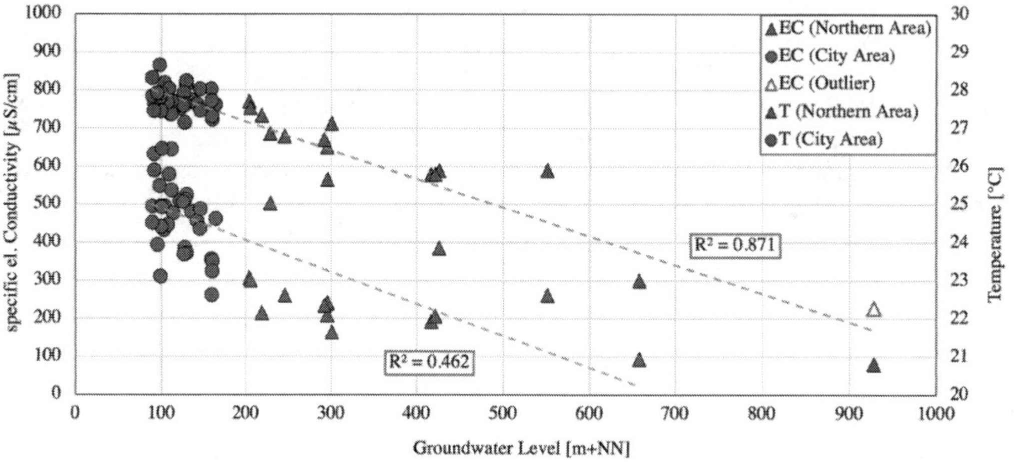

Figure 3. This diagram shows the spatial trend of specific electrical conductivity and temperature of the water samples. The groundwater elevation is highest in the north and decreases southwards to the city area.

Figure 4 displays nitrate concentrations and groundwater level. Samples of the northern area range between no detected nitrate and 57.5 mg/L. The majority of measured nitrate concentrations in the north are below 20 mg/L. The city area shows a high spread of nitrate concentration up to 176 mg/L. This shows very clearly the groundwater pollution in the city area.

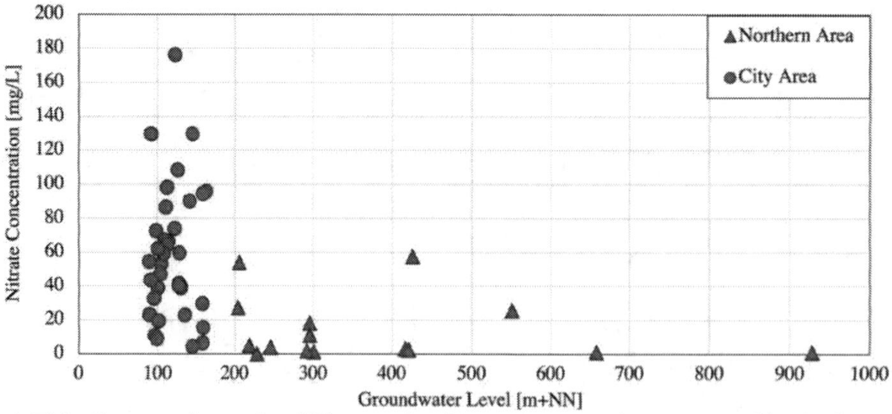

Figure 4. This diagram shows the difference in nitrate concentration measured in the low populated area in the north (high groundwater level) and the city area (low groundwater level).

In figure 5 the concentrations of the major ions measured on groundwater samples are displayed. Cations are shown in a spectrum of red to yellow and anions in blue to green. The presented concentrations are given in mmol(eq)-%. The ion proportion varies clearly within the research area. Characteristic concentrations can be observed in different regions. Bicarbonate is in most springs and wells the dominating anion in western and northern Sleman district. To the south east and in Kota

Yogyakarta the percentage of other anions, especially nitrate, increases and nitrate is higher in concentrations. The most represented cation is Ca^{2+}followed by Na^{+} in the whole research area.

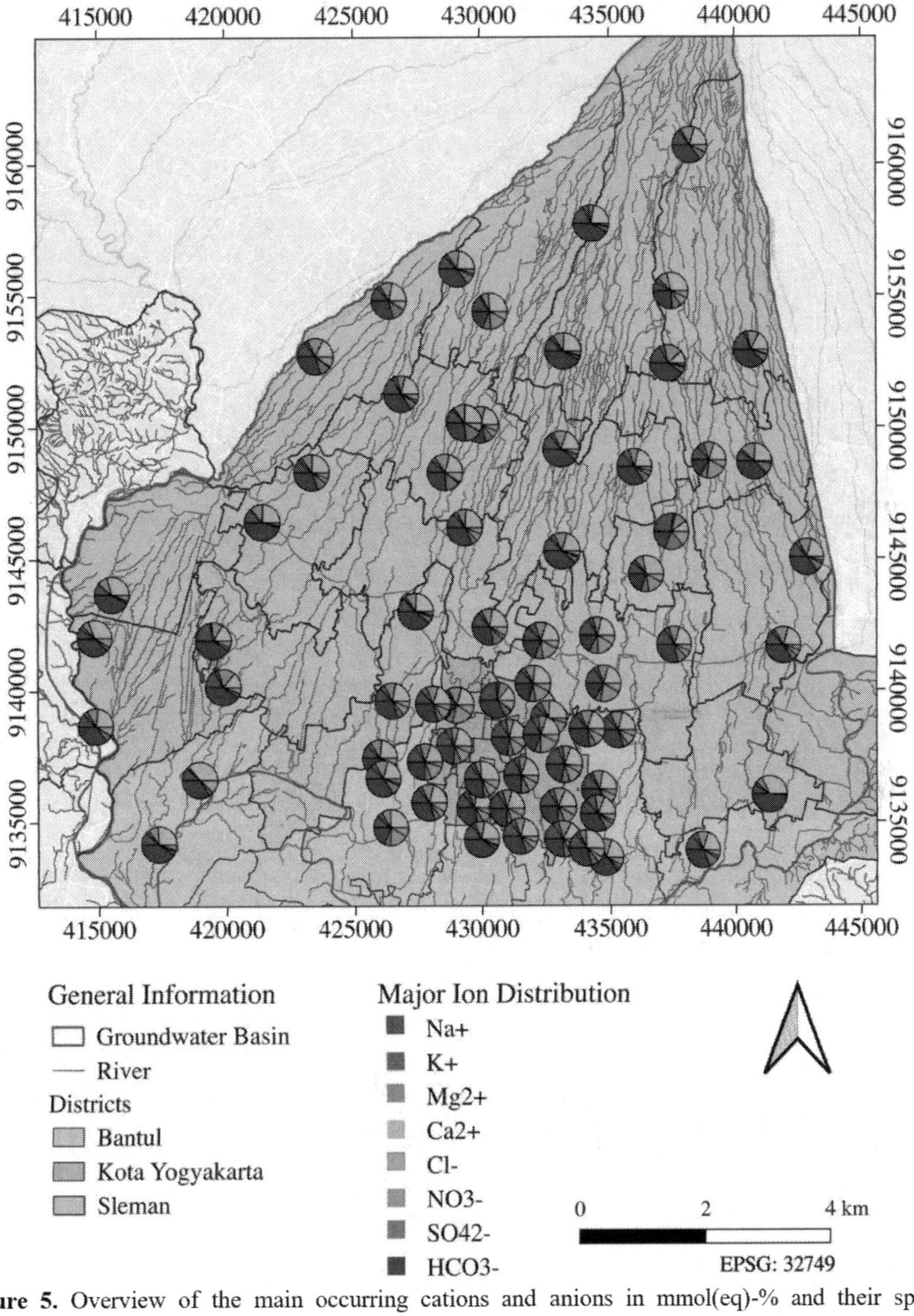

Figure 5. Overview of the main occurring cations and anions in mmol(eq)-% and their spatial distribution in the research area.

6. Discussion

Naturally, each groundwater system has its characteristic, unique water chemistry. It is dependent on several factors such as abundant minerals, soil-water interaction and anthropogenic influence [12]. To describe the water chemistry and distinguish different water types, different diagrams are introduced in the following.

6.1. Water type identification

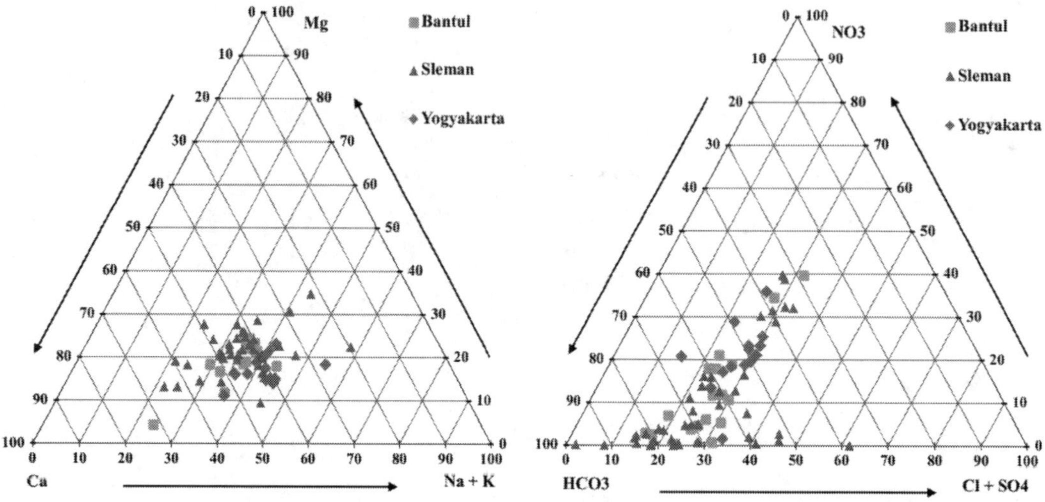

Figure 6. Ternary diagrams of observed cations (left) and anions (right) in mmol(eq)-%.

Figure 6 shows two ternary diagrams displaying each cation (left) and anion (right) distribution in mmol(eq)-%. The cation diagram visualises a similar share for Ca^{2+} and $Na^{+} + K^{+}$ while most water samples show low Mg^{2+} percentages. The right diagram makes clear, that HCO_3^- is the dominating anion for parts of the districts Sleman and Bantul. Most samples of Yogyakarta and some of Sleman and Bantul show higher NO_3^-percentages. It also shows that Cl- and SO4 do not increase with NO_3^-.

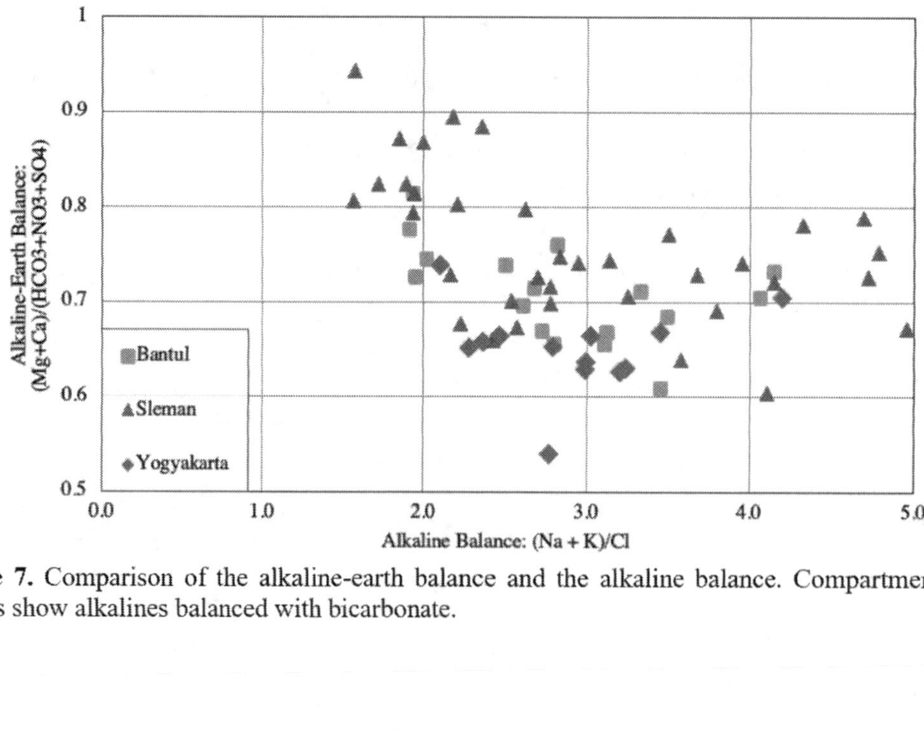

Figure 7. Comparison of the alkaline-earth balance and the alkaline balance. Compartments of all regions show alkalines balanced with bicarbonate.

As indicated before, the water samples show an excess of alkaline elements that are not balanced by chloride. Figure 7 displays the proportion of the alkaline and the earth alkaline balance. Few extreme values with very low chloride concentration lay outside the scale. A part of the water samples including mostly samples from district Sleman show an inverse proportionality. For these samples it can be assumed that the alkalines are balanced by bicarbonate. A majority of water samples can be categorized using Furtak and Langguth as alkaline-earth with higher alkaline mainly bicarbonate [13].

6.2. Identification of possible pollution

In areas of human settlement within a developing society, groundwater pollution is a common issue. There are two different types of sources for groundwater pollution: point sources and non-point sources. Point sources only affect a certain area and can be caused for e.g., by industry, leaking sewage systems or geological anomalies. Non-point sources are not limited to an area and are characterized by a diffusive discharge of pollutant. A typical non-point source is agriculture [14].

Samples from the city area, but also a few samples from north Sleman show over all high nitrate concentrations (Figure 4). To identify possible sources, the nitrate concentrations are compared with the alkaline-earth balance as shown in figure 8. It is obvious that mostly samples from Yogyakarta, but also Bantul and few from Sleman build a coherence of bicarbonate and nitrate. The samples collected in Yogyakarta show a significant correlation with a good coefficient of determination. This correlation may be explained by denitrification and ammonification processes where organic matter is oxidized. Most likely, nitrate is discharged into the saturated zone by sewage and septic tanks [15]. There is a cluster of very low nitrate concentrations in Sleman samples representing water which is rather uninfluenced by human activities.

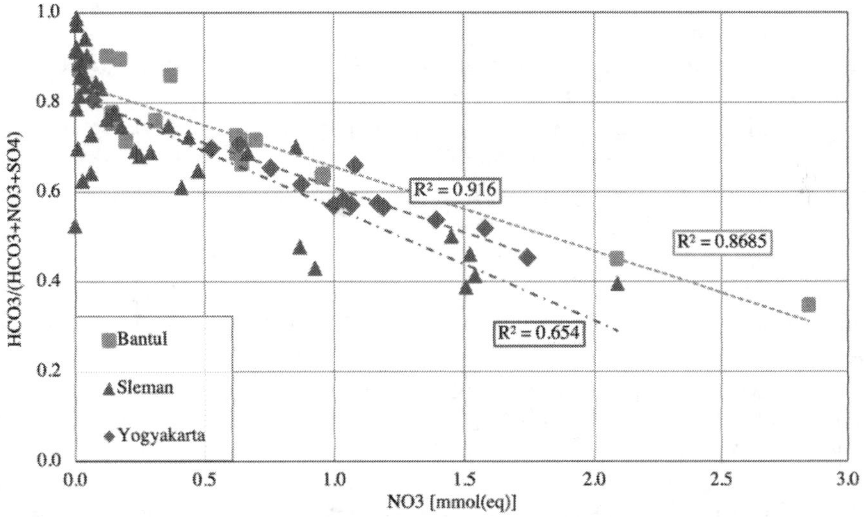

Figure 8. Coherence of bicarbonate and nitrate concentration in the city area and uninfluenced water in parts of Sleman.

In the southern area of Yogyakarta, high phosphate concentrations were observed ranging from 1 mg/L to 4.45 mg/L. High phosphate concentrations often yield from the discharge of washing agent into the groundwater. It is not only used in private households but also in dying processes. A common substance is sodium tripolyphosphate ($Na_5P_3O_{10}$) [15].

7. Conclusion

Sufficient water supply is strongly dependent on water chemistry conditions especially in developing countries. A large number of households in the special region of Yogyakarta still relay on shallow, untreated groundwater from private dug wells. The groundwater chemistry in the Yogyakarta groundwater basin is different throughout the region. Some areas in northern Sleman still show characteristics of uninfluenced groundwater in a good chemical condition. The southern area of Sleman, Kota Yogyakarta and Bantul are more influenced by human activities and very prone to pollution. Especially the city area shows pollution by nitrate and phosphate. These substances are most likely to advance into the groundwater by insufficient sewage water management. Some wells show high DOC concentrations. It was beyond this study to analyse for specific organic contaminants, but such DOC values should be further investigated. To increase groundwater quality for the citizens of Yogyakarta, it would need an intervention and modernisation of a sewage water strategy.

Acknowledgments

One of us (L.K.) likes to express gratitude to the people of Yogyakarta to welcome her so warmly in their homes and to let take samples from their private wells. L.K. would also like to thank the Geological Engineering Department of the Universitas Gadjah Mada for their vast support and for the financial support of the German Academic Exchange Service (DAAD). Analytical work was facilitated by the laboratory of the German-Indonesian GetIn-CICERO cooperation between RWTH Aachen University and Universitas Gadjah Mada. The GetIn-CICERO network is sponsored by the Federal Ministry of Education and Research of Germany.

References

[1] https://jogjaprov.go.id/profil/3-sejarah (visited: 14.01.2021)

[2] Putra D P E, Iqbal M, Hendrayana M and Putranto T T 2013 Assessment of Optimum Yield of Groundwater Withdrawal in the Yogyakarta City (Yogyakarta: *Journal of the Southeast Asian Applied Geology* Volume **5**(1)) pp. 41, 48

[3] Putra D P E and Baier K 2009 The Influence of Uncontrolled Urbanisation on Groundwater Resources Using The Example of The Indonesian Megacity Yogyakarta (online, *Cybergeo : European Journal of Geography* Environment, Nature, Landscape, document 469, Online since 29 September 2009) pp. 7-8 (in German)

[4] https://yogyakarta.bps.go.id/statictable/2020/06/15/88/penduduk-laju-pertumbuhan-penduduk-distribusi-persentase-penduduk-kepadatan-penduduk-rasio-jenis-kelamin-penduduk-menurut-kabupaten-kota-di-d-i-yogyakarta-2000-2010-dan-2019.html (visited: 14.01.2021)

[5] https://www.jogjakota.go.id/pages/geografis (visited: 14.01.2021)

[6] Putra D P E 2011 Evolution of Groundwater Chemistry on Shallow Aquifer of Yogyakarta City Urban Area (Yogyakarta: Journal of the Southeast Asian Applied Geology Volume 3(2)) pp. 116 – 117

[7] Satapona A, Putra D P E and Hendrayana H 2018 Groundwater Flow Modeling in the Malioboro, Yogyakarta, Indonesia (Yogyakarta: Journal of Applied Geology Volume 3(1)) p. 17

[8] Appelo C A J and Postma A 2005 *Geochemistry, Groundwater and Pollution 2nd Edition* (Amsterdam: A.A. Balkema Publishers, Leiden, The Netherlands a member of Taylor & Francis Group plc) pp 15 – 16

[9] Regan S, Hynds P and Flynn R 2017 An overview of dissolved organic carbon in groundwater and implications for drinking water safety *Hydrogeology Journal* 25:959-967 (Berlin: Springer Verlag Berlin Heidelberg) p 960

[10] Hölting B and Coldewey W G 2019 *Hydrogeology* (Berlin: Springer-Verlag GmbH Germany, part of Springer Nature) pp. 172,

[11] https://www.usgs.gov/special-topic/water-science-school/science/nitrogen-and-water?qt-science_center_objects=0#qt-science_center_objects (visited: 15.01.2021)

[12] Singh A K and Kumar S R 2015 Quality assessment of groundwater for drinking and irrigation use in semi-urban area of Tripura, India *Ecology, Environment and Conservation* **21**.1 p. 97
[13] Hölting B and Coldewey W G 2013 Introduction into General and Applied Hydrogeology (Berlin: Springer-Verlag GmbH Germany, part of Springer Nature) pp. 230-231 (in German)
[14] Talabi A O and Kayode T J 2019 Groundwater Pollution and Remediation *Journal of Water Resource and Protection* **11** p. 1
[15] Merkel B and Sperling B 1996 Hydrogeochemical Substances part 1 (Deutscher Verband für Wasserwirtschaft und Kulturbau e.V. DVWK) pp. 125-126, 141 (in German)

IOP Conf. Series: Earth and Environmental Science **851** (2021) 012016 doi:10.1088/1755-1315/851/1/012016

Preliminary conceptual model of the Arjuno Welirang hydrogeological system, and comparison with the Bromo Tengger: An illustration of the hydrogeological systems diversity in volcanic areas

B Baud[1*], P Lachassagne[1], H Jourde[1], V De Montety[1], A Fadillah[2], N Dörfliger[3], H Hendrayana[4] and A Rachmansyah[5]

[1]HSM, Univ. Montpellier, CNRS, IRD, Montpellier, France
[2]AQUA Danone, Jakarta, Indonesia
[3]DANONE WATERS, Water Institute by Evian, France
[4]Universitas Gadjah Mada, Geological Engineering Department, Yogyakarta, Indonesia
[5]Universitas Brawijaya, Department of Civil Engineering, Faculty of Engineering, Indonesia
*E-mail: benjamin.baud@umontpellier.fr

Abstract. While basaltic volcanic aquifers are well described in the literature, andesitic groundwater systems are less studied. Nevertheless, these aquifers supply a large population in all the subduction areas, where this volcanism mainly occurs. In Indonesia, the growing needs of the population induce an increase of the pressure on such aquifers. We present in this paper the case study of the Arjuno-Welirang, and compare it with a well-known hydrogeological system, the Bromo-Tengger. Based on geological, geomorphological, and water chemistry data, this study highlights the diversity of andesitic history inherent to this kind of volcanism, and its strong implications on the groundwater availability downstream. The aim of this paper is also to show that simple investigations can help building a preliminary conceptual model (PCM) of complex volcanic settings. Such PCM is needed to define further detailed hydrogeological investigations required to set-up aquifer exploitation, management and preservation rules.

1. Introduction

In all countries, the socio-economic development and well-being of the population are closely linked to the access to water resources. In Indonesia, a large part of the growing needs is fulfilled by access to groundwater. This country is located on the subduction area of the Indo-Australian plate under the Eurasian plate. Therefore, volcanic aquifers are numerous in this country. This is particularly the case in the Eastern part of the Java Island where this research is focused. Most of these aquifers provide a large quantity of high-quality groundwater, more or less accessible depending on their geological setting.

The regional hydrogeological map of that area (**Figure 1**), even though old, is quite accurate and provides valuable information about the springs, the aquifers productivity and the deep borewells uses at the time of its publication. But this map may lead to think that all the volcanic edifices (a volcanic morphological unit that can be the result of a complex geological history and, as such, is different from a simple "volcano") all have a similar hydrogeological pattern, which contrasts with the geological complexity of the volcanic aquifers' structure and the springs location in the plains around these volcanoes. Our preliminary results on the Arjuno-Welirang complex, and their comparison with the detailed study performed in [1] on the neighbour complex of the Bromo-Tengger illustrate the diversity of the aquifer types, whereas the hydrogeological map suggest an elevation-dependent structure with, from the bottom to the top of the volcanic edifice (**Figure 1**):

- Blue colour: Porous aquifers (coastal plains, intermontane basins and foot of strato-volcanoes), the darker blue the higher productivity.
- Green colour: Fractured aquifers (slope of strato-volcanoes), the darker green the higher productivity.
- Orange Colour: Porous or fractured aquifers of poor productivity, low transmissivity, limited shallow groundwater resources.
- Red Colour: Regions without exploitable groundwater.

These apparent similarities between different volcanic edifices on the hydrogeological map can lead the local authorities to miscalculate while planning water management policies at local scale, as the water availability may be very different from that inferred by the map, and from one volcanic edifice to the others. Thus, our research aims at strongly encourage to build up local conceptual models (CM) of the hydrogeological systems. Indeed, CM are a first step to summarize all the available knowledge on the water cycle at a local scale, necessary to implement numerical modelling and water management scenario testing, and reducing the numerical modelling uncertainties [2–4]. This will help to identify the key missing data to be acquired, in order to address the local water issues like the cities/activities/land use development, and ensure matching human activities and water resources sustainability. To illustrate this, the Arjuno Welirang volcanic complex, located on the Eastern part of the Java Island (**Figure 2**) is studied. We focus on its eastern flank, where the Pandaan city is located, at 40km from Surabaya in the Pasuruan district. The area was studied for its geothermal potential [5–8] on the northern/western part of the volcanic edifice. But on its eastern flank, the city of Pandaan is relying more and more on the less studied groundwater system of the Arjuno-Welirang to sustain its development (industrial, domestic, agricultural).

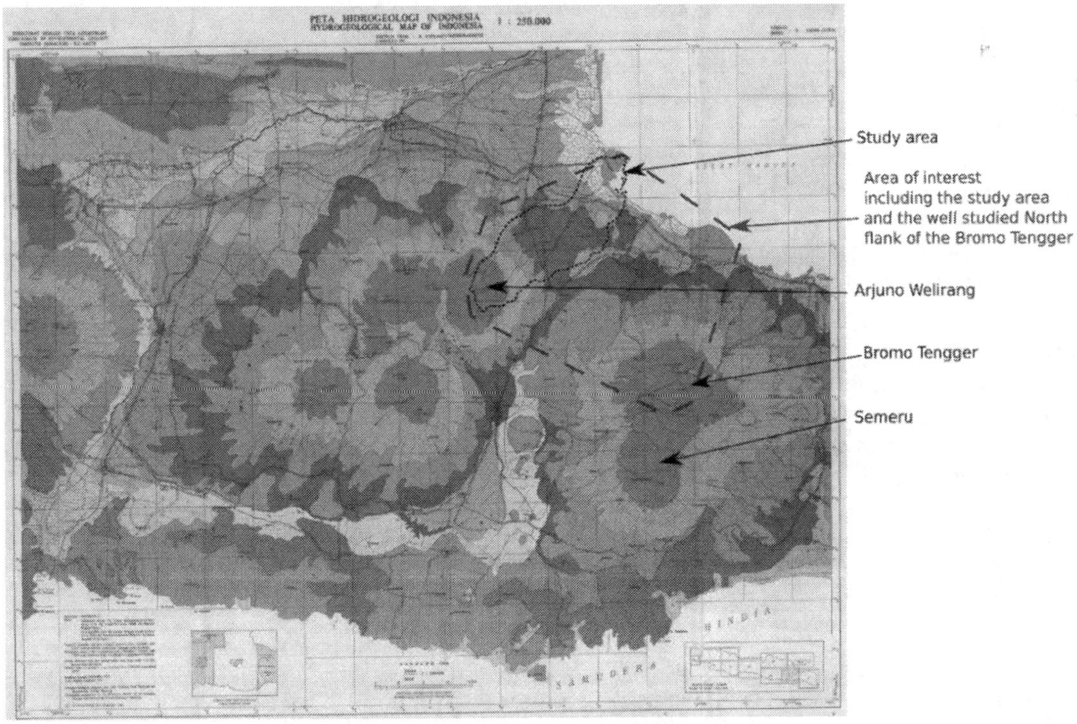

Figure 1. Hydrogeological map of Indonesia [9]– See the explanation of the colour legend in the text.

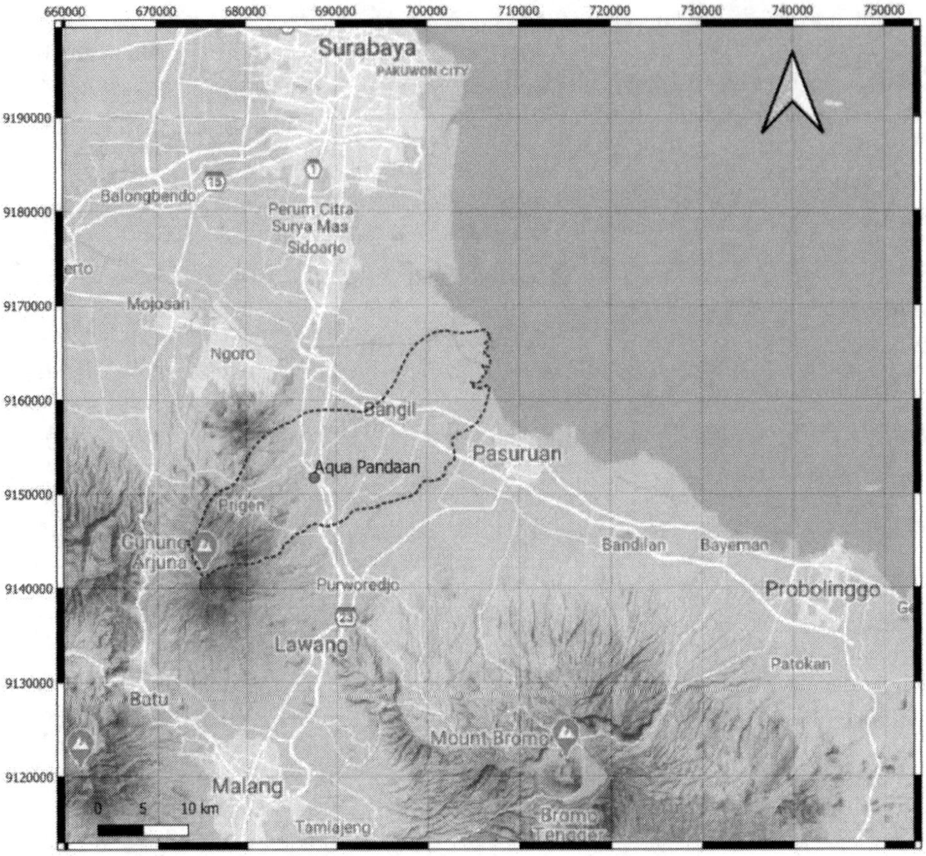

Figure 2. Localisation of the study area. The study area represents an area of 370 km², delineated in black. It is located in Pandaan, in the Pasuruan Regency, East Java.

2. Method and results

As described in [2], there are two types of CM:

- The consensus model, with an iterative integration of all the data into a single CM.
- The multi-model approaches, with several CM tested at the same time.

In such a preliminary stage of study as ours, the PCM approach is more likely to be a consensus model, until there are enough data to raise several hypotheses on the geological settings or water fluxes. The goal of a PCM is to gather all the existing data available to date and define the additional data that can be acquired efficiently during the phases of the project to come. This method is developed on the study case of the Arjuno-Welirang, East Java. A preliminary conceptual model has been built based on a comparative study with the well-studied Bromo-Tengger system [1] on three linked thematic: the geological settings of the volcanic edifice, the measure of the aquifer(s) outflows, and insights on the aquifers' functioning, on the basis of hydrochemical data.

2.1. Structure

To build a preliminary hydrogeological conceptual model, the first step is, as far as possible, to unveil the geological settings of the hydrogeological system.

By settings, we mean the geological units, their geometry, their lithology and, from this last information, at least a qualitative characterization of their permeability (aquifer, aquitard, aquiclude) and of the type of permeability (interstitial, fractures, etc.). For such an identification, the local or regional geological maps provide valuable data. To complete the identification of the geological settings, geomorphological approaches can be easily performed with the use of Google Earth images and the

Digital Elevation Model, that helped delineate broad structural units, and to localize outcrops easy to reach later, once on the field.

The available maps for the study area are the ones of [10], [11], and [12] for the Northern volcano, the Penanggungan (**Figure 3**). These maps present numerous differences, as their scales and their original purposes are different. Therefore, their accuracy for hydrogeological purpose is highly variable, even though all of them provide valuable information. In order to build a stronger preliminary conceptual model, and with field access restriction due to the COVID 19 outbreak during the study (national lockdown, and international travel restrictions), a "geological mapping" through Google maps georeferenced images was performed. On natural touristic sites, plenty of pictures/selfies are taken by visitors and georeferenced on the Google maps. While remote sensing/satellite-based approach allow regional geomorphological studies of the geological bodies, this Google Maps "database" allows to visualize local outcrops, that we broadly described at different levels:

- 1st level: inferred lithology (lava flow, pyroclastic flow/lahar, etc.);
- 2nd level: secondary features such as apparent thickness, cooling fractures, erosion processes, etc.;
- 3rd level: comments about the reliability of these data (number of pictures available, etc.), and also complementary data such as the identification of some springs.

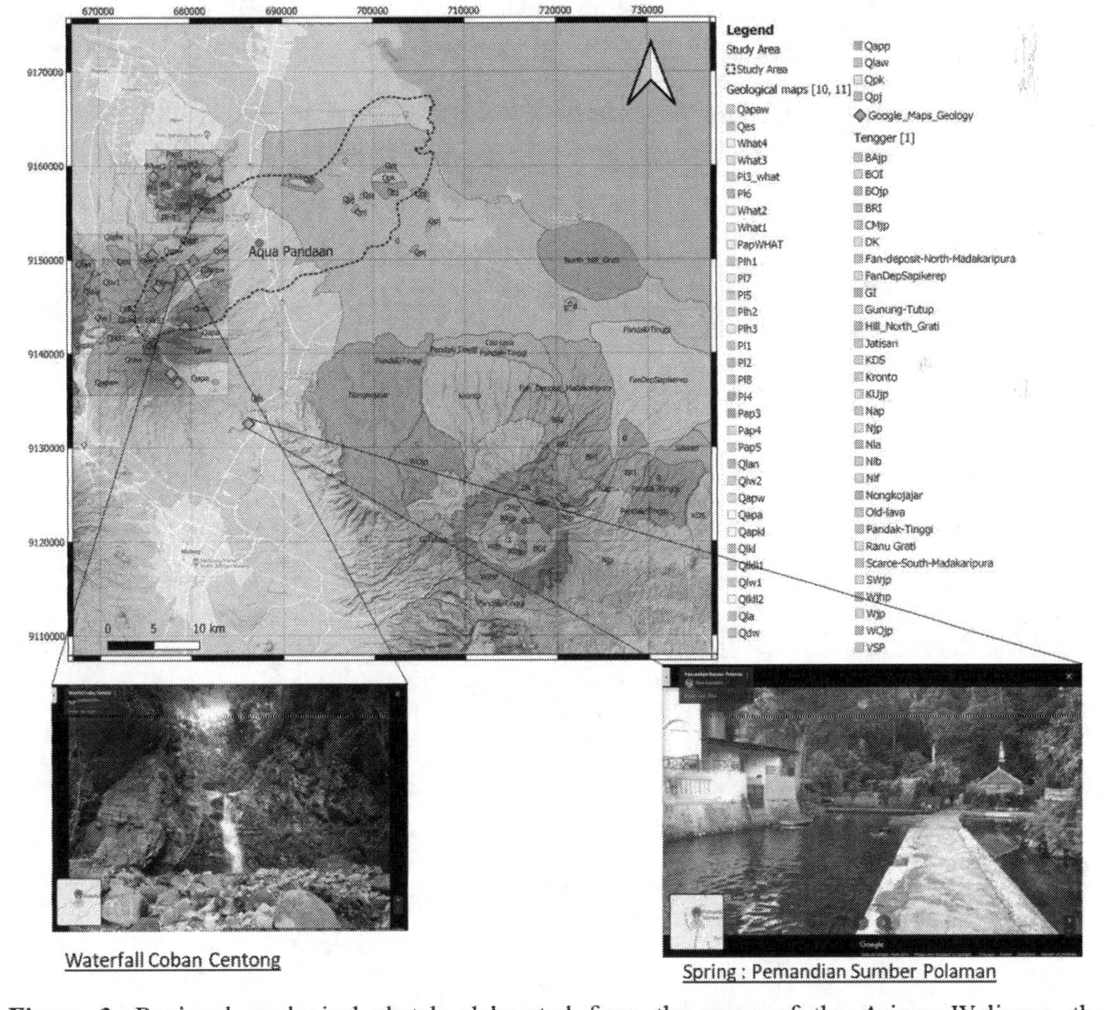

Figure 3. Regional geological sketch elaborated from the maps of the Arjuno Welirang, the Penanggungan and the Tengger Complex, after [1,10,11], and Google photographs. The detailed units description are provided by [1,10,11]. Coordinates WGS84 - UTM 49S.

To ensure that the pictures were correctly referenced, the studied location had to comprise at least 10 pictures with similar outcrops view, consistent altogether from different sources/photographers.
A geomorphological work can also be performed from Google Earth. Of course, the use of a precise Digital Elevation Model (DEM) may enhance the quality of the work. This geomorphological work provides additional quality information on the geological maps. Indeed, in volcanic edifices, the dynamics of construction, erosion, collapses, etc. by various types of eruption and volcanic activity creates complex structures, highly different from an edifice to another. Such processes are well described on Java island, especially on the Bromo-Tengger edifice [1], and on the Merapi [13].

These sites were then compared with the geological maps and their units' descriptions. It turned out that, on the Arjuno Welirang volcanic edifice, the map of [11] seems to be the most accurate to our study purpose, with precise lithological description and location. Thanks to these lithological descriptions, potential aquifers/impervious bodies can be identified, and the kind of aquifers (fractured/porous) as well. But its spatial extension is limited to the volcano; downstream geological information is not available and were then summarized from [10] and confirmed by [1].

2.2. Aquifer outlet identification and outflow measurements

Most of the time, a hydrogeological study is performed in order to evaluate the water budget of an aquifer, for instance to implement water management strategies. The estimation of the water budget might be time consuming, especially to collect input data such as the rainfall, the effective rainfall or the evapotranspiration. But a first step of the water budget easy to carry out on field, or thanks to remote data, is the identification of the aquifer's outflows, natural or anthropogenic, on the study area. The type of outflows and their location already provide valuable information about the groundwater availability, and therefore the number and type of aquifers (unconfined/confined), and sometimes their extent and type.

For instance, in the case of the Arjuno Welirang, a survey of the outflows and a comparison with a similar work on the Bromo-Tengger/Pasuruan plain helped to identify 7 types of outlets on both study areas (**Figure 4**):

- Low discharge springs (< 20 L/s)
- Medium discharge springs (≈ 100 L/s)
- High discharge springs (>>250 L/s)
- Dug wells (shallow unconfined aquifer)
- Pumped wells (deep aquifer)
- Artesian wells (self-flowing wells from artesian confined aquifer)
- Locally, geothermal springs and fumaroles (geothermal systems)

These different types of outflows indicate the presence of several different aquifers, that may be multi layered, which may communicate between each other or not. The relations between these aquifers are to be explored in the future steps of the project.

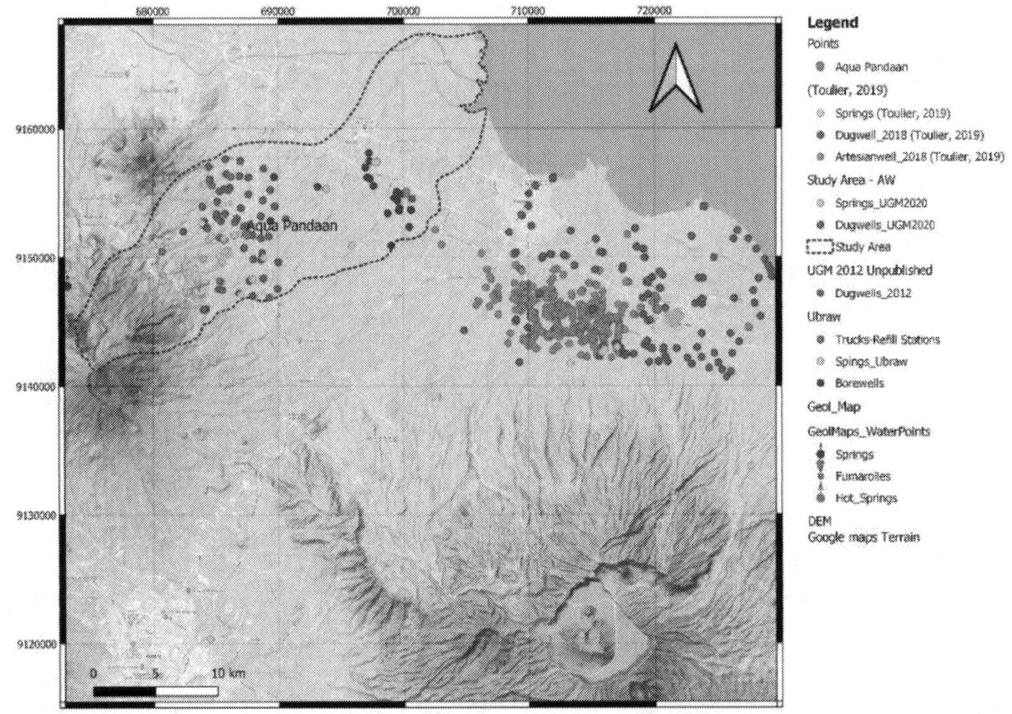

Figure 4. Main types of groundwater outlets on the Arjuno-Welirang, Bromo Tengger and Pasuruan plain, from available data on Dec. 15th, 2020 [1,14]. See a more detailed description in the text. Coordinates WGS84 - UTM 49S.

The main outputs of this comparative study between Arjuno Welirang and Bromo-Tengger were the following:

- High discharge low elevation springs (>>250 L/s) are only found, with numerous self-flowing wells, at the feet of the Bromo-Tengger volcanic edifice where there are identified as natural/anthropogenic outlets from an important and unique artesian volcano-sedimentary aquifer fed by recharge on the volcanic edifice [15]. No such important aquifer seems to exist at the feet of the Arjuno Welirang;
- Additionally, the upstream part of the Bromo-Tengger shows only a few very low discharge springs related to very small perched aquifers, the quite totality of the effective rainfall recharging the above cited main volcanic aquifer [15]. On the opposite, the Arjuno Welirang shows several medium discharge springs (≈ 100 L/s), emerging at medium elevation (**Figure 4** and **Figure 5**), which are interpreted as the outlets of small to medium size aquifers bounded by impervious boundaries, that explain the groundwater outflows. The northern flank of the Arjuno Welirang is thus interpreted as the juxtaposition and/or superposition of several aquifers, pyroclastic and/or lavic, their lithology being identified from the geological maps, "field observations" (see section 3.1.), and boreholes geological logs. The coexistence of artesian wells and dug wells on this same area suggest the existence of superimposed confined and unconfined aquifers;
- The Arjuno Welirang downstream plain mostly shows no self-flowing wells. This is consistent with smaller aquifers than at the feet of the Bromo-Tengger, less recharged than on this latter area (as most groundwater seems to outflow at medium elevation springs) although the recharge elevation still needs to be defined (with 18O data notably).

From these data a sketch of the piezometry may be drawn for the Arjuno Welirang.

2.3. Aquifer Functioning

Along with the outflows field survey described above, an easy quick data gathering is the measurement of basic water chemistry parameters, such as pH, water temperature, and electrical conductivity, as field multiparameter probes are quite affordable, easy to carry on the field, and measurements are quick to perform. More detailed water sampling campaign can be organized after collecting this elementary data, on field or through bibliographic works. This will provide complementary data to the "Outflows measurement", notably the identification of the different aquifer bodies (depth and lateral extent).

Once again on our study area, a first look at the EC data (**Figure 5**) shows that the waters sampled above 350 masl have a low electrical conductivity, as on the Bromo-Tengger volcanic edifice [1], suggesting low water-rock interactions, and probably small extension perched aquifers.

But, between 350 and 100 masl, the electrical conductivity on the Arjuno-Welirang is much higher than on the Tengger at the same elevation (with only a few perched aquifers on this latter), at all the water points (samples, dug wells, borewells), and is closer to the one of the springs and self-flowing wells located at the foothill of the Tengger. This can be explained by different types of water-rock interactions, and/or different residence times, and confirms the conclusions raised from the outflow's location and measurements (section 2.2), and thus the type of identified aquifers: several medium size aquifers. Additionally, the relative homogeneity of the EC data between the springs and dug wells suggest that the confined aquifer may supply the subsurface unconfined ones.

The potential incidence of geothermal processes, as well as potential anthropogenic contaminations will however have to be investigated, notably from further hydrochemical and isotopic data.

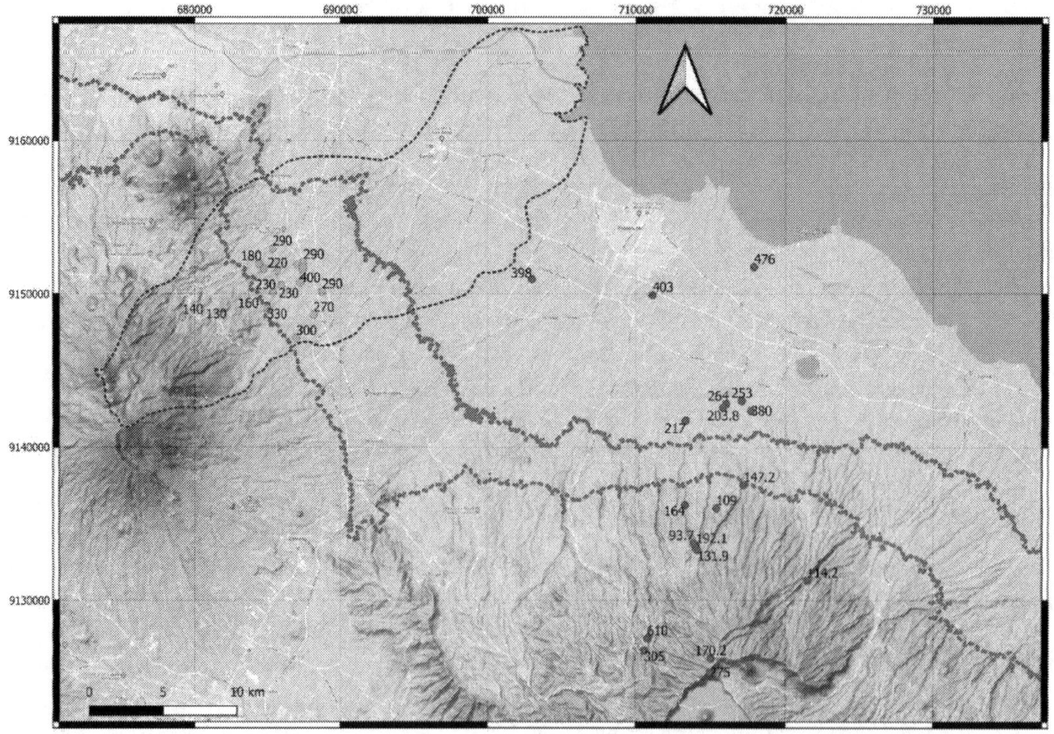

Figure 5. Electrical conductivity (µS/cm) of the springs and unconfined aquifer on Arjuno-Welirang and Bromo-Tengger complex. The 100 – 350 masl area is highlighted by the two dotted contours. Coordinates WGS84 - UTM 49S.

3. Conceptual Model

The building-up of a preliminary hydrogeological conceptual model of the area enables gathering all the collected data (geological settings, first estimates of the water budget, aquifer functioning: *i.e.* notably flow directions in the aquifer from recharge area to discharge areas). Such a conceptual model will be enhanced all along a hydrogeological project, as more and more data will be available to feed it. It will also help to identify the key missing gaps in the overall knowledge. Thus, appropriate research projects can then be launched and their cost evaluated.

[15] provided the complete conceptual model over his study area (**Figure 6**). This provided valuable data to elaborate a numerical modelling tool, which is a pedagogic and precise tool to provide and test water management scenarios [1].

On the Arjuno Welirang complex, our preliminary conceptual model is provided on **Figure 7**, as partly described in the previous section of this paper. First of all, on the upstream part of the study area, the water users in the Pandaan district are relying on several medium size, medium elevation aquifers, exploited with shallow (dug wells) and deeper wells: self-flowing borewells, some of them being – no more -self flowing due to neighbour abstractions, or due to their specific location within the given aquifer. The picture is very different from that of the Bromo Tengger Complex at the same elevation characterized by a unique and deep aquifer with a too deep piezometric level to enable the drilling of acceptable cost wells, and also dug wells. Second, there are no major springs identified at the feet of the Arjuno Welirang volcano, an opposite configuration compared to the Merapi [13], or to the Bromo-Tengger [1]. This leads us to hypothesise that most of the water infiltrated along the flanks of the volcanic edifice outflowed at the above described higher elevation springs, due to local low permeability boundaries/layers/geological bodies. The relatively high electrical conductivity, not observed at such elevations on the Bromo-Tengger, confirms this scheme, with quite long water residence time.

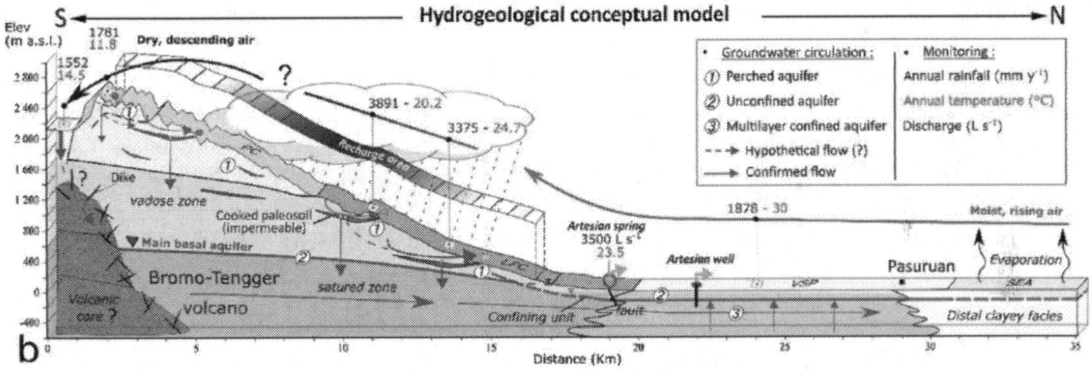

Figure 6. Hydrogeological conceptual model of the Bromo-Tengger [15].

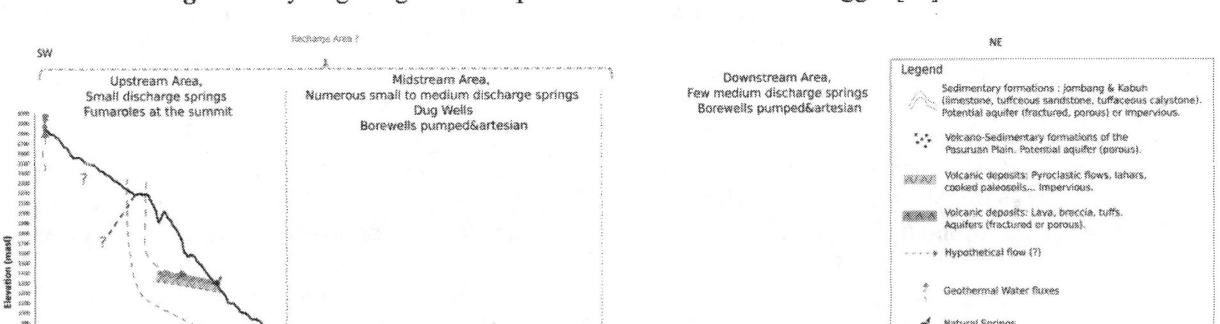

Figure 7. Arjuno Welirang Conceptual model 01/2021, after [10,11].

4. Discussion and perspectives

Our results allowed to elaborate a first hydrogeological conceptual model of the Arjuno Welirang complex, with numerous layered aquifers on the flank of the volcano, with highly diverse productivity. Of course, at this point our study isn't enough to provide usable science for sustainable water management policies. And as studied in [2], our PCM is subject to high uncertainties due to the lack of data. Nevertheless, this PCM gives us a clear view for future research project, and illustrates that our research area has a completely different behaviour than the Bromo Tengger North flank unlike what the regional map (**Figure 1**) may suggest. It highlights the key missing data to build a robust conceptual model that includes all the processes of the water cycle.

In order to have a continuous improvement of the conceptual model, the priority should be given to:

- Acquire geological data, especially in the "midstream" area, where there is the less data, but also the most water extraction (springs, dug wells, borewells pumped & artesian). As stressed earlier in the paper, improving the knowledge about the geological structure of the aquifer will be fundamental, as the geology of these aquifers was poorly characterized in this PCM.
- Acquire piezometric levels for all the different water outlets (springs and wells), in order to build a piezometric map, and get a first idea of the water fluxes, underground sub-watersheds, etc.
- Estimate the water outflows (natural & anthropic) amounts (analysis of various data bases compiling water volume for various water uses (domestic, industrial, …), field survey, remote sensing, …).
- As it appears that there are several multi-layered aquifers with a plenty of natural (and anthropic) outflows, it is expected that the aquifers have limited lateral extend. Therefore, the north east flank of the Arjuno-Welirang would be composed of several small to medium extension aquifers. To differentiate them, and then estimate precise water budget, an extensive water sampling campaign is necessary. This concerns the water major and minor ions, but also the water isotopes (^{18}O and ^{2}H).
- The rivers/groundwaters interaction is still unknow, and shall be checked all over the study area (upstream, midstream, downstream), as it may affect the water budget estimation, and/or potential pollutant transfer. For this, gauging and electrical conductivity measurements will be performed during the dry season in the river beds. This will also complete the geological and hydrogeological characterization of the aquifers and the aquitards/aquicludes.
- In order to precise local geological & hydrogeological parameters, notably in areas without significant outcrops, geophysical campaigns will be performed, and will be calibrated on existing wells and outcrops, as well as on the observations performed along the rivers.
- A rigorous meteorological dataset acquisition will be performed, with collection of existing and new hydrometeorological data (raingauges, temperature, humidity, river discharge, etc.), with in situ station and remote sensing/climate models, in order to compute water budget estimates (rainfall, evapotranspiration, share between recharge and runoff…).

5. Conclusion

Our study shows that despite apparent similarities between volcanic edifices at the regional scale, the hydrogeological patterns of neighbour volcanoes can be very contrasted. Simple investigations enable building-up a preliminary conceptual model (PCM) of complex volcanic settings. Such PCM is needed to define further detailed hydrogeological investigations required to set-up aquifer exploitation, management and protection rules. This is well shown by the comparison of the PCM of the Arjuno Welirang volcanic complex and the CM of the Bromo-Tengger north flank. The Bromo Tengger North flank conceptual model comprises small extension aquifers above 100masl, with an important recharge that feeds an important and productive volcano-sedimentary aquifer at the foot of the volcano. Whereas the Arjuno Welirang PCM suggest small extension aquifers only at high elevation, and an important multi-layered aquifers system a medium-low elevation. The contribution of these aquifers or a deeper infiltration to the volcano-sedimentary/sedimentary plain remains completely unknown for the moment.

Acknowledgements
This research was carried out in the frame of a research project coordinated and funded by AQUA Danone. Scientific supports from Gadjah Mada University, Universitas Brawijaya, and University of Montpellier are gratefully acknowledged.

References

[1] Toulier A 2019 Multidisciplinary study for the characterization of volcanic aquifers hydrogeological functioning: case of Bromo-Tengger volcano (East Java, Indonesia)
[2] Enemark T, Peeters L J M, Mallants D and Batelaan O 2019 Hydrogeological conceptual model building and testing: A review J. Hydrol. **569** 310–29
[3] Anderson M P, Woessner W W and Hunt R J 2015 Modeling Purpose and Conceptual Model Applied Groundwater Modeling (Elsevier) pp 27–67
[4] Bredehoeft J 2005 The conceptualization model problem-surprise Hydrogeol. J. **13** 37–46
[5] Daud Y, Nuqramadha W A, Fahmi F, Sesesega R S, Fitrianita, Pratama S A and Munandar A 2019 Resistivity characterization of the Arjuno-Welirang volcanic geothermal system (Indonesia) through 3-D Magnetotelluric inverse modeling J. Asian Earth Sci. **174** 352–63
[6] Inguaggiato S, Mazzini A, Vita F and Sciarra A 2018 The Arjuno-Welirang volcanic complex and the connected Lusi system: Geochemical evidences Mar. Pet. Geol. **90** 67–76
[7] Sumotarto U, Hendrasto F and Wibagiyo W 2018 GEOHYDROTHERMAL MODEL OF ARJUNO, WELIRANG AND PENANGGUNGAN VOLCANOES EAST JAVA, INDONESIA Penelit. DAN KARYA Ilm. **2** 37
[8] Wardana A, O'Sullivan J and O'Sullivan M 2016 Natural State and Future Production Modelling of Arjuno-Welirang Geothermal Field, Indonesia N. Z. 8
[9] Soekardi P 1984 Hydrogeological Map of Indonesia, Index to sheet location: n° X
[10] Santosa and Suwarti 1992 Geological Map of the Malang quadrangle, Jawa
[11] Martadiastuti 2017 Geological map and water sampling sites of Arjuno Welirang area
[12] Paripurno E T, Pradhana G, Putra W and Hartono G P 2018 New nomenclature of Mt. Penanggungan stratigraphy, Pasuruan and Mojokerto Regency, East Java IOP Conf. Ser. Earth Environ. Sci. **212** 012045
[13] Selles A 2014 Multi-Disciplinary study on the hydrogeological behavior of the eastern flank of the Merapi Volcano, Central Java, Indonesia
[14] Anon 2020 Pandaan Study 2020 - Universitas Gadjah Mada - Universitas Brawijaya - University of Montpellier - UNPUBLISHED
[15] Toulier A, Baud B, de Montety V, Lachassagne P, Leonardi V, Pistre S, Dautria J-M, Hendrayana H, Miftakhul Fajar M H, Satrya Muhammad A, Beon O and Jourde H 2019 Multidisciplinary study with quantitative analysis of isotopic data for the assessment of recharge and functioning of volcanic aquifers: Case of Bromo-Tengger volcano, Indonesia J. Hydrol. Reg. Stud. **26** 100634

IOP Conf. Series: Earth and Environmental Science **851** (2021) 012017 doi:10.1088/1755-1315/851/1/012017

Deformation identification using DInSAR multi temporal analysis method in supporting infrastructure development (case study of the area around the nation's new capital)

A Puspita[1,2*], Y Prihanto[4], S Martha[1], A A Martha[3] and A G Gultom[1]

[1] Sensing Technology Study Program, Faculty of Defense Technology, Indonesian Defense University, Sentul-Bogor, Indonesia

[2] Earthquake and Tsunami Data Center Meteorology Climatology and Geophysics Agency (BMKG), Jakarta, Indonesia

[3] Center of Research and Development Meteorology Climatology and Geophysics Agency (BMKG), Jakarta, Indonesia

[4] Center of Research, Promotion and Cooperation Geospatial Information Agency (BIG), Cibinong-Bogor, Indonesia

*E-mail: aprilpuspita86@gmail.com, joeprihanto3074@gmail.com

Abstract. The relocation of nation's capital is one of the government's flagship programs at this time. The lack of information on soil structures that are prone to deformation can bring an impact on the failure of infrastructure development in the areas of New State Capital (known as Ibu Kota Negara/IKN). Therefore, a significant study is important to find out the deformations that may occur in the region as an initial information in determining the location for safer infrastructure development based on the soil structure conditions in the research area. This research used the Differential Interferometry Synthetic Aperture Radar (DInSAR) method with a multi-temporal approach to identify the deformations in the research area. The data included SAR images of Sentinel 1A type SLC with C band (5.405 GHz) for the period 2015-2019. The results confirmed that deformation had been identified in several areas, both those experiencing subsidence and uplifting. The maximum subsidence was 12.97 cm at Sepaku district on period 2018-2019 and the maximum uplift was 10.01 cm on period 2017-2018. The identified areas with deformation generally take place in areas with a high density of buildings, construction areas, road infrastructure, and river alluvial deposits.

1. Introduction

One of the considerations for choosing the North Penajam Paser and Kutai Kertanegara, East Kalimantan Province as candidates for the New State Capital (known as Ibu Kota Negara/IKN) is that they are the safest areas from disasters, especially geological disasters such as earthquakes and tsunamis [11]. However, based on its geological conditions, the IKN areas do not escape from potential geological disasters caused by ground deformation in the area. Based on the lithology, the research areas are dominated by a fairly thick sedimentary structure, which is part of the Kutai basin. The thick sedimentation is very prone to deformation since it has a relatively weak soil structure.

Based on the geological map, the North Penajam Paser area, which is the site of point 0 of the new IKN development, takes place is included in Balikpapan area. Almost all the rocks in this area had experienced deformation which forms anticlines, synclines, and faults [13]. In the southern area, there is a Holocene age (quarter layer) alluvial deposits with unconformity structure. Based on these geological conditions, the new IKN area has the potential to experience deformation that can trigger geological disasters, such as landslides, subsidence, earthquakes, sink holes, and soil liquefaction. Therefore, it may have a direct impact on the conditions of the infrastructure. Based on the data obtained from the Cent of Volcanology and Geological Hazard Mitigation in the research area, a landslide occurred in the Sepaku district in April 2018, which caused damage to dozens of houses [5].

Therefore, this research used a method to identify and to map the areas that are prone to deformation. This step is one of the attempts to mitigate geological disasters, as well as to provide initial information for planning infrastructure development in the research area. The method used in this research is the Differential Interferometry Synthetic Aperture Radar (DInSAR) multitemporal analysis using SAR Sentinel 1A radar image type Single Look Complex (SLC) with C band to identify the deformation [1][2][4][6][7][10]. The output parameters of the processing results using the DInSAR method was mapped using QGIS software to produce a thematic map of land deformation in the new IKN area.

2. Data And Methods

2.1 Research Area

This research was conducted in the Sepaku and Penajam subdistricts of the Penajam Paser Utara district, East Kalimantan Province. The coordinates of 0.9°S - 1.38°S and 116.48°E - 116.82°E are presented in figure 1. The location of the deformation observation points in this research is shown in the rectangular location with a red dot (figure 1) as presented in Table 2.

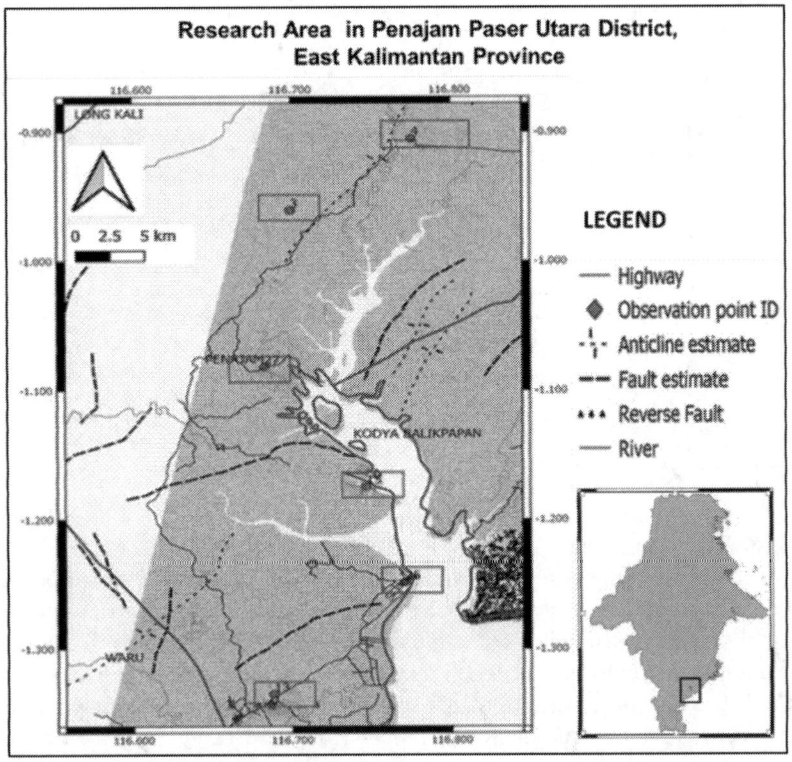

Figure 1. Research area at Penajam Paser Utara District.

2.2 Data

The data used in this study are secondary data obtained from the following sources:

- Sentinel satellite image data set 1-A with C band (working at a frequency of 5.405GHz and a wavelength of 5.6 cm), Single Look Complex (SLC) type 1st level, period 2015-2019. The data can be downloaded from ASF Alaska via https://vertex.daac.asf.alaska.edu (see Table 1).
- Digital Elevation Model Nasional (DEMNAS) data used to validate the DEM model generated from DInSAR processing. The DEMNAS data has a spatial resolution of 0.27-arcsecond. It can be downloaded via http://tides.big.go.id/DEMNAS/.

The software used in this research includes SNAP (Sentinel Application Platform) version 7.0, and QGIS version 3 software.

Table 1. Sentinel 1A Acquisition Data[a].

Processing Year	Mst/Slv	ID Scene	Acquisition	Polarization	Orbit	Abs Orbit
2015-2016	Mst	S1A_IW_SLC__1SDV_201512 13T215045_20151213T215112 009029 00CF3C 47F1	13 Des 2015	VV+VH	Descending	9029
	Slv	S1A_IW_SLC__1SDV_201612 07T215052_20161207T215119 _014279 0171A9 E4E2	7 Des 2016	VV +VH	Descending	14279
2016-2017	Mst	S1A_IW_SLC__1SDV_201612 07T215052_20161207T215119 014279 0171A9 E4E2	7 Des 2016	VV +VH	Descending	14279
	Slv	S1A_IW_SLC__1SDV_201712 02T215051 20171202T215118 019529 021262 E608	2 Des 2017	VV+VH	Descending	19529
2017-2018	Mst	S1A_IW_SLC__1SDV_201712 02T215051 20171202T215118 019529 021262 E608	2 Des 2017	VV+VH	Descending	19529
	Slv	S1A_IW_SLC__1SDV_201812 21T215057_20181221T215124 025129 02C65E 2C74	21 Des 2018	VV+VH	Descending	25129
2018-2019	Mst	S1A_IW_SLC__1SDV_201812 21T215057_20181221T215124 _025129 02C65E 2C74	21 Des 2018	VV+VH	Descending	25129
	Slv	S1A_IW_SLC__1SDV_201912 04T215104_20191204T215131 030204 0373CD 857E	4 Des 2019	VV+VH	Descending	30204

[a]Source: initial inventory of Sentinel 1A data availability

2.3 Processing Method

The processing method used in this research includes the acquisition stage, pre-processing stage, data processing stage, and postprocessing stage. The acquisition process was carried out by downloading secondary data in the form of Sentinel 1A satellite images and DEMNAS data in the research area with the links listed previously. The further details are presented in figure 2, showing the research data processing method. The data processing and analysis techniques of this research were divided into three stages: the data processing technique using the DInSAR method using Snaphu software and the GIS processing using QGIS software.

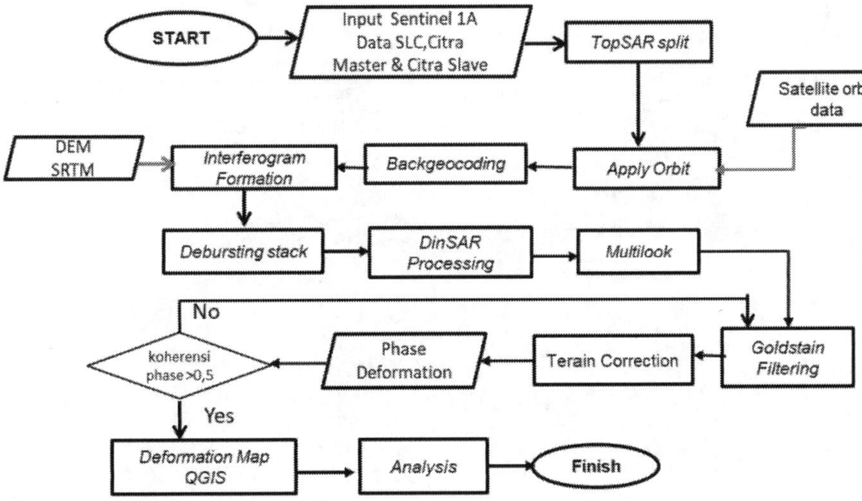

Figure 2. Data processing flow chart on SNAP software.

Data processing using DInSAR method in the initial stages, the TopSar Split process is carried out to divide one satellite image scene into three subwadths (IW) and burst mode (1-9) which covers the research area. This research, the researchers used subwadth IW3 with burst 6 to 9. After the TopSar split process is carried out, the researchers then input the satellite orbit value data. The next process is the formation of an interferogram and input the DEM value, i.e., the SRTM 30 meters 3 seconds [6]. To reduce the noise and to sharpen the fringer on the interferogram, a filtering process was performed, i.e., enhanced spectral diversity. The next step was carried out using TopSar Deburst followed by the process of eliminating the topography phase (Thopo phase removal) and reducing the error phase with the multilooking process. The next process is Goldstain phase filtering to remove noise and increase the coherence value. The final stage is to perform terrain correction. The post processing stage was performed with GIS using QGIS version 3 software to display deformation maps in the research area.

3. Result and Discussion

3.1. Results of Data Processing Using DInSAR Method

Figure 3 shows the phase coherence value resulting from DInSAR processing using SNAP software after the Goldstein filter was carried out. Most of the pixels still have a low coherence value <0.4. The low coherence value indicates that the research area still has a fairly high vegetation density and high noise even though it has gone through the filtering stage. It indicates that the frequency C band (wavelength of 5.6 cm) is not capable enough to penetrate the dense vegetation canopy, so there is considerable noise in the backscatter phase received by the satellite. Therefore, this research only analysed areas that have a coherence value > 0.5 to get the interferogram phase with the best accuracy [4][8].

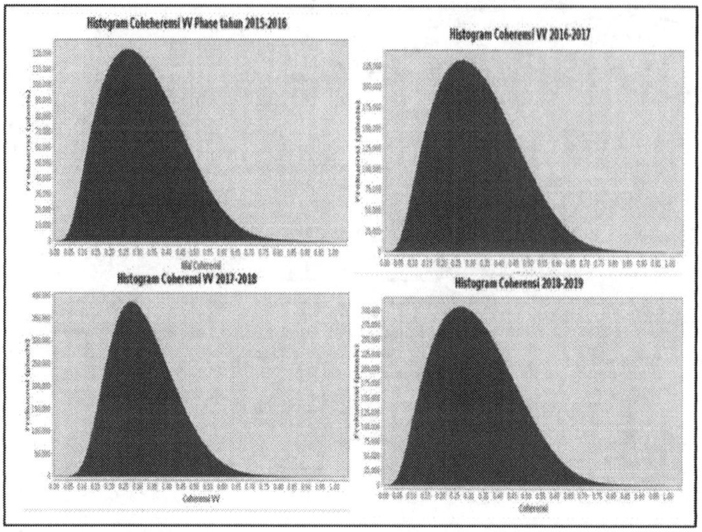

Figure 3. Graph of the coherence value of DInSAR processing result.

In the analysis process, the map of the observation point location is presented in figure 1. The research location, Sepaku district, is shown by point IDs 3,4 and 27. Otherwise, the point samples in Penajam subdistrict are shown by point IDs 13,15 and 26. This research used 6 sample points which became the identified locations where vertical deformation occurred and had a phase coherence value > 0,5. The results of data processing using the DInSAR method at the observation location points are shown in Table 2. On the deformation map (figure 5, 6, 7), the deformation area of the Sentinel 1A image data processing used the DInSAR method with SNAP software. The identified areas getting vertically deformed are indicated by red blocks, showing that the subsidence and the deformation phase is negative, while blue blocks are areas that had experienced an uplift and showed positive phase values. The white areas are generally areas that are relatively stable and do not experience change or deformation. The results of the DInSAR multitemporal processing for the 2015-2019 period showed a maximum subsidence of 12.97 cm and a maximum land increase (uplift) of 10.01 cm (figure 4). The deformation process that occurred in the research area was quite complex. The area where deformation was identified changes every year, whether in the deformation area, the area, or the amount of deformation (cm).

Table 2. Value of data processing results at observation points.

No. ID Point	Subdistrict Name	Coordinate		Magnitude of Deformation (cm)				Deformation Average (cm/year)	Elev. DEMNAS (meter)
		Latitude (°)	Longitude (°)	2015-2016	2016-2017	2017-2018	2018-2019		
3	Sepaku	-0,957	116,693	-2,85	-3,17	-5,06	-5,01	-4,02	55,15
4	Sepaku	-0,905	116,775	-12,42	-5,69	10,01	-12,97	-10,36	9,82
13	Penajam	-1,335	116,688	-7,40	-5,20	-7,40	-5,62	-6,41	26,48
15	Penajam	-1,175	116,746	-9,19	-4,36	-9,19	-7,13	-7,47	8,00
26	Penajam	-1,249	116,770	3,06	2,57	-5,53	-7,48	-6,51	5,27
27	Sepaku	-1,079	116,677	5,80	-5,32	-9,15	-12,04	-8,84	47,00

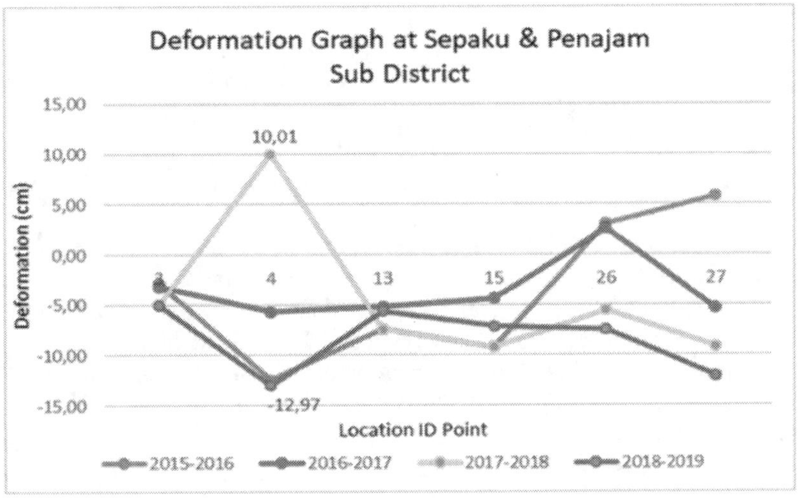

Figure 4. Deformation chart in Penajam and Sepaku district.

3.2. Discussion

The location of point 13 (figure 5) is the terminal area of PT. Pertamina's oil refinery in the Penajam district. In the 2015-2019 period, the location of this point experienced an average land subsidence of 6.41 cm per year. The subsidence occurred at point 13 due to the continuous load from buildings and oil refineries. Meanwhile, the soil structure at the location of that point was not strong enough to withstand the loads of the building above it. The point 26 was a densely populated port city with a fairly high building density. This area was identified as having deformations in the form of subsidence and uplift. The subsidence that occurred at point 26 had a speed of 6.51 cm per year. The subsidence occurred in this area due to the load of buildings with high density and human activities such as groundwater extraction. Based on geomorphology, the uplift that occurred in this area is closely related to alluvial material deposits, because this area is located at the mouth of rivers and the sea.

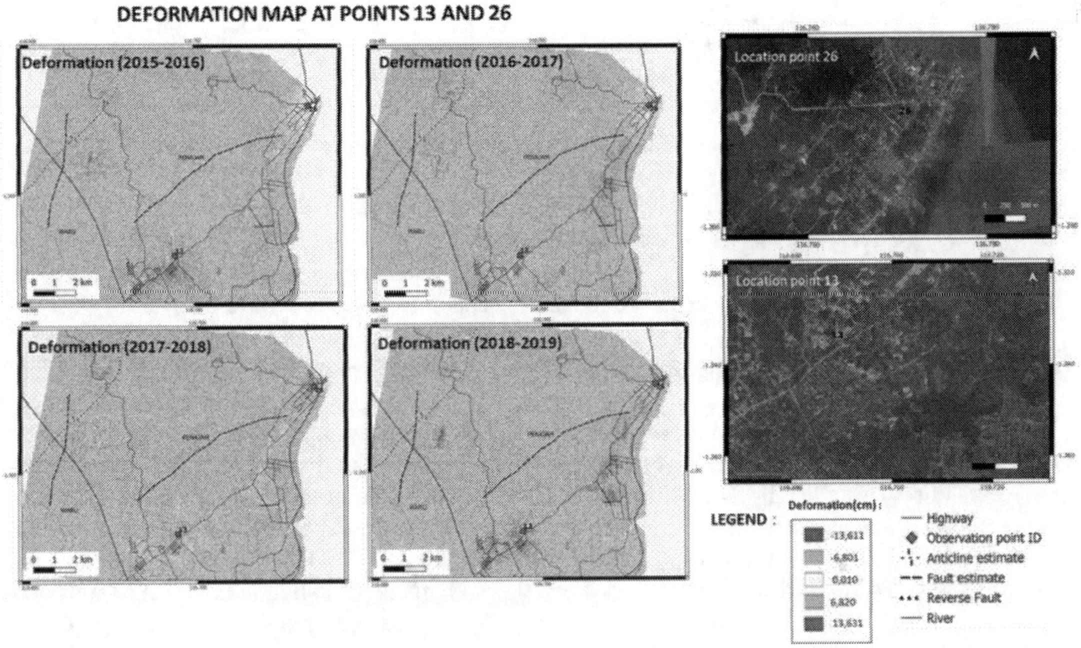

Figure 5. Deformation Map at Point 13 and 26, Sepaku District for Period 2015-2019

IOP Conf. Series: Earth and Environmental Science **851** (2021) 012017 doi:10.1088/1755-1315/851/1/012017

Based on figure 6, the location of point 3 is the location of point 0 IKN, where it is planned to be the centre of government. This location is located at an elevation of 55 meters in a hilly area. The land subsidence that occurred in this area is still classified as the lowest area compared to the location of point 4 and point 27 of about 4.02 cm per year. This could happen since there had been no development of supporting infrastructure for the capital city, while the deformation occurred due to the deforestation activities in the region.

The location of point 4 (figure 6) is the main provincial road in the Sepaku district with an elevation of about 9-10 meters. Based on figure 4, the location of point 4 experienced the highest land subsidence in the period 2018-2019 of 12.97 cm and the average land subsidence per year was 10.36 cm per year. The deformation that occurred in this area is closely related to development activities in the form of road infrastructure and buildings around point 4. The location of point 4 is the main access that connects the regions. The soil structure in the area of point 4 had a weak structure compared to that of point 3, making it more prone to deformation. In 2017-2018, this location was identified as uplift 10 cm increase in land due to development activities, such as elevating the road body at location 4.

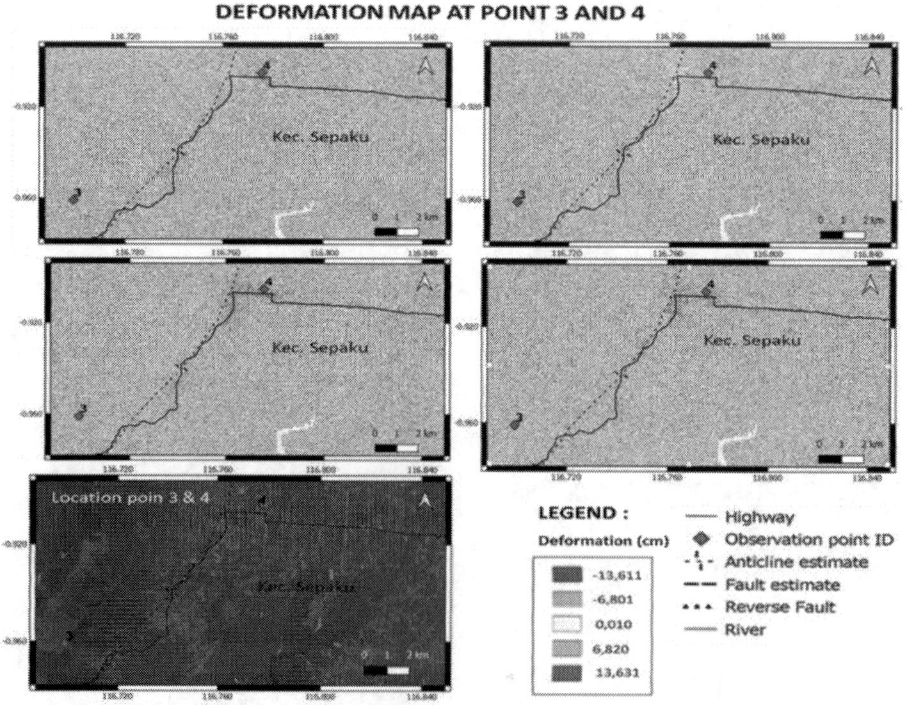

Figure 6. Deformation Map at Point 3 & 4, Sepaku District for Period 2015-2019

Land subsidence at observation point 15 (figure 5) was influenced by building density and alluvial material deposited in the area around Balikpapan Bay. The presence of sedimentary material deposits is a factor in the area around point 15 experiencing an uplift. The location of point 27 is an area that had experienced landslides in Sepaku district in April 2018 [5] as shown in figure 7. This area is a hilly area with steep slopes with a height of 47 meters. The DInSAR multitemporal processing results confirms that this area had been identified as experiencing subsidence in the time span before the landslide disaster of 9.15 cm [4][9]. The average land subsidence value in this area is 8.84 cm per year. Hence, the underlying cause for landslides at point 27 based on this study is deformation in the form of subsidence due to the loading of buildings with a high enough density in the area around the landslide, the topographical conditions, and the presence of high rainfall.

IOP Conf. Series: Earth and Environmental Science **851** (2021) 012017 doi:10.1088/1755-1315/851/1/012017

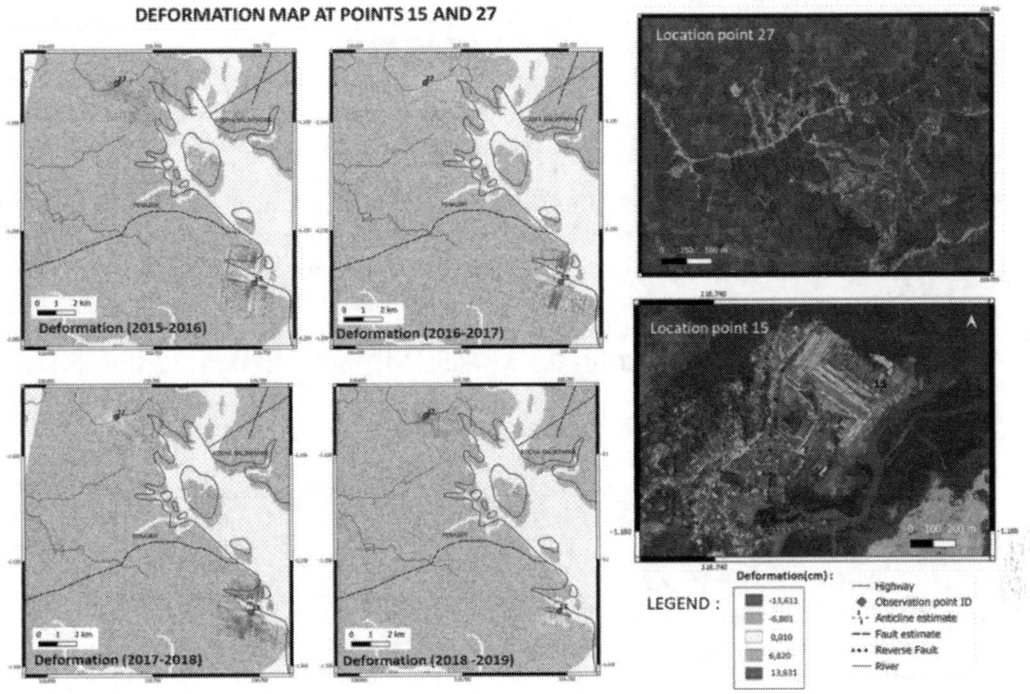

Figure 7. Deformation Map at Point 15 and 27, Penajam and Sepaku District for Period 2015-2019

The observational areas generating a deformation process can be used to monitor the developments in an area. Areas under development can generally be identified from the deformation contrast values that have changed significantly, compared to areas that are not under development [6][12]. This can be seen from areas that experience deformation in the form of subsidence with high building density, and this is closely related to the loading of big building volumes on the soil structure. Land subsidence may also occur due to the extraction of big amount of groundwater, particularly in densely populated urban areas [2].

4. Conclusion

The deformations that occur in the research area can be identified properly using the DInSAR method. Multitemporal analysis using DInSAR on Sentinel 1A images in the research area is able to detect changes of ground deformation, both subsidence and uplift from time to time. This method can identify areas that experience landslides caused by subsidence that have occurred in the research area. Based on the results of data processing, the area with the greatest land subsidence rate is the location of point 4 at Sepaku Distrct, with an average annual subsidence of 10,75 cm per year. This location also experienced a maximum uplift of 10 cm which occurred in the 2017-2018 period. Meanwhile, the area with the lowest land subsidence is the location of point 0 IKN (point 3) at Sepaku subdistrict with a land subsidence rate of 4,02 cm per year. The identified areas with deformation generally take place in areas with a high density of buildings, construction areas, road infrastructure, and river alluvial deposits.

References

[1] Agustan, Hanifa R H, Anantasena Y, Sadly M and Ito T 2019 Ground Deformation Identification related to 2018 Lombok Earthquake Series based on Sentinel-1 Data, *IOP Conf. Ser.: Earth Environ. Sci. 280 012004.*

[2] Al-Akbar T O, Prasetyo Y and Wijaya A P 2015 Analisis Dampak Penurunan Muka Tanah Terhadap Tingkat Ekonomi Menggunakan Kombinasi Metode DINSAR dan SIG (Studi Kasus: Kota Semarang). *Jurnal Geodesi Undip*, **4(4)**, 136-143.

[3] Azkiya dan Jatmiko 2015 Studi Perbandingan Dua Algoritma Phase Unwrapping (Region Growing dan Minimum Cost Flow) pada Teknik Interferometric Synthetic Aperture Radar (INSAR) dalam Menghasilkan Digital Surface Model (DSM*), Jurnal Bumi Indonesia*, **4(1**).

[4] Bonì R, Bordoni M, Colombo A, Lanteri L and Meisina C 2018 Landslide state of activity maps by combining multi-temporal A-DInSAR (LAMBDA), *Remote Sens. Environ*, **217**, 172-190.

[5] Badan Geologi (PVMBG) 2018 Laporan Singkat Pemeriksaan Gerakan Tanah Di Desa Telemow Kec. Sepaku, Kab. Penajam Paser Utara Provinsi Kalimantan Timur, https://vsi.esdm.go.id/index.php (accessed on 12 November 2020)

[6] Blasco J M D, Foumelis M, Stewart C and Hooper A 2019 Measuring Urban Subsidence in the Rome Metropolitan Area (Italy) with Sentinel-1 SNAP-StaMPS Persistent Scatterer Interferometry, *Remote Sensing*, **11**, 129.

[7] Even M and Schulz K 2018 InSAR Deformation Analysis with Distributed Scatterers: A Review Complemented by New Advances, *Remote Sensing*, **10**, 744.

[8] Hanssen and Ramon F 2001 Radar Interferometry Data Interpretation and Error Analysis. *Netherlands: Kluwer Academic Publisher*, Delft University of Thechnology

[9] Huang J et al. 2017 DInSAR Technique for Slow-Moving Landslide Monitoring Based on Slope Units, *Survey Review*, **51(364)**, 70-77

[10] Kurniawan R and Anjasmara I M 2016 Pemanfaatan Metode Differential Intermerometry Synthetic Aperture Radar (DInSAR) untuk Pemantauan Deformasi Akibat Aktivitas Eksploitasi Panasbumi, *Jurnal Teknik ITS*, **5(2)**, doi: DOI: 10.12962/j23373539.v5i2.17361.

[11] Kurniadi, A 2019 Pemilihan Ibukota Negara Republik Indonesia Baru Berdasarkan Tingkat Kebencanaan, *Jurnal Manajemen Bencana*, **5(2)**, 1-12.

[12] Miky Y H 2019 Multitemporal Sentinel-1 SAR Interferometry for Surface Deformation Monitoring near High Dam in Aswan, Egypt, *American Journal of Geographic Information System*, **8(2)**, 96-102.

[13] Makatia W D C, Pratiknyo P and Nugroho A R B 2014 Geologi dan Karakteristik Akuifer Daerah Mutiara, Kecamatan Semboja, Kabupaten Kutai Kertanegara Provinsi Kalimantan Timur, *Jurnal Ilmiah Geologi Pangea*, **1(2)**, 1-10.

[14] Morishita Y and Hanssen F 2015 Deformation Parameter Estimation in Low Coherence Areas Using a Multisatellite InSAR Approach, *IEEE Trans. Geoscience Remote Sensing*, **(53)**, 4275–4283

[15] Sunu H A, Yuwono B D and Suprayogi A 2019 Analisis Ketelitian DSM Kota Semarang Dengan Metode INSAR Menggunakan Citra Sentinel-1, *Jurnal Geodesi Undip*, **8(3)**, 17-26.

Estimation of near-surface temperature in Suwawa Geothermal Prospect, Gorontalo, Sulawesi, Indonesia, based on magnetotelluric and artificial neural network

M Maryadi[1,3*], P Bramanthyo[1], A Zarkasyi[2] and H Mizunaga[3]

[1]Undergraduate Program of Geophysics, Faculty of Mathematics and Natural Sciences, Universitas Indonesia, Depok, 16424, Indonesia.
[2]Center for Mineral, Coal, and Geothermal Resources, Geological Agency, Bandung, 40254, Indonesia.
[3]Department of Earth Resources Engineering, Graduate School of Engineering, Kyushu University, Fukuoka, 819-0395, Japan.
*E-mail: maryadi@sci.ui.ac.id

Abstract. A geophysical survey using broadband magnetotelluric (MT) technology was carried out in Suwawa Geothermal Prospect Area, Gorontalo Province, Sulawesi Island, Indonesia. The target of that research is to evaluate the geothermal potential hidden below the surface, based on underground resistivity distribution. However, the information about resistivity alone is not enough to get a proper understanding of the geothermal system in this area. Another important subsurface feature that could be useful for the evaluation is temperature. In this study, an attempt to predict the subsurface temperature using resistivity and limited information from a shallow borehole thermogram was carried out. Employing the dependency between resistivity and temperature an indirect temperature estimator was built, thanks to the applicability of artificial neural network (ANN) to learn the pattern connecting both parameters. Comparing some neural network training data shows that the predictive powers of the calibrated neural network highly influenced by the geological difference between the location of borehole and MT station. The best trained ANN was then used to predict the temperature below the other MT stations. The result shows that a proper ANN architecture is important to improve the deeper temperature estimation. The best ANN estimator was obtained from the BT01 and AMT39 data pair, which has the highest correlation as well. This preliminary study gives useful insight into how resistivity could be an alternative tool to delineate the near-surface temperature profile, in order to get a more comprehensive image of the subsurface geothermal system.

1. Introduction

Geothermal is one of the most promising energy resources preserved in Indonesia, with approximately 23.9 GW of geothermal energy potential [1].The utilization of this energy is highly related to the effort that have been given during the exploration stage. One of the geothermal exploration activities was carried out in Suwawa, Gorontalo, Sulawesi Island, Indonesia by the researchers from the Center of Minerals, Coal, and Geothermal Resources back in 2012 using magnetotelluric (MT) method. The magnetotelluric method is used to image the subsurface condition based on its resistivity value [2–4]. However, in order to get a reliable exploration result, one would need other subsurface parameter, such as temperature. Temperature is an important subsurface feature which directly related to geothermal system [5,6]. Knowledge about its distribution would be useful to gain a deeper understanding about the geothermal system.

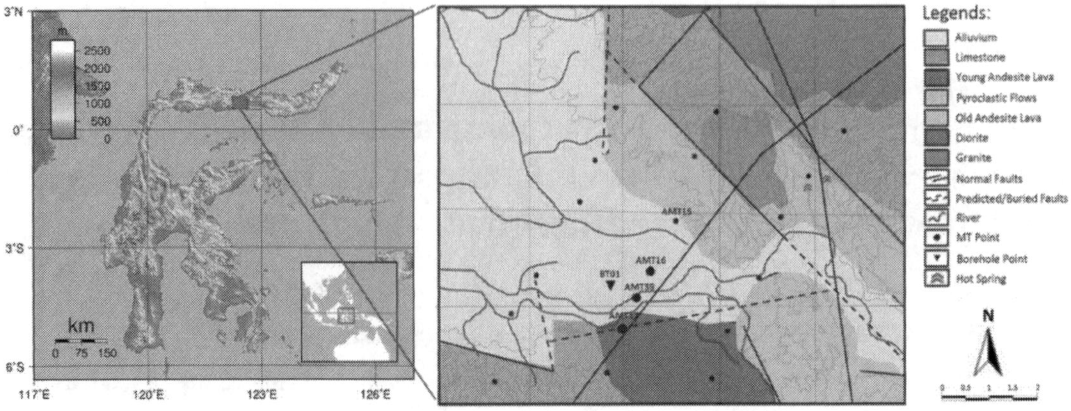

Figure 1. AMT stations and shallow borehole location on regional geological map of Suwawa region, Gorontalo province, Indonesia (modified from [7]).

In this study, the attempt to estimate the shallow temperature in Suwawa Geothermal Prospect was carried out. Rather than using direct measurement using temperature logs, temperature profile was predicted from resistivity value below an MT station using artificial neural network (ANN). Resistivity is a temperature-dependent parameter [8–11], therefore predicting the temperature from resistivity is possible if the relationship between both parameters was recognized [12]. The resistivity data was obtained from MT sounding, while a limited number of temperature data was collected from a shallow borehole thermogram. The degree of correlation between temperature and resistivity information, and the reliability of the ANN technique was firstly evaluated before finally being used to estimate the temperature below MT stations nearby.

2. Field Data

The geothermal prospect is situated in between 0°28'13.7" N – 0°36'54.8'' N and 123°06'00'' E – 123°15'00" E. Lithological condition of Suwawa Geothermal Prospect area consists of tertiary to quaternary rocks. The oldest rock is Bilungala Volcanic which includes breccias, tuffs, and andesitic lavas from Early to Late Miocene ages. Diorite Bone, located in the middle of the area, consists of quartz diorite, diorite, granodiorite, and late Miocene granite. In the south-eastern part, there is Pinogoe volcanic group which comprises of tuffs, lapilli tuffs, breccias, and lava from Pliocene to Plistocene ages. The youngest rocks are lake sediment and alluvium [7].

There are two kinds of data sets that were used in this preliminary work. These data were collected from Suwawa geothermal prospect area in Gorontalo Province, Indonesia. One data set needed is shallow borehole thermogram data, and the other is audio-magnetotelluric data. This field has only two borehole thermogram data recorded up to this time: BT01 and BT02. However, the latter does not have any AMT station nearby, so it cannot be used. For BT01, there are 3 surrounding AMT stations: AMT16, AMT17, and AMT39. The location of the borehole point and all AMT stations are shown in figure 1. As seen on the map, these AMT points are situated in different geological conditions and at different distances relative to the exploratory well. The AMT16 station is located on an alluvium layer about 900 m from the borehole. In the same surface geological condition, AMT39 lies 680 m to the west from the borehole. The last station, AMT17 is located at around 700 m from the borehole and sits on an andesitic body. This different condition might affect the correlation between resistivity derived from AMT and borehole temperature value.

2.1. Borehole thermogram data

Borehole BT01 is located on an alluvial overburden. The well log of BT01 is shown in figure 2. From the drilling results, it is known that lithology below the well site consists of alluvial deposits (Al) from surface to 34 m depth, unaltered to weakly altered polymictic breccias (BX) at a depth of 34 - 120 m, and a moderate to very strong altered polymictic breccia layer (BXT) at a depth of 120 - 250 m. Subsurface temperature was measured up to 250 m deep at 2 m intervals. The profile shows that the temperature is mainly increasing with the depth. In parallel with the heat measurement, investigation of lithology, as well as alteration intensity, was also carried out.

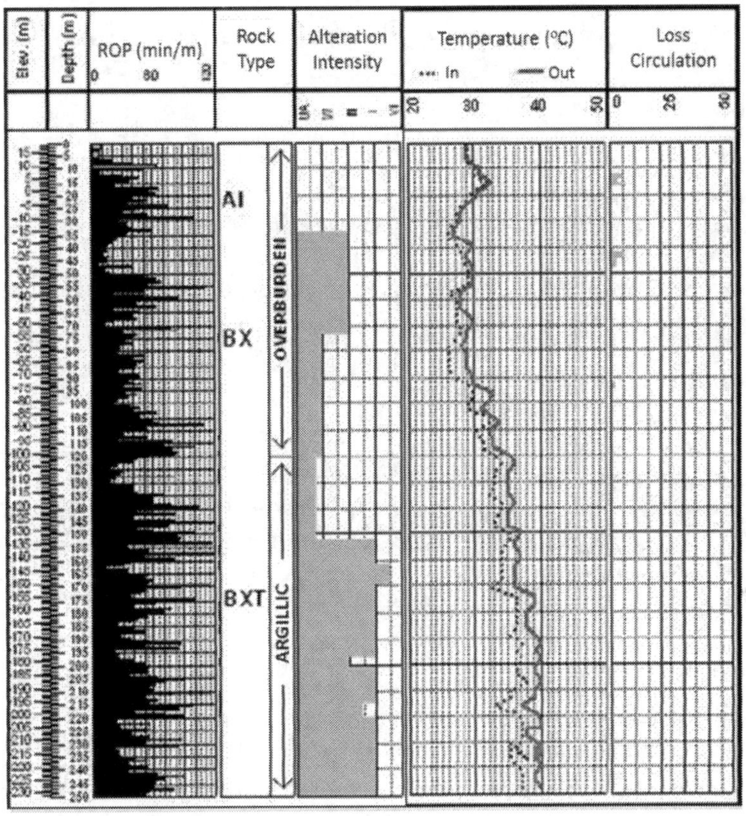

Figure 2. Shallow borehole thermograms data up to 250 m depth at BT01 (after [13]).

The existence of the altered breccia also shows a high alteration at the respective depth, which is caused by mineralization of geothermal fluids. This argillic type of alteration rock functions as the clay cap of the geothermal system below. The increasing temperature verifies that this borehole penetrates the up-flow zone [13].

2.2. Audio-magnetotelluric data

Three AMT impedance data sets from the nearest sites from borehole BT01 were used for calculating the relation between resistivity and subsurface temperature. These data sets were actually derived from a combined AMT-MT survey. The frequency range of the signal is recorded from a high of 8,192 Hz to a low of 0.117 Hz in average. These data sets were collected by using Zonge Geophysical Receiver GDP-32II. For balancing the comparison, the electromagnetic data that was used for correlation was only the high-frequency part (AMT), which also has better resolution for the shallower part [14,15]. The AMT responses (apparent resistivity and impedance phase) at each station are shown in figure 3.

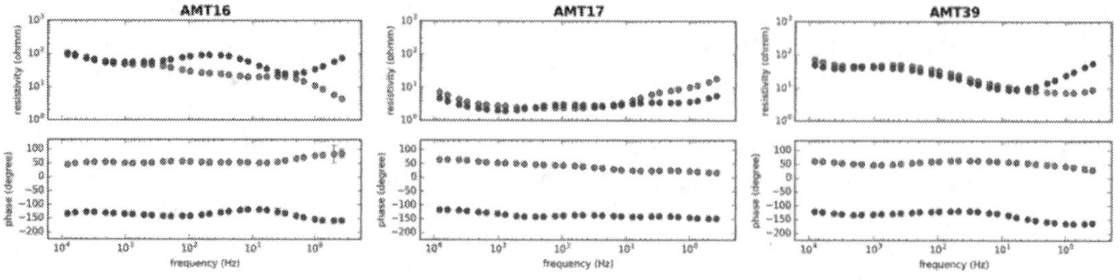

Figure 3. AMT responses, apparent resistivity (upper part) and phase (lower part) at AMT16, AMT17, and AMT39.

From both measurement modes (TE and TM), the average one-dimensional sounding profile is calculated as the determinant. In this work, the audio-magnetotelluric sounding responses, resistivity-phase versus frequency, were inverted by using Bostick transformation for a quick calculation. This simple inversion scheme produces an approximate curve of resistivity versus depth [16,17]. Resistivity values at every 10-meter depth were then calculated by employing the linear interpolation technique. Tens of data pairs of resistivity and temperature at the same depth were examined before finally being used as the training basis.

3. Dependency analysis

An attempt was made to determine the relationship between resistivity and temperature at the same depth below the surface. Bostick resistivity values from all AMT stations were first plotted against temperature at the same depth, then the dependency between both variables was approximated by using several types of regression formulae to see which one was best for this case. The quality of fitness was analysed by the coefficient of determination value (R-squared). This coefficient gives information about how well a regression equation represents the data distribution. The R-squared value for each regression equation is listed in Table 1.

Table 1. Coefficients of determination value for each regression type for each respective data pair.

Regression Type	BT01 and AMT16	BT01 and AMT17	BT01 and AMT39
Linear	0.2370	0.3526	0.3105
2nd Order	0.2431	0.3745	0.3382
Power	0.2920	0.4166	0.5346
Logarithmic	0.2393	0.3454	0.3103
Exponential	0.2883	0.4245	0.5526

The value itself is the ratio between explainable and unexplainable data values in a set of data pairs. Based on the regression result, it can be roughly said that temperature was more strongly correlated to resistivity in an exponential equation than in other regression types. Due to low R-squared value, the result is not so satisfying. Nonetheless, this kind of analysis remains beneficial to explain how the variables are related one to another.

4. Effect of Resistivity-Temperature Dependency

By using some training data, which was extracted from the shallower part of the data pairs, we tried to estimate the rest. This analysis was carried out to evaluate the effect of correlation value obtained in the previous section, to the prediction accuracy. The result is shown in figure 4 along with the observed temperature value for comparison, with the root mean square error (RMSE) value.

From this result, the calculated temperature profile produced from the BT01 and AMT39 pair has the best fit among all the results, as it has the smallest RMSE of the samples. This result also gives information about how strongly the geological condition and distance affects the relation between resistivity and temperature. Having the least error, the BT01 and AMT39 data pair can be used in the network training phase, before being employed to estimate the deeper temperature values below the surrounding AMT stations. However, the predicted values from the BT01 and AMT16 data pair does not give poor result. It could still be utilized, but with a lower degree of confidence. On the other hand, the BT01 and AMT17 data pair gives quite rough predictions even visually. Therefore, it is unwise to choose the pair as the neural network basis.

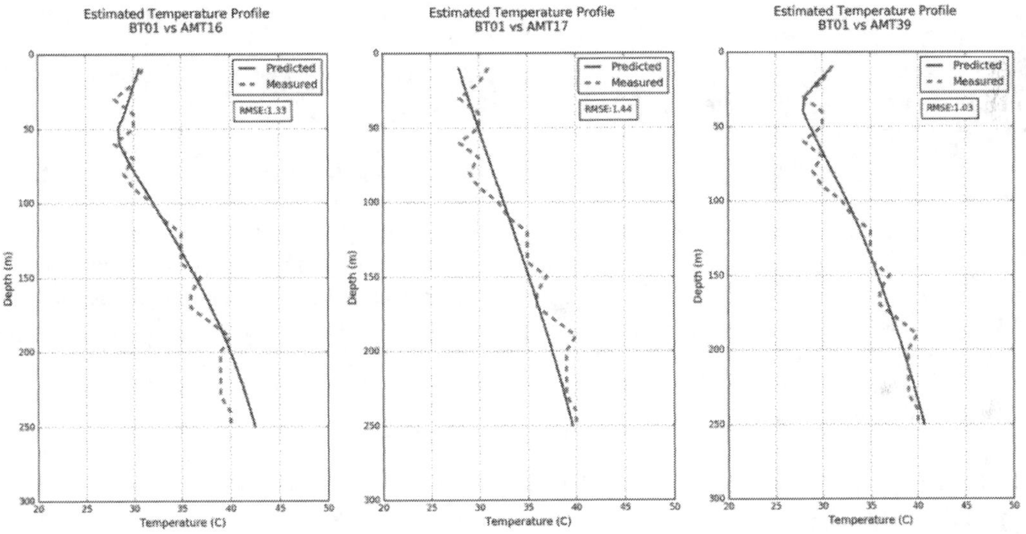

Figure 4. Measured (dashed line) and calculated (solid line) temperature profiles from three different data pairs. The result of BT01-AMT39 has the better prediction than the others, as this pair has the highest R-squared value as well.

5. Result and Discussion

From the comparison result, the deeper portion of temperature values was predicted using the transfer function based on the ANN training of BT01 and AMT39 data pair. With this transfer function, the temperature below the additional AMT station AMT15, AMT16 and below the AMT39 itself were then tried to estimate. The AMT15 station lies 1 km northeast from AMT16 as shown earlier in the map earlier in figure 1. Although it is unfair to forecast the temperature below the other AMT stations with only one training data using the proposed ANN model (which roughly means that the effect of relative distance from the AMT station to the heat source is neglected), the estimation was still carried out in order to test the impact of resistivity differences between the AMT stations. The results are shown in figure 5.

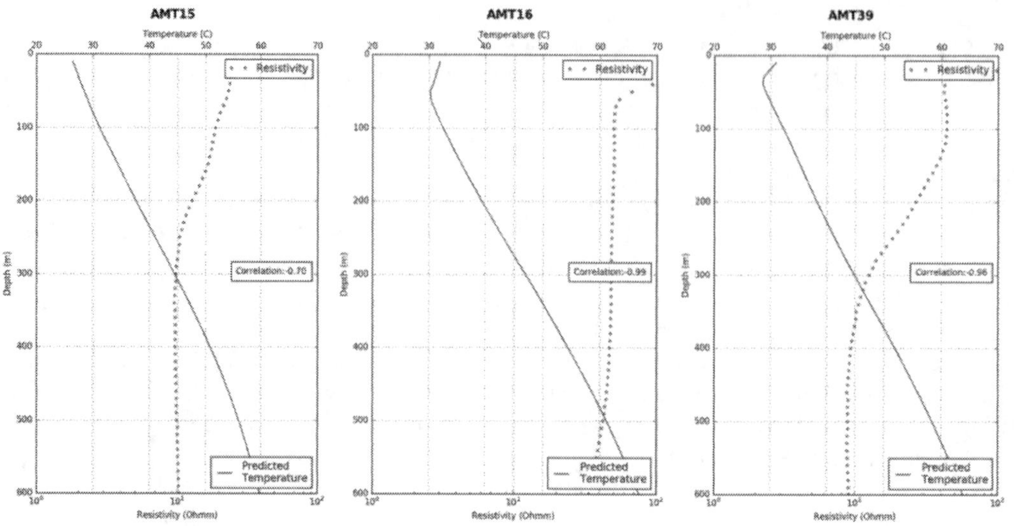

Figure 5. Predicted temperature profile after temperature forecasting under AMT sites nearby, and its updated correlation coefficient with the Bostick resistivity value.

Due to the limited training data, this prediction can only be performed for a limited depth, up to 600 m, to avoid over-estimated values. As shown in the results, the estimated temperature values are different for each resistivity profile, even though the differences are small compared with the resistivity changes. The correlation coefficient was also recalculated. The numbers mainly show that the resistivity and the estimated temperature are associated. In this result, the temperature below the AMT39 station is assumed to be the same as the temperature measured from borehole BT01, because the resistivity from this point is calibrated with the temperature from the borehole. Meanwhile, the temperature values in degree Celsius below the AMT15 and AMT16 stations are calculated from the transfer function of the calibrated ANN, assuming that they are lying at the nearly same geological condition and at about equal relative distance to the heat source with AMT39. Predictions below other points that lie in a different geological condition were not carried out because of the probability that the transfer function is only valid for one certain environment, although this behaviour needs to be further studied.

From the estimated temperature value, regardless of the large difference in the AMT resistivity responses between all stations, the calculated temperature values have only slight variation. From this result, the estimated temperature is increasing with depth at a rate of approximately 70 °C/km of geothermal gradient. However, the temperature profile is not in linear form like the constant geothermal gradient due to the effect of subsurface resistivity values. This estimated gradient is reasonable for a geothermal gradient value in a geothermal field for near-surface temperature.

Then, a minimum curvature gridding technique was used to interpolate the grids between AMT15, AMT16, and AMT39 in order to test the appropriateness of representing subsurface temperature distribution in a two-dimensional vertical cross section in this introductory study. Due to the limited number of data, the horizontal grid is minimized to prevent error and over-estimating value. The result is quite satisfying and plausible as shown in figure 6.

In the shallow and middle parts, subsurface temperature is distributed almost uniformly. While, in the deeper zone, the contour starts bending with the temperature below AMT16, and is slightly higher than its surrounding. The minimum curvature method minimizes the bends of the interpolated grid. So, the curve seen in the graph mainly comes from the actual data. Despite the limited displayed data in this study, it is possible and promising to create a larger temperature distribution map, if some additional borehole data are at hand in the future.

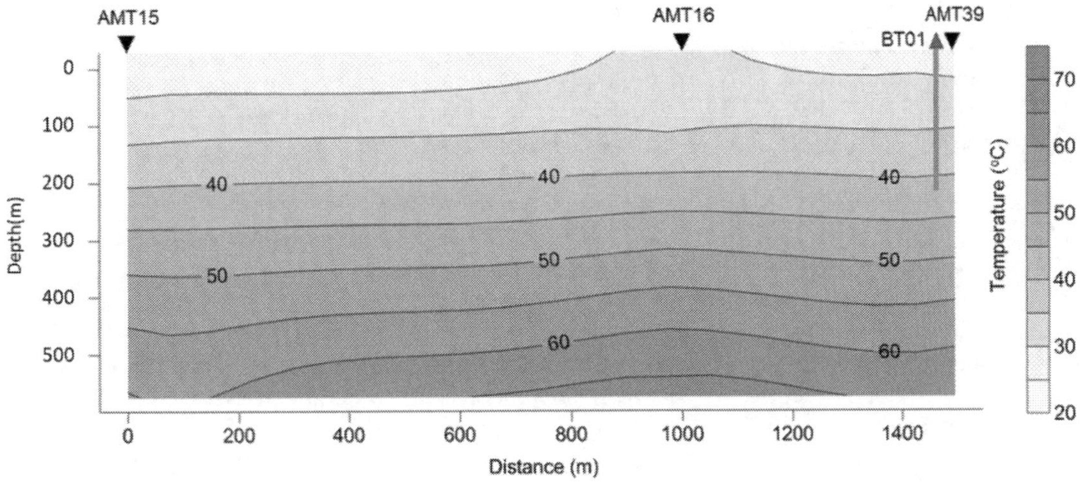

Figure 6. Predicted temperature distribution below AMT15, AMT16, and AMT39 up to 600 m based on the result of artificial neural network using BT01 and AMT39.

6. Conclusions

In the subsurface, resistivity and temperature at certain depths are most likely correlated exponentially from the coefficient of determination analysis, despite the unsatisfactory value due to the far distance between borehole and AMT points. In order to estimate subsurface temperature, the ANN technique was applied using resistivity data from the AMT method and temperature data from a borehole. A reasonable ANN model is needed to sharpen the temperature estimation for the deeper subsurface. In this case, prior knowledge about the factors that affect the subsurface resistivity value is needed. As a result, deeper temperatures were able to be estimated by resistivity and depth information. From the comparison of ANN test results, the best temperature estimation was obtained from the BT01 and AMT39 data pair, which has the highest correlation coefficient as well. This shows that the geological condition, as well as distance, should be wisely considered. In other word, AMT-Temperature data pairs that come from the same geological location and/or close in distance would be a better ANN estimator training basis. Therefore, the data pair that will be used in this indirect geothermometer method should be chosen carefully.

The predicted temperature profiles below each AMT stations are slightly different, regardless of the large difference in resistivity values. It shows that the influence of resistivity in calculating the temperature value is small compared with the effect of depth or other influential factors. A two-dimensional cross section of subsurface temperature is possible and could clearly describe the temperature distribution below the AMT stations. Nevertheless, more training data necessary for obtaining a better training result. So, deeper and larger sets of borehole data will surely provide a better result.

Acknowledgments

Authors wishing to acknowledge the supports from Universitas Indonesia through PUTI Saintekes Program 2020 (contract no. NKB-4907/UN2.RST/HKP.05.00/2020). We would also thank the Center for Mineral, Coal, and Geothermal Resources for giving permission to the authors to use the data for this study.

References

[1] Indonesian Ministry of Energy and Mineral Resources 2019 *Report of Indonesian Geothermal Energy Potential* (Jakarta)
[2] Simpson F and Bahr K 2005 *Practical magnetotellurics*
[3] Chave A D and Jones A G 2012 *The magnetotelluric method: Theory and practice*
[4] Naidu G D 2012 Magnetotellurics: Basic Theoretical Concepts
[5] Ranalli G and Rybach L 2005 Heat flow, heat transfer and lithosphere rheology in geothermal areas: Features and examples *J. Volcanol. Geotherm. Res.*
[6] Gupta H and Roy S 2007 *Geothermal Energy: A alternative resource for the 21st century*
[7] Rezky Y and Hasan A R 2005 Penyelidikan Geologi Daerah Panasbumi Suwawa Kabupaten Bone Bolango, Gorontalo *Prosiding Pemaparan Hasil Kegiatan Lapangan dan Non-lapangan* (Bandung)
[8] Ussher G, Harvey C, Johnstone R and Anderson E 2000 Understanding the resistivities observed in geothermal systems *World Geothermal Congress 2000* (Kyushu-Tohoku, Japan) pp 1915–20
[9] Llera F J, Sate M, Nakatsuka K and Yokoyamas H 1990 Temperature dependence of the electrical resistivity of water-saturated rocks *Geophysics* **55**
[10] Ruhaak W, Bar K and Sass I 2014 Combining numerical modeling with geostatistical interpolation for an improved reservoir exploration *Energy Procedia*
[11] Nouveau M, Grandjean G, Leroy P, Philippe M, Hedri E and Boukcim H 2016 Electrical and thermal behavior of unsaturated soils: Experimental results *J. Appl. Geophys.*
[12] Spichak V V., Zakharova O K and Rybin A K 2011 Methodology of the indirect temperature estimation basing on magnetotelluric data: Northern Tien Shan case study *J. Appl. Geophys.* **73** 164–73
[13] Nanlohi F and Dikdik R 2006 Pemboran Sumur Landaian Suhu SWW-1 Lapangan Panasbumi Suwawa, Kabupaten Bone Bolango, Gorontalo (In Bahasa Indonesia) *Prosiding Pemaparan Hasil Kegiatan Lapangan dan Non-lapangan* (Bandung)
[14] García X and Jones A G 2005 A new methodology for the acquisition and processing of audio-magnetotelluric (AMT) data in the AMT dead band *Geophysics*
[15] Yang B, Zhang A, Zhang S, Liu Y, Zhang S, Li Y, Xu Y and Wang Q 2016 Three-dimensional audio-frequency magnetotelluric imaging of Akebasitao granitic intrusions in Western Junggar, NW China *J. Appl. Geophys.*
[16] Bostick F X 1977 A simple almost exact method of magnetotelluric analysis, in workshop on electrical methods in geothermal exploration *US Geol. Surv.*
[17] Goldberg S and Rotstein Y 1982 A simple form of presentation of magnetotelluric data using the bostick transform *Geophys. Prospect.*

Estimation of the subsurface sedimentary structure of Amarapura township (Mandalay Region) based on microtremor survey

P S Thein[1], K M O Kyaw[2], J Kiyono[3], S Pramumijoyo[4], T T Nu[5], K H Khaing[6], W M Than[7], T T Win[8] and M Thant[9]

[1]Department of Geology, University of Magway, Myanmar
[2]Department of Geology, University of Mandalay, Myanmar
[3]Graduate School of Global Environmental Studies, Kyoto University, Japan
[4]Geological Engineering Department, Gadjah Mada University, Indonesia
[5]Department of Geology, University of Mandalay, Myanmar
[6]Department of Geology, Defence Services Academy, Myanmar
[7]Department of Geology, Sittwe University, Myanmar
[8]Department of Geology, Kyaukse University, Myanmar
[9]Department of Geology, University of Yangon, Myanmar
*E-mail: pyisoethein@yahoo.com

Abstract. In this study, we investigated the underground structure for Amarapura township, Mandalay region. One of the major structures in Myanmar is the Sagaing Fault system. Several powerful earthquakes have struck along the Sagaing Fault during recent years, one of the largest of which was an M 7.0 event that occurred on July 16, 1956 and caused several casualties. Following the event, we conducted a microtremor survey to estimate the shaking intensity distribution during the earthquake. Seven boreholes were drilled throughout the basin, especially in Amarapura township to evaluate the geotechnical properties of subsurface soil layers. The H/V analysis of the noise measurements reveals frequency peak values ranging from 0.5 to 3 Hz on the Amarapura township. The shear wave velocity becomes less than 190 m/s in some areas, which causes severe damage for buildings in high probability. The sediment thickness values range from less than 50 m to greater than 320 m on Amarapura township, Mandalay region. In this way, the accomplished work has contributed to the characterization of the underground structure and the identification of sites prone to amplify soil shaking during a possible strong earthquake.

1. Introduction

Mandalay is the second largest capital city in Myanmar and the population is 1.727 million. This city is located very closed to the most active Sagaing fault in Myanmar [1],[2]. Mandalay region, where a destructive earthquake is expected in the near future, is considered to be a target area. The geometry of the underground structure and plate tectonic movement are at the origin of ground vibration during destructive earthquakes [3]. We carried out geophysical and geotechnical survey in Amarapura township, Mandalay region. Based on the survey data, we calculated the distribution of the sediment thickness and Vs30 of Holocene deposits.

2. Seismic Sources of Myanmar

Tectonically, Myanmar lies in the frontier zone which consists of two major plates, the India Plate and the Eurasia Plate. Numerous historic earthquake events recorded throughout Myanmar exhibit the seismic nature of the country. Since the 18th century, at least 15 major earthquakes including events in 1762 (north Rakhine Coast) and 1839 (Innwa Ancient Capital) occurred in Myanmar territory. Recent earthquakes occurring in Central Myanmar with notable magnitudes include the 1975 Bagan Earthquake (Magnitude 6.5), the 2003 Taungdwingyi Earthquake (Magnitude 6.7) and the 2011 Tarlay Earthquake (Magnitude 6.8).

2.1. Historical Seismicity around Mandalay Region

2.1.1. Innwa Earthquake (1839)

The Innwa earthquake occurred at about 4 am on 23 March 1839. Its epicenter was estimated at 22° North and 96° East with the maximum intensity of XI (Modified Mercalli Intensity Scale). The shock was accompanied by a tremendous roar and lasted about 20 seconds. While the damage was experienced in neighboring areas such as Amarapura, Innwa sustained extensive destruction. It is estimated that about 300 to 400 people were killed in the event [4].

2.1.2. Maymyo Earthquake (1912)

A disastrous intraplate earthquake occurred in 1912 at 97°E and 21°N with a moment magnitude of 8.0. The damage was severe in Maymyo (Pyin Oo Lwin). Railway lines were bent where they crossed the fault and massive landslides occurred. Aftershocks occurred for six months following the event.

2.1.3. Kyaukse Earthquake (1931)

Occurring on 19 August 1931, the Kyaukse earthquake was strong enough to cause minor cracking to buildings in Mandalay and to knock light articles over in Kalaw.

2.1.4. Sagaing Earthquake (1956)

The epicenter of the Sagaing earthquake was at 24.6°N and 99°E near Sagaing in central Myanmar with the depth between 8 and 10 km. It occurred at 15:07 hours GMT on 16 July 1956 and had a magnitude of 7.0 on the Richter scale. The earthquake was destructive in Upper Myanmar causing property damage at Mandalay and damaging 80% of the houses in Sagaing (22 km from Mandalay) as well as causing up to 50 casualties and 50 injuries [4].

2.1.5. Bagan Earthquake (1975)

The Bagan earthquake on 8 July 1975 was the most memorable in recent Myanmar history. It had a magnitude of 6.8, but its destruction horrified the people of Myanmar. The epicenter lies at latitude 21.48° N and longitude 94.70°E which is located in the northern part of Tankgyi Taung. Tremors of three aftershocks with moderate intensity were felt in Bagan and its surrounding area.

3. The GIS-Based Model proposed for Strong Ground Motion Analysis

The GIS based model involves three types of objectives. These are interpretation of vector, tabular and raster database. The imageries such as aerial photographs and satellite images were used for the generation of stereo-pairs in order to generate geomorphological map of Mandalay region. The necessary analysis was carried out for the generation of GIS layer models and lithological cross section. The strategy for earthquake risk assessment is shown in figure 1.

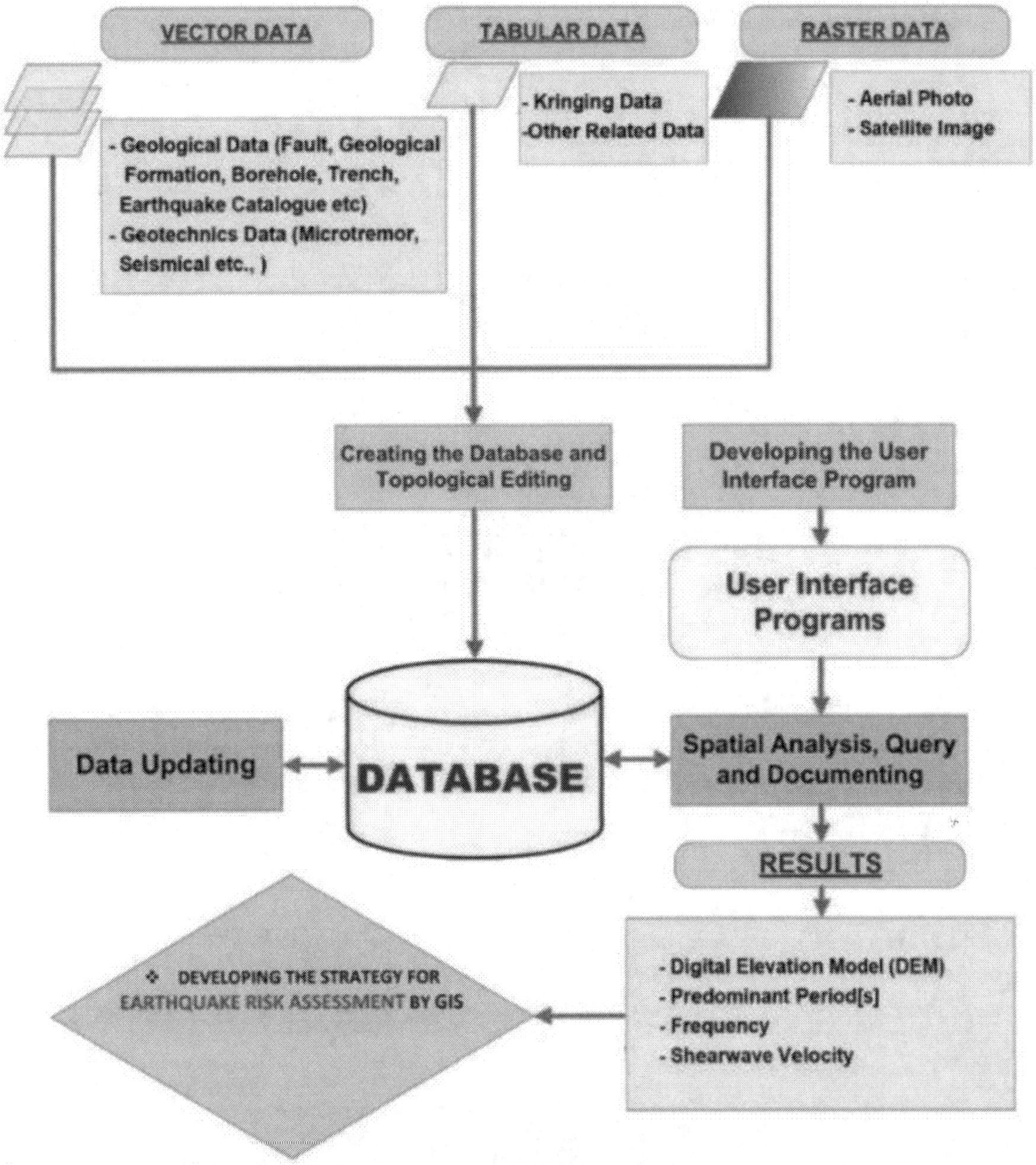

Figure 1. Data workflow diagram of earthquake risk assessment.

4. Regional Geology

Mandalay region is tectonically bounded by the Shan Plateau in the east, in which several lateral strike-slip faults are included such as Moemeik fault, Shweli fault, and Namma fault etc., and the subduction zone of Indian plate beneath Burma plate in the west and these tectonic environs are the main seismic sources of causes rather than Sagaing fault [5]. The Holocene sediments were deposited on Mandalay City. Geological map of Mandalay region is shown in figure 2 [6].

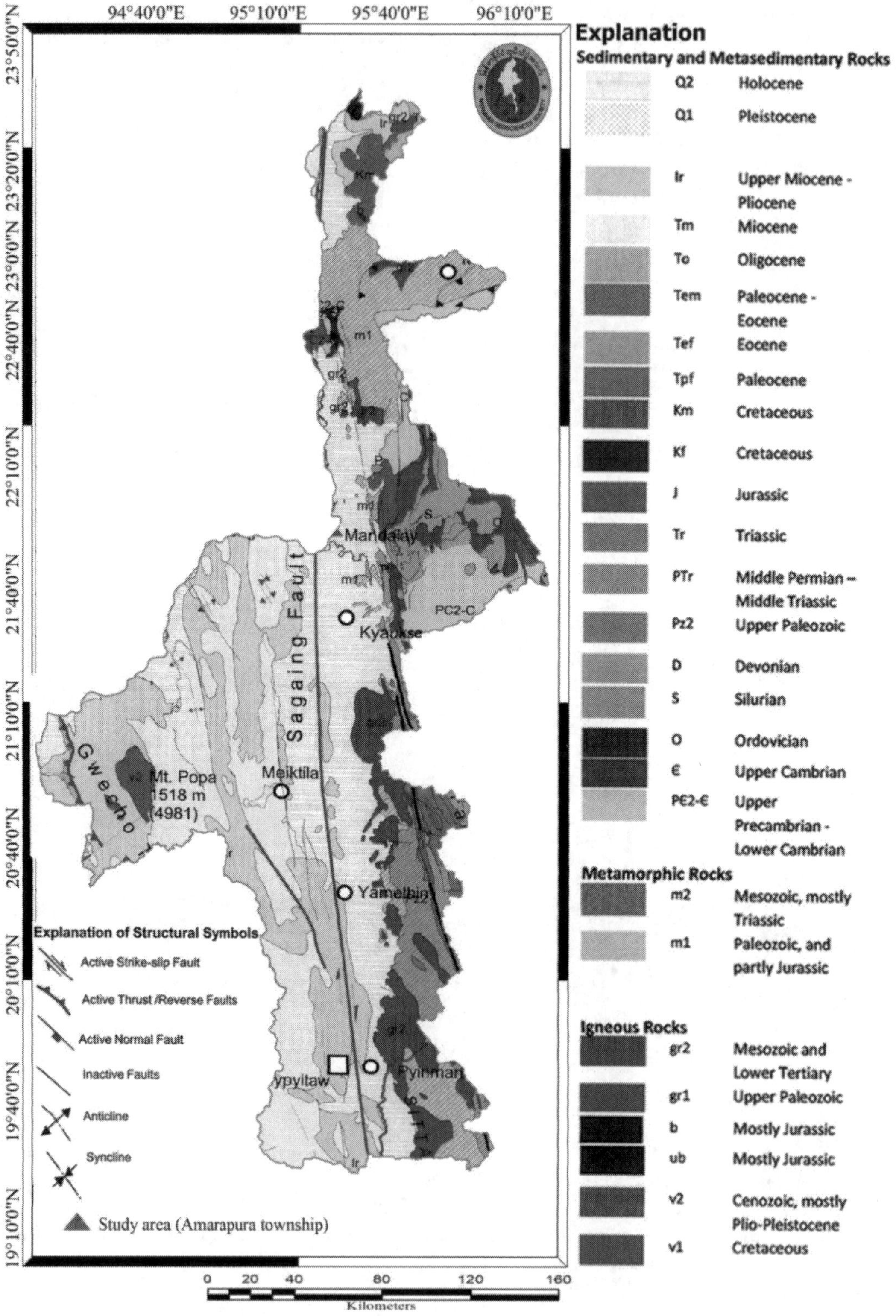

Figure 2. Geological map of Mandalay region [6].

5. Microtremor Survey

Microtremor instrument developed by Akashi Corporation, Model SMAR 6A3P was used in our site investigation. Microtremor measurements were taken in the Amarapura township in March 2020. In this research, 21 microtremor measurements were performed that sampled every district of Amarapura Township (figure 3).

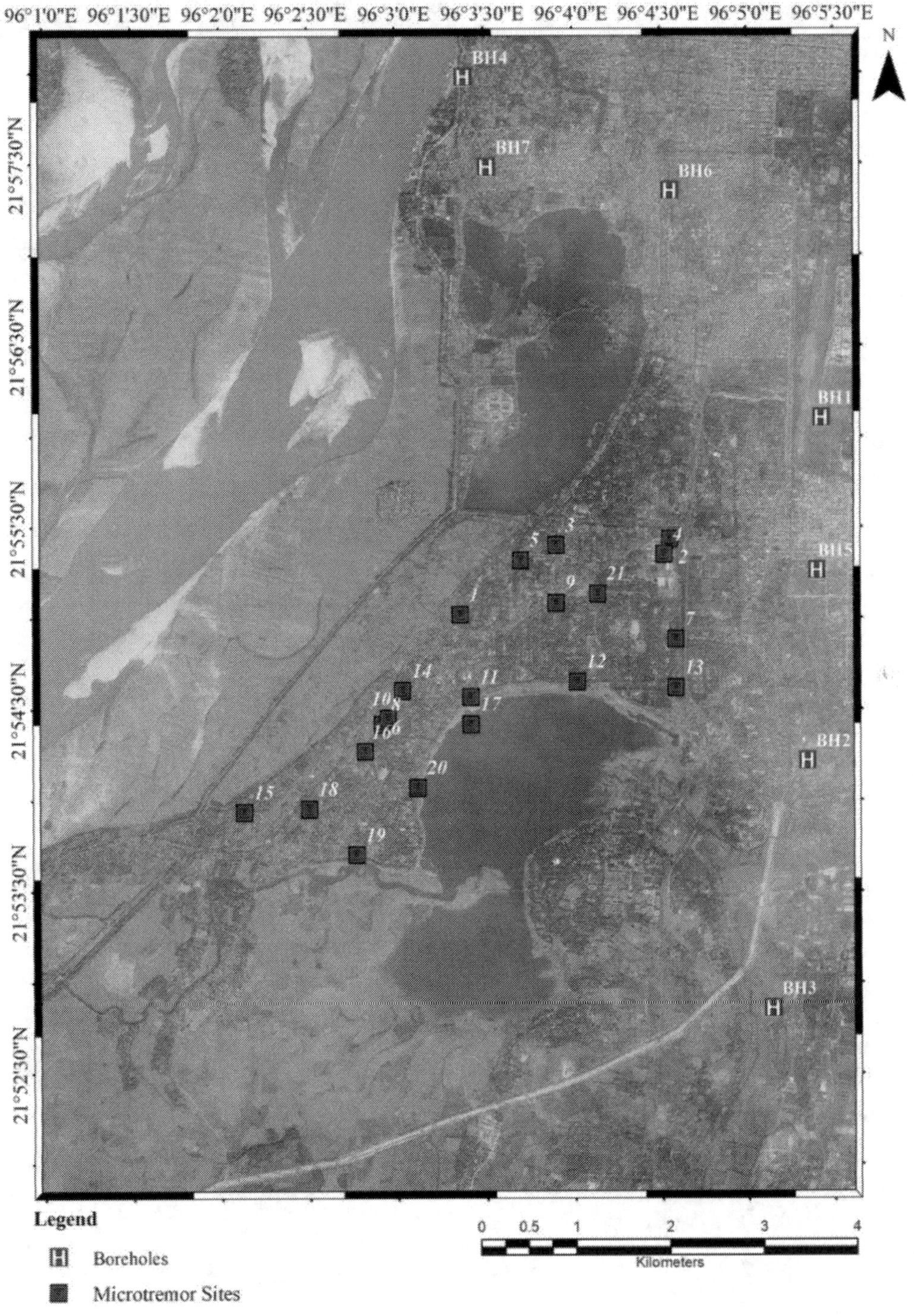

Figure 3. Location map of the boreholes and microtremor measurements [7].

6. Evaluation of Subsurface Soil Profile from Borehole data

Geotechnical site investigation is carried out to determine the subsurface soil structure and rocks. This investigation consisted of seven bore holes which were drilled up to 30m depth (Figure 3). In Mandalay region, bore hole drilling is carried out on 27 November 2020 to 9 January 2021. The subsurface soil profile of bore hole site no.1 is shown in figure 4. Shear wave velocity is an important parameter for estimation of the soil dynamic properties in the shallow subsurface. Vs^{30} values are computed for each target depth as equation 1 [8]

$$Vs^{30} = \frac{\sum d_i}{\sum t_i} = \frac{\sum d_i}{\sum(\frac{d_i}{vs_i})} \tag{1}$$

where, v_{si} = shear wave velocity, d_i= thickness of i layer and t_i = one-way travel time in i[th] layer. According to subsurface investigation, seven standard penetration test (SPT) had been carried and the correlation result of Vs profile and depth is shown in figure 5. These soil profiles and properties are fundamental input parameters for ground response analyses.

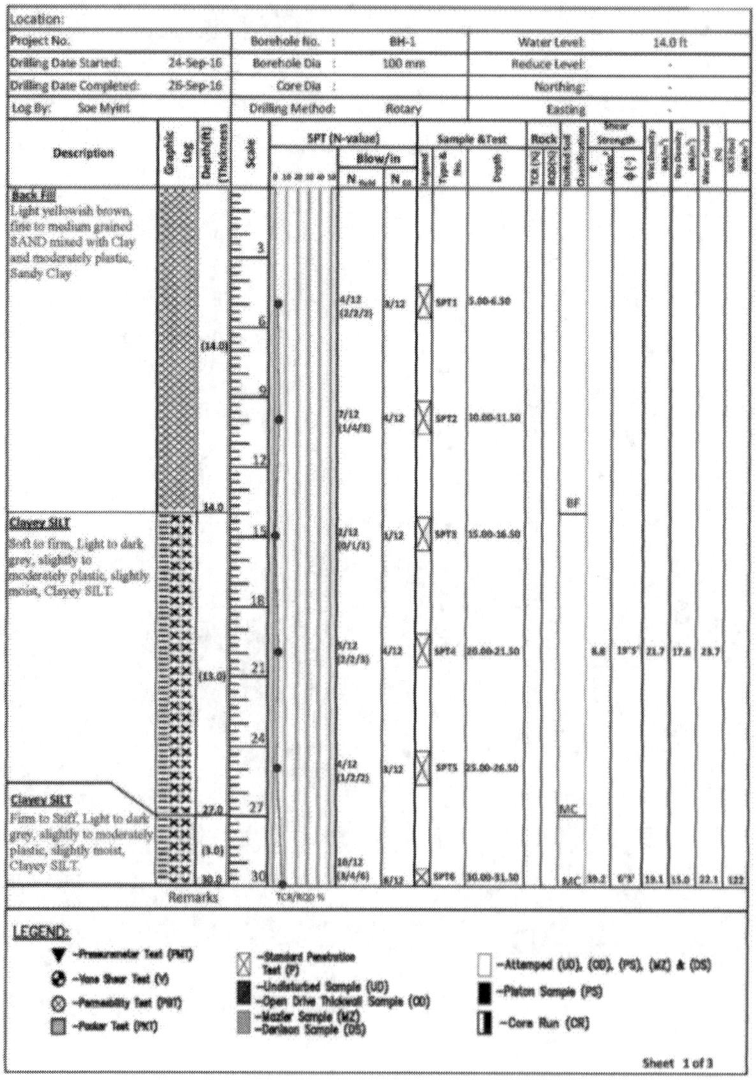

Figure 4. Subsurface soil profile at BH site no. 1.

IOP Conf. Series: Earth and Environmental Science **851** (2021) 012019 doi:10.1088/1755-1315/851/1/012019

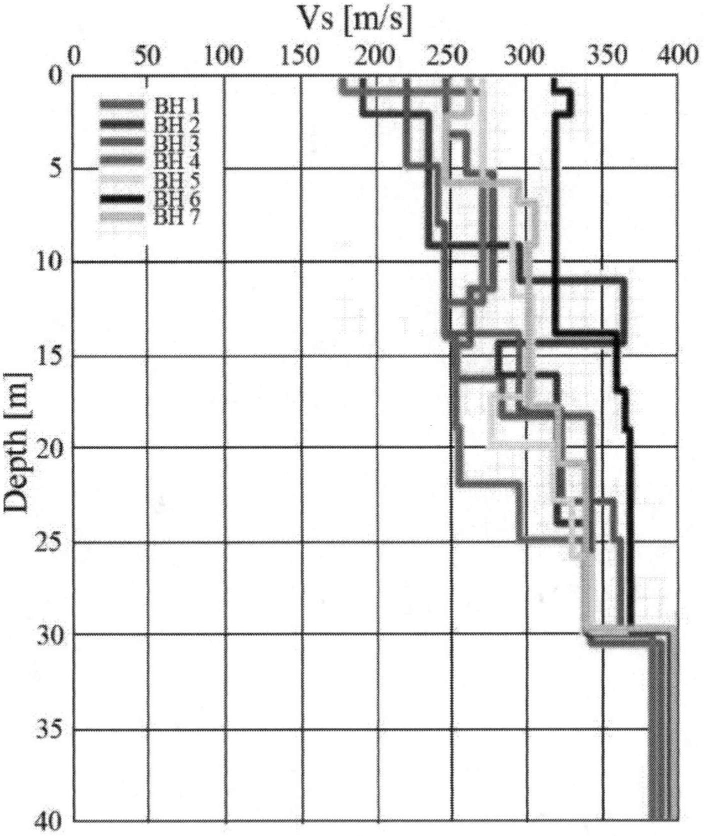

Figure 5. The evaluated Vs profiles and depth at Amarapura Township.

7. Modelling of Sedimentary Layered Media

The multiple reflection analysis was used to calculate the transfer function, which expresses the relation between the period and the corresponding magnification factor. Calculation of predominant period by using borehole data and the ground model profile is done according to the multiple reflection analysis. The governing equation is

$$\rho \frac{\delta^2 \mu}{\delta t^2} = G \frac{\delta^2 \mu}{\delta Z^2} + \eta \frac{\delta^3 \mu}{\delta Z^2 \delta t} \qquad (2)$$

Where, μ is the displacement of horizontal S-wave (SH), Z the direction of wave propagation (up-down), *t* the time, ρ the density, *G* the shear modulus and η the coefficient of visco-elasticity. The soil damping is considered by giving the complex value to the shear modulus and solving equation 2. The damping constant is 5% of critical damping for each layer [9]. The H/V spectral ratio of microtremor observation is frequently used for estimating the predominant period of the surface ground [10]. We estimate here the predominant period by calculating the transfer function of model ground based on the borehole data. The multiple reflection analysis is the linear analysis, however, above H/V ratio results also obtained as linear vibration phenomena. Therefore, we adopted this method for the determination of the ground motion parameter. Figure 6 shows examples of transfer functions.

IOP Conf. Series: Earth and Environmental Science **851** (2021) 012019 doi:10.1088/1755-1315/851/1/012019

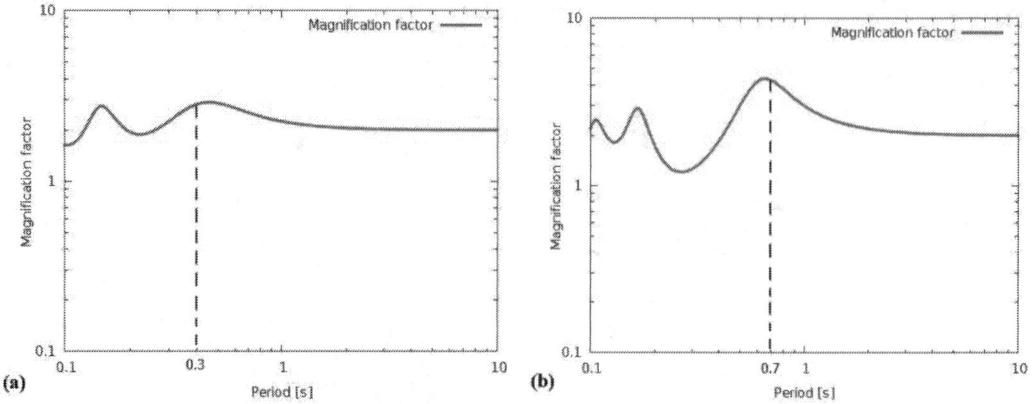

Figure 6. Example of the predominant periods. (a) Predominant period appears in a shorter period (b) predominant period appears in a longer period.

We could obtain Vs structures at microtremor survey. We proposed three-layer model with inversion methods in Amarapura township by using the microtremor sites (Figure 7). Periods parameters are computed for each target depth as equation 3 [11].

$$T = 4H / Vs \tag{3}$$

where, H is a thickness of a layer, Vs shear wave velocity and T predominant period. The distribution of sediment thickness map is shown in figure 8.

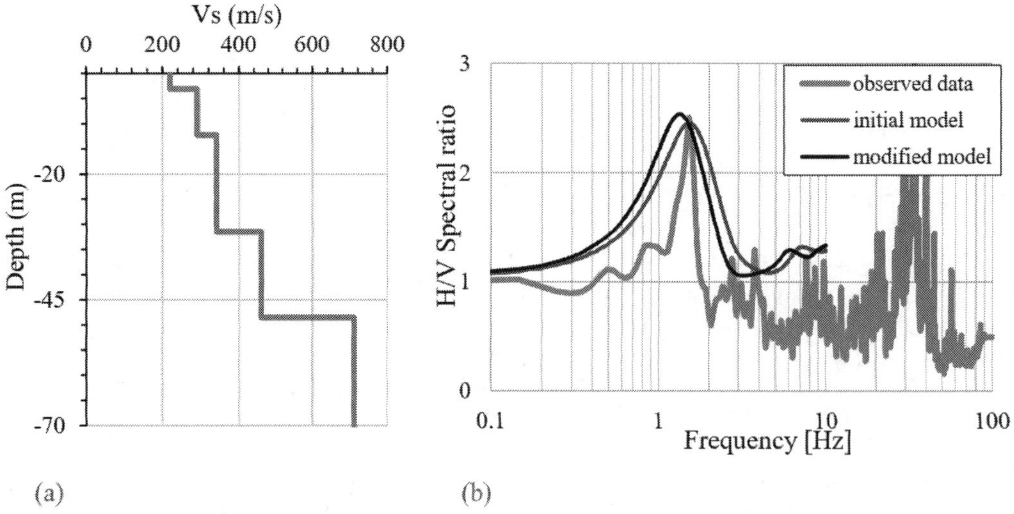

Figure 7. (a) The evaluated Vs profile from inversion of microtremor site no. 3, and (b) H/V Spectral ratio at microtremor site no.3.

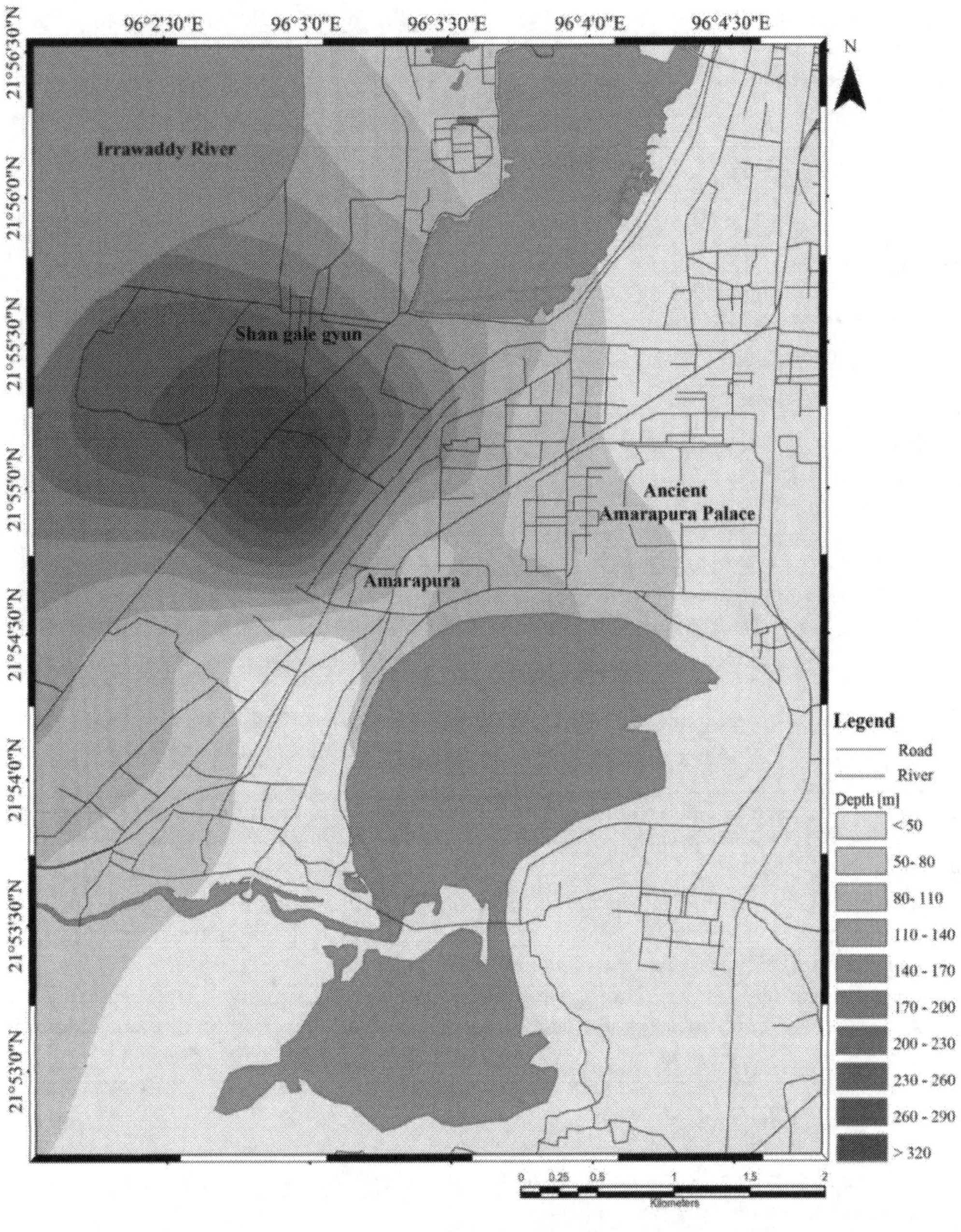

Figure 8. Sediment Thickness map of Amarapura Township, Mandalay region.

8. Conclusions

This research mainly focuses on the results of seismic microzonation based on 7 boreholes and 21 microtremor data, which covered almost the whole township area of Amarapura. Shear wave velocity (Vs^{30}), frequency distribution and sediment thickness are produced as the main results for seismic microzonation in Amarapura township. The kriging interpolation method was used for subsurface information such as predominant periods, frequency, shear wave velocity and sediment thickness with the GIS software [12]. The frequency distribution events ranged from less than 0.5 Hz to greater than 3 Hz and the Vs^{30} of the top layer is $Vs \leq 190$ m/s. The sediment thickness values ranged from less than 50 m to greater than 320 m. The deepest layer of sediment belongs to the Eastern parts of the Irrawaddy River and the thickness value is 110 – 320 m. This portion includes the Southern part of Shan gale gyun, the western parts of Amarapura and the ancient Amarapura palace. The outputs of this research would be very applicable for the seismic risk assessments, urban land use planning, retrofitting the various sorts of building after assessing the vulnerability, in designing the future construction of the various sorts of structures and earthquake preparedness, thereby reducing earthquake risk in Amarapura.

Acknowledgments
This research project is supported by Department of Higher Education, Upper Myanmar, Ministry of Education. We would also like to thank to all researchers who provided advice regarding the writing of this research paper.

References
[1] Swe W 1970 Rift-features at the Sagaing-Tagaung Ridge *Abstract, Burma Research Congr Rangoon* **101**.
[2] Thein M, Tint K and Aung A K 1991 On the lateral displacement of the Sagaing fault, *Georeports* **1** pp 23–34
[3] Kramer S L 1996 Geotechnical Earthquake Engineering Prentice-Hall International Series in Civil Engineering and Engineering Mechanics p 653
[4] Chibber H L 1934 *The Geology of Burma* Macmilland Com. and Limited, London p 538.
[5] Maung H 1987 Transcurrent movements in the Burma-Andaman Sea region *Geology* **15** pp 911–912
[6] Myanmar Geoscience Society 2014 *Geological map of Burma* Partly based on Geological Map of the Socialist Republic of the Union of Burma (1977 (Scale 1:1,000,000))
[7] http://www.esri.com/metadata/esriprof80.dtd
[8] Bernard R W, Jason T D and Thomas S 2012 Guidelines for Estimation of Shear Wave Velocity Profiles *Pacific earthquake engineering research center, PEER Report* 2012/08
[9] Thein P S, Kiyono J, Win T T, Nu T T and Aung D W 2018 Seismic microzonation of Mandalay City, Myanmar, *Journal of Geological Resources and Engineering* **6**, pp 1–13
[10] Nakamura Y 2000 Clear identification of fundamental idea of Nakamura's technique and its application *World Conference of Earthquake Engineering*
[11] Thein P S, Pramumijoyo S, Brotopuspito K S, Wilopo W, Kiyono J and Setianto A 2013 Investigation of subsurface soil structure by microtremor observation at Palu, Indonesia *the 6th ASEAN Civil Engineering Conference (ACEC) & ASEAN Environmental Engineering Conference (AEEC), Civil and Environmental Engineering for ASEAN Community, Bangkok, Thailand* p C–6
[12] Kiyono J and Suzuki M 1996 Conditional Simulation of Stochastic Waves by Using Kalman Filter and Kriging Techniques *Proc. of the 11th World Conference on Earthquake Engineering Acapulco Mexico* **1620**

IOP Conf. Series: Earth and Environmental Science **851** (2021) 012020 doi:10.1088/1755-1315/851/1/012020

Analysis of the distribution of seawater intrusion using electrical conductivity and total dissolved solid data and intelligence for residents in Sayung District, Demak Regency

K L Ariwibowo[1*], A M Riski[1], A S Fajarulloh[1] and T T Putranto[1]
[1]Department of Geological Engineering, Diponegoro University, Semarang, Jawa Tengah, Indonesia
*E-mail: kartikabasymeleh@gmail.com

Abstract. Massive groundwater extraction from production wells can trigger seawater intrusion, such as what happened in Sayung District, Demak. Seawater intrusion can be identified using quality analysis in the form of the electrical conductivity of each water sample taken. The research area is located in Sayung District, Demak Regency and is on the north coast, and is composed of alluvium quaternary deposits in the form of clay, sand sediments, and organic material of the plant. The purpose of this study is to determine the distribution of seawater intrusion in Sayung District and to know the prediction of the direction of seawater intrusion to provide an overview of groundwater quality recommendations in the study area to residents. To determine the distribution of seawater intrusion, you can use Electrical Conductivity and TDS (Total Dissolved Solids) data, from the results of the analysis there is a decrease in groundwater quality as indicated by the electrical conductivity value in the range 2,800-3,265 µmhos/cm with slightly brackish salinity and damage levels. categorized as critical and has a relatively high TDS value in the range 1,904-2,220.2 mg/L which is spread in the northern part near the coast including Timbulsloko and Sidorejo villages.

1. Introduction

Water is one of the most indispensable needs of every resident. With an increasing population every year, the increase in water demand will tend to increase. The dependence of the population on the use of groundwater is very high, this is due to the limited supply of groundwater which depends on the quality of groundwater, and the lack of interest in the population to use water by the PDAM (Local Water Supply Utility). It is feared that the increasing use of groundwater will exceed the limit of groundwater availability. Sayung District, Demak Regency is one of the coastal areas in the Central Java region, Indonesia (Figure 1.). This area has experienced events such as land subsidence as indicated by the relatively submerged houses of residents and there are cracks in road access and frequent tidal flooding with certain weather conditions (Figure 2.). The area is crossed by the sea so that the area is very potential for intrusion from seawater. For its use, the area has used groundwater from drilled wells, so that the water comes from a confined aquifer which can still be used for daily needs. One of the disasters that often occurs in coastal areas is seawater intrusion. Where in that area, groundwater should be used for community use but cannot be used and is not suitable for use because the composition of groundwater has been changed by seawater intrusion. Seawater intrusion often occurs in areas close to the sea, which is caused by several factors, which can occur due to excessive groundwater extraction or

IOP Conf. Series: Earth and Environmental Science **851** (2021) 012020 doi:10.1088/1755-1315/851/1/012020

empty voids in a lithology. Sayung sub-district is a coastal area where one of the potential disasters is seawater intrusion. In this area, the sea level is almost approaching the plains in the sub-district. The purpose of the research conducted in Sayung sub-district is to determine the zoning of seawater intrusion in Sayung sub-district and provide recommendations to minimize the occurrence of intrusions. From this goal, it is hoped that it can be useful for something that may happen in the future.

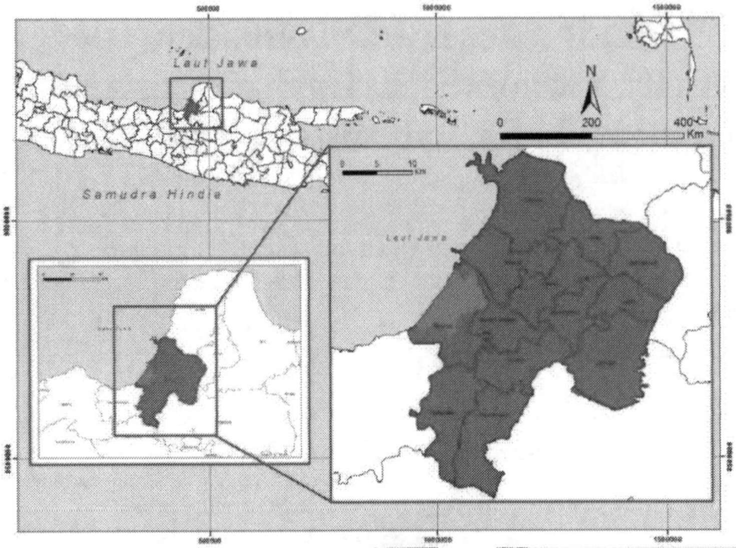

Figure 1. The location of Sayung District, Demak Regency, Central Java, Indonesia.

Figure 2. (A) The condition of the house which is affected by land subsidence; (B) The condition of the research area affected by tidal flooding.

2. Regional Geology

The regional geological conditions of the study area are included in the Magelang Semarang sheet [1] and the Kudus sheet [2]. Based on this, the oldest to youngest regional geological conditions are composed of two types of rock, namely Tertiary sedimentary rocks, namely the Kerek and Kalibeng Formations, then the Damar Formation and the Kaligetas Formation which are Quaternary age and surface sediments (alluvium) which are Holocene age. The Kerek and Kalibeng Formations are predominantly composed of claystone, marl, and local limestone. The formation is located on low to medium slope morphology. The Damar Formation is composed of conglomerates, tuff sandstones. Volcanic breccias are in the form of lava and are located in morphology with moderate slopes, while the Kaligetas Formation is composed of dominant volcanic products such as volcanic breccias, lava, and tuff sandstones, and local claystone. This formation is spread over morphology with moderate to high slopes. The plains area is dominated by coastal alluvium deposits, rivers, and lakes, consisting of clay, silt, sand, and gravel [3].

3. Methodology

In doing research on the distribution of seawater intrusion that occurs in Demak, some data is needed including geological data of the research area, measurement of elevation for taking wells, and depth data of boreholes to calculate the value of the groundwater level using MS Excel 2013 software. In addition, water samples were taken to test their electrical conductivity using an electrical conductivity meter. Then the data that has been obtained will be processed using Surfer 13 to find out 3D conditions under the surface and using ArcMap 10.5 software to determine the distribution of each parameter needed in determining the distribution of intrusion, including geological maps, electrical conductivity distribution map, total dissolved solid distribution map, groundwater map. To analyze the distribution of each map using the IDW (Inverse Distance Weighted) data interpolation method is a simple method of determination by considering the surrounding points [4], where the interpolation value will be more similar to the closer sample data [5].

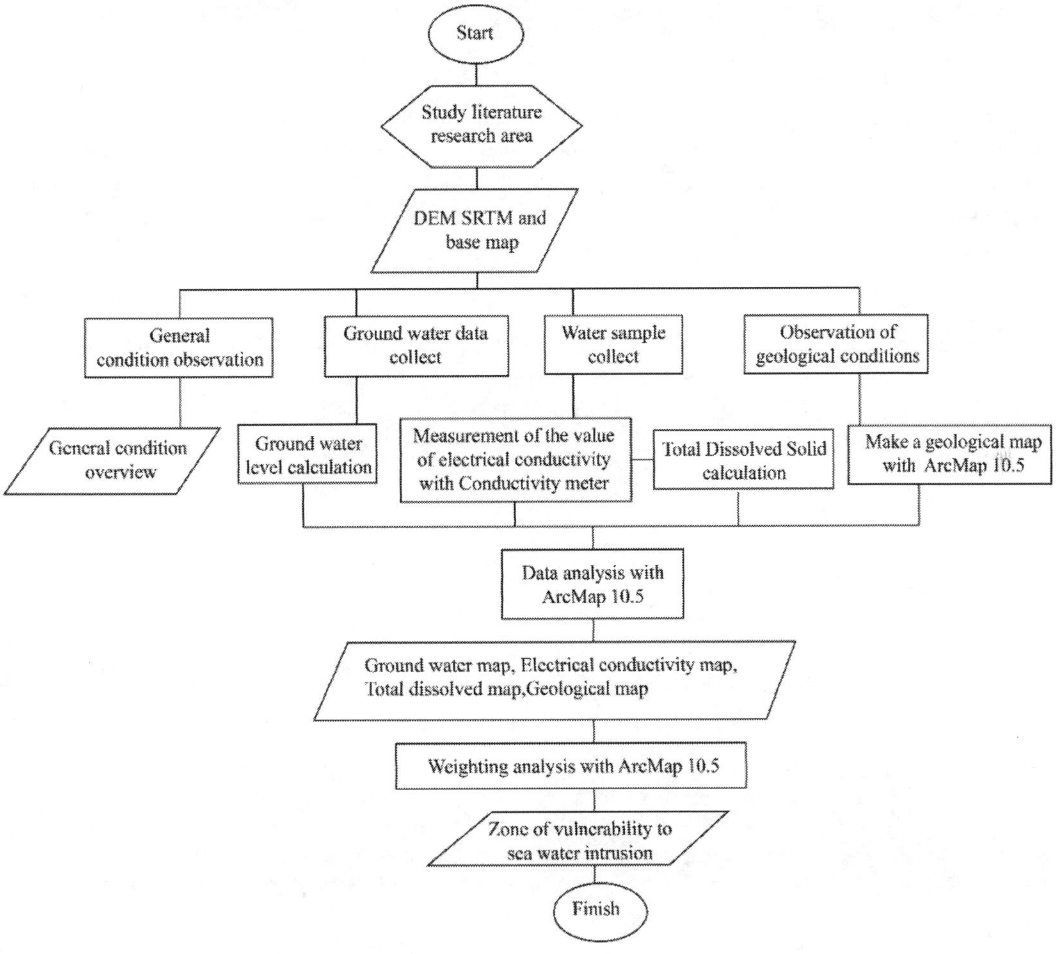

Figure 3. Flowchart of the research.

4. Result and Discussion

4.1. Geological Conditions

The state of lithology in Sayung sub-district is an alluvium lithology. This area is a coastal area of the Central Java region. The lithology is a collection of several materials transported from the area.

Figure 4. The research area there are alluvium deposits.

Based on the regional geology, the area is included in the alluvium formation which is in the north and approaches the Java Sea. The lithological composition in the area is the presence of clay minerals with the size of clay, sand grains, clastic sedimentary material, and plant deposits as well as the presence of organic material because the area is close to the swampy area of origin.

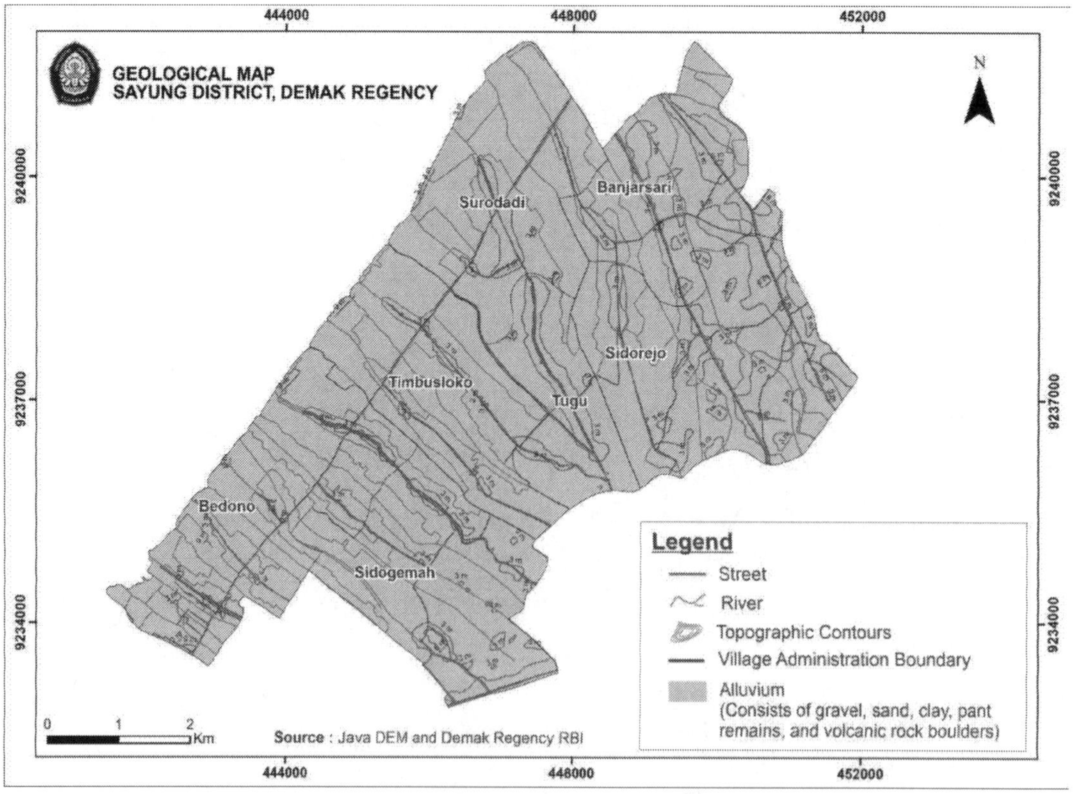

Figure 5. The research area there are alluvium deposits.

4.2. Groundwater Elevation Map

The state of groundwater elevation in the research area is by examining the depth of the borewell because the research area has not used dug wells for its water source so that it does not find dug wells to determine the groundwater level. Groundwater elevation data was collected directly using field survey methods and interviews with residents who owned boreholes. So that it can be estimated about the depth of the wells in the study area. In the study area, it was found that the depth range of the groundwater elevation table was 60 to 120 meters below the surface (Figure 6B).

At the groundwater level, it forms a flow pattern leading to the respective wells (Table 1.). In the southwest part of the research area, the direction of the flow of the groundwater elevation is towards the sea, while in the eastern area of the study area the direction of flow is dominant towards the northeast to the north of the study area. Based on the area, the direction of groundwater flow in the confined aquifer is towards Bedono and partly towards the Banjarsari sub-district (Figure 6A). A 3-dimensional illustration of the high and low groundwater level conditions can be seen in Figure 6A. If the red area shows that the groundwater level is high or relatively close to the surface and will flow to the lower area, it is shown in blue The lowest contour value is -130 meters while the contour value the highest is -50 meters (Figure 6A).

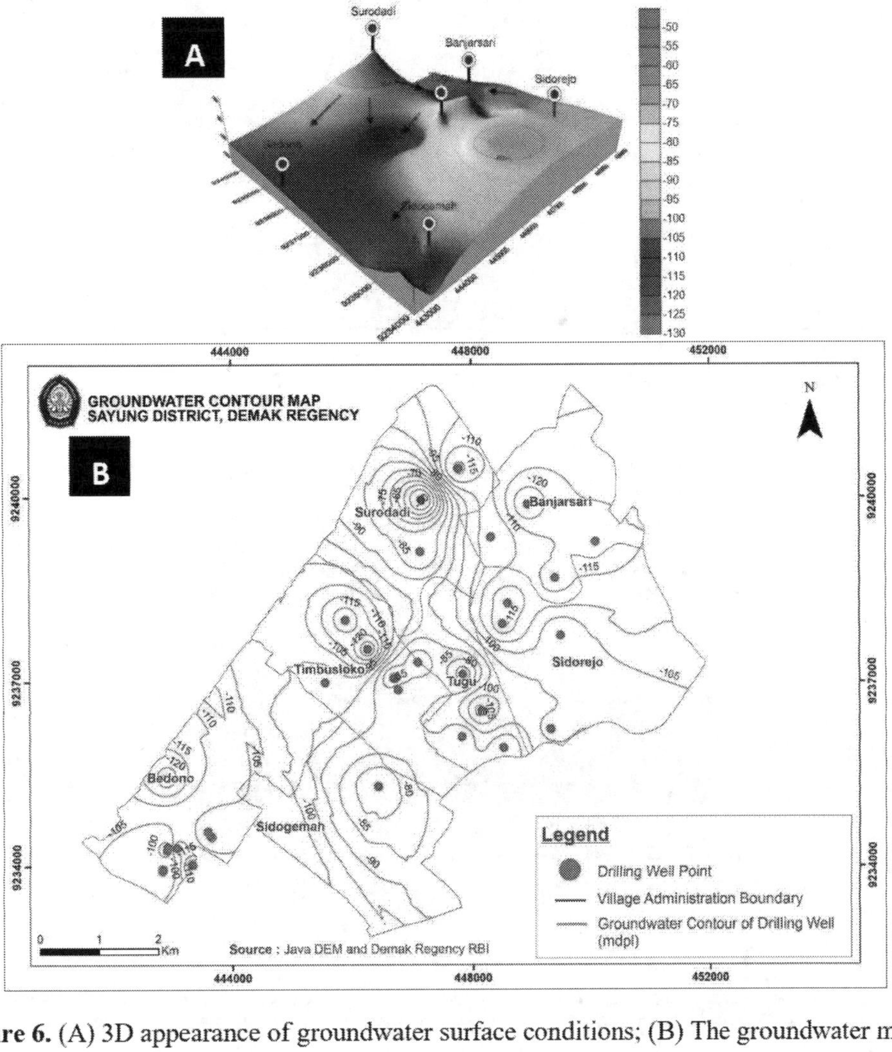

Figure 6. (A) 3D appearance of groundwater surface conditions; (B) The groundwater map of the research area.

Table 1. Groundwater surface coordinate point in the study area.

Coordinate X	Coordinate Y	Groundwater Level (m)
447818	9236106	90
446770	9236873	90
445895	9238004	120
443071	9234311	120
442922	9234320	100
443599	9234572	110
443641	9234502	100
442905	9235442	125
442908	9234265	100
442842	9233944	100
443335	9234028	125
448531	9235928	90
448196	9236523	100
447846	9237126	82
447101	9237319	85
449488.6	9237752	102
449324.8	9236230	105
448160.7	9236529	120
448516.2	9237942	120
448942.5	9239884	125
450071.7	9239277	120
449400.8	9238690	120
448604.5	9238282	125

Figure 7. One of the drilled wells in the study area.

4.3. Electrical Conductivity and Total Dissolved Solid

Based on the value of Electrical Conductivity and Total Dissolved Solid (TDS), the research area includes groundwater quality in the form of fresh and slightly brackish water [6]. The classification is based on the value obtained from the measurement results of water samples in boreholes in the study area. Based on the results (Table 2) which have been presented, it can be seen that the distribution of groundwater content with a high electrical conductivity value is shown in dark green (Figure 8B) and a high TDS value is shown in dark red (Figure 9B), the distribution of these conditions is In Sidorejo Village, Tugu Village and Timbusloko Village, this is interpreted by the quality content of groundwater that has been mixed with seawater or it can be indicated that the area has experienced seawater intrusion and is supported by the groundwater content in the area which is included in the type of groundwater quality which is rather brackish. In addition, in several villages, namely Banjarsari Village, Sidodadi Village, Bedono Village, and Sidogemah Village, they tend to have a low electrical conductivity value content as shown in light green (Figure 8B.) and have a low TDS value content as shown in pink. (Figure 9B.), this shows that in these areas the quality of groundwater contained is still not many signs of seawater intrusion and based on its quality it is still included in freshwater.

Table 2. TDS coordinate points and electrical conductivity in the study area.

Coordinate X	Coordinate Y	TDS (mg/L)	Electrical Conductivity (µmhos/cm)
443071	9234311	465.8	685
442922	9234320	541.28	796
443599	9234572	469.88	691
443641	9234502	999.6	1,470
442905	9235442	553.52	814
442908	9234265	394.4	580
442842	9233944	425.68	626
443335	9234028	427.04	628
448531	9235928	705.16	1,037
448196	9236523	577.32	849
447846	9237126	687.48	1,011
447101	9237319	1904	2,800
449489	9237752	2220.2	3,265
448161	9236529	490.96	722
448516	9237942	2097.8	3,085
448943	9239884	732.36	1,077
450072	9239277	867	1,275
449401	9238690	592.28	871
448605	9238282	636.48	936

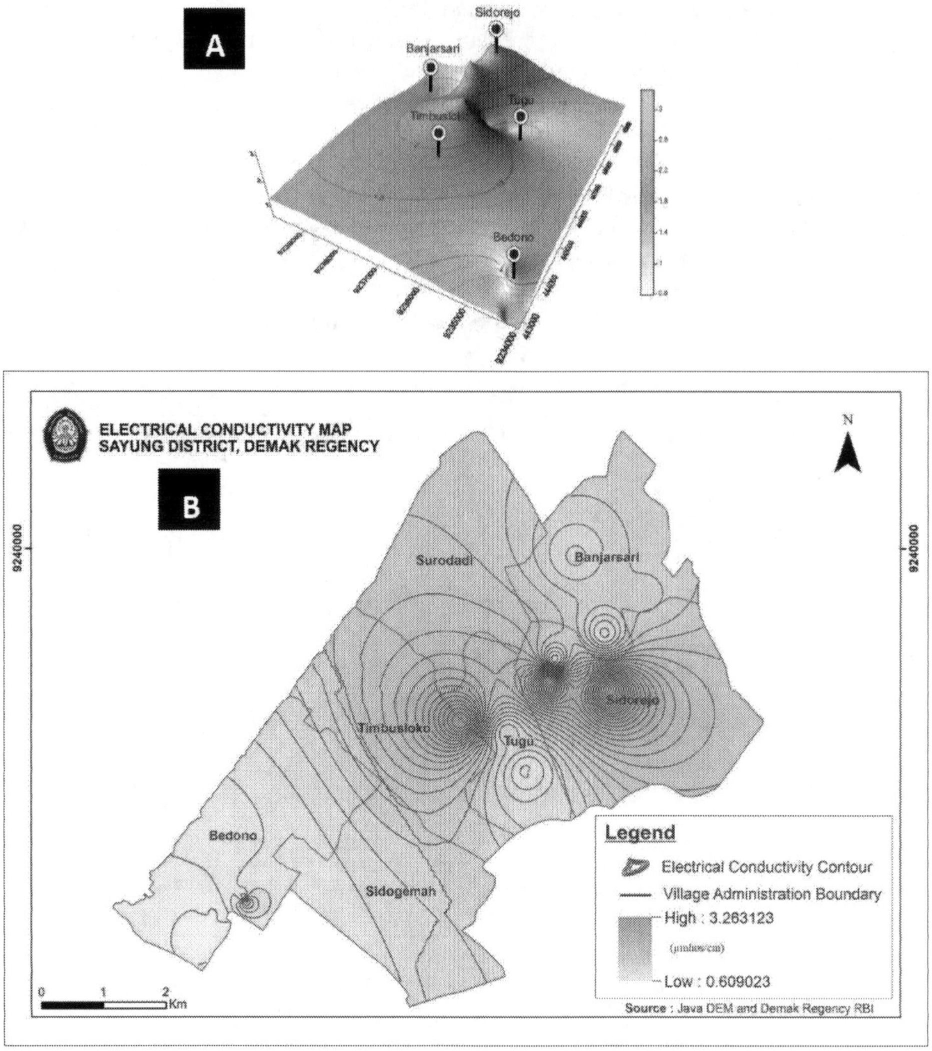

Figure 8. (A) 3D appearance of electrical conductivity value; (B) The electrical conductivity map of the research area

4.4. Seawater Intrusion Vulnerability Zone Analysis

Seawater intrusion is the infiltration of seawater into the ground or the process of entering seawater into groundwater. This intrusion can be identified using data from electrical conductivity and TDS (total dissolved solid). If the values of electrical conductivity and TDS are in the high category, there will be a decrease in groundwater quality. Usually, seawater intrusion is found in wells near the coast which is triggered by large amounts of groundwater withdrawal [7]. Analysis of the seawater intrusion vulnerability zone uses weighting analyzed by ArcMap 10.5.

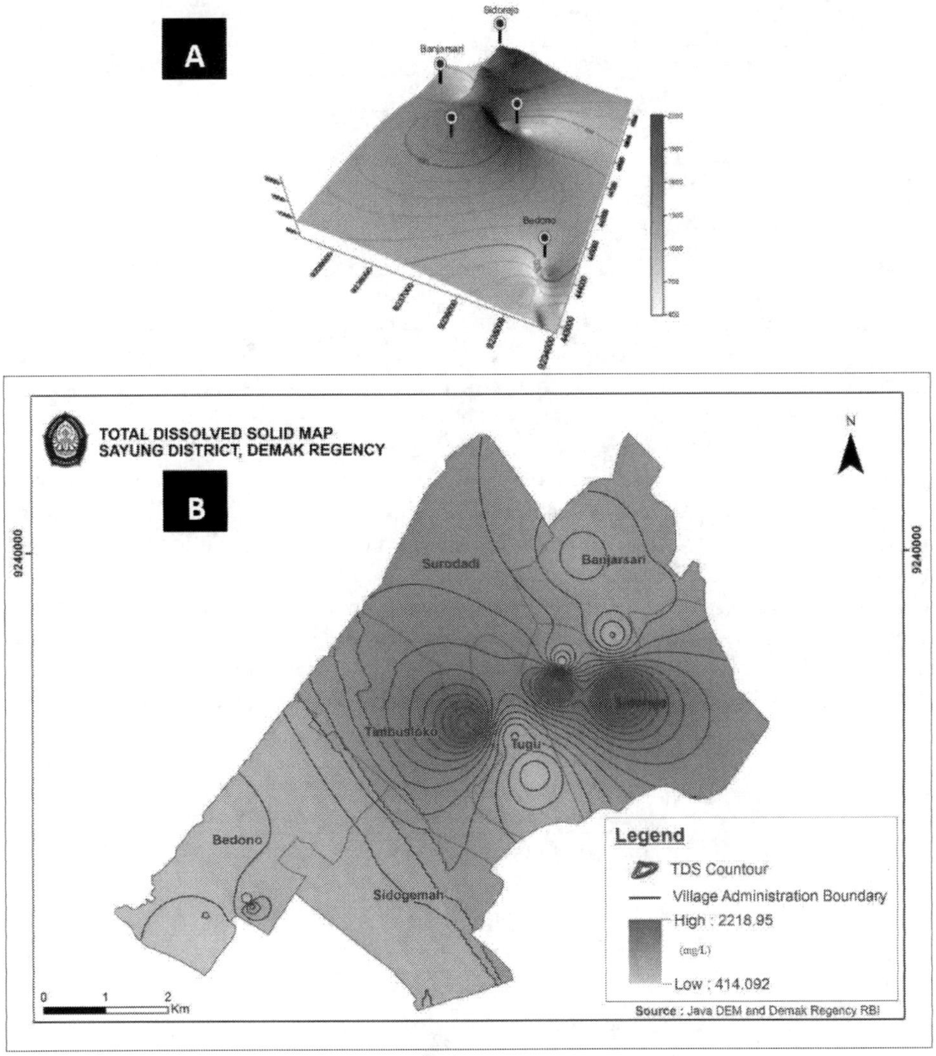

Figure 9. (A) The total dissolved solid map of the research area; (B) 3D Appearance of total dissolved value

The parameters used in the weighting include the distribution of lithology, distribution of electrical conductivity values in the study area, and distribution of TDS values. The values of electrical conductivity and TDS measured in the field are in class 2 and class 1 (Table 3.). The measured electrical conductivity values ranged from 580 – 3,265 µmhos/cm, while the TDS ranged from 394.4 to 2,220.2 mg/L. So, that is categorized into 2 classifications of vulnerability, namely class 1 which is categorized as safe, and class 2 which is categorized as vulnerable. Meanwhile, the lithological condition in the form of alluvium is very influencing so that it has the highest value. To produce a visualization map of the intrusion distribution, the map that has been classified into 2 classes of each parameter is overlaid using a weighted percentage.

Table 3. Weighted value of electrical conductivity and Total Dissolved Solid

Electrical Conductivity (µmhos/cm)	TDS (mg/L)	Value of weighting calculation
<1,500	<1,000	Class 1
1,500-5,000	1,000-3,000	Class 2
5,000-15,000	3,000-10,000	Class 3
15,000-50,000	10,000-35,000	Class 4
>50,000	>35,000	Class 5

The highest percentage of weighting (Table 4.) is 40% for lithology maps, 30% for electrical conductivity maps, and 30% for TDS maps. The lithology map has the greatest influence because the lithology of the study area consists of alluvium deposits which form swampy areas that are inundated by seawater, brackish water, and freshwater so that it further accelerates the aquifer layer to be entered by water with high salinity. The weighted value of electrical conductivity and TDS has the same value, namely 30%, the two values are related to each other, water with high salinity will more easily flow electric current so that the electrical conductivity is high. While the TDS value is obtained from the calculation of the electrical conductivity value formula. From the calculation analysis using ArcMap 10.5, it is obtained 3 classifications that identify that there are 3 seawater intrusion zoning in the study area including safe, moderate, and trouble zones. The safe zone is marked with a green area covering the villages of Banjarsari, Tugu, Sidogemah, and Bendowo. The zone being shown is a yellow area covering the northern part of Surodadi Village. Meanwhile, the vulnerable areas marked with red residents include the villages of Surodadi, Timbulsloko, and Sidorejo (Figure 10.).

Table 4. Weighted value for sea water intrusion zoning

Parameter	Percentage
Lithology	40%
Electrical Conductivity	30%
Total Dissolved Solid (TDS)	30%

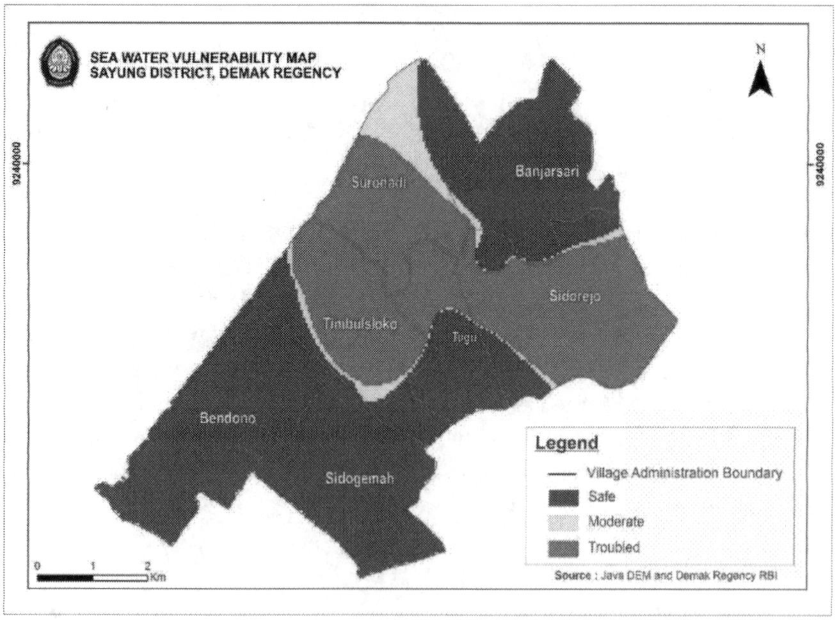

Figure 10. Seawater Intrusion Vulnerability Zone Map

4.5 Simple Water Treatment Intelligence

The condition around the research area is an area close to the north coast and areas that are mostly surrounded by swamps or areas of brackish water and freshwater. Residents' houses are built on areas that are inundated by water and residents build houses using the landfill method. However, this area is an area of tidal flooding or seawater entering the land due to the high sea level [8]. Geologically, this area is composed of alluvial deposits so that the soil conditions in the Sayung area are quite unstable and land subsidence is marked by several resident's houses that are abandoned due to drowning, as well as several residents' houses that have to be landfilled every year.

Based on these conditions, Sayung residents can only get water sources from confined aquifers or use artesian wells to get clean water. In fact, there are several wells that produce water with the physical characteristics of the water being slightly yellow and tasting salty. Based on interviews with the Village Head of Sidorejo, some residents often find it difficult to find water for drinking and require buying water with unstable economic conditions, this is quite a problem in several villages in Sayung District.

After studying social problems in the research area, a simple right solution that can help residents around the research area is considered. Identified groundwater is water with brackish characteristics in Tugu Village and Sidorejo Village based on the PAHIAA classification. Brackish water tends to have a high solute content, the chemical content of brackish water includes sodium (Na^{+}), potassium (K^{+}), calcium (Ca^{2+}), magnesium (Mg^{2+}), chloride (Cl^{-}), sulfate (SO_4^{2-}), and bicarbonate (HCO_3^{-}) in water [9]. The salty taste of brackish water is caused by the high salinity of the water, if water is consumed daily it can cause health problems in the future. So it is necessary to handle it so that water can still be used for drinking needs. In addition, the existence of this simple filter reactor can help residents understand that surface water that has not been contaminated by chemical waste can still be used for bathing needs can be filtered so that this recommendation can reduce water intake through deep wells (bores) to suppress seawater intrusion. This treatment can take advantage of the adsorbent properties of natural zeolites. Zeolite has a high adsorbing ability because it has pores that can exchange high cations, which are hydrated alumino-silicates with cations of sodium, potassium, and barium [10]. The filter reactor to filter brackish water requires simple materials using activated charcoal, zeolite, silica sand, and gravel (Figure 11D.). These materials are made for filtration prototypes with a ratio of charcoal, zeolite, sand, gravel is 4: 4: 3: 2 [11].

Socialization in an effort to educate residents in the research area is descriptive of the results of data processing carried out by the delivery method in the form of door-to-door with the aim of implementing health protocols during the COVID-19 pandemic. In addition, this effort is very helpful in providing insight to residents regarding the condition of the research area that has been affected by the seawater intrusion process so that it can provide recommendations in order to minimize the use of boreholes which are interpreted as sustainable intrusions in the future (Figure 11.).

5. Conclusion

The research area which is included in the coastal area of Sayung District, Demak Regency generally has lithological conditions in the form of alluvium deposits and is supported by field conditions that experience subsidence of land level, tidal flooding, and roads that have cracks. Based on field data processing, the value of electrical conductivity, and total dissolved solids, there are three well points that have been indicated to experience seawater intrusion, namely around the villages of Sidorejo, Tugu, and Timbulsloko. To prevent and minimize the occurrence of sustainable seawater intrusion, intelligence and understanding of seawater intrusion and recommendations for the use of reactor filters are carried out.

Figure 11. Socialization with the door to door method in the research area. (A) The process of water filtration using a simple reactor filter, (B) The process of intelligence and socialization of sea water intrusion to residents of Sayung District, (C) The process of intelligence and socialization of sea water intrusion to residents of Sayung District, (D) Design of a filter or simple reactor filter.

Acknowledgments
We would like to thank the lecturers of the Geological Engineering Department, Faculty of Engineering, Diponegoro University who have guided the authors to carry out this research.

References

[1] Thanden and Sumadirja 1996 *Peta Geologi lembar Magelang dan Semarang* (Bandung: Direktorat Badan Geologi ESDM)
[2] Suwarti T and Wikarno R 1992 *Peta Geologi Lembar Kudus Skala 1 : 100.000* (Bandung: Pusat Penelitian dan Pengembangan Geologi)
[3] Marfai M A Cahyadi A Krisnantara G Gustiar G G 2015 *Karakteristik Hidrogeokimia Airtanah di Pesisir Kabupaten Demak, Jawa Tengah* (Seminar Nasional Pekan Ilmiah)
[4] NCGI 2007 Interpolation: Inverse Distance Weighting www.ncgia.ucsb.edu/pubs/
[5] Pramono G H 2008 Akurasi metode IDW dan kriging untuk interpolasi sebaran sedimen tersuspensi. *In Forum geografi* **22** L 97-110
[6] PAHAAI 1986 *Klasifikasi Keasinan Perairan* (Jakarta: Panitia Ad Hoc Intrusi Air Asin Jakarta)
[7] Waspodo R S B Kusumarini S Dewi V A K 2019 Prediksi Intrusi Air Laut Berdasarkan Nilai Daya Hantar Listrik dan Total Dissolved Solid di Kabupaten Tangerang. *Journal of Agricultural Engineering*, **8** L 243-250
[8] Rahmawan L E Yuwono B D Awaluddin M 2016 Survei Pemantauan Deformasi Muka Tanah Kawasan Pesisir Menggunakan Metode Pengukuran GPS Di Kabupaten Demak Tahun 2016 (Studi Kasus: Pesisir Kecamatan Sayung, Demak). *Jurnal Geodesi Undip*, **5(4)**L 4455

[9] Aziza F N Latifah L Kusumastuti E 2014 Pemanfaatan Zeolit Alam Teraktivasi Ammonium Nitrat (NH4NO3) Untuk Menurunkan Salinitas Air Sumur Payau. *Indonesian Journal of Chemical Science*, **3(3)**

[10] Khiqmah N V 2015 *Pengembangan Teknik Adsorpsi dengan Menggunakan Ion Exchanger Berbasis Zeolit-Karbon Aktif untuk Produksi Air Sanitasi (Development of Adsorption Techniques Using the Ion Exchanger Zeolite-Based Activated Carbon for the Production Sanitation Water)* (Semarang: Doctoral dissertation, Diponegoro University)

[11] Hamidah L N Rahmayanti A 2018 Pemanfaatan Zeolit dan KarbonAktif dalam Menurunkan Jumlah Bakteri pada Filter Pengolah Air Payau. *In Conference Proceeding on Waste Treatment Technology* **1** L 113-118

IOP Conf. Series: Earth and Environmental Science **851** (2021) 012021 doi:10.1088/1755-1315/851/1/012021

Soil stiffness identification using fuzzy logic based on seismic tomography and its relationship with dynamic elastic moduli

L Yufajjiru[1], M Maryadi[1,2*]

[1]Undergraduate Program of Geophysics, Faculty of Mathematics and Natural Sciences, Universitas Indonesia, Depok 16424, Indonesia
[3]Department of Earth Resources Engineering, Graduate School of Engineering, Kyushu University, Fukuoka, 819-0395, Japan.
*E-mail: maryadi@sci.ui.ac.id

Abstract. Either P-wave or S-wave tomography has been used to interpret soil stiffness zone to extend subsurface conditions' visibility. The dynamic elasticity moduli have been calculated to connect both seismic wave velocities to geotechnical rationale. However, the interpretation has to be more integrated and less subjective. Fuzzy logic has been applied for both velocities fusion to generate geotechnical cross-sections to diminish the subjectivity. Nevertheless, this is a new application, and it is not known whether this result has a relation to the dynamic elasticity moduli or not. Therefore, this research uses Fuzzy Logic to identify the soil stiffness zone and its relationship to the dynamic elasticity moduli. Meta-hypotheses are produced by possibility function in the early phase of the fusion. The possibility function is a fuzzy set that describes each seismic wave velocities to a geotechnical rationale in a possibility scale. Meta-hypotheses fused into hypotheses: the possibility of solid-state behavior and plastic-state behavior. It is found that solid-state behavior is analog to Bulk modulus, and the plastic state is analog to the elastic modulus. Both hypotheses' extreme values are highlighting the target based on the geotechnical rationale from the possibility function.

1. Introduction

Soil investigation has been being used in the early phase of civil works. It needs plenty of either a standard penetration test (SPT) or the other penetration test points in an extensive area. These kinds of jobs are costly. We need to map at a larger scale to makes penetration tests are more targeted.

Seismic refraction is the most common geophysical tool to maps the P-wave velocity in the geotechnical assessment. However, this methodology is not directly indicated the soil stiffness. The S-wave velocity is meant to be the best indicator seismically [5]. The shear modulus is directly linked to the material stiffness. Multichannel Analysis Surface Wave (MASW) has been used to maps the S-wave velocity by using the Rayleigh wave dispersion.

The dynamic elasticity moduli have been calculated to interpret both seismic wave velocities for the geotechnical rationale. These moduli required a density model to produce. However, the interpretation has to be more integrated and less subjective. Fuzzy logic has been applied to diminish the subjectivity by fusing both seismic wave velocities. The first application has shown that it generates a geotechnical cross-section without creating any initial model [2]. Nevertheless, it is not known whether this result has a relation to the dynamic elasticity moduli or not. Therefore, this research uses Fuzzy Logic to identify soil stiffness zone from geotechnical cross-section and find its relationship to the dynamic elasticity moduli. A complete workflow of this study is described in figure 1.

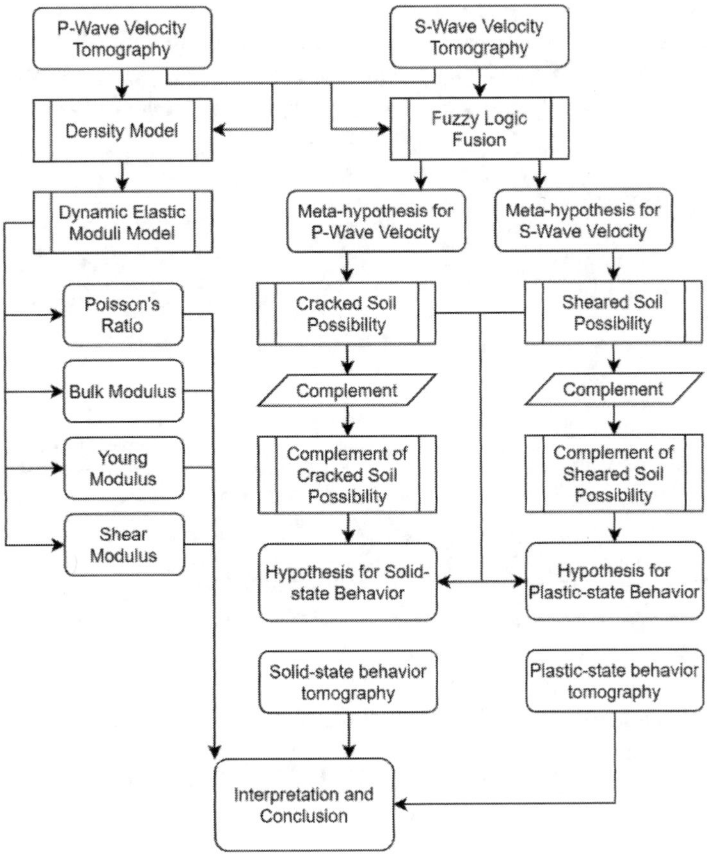

Figure 1. Workflow diagram.

2. Seismic Refraction

Seismic refraction is an active seismic method that used critically refracted waves for analysis. Head wave is a wave that returns to the surface when a seismic wave is being critically refracted. The critical angle is an incoming angle that made the refraction angle reach 90° when the wave hits a new medium with speed more significant than the previous medium. The medium which has a lower velocity or slightly different than a previous medium is not detected or called a blind zone [6]. The P-wave velocity tomography is generated by the general reciprocal method (GRM). The GRM method has excellent lateral difference detection, which good for geotechnical investigation [6].

3. Multichannel Analysis Surface Wave

Multichannel analysis surface wave is a methodology that generates S-wave velocity from surface waves, particularly the Rayleigh wave. The Rayleigh wave has the most considerable energy recorded in seismic traces. It could 67% for Rayleigh wave, 23% for S-wave, and 7% for P-wave [5]. Thus Rayleigh wave could serve as an excellent signal to noise ratio. Another surface wave velocity could invert to S-wave velocity. The Love wave velocity has equal to the S-wave velocity, and the Rayleigh wave velocity is about 91% of the S-wave velocity. However, the Love wave velocity characteristic made it does not always have been recorded in vertical geophones, and the Rayleigh wave has been

IOP Conf. Series: Earth and Environmental Science **851** (2021) 012021 doi:10.1088/1755-1315/851/1/012021

consistently appearing in any seismic recordings as a ground roll. Thus still, the Rayleigh wave is favorable to obtains S-wave velocity in this research.

The Rayleigh wave dispersion analysis has produced the S-wave velocity. The Rayleigh wave is dispersed in a non-homogeneous half-space. Three parameters defined the dispersion property, such as S-wave velocity, Poisson's ratio, and density. The S-wave velocity is the most affecting the dispersion up to 98% dependency [5]. The theoretical dispersion curve models the S-wave velocity over a thousand iteration that matches the extracted dispersion curve.

4. The Dynamic Elastic Moduli

The dynamic elastic moduli consist of an elastic modulus, Bulk modulus, Shear modulus, and Poisson's ratio derived from wave velocity.

$$\sigma = \frac{vp^2 - 2vs^2}{2(vp^2 - vs^2)} \tag{1}$$

$$G_{\max=\rho} \times V_s^2 \tag{2}$$

$$E = G\left(\frac{3V_p^2 - 4V_s^2}{V_p^2 - V_s^2}\right) \tag{3}$$

$$K = \rho\left(V_p^2 - \frac{4}{3}(V_s^2)\right) \tag{4}$$

σ = The Poisson's ratio,
G_{max} = Maximum shear modulus,
E = Elastic modulus,
K = Bulk modulus,
vp = P-wave velocity, and
vp = S-wave velocity.

5. Fuzzy Set

The set is a collection that defining an object. Thus every object could be differentiated from each other from the set. A classical set has a crisp boundary. This classical set could not recognize an object if we have a lack of information. It does not seem enough to satisfy the object definition. This lack of information condition made a "fuzzy" condition.

A fuzzy set defines how "fuzzy" the object is. An example is when we talk about how "tall" is someone, so we need a reference for the tallness. This tallness does not have an absolute value. Thus it is defined by a degree of tallness from the reference. A fuzzy set defines an object through the fuzzy membership. There are many fuzzy membership functions. In this research, the trapezoidal function is used, as shown in Equation 5.

$$\mu_x \begin{cases} \frac{x-a}{m-a}, & a < c \leq m \\ 1, & m < x \leq n \\ \frac{b-x}{b-m}, & n < x \leq b \end{cases} \tag{5}$$

x = The condition's universe,
μ_x = Fuzzy set in x universe,
a = The first boundary of the lowest value of the degree,
m = The first boundary of the highest value of the degree,
n = The end boundary of the highest value of the degree,
b = And the end boundary of the lowest value of the degree.

6. Fuzzy Operator

6.1. And/Intersection

$$(A \cap B)(x) = min\ (A(x), B(x)) \tag{6}$$

The standard intersection has shown in Equation 6. The $(A \cap B)(x)$ is the notation of the intersection of set A and set B in the x universe. The illustration of an intersection is shown in figure 2a. There are the non-standard intersections, as described in Equation 7 – Equation 9.

Algebraic Product : $\mathrm{I(a,b)} = a.b$ (7)

Bounded Difference : $\mathrm{I(a,b)} = \max(0, a + b - 1)$ (8)

Drastic Intersection : $\mathrm{I}_{min}(a.b) = \begin{cases} a\ when\ b = 1 \\ b\ when\ a = 1 \\ 0\ others \end{cases}$ (9)

6.2. Complement

Complement is an inverse of an element in a set. The complement operation is shown in Equation 10. The illustration of complement is shown in figure 2b.

$$\bar{A}(x) = (1 - A(x)) \tag{10}$$

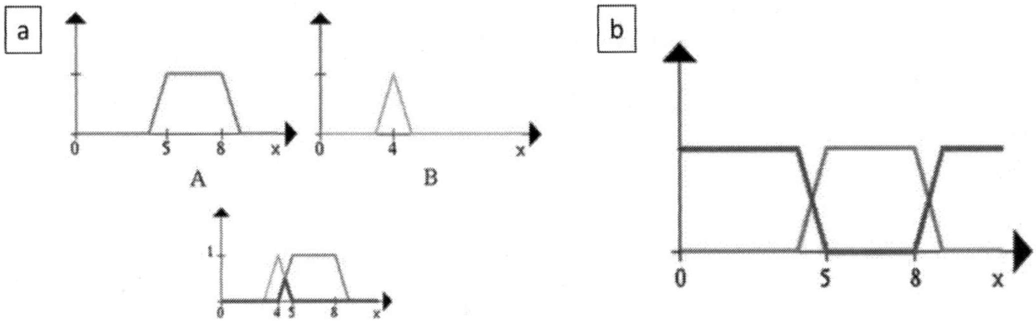

Figure 2. (a) Intersection of fuzzy sets and (b) operation of complement.

7. Fusion Operator

$$\pi_{(x)} = \pi_{1_{(x)}} \otimes \pi_{2_{(x)}} = \frac{\pi_{1_{(x)}} \wedge \pi_{2_{(x)}}}{\sup\left(\pi_{1_{(x)}} \wedge \pi_{2_{(x)}}\right)} \tag{11}$$

A fusion operator is an operator that fusing several pieces of information into new information.

$\pi_{(x)}$ = A result set,
$\pi_{1_{(x)}}$ = First set,
$\pi_{2_{(x)}}$ = Second set,
$\otimes$ = The fusion operator,
$\wedge$ = The intersection operator,

sup = And supremum.
Supremum is the maximum of the list of intersection sets, which is used for normalizing the results. The maximum value in the set turns into "1" and references the other value.

8. Meta-hypotheses and Hypotheses

Meta-hypotheses define each velocity tomography to the geotechnical definition. We define that when the velocity below 500 m/s, the soil has been cracked. When it exceeds 1500 m/s, the soil has not been cracked. We referenced these definitions as a possibility of cracked soil, as shown in figure 3a. The geotechnical definition for S-wave velocity is when it goes below 500 m/s, the soil has been sheared, and when it surpasses 1000 m/s, the soil has not been sheared and stated as a possibility of sheared soil as shown in figure 3b. These meta-hypotheses posed as a first step calculation from physical value-form into fuzzy logic operation. It is behaved as a translator from seismic wave velocity into physical reason in geotechnical properties.

The hypotheses are the form of physical inference of our fusion through the formula. The formulas of those hypotheses are a primary key in this fusion process. It drives away the inputs into a new form of tomography corresponding to the formula's physical rationale. The hypotheses serve a new form of tomography from the fusion of several geophysical data. In this case, we are using P-wave velocity and S-wave velocity tomography. It combines those meta-hypotheses using the following general equation [3]:

$$\pi = \pi_1 \otimes \overline{\pi_2} \tag{12}$$

π = hypothesis
π_1 = First meta-hypothesis
$\overline{\pi_2}$ = Second complement meta-hypothesis
$\otimes$ = Fusion operator

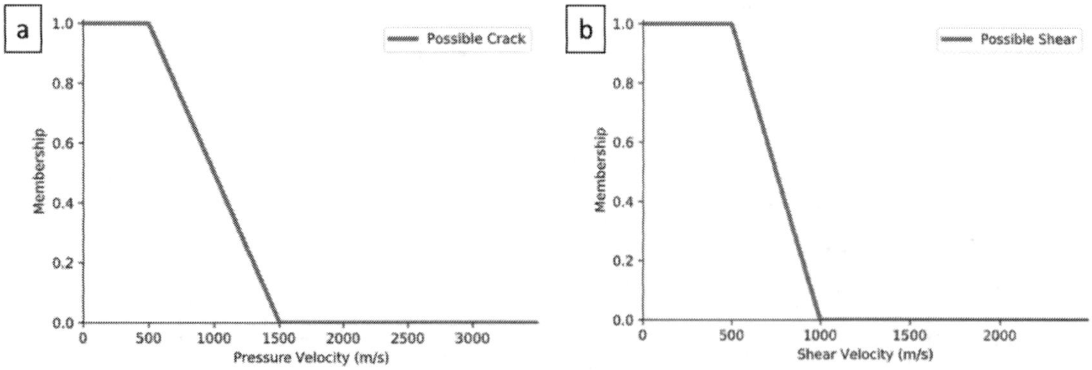

Figure 3. Fuzzy set for possibility of (a) cracked soil and (b) sheared soil meta-hypothesis.

8.1. Solid State Behavior Hypothesis

Solid-state behavior is a hypothesis in this research that delivers a geotechnical characteristic that the rocks are vulnerable to the compressional force. When the soil is likely cracked (π_1) or not likely sheared ($\overline{\pi_2}$), then the soil is likely vulnerable to compressional force. That condition's possibility is inherited from the minimum value between the possibility of being cracked or not being sheared. When the soil is not cracked, or soil is sheared, the solid-state behavior possibility is zero. In other words, when neither the soil is not cracked nor sheared, the soil is in solid-state behavior.

IOP Conf. Series: Earth and Environmental Science **851** (2021) 012021 doi:10.1088/1755-1315/851/1/012021

8.2. *Plastic State Behavior Hypothesis*

Plastic-state behavior is a hypothesis in this research that delivers a geotechnical characteristic that the rocks are vulnerable to the shear force. When the soil is likely sheared (π_1) or not likely cracked $(\overline{\pi_2})$, then the soil is likely vulnerable to shear force. That condition's possibility is inherited from the minimum value between the possibility of being sheared or not being cracked. When the soil is not sheared, or the soil is cracked, the possibility of plastic-state behavior is zero. In other words, when neither the soil is not sheared nor cracked, the soil is in plastic-state behavior.

9. Geological Overview and Methodology

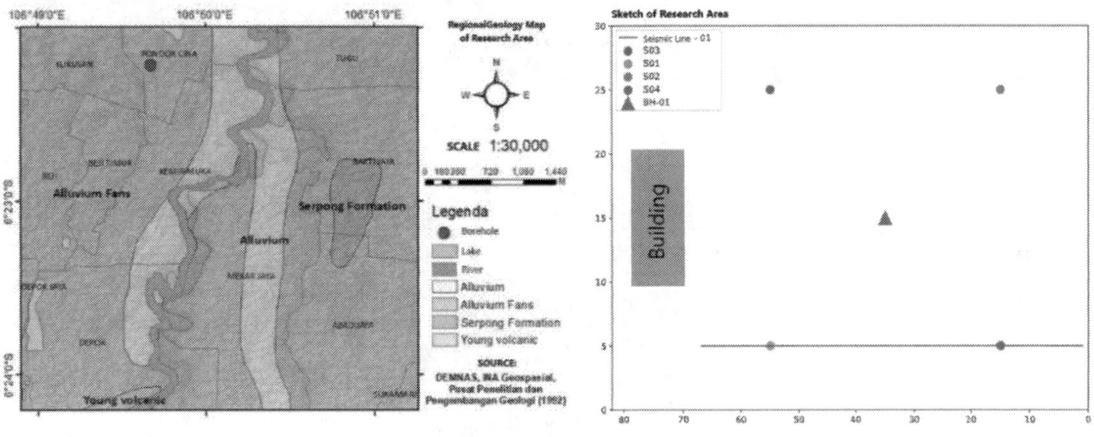

(a) (b)

Figure 4. Research Area [7].

The geology of the research area is a sedimentary environment. The research area's surface is covered by alluvium fans, which has poorly sorted grains presumed that produced by landslides, as seen in figure 4a. It is shown that at the east of the area is covered by alluvium and has an active meandering river from southwest to northeast. The alluvium is considered to be a past deposit from the active river. It is shown that the seismic line lies along 67.5 m in the west-east direction and crossing the two cone penetration test points in figure 4b.

The seismic recording was done using the seismic refraction method and the MASW method. The 24 geophones had planted along 67.5m, 2.5m between each geophone, 10m offset, and 14 shots from the west side with a 5m increment for the seismic refraction method. The distance between each geophone and the offset is the same. However, it only uses 16 actives geophones and moves by one geophone from the first configuration sequentially.

The P-wave velocity tomography was obtained from Easyrefract software, which uses the general reciprocal method for time travel interpretation. The S-wave velocity tomography was obtained from EasyMASW, which uses the phase-shift method for the dispersion curve transformation. The dynamic elastic modeling and fuzzy logic fusion was performed using Python 3.7, employing Numpy and Scikit-Fuzzy modules.

10. Results and Discussion

The P-wave velocity has shown that there is a possibility that the research area has a channel-fill with the 10m – 20m thick at 32.5m, as shown in 5a. This is agreed with geological interpretation in figure 4a which shows that the current river moves from west-east direction in the past and changes to east-west in the current era [4]. The bedrocks lie in 20 m – 23 m depth and >25 m depth, as shown by S-wave velocity and density model in figure 5b and 5c. This finding accords with the previous CPT investigation as shown in figure 5d.

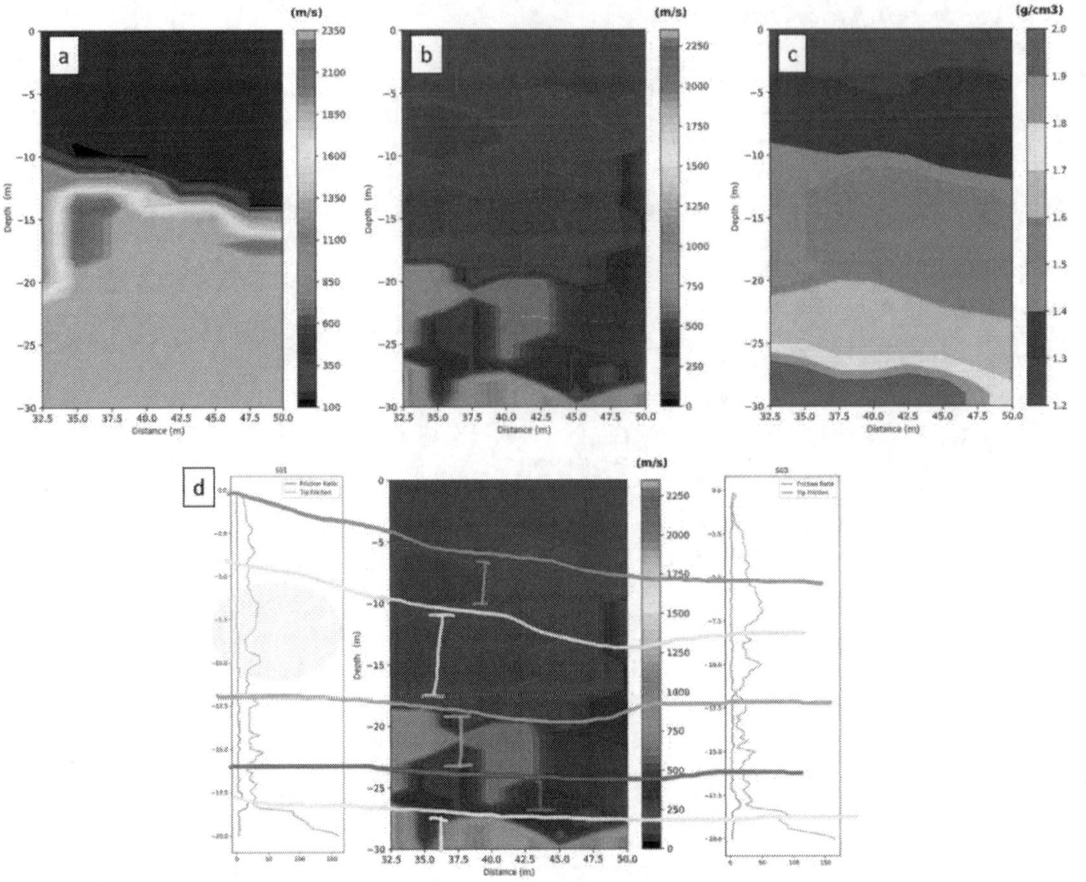

Figure 5. Tomography of (a) P-wave velocity, (b) S-wave velocity, and (c) density model, and integrated interpretation of S-wave velocity tomography and Cone Penetration Test (CPT) data.

The dynamic Bulk modulus illustrates the subsurface in more detail, as shown in figure 6. The Bulk modulus physically explains the compression of the rocks, which relies more on P-wave velocity. The rocks are more fragile to the pressure crack, or the P-wave tomography could not detect the layers in that depth. Thus there are no velocity changes in this depth that made this situation.

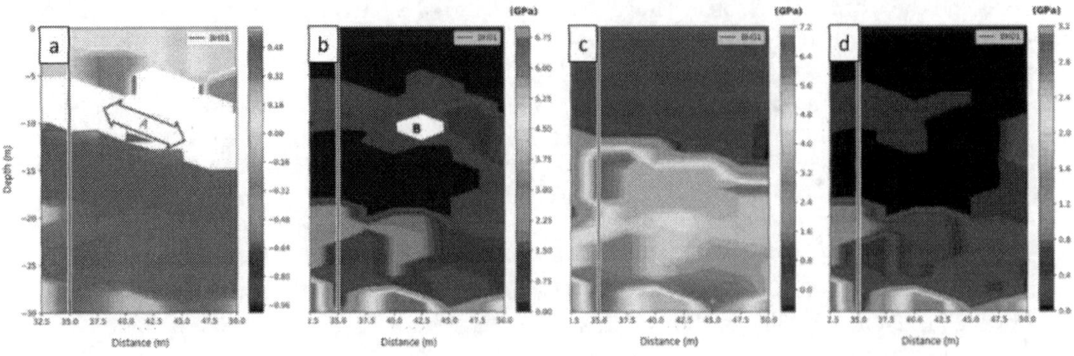

Figure 6. Subsurface dynamic elastic moduli: (a) Poisson's ratio, (b) Young's modulus, (c) bulk modulus, and (d) shear modulus.

The dynamic elastic moduli unit in the GPa should not make sense in term of the stiffness of the soil, this because the elastic moduli are dynamic and should be corrected with static testing data. Unfortunately, those static data are not in the range of the tomography zone. The anomaly in dynamic Poisson's Ratio could be possible by the blind zone. The ambiguity of these results could be minimized by doing fusion with Fuzzy Logic fusion.

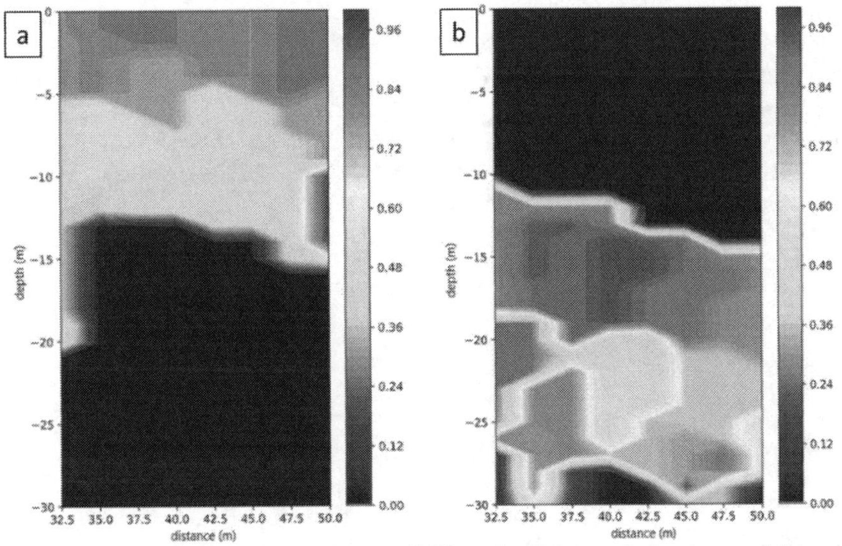

Figure 7. Results of fuzzy logic fusion: (a) possibility of solid-state behavior, and (b) possibility of plastic-state behavior.

It is shown from figure 7a and 7b that below 18m - 24m and > 26m depths are hard layers, while 0m - 10m are layers that can be deformed by stress or cracking forces. The degree of cracking at a depth of 0m - 5m not greater than in depth of 5m - 10m. This finding coincides with anomaly A at the Poisson's ratio, shown earlier in figure 6a. The seam is thought to be an unconsolidated sand layer. At a distance of 32.5m and a depth of 20m, the object is shown to have a Poisson's ratio value of 0.36, and a high value of cracked soil possibility as illustrated in figure 7. The object also has a similar shape between the Poisson's cross-section and the possibility of solid-state behavior. The possibility of the plastic state behavior has an anomalous geometric shape, analogous to Young's modulus of elasticity and dynamic shear modulus. Physically, the cross-section has the same substance, namely Young's modulus, and shear modulus, which describe the resistance of rock or soil to shear forces.

11.Conclusions

The combined result of the P wave velocity and S wave velocity with the concept of fuzzy logic provides ease of interpretation because it is following the analytical calculation of dynamic modulus of elasticity and is more comfortable and more intuitive to use. However, this study's results are subject to improvement because there could be other influential data that need to be considered, e.g., resistivity, and each data's likelihood is not accounted yet.

Acknowledgment

This work was financially supported by Universitas Indonesia through PUTI Saintekes Program No. KB-4883/UN2.RST/HKP.05.00/2020. We acknowledge the Division of Facility Management and Maintenance of Universitas Indonesia to cooperate and support this study.

References

[1] Ekinci B 2012 The Determination of The Dynamic Elastic Moduli by MASW and Seismic Refraction Method 12th International Multidisciplinary Scientific GeoConference and EXPO - Modern Management of Mine Producing, Geology and Environmental Protection SGEM 649–54

[2] Grandjean G, Malet J-P and Bitri A 2007 Geophysical data fusion by fuzzy logic for imaging the mechanical behaviour of mudslides *Bulletin de la Société Géologique de France* **178** 127–36

[3] Hajian A and Styles P 2019 *Application of Soft Computing and Intelligent Methods in Geophysics* (Springer) p 533

[4] Nichols G 2009 *Sedimentology and Stratigraphy* (John Wiley & Sons Ltd) p 432

[5] Park C B, Richter J, Rodriguez R and Cirone A 2018 MASW applications for road construction and maintenance *The Leading Edge* **27** 724–30

[6] Sheriff R E and Geldart L P 1995 *Exploration Seismology* (Cambridge: Cambridge University Press) p 592

[7] Sukardi T 1992 Peta Geologi Lembar Jakarta (Bandung: Pusat Penelitian dan Pengembangan Geologi)

Regression kriging analysis for predicting the shallow depth water from Sentinel-2 satellite multi-spectral images, study area: Coastline of Florida, USA.

B G Dewanto [1,4], I Arifianto [2,4*], C Suhendi [3,4] and M Fittipaldi [4]

[1] Department of Geodetic Engineering, Faculty of Engineering, Universitas Gadjah Mada (UGM), Yogyakarta 55281, Indonesia.
[2] Department of Geological Engineering Faculty of Engineering, Universitas Gadjah Mada (UGM), Yogyakarta 55281, Indonesia.
[3] Geophysical Engineering, Institute Technology of Sumatera (ITERA), Lampung Selatan 35365, Indonesia.
[4] Physical Science and Engineering Division, King Abdullah University of Science and Engineering, Thuwal 23955-6900, Kingdom of Saudi Arabia.
*E-mail: indra.arifianto@mail.ugm.ac.id

Abstract. The shallow depth water mapping has become important to the study of morphology and resources management of the coastal area. Moreover, for city or urban planning, it can be used for determining the proper location for seaport or tourist destination areas such as diving spots, coral reef monitoring, etc. However, the acquisition of shallow depth water data in a large area is somehow costly. In this paper, we represent an approach of mapping the high accuracy bathymetry with input from open satellite access data and some point measurement of bathymetry in the field (can be from LiDAR, Fathometer, Single Beam, or Multi-Beam). The study area is located in the Florida coastline, the USA that has satellite data from Sentinel-2 while the bathymetry is from a single beam survey. The method is combining satellite-derived bathymetry (SDB) with the regression kriging analysis, which shows a better depth water prediction compared to the SDB alone or the ordinary kriging method. The statistical result of the bathymetry shows the regression kriging has a better mean value, standard deviation and coefficient correlation compared to the true bathymetry value. Thus, this method can be utilized as an alternative method to map shallow depth water.

1. Introduction

Shallow depth water mapping is one of the important surveys to be conducted in the scientific study of the marine environment, geomorphology, coastal, and marine planning management [1]. Bathymetric information also very useful in the nautical navigation for shipping traffic safety in the seaport, this application is necessary to map the underwater sand bars, rock, shoals, reefs, and other marine feature that can be dangerous for a big ship in the shallow water docking [2]. In marine science, bathymetric data is important for modelling coastal storm surge, flooding, circulation of ocean water, and tsunami wave propagation. There is global bathymetric data derived from satellite marine gravity measurements like Geosat and ERS-1 satellite radar altimeters that can be accessed and downloaded free online such as from GEBCO, NOAA, or TOPEX website. This global bathymetric is cover shallow to the deep-water area, but the spatial resolution is from 400 meters to 4 km. This global bathymetric data resolution is not sufficient in many coastal applications that need detailed bathymetry information.

The traditional bathymetry acquisition is by shipborne echo sounding may not be feasible for several cases of shallow waters due to sound saturation, narrow survey swath, and inaccessible path for a big vessel [3]. The other popular shallow water bathymetry survey is the airborne survey by light detection and ranging systems (LiDAR). The airborne survey has the advantage of rapid and high accurate bathymetric measurement (4 m spatial and 20 cm vertical accuracy) in the shallow water area, but this survey is a relatively high-cost operation for a large area [4]. This condition aims for another approach that more cost-effective alternative, various passive remote sense multispectral images (i.e., satellite image) have been used to obtain water depth estimation in the shallow water area. The methodology applying satellite data known as satellite-derived bathymetry (SDB) either linear or log-linear regression have been started since 1980 such as [5], [6], [7], [3], [4], and [8]. Although this method has a good spatial resolution (4 m to 10 m), the vertical accuracy has a root mean square errors of around 2 meters [4] [9] [10].

A new approach to minimize the vertical accuracy based on the SDB method has been applied by some researchers from machine learning algorithms [10] to a simple prediction by kriging method as Best Linear Unbiased Estimator (BLUE) methods [9]. In this research, we want to apply regression kriging as the better prediction tool for shallow depth estimator using the multi-spectral image from satellite data. The idea is by combining deterministic approach from SDB with the stochastic variable is acquired from residual value from SDB and field samples, it is similar to [9] but instead of using IKONOS satellite and LiDAR as true data, we use Sentinel-2 satellite and single beam and fathometer survey as true data. The reason of using the Sentinel-2 is because it is an opensource data and it also has the SAR image which one or both reason is not available in the other platform. Since this method requires field samples, we choose to perform this method in the area that has open bathymetric data as true depth water for calibration. Florida coastline the USA has complete bathymetry survey from Fathometer, single beam to LiDAR data (figure 1). Thus, the objectives in this research; firstly, we want to implement the regression kriging in the Florida coastal area to estimate the bathymetry from multi-spectral satellite Sentinel-2 image. The more important thing, we want to compare the water depth estimation from regression kriging with the ordinary kriging and log-linear regression SDB method.

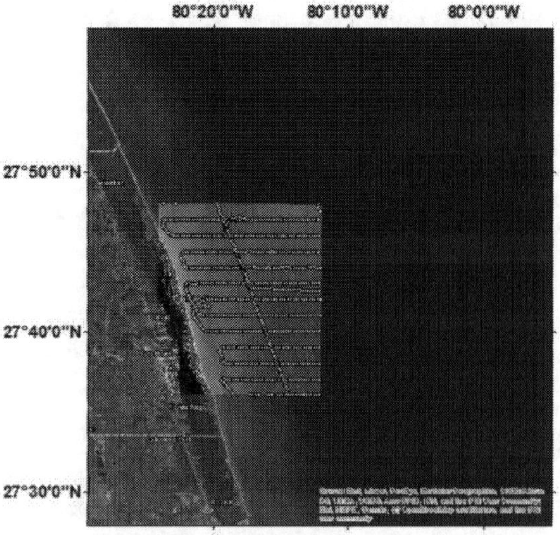

Figure 1. Florida USA coastline area with the bathymetric data and Sentinel-2 satellite image.

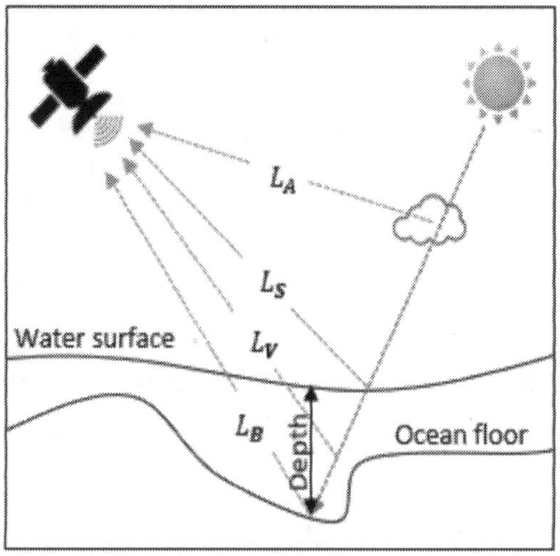

Figure 2. SDB concept (Su et al., 2008).

2. Methodology

Satellite observations include removing the bottom radiance from the measured water-leaving reflectance for bathymetry calculation [11]. The ratio of the radiance leaving the water surface to the downwelling irradiance just above the water surface is known as the reflectance R. The core principle of bathymetry derivation using satellite measurements is based on the dynamics of water column light pass attenuation, bottom reflection, and refraction (figure 2). The mathematical equation can be written as Equation 1:

$$L_{TOA} = L_B + L_V + L_S + L_A \quad (1)$$

where the radiance L_{TOA} located at the sensor. L_A is the atmospheric scattering, L_S is the reflection of optical energy from the water surface, L_V is the subsurface volumetric radiance, and L_B is the result of energy reflection from the seafloor (gives the information about bottom scattering water depth). Disaggregating the bottom and volumetric radiance from the total radiance is very important for deriving the depth of seafloor from satellite observation.

From the Sentinel 2 satellites images, we can get the information about Multi-Spectral Instrument (MSI) with 12 spectral bands. The main bands for the SDB consist of red, green, and blue bands with 5 m, 15 m, and 30 m as the resolutions, respectively. In this research, we only use the green and blue bands. Then, in the pre-processing step, the conversion from the reflectance to radiance (L) is needed which can be seen in Equation 2:

$$L = \frac{pixelValueBandY\ x\ cos(incidenceAngle)\ x\ solarIrradianceBandY}{phi\ x\ d2}\ x\ 10000 \quad (2)$$

where d2 is 1 divided by reflectance conversion coefficient. The figure 3 shows us about the conversion process in this study.

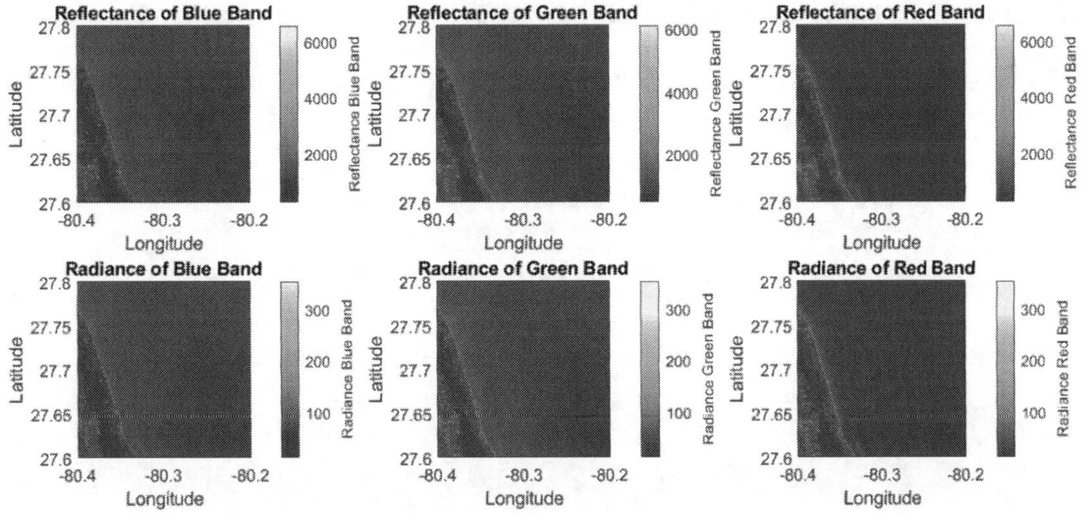

Figure 3. Conversion from the reflectance to radiance.

Kriging is often favoured by different spatial interpolation techniques because it enables one to calculate and manipulate the spatial autocorrelation and joint dependency between neighbouring observations to get an optimized approximation with the corresponding uncertainty estimate for the variable values at unsampled locations. The most widely used kriging variant, ordinary kriging (OK), is used in this analysis as the benchmark system for comparison.

$$\tilde{z}(s) = \sum_{j=1}^{n} w_j(s) z(s_j)\,; \quad \sum_{j=1}^{n} w_j = 1 \tag{3}$$

where $\tilde{z}(s)$ is the the estimated water depth at location s, $z(s_j)$ is the observed water depth at location s_j, n is the number of water depth sample points used in the calculation, and w_j is the weight assigned to the sample point at s_j. In conjunction with the secondary variable (auxiliary information), the spatial autocorrelation of the primary variable (densities) is sampled with the primary variable. In the case of RK, the water depth at a new unsampled location (s) is predicted by summing the predicted drift and residual (see figure 5).

$$\tilde{z}(s) = \tilde{m}(s) + \tilde{e}(s) \tag{4}$$

where the drift $\tilde{m}(s)$ is estimated from the log-linear inversion model. Lyzenga (1978,1981) derived water depth from inverting multispectral satellite images by using a log-linear inversion model (Equation 5), that assuming the ratio of the bottom reflectance between two spectral bands is constant for all bottom types within a given scene. $L(\lambda_i)$ is the water-leaving radiance measured by a remote sensor for spectral band λ_i, $L_\infty(\lambda_i)$ is is the deepwater radiance for spectral band λ_i, $\alpha_i(i = 0, 1, \ldots, N)$ are the model parameters to be determined, and N is the number of spectral bands to be used.

$$\tilde{m}(s) = \alpha_0 + \sum_{i=1}^{N} \alpha_i + \ln\left[L(\lambda_i) - L_\infty(\lambda_i)\right] \tag{5}$$

and the residual $\tilde{e}(s)$ are interpolated using the ordinary kriging method. $e(s_j)$ explains the regression residuals from the log-linear inversion model and $w_j(s)$ tell us about the weights. The histogram residual of measured bathymetry and residual error can be seen in figure 4a and 4b, respectively. Then, 4c shows us about the covariogram vs variogram of residual. Our research's workflow of can be seen in figure 6.

$$\tilde{e}(s) = \sum_{j=1}^{n} w_j(s) e(s_j)\,; \quad \sum_{j=1}^{n} w_j(s) = 1 \tag{6}$$

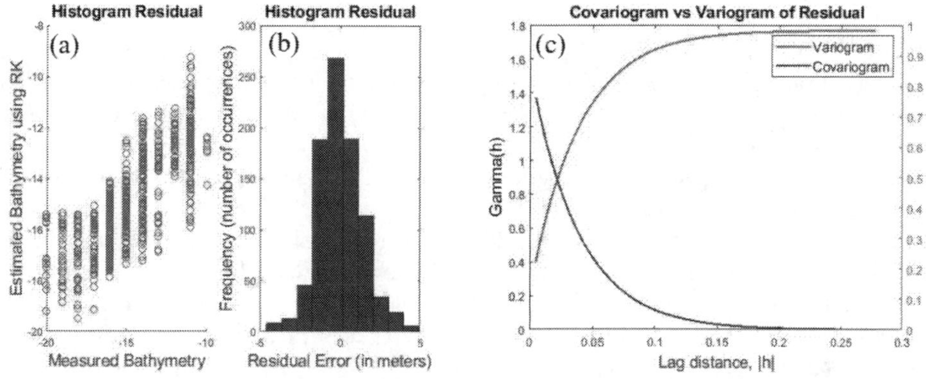

Figure 4. Histogram residual of measured bathymetry (a), histogram residual of residual error (c), and covariogram vs variogram of residual (c).

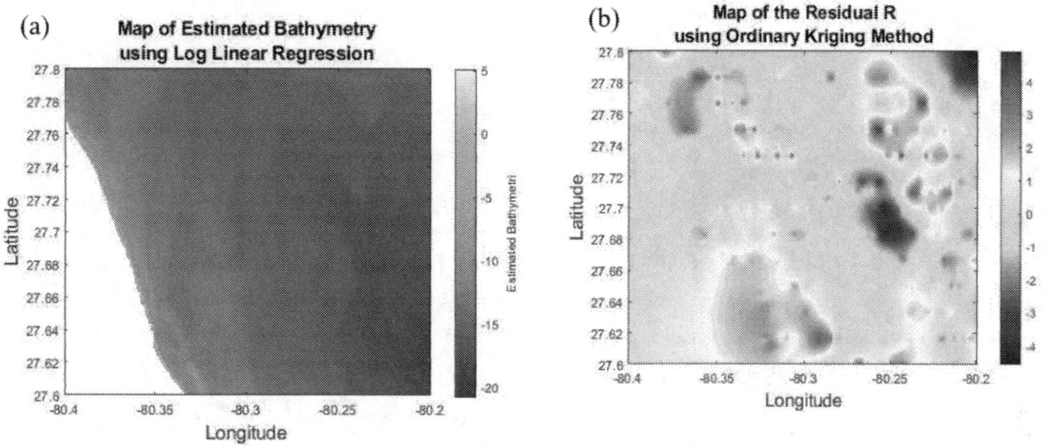

Figure 5. Map of estimated bathymetry using log linear regression (a) and map of the residual R using OK method (b).

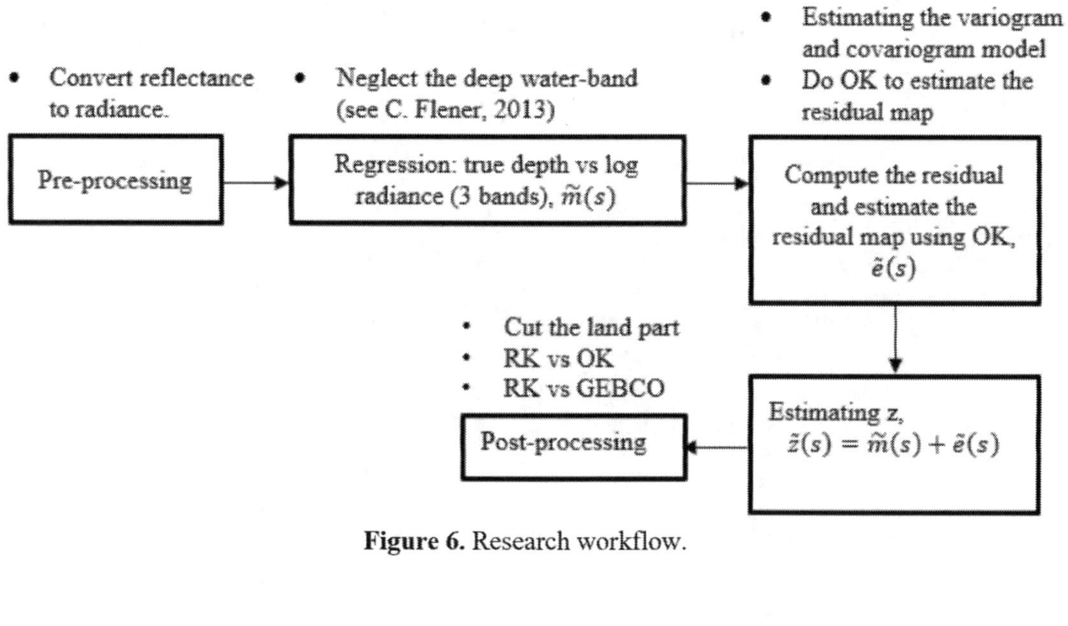

Figure 6. Research workflow.

3. Result

In our research, the radiance value from the land is not used. In the other word, we only use the radiance value from the water area. Jagalingam (2015) used the NIR for separating the radiance of water from the land. From their study, the NIR value anomaly detected in land area, while the smooth result shown in water area.

For comparison to our result, we use grid data from 15 arc-second GEBCO from various method and NOAA Bathymetry data acquired from Fathometer method. The GEBCO data comes from the combination between single beam echosounder, lidar, and satellite images surveys that acquired in 1983. Then, the fathometer data was coming from which was collected in 1930.

In the GEBCO and NOAA data, we also did the gridding process by using OK method. The results of these two processes can be seen in figure 7. The water depth resolution of GEBCO is 400 m, then for the fathometer survey is 150 m and 500 m.

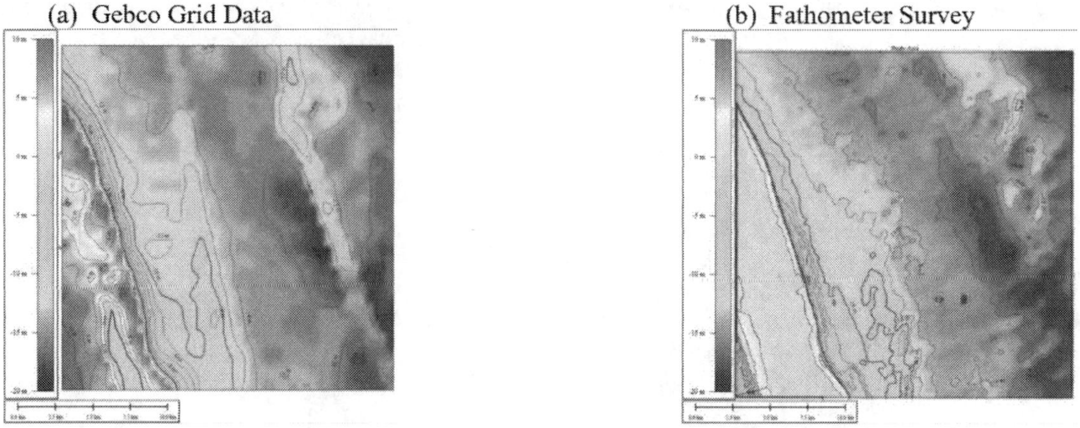

Figure 7. Water depth data acquired from GEBCO (a) and fathometer survey NOAA (b). The water depth calculation from Sentinel 2 satellite image by using log linear method can be seen in figure 8a. It has 22 m as the resolution and compared to the GEBCO and fathometer data, it has better comparison in shallow depth water. Then, we also tried to use the OK method for estimating the water depth from the satellite image. The result shows us that the water depth estimation is much better than by using the log linear, especially in deeper water. The OK method result is shown in figure 8b.

IOP Conf. Series: Earth and Environmental Science **851** (2021) 012022 doi:10.1088/1755-1315/851/1/012022

(a) Satellite Derived Bathymetry b) Ordinary Kriging c) Regression Kriging

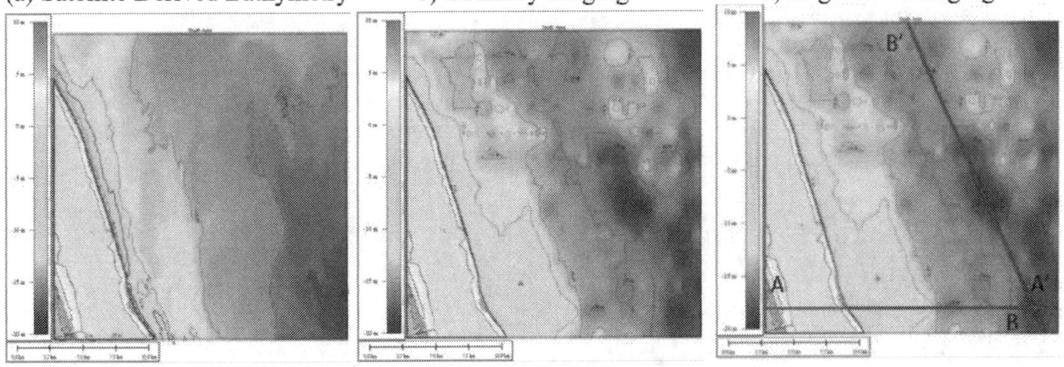

Figure 8. Water depth estimation from the log linear (a), ordinary kriging (b), and regression kriging methods (c).

In the next step, we used the result from the log linear and OK estimations to calculate the water depth using the RK method (see figure 8c). From the result, we can see that it looks similar with the water depth detection from OK. We will explain more details about the difference between log linear, OK, and RK estimations results in the discussion section.

Overall, ordinary kriging and regression kriging maps has high similarity to the GEBCO and fathometer data. These two methods show more detail maps compare to the log linear map. The water depth is explained by the blue to red colours that means deep to shallow bathymetry, respectively.

4. Discussion

Based on the bathymetric profile (figure 9), the East-West profile clearly shows that the ordinary kriging cannot capture the slope of the beach or the near shore it is due to no data constrain in the land area, more over the range of variogram is too far from the nearest data point. While from the satellite derive bathymetry (SDB) profile shows that the slope is still capture although it steeper than the real data from NOAA fathometer or GEBCO data where the inaccuracy of SDB is caused by no bathymetry data in the near shore therefore we do not have good gradient at the near shore area. The regression kriging method profile show the best estimation in the near shore area, where it accounts for the SDB bathymetry plus the residual error from the regression of the OK method although the result still quite steep [9].

The other interesting feature in the SDB is that the SDB failed to estimate a basinal area in the middle of the area (figure 8) where the profile also shows that the bathymetry just dipping toward the east or left side (figure 9). This is maybe due to several problem such as atmosphere condition, sunlight correction, deep water correction or there is habitat or substrate reflectance which cause the SDB cannot estimate the water depth accurately [12]. In this case we are neglected the atmosphere and deep-water correction because choose the clear satellite image and relatively shallow water less than 30 meter where the blue light still can penetrate clear down to 30 m depth. The possible reason for this is we do not perform most of the corrections, however the result is still acceptable, and we argue that in this basinal area error might be due to bottom radiance which integrates the information about ware depth and bottom characteristics including habitat and substrate.

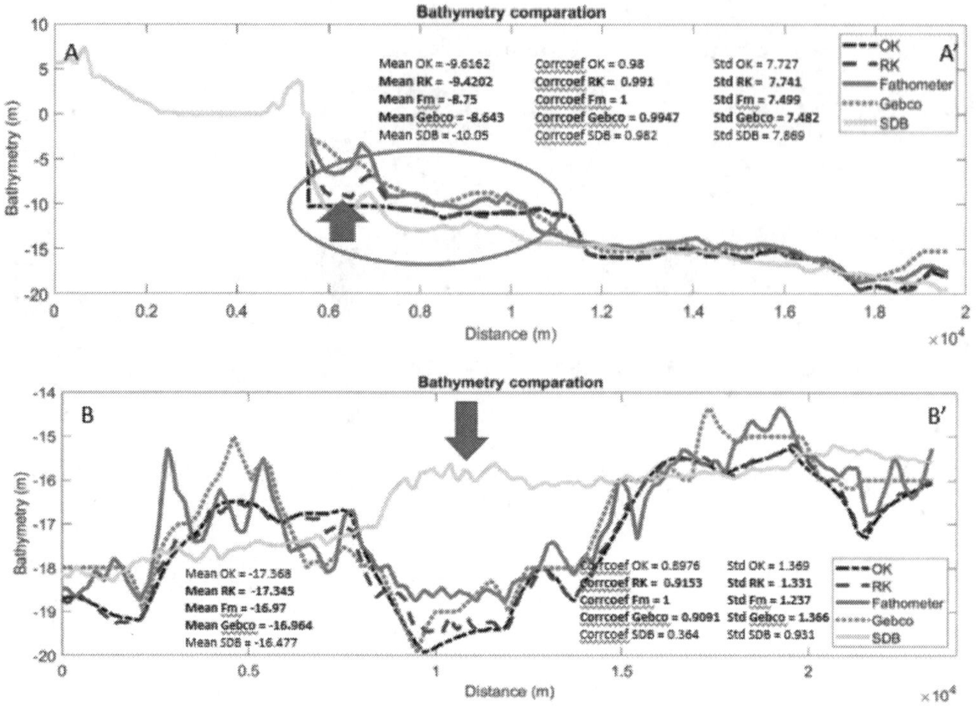

Figure 9. Water depth cross section of the GEBCO, fathometer, log linear, ordinary kriging, and regression kriging, for the section map see figure 8(c).

Beside from qualitative analysis, we also perform statistical analysis to convincing that the bathymetry estimation result has good accuracy. We use the NOAA Fathometer acquired in 1930 as the real bathymetry data because it has close direct measurement data point, and with assumption there is no significant change in seafloor surface due to current or storm activity since it has depth below the fair-weather wave base (figure 9). The other reason is GEBCO data has resolution of 400 meter which is bigger than the NOAA data set, moreover it also has some North-South trace or relict in the bathymetry data due to many data sources accounted in the gridding process (figure 7a). The statistic we performed is the mean of the profile, the correlation coefficient that we compare to the true data (NOAA Fathometer) and the standard deviation of the bathymetry. As we can see that the RK method has better accuracy from SDB and OK method. However, the OK and RK have statistically close value which mean if we increase the number of the bathymetry data the result of OK and RK will be the same. In the other side, the OK can perform better than RK for the data set with strong spatial autocorrelation in the primary variable even when the correlation with a strong second variable. The RK method may be more suitable for water depth estimation with high variable in water depth (low spatial autocorrelation) and might be not useful for the area without too much variation in the water depth (high spatial autocorrelation).

5. Conclusions

We successfully perform bathymetry estimation with good accuracy using Regression Kriging method (RK) compiled with Satellite Derived Bathymetry method (SDB) from an open online database. The accuracy of the bathymetry estimation from SDB-RK is dependent on the number and the distribution of the calibration depth point where the SDB suppose not to be standalone without calibration data or the result will have very low accuracy. The Bathymetry acquired from SDB-RK is qualitatively and quantitatively better than bathymetry prediction only from Ordinary Kriging (OK) or SDB where the more real bathymetry data the result of OK will be similar to RK and the accuracy of SDB is dependent on the heterogeneity spatial of the shallow water condition, such suspension clay-sand material, pelagic micro algae, also weather when the image is taken. Furthermore, SDB-RK result in our area is comparable with direct measurement (LiDAR or Multibeam) survey.

Acknowledgments
We would like to say thank you to Prof. Sigurjon Jonsson and Dr. Ibrahim Hoteit as the instructors in the Data Analysis course in Earth Science and Engineering Department, King Abdullah University of Science and Technology so that this paper can be done as the final project of the course. We also would appreciate NOAA and GEBCO organization for opening the bathymetry data into public and providing the dataset available online.

References

[1] Finkl C W, Benedet L and Andrews J L 2005 Submarine geomorphology of the continental shelf off Southeast Florida based on interpretation of airborne laser bathymetry. *Journal of Coastal Research,* **21**:1178–1190.

[2] Jupp D L B 1989 Background and extension to depth of penetration (DOP) mapping in shallow coastal waters. *Conference on Remote Sensing Application Projects Using microBRIAN Image Processing System*. pp.IV.2.1-19, 12 refs, illus.

[3] Stumpf R P, Holderied K and Sinclair M 2003 Determination of water depth with high-resolution satellite imagery over variable bottom types. *Limnology and Oceanography* **48**:547–556.

[4] Su H, Liu H, and Heyman W D 2008 Automated Derivation of Bathymetric Information from Multi-Spectral Satellite Imagery Using a Non-Linear Inversion Model, *Marine Geodesy*, **31:4**, 281-298.

[5] Lyzenga D R 1981 Remote sensing of bottom reflectance and water attenuation parameters in shallow water using aircraft and Landsat data. *International Journal of Remote Sensing* **2**:71–82.

[6] Lyzenga D R 1985 Shallow-water bathymetry using combined lidar and passive multispectral scanner data. *International Journal of Remote Sensing* **6**:115–125.

[7] Philpot W 1989 Bathymetric mapping with passive multispectral imagery. *Applied Optics* **28**:1569–1578.

[8] Jagalingam P, Akshaya B J and Hegde A V 2015 Bathymetry Mapping Using Landsat 8 Satellite Imagery. *Procedia Engineering* 116: 560 – 566.

[9] Su H, Liu H and Wu Q 2015 Prediction of Water Depth from Multispectral Satellite Imagery—The Regression Kriging Alternative, *IEEE Geoscience and Remote Sensing Letters, Vol. 12.*

[10] Misra A, Vojinovic Z, Ramakrishnan B, Luijendijk A and Ranasinghe R 2018 Shallow water bathymetry mapping using Support Vector Machine (SVM) technique and multispectral imagery, *International Journal of Remote Sensing*, **39**:13, 4431-4450

[11] Jensen J R 2007 Remote sensing of the environment: An earth resource perspective, 2nd ed. Upper Saddle River, NJ: Prentice Hall.

[12] Lyzenga D R, Malinas N P and Tanis F J 2006 Multispectral bathymetry using a simple physically based algorithm. *Geoscience and Remote Sensing, IEEE Transactions on* **44**:2251–2259.

IOP Conf. Series: Earth and Environmental Science **851** (2021) 012023 doi:10.1088/1755-1315/851/1/012023

Groundwater potential zones identification in Purwobinangun, Pakem, Sleman, D.I. Yogyakarta using vertical electrical sounding (VES)

S B Samodra[1*], R T Sembiring[1], G R Ramadhan[1], H A Kharis[1] and T R Siregar[1]
[1] Department of Geological Engineering, Faculty of Engineering, Universitas Gadjah Mada, Yogyakarta 55281, Indonesia
*E-mail: sbsamodra@ugm.ac.id

Abstract. Water is crucial part in tourism development. However, not all locations have good and accessible groundwater resources. Ngepring Vineyard in Purwobinangun Village, Pakem District, Sleman, Yogyakarta face this problem. Ngepring vineyard is a developing tourism area at southern mountainside of Merapi which is one of geotourism destination in Indonesia. In its development, there were obstacles to fulfill the water supply for tourism and locals' daily use. This research aims to determine groundwater potential zone and recommend a drilling point for a new well. Geoelectric resistivity survey conducted to evaluate subsurface condition by measuring and interpretating rock resistivity properties. Schlumberger Vertical Electrical Soundings (VES) configuration used with a maximum spacing from the center point as far as 250 meters with three VES were measured in study area along two N-S profiles perpendicular to one E-W profile. As the results, there is a sand layer at approximately 80m depth interpreted as aquifer. This result can be used as a preliminary study used to determine potential location for a new water resource in Ngepring vineyard.

1. Introduction

Water has vital role in human activities as well as for tourism development. In dried area where there is lack of water, water consumption supported by using ground water production [1]. In groundwater exploration, geophysical assessment usually conducted to evaluate and predict possibility position of aquifer in subsurface. Geoelectric resistivity is one of common geophysical assessment in groundwater exploration. Geoelectric survey measures rock resistivity properties by flowing electric current in specific value to determine the rock by its properties [1].

Ngepring Vineyard, Purwobinangun Village, Pakem District (as shown in figure 1), is a developing tourism area in Yogyakarta Province. This location situated at southern mountainside of Merapi. Ngepring vineyard does not have enough water resources since it is located above Merapi's spring belt. As the result of this condition, Ngepring vineyard is experiencing drought during dry season because most of the groundwater layers are deep aquifers (>40 meter). This condition makes difficulties for people in surroundings area to utilize water for tourism and daily use. Based on that condition, Ngepring vineyard needs a new water resources to fulfill water needs. This survey aims to determine a new well location based on interpretation from geoelectric assessment.

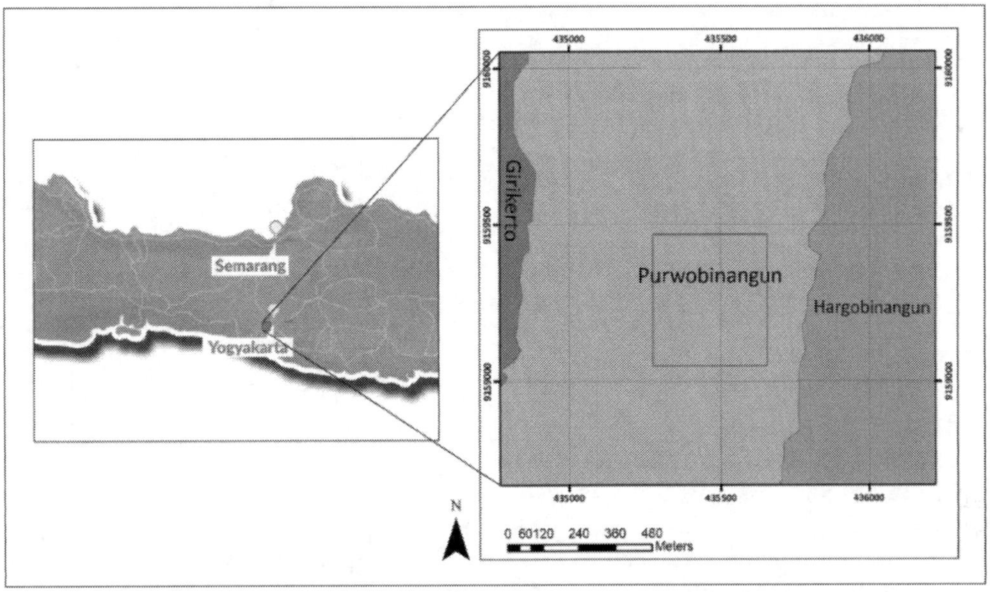

Figure 1. Location of the study area with area of interest marked with red square.

2. Geological Settings

Study area located in Geologic Map of Yogyakarta Quadrangle [2]. Yogyakarta quadrangle geologic map composed by 24 formations. Of these 24 formations in Yogyakarta Quadrangle, only one formation represented in study area namely Young Volcanic Deposit of Merapi Volcano. This formation consists of tuffs, ash, breccia, and agglomerates [2]. Based on Yogyakarta Quadrangle Map, there is not any geological structure found in study area [2].

3. Hydrogeological

Research area is located on the Yogyakarta – Sleman groundwater basin. Yogyakarta – Sleman groundwater basin divided into 2 main formation, Old Merapi and Young Merapi Volcanic Deposits [3,5]. Young Merapi volcanic deposit consists of Yogyakarta Formation and Sleman Formation [3,4]. Sleman Formation extends from the upper slope of Merapi Volcano to the southern of Bantul Regency and features mostly sands, gravels with interspersed boulders and derivations from volcanic ejecta with a varying thickness of 38 to 120 m [4]. Yogyakarta Formation covers most of Yogyakarta – Sleman groundwater basin. This formation formed by sand, gravel, silts, clays up to 45 m thickness and lying unconformably upon the Sleman Formation [4]. Both volcanic formations form aquifer system known as Merapi Aquifer System.

These aquifers in Yogyakarta – Sleman groundwater basin can be classified into three aquifers: upper, lower, base aquifer [3,6]. Upper aquifer formed by Yogyakarta Formation which is consist by andesite lava, breccia, and sand. This aquifer lied in northern side of Yogyakarta – Sleman groundwater system. Lower aquifer dominated by Sleman Formation which is consists of coarse sand, coarse gravel, and lenses andesite lava [6]. Base aquifer located in north part of Yogyakarta – Sleman groundwater. This aquifer formed by volcanic breccia and andesite lava from Old Merapi Volcanic Deposits [3,6]. Study area is situated as shown as figure 2.

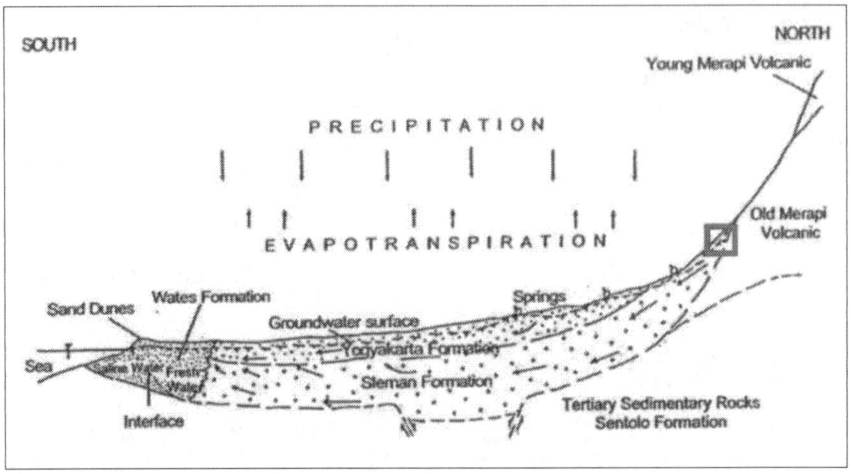

Figure 2. South-North cross section of hydrogeological concept of aquifer in the Yogyakarta - Sleman groundwater basin. The approximate location of the study area is marked with the red square [3].

4. Methods

Geophysical investigation was carried out to locate groundwater potential area. Geoelectrical resistivity as one of geophysical method used in this study because considered as a promising and most suitable method for groundwater exploration [5]. Vertical electrical resistivity (VES) applied in this geoelectric survey. Schlumberger electrode configuration was used, with four electrodes that are moved sequentially as shown in figure 3.

Figure 3. Installation of electrodes at point GL-02, using four electrodes in a E – W direction.

Three VES were measured in study area, two N – S profiles perpendicular to one E – W profile. Resistivity data for research area is taken from three measurement points as shown in figure 4 with the coordinates of each measurement point shown in table 1.

IOP Conf. Series: Earth and Environmental Science **851** (2021) 012023 doi:10.1088/1755-1315/851/1/012023

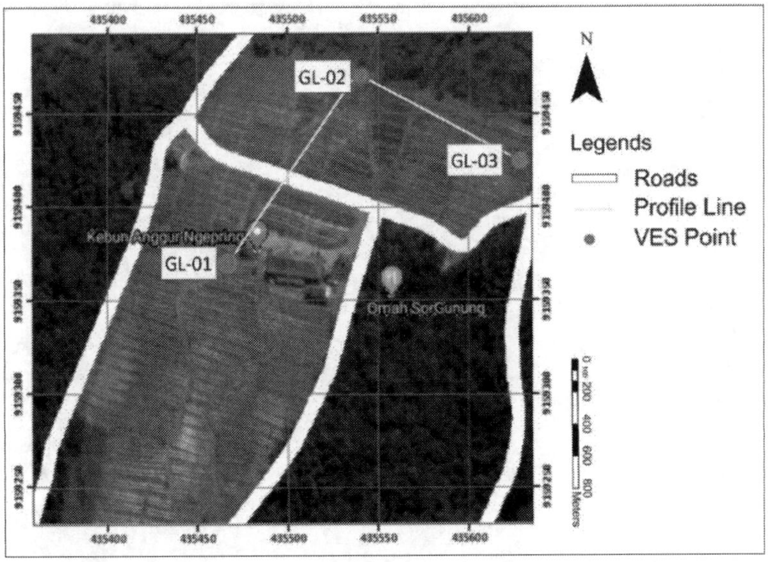

Figure 4. Research location map. Points GL-01, GL-02, and GL-03 (in red points) are located in the Ngepring Vineyard area, Purwobinangun Village, Pakem District.

Table 1. VES points coordinate.

Point	Coordinate	Azimuth
GL-01	7°36'16,4" S/ 110°24'54,2" E	N 22° E
GL-02	7°36'13.1" S/ 110°24'56.4" E	N 82° E
GL-03	7°36'14.2" S/ 110°24'59.5" E	N 186° E

Measurement points not selected based on geological and geomorphological condition, because in study area only consists of one formation and no geological structure found. The distribution of measurement points determined by the availability of land for further construction of drilling well. The measurement points are not influenced by external factors such as irrigation pipes, due to the lack of infrastructure being built around the vineyard. Measured vertical electrical soundings (VES) calculated to get true resistivity (ῤ). True resistivity value then interpreted to get subsurface overview which includes type, depth and thickness, distribution of aquifers both vertically and laterally. All the available data about the geologic and geophysical settings were used in this research to initiate the geoelectrical model. Acquired geophysical data processed by using PROGRESS software. Data processing results (layer resistivity and thickness) analyzed to determine groundwater potential zone in area of study.

5. Results

Based on the resistivity data acquisition, the measurement results from the three VES points are as follows:

a. VES point GL-01

Point GL-01 is located west of the gazebo, on the downhill bike trail, in the middle of the vineyard. The direction of the electrode stretches relatively North - South with the outer electrode distance of 250 m each from the center point. The measurement results can be seen in figure 5.

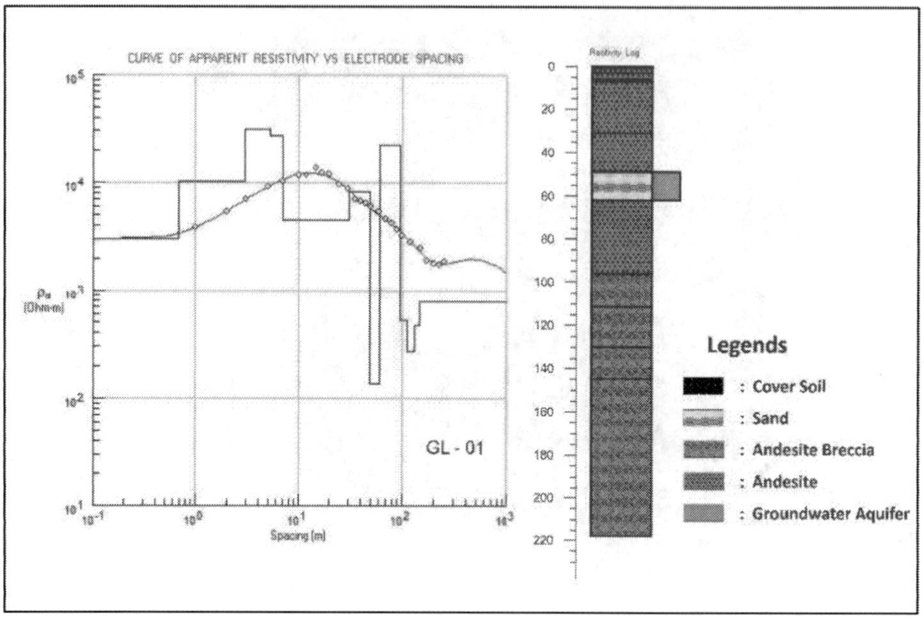

Figure 5. The results of the VES point GL-01.

b. VES point GL-02
Point GL-02 is located on the northern edge of the vineyard. The direction of the electrode stretches relatively West - East with the outer electrode distance of 250 m each from the center point. The measurement results can be seen in figure 6.

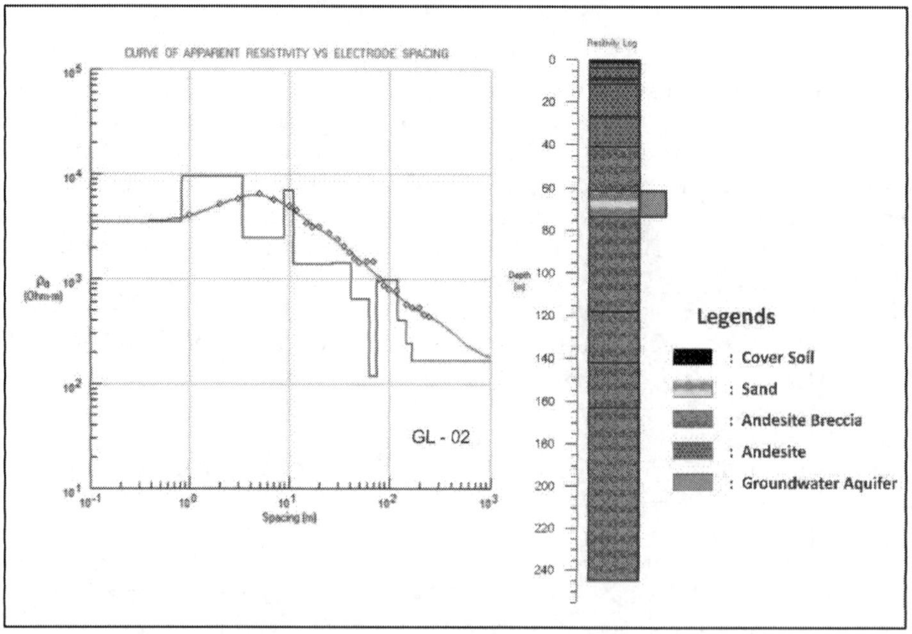

Figure 6. The results of the VES point GL-02.

c. VES point GL-03

Point GL-03 is located at the northeast of the vineyard, to the east of the newly constructed reservoir. The direction of the electrode stretch is relatively North - South with the outer electrode distance of 250 m each from the center point. The measurement results can be seen in figure 7.

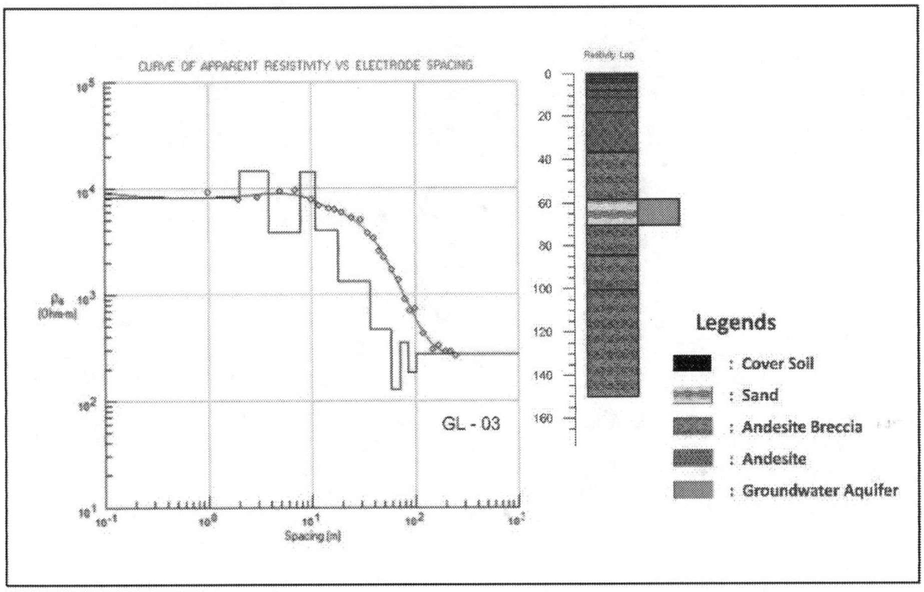

Figure 7. The results of the VES point GL-03.

From interpreted VES data, there is a sand layer at approximately 80 meters depth. This sand layer interpreted as an aquifer at all measurement points. Thus, drilling can be carried out at all locations. Considering the possibility of aquifer presence and proximity to the constructed reservoir, it is recommended to drill wells around point GL-02 at a depth of approximately 80 meters. The results of the geoelectric interpretation correlation of the measurements at the GL-01, GL-02, GL-03 points are as shown in figure 8.

IOP Conf. Series: Earth and Environmental Science **851** (2021) 012023 doi:10.1088/1755-1315/851/1/012023

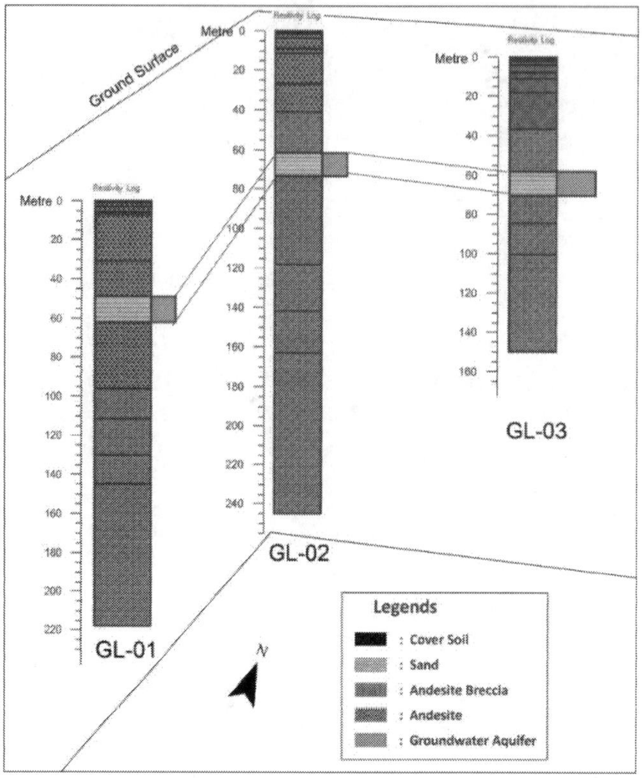

Figure 8. Perspective illustration of VES point correlation GL-01, GL-02, and GL-03.

6. Discussion

Based on three available profiles, there is a difference between GL-01 profile and GL-02 & GL-03. At GL-01 profile, there is a layer of sand between andesite. This condition is possible because deposition time gap between sand layer and andesite at point GL-01. The first andesite layer deposited, then the sand layer covered, before the andesite layer reappears. It achieved by andesite occurrence in research area present as sill and can reappears at any time in stratigraphy. The existence of this sand layer is probably the result of deposition from fluvial environment which flank the study area, namely Boyong River (east side) and Gendol River (west side).

However, the presence of aquifer layer does not necessarily mean this area is potential to fulfill their water need. We need to consider whether existing recharge supply will always be available in the aquifer. Groundwater recharge in this location is characterized by high magnitude supply up to 4270 mm/year with average rainfall 2801.5 mm/year based on the geomorphological model and calculations using the Water Table Fluctuation method [7]. It is known that the water supply at the research location is quite good, which resistant to changing seasons in Indonesia [7]. Based on recharge supply data, we can predict how much water we can utilize. It suggested maximum 40% of recharge supply can be extracted based on Regulation of The Minister of Energy and Mineral Resources the Republic of Indonesia [8]. So, that it can be used sustainably to fulfill the water supply for tourism and locals' daily use.

7. Conclusions

Based on the results and discussion above, the following conclusions can be drawn:

1. The research area is located on the southern slopes of Mount Merapi and above the Merapi spring belt.
2. Geologically, the research area is composed of layers of andesite breccias, andesite lava, and sand which is a product of Young Merapi volcanic deposit.

3. There is a layer of sand at approximately 80 meters below the surface that interpreted as aquifer in the research area.
4. Andesite layer in the study area interpreted as a sill, so it can be formed any time and existence of sand layer between andesite seems possible.
5. Based on previous researcher, the groundwater recharge is quite good and resistant to changing seasons in Indonesia.
6. We suggest maximum 40% of recharge supply can be extracted based on Regulation of The Minister of Energy and Mineral Resources the Republic of Indonesia to conserve ground water sustainability.

Acknowledgement
Thanks to the owner of the Ngepring Vineyard for allowing the use of his land as research location.

References
[1] Wintolo D, Arifianto I, Wibowo J P Resistivity Survey for Groundwater Exploration in Tiron Village, Kediri, East Jawa *Proceeding Seminar Nasional Kebumian Ke-10*
[2] Rahardjo W, Sukandarrumidi, Rusidi H 1977 *Geological Map of Yogyakarta Quadrangle* (Bandung: Geological Surveys of Indonesia)
[3] Putra D and Indrawan I 2014 Assessment of Aquifer Susceptibility due to Excessive Groundwater Abstraction: A Case of Yogyakarta-Sleman Groundwater Basin *ASEAN Engineering Journal Part C* vol 3(2), pp 105
[4] Manny L, Atmaja R, Putra D 2016 Groundwater Level Changes in Shallow Aquifer of Yogyakarta City, Indonesia: Distribution and Causes *Journal of Applied Geology* vol 1(2), pp 89-99
[5] Ibraheem I, El-Qady G, and ElGalladi A 2016 Hydrogeophysical and structural investigation using VES and TDEM data: A case study at El-Nubariya–Wadi El-Natrun area, west Nile Delta, Egypt NRIAG Journal of Astronomy and Geophysics, vol 5(1), pp 198-215
[6] Mutianto H 2008 Studi Konservasi Air untuk Pemanfaatan Airtanah yang Berkelanjutan pada Recharge Area Lereng Gunungapi Merapi Kabupaten Sleman Yogyakarta *Jurnal Geografi* vol 8(2), pp 1-15
[7] Tjahyo, A 2008 Application of Water Table Fluctuation Method to Quantify Spatial Groundwater Recharge Within the Southern Slope of Merapi Volcano, Indonesia *Indonesian Journal of Geography* vol 39, pp 157-171
[8] The Minister of Energy and Mineral Resources the Republic of Indonesia Number 31 of 2018 *Guidelines for Determining Groundwater Conservation Zone*

Mass transport modelling of total organic carbon in groundwater to determine the location of petroleum fuel pollutant sources in the area of Jlagran, Yogyakarta City, Indonesia

E G Manek[1], D P E Putra[1*], H Hendrayana[1]

[1]Geological Engineering Departement, Faculty of Engineering, Universitas Gadjah Mada, Jl.Grafika No.2 Bulaksumur, Yogyakarta, Indonesia
*E-mail: putra_dpe@ugm.ac.id

Abstract. Hydrocarbon pollution in the Jlagran area's groundwater, Yogyakarta City, has been going on for more than 20 years. So far, the pollution source has been inferred from a leak in the diesel fuel storage tank from near Train Station. This study aims to prove the claim that the source of hydrocarbon pollutants comes from the location of the fuel storage tank or not. This research was conducted by collecting data on hydrogeological aspects and total organic carbon (TOC) concentration in groundwater, followed by modeling flow and mass transport using Visual Modlfow MT3DMS. The modeling results show that the contamination plume distribution in groundwater at the recent time cannot be caused by only one point source, namely the fuel oil storage tank's location, but also may come from the train maintenance area. It is estimated on this area that the disposal of train maintenance waste of hydrocarbon as a result of poor practices for at least two decades has been developing a residual hydrocarbon zone in the unsaturated zone, and now potentially seeped into the aquifer and also contribute as a source of hydrocarbon pollutant to groundwater.

1. Introduction

In the year between 2000-2001, groundwater in the dug wells of residents around Jlagran and Jlagran Lor, Kapanewon Gedongtengen, Yogyakarta City, was contaminated by diesel fuel, and the contamination was at that time due to a leak in the fuel storage tank at the Train maintenance, Yogyakarta Station (see figure 1) [1]–[3]. The petroleum/hydrocarbon contamination in groundwater is a typical fuel tank leakage [4]–[6]. Diesel fuel is one type of organic carbon contaminant, which is composed of approximately 75% aliphatic hydrocarbons ($C_{10}H_{20}$–$C_{15}H_{28}$), more or less 25% aromatic hydrocarbons (benzene, styrene), and a minor fraction of sulfur depending on crude oil source and cleaning quality [7]. The characteristics of the liquid phase organic carbon contaminants are divided into two, namely NAPL and plume. NAPL (nonaqueous-phase liquid) is a phase of organic carbon contaminants that forms its liquid phase. NAPL is divided into two types based on its density: LNAPL (light nonaqueous-phase liquid) and DNAPL (dense nonaqueous-phase liquid). LNAPL, such as fuel, has a lower density than water, so the movement of LNAPL is strongly influenced by the direction of groundwater flow [8]. On the other hand, plume is a phase of organic carbon contaminants that is dissolved in groundwater. Additional components in fuel (additive oxygenate ether) such as methyl

IOP Conf. Series: Earth and Environmental Science **851** (2021) 012024 doi:10.1088/1755-1315/851/1/012024

tertiary butyl ether (MTBE) or ethanol have a high level of solubility in water, so that some of the LNAPL phase can be dissolved. The extent of the plume coverage varies depending on the natural attenuation process. Natural attenuation is a variety of physical, chemical, or biological processes that occur naturally, reducing the mass, toxicity, mobility, volume, or concentration of contaminants in groundwater [9].

The spreading of hydrocarbon contamination plumes in the study area (see figure 1) has been studied by some researchers. It was found that the dimensions of hydrocarbon contaminant plumes are 350 m long and 270 m wide [10]. A similar study was conducted [11], and it seems that there was no significant change in the contaminant plume. The plume length was still around 338 m and 325 m wide, with the highest oil and fat concentration, reported to be 55 mg/l. Theoretically, the length of the plume condition, which tends to be constant, shows a balance between the attenuation process and the continuation of pollutant spilling. This condition certainly raises a question; if the pollutant source is a leak in a fuel storage tank where the source was stopped in the early 2000s, then the plume's levels and distribution should shrink over time. One simple way to simulate this may be done by performing numerical modeling of the mass transport, such as MT3DMS. The MT3DMS is a modular three-dimensional, multispecies solute transport model in groundwater flow systems and is intended to simulate changes in concentrations of soluble contaminants in groundwater flow systems taking into account advection, dispersion, molecular diffusion, and some essential chemical reactions, including non-linear absorption, and first-order radioactive decay or biodegradation [12]. Many researchers have carried out the application of MT3DMS for modeling hydrocarbon pollution in groundwater. MT3DMS was used to model and predict groundwater pollution due to leakage of petroleum pipes [13]. Xu et al. [14] and Colombo et al. [15] clearly states that Modflow / MT3DMS can simulate organic chemical contamination of PCE (perchloroethylene) in groundwater. On the other hand, Gao et al. [16] utilized MODFLOW / MT3DMS for modeling migration of organic pollutants in shallow groundwater conditions validated by electrical resistivity tomography results.

One of the manifestations of hydrocarbon pollution in groundwater is the increase in the level of Total Organic Carbon (TOC) concentration of groundwater [17], [18]. TOC is a parameter that describes the total concentration of organic carbon in groundwater. The average concentration of natural organic carbon in the groundwater system is only 0.7 mg/l [7]. Of course, a high TOC concentration will be associated with the presence of either natural or non-natural sources of hydrocarbons. Based on the above, in this study, the TOC level in groundwater is used as a species parameter in MODFLOW / MT3DMS to simulate whether the source of hydrocarbon pollutants in the study area is a point source or an area source.

2. Study Area

The research area is located in Gedongtengen District, Yogyakarta City, with a surface elevation ranging from 106-112 m above sea level (see figure 1). Geomorphologically, Yogyakarta is located at the foot of the volcanic landscape of Merapi Volcano. Yogyakarta City's regional stratigraphy comprises Tertiary-aged rock formations that form basement and Quaternary deposits [19]. In Tertiary rock formations, the regional geological structure in Yogyakarta is in the form of faults that form a graben morphostructure called the Yogyakarta graben and Quaternary deposits as filling material [19]. Based on the Yogyakarta sheet's geological map [20], the Quaternary deposits of Yogyakarta City are Quaternary sediments resulting from a fluvio-volcanic process known as the Young Merapi Sediment Formation. This formation is divided into the Sleman Formation and the Yogyakarta Formation [19]. In general, these two formations are composed of gravel, sand, silt, and clay deposits and form the primary aquifer in the Merapi Aquifer System [21].

There are four dominant types of lithology between the bedrock and Yogyakarta City's surface [22]. The lithology is fine sand, coarse sand, silt sand, and loamy sand, interpreted in a stratigraphic map (See figure 2). Groundwater in Yogyakarta City is part of the Merapi Aquifer System. In general, in Yogyakarta City, the groundwater level's depth is less than 20 m from the ground surface, with the regional groundwater flow direction from north to south. Local groundwater also flows out into rivers that divide Yogyakarta city [23]. The Merapi Aquifer System's hydraulic conductivity is known to range from 9.26 x 10^{-6} m/s to 1.1 x 10^{-3} m/s, with storativity values ranging from 0.001 to 0.2 [24]. In Yogyakarta, Wijaya [11] states that the study area's permeability value ranges from 4.6 x 10^{-5} m/s to 8.2 x 10^{-4} m/s, and the average permeability value is 3.8 x 10^{-4} m/s. In contrast, the average aquifer porosity value in the study area is 35%. The groundwater recharge in Yogyakarta ranges from 400-600 mm/year [25]. Concerning the TOC level in groundwater in the Yogyakarta-Sleman area, Kaempfner [26] states that the initial environmental baseline conditions for TOC in groundwater in this area are 3.7 mg/l, meaning that values higher than this are categorized contaminated.

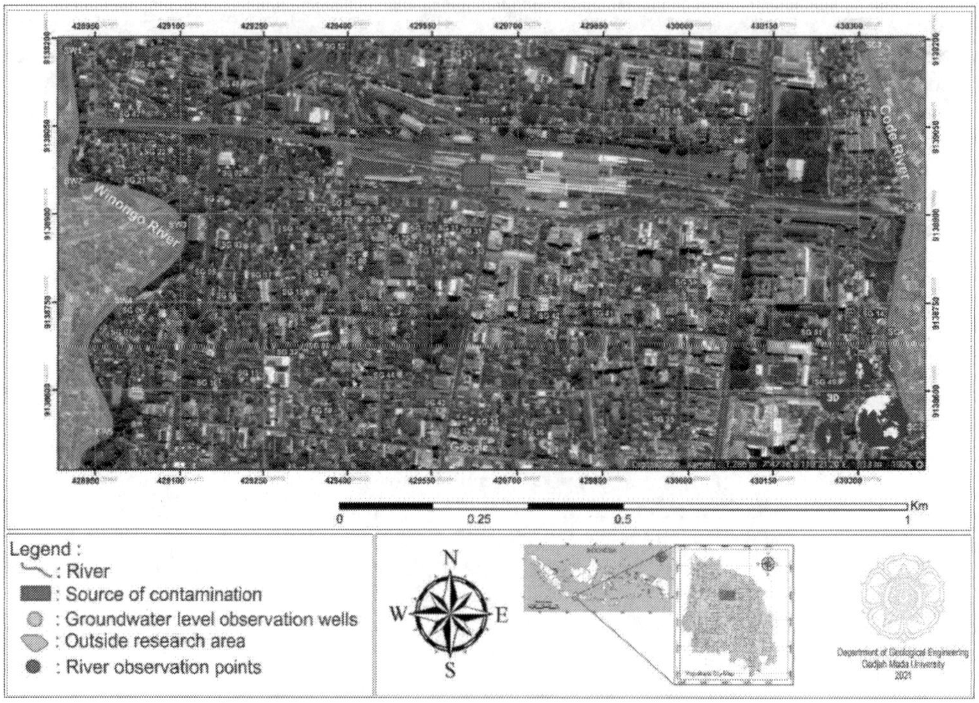

Figure 1. Study area in Yogyakarta City of Indonesia.

Based on the stratigraphic conditions shown in figure 2, in general, the aquifer unit of the study area can be divided into two main hydrostratigraphic layers; namely, the upper hydro-stratigraphic unit, which is composed of alternating layers of sandstones with loam-clay sand and lower hydrostratigraphic units composed of alternating clay sandstones with clay. The boundary between the two hydrostratigraphic layers is a layer of claystone that is assumed to be continuous. This limiting clay layer is considered as basement of the aquifer for upper hydrostratigraphic unit. It is located at a depth of 50-55 m from the ground surface. Groundwater recharge through rain enters directly into the upper hydrostratigraphic unit.

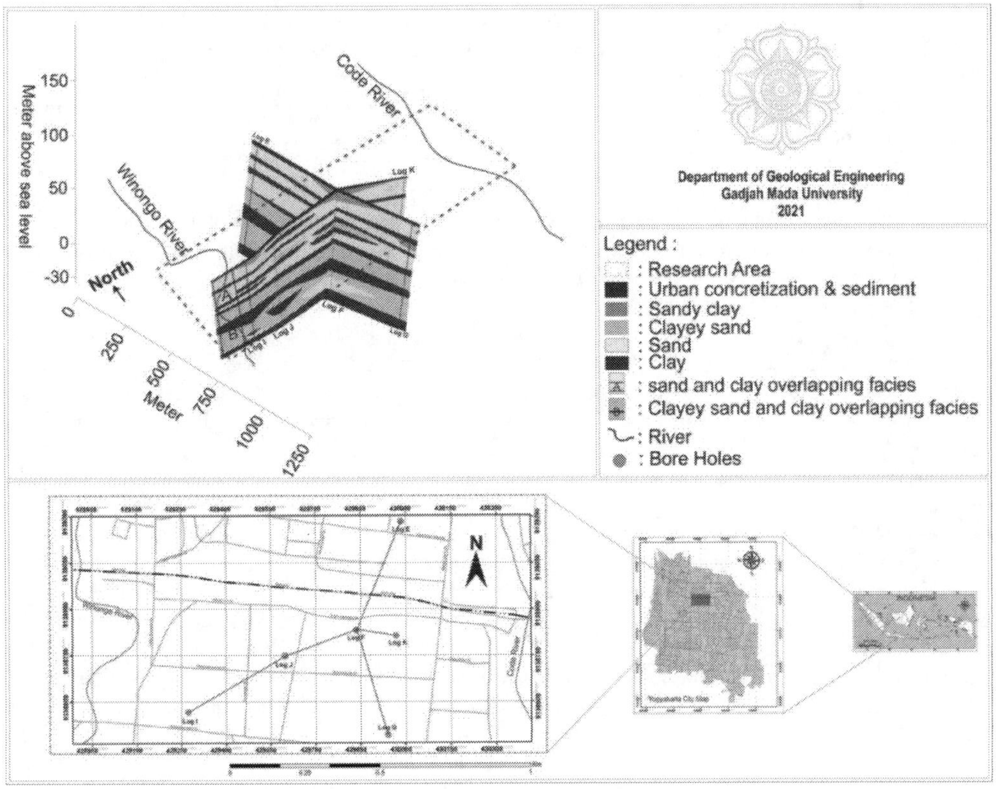

Figure 2. Fence diagram of subsurface conditions around the study area [21].

3. Methodology and Conceptual Model

3.1 Groundwater Sampling and Laboratory Works

This study aims to determine the level of TOC in groundwater as the basis for numerical modeling of groundwater pollution. In this study, TOC groundwater samples were taken and collected using dark glass bottles and stored in an icebox to keep groundwater samples associated with volatile pollutants in temperatures below room temperatures. Because the distance between the study area and the Get in Cicero laboratory is relatively close, about 6 km, groundwater samples can be analyzed immediately after being taken to the laboratory.

The tool used to analyze the TOC concentration of the groundwater sample in this study was the Vario TOC Analyzer. This tool provides measurement results in the form of TOC concentration in mg/l units. TOC standard solution (20 mg/l) is needed to calibrate the Vario TOC Analyzer tool's sensitivity, to get the accuracy of measuring TOC levels in groundwater samples. A TOC standard solution of 20 mg/l was used, obtained from a 500 mg/l dilution of the TOC standard solution that was made before analyzing groundwater samples. This 500 mg/l TOC standard solution is prepared by mixing 531.25 mg KHP (potassium hydrogen phthalate), 2206.1 mg Na_2CO_3 (sodium carbonate), 758.95 mg $NaNO_3$ (sodium nitrate), and 477.7 mg NH_4Cl (ammonium chloride) with ultrapure water as much as 500 ml. It is necessary to calibrate by analyzing the TOC level in the TOC standard solution (20 mg/l) that has been made. Suppose that reading the TOC level in the Vario TOC Analyzer tool against the TOC standard solution (20 mg/l) matches the TOC level (20 mg/l) in the TOC standard solution made. In that case, the Vario TOC Analyzer can be used to analyze groundwater samples that have been collected from the study area. The ultrapure water is also used to test the Vario TOC Analyzer's sensitivity to ensure that the instrument was sensitive to changes in TOC levels in the sample measurement. Ultrapure water has a TOC level of 0 mg/l. It is expected that the results of the Vario TOC Analyzer reading on the ultrapure water will show a TOC level of 0 mg/l. Suppose that reading on the results of the ultrapure water analysis is 0 mg/l. In that case, the Vario TOC analyzer can be declared sensitive to changes in the TOC concentration in the samples can be analyzed being tested so that the collected groundwater sample. After the sensitivity test of the TOC standard solution (20 mg/l) and ultrapure water, shows the yield levels of the Vario TOC Analyzer according to the TOC levels in the standard solution (20 mg/l) and ultrapure water (0 mg/l), analyse of groundwater samples may conducted.

3.2 Numerical Model

In this study, numerical modeling of groundwater flow and mass transport was carried out only for the upper hydrostratigraphic unit using Visual Modflow and MT3DMS. The aquifer is assumed to be isotropic and homogeneous. Groundwater flow modeling was carried out at a steady-state, while mass transport was carried out transiently.

The research area with an area of 1.2 km^2 is discretized into 60 x 40 grid cells, where each grid cell's size is approximately 26.2 m x 19.6 m. Vertically, this model is divided into two layers, where layer 1 represents the upper hydrostratigraphic unit / unconfined aquifer and layer 2 represents the clay layer as the base of the unconfined aquifer. The hydraulic conductivity value in the hydrostratigraphic unit was applied at 3.8 x 10^{-4} m/s, while the limiting clay layer was 1 x 10^{-9} m/s. The first surface elevation layer is 120-113 masl to a depth of 49-50 masl. Furthermore, the lower level of the boundary between the shallow aquifer and the model's lower elevation limit is entered. The lower limit of this model's elevation is at an elevation of 50-55 masl, representing a continuous layer of clay.

The modeling boundary conditions applied to the model are river boundary, constant head boundary, specified flow boundary, and no-flow boundary. Input data from the river boundary were obtained from observations of the Winongo river in the western part of the study area and observations of the Code River in the eastern part of the study area. Data relating to the river boundary include river water level elevation, river sediment base elevation, river sediment thickness, and the value of the hydraulic conductivity of river sediments and river width. The river observation data are shown in Table 1. The model's constant head boundary conditions are placed on the model area's northern and southern boundaries. The constant head value was applied in the range 99-103 meters above sea level in the northern boundary. In the southern part of the model, the constant head boundary value was defined 95-100 meters above sea level. Groundwater level and groundwater flow patterns were obtained from shallow wells' elevation measurements in the study area (see Table 2 and figure 3). Recharge is applied as the specified flow boundary and has value of 500 mm/year. In contrast, the no-flow boundary was placed on the model's western and eastern boundaries as inactive cells (see figure 4).

Table 1. Observations of the Winongo and Code rivers.

Site	River stage (m)	River width (m)	The thickness of water (m)	The thickness of sediment (m)	River base elevation (m)	Type of sediment	Hydraulic conductivity of sediment*(m/s)
SW1	92	8.7	0.49	0.3	91.21	Medium sand	9 x 10^{-7}
SW2	90	10.8	0.19	0.11	89.7	Medium sand	9 x 10^{-7}
SW3	90	13.5	0.27	0.15	89.58	Medium sand	9 x 10^{-7}
SW4	90	10	0.94	0.25	88.81	Medium sand	9 x 10^{-7}
SW5	88	11.6	0.3	0.16	87.54	Medium-fine sand	2 x 10^{-7}
SC1	98	13	0.14	0.12	97.74	Medium-fine sand	2 x 10^{-7}
SC2	95	10.4	0.26	0.18	94.56	Medium-fine sand	2 x 10^{-7}
SC3	100	11.1	0.18	0.16	99.66	Medium-fine sand	2 x 10^{-7}
SC4	96	11.4	0.3	0.2	95.5	Medium sand	9 x 10^{-7}

SW: Winongo River, SC: Code River *K average value based on sediment grain size [27]

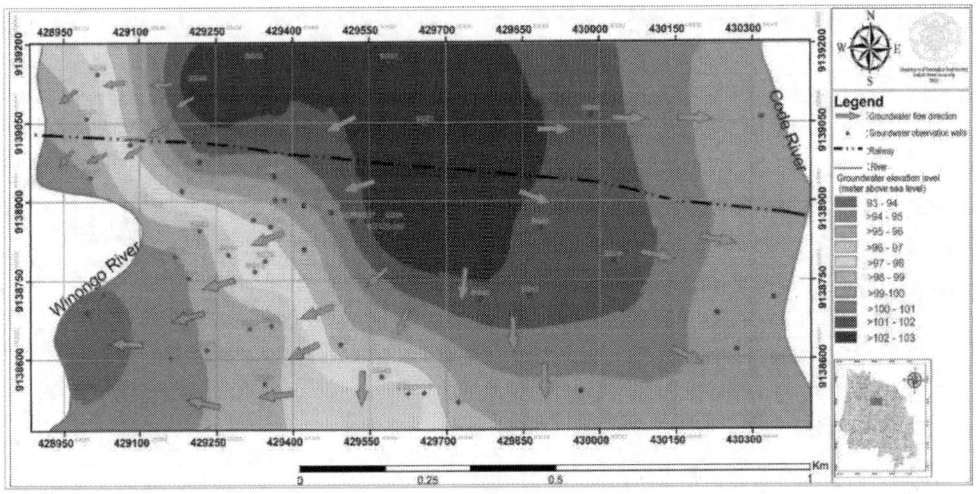

Figure 3. The contour of groundwater level and groundwater flow patterns measured in the study area.

Table 2. Observations of the depth and groundwater level in the study area.

Obs.Site	Elevation (masl)	Water table elevation (masl)	Obs.Site	Elevation (masl)	Water table elevation (masl)	Obs.Site	Elevation (masl)	Water table elevation (masl)
SG 01	115.8	102.8	SG 19	113.5	100.3	SG 37	109	96.35
SG 02	111	100.8	SG 20	107	98.45	SG 38	110	97.45
SG 03	110.5	95.8	SG 21	105	95	SG 39	114.5	102
SG 04	110	94.9	SG 22	112	98	SG 40	114	101.55
SG 05	106.9	94.9	SG 23	111	100	SG 41	114.5	101.6
SG 06	104	93.9	SG 24	111	97.3	SG 42	114	101.8
SG 07	106	93.3	SG 25	111	98.7	SG 43	113	101.8
SG 08	110	98.4	SG 26	114	103.1	SG 44	110.5	98.55
SG 09	110	98	SG 27	115	103.1	SG 45	114.5	101.5
SG 10	111	99.4	SG 28	111	97.5	SG 46	113	97
SG 11	114	102.9	SG 29	115	103.45	SG 47	111	96
SG 12	114	102	SG 30	114.95	103	SG 48	118	103
SG 13	101	99	SG 31	114	101.65	SG 49	114	100
SG 14	102.6	99.6	SG 32	108	95.2	SG 50	108	95.7
SG 15	110	95.1	SG 33	108	95.3	SG 51	114	99.4
SG 16	110	94.55	SG 34	112	100.6	SG 52	116.5	102.2
SG 17	109	95.85	SG 35	112	99.7	SG 53	115.5	102.2
SG 18	111	97.8	SG 36	111	98.9			

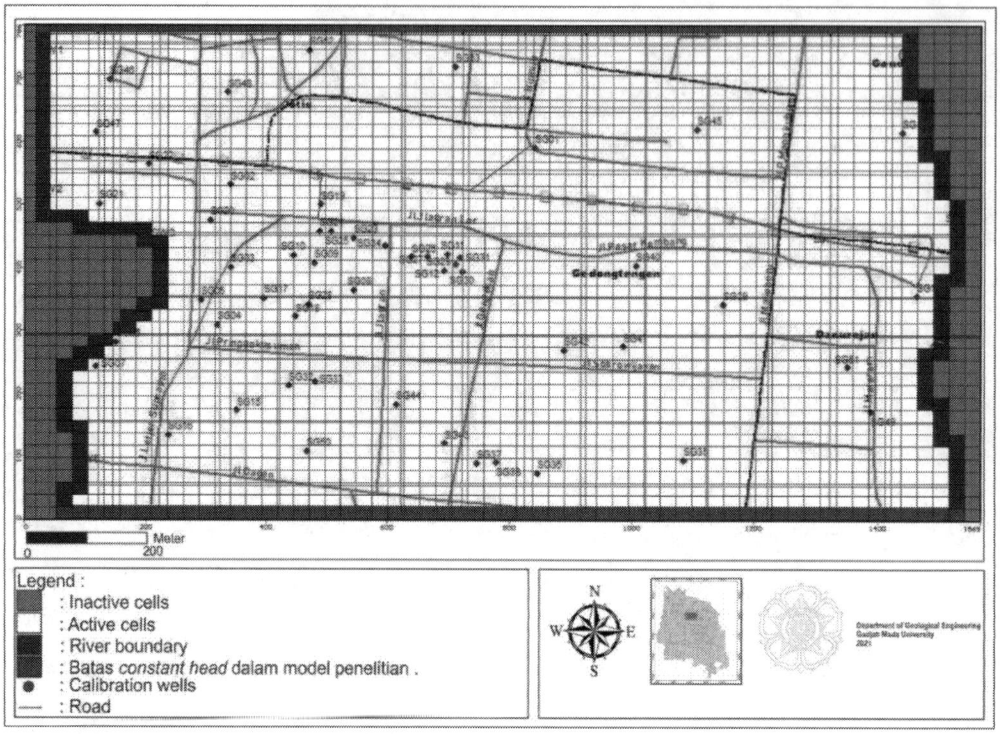

Figure 4. Boundary conditions applied to the groundwater flow model.

In the mass transport model, in addition to the input data above, data are required in the form of pollutant source points, species parameters, bulk density, dispersion, and initial concentration and particle tracking. It is known that organic pollutants undergo biodegradation, so the input data for species parameters in the form of decay value or first order of decay is required. Pollutants in the form of fuel have a half-life of decay ranging from 3-200 days with an average of 34 days [28] or converted to a decay rate around 0.02 /day. The movement of fuel pollutants in groundwater will experience sorption, which is called retardation and the influencing factor is the bulk density of aquifer material. The input bulk density in the model was 1.8 kg/cm^{3,} where the value is taken from the range of bulk density values for sand-sandstones between 1.6 - 2.2 kg/cm^3 [29]. The longitudinal dispersion coefficient value of materials similar to the aquifer material ranges from 1.31×10^{-7} m^2/s to 1.48×10^{-5} m^2/s so that the average longitudinal dispersion coefficient is 2.20×10^{-6} m^2/s [1]. The transverse dispersion value is assumed to be 1/10 of the longitudinal dispersion. The initial concentration of represented fuel polluting groundwater (TOC) in all model areas are considered to be 3.7 mg/l, according to the initial environmental baseline conditions of TOC of groundwater in the Yogyakarta-Sleman Groundwater Basin [26]. The initial concentration of pollutant sources is estimated to be in the range of 300 - 30000 mg/l, where the pollutant sources are in the form of fuel storage tank point and the entire train maintenance location [2]. The TOC level test was carried out using the Vario TOC Elementar tool at the GetIn-Cicero Laboratory. Data from physical observation and TOC levels in community dug wells are shown in Table 3. The TOC levels representing dissolved hydrocarbon at a depth of 1 m below the groundwater level as the LNAPL pollutants are still in their original form or their derivatives [30].

Groundwater flow model simulations were carried out in a steady-state assumption, and groundwater flow model calibrations were based on 53 groundwater level elevations obtained from field observations. The results of the calibrated model were used for transient modeling of MT3DMS mass transport in groundwater. Mass transport modeling was carried out to simulate the incidence of pollution occurring from 2000 to 2001. The time discretization was divided into 0-180 days when the leak occurred 180-8030 days when the fuel storage tank was cleaned and has been no longer used until now. The mass transport model running was carried out in two cases. The first case was carried out when the pollutant source was determined to be the fuel storage area that leaked. Meanwhile, in the second case, the source of pollution was the train maintenance facility's entire area. The mass transport modeling calibration with TOC as its parameter was carried out only for the 2020 simulation results or 8030 days since the fuel storage tank leak occurred.

Table 3. Observation data and TOC levels in dug wells of residents in the study area.

No	Well Code	TOC (mg/l)	Description
1	SG 08	5.02	The surface is cloudy, smells like diesel
2	SG 09	5.35	Clear, no diesel smell
3	SG 10	6.28	Clear, no diesel smell
4	SG 11	25.49	Slightly cloudy, Very smelly diesel, there is some trash
5	SG 12	10.42	Cloudy, smelled a little diesel.
6	SG 19	5.89	Clear, no diesel smell
7	SG 23	9.35	Clear, a little smelling of diesel
8	SG 25	13.41	Slightly cloudy, smelled a bit diesel, not used
9	SG 26	18.10	The surface is LNAPL cake, thick & smells very badly of diesel fuel
10	SG 27	17.69	It smells of diesel fuel, is cloudy yellowish
11	SG 28	5.35	Clear, a little smelling of diesel
12	SG 29	9.42	Clear, the smell of diesel
13	SG 30	4.88	Clear, a little smelling of diesel
14	SG 31	5.97	Clear, a little smelling of diesel
15	SG 34	19.68	A little cloudy yellowish color smells of diesel fuel.

** Water samples for TOC were taken 1 m below the groundwater level*

4. Results and Discussion

4.1 Groundwater flow model

The results of the calibrated steady groundwater modeling with the slightest error are shown in figure 5. The groundwater flow model calibration was performed by zoning the hydraulic conductivity parameters of layer 1 into zones with different hydraulic conductivity, as shown in figure 6. In contrast, in layer 2, the hydraulics conductivity was not changed. In the calibration process, it is known that changes in the recharge value of groundwater, either increasing or decreasing the value, cause a more significant error. Therefore the recharge value was not changed. The plotting diagram between the measured and calculated groundwater level elevation values from the calibrated groundwater flow model results is shown in figure 7. When compared between figure 3 and figure 5, it can be seen that there is a visual correspondence between these two figures, where the groundwater flow was generally divided into three flow directions, namely to the south, to the west as inflow into the Winongo River, and to the east as inflow into the Code River. The groundwater flow pattern towards the rivers shows that the two rivers are a gaining stream type. This gaining stream can be seen as seepages or springs coming out of the river wall. In several locations, the seepage/spring on the river's edge is used by the community as a water source. Based on the calibration diagram and the suitability of groundwater flow patterns with actual conditions, this calibrated groundwater flow model meets the requirements for input data in MT3DMS mass transport modeling. The result of groundwater flow modeling in this study is also in accordance to previous groundwater flow models carried out in Yogyakarta City [22], [23].

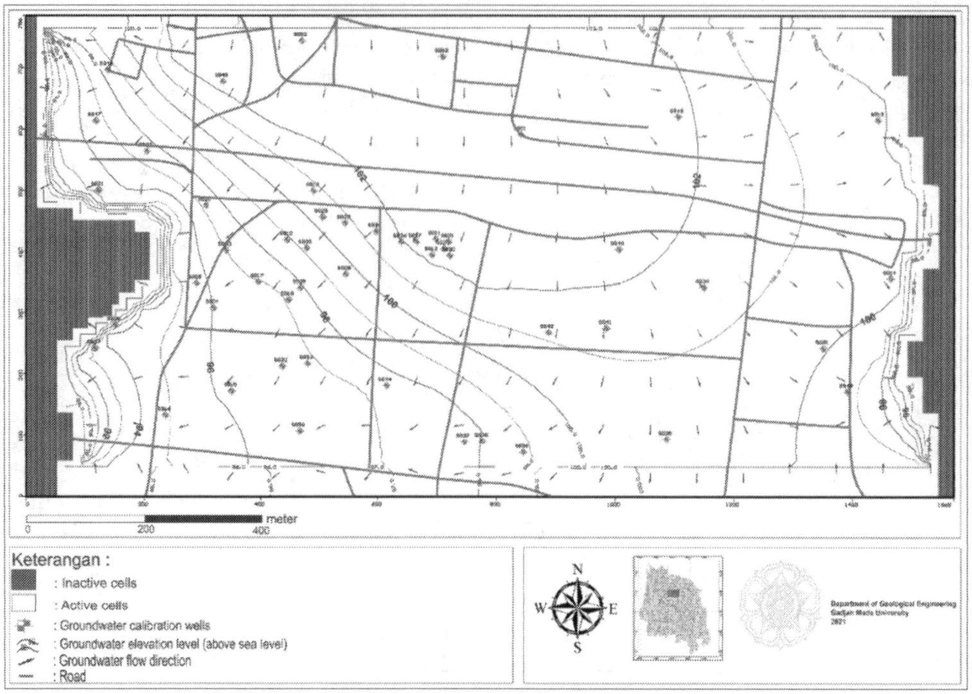

Figure 5. Calculated groundwater levels and flow patterns.

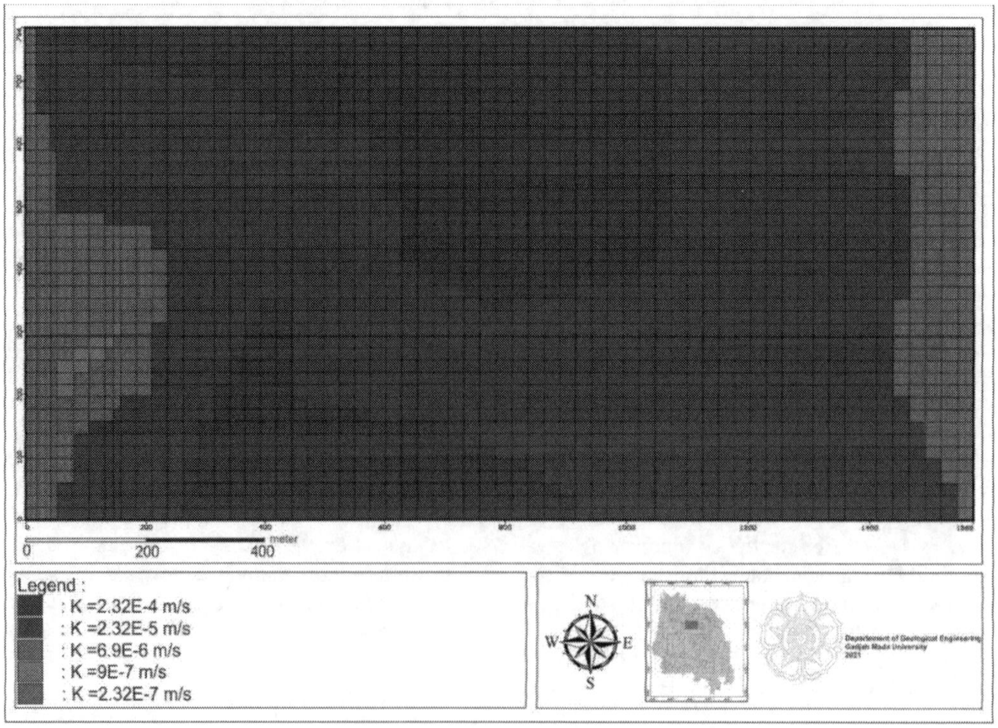

Figure 6. Zoning of hydraulic conductivity in the groundwater of flow model calibration process.

IOP Conf. Series: Earth and Environmental Science **851** (2021) 012024 doi:10.1088/1755-1315/851/1/012024

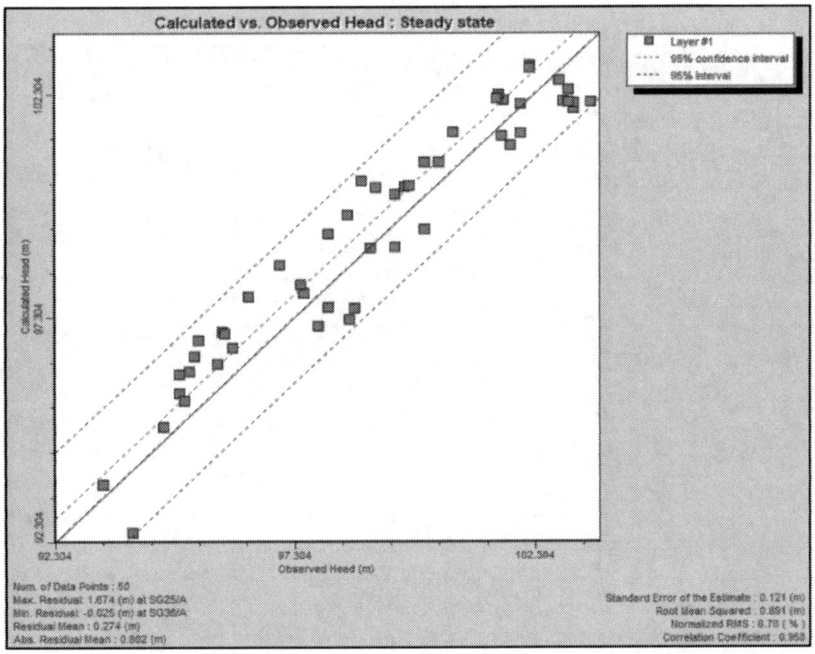

Figure 7. Calibration diagram of measured versus calculated groundwater levels.

4.2. Mass Transport Model of MT3DMS – Case 1

As explained in the methodology, the Case-1 mass transport model was carried out by entering a pollutant source in the area indicated as a fuel storage area of about 20 years ago (see figure 8) as a recharge concentration, with TOC concentration in groundwater is applied as 300 mg/l at a time interval of 0 - 180 days, but after 180 days to 20 years the pollutant source is not activated. The results of MT3DMS case 1 mass transport modeling are shown in figure 9. In figure 9, it can be seen that the pollution plume forms a plume that extends south-southwest about 200 m with a maximum plume width of approximately 70 m to the west-east. In this Case-1, it is clear that the plume of contamination in Case-1 does not affect several dug wells that are indicated to be contaminated, which are analyzed either directly or in the laboratory and have a distinctive fuel odor, turbidity and have high TOC levels such as SG08, SG09, SG10, SG23, SG28, SG29, SG30, SG31, and SG34 wells.

Despite, for the past 20 years, it has been believed and determined that the cause of groundwater fuel pollution in the study area has been only caused by leakage of subsurface fuel storage tank. The Case-1 mass transport modeling results indicate that the possible pollutant source as a point caused by a fuel storage tank leak is not quite right. The result of Case-1 was contrary to the point pollution source used by previous studies on fuel pollution in groundwater of the research area, as reported by Setyaningsih [10], Wijaya [11], and Rahman, et al. [31].

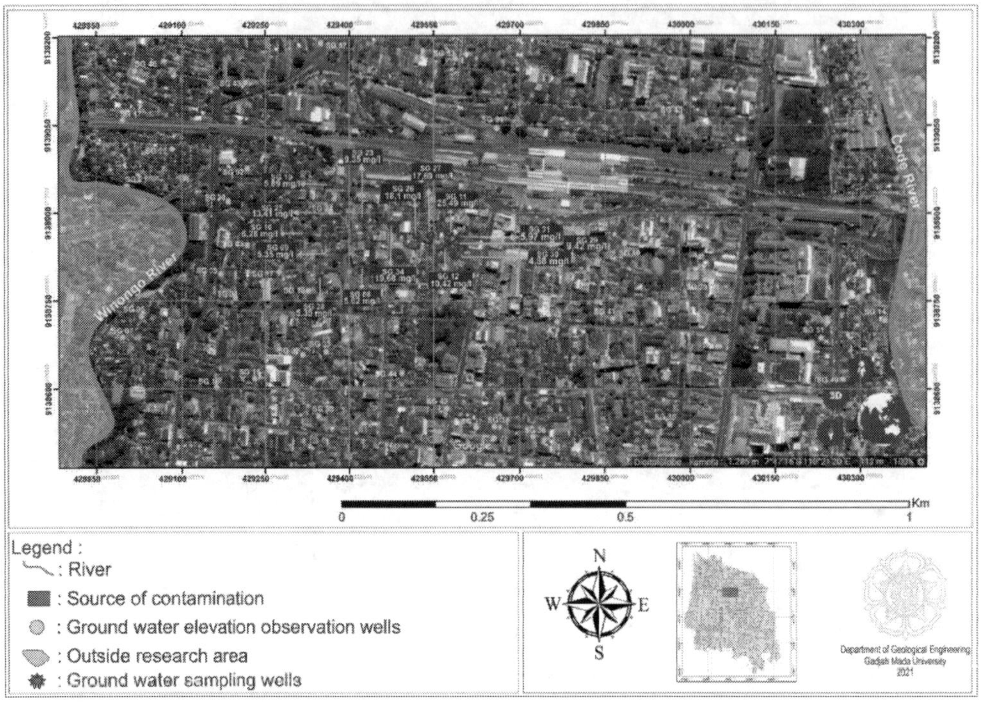

Figure 8. Estimated locations of Case-1 pollutant sources and TOC levels in observation wells.

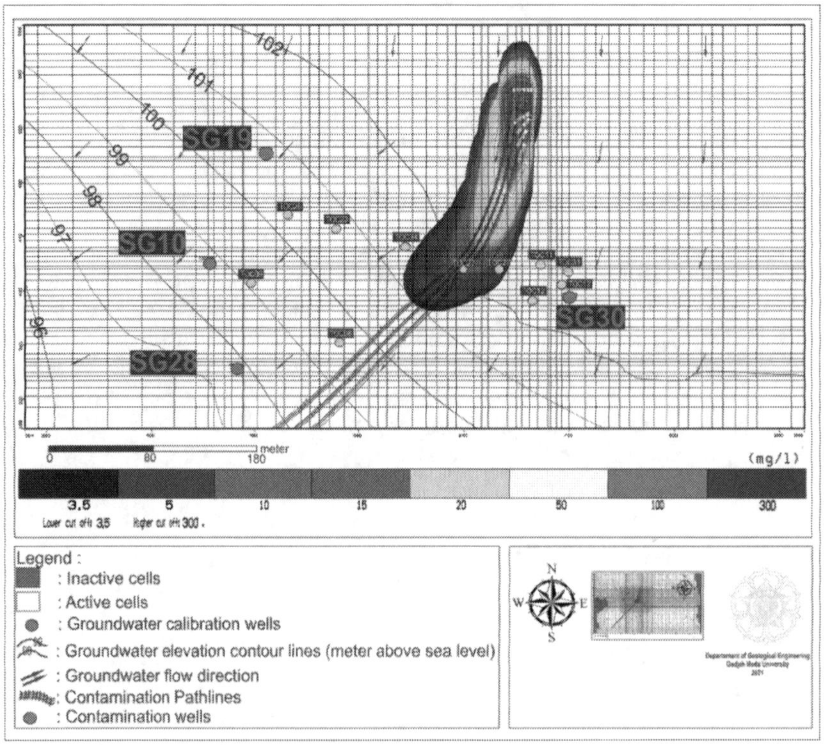

Figure 9. Modeling results of mass transport case-1.

IOP Conf. Series: Earth and Environmental Science **851** (2021) 012024 doi:10.1088/1755-1315/851/1/012024

4.3. Mass Transport Model of MT3DMS – Case 2

Based on the results of the Case-1 simulation, where the plume distribution does not match the current plume distribution, the pollutant source is thought not only from the tank leak (point source) but also from other sources. The best other probable source is the maintenance workshop area. The suspicion area was made possible by the existence of non environmentally practices in disposing of waste in the form of oil or fuel in the past 20 years. It can be seen in the results of research by Priambodo [2], which states that the distribution of TOC levels in the unsaturated zone in the train maintenance area are very high, some even exceeding 30000 mg/l levels and smells of fuel. Priambodo [2] also reports the closer to the water table, the TOC level in the unsaturated zone will decrease. In the Case-2 mass transport model, the facts above were used as the basis for allegations that the source of fuel pollutants has been the source area, as shown in figure 10.

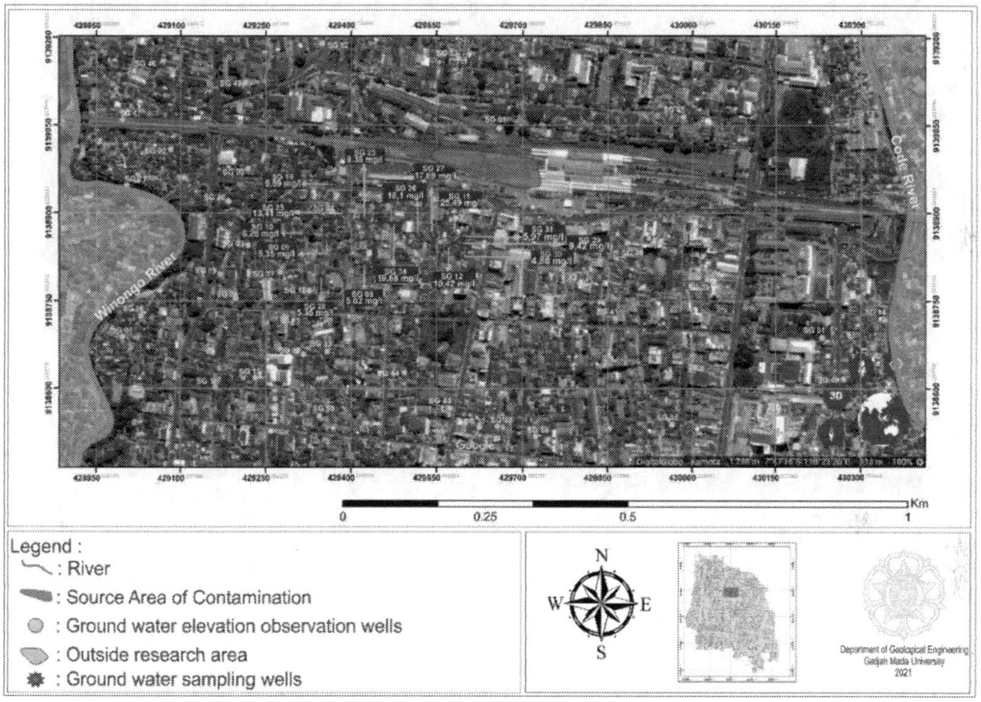

Figure 10. Estimated locations of case-2 pollutant sources.

The Case-2 mass transport model's simulation results by applying a broader pollutant source with the same concentration of 300 mg/l and the same time-division for pollution events can be seen in figure 11. Figure 11 shows that fuel pollution in groundwater is moving according to the direction of groundwater flow, where the plume moves to the south and southwest of the model area. The contamination plume is elliptical and has approximately 370 m and a maximum width of 270 m. For the calibration process, not all of the TOC sampling wells were used for calibration. Since the case of petrol contamination occurred until now, cleaning of community dug wells by pumping was carried out by PT. KAI (Indonesian Railways Company). However, the pumping time, pump discharge rate, and pumping duration to reduce pollutant levels in groundwater around the affected area are not recorded. Due to the pumping activity, the actual results of the pollutant level in most of observation wells cannot describe the actual situation. Based on the interview information obtained during field observation, only SG10, SG19, SG28, and SG30 were never been pumped so that only these four wells were used as a reference for calibrating the distribution of LNAPL pollutant plume concentrations in groundwater. The measured and calculated TOC data calibration at wells SG10,

SG19, SG28, and SG30 reveals good calibration results with a small error value and a correlation coefficient of up to 0.992 (see figure 12). The results of this Case-2 modeling prove that the source of the case of fuel pollution in groundwater in the study area does not only comes from the leakage of the fuel storage tank as a point source, but also from the entire train maintenance (area source) due to the possibility of poor waste management practices in the train maintenance in the past.

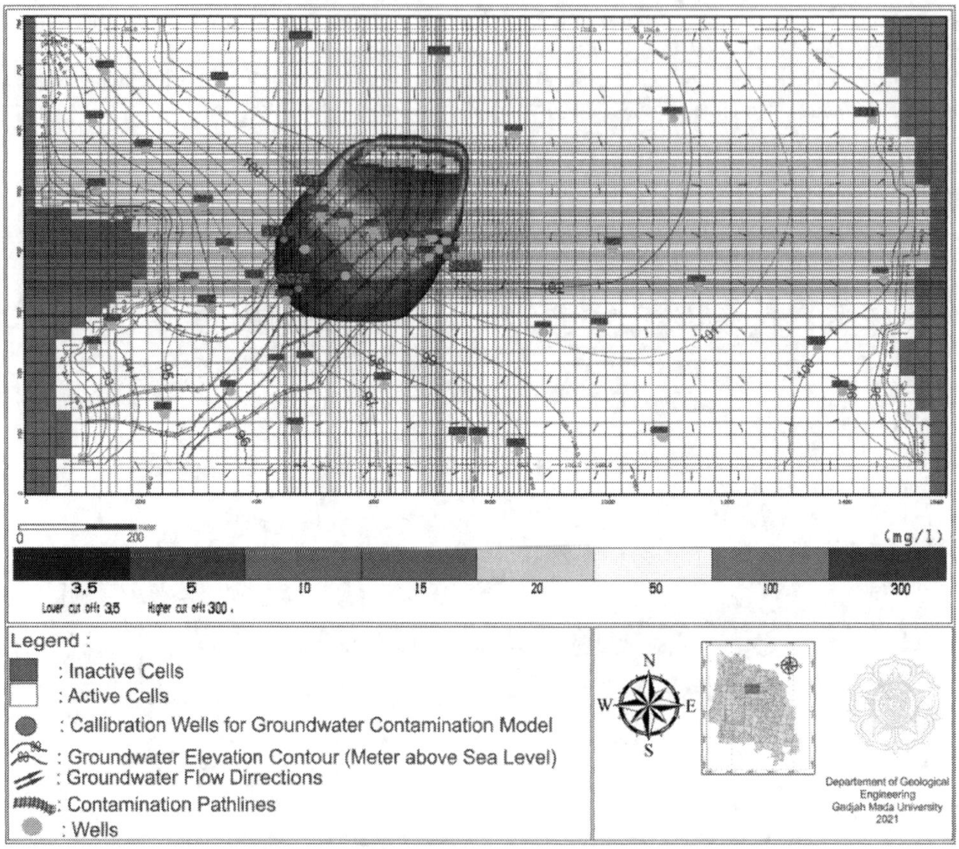

Figure 11. Mass transport modeling results of case-2.

The length and width of the pollution plume produced by the MT3DMS mass transport model are not much different from Setyaningsih [10] and Wijaya [11]. In theory, this indicates that the pollution plume has tended to be stable since 2010, which means a balance between the pollutant input (mass loading) and the pollutant attenuation process in groundwater. Concerning attenuation, in addition to natural aspects in the form of sorption and biodegradation, it may also be supported by the pumping activities of the dug wells of the affected population to clean LNAPL, which carried out so far by PT. KAI. According to PT. KAI, although the pumping is not regular, these activities have continued to be carried out since groundwater pollution occurred. Based on these findings, groundwater remediation from this fuel pollution must be carried out immediately, including remediation of the pollution plume and remediation actions for the presence of hydrocarbon residuals in the underground water-unsaturated zone of the train maintenance area.

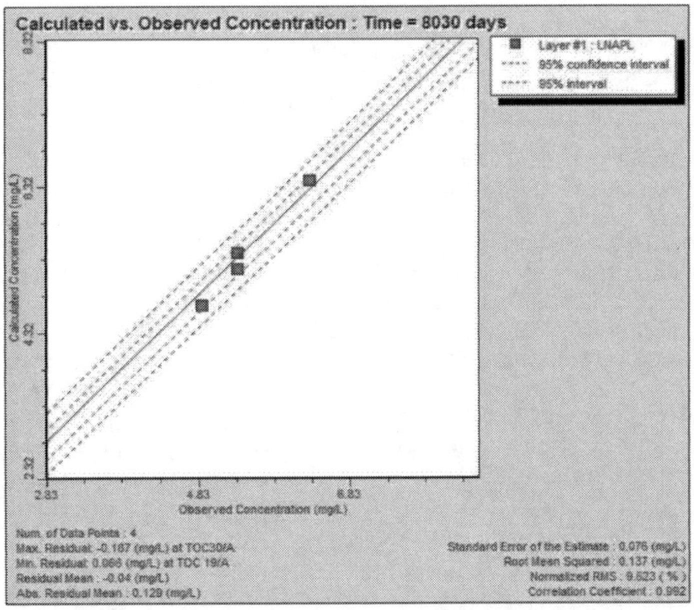

Figure 12. Calibration diagram of measured TOC concentration vs. calculated TOC concentration in the Case-2 mass transport model.

5. Conclusion

The study of mass transport modeling of fuel hydrocarbon pollution in groundwater of Jlagran, Kapanewon Gedongtengen, Yogyakarta City using TOC as the pollutant and the application of Visual Modflow / MT3DMS software can show the actual possible sources of hydrocarbon pollutant. It is already two decades that the source of pollution was believed to come from the fuel storage tank's leaks. However, the recent study proves that the train maintenance area is also the source of pollutants. In the mass transport model, the actual plume of pollution based on the TOC concentration in groundwater only matches if the pollutant source was determined as a whole area of the train maintenance area. Despite the negative findings, the modeling also contributed to non-continuous operational pumping to clean LNAPL in community dug wells carried out by the train company. Moreover, these pumping activities can reduce pollutant levels in dug wells. However, the pumping indeed has no significant effect on the spreading of the pollution plume. To accelerate the restoration of the groundwater environment in this area, active and passive remediation measures are required to reduce the length, width, and concentration of pollutants in the groundwater.

Acknowledgments

The authors would like to thank PT. KAI which provides access to the affected locations and to Department of Geological Engineering, Universitas Gadjah Mada for supporting the research budget. Last but not least, thanks also to the GetIn-Cicero Laboratory for access and support the TOC analysis.

References

[1] C. Destunawati, 2020. *Kajian Penyebaran Pencemaran Organic Carbon Pada Air Tanah Di Kecamatan Gedongtengen, Kota Yogyakarta, Daerah Istimewa Yogyakarta.* Skripsi, Universitas Gadjah Mada (*Unpublished*).

[2] Y. Priambodo, 2020. *Kajian Kadar Total Organic Carbon (TOC) di Zona Tidak Jenuh Air Kelurahan Pringgokusuman, Kecamatan Gedongtengen, Kota Yogyakarta.* Skripsi, Universitas Gadjah Mada (*Unpublished*).

[3] Pusat Kajian LKFT UGM, 2019. *Laporan Survey dan Pengukuran Nilai Parameter Pencemaran Air Tanah di Stasiun Tugu Yogyakarta dan Sekitarnya, pada Tahun Anggaran 2019.* Yogyakarta (*Unpublished*).

[4] C. Y. Chiang, J. P. Salanitro, E. Y. Chai, J. D. Colthart, and C. L. Klein, 1989. *Aerobic biodegradation of benzene, toluene, and xylene in a sandy aquifer-data analysis and computer modeling.* Groundwater, vol. 27, no. 6, pp. 823–834. https://doi.org/10.1111/j.1745-6584.1989.tb01046.x.

[5] G. B. Davis *et al.*, 1999. *The variability and intrinsic remediation of a BTEX plume in anaerobic sulfate-rich groundwater.* Journal of Contaminant Hydrology, vol. 36, no. 3–4, pp. 265–290. DOI: 10.1016/S0169-7722(98)00148-X.

[6] P. D. Franzmann, B. M. Patterson, T. R. Power, P. D. Nichols, and G. B. Davis, 1996. *Microbial biomass in a shallow, urban aquifer contaminated with aromatic hydrocarbons: Analysis by phospholipid fatty acid content and composition.* Journal of Applied Bacteriology, vol. 80, no. 6, pp. 617–625. DOI: 10.1111/j.1365-2672.1996.tb03266.x.

[7] M. Huth and A. Helios, 2013. *Fuel flexibility in gas turbine systems: impact of burner design and performance,* Modern Gas Turbine Systems, Woodhead Publishing, pp.635-684.

[8] D. W. Tomlinson, M. O. Rivett, G. P. Wealthall, and R. E. H. Sweeney, 2017. *Understanding complex LNAPL sites: Illustrated handbook of LNAPL transport and fate in the subsurface.* Journal of Environmental Management, vol. 204, pp. 748–756. DOI: 10.1016/j.jenvman.2017.08.015.

[9] B. Bekins, B. E. Rittmann, and J. A. MacDonald, 2001. *Natural attenuation strategy for groundwater cleanup focuses on demonstrating cause and effect. Eos*, vol. 82, no. 5, pp. 57–58. DOI: 10.1029/01EO00028.

[10] W. Setyaningsih, 2010. *Model Pergerakan Bahan Pencemar Minyak Desel Pada Akuifer Batupasir Formasi Volkanik Merapi Muda.* Jurnal Geografi, pp. 75–87.

[11] S. Y. C. Wijaya, 2015. *Pencemaran airtanah oleh bahan bakar minyak di kelurahan jlagran, kecamatan gedongtengen, kota yogyakarta*. Universitas Gadjah Mada.

[12] C. Zheng, M. C. Hill, G. Cao, and R. Ma, 2012. *MT3DMS: Model use, calibration, and validation.* Transactions of the ASABE, vol. 55, no. 4, pp. 1549–1559.

[13] H. Liang, L. Guodong, and L. Xinfeng, 2009. *Modeling and predicting groundwater pollution caused by oil pipeline leakage using MODFLOW and MT3D.* 3rd International Conference on Bioinformatics and Biomedical Engineering, iCBBE 2009, pp. 1–5. doi:10.1109/ICBBE.2009.5163071.

[14] Z. Xu, Y. Wu, and F. Yu, 2012. *A Three-Dimensional Flow and Transport Modeling of an Aquifer Contaminated by Perchloroethylene Subject to Multi-PRB Remediation*. Springer, vol. 91, pp. 319–337.

[15] L. Colombo, L. Alberti, P. Mazzon, and G. Formentin, 2019. *Transient flow and transport modeling of a historical CHC source in North-West Milano.* Water (Switzerland), vol. 11, no. 9, pp. 1–18. DOI: 10.3390/w11091745.

[16] C. Gao, X. Guo, S. Shao, and J. Wu, 2020. *Using MODFLOW/MT3DMS and electrical resistivity tomography to characterize organic pollutant migration in clay soil layer with a shallow water table.* Environmental Technology (United Kingdom), pp. 1–24. DOI: 10.1080/09593330.2020.1767699.

[17] N. Marić, S. Jana, Kurilic, Beskoski, N. Zoran, Ignjatovic and Malbasic, 2020. *Hydrochemistry of groundwater contaminated by petroleum hydrocarbons: the impact of biodegradation (Vitanovac, Serbia).* Environmental Geochemistry and Health, vol. 42, no. 7, pp. 1921–1935. DOI: 10.1007/s10653-019-00462-9.

[18] T. B. Spruill, 1988. *Use of Total Organic Carbon as an Indicator of Contamination from an Oil Refinery, South-Central Kansas.* NGWA, vol. 8, no. 3, pp. 76–82. DOI: https://doi.org/10.1111/j.1745-6592.1988.tb01088.x.

[19] Sir MacDonald and Partners, 1984. *Greater Yogyakarta Groundwater Resources Study,* Volume 3: Groundwater, Directorate General of Water Resources Development Project (P2AT), Ministry of Public Works, Government of The Republic of Indonesia, 116p.

[20] W. Rahardjo, Sukandarrumidi and H.M.D. Rosidi, 1977. Geological map of Yogyakarta Quadrangle, Geological Survey of Indonesia, Bandung.

[21] H. Hendrayana and D. P. E. Putra, 2003. *The Improvement of Yogyakarta Groundwater Basin Concept,* Proceeding of 32nd IAGI & HAGI annual convention, December 15-17, 2003, Jakarta.

[22] A. Satapona, D. P. E. Putra, and H. Hendrayana, 2018. *Groundwater Flow Modeling in the Malioboro, Yogyakarta, Indonesia*. Journal of Applied Geology, vol. 3, no. 1, pp. 11–22.

[23] D. P. E. Putra, M. Iqbal, H. Hendrayana, and T. Putranto, 2013. *Assessment of Optimum Yield of Groundwater Withdrawal in Yogyakarta City*. Journal of Southeast Asian Applied Geology, vol. 5, pp. 41–49.

[24] D. P. E. Putra, L. Manny, and R. R. S. Atmaja, 2017. *Groundwater Level Changes in Shallow Aquifer of Yogyakarta City, Indonesia: Distribution and Causes*. Journal of Applied Geology, vol. 1, no. 2, p. 89. DOI: 10.22146/jag.27584.

[25] D. P. E. Putra, 2007. *The Impact of Urbanization on Groundwater Quality, A Case Study in Yogyakarta City–Indonesia*, Mitteilungen zur Ingenieurgeologie und Hydrogeologie, Heft 96, Okt 2007, Herausgegeben vom Lehrstuhl fuer Ingenieurgeologie und Hydrogeologie, Univ. Prof. Dr. R. Azzam, RWTH Aachen, Germany, 148p.

[26] L. Kaempfner, 2019. *Redox State and Zonation of Groundwater In the City of Yogyakarta (Java, Indonesia)*. Thesis, Lehr- und Forschungsgebiet Hydrogeologie Faculty of Georesources and Materials Engineering, RWTH Aachen and Universitas Gadjah Mada *(Unpublished)*.

[27] P.A. Domenico and F.W. Schwartz, 1998, Physical and Chemical Hydrogeology, 2nd Edition, John Wiley & Sons, Inc. 506p.

[28] P. Noble and P. Morgan, 2002. *The Effects of Contaminant Concentration on the Potential for Natural Attenuation*. Environment Agency (for England & Wales) R&D Technical Report P2-228/TR. ISBN: 1 85705 5993. Bristol

[29] R. . Freeze and R. Cherry, 1979. *Groundwater*. Edgewood Cliffs, NJ: Prentice Hall Inc. 604p.

[30] P. B. Bedient, H. S. Rifai, and C. J. Newell, 1999. *Groundwater Contamination : Transport and Remediation*. USA: PTR Prentice Hall, 604p.

[31] H. A. Rahman, D. P. E. Putra, and H. Hendrayana, 2018. *Pemodelan Pergerakan Pencemar Airtanah oleh Hidrokarbon di Kelurahan Jlagran Kota Yogyakarta*. Jurnal Geosains Dan Teknologi, vol. 1, no. 3, pp. 81–88.

Study of negative skin friction on floating piles foundation due to long-term groundwater extraction in Semarang, Indonesia

M S Tazakka[1], B A Tirta[2], M A Popang[2] and Y A Sariska[2]

[1]Sandy Soilindo, Yogyakarta, Indonesia
[2]PT. Witteveen Bos Indonesia, Jakarta, Indonesia
*E-mail: sanditazakka@hotmail.com

Abstract. The northern Semarang area is prone to a land subsidence problem as the underlying soil consists of a thick layer of normally consolidated clay which also exacerbated by groundwater extraction to meet water demand. The downdrag impact on a pile structure due to the water extraction from two major aquifers that have been used for domestic consumption was investigated. A 3D-finite element analysis was performed to a raft foundation with steel pipe piles. The study used Banger polder data and soil characteristics based on previous investigations. The evaluation was done by comparing the results of negative skin friction (NSF) and total settlement within 50 years period under continuous extraction and one without. From the result, the corner pile located farthest from the pump mobilized NSF of 64% of the total negative skin friction which was highest than middle and interior-corner piles of 28% and 20%, respectively. The groundwater extraction was also found to contribute to almost double the settlement in 50 years in the non-piled area, from 0.24 m without extraction to 0.42 m accounted for. Also, the empirical $0.75L_{pile}$ for the maximum neutral point of the pile is in agreement in the result of this study.

1. Introduction

Semarang is one of the biggest cities located in Java Island Indonesia. From a recent survey in 2017, The city encountered one of the biggest increases in the population in its history as the city population increased from 1.5 million to almost 1.7 million in 2017 to 2019 [1, 2]. The growing demands of basic needs are inevitable, especially the domestic water demands for daily consumption. The water supply in Semarang is provided and is under PDAM (municipal water supply company) authority for both surface and groundwater sources. The water demand in the area increased yearly with the largest percentage of 83% or 37.50 million m^3 that had been supplied by PDAM for domestic purposes [3].

According to DESDM (government local office of energy and mineral resources), in Semarang, 96% of the groundwater extraction is used for industrial purposes and the remaining 4% is for domestic purposes (e.g., public housing). In this project located in Banger, northeastern of Semarang, there are two confined aquifers found within the first 100 m below the ground (upper and lower aquifer). From both aquifers, the groundwater that is extracted via a deep well on the lower aquifer of the area is mainly used for industrial purposes and the upper aquifer is used via a shallower well dedicated for domestic purposes. There has been groundwater monitoring at this location conducted by DESDM [4]. From time to time, as recorded on the monitoring well, the groundwater piezometric head on the monitoring well becomes lower due to this water extraction on both aquifers.

Groundwater extraction is known as one of the causes of land subsidence in Semarang especially in this project location in particular [5]. In this paper, the effect of groundwater extraction on both of these aquifers was studied. The study was focused on the impact of long-term groundwater extraction on the piles' foundation of building on the Banger polder based on finite element analysis. Finite element analysis can be used as a method to predict the impact of land subsidence due to groundwater extraction [6, 7]. Nguyen, Small and Poulos [7] showed the groundwater extraction can induce axial and lateral deflection on piles increasing piles' bending moment. In the long-term run of groundwater pumping, the piles may reach critical condition and get damaged. In this paper, the finite element model of land subsidence due to groundwater extraction was modelled using software called Plaxis 3D 2020 and the scope was limited to study the negative skin friction on the piles. In Banger polder, there has been a study on the impact of the landfill that includes groundwater extraction. It is mentioned that land subsidence in the area is significantly caused by the landfill instead (e.g., past and new reclamation) [8]. The groundwater extraction impact is rather small if compared to the overburden compressive load caused by the landfill. This paper discusses the extension of the research on the impact of the land subsidence in Banger to piled building solely from groundwater extraction and natural consolidation of the local area.

1.1. Land subsidence from a geotechnical engineering perspective

Land subsidence occurs worldwide and becomes one of the most scrutinized geohazards in an urban or rural area. Land subsidence is the occurrence of a vertical settling or the sinking of the ground surface [9]. once the ground is settled, it is a non-reversible situation. It can occur naturally for instance due to the degradation of organic material in soils, or it can be induced by human activities such as underground mining or in combination [10, 11].

In civil engineering design, land subsidence can be taken into account for design consideration. For example is Jakarta Bay artificial islands reclamation design in which the land design elevation did consider the land subsidence from the nearby location which was recorded from digital measurement (e.g., GPS, InSAR) incorporated with consolidation settlement [12]. Full Modelling of land subsidence can be quite a challenge since many factors can influence this phenomenon. When a soil stratum or soils strata below the ground are compressed, the surface area is facing land subsidence. The discussion of how and why compression of soil occurs is widely discussed in the geotechnical engineering world. In the other words, land subsidence can be approached with a geotechnical engineering model. Bootsma, Kooi and Erkens [13] illustrated to model land subsidence in Netherland's peat by taking into account three main causes of compression on soils (compression by the degradation of organic materials due to oxidation, compression by shrinkage, and compression by loadings). A compression by loading in this matter is due to the change of effective stress in soil. For instance, an increase of effective stress due to embankment construction above clayey soil will cause settlement that consists of initial settlement and consolidation settlement (primary). Eventually, under constant loading, the creep settlement (secondary) on the ground will take place.

In theory, either primary settlement or secondary settlement is a result of a decrease in soil void ratio. Bjerrum [14] also describes how the primary consolidation can occur in nature as a result of an increase of effective stress (e.g., overlying natural sedimentation soils). Under a constant loading, clay deposit will encounter what is called creep settlement or secondary settlement at any time. Creep term here is limited to a time-dependent vertical directional settlement, as closely related to the primary consolidation of clays soil from a geotechnical perspective. In the simplest words, creep is the rearrangement of the soil grains over time resulting in a volume-change in the soil body. Creep theory was pioneered by Buisman [15] and Terzaghi [16]. The approach to explain the natural creep phenomenon in fine-grained and organic soil has been researched for the past decades [17-26].

Another example of soil compression by loading is groundwater extraction. The water that is extracted from the ground will reduce the pore water pressure on the extracted stratum increasing effective stress below the remaining submerged soils. Hence the consolidation can take place in the first place at the overburdened soils. The more complicated case can also happen when the aquifer itself becomes compacted due to this water extraction. The famous land subsidence example is in San Joaquin Valley where the land subsidence was induced by continuous groundwater pumping over decades [27]. Water extraction in porous soil reduces the water amount in the soil void, and it will increase the soil compressibility resulting in the settling of soils above the aquifer.

1.2. Piles design subjected with negative skin friction
Piles foundation may be used for building in the location where the soft soils exist or when the load of the building is high (e.g. high-rise buildings). Designing a pile foundation in the soil where the surrounding soil settles or subsides requires extra attention to the bearing capacity of the foundation. Piles will experience downdrag as the soil surrounds settle.

1.2.1. Negative skin friction (NSF) on piles
Piles foundation needs to be designed to bear the building above the ground. To mobilize pile bearing capacity, relative movement of the pile to the soil is needed. Pile bearing capacity is mobilized through the positive skin friction (PSF) whereas piles are needed to be loaded from the top hence shear resistance and end bearing, respectively, on the shaft and pile tip are developed due to the relative movement of the piles to the soil (e.g., [28, 29]).

The opposite of positive skin friction (PSF) is known as negative skin friction (NSF). NSF is generated when the soil moves down relative to the pile. NSF will produce additional loading on piles as drag load develops on piles. Drag load on piles is caused by downdrag on the piles. Downdrag is a settlement of pile due to the settling of the surrounding soil. The effect of negative skin friction has been extensively studied over the past decades whether for single pile or piles in a group [30-32].

Consolidating soft soils or settling soils surround the structure's piles will induce drag load on piles. The drag load that is related to NSF is much more generated on the pile located in the outer corner compared to the pile located in the interior (inside the perimeter). This is known as the "shielding effect" in which piles located in the perimeter will act as a shield that reduces the amount of drag load and settlement that occurred on piles interior [33].

1.2.2. The practical design of pile subjected with NSF
When a pile is subjected to NSF, no bearing capacity or positive skin friction is developed along the pile skin until the neutral transition point between NSF to PSF is reached. This means, that the piles tend to experience settlement issues and hence piles need to be designed to rectify this. The transition point or neutral point plays an important role to calculate the pile bearing capacity. Axial load from pile top needs to be transferred to the soil via pile skin friction contact. This pile bearing mechanism is illustrated in figure 1. Indraratna [32] showed that the neutral point of piles shifted downward as the soil surrounding the piles settled/consolidated. At the early stage of construction, the neutral point position is relatively higher compared to the later time with a low degree of negative skin friction mobilized. After some time, eventually, the mobilized skin friction reaches critical value with the most minimum neutral point on the piles. This critical value means the highest possible mobilized negative skin friction that occurs on the pile's system and can be measured via in-situ tests such as static pile loading test. However, conducting a static pile loading test is very expensive and highly time-consuming and may only be cost-effective for big projects. When designing piles subjected to NSF, a designer may consider following international standards such as SS CP-4 2003 [34]. For floating piles or friction piles on soft clays, SS CP-4 2003 suggests that the position of neutral point is 0.6 of the entire pile lengths (0.6 L_{pile}) and the degree of mobilization (η) of negative skin friction (Q_{NSF}) that resulting the down drag load is typically 0.67 or 1 out of the total Q_{NSF} (0.67Q_{NSF} or 1 Q_{NSF}). As another rule of thumb, the neutral point may also be taken as 75% from the total pile length or 0.75 L_{pile} [35].

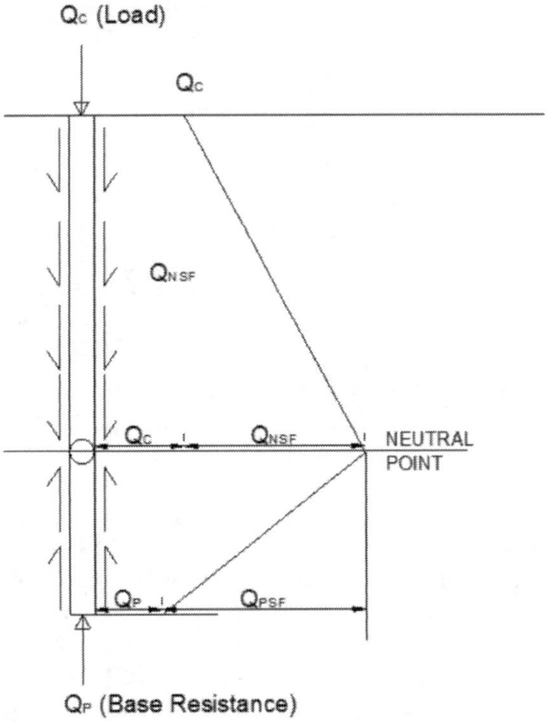

Figure 1. Scheme of axial forces on pile [29].

2. Geological setting on Banger

Semarang stratigraphy, in general, comprises three underlying formations, surficial deposit and sedimentary rocks, volcanic rocks, and intrusive rocks [36, 37]. From a geotechnical perspective, the sedimentary, volcanic, and intrusive rocks categorized as Pleistocene, Pliocene, and Miocene formations, respectively, are assumed to be incompressible layers. The surficial deposit is categorized as Holocene formation which is considered compressible. The surficial deposit is widely represented by river channel and beach deposit which is situated in a low-lying coastal area in northern Java, including also Semarang [38].

2.1. Banger polder's subsurface condition and hydrogeological setting

Banger polder is on the Semarang's polder located in the northeastern part of Semarang. Banger area is categorized as Java's surficial deposit formation. It is given the estimated soil profile which includes Banger area in figure 2, that this location is mainly dominated by clay and silt soils for the first 100 m below the ground surface. The estimated cross section is applied only for the natural formation on the area, the artificial landfill (e.g., past, and present reclamation) is not included.

In Banger, a total of 2 aquifers has been discovered from some boreholes executed [8, 39, 40]. Figure 2 shows two major confined aquifers shallow and deep aquifer called alluvium and Delta Garang deposits aquifer, respectively. The first aquifer, the alluvium aquifer, is located at depth -19 to -25 below the ground surface. This first aquifer is confined by two aquitards in between this shallow aquifer layer. On the deeper part of the area, the Delta Garang aquifer is located at the depth -65 to -75 below the ground surface and is confined by two aquitards as well. Both aquifers are categorized as sandy soil.

Across Central Java Province, every groundwater extraction/exploration needs to have permission from a related authority. Information such as the location of the wells, extraction purpose, and volume of the extracted water is under DESDM. According to DESDM, the smallest and largest extraction discharge registered was 2 up to 200 m^3/day, respectively [8]. As of 2020, the actual number of wells remains unknown.

Though the actual operating number of wells is unknown, groundwater monitoring is still being conducted. This groundwater monitoring is important to evaluate the water availability and the condition below the ground. This groundwater monitoring data is discussed further in section 4.

2.1.1. Banger Polder land subsidence

Besides due to its natural soil characteristics (e.g., consolidation and creep settlement of fine-grained soil), land subsidence in Semarang is induced by man-made activity such as groundwater pumping. The aquifers are compressed slowly as the water-extraction is commencing. Efforts of measuring land subsidence in Semarang has been started in early 1996 and has become national concerns since then [41].

Some previous studies have shown the common impacts of land subsidence in Semarang [5, 42]. Other than the apparent 'sinking' of the buildings, coastal flooding called 'rob' (in Javanese) is the widely known impact of subsidence occurrence on the northern coast of Java Island. The impacted area of the flooding is reportedly increased each year. The inundation due to the flooding deteriorates buildings and other infrastructures, such as roads, dikes, and bridges, in the form of cracking, tilting, and some level of differential subsidence. Several public facilities such as dikes and bridges no longer fulfil their functional requirement, being also slowly inundated. Additionally, these resulted in temporary or even permanent displacement of the affected population. The economic impact is also high, not only due to cover the costs of structural repair and/or rehabilitation of these facilities, but also losses from economic activities at the inundated industrial zone. The indirect impacts also include the living condition of the affected population, such as health and sanitation problems that are caused by poor drainage due to malfunction of the drainage system, increased seawater intrusion, and also substantial damaged and/or loss of wetland habitat.

The subsidence rate in the northeast part of Semarang varies, ranging from 0-2 cm/year to more than 10 cm per year. As given in figure 2, the Banger polder location is considered a yellow to red zone with a land subsidence rate of more than 5 cm per year.

2.1.2. Comparison of Banger Polder's clay with other location

In construction practice, especially embankments, area dominated with clay soil needs to be treated extra since clay material is considered as highly compressible material. From a geotechnical perspective, high compressible soil such as clays and peats contain micro voids which are reflected as high porosity (n) or high void ratio (e) value [43]. Clay material is also able to retain a high amount of natural water content in its void. Natural water content (w_n) of soil is expressed as the ratio of water weight over the weight of the soil's solid component in a volume of soil. In some areas in the north coastline of Java, the first few meters from the ground surface is usually clay soil. This clay soil deposited in the coastal region is often called marine clay which is characterized by ultra-soft soil stiffness and high water content [44].

The natural water content of near surface clay on the Banger polder area lies in the same trends if compared with the famous area such as Jakarta Bay or Gresik regardless of the variation of the water contents as shown in figure 3 [8]. In the Banger area, the clay generally has lower natural water content compared to Jakarta Bay's and Gresik's clay. This leads to higher undrained shear strength (c_u) of clays and may be proven by higher N-SPT value, as long as the plasticity index (I_P) value at the same comparable range. With the same stiffness consistency, Banger polder's clays also have lower compressibility parameter of the clays compared to Jakarta Bay's.

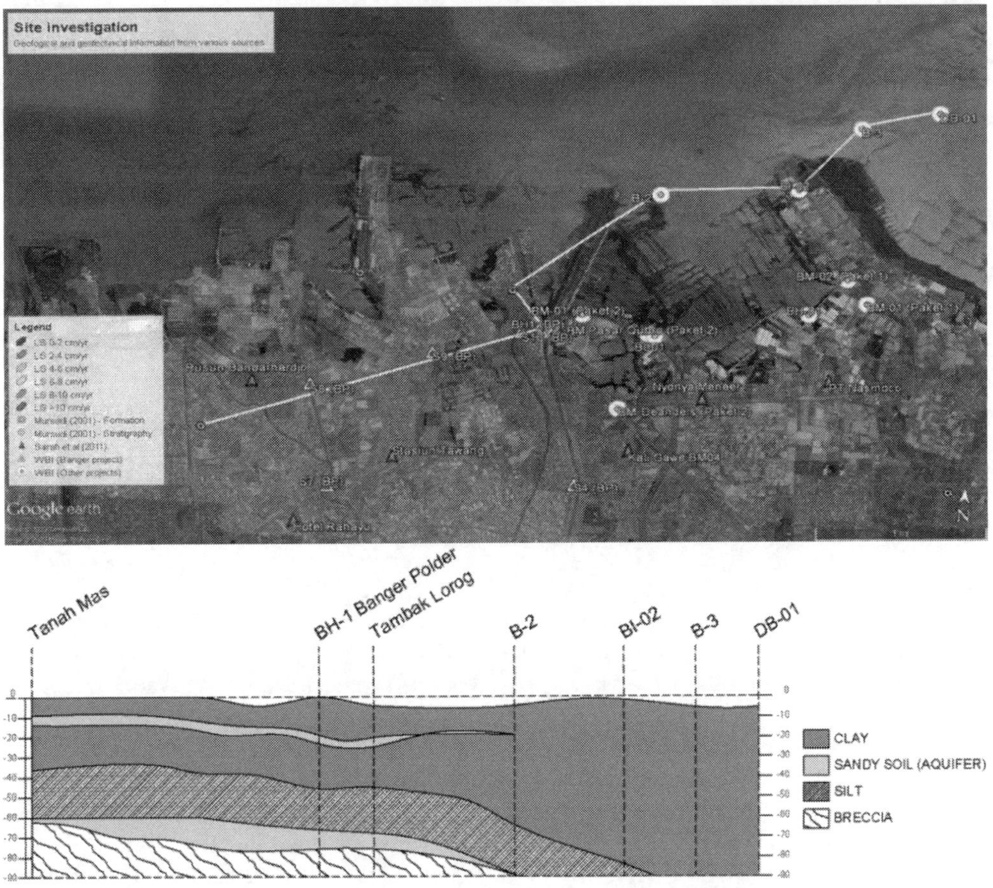

Figure 2. Land subsidence rate on the northeastern part of Semarang (top) and estimated soil profile at Banger polder area (bottom) [8, 39, 40].

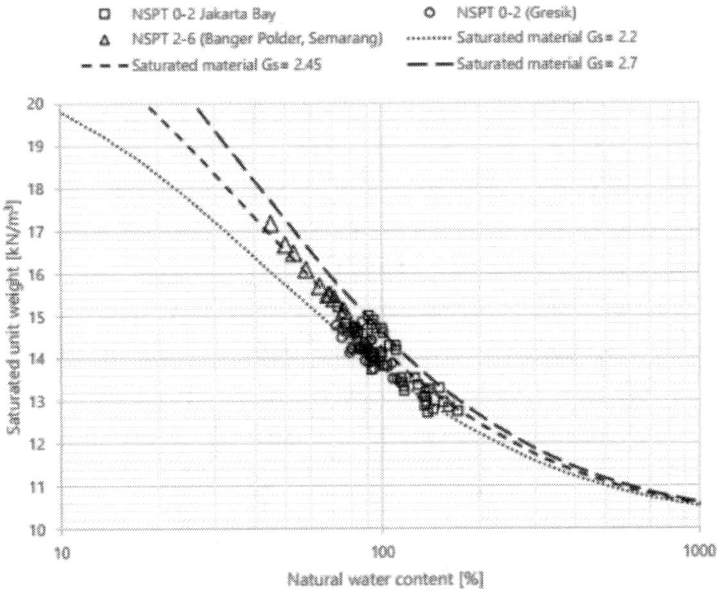

Figure 3. Comparison of near surface natural water content of Banger's clay with Jakarta Bay's and Gresik's clay [8].

IOP Conf. Series: Earth and Environmental Science **851** (2021) 012025 doi:10.1088/1755-1315/851/1/012025

3. Basis of analysis

3.1. Overview of analysis

The effect of groundwater extraction can lead to aquifer compression in which can trigger subsidence above the compressed aquifer. This simulation of groundwater extraction with pumps was carried out to study the effect of negative skin friction on piles foundation on building nearby as the land subsidence occurs. Pumps were installed into the middle of two existing aquifers simulating 50 years of continuous pumping.

3.2. Model boundary condition and geometry

To conduct FEM analysis, some geotechnical and hydraulic parameters are essentially needed. The finite element model geometry and the related boundary condition for analysis are discussed in the following sections.

3.2.1. Soil profile for finite element analysis

The soil profile for finite element analysis was generalized from local boreholes executed in Banger polder [8]. A total of 6 soil layers consist of clays with stiffness consistency soft to stiff and two aquifer layers were defined for analysis. The soil profile for analysis is presented in figure 4.

3.2.2. Soil parameter for finite element model

The geotechnical parameters of Semarang soil were derived as presented in table 1. A typical ratio of Cc/Cr for soft (5.0-6.0) to stiff (7.0-8.0) clay was considered. A similar procedure in determining Cα, the typical ratio Cα/Cr=0.05-0.06 (soft) to 0.02-0.03 (stiff) were used. This ratio was based on engineering judgment and experience in the other projects in dealing with soft soil. All the layers were assumed to be normally consolidated (NC) as there was no indication otherwise. However, considering the q_c value obtained from CPT for layer deeper than -25 m MSL is very high for an NC condition, which typically comparable for stiff to very stiff clay consistencies. Therefore, for those layers, overconsolidation ratio (OCR) of 1.3-1.4 was used. For the layers above -25 m MSL or of soft to medium stiff consistencies OCR=1.1 was used, which is reasonable for the top layer developed from sedimentation [8].

Depending on the location, the thickness of each layer varies spatially. Considering the intrinsic parameter used for the initial check of consistency can be used to determine the thickness of those layers or to identify a new layer based on any borehole information at the site of interest.

Table 1. Conventional soil parameter [8].

Layer	γsat/γ	Cc	Cr	Cα	e_0	OCR	Cv
	[kN/m^3]	[-]	[-]	[-]	[-]	[-]	[m^2/s]
Very soft clay	14.0/14.0	0.65	0.15	0.015	1.7	1	7.4E-08
Very soft to soft clay	15.5/15.5	0.5	0.1	0.0125	1.4	1.1	6E-08
Soft clay	15.5/15.5	0.5	0.08	0.01	1.2	1.1	5.5E-08
Very stiff clay A	17.5/17.5	0.5	0.05	0.004	1.15	1.3	8.00E-08
Very stiff clay B	17.5/17.5	0.35	0.045	0.003	1.2	1.4	6.50E-08
Deep stiff clay	16.8/16.8	0.5	0.07	0.0065	1.3	1.4	6.00E-08

To identify and to quantify the causes of land subsidence in Semarang, finite element method software Plaxis was used to model a two-dimensional soil layer in Banger polder. After shortlisting the causes in the previous section, the Plaxis model is mainly used to explore the effect of groundwater extraction and consolidation. Plaxis allowed the user to perform a couple of flow-deformation method analysis that simulates deformation due to loading and groundwater extraction due to pumping. Thus, in this section, the term subsidence is to represent the consolidation (primary) and creep settlement (secondary) and used interchangeably in later text. Table 2 presented soil parameters that were Plaxis-specific in the model. The compressible soil layer is assigned as soft soil creep model, whereas the sandy layers that were presumed as aquifers are modeled as linear elastic (E'=8000 kPa) [8].

Table 2. Soil parameter for finite element Soft Soil Creep (SSC) model [8].

Soil layer	$\gamma sat/\gamma$ [kN/m^3]	λ^* [-]	k* [-]	μ^* [-]	K_x=K_y [m/day]	K_z [m/day]	$K_{x,y}/K_z$ [-]
Very soft layer	14.0/14.0	0.103	0.0412	0.00229	8.50E-05	5.67E-05	1.5
Very soft to soft clay	15.5/15.5	0.087	0.0348	0.00217	5.19E-05	3.46E-05	1.5
Soft clay	15.5/15.5	0.099	0.0329	0.00225	4.20E-05	2.80E-05	1.5
Very stiff clay A	17.5/17.5	0.093	0.0233	0.00087	1.02E-04	6.78E-05	1.5
Very stiff clay B	17.5/17.5	0.071	0.0178	0.00065	6.28E-05	4.19E-05	1.5
Deep stiff clay	16.8/16.8	0.094	0.027	0.00123	5.19E-05	3.46E-05	1.5

3.2.3. Pile foundation and finite element model geometry

The floating piles foundation model was designed using a simplified loading condition of 15 kPa area-load on the raft. This load corresponds to 1 to 2 story buildings. The raft sits on low-displacement piles of steel pipe pile (SPP) foundation with 610 mm diameter (D) and 3.5 mm thickness. These very low load and low-displacement piles were intended to produce a low strain on the soil-pile to minimize the undisturbed zone in the soil. In short term condition, the raft-pile foundation was designed to comprise the 25 mm maximum settlement from structural loading solely not included with the global area consolidation and creep settlement. In the long-term run, the consolidation and creep settlement will take place and it was modeled in automatic manners on finite element analysis with Plaxis. Raft thickness of 300 mm was used with a total of 9 SPP piles installed, altogether with 1 m excavation on the topsoil. The center to center of pile (S) distance is 2 m and hence S/D is equal to 3.3. The soil model geometry was set to 50 x 50 x 120 m (length x width x depth) and the raft geometry was defined 6 x 6 x 0.3 m (length x width x depth). In the model, the raft-pile was positioned on the edge of soil model geometry. The geometry of the model can be seen in figure 5. This position was intended to study the effects of groundwater extraction and consolidation with creep settlement between the outer corner pile (C) and the interior piles (middle, B, and interior-corner, A). The groundwater pumps were installed in the middle of the soil model with the installation level at each corresponding aquifers both for the upper and lower aquifer. The first pump is located at X=25 Y=25 with Z_{top} at -19 and Z_{bottom} -25. However, the mentioned geometry is not related to the actual geometry of the pumps. This geometrical parameter was intended to produce a linear groundwater head lowering within the top to bottom of an aquifer. The same principle was applied for the bottom aquifer in which the pump was positioned at X=25 and Y=25 with Z_{top} and Z_{bottom} at -65 and -75, respectively. The discussion on how the groundwater pump was modeled is further discussed in section 3.2.4.

On the software, the piles were modeled as "embedded beam" with the predefined circular tube. In Plaxis 3D, the piles can be modeled as volume piles instead of embedded beam piles as an alternative. However, it is not simple to do the volume piles modeling, and hence embedded piles type of piles were chosen for simplicity and most efficient way to get structural and geotechnical internal forces results (e.g., quick interpretation of bending moment, axial forces, shaft friction, and foot forces on piles). The piles were installed at a tip level of -18, with the pile head elevation of -1 (17 m piles length). The piles were designed to act as floating piles penetrating the soft clay layer until 1 m above the upper aquifer. With this foundation system, the bearing capacity significantly relies on the shaft resistance instead of end bearing resistance. The shaft resistance is modeled as a "layer dependency" model with the maximum mobilized shaft resistance is calculated about 25 kN/m for all piles in all depth and the base resistance is calculated about 65 kN only.

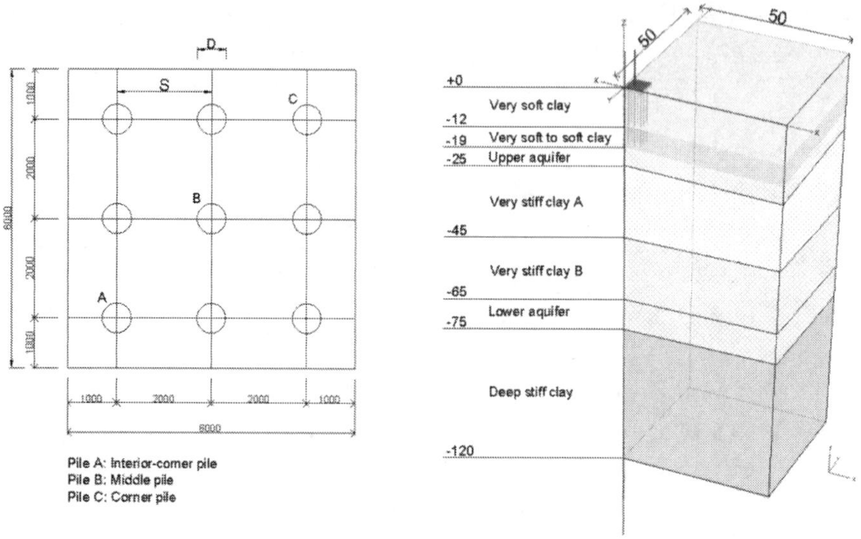

Figure 4. Piling plan in millimetre (left) and finite element model geometry in meter (right).

Subsidence can occur as a result of the loss of soil organic matter (SOM) due to soil oxidation [45]. This type of subsidence factor occurs for organic soils such as peat soil. However, in this project, the biological complexity is not taken into account in the finite element model.

3.2.4. Groundwater head in the finite element model

Figure 5 presents the measurement result from the Anjasmoro station. The distance between the monitoring well and the Banger polder is up to 6 km (this station is one of the nearest stations and has the best data integrity than the others). Besides that, this station is chosen for analysis based on the similar subsidence rate on Anjasmoro to Banger polder as shown in the figure which is 6-8 cm/year. The effect of seasonal recovery in measurement was simplified with a linear trend of 0.4 m/year (4 kPa/year) head decrease rate. This head lowering rate was used to model the groundwater lowering as the impact of continuous extraction at the deep aquifer layer (Delta Garang aquifer). Additionally, groundwater head lowering at the upper aquifer (alluvium aquifer) was modeled 2 times lower at 0.2 m/year (2 kPa/year) based on experimental condition. These groundwater head lowering was applied as a water pump model in Plaxis for both upper and lower aquifer.

The pumps were positioned in the middle of the 3D model (X=25, Y=25) to simulate the groundwater extraction. Although different water head in every stratigraphy layer can be specifically modeled in Plaxis (e.g., entire interested area from edge-to-edge model X=0 Y=50), though it requires a lot of time and computer resources to conduct such coupled flow-deformation analysis. Hence, the water head lowering was simplified into groundwater extraction with pumps with the same water head declines rate as measured in Anjasmoro station for the lower aquifer (0.4 m/year) and for the upper aquifer 0.2 m/year was defined.

In Plaxis, the decreases of groundwater head by pumping that resulting in a decrease of active hydrostatic pressure (in kPa) is produced by inputting Q(m^3) or pumping discharge, h_{min} (m) or the lowest elevation of the funnel, and pump filter length (m). Trial and error of Q, h_{min}, and filter length values were conducted to produce groundwater head lowering at X=25 and Y=25 of pump locations as presented in figure 6. The horizontal distance between pumps to the raft is about 20 m.

This model requires two types of water head, first water head is the initial water head of all soil layers before pumping commencement and the second is the groundwater head lowering of the aquifers. The default of -1 m piezometric head was defined as the initial piezometric head which was obtained from the local boreholes. As the pumping in the aquifers executed with constant discharge, the head in aquifers will decrease over time. The pumping process will reduce the current active pore pressure from initial time (t=0 year) to 50 years (t=50 year) as illustrated in figure 6. In 50 years, the active pore water pressure reduction on the upper aquifer is about 100 kPa (50 x 2 kPa) and for the lower aquifer is about 200 kPa (50 x 4 kPa) reduction of active pore water pressure.

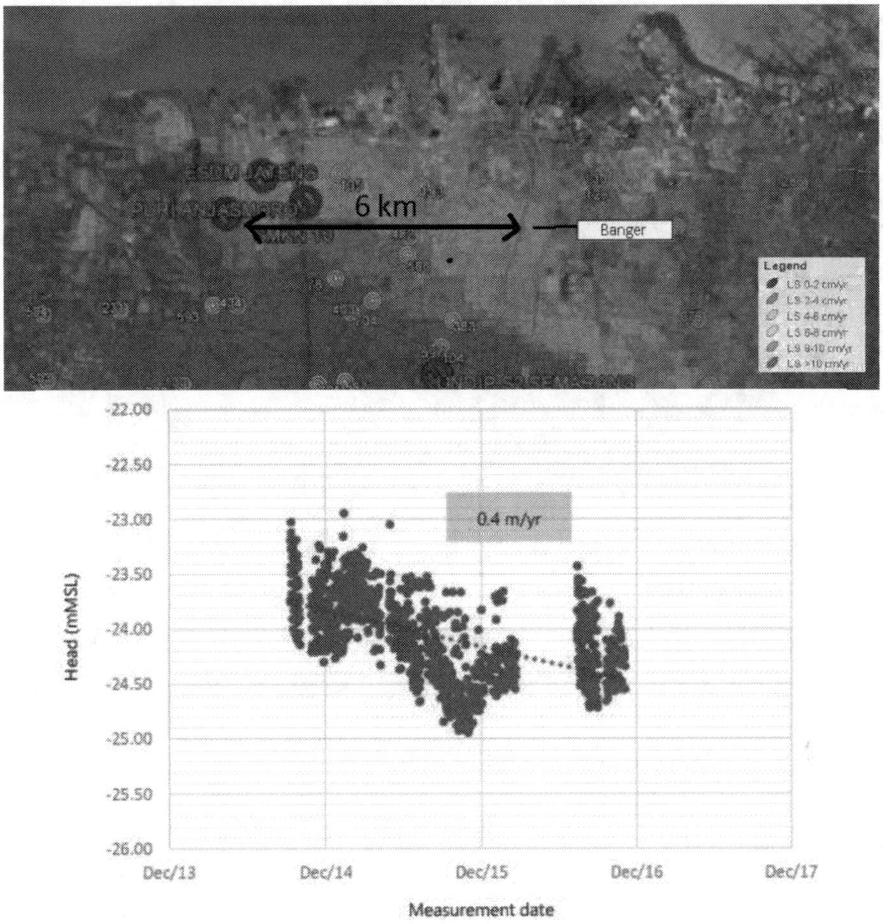

Figure 5. Distance between Anjasmoro to Banger polder (top) and well measurement from Anjasmoro station (bottom) [8].

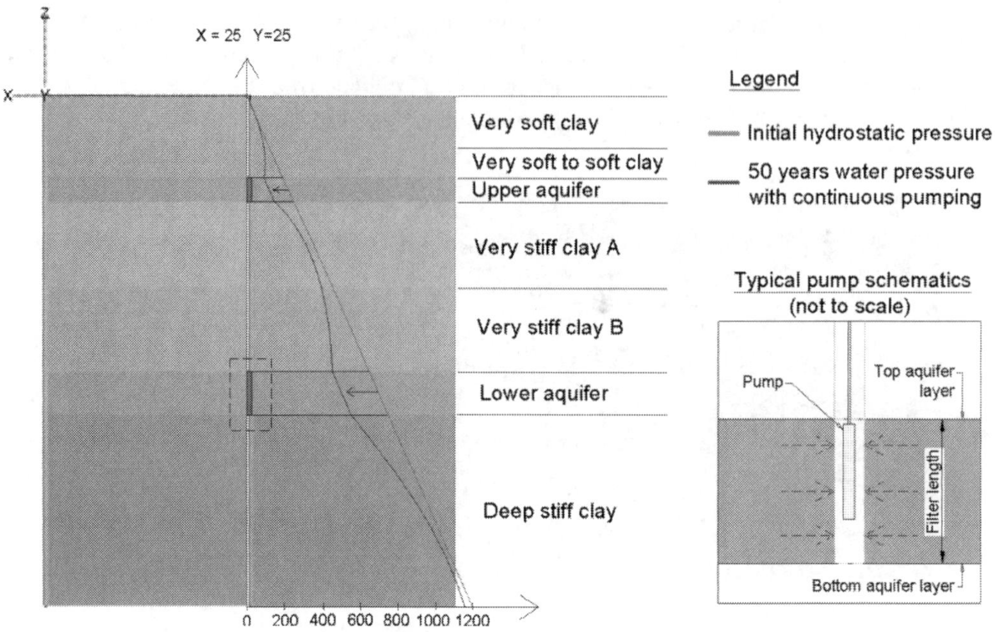

Figure 6. Idealized groundwater head profile in from initial to 50 years condition due to groundwater extraction.

4. Analysis results and discussion

The analysis was conducted with two analysis scenarios. The first model is foundation calculation within 50 years projection without the groundwater extraction. This is the best scenario since the settlement in the area is only affected by the consolidation and creep settlement that will take place for 50 years period. The second model is a calculation with groundwater extraction in 50 years. These two scenarios' results are compared and discussed in the following sections.

4.1. Active pore pressure profile initial to 50 years condition

The pumps were positioned in the middle of each aquifer layer as shown in figure 6. To produce groundwater head declines similar as discussed in section 3.2.4, a constant head of 350 m^3/ day and 1000 m^3/day need to be applied for the upper aquifer and lower aquifer, respectively. The pumps' filter length was defined to be equal to the thickness of each aquifer. The h_{min} (m) or the lowest elevation of the funnel was defined in Plaxis as similar to the top elevation of each aquifer. These simplifications were introduced to produce a constant reduction of active pore pressure profile on each aquifer. As illustrated in figure 7, it is depicted an initial active pore water pressure and 50 years active pore water pressure during pumping at the middle of the model. With the constant discharge of the pumps, this will result in steady-state condition starting at a particular time somewhen between t=0 years and t=50 years. On the active pore water pressure chart, it can be seen that the pumping has also an effect on the changes of active pore water pressure in soil layers whether on top or bottom of each aquifer. This means that the reduction of active pore water pressure will lead to compression not only in the aquifers' thickness but also on the influenced zone due to pumping. However, the effect of water pressure reduction on the overlying and underlying soils on settlement amount is small. The settlement amount due to groundwater extraction is discussed in the next section.

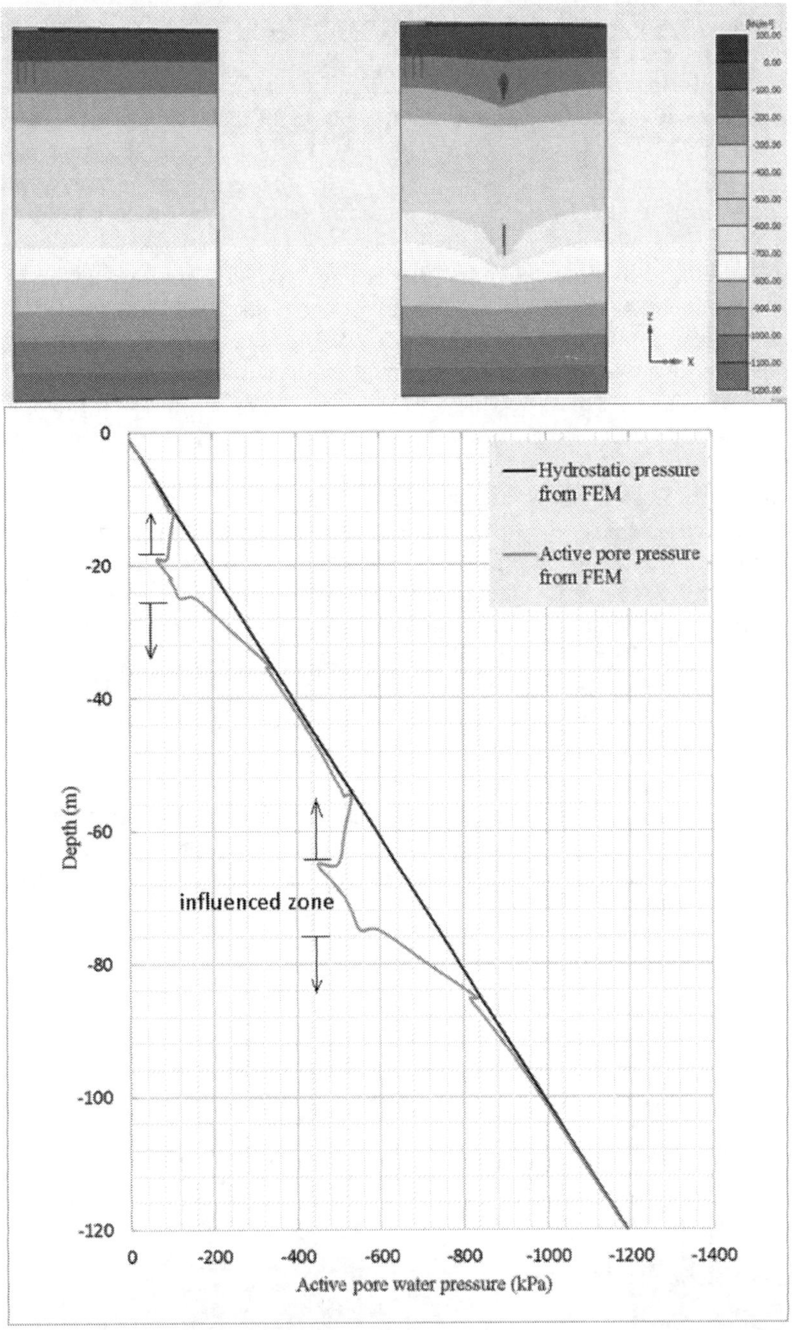

Figure 7. Active pore water pressure changes at initial and 50 years condition due to groundwater extraction from finite element analysis (top) and plot of initial hydrostatic pressure with 50 years water pressure profile at the middle of the model (bottom).

4.2. Settlement comparison in 50 years

Natural settlement in Banger polder is mainly due to the nature of clay soil. Clay soil tends to have vertical creep settlement over time. In 50 years, from predictive analysis, the Banger polder will encounter about 0.24 m settlement and it mainly occurs on the surface. From figure 8, the creep settlement is dominant in very soft and soft clay above the upper aquifer. If the groundwater extraction is commenced for 50 years, the settlement in the area will increase to 0.42 m. This is almost double the increase in settlement from 0.24 m to 0.42 m. The groundwater extraction will increase the settlement of about 0.18 m from the result of 3D FEM.

The increase of settlement of groundwater extraction is considerably low in number. Although, the actual land subsidence rate on the field is very high about more than 5 cm/year as discussed in 2.1.1. This actual land subsidence rate is a product of many factors, one of the most impacting is the landfilling on the surface. As illustrated by PT WBI [8], simulation of 1 to 3 landfill with groundwater extraction was performed to study the effect of a landfill on the subsidence rate in the Banger area conducted with Plaxis 2D axis symmetry model. With a 1 - 3 m landfill, the total settlement is increased to a range of 1 m to almost 3 m in 50 years, respectively. Without taking into account the landfill, it is mentioned that the land subsidence with groundwater extraction only was 0.48 m and it is comparable with the result in this study (0.42 m). It was also conducted a sensitivity analysis of the aquifer's stiffness to evaluate the amount of settlement on the surface. The compression of an aquifer is sensitive with the defined stiffness for the finite element analysis, thus 8000 kPa stiffness for the upper aquifer layer for this area is the most realistic value for this type of soil based on engineering judgment and NEN 6740 [8, 46].

Taking cross-sections from figure 8, it is depicted in figure 9 that the settlement on the raft surface area is reduced. The settlement on the raft surface is apparent to be lesser. The piles are limiting the land subsidence rate on the raft creating differential settlement between raft area and surroundings (e.g. point 1A vs. 1B and 2A vs 2B). Comparing point 1A vs. 1B, it has a differential settlement value of 150 mm (subtracting 250 mm with 100 mm) and 2A vs 2B has a similar value of 150 mm (subtracting 325 mm with 175 mm). The soil beneath the raft looks like hanging on the bottom raft. The reason is that in this model the pile-soil interface is set to R=1.0 to simplify the analysis. However, in reality, this is not always the case since soils beneath the raft can still settle creating a gap beneath the raft to the soil. The groundwater extraction from the upper aquifer will have a high impact on the surface settlement since the upper aquifer layer is compressed as the water from the aquifer soil's pore is gone. Water in the soil pores increases the stiffness of soil, when the water is reduced, the soil stiffness, in reality, is reduced as well. With the increase or even sustain overburden soil weight above the aquifer, the aquifer will have a tendency to compact. The extraction from the lower aquifer from the analysis also has an impact on the area. From depth -25 to -65 the settlement rate is 25 mm to 75 mm.

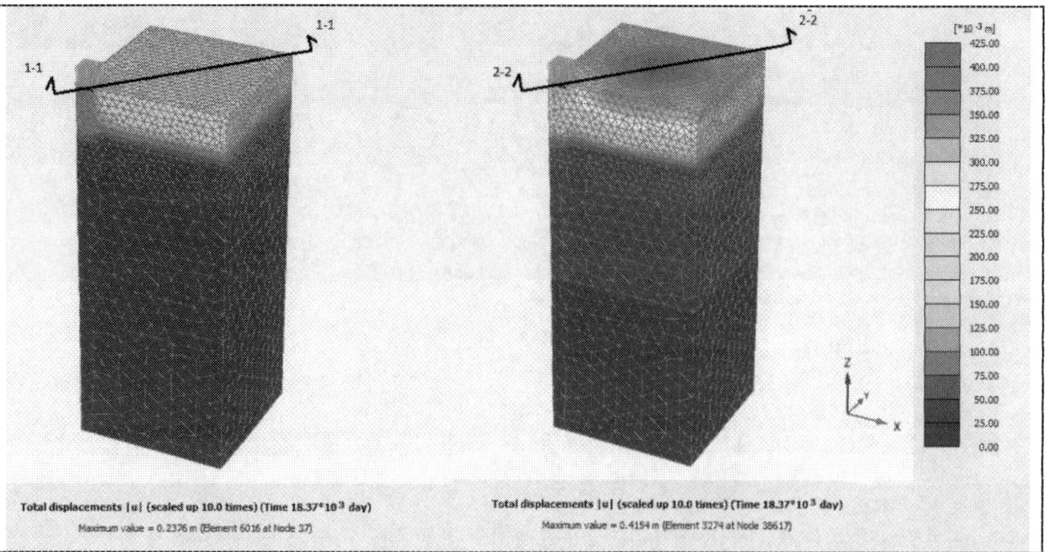

Figure 8. Settlement calculation without groundwater extraction (left) and with groundwater extraction (right) in 50 years.

4.3. Comparison of axial forces with vs without groundwater extraction in 50 years

Reduction of settlement rate as shown in figure 9 on the raft-piles area cost drag loads on the piles in which the settling soil surrounds the piles are dragging down the soil producing negative skin friction (NSF) on piles.

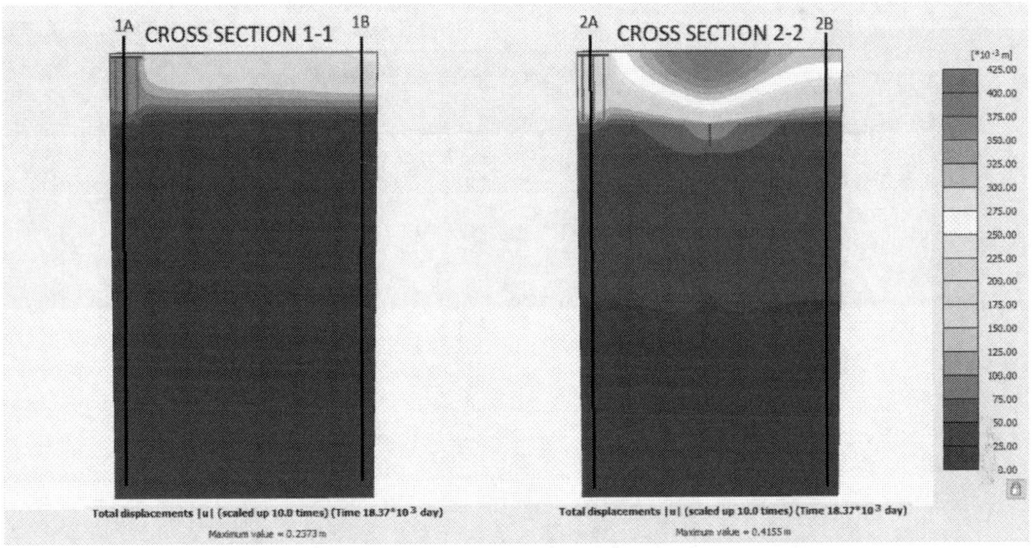

Figure 9. Cross-section of soil model showing piles structure.

Three piles were inspected: interior-corner pile (A), middle pile (B), and corner pile (C) from the results from two scenarios analysis. Figure 10 shows Axial loads plotted altogether for those 3 corresponding piles in the two scenarios. It appears that the raft affects the distribution of axial load on the top piles with pile (C) in the corner suffer the lowest axial load and the middle pile (B) suffers the highest axial load among others. This is reasonable since the surface area of 15 kPa will concentrate in the middle of the raft.

Overall, the groundwater extraction will increase the amount of drag load generated on all piles. The total drag load experienced by the pile C is about 190 kN without groundwater extraction and 210 kN with the commencement of groundwater extraction. Comparing interior-corner pile (A) and middle pile (B), the middle pile's axial force is about 10 percent higher than the interior-corner pile's for the first 10 m depth. Going deeper after -10 depth, the pile axial forces between two piles lies on the same axial forces level for both with and without groundwater extraction. Eventually, the two piles' axial loads reduced at depth -13.5 m, which means that positive skin friction is already developed.

From both scenarios, the pile C in the corner encounters the highest drag load among others. This corner pile and piles located on the perimeter will act as a shield to the other piles, and hence the pile A and B suffer lower drag load. The neutral point of pile C is located about depth -12 and the other piles' neutral point located about at depth -13.5. The pile in the corner suffers the highest rotation due to the drag load or raft surface load, this may be the reason why it has a higher neutral point among others. Comparing those values with the neutral point value obtained from the rule of thumb method 0.6 and 0.75 L_{pile}, -12 and -13.5 are still within approximation value. Recall that Pile top-level is -1 and tip level is -18, the pile length is 17 m. Using the rule of thumb of 0.6 L_{pile} and 0.75 L_{pile} to calculate the neutral point, the piles' neutral points are 10.2 m (depth -11.2 m) and 12.75 m (depth -13.75 m), respectively. Using 0.6 L_{pile} is on the more optimistic side in this case, since 0.6 L_{pile} may only represent the neutral point for the corner pile (-11.2 m vs. -12 m). The 0.75 L_{pile} is considered the more reliable method to simplify the design process to represent all piles, as the hand calculated value of -13.75 is comparable with -13.5 from the result of finite element analysis.

The increases in drag loads after the commencement of the groundwater extraction is related to the degree of mobilization of the skin friction. The axial loads distribution curve for each pile for both scenarios has a similar trend or shape. This means that the mobilized negative skin friction development is more or less have a similar value. The degree of mobilization issue is discussed in the next section.

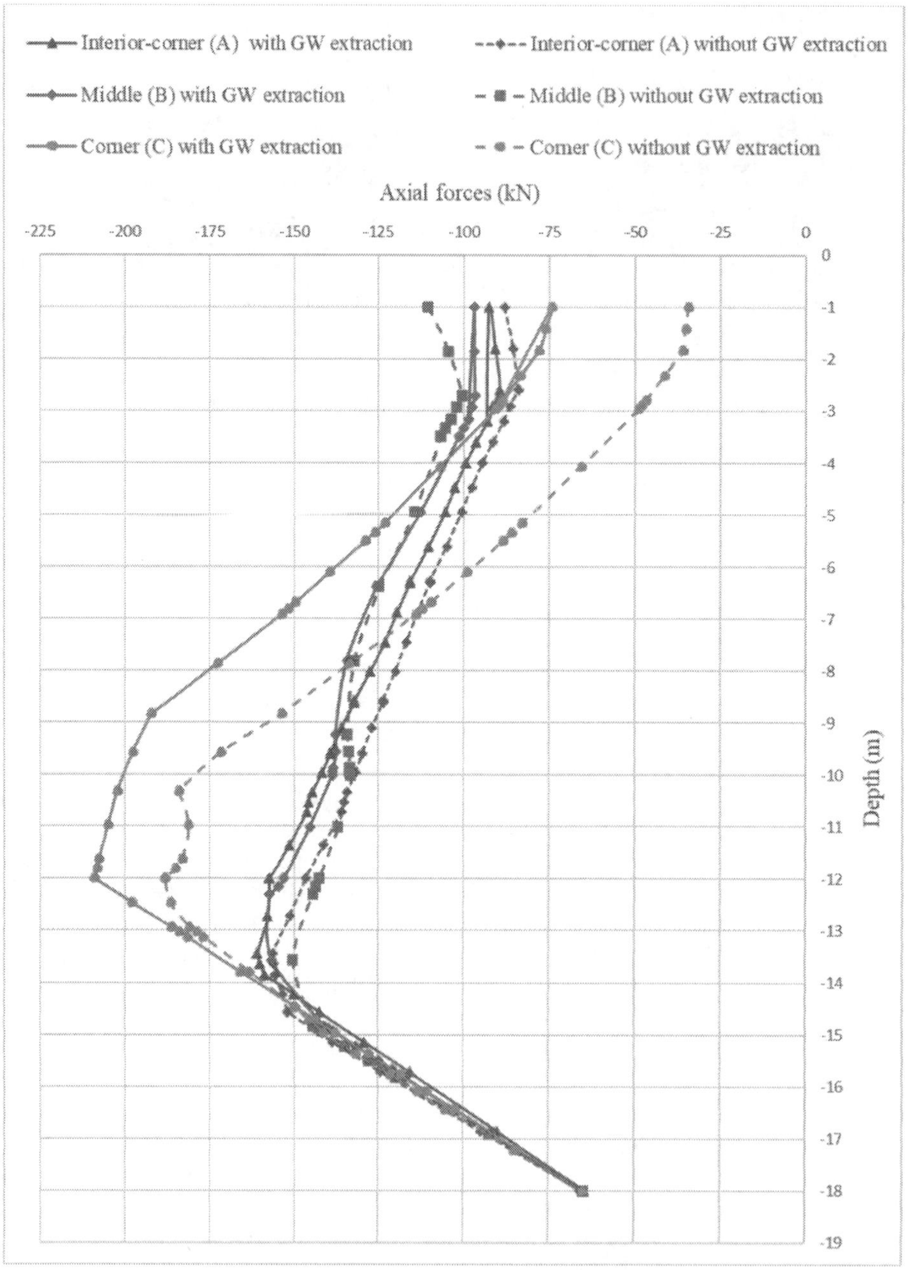

Figure 10. Distribution of axial forces over depth with and without groundwater extraction of three observed piles in 50 years.

4.4. Degree of mobilization between observed piles

In Plaxis, the analysis of axial load on piles can be broken down into a smaller fraction of skin friction mobilization. The positive or negative skin friction development can be obtained and plotted per meter run along the pile length as shown in figure 9. From the figure, each pile has the same trend and shape of skin friction mobilization between two scenarios. The maximum skin friction is limited to 25 kN/m. Corner pile (C) encounters overall higher axial loads, in the other words, it encounters the highest mobilized negative skin friction. The average mobilized skin friction for pile C is about 16 kN/m, with an average degree of mobilization (η) of 0.64 or 64 % (calculated from 16/25 x 100%). The value complies well with the rule of thumb method in which for floating piles at least $0.67Q_{NSF}$ will be mobilized. Piles A encounters the lowest average mobilized negative skin friction of 5 kN/m with the average degree of mobilization of 0.2 or 20% calculated from 5/25 x 100%. Pile B has a little bit higher degree mobilization than pile A, with an average mobilized negative skin friction of 7 kN/m and hence the average degree of mobilization of 0.28 or 28% calculated from 7/25 x 100%. This means that the average degree of mobilization is getting higher on piles towards to outer corner or perimeter. Piles located on the perimeter will act as a shield for the inner piles resulting in a lower degree of negative skin friction mobilization for the shielded piles.

The mobilized degree of mobilization may be related to the spacing of piles and the type of piles that are used on the construction. Thus, the degree mobilization for the interior piles (Pile A and B) may not be used for different conditions. Simply taking all piles as $0.67Q_{NSF}$ will be much more reasonable as an act of a more conservative approach to simplify the design process.
Although in designing "floating" piles designer can assume that NSF will be $0.67Q_{NSF}$ for all piles, it is safer to assume that the mobilized NSF will have 100 percent mobilization unless a static pile loading test is conducted on-site for verification.

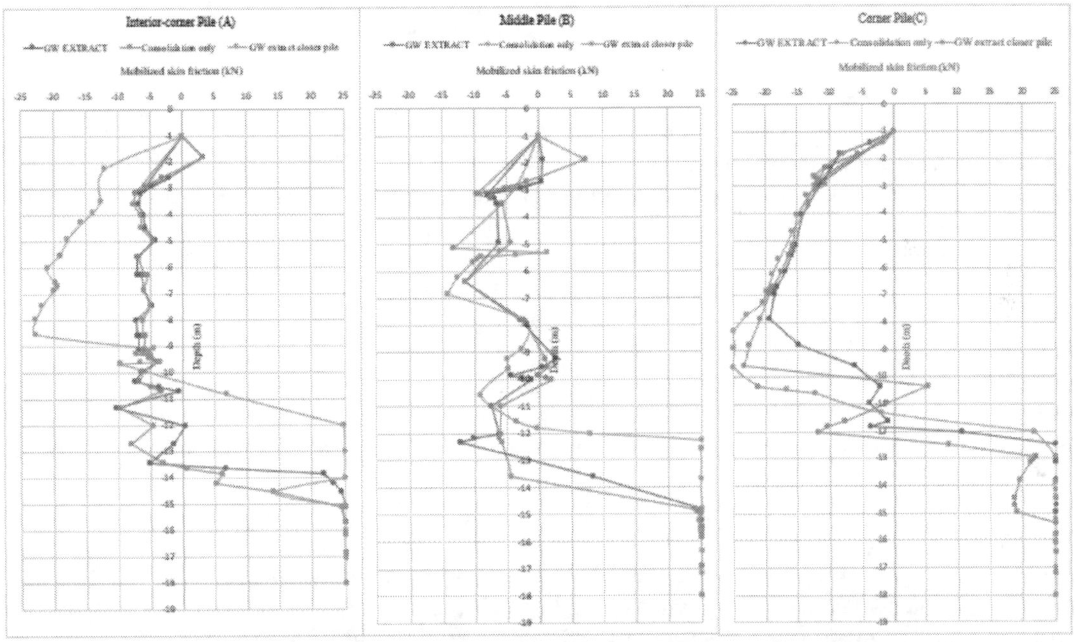

Figure 11. Skin friction mobilization for each observed piles.

The discussed results were modelled by using the corner pile (Pile C) horizontal distance about 20 m to the pump. This means that pile A is completely shielded by the piles located on the perimeter. Another FEM was conducted to put the piles formation closer to the pump with the horizontal distance to pump about 5 m. With this situation, pile A became unshielded, and hence it experiences higher mobilized negative skin friction as added in figure 11. As pile A becomes unshielded, therefore the mobilized negative skin friction is apparent to be similar to Pile C. The other piles, Pile B and C have about the same degree of mobilization although the piles are closer to the pump. Given in figure 12 the settlement profile difference between original piles vs. piles with closer distance, the settlement occurred on the raft is still reduced by the installed piles but the settlement on the raft surface is higher. The closer the building location to the pump, the higher the building settlement.

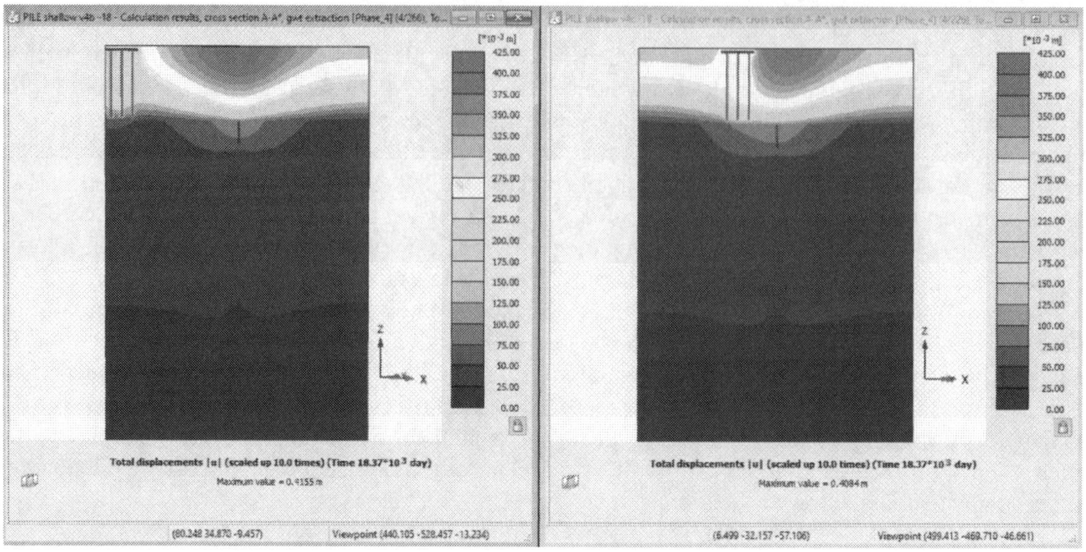

Figure 12. Reduced settlement in piles area.

5. Conclusion and recommendation

The Banger polder area has two aquifers that are still being used for water consumption by the local as of 2020. The groundwater extraction into those two aquifers has triggered land subsidence in the area. The effect of land subsidence induced by groundwater extraction in Banger was studied for long-term situations of 50 years, particularly about the effect of drag loads induced by settling soils surround the piled building.

A finite element model was executed to model and evaluate a foundation building with two scenarios analysis. The first scenario was to evaluate the NSF values of piles in 50 years without the groundwater extraction and the second scenario was to evaluate the NSF values of piles in 50 years with continuous groundwater extraction. The groundwater extraction was modelled as a water pump model by using software called Plaxis 3D 2020. The foundation type was floating piles underlying a concrete raft with 300 mm thickness. The pile used for analysis was steel pipe pile (SPP) type with the diameter and thickness of 610 mm and 3.5 mm respectively. The piles were modelled as embedded beams in Plaxis.

Three piles located on the corner (Pile C), middle (Pile B), and interior-corner (Pile A) were observed from the result of a finite element model. In 50 years with groundwater extraction, the settlement on the area is almost double the value from 0.24 m without water extraction to 0.42 m. The settlement amount may not be accurate in reality (e.g., the calculated settlement vs. direct land subsidence measurement) since the model is dedicated only to study the effect purely from the water extraction and natural subsidence of the area. The effect of actual landfilling (e.g., heightening of the land for housing and roads) was not taken into account in the model.

The pile on the outer corner (Pile C) experiences the highest mobilized NSF for both cases with and without groundwater extraction. Mobilized negative skin friction for pile in the outer corner is about 64 percent out of the total skin friction capacity while the middle and interior-corner experience less mobilization. Therefore, the pile located in the corner is the most critical pile to be concerned when designing piles foundation subjected with downdrag induced by the groundwater pumping. The middle and interior-corner piles experience less downdrag as the corner pile acts as a shield for piles located in the interior.

The neutral points of piles obtained from FEM were also compared with the rule of thumb method when designing piles subjected to NSF. The neutral point of floating piles obtained from FEM shows good agreements with 0.75 L_{pile} to simplify the design process.

This coupled flow-deformation finite element model was produced with a medium mesh system size. The finite element modelling result using Plaxis is very mesh sensitive. The finer the mesh, the precise the analysis. Conducting such analysis with a very fine mesh system is recommended, although refining the meshing for analysis will cost time to finish the analysis. Modelling groundwater extraction with a pump produces the highest settlement magnitude only in the middle of the model. Lowering the water head from edge to edge in the model in a specific aquifer will have better insight to model total subsidence due to the compaction of the whole aquifer. However, doing that will significantly increase the time to run the model. Moreover, land subsidence is a world-wide geohazard problem that can cause issues to the building structures, and hence discussing and compiling geotechnical and structural failures of infrastructures caused by land subsidence can be very interesting and important for future research.

References

[1] Semarang B P S K 2016 Kependudukan: Laju Pertumbuhan Penduduk Kota Semarang 2003 - 2016 (in Indonesian; accessed in November 2019) https://semarangkota.bps.go.id/

[2] Semarang D K d P S K 2018 Statistik Jumlah Penduduk 2018 dan 2018 (in Indonesian, accessed in January 2020) http://www.dispendukcapil.semarangkota.go.id/statistik/jumlah-penduduk-kota-semarang/2018-12-02

[3] Semarang D P R D K 2017 Peraturan Daerah Kota Semarang No 11 Tahun 2017 tentang Perubahan Atas Peraturan Daerah Kota Semarang Nomor 6 Tahun 2016 Tentang Rencana Pembangunan Jangka Menengah Daerah Kota Semarang Tahun 2016-2021, Semarang (in Indonesian) *11*

[4] Mineral D E d S - Bab VII Spesifikasi Teknis dan Gambar Sumur Pemantauan (in Indonesian). (Semarang: DESDM)

[5] Abidin H Z, Andreas H, Gumilar I, Sidiq T P and Fukuda Y 2013 Land subsidence in coastal city of Semarang (Indonesia): characteristics, impacts and causes *Geomatics, Natural Hazards and Risk* **4** 226-40

[6] Hsi J P, Carter J P and Small J C 1994 Surface subsidence and drawdown of the water table due to pumping *Géotechnique* **44** 381-96

[7] Nguyen V D, Small J C and Poulos H G 2001 *Computational Mechanics–New Frontiers for the New Millennium,* ed S Valliappan and N Khalili (Oxford: Elsevier) pp 433-8

[8] Witteveen+Bos 2020 Water Management Semarang In: *Component 1 report: Land Subsidence,* (Jakarta: PT WBI)

[9] Galloway D L, Jones D R and Ingebritsen S E 1999 Land subsidence in the United States. In: *Circular,*

[10] Erkens G, van der Meulen M J and Middelkoop H 2016 Double trouble: subsidence and CO2 respiration due to 1,000 years of Dutch coastal peatlands cultivation *Hydrogeology Journal* **24** 551-68

[11] Holzer T L 1984 *Man-Induced Land Subsidence*: Geological Society of America)

[12] Witteveen+Bos 2010 Kapuknaga Indah Land development Detailed Design Report (Jakarta: PT WBI)

[13] Bootsma H, Kooi H and Erkens G 2020 Atlantis, a tool for producing national predictive land subsidence maps of the Netherlands *Proc. IAHS* **382** 415-20

[14] Bjerrum L 1973 Problems of soil mechanics and construction on soft clays and structurally unstable soils (collapsible, expansive and others).

[15] Buisman A S 1936 Results of long duration settlement tests *Proc. 1st ICSMFE* **1** 103-7

[16] Terzaghi K 1941 *Undisturbed clay samples and undisturbed clays*: Harvard University)

[17] Bjerrum L 1967 Engineering geology of Norwegian normally-consolidated marine clays as related to settlements of buildings *Geotechnique* **17** 83-118

[18] Mitchell J K, Campanella R G and Singh A 1968 Soil creep as a rate process *Journal of Soil Mechanics & Foundations Div*

[19] de Jong G d J 1968 Consolidation models consisting of an assembly of viscous elements or a cavity channel network *Geotechnique* **18** 195-228

[20] Janbu N 1969 The resistance concept applied to deformations of soils. In: *Proceedings of the 7th international conference on soil mechanics and foundation engineering, Mexico city,* pp 191-6

[21] Garlanger J E 1972 The consolidation of soils exhibiting creep under constant effective stress *Geotechnique* **22** 71-8

[22] Mesri G and Godlewski P M 1977 Time and stress-compressibility interrelationship *ASCE J Geotech Eng Div* **103** 417-30

[23] Butterfield R 1979 A natural compression law for soils (an advance on e–log p') *Géotechnique* **29** 469-80

[24] Den Haan E J 1994 Vertical compression of soils

[25] Mesri G 2003 *Soil behavior and soft ground construction,* pp 122-66

[26] Zhang L and O'Kelly B C 2013 Constitutive models for peat-a review. In: *COMPLAS XII: proceedings of the XII International Conference on Computational Plasticity: fundamentals and applications*: CIMNE) pp 1294-304

[27] Poland J F, Lofgren B E, Ireland R L and Pugh R G 1975 Land subsidence in the San Joaquin Valley, California, as of 1972. In: *Professional Paper,*

[28] Fellenius B H 1972 Down-drag on Piles in Clay due to Negative Skin Friction *Canadian Geotechnical Journal* **9** 323-37

[29] Fellenius B 2017 *Basics of foundation design*: Lulu. com)

[30] Shibata T, Sekiguchi H and Yukitomo H 1982 Model Test and Analysis of Negative Friction Acting on Piles *Soils and Foundations* **22** 29-39

[31] Briaud J L, Jeong S and Bush R 1991 Group effect in the case of downdrag. In: *Proceedings of the Geotechnical Engineering Congress*: ASCE) pp 505-18

[32] Indraratna B, Balasubramaniam A S, Phamvan P and Wong Y K 1992 Development of negative skin friction on driven piles in soft Bangkok clay *Canadian Geotechnical Journal* **29** 393-404

[33] Lam S Y, Ng C W W and Poulos H G 2013 Shielding Piles from Downdrag in Consolidating Ground *Journal of Geotechnical and Geoenvironmental Engineering* **139** 956-68

[34] Councils S 2003 SS CP4: 2003. Code of Practice for Foundations. (Singapore: Singapore Standard Councils)

[35] Jeong S and Briaud J L 1994 Nonlinear three dimensional analysis of downdrag on pile groups. In: *Proceedings of the Conference on Vertical and Horizontal Deformations of Foundations and Embankments. Part 2 (of 2)*: ASCE) pp 1366-84

[36] Thaden R E, Sumadirdja H, Richards P W, Indonesia and Direktorat G 1975 Peta geologi lembar Magelang dan Semarang, Java = Geologic map of the Magelang and Semarang quadrangles, Java. (Bandung: Direktorat geologi)

[37] Putranto T and Rüde T 2016 *Hydrogeological Model of an Urban City in a Coastal Area, Case study: Semarang, Indonesia* vol 3

[38] Bemmelen R W v and Indonesia 1949 The geology of Indonesia

[39] Sarah D, Soebowo E, Lubis R F, Murdohardono D and Mulyono A 2011 Identifikasi faktor geoteknik penyebab amblesan di kota Semarang *PROSIDING Pemaparan Hasil Penelitian Puslit Geoteknologi – LIPI (in Indonesian)*

[40] Marsudi 2001 Prediksi laju amblesan tanah di dataran alluvial Semarang Provinsi Jawa Tengah (Ph.D. Dissertation, in Indonesian). (Bandung: Institut Teknologi Bandung)

[41] Marfai M A and King L 2007 Monitoring land subsidence in Semarang, Indonesia *Environmental Geology* **53** 651-9

[42] Andreas H, Zainal Abidin H, Pradipta D, Anggreni Sarsito D and Gumilar I 2018 Insight look the subsidence impact to infrastructures in Jakarta and Semarang area; Key for adaptation and mitigation *MATEC Web Conf.* **147** 08001

[43] Waltham T 2009 *Foundations of engineering geology*: CRC Press)
[44] Raju V R and Sondermann W 2015 Ground Improvement Using Deep Vibro Techniques. In: *Ground Improvement Case Histories*: Elsevier) pp 175-213
[45] Zain N 2019 Effect of Oxidation on the Compression Behaviour of Organic Soils. Delft University of Technology)
[46] NEN 2006 Geotechnics - TGB 1990 - Basic requirements and loads. (Netherland: NEN)

IOP Conf. Series: Earth and Environmental Science **851** (2021) 012026 doi:10.1088/1755-1315/851/1/012026

Urban soil assessment caused by heavy metals contamination in Yogyakarta City, Indonesia

A T Sekarningsih[1], W Budianta[1*], I W Warmada[1], H Hinode[2]

[1]Department of Geological Engineering, Universitas Gadjah Mada
Jl. Grafika 2 Kampus UGM, Yogyakarta, Indonesia
[2]Tokyo Institute of Technology, Japan
*E-mail: wbudianta@ugm.ac.id

Abstract. The effect of urbanization and industrialization in the urban city is soil contamination by heavy metals. This study was conducted to assess Pb, Cu, Zn, and Cd, in the soil of Yogyakarta city and its surrounding, Indonesia. The assessment was done by analyzing 45 surface soil samples in the study area, divided into three-zone. They are divided based on the distance of each zone to the center of the city. The zone III is located in the outermost of the study area, and zone I is inside the city. The results of the study showed that generally, the highest concentration of metals was located in zone I, which is located near or directly situated in a city center. The result indicated that Pb and Cd had the highest pollution index compared to Cu and Zn. The pollition load indeks *(PLI)* and geoaccumulation indeks (I_{geo}) calculations in the whole study area showed that the values demonstrated a moderate class in average. Special attention was needed to be given to the zone I, which has a higher PLI and I_{geo} index to reduce the source of emission for Pb and Cd.

1. Introduction

The geochemistry of soil contamination by heavy metals is a common problem that has been investigated in several countries around the world in the last decades [1][2][3][4]. In the urban area, the degradation of soil quality is caused by industrialization and urbanization [5]. Heavy metal contamination in surface soil in the urban area is an important issue since this phenomenon will affect the environment and human health. Several heavy metals have a toxic behavior, such as Pb, Cu, Zn, and Cd, and have a carcinogenic characteristic that seriously affects human health [6]. Today, about 55% of the world population lives in urban areas, and 54% of the world's urban population is located in Asia. The urban area has rapid population growth, especially in developing countries [7]. This growth usually is located in the peripheral area outside of the city or in the area projected for city development [8].

Yogyakarta city is one of the developing cities in Indonesia, and a world tourist destination, hence the urbanization is an important issue in this city today [9]. As the implication of urbanization in Yogyakarta city, soil quality degradation is one of the environmental issues. Several studies on the soil contamination in Yogyakarta city, especially for Pb soil contamination, show that the concentration of Pb in surface soil in Yogyakarta city was exceeding the normal concentration in general [10][11][12]. This study aims to conduct the assessment of Pb, Cu, Zn, and Cd contamination in Yogyakarta city and its surrounding area.

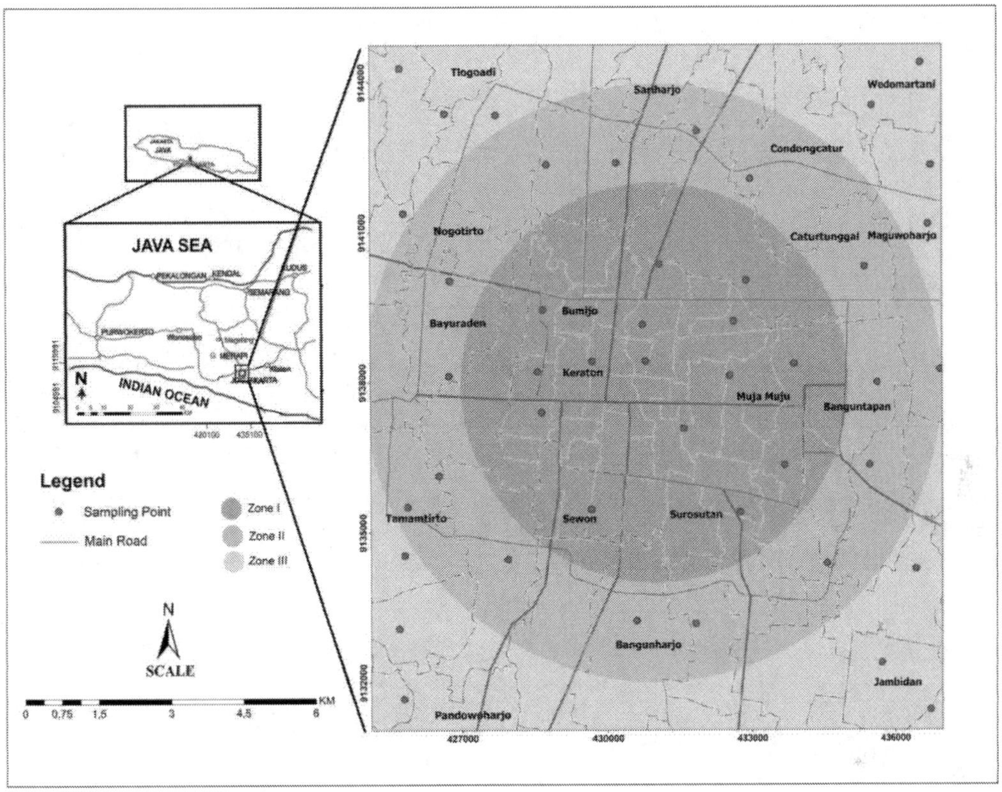

Figure 1. Location of study area and sampling.

2. Material and methods

2.1 Sampling andanalysis

The study area is located in Yogyakarta city, Indonesia, and the location is shown in Figure 1. In this study, Yogyakarta city and its surrounding area were divided into three-zone, as shown in Figure 1. This dividing is based on the distance of each zone to the center of the city. Zone III was located in the outside of the study area, and zone I was inside the city.

Soil samples were obtained in each zone (Figure 1), in which 15 samples were taken in each zone for 5 cm depth from the surface by using a hand auger. About 0.5 kg of soil samples were obtained at each sampling point and put in a plastic bag for laboratory analysis. The samples were dried at room temperature for about two weeks before sending to the laboratory. Laboratory analysis of heavy metals was conducted by preparing a 1.5-gram soil sample that was crushed by agate mortar and pestle and sieved to obtain grain size for <2 mm. The sample digestion was conducted refer to Bettinelli et al, 2000 [13], by adding with strong acid (HF-HCl and HNO_3 mixture) followed by microwave digestion. The heavy metal concentrations in the solution were analyzed by using inductively coupled plasma - atomic emission spectrometry (ICP-AES).

As a representative of each zone (zone I, II, and III), the samples also were taken to analyze for heavy metal analysis. Organic content analysis that was measured by Walkley and Black (1934) method [14] was included. The pH was also measured referring to Conklin (2013) [15].

2.2 Heavy metal assessment in contaminated soil

In this study, the enrichment of Pb, Cu, Zn, and Cd in contaminated soil samples was assessed by pollution index (PI) calculation and compared to the background concentrations in soil. The background concentration was obtained by the previous study, which studied the level of metals in natural unpolluted soil in the study area (Table 1) [16]. The PI was calculated with the formula below [17]:

$$PI = C_s/B_n \qquad (1)$$

where C_s is heavy metal concentration in contaminated soils, B_n is heavy metal background in uncontaminated soils.

Table 1. Background environmental value of the soil in study area.

Parameter	Value (mg/kg)				References
	Pb	Cu	Zn	Cd	
B_n	25.5	23.6	86.1	0.10	Wulaningsih, 2009 [16]

After calculation, the values of the *PI* were classified as followed: $PI < 1$ for low; $1 \leq PI < 3$ for moderate; $3 \leq PI < 6$ for considerable high, and $PI \geq 6$ for very high contamination [18].

A pollution load index *(PLI)* for each sample was calculated to demonstrate the level of pollution of heavy metals [19]:

$$PLI = \sqrt[4]{PI_{Cu} \times PI_{Zn} \times PI_{Cd} \times PI_{Pb}} \qquad (2)$$

The rating of pollution was then classified on four levels based on *PLI* value as followed: no pollution with value $PLI < 1$; moderate with value $1 \leq PLI < 2$; high with value $2 \leq PLI < 3$; and very high pollution with $PLI \geq 3$ [19].

Geoaccumulaition index *(I_{geo})* calculation was also used in this study to measure the degree of pollution. This index was calculated as followed [20]:

$$I_{geo} = \ln C_n/1.5{*}B_n \qquad (3)$$

C_n is the measured concentration in the soil samples, B_n is the background values, and 1.5 is a constant to allow for natural fluctuations analysis in the environment. In this study, the classes of I_{geo} were simplified by the classification given by Muller (1969) [20]. The simplified classes were as followed : <0-1 is low class; >1-3 moderate class; >3-5 is moderate high class; and >5 is high class.

3. Result

3.1 Heavy metals in soil

The results of the heavy metal analysis are shown in Table 2. The soil samples taken from the study area had variable values of pH and organic content. The value of pH from 5.5 to 6.5, and this value was slightly lower than normal. This lower value was influenced by organic content in soil samples, ranging from 4.1% to 5.8%. The organic content value had no significant difference in each zone in the study area.

In general, the concentrations of heavy metals (Pb, Cu, Zn, and Cd) showed that the zone I had the highest value which was compared to the other two zones in the study area. Especially for Pb and Cd, the value was higher than the natural background in study area. For Cu and Zn, the value was in the range of the natural background. The results showed that there was no significant enrichment in these two metals. By giving an average value for each zone, it can be easier to determine that the zone with a high value has a high risk and requires special attention for the soil remediation strategy. Then, the locations that have a maximum heavy metal concentration can be affected by land use. This is because the increase in the number of settlements has an impact on increasing the concentration of heavy metals in the soil.

3.2 Heavy metal pollution index

The results of the calculation of the metals index are shown in Table 3. Generally, the highest *PI* value was located in zone I. *PI* values for Pb and Cd were higher than Cu and Zn in general. They showed that about 96% of the total samples were included in the considerable high class, and that Pb and Cd occupied 69% of total samples for considerable high class.

The results of the *PLI* calculation showed that the values rangedg from 1.16 to 3.13. Generally, the highest *PLI* value was located in the zone I; with an average value of *PLI* of 2.72. Table 3 also showed that the value of I_{geo} in the zone I was the highest compare to the other two zones. Pb and Cd had higher values than Cu and Zn. It seems that Pb had the highest number in the moderate class of I_{geo}. However, Cd had 11% of the total sample included in the considerable high class of I_{geo}. The low values of I_{geo} for Cu and Zn with minus values were similar to the finding of several studies [21][22][23]. These results indicated that there was very low or no pollution in the soil caused by these metals.

Table 2. Heavy metals and soil properties.

Zones[a]	I (n=15) Mean±SD	II (n=15) Mean±SD	III(n=15) Mean ±SD	All samples n=45 Mean ±SD	All samples n=45 Range
Soil heavy metal concentration (mg/kg)					
Pb	105±12.4	85±14.4	59.8±24.7	83.3±6.59	22-129
Cu	46.9±2.83	45.4±9.34	41.8±3.34	44.7±3.62	34-67
Zn	76.3±4.06	70.6±10.3	75.3±4.95	74.1±3.39	49-81
Cd	1.19±0.44	0.54±0.31	0.19±0.12	0.64±0.16	0.1-1.9
Soil characteristics					
Organic content(%)	4.8±1.2	4.9±1.3	4.2±1.1	4.6±0.9	4.1-5.8
pH	5.9±0.9	6.2±0.8	6.1±0.2	6.1±0.4	5.5-6.5

[a]see Figure 1 for zone I, II and III

Table 3. The value of pollution index *(PI)*, pollution load index *(PLI)*, and geoaccumulation index *(I_{geo})*

Zones[a]	I (n=15) Mean±SD	II (n=15) Mean±SD	III (n=15) Mean±SD	All samples (n=45) Mean±SD	All samples (n=45) Range	Low	Moderate	Considerable High	High
						Proportions of total samples (%)			
Pollution index (PI)									
Pb	8.92±1.05	7.20±1.22	5.07±2.09	7.06±2.18	1.86-10.9	0	4	96	0
Cu	1.35±0.08	1.30±0.27	41.8±0.10	1.28±0.18	0.98-1.93	2	98	0	0
Zn	1.10±0.06	1.02±0.15	1.09±0.07	1.07±0.11	0.71-1.17	20	80	0	0
Cd	11.9±4.42	5.4±3.09	0.83±0.40	1.87±1.25	1-19	0	31	69	0
Pollution load index *(PLI)*									
	2.72±0.23	2.14±0.39	1.59±0.30	21.5±0.56	1.16-3.13	0	44	51	4
Geoaccumulation index *(I_{geo})*									
Pb	2.56±0.17	2.24±0.26	3.05±2.06	1.62±0.68	0.31-2.87	7	93	0	0
Cu	-0.49±0.08	-0.23±0.26	-0.33±0.12	-0.24±0.18	-0.62-0.36	100	0	0	0
Zn	-0.45±0.08	-0.57±0.23	-0.47±0.10	-0.50±0.16	-1.09-0.36	100	0	0	0
Cd	2.90±0.51	1.60±0.89	0.08±0.82	1.53±1.38	-0.58-3.66	40	27	11	0

4. Discussion

4.1 Heavy metals enrichment in soils

Several studies have been done on the concentration of heavy metals in surface soils in urban cities. The results of this study showed that most Pb had a higher concentration than normal compare to several references. Cd also showed a similar pattern to Pb, but Cu and Zn had a lower than normal values in general. The high Pb and Cd concentration in soil in the study area are suspected from human activities such as vehicle emission and industry. Enrichment of heavy metals in soils may be derived from the host rock. However, the host rock in the study area is a volcanic rock, which has very low contents of the heavy metals. This, the enrichment may be derived from human activities. Another factor influenced may be from pH and organic content in the soil.

Vehicle emission gives the most contributions for Pb concentration in the surface soil in the urban city, as stated by several researchers [24][25]. The vehicle emission is generated by carbon monoxide and the burning of lead in the vehicle engine. The emission comes from engine friction, vehicle rusty, and gas emission due to incomplete combustion. These substances then spread with air and deposit in the surface soil. According to another study, the concentration of heavy metals in the air in several big cities in Indonesia contain Pb (0.2-2664), Zn (2.9-913), Cu (0.05-18.79) in mg/m^3 [26].

4.2 Environmental impact assessment

The heavy metals in the soil in urban areas were a serious consideration in several studies [2][3][4]. The result of the calculation of the assessment in this study showed that the heavy metals in the soil in the study area had high Pb and Cd, and the concentration is above normal. The *PI*'s value in Table 3 showed that the values were on the moderate to high condition, except for Cu and Zn. Generally, the *PI* value in the soil in the zone I was higher than the other two zones. The I_{geo} values in zone I had the highest value in the study area due to the closest location to the center of the city. Indeed, the location in zone I had the highest risk for the environment.

The calculation of *PLI* and I_{geo} showed a difference in the prediction of pollution level in the soil in the study area. According to *PLI*, the soil samples are in the range of moderate to considerable high; however, according to I_{geo} the samples are included in the low to considerable high level. Overall, compared to the natural background level (Table 1), the Pb and Cd had higher values. On the other hand, Cu and Zn had lower values. This difference may be due to the metals' source because the human activities generating Pb and Cd are more intensive in the study area. Another factor that may influence is the soil characteristic in the study area, with strong bonding for Cu and Zn in the soil particle. This phenomenon needs to be investigated more in-depth in the future.

5. Conclusions

The investigation of the study concluded that the concentrations of heavy metals (Pb, Cu, Zn, and Cd) in the soil in Yogyakarta city and its surrounding area varied according to the distance from the city center. The highest concentration of Pb and Cd in the soil in the study area was found in zone I due to the intensive traffic and industrial area. Cu and Zn in the soil in this study showed that the level of pollution in the study area was in a moderate category. There was a different value of pollution index depending on heavy metal and a potential effect on migrating in the soil profile. Generally, the concentration of heavy metals in the soil in the study area were moderate; however, it still needs more attention for zone I especially for human activities in that zone. Several policies need to be made in order to reduce Pb and Cd emission in the area.

Acknowlegment

Authors thanks to Department of Geological Engineering Universitas Gadjah Mada for the research grant. Authors thanks to International Development Engineering Department, Tokyo Institute of Technology, Japan, for the facility of laboratory on the analysis of the samples.

References

[1] Argyraki A and Kelepertzis E 2014 Urban soil geochemistry in Athens, Greece: the importance of local geology in controlling the distribution of potentially harmful trace elements. *Sci. Total. Environ.***482** 366-377

[2] Gaberšek M and Gosar M 2018 Geochemistry of urban soil in the industrial town of Maribor, Slovenia. *J. Geochem. Explor.***187** 141-154

[3] Andersson M, Ottesen R T and Langedal M. 2010 Geochemistry of urban surface soils - monitoring in Trondheim, Norway. *Geoderma.***156**(3-4) 112-118

[4] Flem B, Eggen O A, Torgersen E, Kongsvik M K and Ottesen R T 2018 Urban geochemistry in Kristiansand, Norway. *J. Geochem.Explor.***187** 21-33

[5] Tiller K G 1992 Urban soil contamination in Australia. J. *Soil. Res.***30**(6) 937-957

[6] Li G, Sun G X, Ren Y, Luo X S, and Zhu Y G 2018 Urban soil and human health: a review. *Eur. J. Soil. Sci.***69**(1) 196-215

[7] United Nations 2018 *World urbanization prospects*: The 2018 revision, United Nations

[8] Inostroza L 2017 Informal urban development in Latin American urban peripheries. Spatial assessment in Bogotá, Lima and Santiago de Chile.*Landsc. Urban Plan.***165** 267–279

[9] Pradoto W 2011 Dynamics of Peri-urbanization and socioeconomic transformation: Case of Metropolitan Yogyakarta, Indonesia. *Intern. J. Arts. Sci.***4**(27) 19

[10] Renata, P 2009 *Studi Pencemaran Tanah oleh Logam Berat di Kota Yogyakarta dan Sekitarnya*, Undergraduate Thesis, Department of Geological Engineering, Universitas Gadjah Mada (in Indonesia language - unpublished)

[11] Budianta W 2012 Lead Contamination in Soil of Yogyakarta city, Indonesia. *J. South East Asia* Appl. Geology. **4**(2) pp 90-98

[12] Aye Z A 2016 *Lead Contamination and Behavior in Soil of Unsaturated Zone, Yogyakarta City and Surrounding Area, Indonesia*, Doctoral Thesis,Department of Geological Engineering, Universitas Gadjah Mada (in Indonesia language - unpublished)

[13] Bettinelli M, Beone G M, Spezia S, and Baffi C 2000 Determination of heavy metals in soils and sediments by microwave-assisted digestion and inductively coupled plasma optical emission spectrometry analysis. *Anal. Chim. Acta.***424**(2) 289-296

[14] Walkley A and Black I A 1934 An examination of the Degtjareff method for determining soil organic matter, and a proposed modification of the chromic acid titration method. *Soil Sci.* **37**(1)29-38

[15] Conklin A R 2013 *Introduction to Soil Chemistry: Analysis and Instrumentation*. John Wiley and Sons.

[16] Wulaningsih T 2009 *Karakteristik Abu Vulkanik Gunung Merapi dan Pengaruhnya terhadap Kesehatan Masyarakat di Desa Kaliadem dan Sekitarnya, Kecamatan Cangkringan, Kabupaten Sleman Propinsi Daerah Istimewa Yogyakarta*, Undergraduate Thesis, Department of Geological Engineering, Universitas Gadjah Mada (in Indonesia language - unpublished)

[17] Hakanson L 1980 An ecological risk index for aquatic pollution control: A sedimentological approach. *Water Res.* **14**(8) 975–1001

[18] Loska K, Cebula J, Pelczar J, Wiechula D and Kwapulinski J 1997 Use of enrichment, and contamination factors together with geoaccumulation indexes to evaluate the content of Cd, Cu, and Ni in the Rybnik water reservoir in Poland. *Water Air Soil Pollut.* **93**(1) 347– 365.

[19] Wang, X. Q., He, M. C., Xie, J., Xi, J. H., and Lu, X. F. 2010. Heavy metal pollution of the world largest antimony mine-affected agricultural soils in Hunan province (China). *J. Soil. Sediment.* **10**(5) 827–837

[20] Muller G 1969 Index of geoaccumulation in sediments of the Rhine River. *GeoJournal.***2**108-118

[21] Li H, Lin L, Ye S, Li H and Fan J 2017 Assessment of nutrient and heavy metal contamination in the seawater and sediment of Yalujiang Estuary. *Mar. Pollut. Bull.* **117** 499–506

[22] Hieu H O, Swennen R, and Van Damme A 2010 Distribution and contamination status of heavy metals in Estuarine sediments near CuaONg Harbor, Ha Long Bay, Vietnam. *Geol. Belg.* **13** 37-47.

[23] Hwang D W, Kim S G, Choi M, Lee I S, Kim S S, and Choi H G 2016 Monitoring of trace metals in coastal sediments around Korean Peninsula. *Mar. Pollut. Bull.* **102**(1) 230-239

[24] Yulius U and Afdal A 2014 Identifikasi Sebaran Logam Berat Pada Tanah Lapisan Atas dan Hubungannya Dengan Suseptibilitas Magnetik di Beberapa Ruas Jalan di Sekitar Pelabuhan Teluk Bayur Padang. *J. Fisika* Unand, **3**(4) 198-204

[25] Thomaidis N S, Bakeas E B, and Siskos P A 2003 Characterization of lead, cadmium, arsenic and nickel in $PM_{2.5}$ particles in the Athens atmosphere, Greece. *Chemosphere*.**52**(6) 959-966

[26] Mukhtar R, Wahyudi H, Panjaitan E H, Lahtiani S, Santoso M, Lestiani D D, and Kurniawati S 2013 Kandungan logam berat dalam udara ambien pada beberapa kota di Indonesia. *Ecolab*.**7**(2) 49-59

Petrography and geochemistry preliminary study of Pucangan Serpentinite Geosite Karangsambung – Karangbolong Geopark Central Java, Indonesia

Isyqi[1*] and C Ansori[1]

[1]Research & Development Division for Earth conservation and Information – LIPI Karangsambung Highway Km. 19, Kebumen, Indonesia.
***E-mail:** Isyqi@Karangsambung.lipi.go.id

Abstract. Geosite is a Geological Heritage object in the Geopark area with specific characteristics and is an integral part of the evolutionary history of forming an area. Serpentinite Pucangan is one of the Geosite located in Karangsambung – Karangbolong geopark. This study aims to find out the characteristics of petrography and geochemistry of Serpentinite Pucangan Geosite. The methods used in this study include literature studies, fieldwork research, petrography analysis, ore microscopy analysis, whole rock analysis consist of major element analysis using fusion inductively coupled plasma (Fus ICP) methods, trace element analysis, and rare earth element analysis using inductively coupled plasma mass spectrometry (ICP-MS) methods. Serpentine minerals with antigorite type dominate the petrography composition of Pucangan serpentinite with a little magnetite mineral. SiO_2, MgO, and Fe_2O_3(t) are the major elements with the largest percentage of 37.93%, 5.78%, and 9.73% that have an agreement with the mineralogical composition of the rock that dominated by ferromagnesian minerals. This Geosite is interpreted to be formed due to the metasomatic process when the subduction process is happening in this area in the Pre-tertiary period.

1. Introduction

Geopark is a geographical area with valuable Geosite and landscapes related to Geoheritage, Geodiversity, Biodiversity, and Cultural Diversity [1]. Karangsambung - Karangbolong Geopark is located in Kebumen Regency, Central Java, has 41 Geosites spread across northern Kebumen, central Kebumen, to Southern Kebumen. The entire Geosites reflect the geological history on the Java Island from the Pre-Tertiary until the Quarternary period that begins with java island formation due to plate subduction. These tectonic events produce a diversity of igneous, sediments, and metamorphic rocks that we can find in the northern Kebumen. One of the subduction process evidence is serpentinite rock that was well outcropped in the Pucangan area.

Serpentinite is a greenish-black metamorphic rock dominated by serpentine minerals: antigorite, chrysotile, or lizardite (allomorphs of $Mg_3Si_2O_5(OH)_4$), which are formed from the hydrous alteration of olivine minerals, orthopyroxene, and sometimes, clinopyroxene [2]. Serpentinite generally formed beneath the earth's surface and are rarely exposed to the surface. Therefore, the serpentinite outcrop in the Pucangan area, North Kebumen, is significant to be conserved to maintain the rocks' educational value. One way to conserve a geological heritage is to make it a Geosite within the Geopark area. Geosite is a Geological Heritage object (Geoheritage) in the Geopark area with specific characteristics and is an integral part of forming a region's evolutionary history. This study aims to determine the attributes of petrography and geochemistry of serpentinite rocks in the Pucangan area, one of the Geosite in Karangsambung – Karangbolong Geopark so that the history of Geosite formation can be known.

2. Geological setting

This research is located at coordinates E 0355735 N 9168547, Pucangan Village, Sadang Sub-District, North Kebumen (Figure 1). This area is included in the Luk Ulo Melange Complex (KMLU), which consists of a varied block (igneous, sediment, and metamorphic) that mixed tectonically within the black shale matrix [3]. The components of KMLU, in general, include:

- Metamorphic rocks are the oldest rocks found and consist of Gneiss, Green Schist, Mica Schist, Blue Schist, phyllite, amphibolite, eclogite, and marble. Rb – Sr dates of the Garnet-Phengite Schists indicate the age of 118 and 114 Ma [4].
- Igneous rocks: a series of ophiolite rocks exposed in this area consisting of peridotite, serpentinite, gabbro, diabase, and basal rocks that commonly form pillow structures. Pillow-structured basalt is generally associated with deep-sea sediment.
- Deep-sea sediment: consists of a Chert interlayered with red clay or red limestone clay. Sedimentary rocks generally consist of layered pelitic rocks with sandstone, greywacke, and metagreywacke, which generally form boudinage structures.

The research area belongs to the South Serayu Mountains physiography zone [5] that dominates the structural hills morphology. The geological structure found in the research area is strike-slip faults with directions almost north-south and northwest-southeast [3].

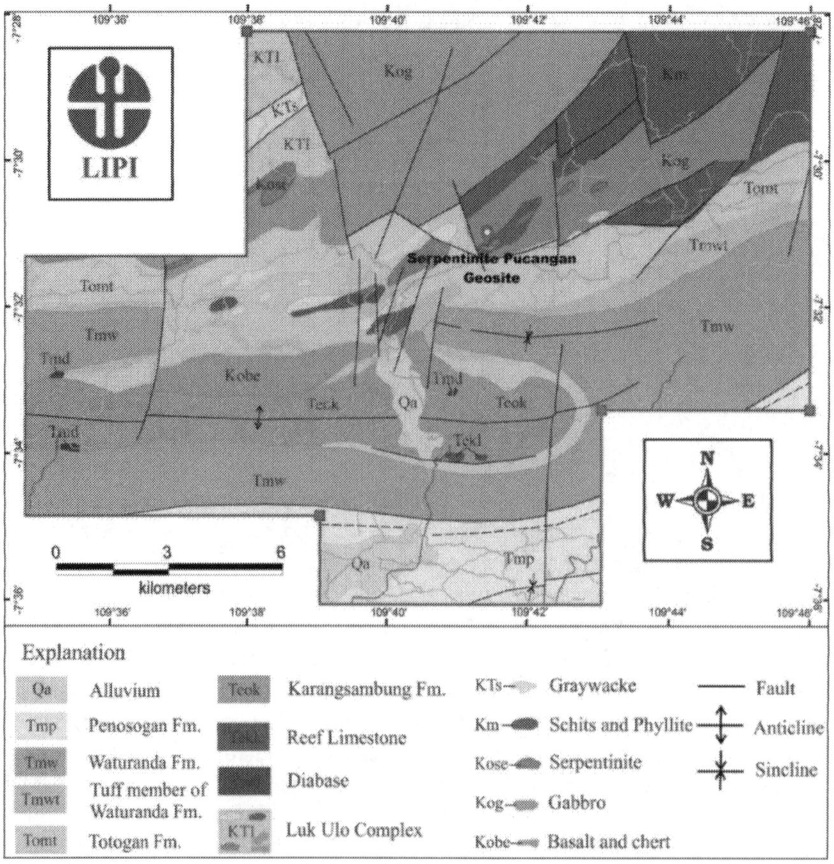

Figure 1.Geological map of the research area (modified after [3]).

3. Methods

The method used to achieve this research's objectives includes three stages, the literature study stage, the fieldwork stage, and the laboratory analysis stage. The first stage is to study the research area's regional geological conditions and a review of previous research. The fieldwork activities carried out are the plotting of outcrop location, lithology description in the research area, rock sampling, and documentation. The last stage is laboratory analysis, including petrography analysis, ore microscopy analysis, XRD (X-Ray Diffraction) analysis, and bulk rock chemical analysis.

A sample from the research area with code SR – 41 was carried out for the petrography analysis, ore microscopy analysis, XRD (X-Ray Diffraction) analysis, and bulk rock chemical analysis. The thin section and polished surfaces were made in the optical laboratory of Gadjah Mada University. Subsequently, petrography observations were carried out in the optic laboratory of the Research & Development Division for Earth conservation and Information – LIPI. Preparation and Analysis of XRD were done in the laboratory of the Geological Survey Center, Bandung, while the bulk rock chemical analysis was carried out at Activation Laboratories LTD., Canada. Bulk rock chemical analysis consists of the analysis of the major elements using fusion inductively coupled plasma (Fus ICP) method with units of measurement in the form of a percentage (%), analysis of trace elements, and analysis of rare earth elements use the method of inductively coupled plasma mass spectrometry (ICP-MS) with units of ppm (part per million).

4. Result

Based on field research, it is known that Pucangan serpentinite outcropped in Karangsambung - Sadang highway precisely at coordinates E 0355735 N 9168547. The outcrop has dark green - blackish colors with quite large dimensions covering a hill (Figure. 2). The condition of the outcrop is not damaged, but it is intensively fractured. Some sides of the outcrop were weathered until a layer of soil forms. When observed in more detail, fresh rocks are green, and some parts still show the original rock's texture.

Figure 2. Condition of the Pucangan serpentinite outcrop (a) the dimension is covering a hill; (b) located on the side of Karangsambung – Sadang highway; (c) the color of fresh rock; (d) intensively fractured.

IOP Conf. Series: Earth and Environmental Science **851** (2021) 012027 doi:10.1088/1755-1315/851/1/012027

The thin sections from the research area showed the rocks' composition was dominated by serpentine, with minerals size 0.05-0.5mm, subhedral - anhedral shape as presented in figure 3. Another mineral found in the thin section is the opaque mineral. Mineral serpentine (96%), colorless - brownish-green, has low relief, flaky shape, and weak pleochroism. Opaque mineral (4%), color black, has high relief, and having size 0.02 – 0.15mm. The opaque mineral is magnetite with brownish-grey color based on ore microscopy analysis, does not show pleochroism, euhedral shape, size <10-75 μm, and dark interference color (figure 4). XRD analysis shows that the type of serpentine mineral in rock sampling is an antigorite type.

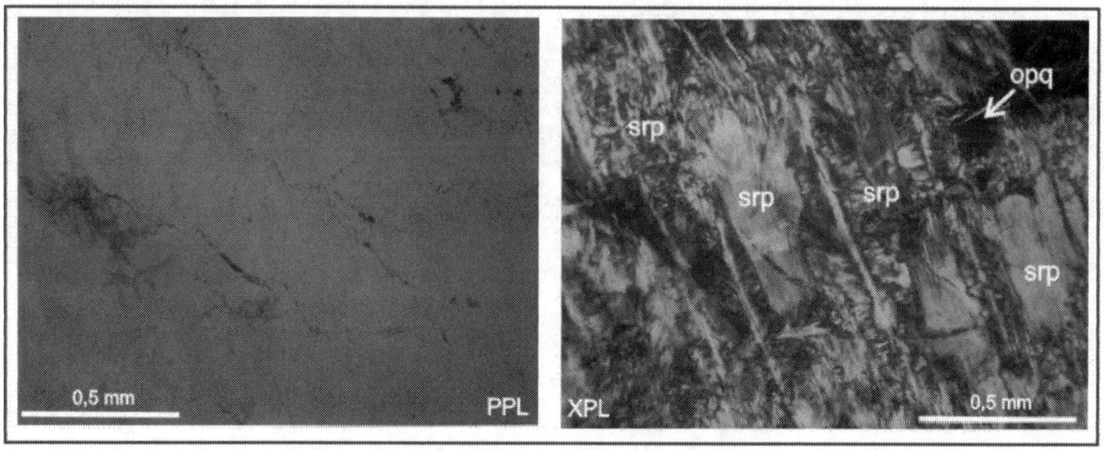

Figure 3. Thin section of serpentinite Pucangan geosite showing serpentine (srp) and opaque (opq) minerals.

Figure 4. Ore microscopy of serpentinite Pucangan geosite showing magnetite (mag) mineral.

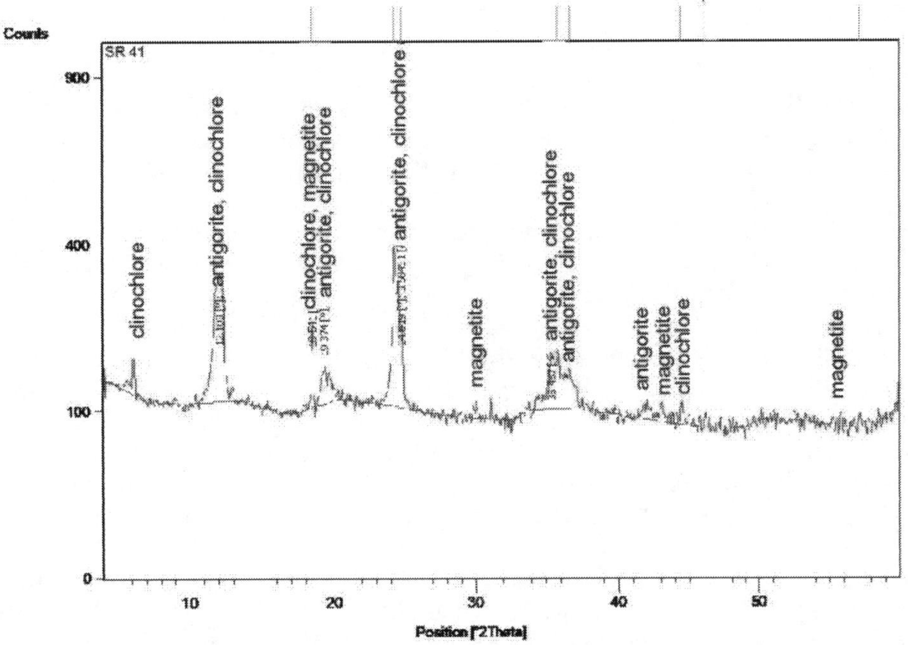

Figure 5. Spektrum XRD of Serpentinite Pucangan.

Table 1. Serpentinite Pucangan minerals composition based on XRD analysis.

Ref. Code	Score	Compound Name	Displacement [°2Th.]	Scale Factor	Chemical Formula
*00-052-1572	30	**Antigorite**	0.000	0.406	$Mg_3Si_2O_5(OH)_4$
*00-029-0853	29	**Clinochlore**	0.000	0.192	$Mg_5Al(Si_3Al)O_{10}(OH)_8$
*00-001-1111	31	**Magnetite**	0.000	0.057	Fe_3O_4

The major elements of serpentinite Pucangan are presented in table 2, whereas all of trace elements and rare earth elements are below to detection limit except Ba = 8 ppm.

Table 2. The concentration of major elements in Serpentinite Pucangan.

Major Elements (wt%)	SR-41
SiO_2	37.93
Al_2O_3	0.22
Fe_2O_3(t)	9.73
MnO	0.108
MgO	35.78
CaO	0.02
Na_2O	0.01
K2O	0.009
TiO_2	0.009
P_2O_5	0.009
LOI	13.88
Total	97.68

5. Discussion

Petrography and XRD analysis of rock samples from the research area showed antigorite-type serpentine minerals' dominance. Serpentine minerals are formed due to the hydration process of Mg-rich minerals such as olivine, orthopyroxene, and clinopyroxene at various temperatures, including at hydrothermal temperatures and sub-surface conditions [6]. According to [7], antigorite is a type of serpentine mineral stable at high temperatures and high pressures. The process of serpentinization of antigorite minerals generally occurs at a temperature of 400 – 600 °C [8], usually occurs in partially recrystallized and metamorphosed serpentinite [6]. [9] reported that antigorite serpentinite was found in three paleo-subduction zones: The Himalayas, the Alps, and the Cub and associated with High Pressure to Ultra-High Pressure (HP-UHP) metamorphic rocks. This theory agrees with the geological conditions of the research area: a Pre-Tertiary subduction zone that can form serpentinite antigorite due to the metamorphosis process. Moreover, near the research area precisely in Muncar and Loning rivers were found high-pressure metamorphic rocks such as eclogite-, blueschist-, and amphibolite-facies [10]. Serpentinite is believed to have an important role in the outcropping of high pressure to ultra-high pressure (HP-UHP) metamorphic rocks due to their lighter physical properties (buoyant) than other rocks [9].

Another mineral found in the rock sample is magnetite mineral (Fe_3O_4). In the serpentinization process, the content of #Mg in origin minerals such as olivine, orthopyroxene, and clinopyroxene will be distributed in serpentine minerals. In contrast, the content of #Fe will precipitate into magnetite minerals [6]. Furthermore, [8] explained that the higher the serpentine mineral formation's temperature, the less magnetite mineral formation would be. That's why lyzardite serpentinite, in general, have more abundant magnetite content than antigorite serpentinite. Petrographic analysis on serpentinite Pucangan from the research area shows results in agreement with the theory that magnetite content is not abundant, only about 4%.

The chemical composition of Serpentinite Pucangan is presented in Tables 2. The composition of the major elements of Serpentinite Pucangan is dominated by SiO_2 37.93%, MgO 35.78%, and Fe_2O_3(t) 9.73%. The value agrees with the results of the petrography analysis that shows the dominance of ferromagnesian minerals. Other major elements are Al_2O_3 0.22%, MnO 0.108%, CaO 0.02%, Na_2O 0.01%, K_2O 0.009%, TiO_2 0.009%, and P_2O_5 0.009%. The LOI (lost of ignition) value of 13.88% indicates the high number of hydrous minerals.

The result of most trace elements and rare earth elements analysis of Pucangan serpentinite are below detection limits. Therefor advanced research is needed with another geochemical analysis method so that the trace element value in this rock can be detected and the origin can be known.

6. Conclusion

Serpentinite Pucangan Geosite is composed by the dominance of antigorite type serpentine minerals (956%) and a little magnetite mineral (4%). The rocks are interpreted to be formed due to the metamorphism process at the time of pre-tertiary Subduction. The major element content of Serpentinite Pucangan is dominated by SiO_2 37.93%, MgO 5.78%, and Fe_2O_3 9.73% that agrees with the mineralogical composition that is dominated by ferromagnesian minerals.

Limited data, especially about trace elements in Pucangan serpentinite, interpret the rock genesis has not suffered yet. Therefore, further research is needed so that the rock formation can be known. The increase in the number of rock samples is recommended for where further research is. Besides, it considers choosing other geochemical methods with smaller detection limits so that traces and rare earth elements in Pucangan serpentinite can be detected.

Acknowledgments
Thank you to the Head of Research & Development Division for Earth conservation and Information LIPI, who has allowed us to run this research. A sincere thank you to the local government and all those who have helped in this research. This research was funded with DIPA LIPI funds in the fiscal year 2018.
Author Contributions: For this research article, Isyqi is the main contributor.

Reference
[1] Cabinet Secretariat of the Republic of Indonesia, 2019 *Peraturan Presiden No. 9 tahun 2019 tentang Pengembangan Taman Bumi (Geopark)* Indonesia.
[2] Park A F, 1990, Serpentinite, in *Petrology*, (Dordrecht: Kluwer Academic Publishers), p. 524–525.
[3] S. Asikin, A. Handoyo, A. Busono S G, 1992, Geologic Map og the Kebumen Quadrangle, Jawa, Scale 1 : 100.000. Geological Research and Development Center of Indonesia, Bandung.
[4] Julian Alfing J Michael Bröcker M and Setiawan N I, 2021 Rb–Sr geochronology of metamorphic rocks from the Central Indonesian Accretionary collision complex: Additional age constraints for the Meratus and Luk Ulo complexes (South Kalimantan and Central Java) *Lithos* **389** p. 105971.
[5] Bemmelen R W Van, 1949 *The Geology of Indonesia* **1A**, 1 Hague: Goverment Printing Office Hague.
[6] Evans B W Hattori K and Baronnet A, 2013 Serpentinite: What, why, where? *Elements* **9**, 2 p. 99–106.
[7] Guillot S Schwartz S Reynard B Agard P and Prigent C, 2015 Tectonic significance of serpentinites *Tectonophysics* **646** p. 1–19.
[8] Evans B W, 2010 Lizardite versus antigorite serpentinite: Magnetite, hydrogen, and life(?) *Geology* **38**, 10 p. 879–882.
[9] Hattori K H and Guillot S, 2007 Geochemical character of serpentinites associated with high- to ultrahigh-pressure metamorphic rocks in the Alps, Cuba, and the Himalayas: Recycling of elements in subduction zones *Geochemistry, Geophys. Geosystems* **8**, 9.
[10] Setiawan N, 2013 An overview of metamorphic geology from central Indonesia: Importance of South Sulawesi, Central Java and South-West Kalimantan metamorphic terranes *Bull. Grad. Sch. Soc. Cult. Stud. Kyushu Univ.* **19** p. 39–55.
[11] Closs L G, Mar. 1983 Book Review: Principles of Geochemistry (Fourth Edition). Brian Mason and Carleton B. Moore, John Wiley, New York, 1982, 344 p., 29.95. (also available in paperback in the Wiley International Edition Series at {\textsterling}7.25/12.35.) \g*ca* **47**, 3 p. 661–662.

IOP Conf. Series: Earth and Environmental Science **851** (2021) 012028 doi:10.1088/1755-1315/851/1/012028

The relationship of physical and mineralogical properties and geochemical compositions of limestone and its implications to the quality as cement raw material in Gunem District, Rembang Regency, Central Java

T Winarno[1*], J Marin[1] and Nurjayanti[1]

[1]Geological Engineering Department, Universitas Diponegoro, Jalan Prof. Soedarto SH, Tembalang, Semarang, Central Java, Indonesia 50275.
*E-mail: tri.winarno@live.undip.ac.id

Abstract. The increasing development in Indonesia today will require higher portland cement consumption. Good quality cement comes from good quality raw materials. A research was conducted in the Gunem area, Rembang, Central Java which is a limestone quarry of PT Semen Indonesia. The research aims to determine physical properties of the limestone and determine the percentage of CaO and MgO and determine the quality of limestone as raw material for Portland cement. This research uses several methods, which are megascopic analysis, petrographic analysis and XRF (X-Ray Fluorescence) analysis. The results shows that wackestone has the lowest content of CaO and the highest content of MgO, while packstone and grainstone have the highest content of CaO and the lowest content of MgO. Packstone and grainstone are classified as high grade for cement raw material, while wackestone is classified as low grade for cement raw material.

1. Introduction

Limestone is one of the raw materials for the cement manufacture, apart from claystone, gypsum, silica, and iron as additives. In the current development era, cement demand has always increased in line with the pace of development throughout Indonesia [1]. Indonesia has enormous limestone potential, which is spread almost throughout the Indonesian archipelago. In general, the potential of limestone in Indonesia based on its geological map is estimated at around 28.678 billion tons [2].

One of the cement factories in Indonesia is located in Rembang, Central Java, i.e., PT Semen Indonesia Tbk. The cement factory is located in the limestone quarry, which is a member of the Paciran Formation. The Paciran Formation consists of brownish and yellowish white limestone, coral texture and has a layer thickness of ± 25 m [3].

The purpose of this research is to determine the physical characteristics and components of limestone in the limestone quarry based on megascopic and petrographic features. This research also aims to determine the chemical composition of limestone in the limestone quarry based on XRF analysis and to determine the relationship between physical and chemical properties of rock samples to determine the limestone's quality in cement manufacture.

2. Literature Review

2.1. Regional Geology

The research location, which is the limestone quarry of PT Semen Indonesia, is at Tegaldowo Village, in Gunem District, Rembang Regency. The research location is included in the Paciran Formation (Figure 1). The original Paciran Formation was referred to as Karren Limestone [4]. This formation is found only in the northern part of the Rembang Zone. This formation is characterized by coral limestone, which is grayish-white, massive, often dolomitic, consisting of algae, corals, large foraminifera, and other reef-forming organisms. The distinctive feature of this limestone is the disappear of the fossil content of Lepydocyclina, so that it can be distinguished from the limestone underneath this formation [3].

The age of the Paciran Formation is Pliocene - Pleistocene, based on the discovery of *Alveolinella quoyi* which is a fossil that often appears in the Pliocene. The depositional environment of this formation is in shallow seas, near the coast, warm climates, clear, depth <50 m, the littoral zone - sublittoral edge where the environment allows reef growths [4].

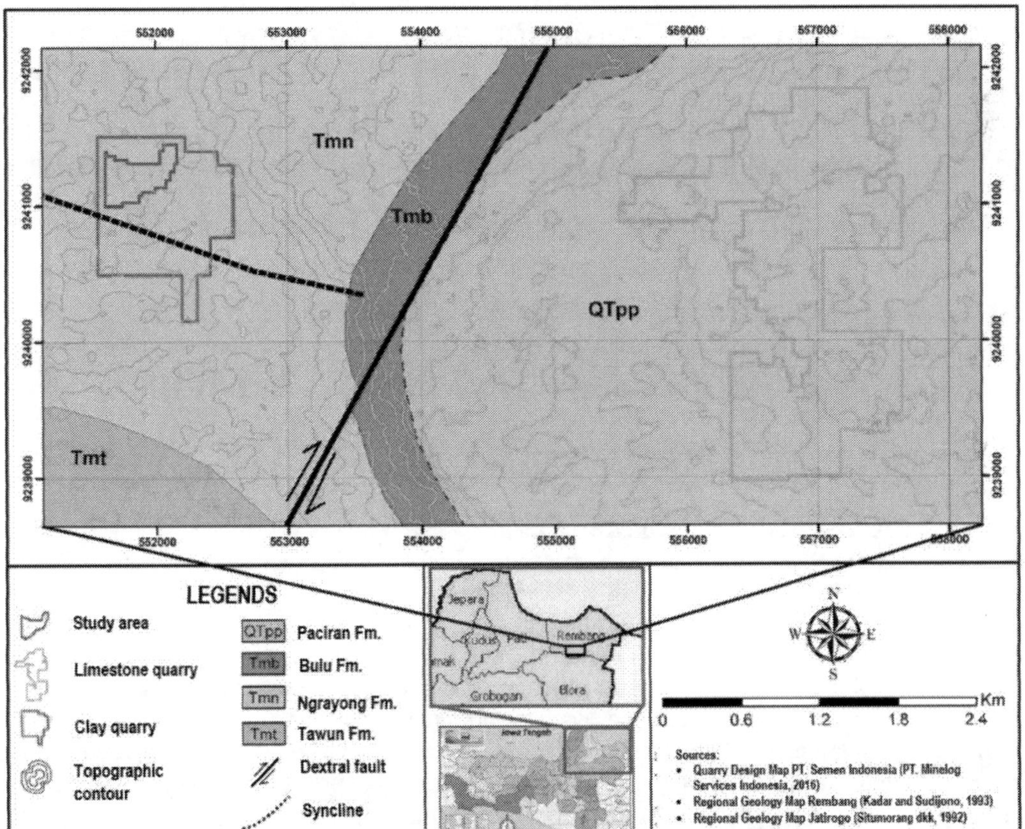

Figure 1. Regional geology map.

2.2. Limestone as Cement Raw Material

Cement is a crystalline compound of calcium silicates and other calcium compounds having hydraulic properties [5]. The raw material for artificial cement is a mixture of the following materials: limestone, clay, silica, iron or copper slag, gypsum and trass (feldspar) [6] [7].

Limestone is the main raw material in the manufacture of cement. In the exploration of limestone as a raw material for cement, it is necessary to analyze the distribution of high-quality limestone. The high and low quality of limestone as a raw material for cement is largely influenced by its CaO content. On the other hand, limestone has different physical characteristics depending on the facies of the limestone where it found. So, it is important to study the relationship between the physical and mineralogical properties and geochemical of limestone [8].

Based on the standards of PT Semen Indonesia Rembang, limestone which meet the requirements for cement raw materials (high grade class) must have > 50% CaO and <3% MgO content [9].

3. Methods

The methods used in this study are field observations, X-Ray Fluorescence (XRF) analysis, and petrographic analysis.

a. Field observations were carried out to determine the geological conditions of the study area, such as the type of lithology and geological cross section and to collect the rock samples to be analyzed in the laboratory.
b. Megascopic observations were carried out to determine the physical properties of limestone.
c. Petrographic analysis was carried out to determine the rock description optically. Thin section samples were prepared to identify mineralogical and petrographical characteristics under polarizing microscope. The rock samples from the research area were polished to make the thin sections. The thin sections were analyzed using polarization microscope. The petrographic analysis was carried out at the Laboratory of Paleontology, Remote Sensing and Optical Geology, at Geological Engineering Department, Universitas Diponegoro.
d. The geochemical analysis is carried out by using XRF analysis. XRF analysis is carried out to analyze the main elements and trace elements along with their concentrations in the rocks deposits using spectrometric methods. Currently, XRF is the most common method of analysis used in the determination of major elements and traces elements on rock samples. Compared to other geochemical analysis methods, the XRF analysis can detect the lowest elements more sensitive. However, it cannot detect elements lighter than Na, which have an atomic number of 11. The XRF analysis was carried out at Laboratory of Quality Assurance of PT Semen Indonesia Rembang.

4. Result

4.1. Lithology

Based on the field observation in the limestone quarry, three lithologies were found in the study area: wackestone, packstone and grainstone. The distribution of the lithology units can be seen in Figure 2.

4.1.1. Wackestone

This limestone is found in the northwest part of the research area. From the observation of its physical properties, the rock has a moderate level of weathering, white color, massive structure, silt-sized grain texture, and the composition consists of bioclasts (fragments of fine foraminifera test). Based on petrographic observations, the thin section of the rock shows the texture of the rock is mud-supported, grain size <1 mm, with subangular grain shape and poorly sorted. The composition of limestone consists of skeletal grain, including foraminifera and algae (15%), the matrix is microsparite (80%) and calcite cement (5%) (Figure 3) and then the rock can be classified as wackestone [10].

IOP Conf. Series: Earth and Environmental Science **851** (2021) 012028 doi:10.1088/1755-1315/851/1/012028

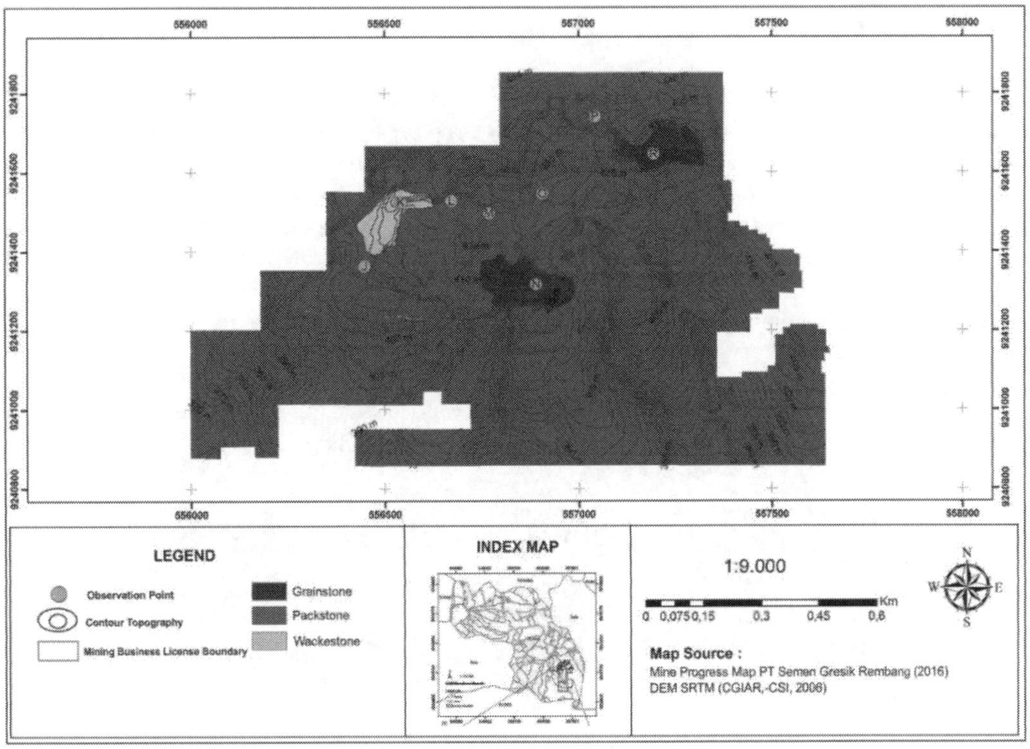

Figure 2. Geological map of research area.

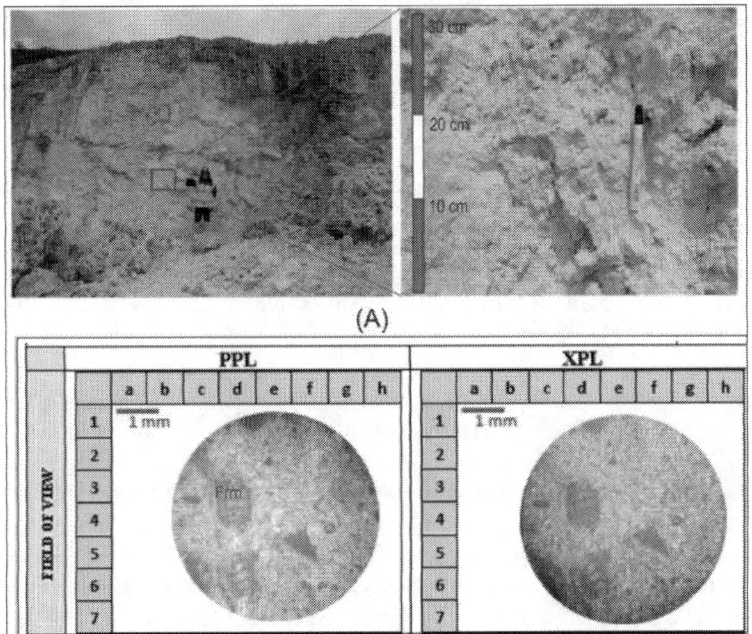

Figure 3. Outcrop of wackestone (A) and its petrographic thin section (B). Note: Frm (foraminifera), Alg (algae).

4.1.2. Packstone

This limestone is found in almost all parts of the research area. From the observation of its physical properties, this limestone has a yellowish to brownish-white color, has medium to coarse sand grain size (0.25 - 1 mm), and the composistion consists of bioclasts (fine shell fragments) and calcite. Based on the petrographic observation, the thin section of the rock shows grain size <1mm, subangular grain shape and poorly sorted. The composition of the limestone consists of skeletal grain of foraminifera (50%), calcite crystal (10%), the matrix is microsparites (30%) and also calcite cement (10%) (Figure 4). Based on the compositions and classification of [10], the rock can be categorized as packstone.

Figure 4. Outcrop of packstone (A) and its petrographic thin section (B). Note: Frm (foraminifera), Alg (algae), Pr (porosity), Ms (microsparite).

4.1.3. Grainstone

This limestone is found in the middle and northeast part of the research area. From the observation of its physical properties, this limestone has a low level of weathering. This limestone has a darker color than packstone and wackestone, which is yellowish-white to reddish-brown. It has a grain size of coarse sand (1-2 mm) and consists of bioclast (foraminifera test). Based on the petrographic observation, the thin section of the rock shows grain size <1mm-2 mm, subangular grain shape and poorly sorted. The composition of the limestone consists of skeletal grain of foraminifera and algae (60%), calcite crystal (20%), the matrix is microsparites (10%) and also calcite cement (10%) (Figure 5). Based on the compositions and classification of [10], the rock can be categorized as grainstone.

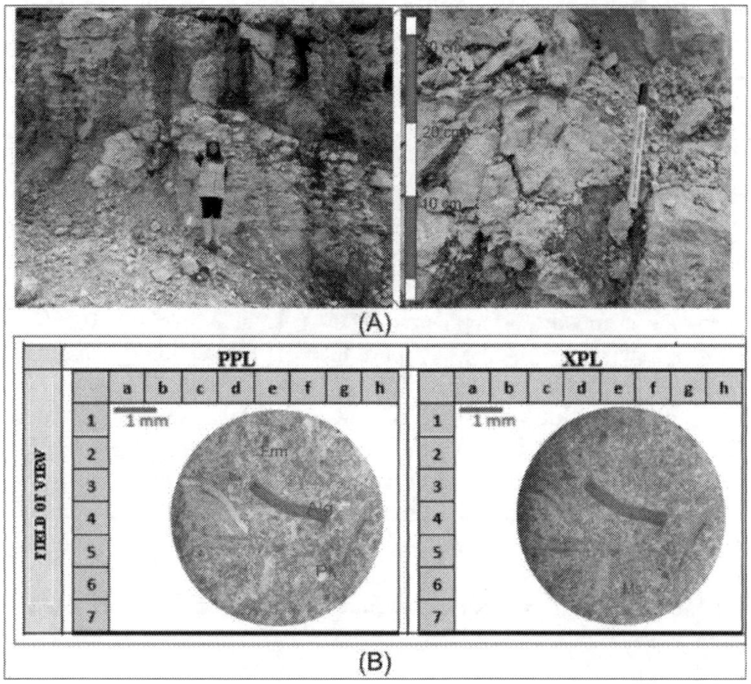

Figure 5. Outcrop of grainstone (A) and its petrographic thin section (B).
Note: Frm (foraminifera), Alg (algae), Pr (porosity), Ms (microsparite).

4.2. Geochemical Analysis

The geochemical analysis was carried out by using XRF analysis method to determine the major oxide compounds of limestone. XRF analysis was carried out on nine rock samples of different lithology types. The XRF analysis result shows that each lithology has a different chemical composition (Table 1). There is no LOI data available from Laboratory of Quality Assurance of PT Semen Indonesia Rembang.

Table 1. Chemical composition of limestone.

Lithology	SiO_2 (%)	Al_2O_3 (%)	Fe_2O_3 (%)	CaO (%)	MgO (%)	K_2O (%)	Na_2O (%)	SO_3 (%)
Wackestone	0.56	0.30	0.17	49.20	4.19	0.02	0.09	0.02
Packstone	0.80	0.48	0.24	55.80	0.50	0.02	0.04	0.01
Packstone	1.47	0.76	0.22	55.38	0.68	0.05	0.05	0.03
Packstone	0.38	0.23	0.13	55.56	1.05	0.01	0.05	0.00
Packstone	0.99	0.61	0.25	55.25	0.74	0.04	0.04	0.00
Packstone	1.11	0.67	0.30	55.14	0.68	0.04	0.05	0.00
Packstone	1.26	0.71	0.30	55.38	0.52	0.04	0.04	0.02
Grainstone	2.03	1.24	0.49	54.38	0.57	0.03	0.04	0.07
Grainstone	1.19	0.69	0.27	55.33	0.57	0.04	0.04	0.01

Based on the XRF analysis, it shows that wackestone has a lower CaO content of 49.20% and higher MgO content (4.19%), compared to other samples (packstone and grainstone) which have Cao content> 50% and lower MgO content (0.50-1.05%).

4.3. The relationship of physical properties and chemical composistion of limestone
The wackestone has less than 50% CaO content and 4.19% MgO content. The wackestone has white color, very fine grain size and contains less fossil composition. The composition of wackestone is mostly micrite (80%), skeletal grain (15%) and calcite cement (5%). It can be concluded that the fine-grained limestone with low content of skeletal grain and calcite mineral has low content of CaO.

Packstone generally has more than 50% CaO content and 0.5-1.05 % MgO content. The limestone has a yellowish to brownish-white color, and medium to coarse sand grain size (0.25 - 1 mm). The composition of packstone is mostly skeletal grain (50%), calcite crystals (10%), micrite (30%) and calcite cement (10%). Likewise, grainstone has more than 50% CaO content and 0.57% MgO content. The physical properties of this limestone are yellowish-white to reddish-brown, and coarse sand grain size (1-2 mm). The composition of grainstone is mostly skeletal grain (60%), calcite crystals (20%), micrite (10%) and calcite cement (10%). It can be concluded that the coarse-grained limestone with high content of skeletal grain and calcite mineral has high content of CaO.

The above discussion shows that the fine-grained limestone has low CaO content and high MgO content, while the coarse limestone has high CaO content and low MgO content. This is because the fine-grained limestone contains less bioclastic and calcite minerals, while the coarse limestone contains more abundant fossil shells (bioclastic) and calcite minerals. The fossil shells and calcite minerals contain high levels of CaO.

Based on the limestone classification for cement raw material from PT Semen Indonesia Rembang, wackestone is classified as a low grade because it contains low CaO content but high MgO content, which is not suitable for cement raw material [9]. While packstone and grainstone are classified as a high grade because they contain high Cao content and low MgO content. Thus, packstone and grainstone are recommended to be used as cement raw material, while wackestone must be mixed with other limestones with high CaO content to be used as cement raw material.

In comparison, the same result also occurred in PT Indocement Tunggal Prakarsa, Palimanan, Cirebon, West Java, Indonesia. The results in this area showed that grainstone, rudstone, bafflestone and packstone (coarse-grain limestone) tend to have higher CaO content and lower MgO content, while wackestone and mudstone (fine-grain limestone) tend to have lower CaO content and higher MgO content. The high level of CaO and low level of MgO in grain-supported limestones are interpreted to occur because the coarse-grain limestone has a higher grain percentage than the matrix. The CaO content in limestone comes from limestone grains consisting of fragments of organisms (foraminifera, algae, molluscs, corals, etc.) or cement consisting of aragonite or calcite with the chemical formula $CaCO_3$ [8].

5. Conclusion

1. The physical properties of wackestone have fine grain and contains less skeletal grain and calcite mineral, while packstone and grainstone have coarse grain and contain higher skeletal grain and calcite mineral.
2. Wackestone has lowest content of CaO and highest content of MgO, while packstone and grainstone have highest content of CaO and lowest content of MgO.
3. Packstone and grainstone are classified as a high grade for cement raw material, while wackstone is classified as a low grade for cement raw material.

Acknowledgments
The authors wish to thank Universitas Diponegoro, especially to Faculty of Engineering that has supported the research fund. The author also thank to the support of PT Semen Indonesia (Persero) Tbk Rembang for allowing the authors to get the data.

References

[1] Sukandarrumidi 1998 *Bahan Galian Industri* Universitas Gadjah Mada (Yogyakarta)

[2] Madiadipura T 1990 *Bahan Galian Industri di Indonesia* Direktorat Sumber Daya Mineral (Bandung)

[3] Pringgoprawiro H 1983 *Biostratigrafi Dan Paleogeografi Cekungan Jawa Timur Utara:* Suatu Pendekatan Baru Institut Teknologi Bandung

[4] Van Bemmelen R W 1949 *The Geology of Indonesia, Vol. IA: General Geology of Indonesia* (The Hague Netherlands)

[5] Vipulanandan C and Liu J 2012 *Polymer Coatings for Concrete Surfaces: Testing and Modeling*, Center for Innovative Grouting Materials and Technology (CIGMAT) Department of Civil and Environmental Engineering University of Houston Houston Texas

[6] Macfadyen J D 2006 *Cement and cement raw materials* Pages 1121-1136 in Industrial Minerals and Rocks 7th edition Littleton Colorado

[7] Fithri P and Sindikia A 2014 Pengendalian Persediaan Pozzolan di PT Semen Padang. *Jurnal Optimasi Sistem Industri,* Vol. **13** No. 2, Oktober 2014: 665-686

[8] Hidayatillah A S, Winarno T and Khasanah R 2020 Hubungan Antara Fasies Batugamping Terhadap Kualitasnya Sebagai Bahan Baku Semen Portland Menurut Kadar CaO dan Senyawa Terkait di Kuari B dan C, PT Indocement Tunggal Prakarsa Tbk Unit Palimanan, Cirebon. *Jurnal Geosains dan Teknologi,* Vol. **3** No. 1, Maret 2020:1-11

[9] PT Semen Indonesia (Persero) Tbk Rembang 2016 Peta Detail Kemajuan Tambang (unpublished)

[10] Embry A F and Klovan J E 1971 A Late Devonian reef tract on northeastern Banks Island. *NWT Bulletin of Canadian Petroleum Geology* **19** (4): 730–781

Petrogenesis of Botorubuh igneous rocks at Gunungkidul: A preliminary study for paleomagmatism phenomenon in the southern beach of Yogyakarta - Indonesia

Y K Sari, A D Titisari, I W Warmada and F Hakim

Geological Engineering Department, Faculty of Engineering, Universitas Gadjah Mada, Jalan Grafika Bulaksumur 2, Yogyakarta 55284, Indonesia
*E-mail: adewititisari@ugm.ac.id

Abstract. Botorubuh Beach is located at Southern Beach of Gunungkidul, in the Southern Mountain of Yogyakarta - Indonesia and is a promontory composed of Middle - Late Miocene igneous rock with a columnar jointing structure. This isolated igneous rock area is surrounded by limestone. Therefore, the regional geological map of Surakarta Giritontro classifies this area as the Punung-Wonosari Formation which is dominated by limestone. Because of the geological phenomenon of the isolated igneous rock area, it is necessary to study the petrogenesis of the igneous rocks at the area. The petrogenesis research is based on a preliminary study of the petrographical and geochemical characteristics of this igneous rock samples. The petrographic identification of andesite samples shows porphyritic, trachytic, oscillatory zoning, and sieve textures. The results of geochemical analysis (major and trace elements) show that the rock samples are are classified to andesite rocks and calc-alkaline suites. These rocks are enriched in LILEs (Rb, Ba, K) and depleted in HFSEs (Nb, Ti, Ce). Additionally, REE shows a slight enrichment of light-REE and a slight negative anomaly of Eu. The patterns of the trace elements including REE show a typical pattern of calc alkaline arc. Petrographical and geochemical characteristics suggest evidence of magma differentiation process, that is by a mechanism of crystallization fractionation. The andesite was formed in relation to a Middle – Late Miocene paleomagmatism and the Late Miocene-Pliocene subduction zone.

Keywords: petrogenesis, Botorubuh, andesite, calc-alkaline, columnar jointing structure

1. Introduction

A cape is called Botorubuh Beach which is located at the Southern Beach of Gunungkidul, in the Southern Mountain of Yogyakarta - Indonesia, exposes andesitic rocks with a columnar jointing structure. However, on the regional geological map of Surakarta - Giritontro, with a scale of 1: 100,000 compiled by [1], the rocks are grouped into the Wonosari Formation whose constituents are dominated by limestone because the isolated area is surrounded by limestone. In contrast, the Mount Batur composed of andesitic igneous rocks at Wediombo Beach which is located around 2 km to the northwest of Botorubuh Beach area, is classified as the Wuni Formation on the regional geological map. The Wuni formation is composed of andesite to basalt breccias, lava breccias, andesite lava and tuffaceous sandstone of the Middle – Late Miocene Epoch [1]. Meanwhile, the Wuni Formation represents one of the formations in the Southern Mountain whose location is far away in the Pacitan area, East Java, around 60 km to the southeast from Botorubuh area. Furthermore, several studies that have been conducted around the area such as [2], [3], [4], [5] and [6] are more focused on studying the Mount

Batur. According to Hartono (2000) in [2], the Mount Batur is identified as an ancient volcanic center. This statement is supported by [6] who explain that andesite lava of Mount Batur which was formed in the initial magmatisme phase, while the diorite intrusion on the area was formed in the next phase with relation to Calc-Alkaline Arc. Meanwhile [7] who conducted a research at Botorubuh Beach investigated more on the topic of geotourism. A preliminary study carried out in this research area has identified andesite rocks with a columnar jointing structure, which have similarities with rocks at the Mount Batur. For the reason that there are similarities in the field characteristics of the igneous rocks distributed on Mount Batur (Wediombo Beach) which are classified as the Wuni Formation and identified as an ancient volcanic center, however the igneous rocks of Botorubuh Beach which are categorized as the Wonosari Formation, it is necessary to conduct a petrogenesis research to determine the occurrence of the lithology at Botorubuh Beach. This research is expected to explain better understanding of andesite petrogenesis on studied area as a part of paleomagmatism phenomenon at the Southern Beach of Yogyakarta, in the Southern Mountain of Java Island at Late Miocene – Pliocene subduction. Discussion of the petrogenesis is confined to rock characteristics, magma suites, and patterns of trace elements which is based on petrographic observations and whole-rock geochemical data (concentrations of major oxide compounds and trace elements).

2. Regional Geology

In the southern part of the regional geological map the Surakarta-Giritontro sheet [1] shows several rock formations from the oldest to the youngest, namely the Mandalika Formation, Semilir Formation, Nglanggran Formation, Wuni Formation, Wonosari-Punung Formation, and Baturetno Formation (Figure 1). Lithological formations which are associated with the study area are the Wuni Formation which is at the Middle - Late Miocene Epoch, and the Wonosari-Punung Formation. More regionally, magmatic arcs that formed mountainous belts on the Java Island have undergone three phases, namely during the Late Eocene-Early Miocene, Late Miocene-Pliocene and at this time what is often referred to as the Quaternary Age [8]. On the other hand, [9] has reconstructed Tertiary volcanic centers in Java Island and Mount Batur was identified as one of the Lower Miocene volcanic center.

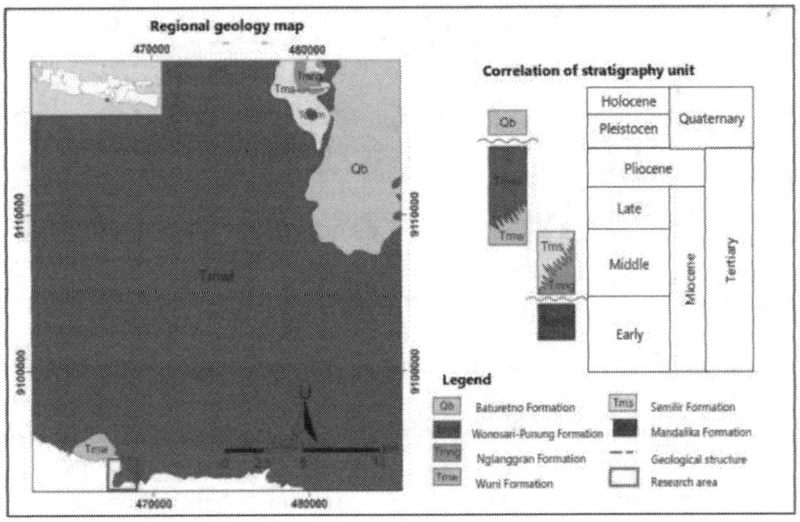

Figure 1. Regional geological map of the southern part of Surakarta-Giritontro geological map, simplified from [1] and the location of studied area.

IOP Conf. Series: Earth and Environmental Science **851** (2021) 012029 doi:10.1088/1755-1315/851/1/012029

3. Research method

More detailed geological mapping was carried out to determine the geological condition and distribution of lithologies in the study area. Rock sampling was collected at several points which represents lithology in the study area. Laboratory analysis was conducted for petrographical observations and geochemical analysis. Petrographic observation for 4 rock samples using a polarizing microscope was carried out at the Optical Geology Laboratory, Department of Geological Engineering, Universitas Gadjah Mada. Geochemical analysis to determine the percentage of major oxide compounds and the concentration of trace elements for representative rock samples using the ICP - AES (Inductively Coupled Plasma - Atomic Emission Spectrometry) and the ICP – MS (Inductively Coupled Plasma - Mass Spectrometry) methods was conducted at ALS Minerals Laboratory, Vancouver, Canada. This study uses additional geochemical data of igneous rocks from previous studies [4], [5] and [6]. The sampling locations and geochemical data for this study and previous researches are summarized on Figure 2 and Tables 1-2, respectively. The types of rocks in the previous studies are andesitic lava, andesitic intrusion [4,5] which is identified as dioritic intrusion by [6].

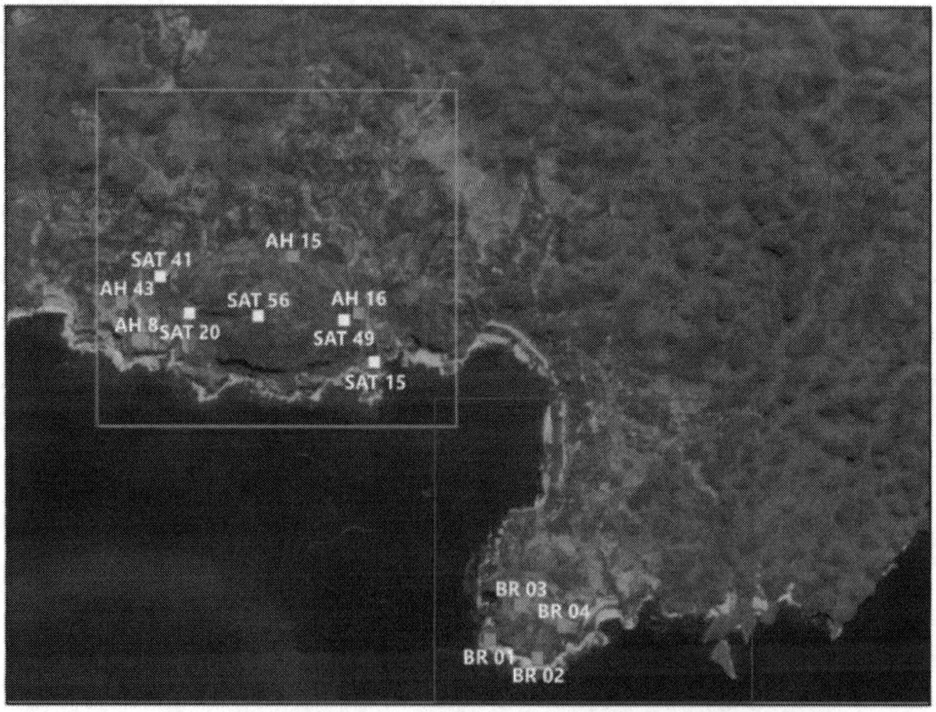

Figure 2. Location of this study (red box) and previous studies of [4], [5] and [6] indicated by green box, and their distribution of geochemical sampling locations. Note: solid orange squares indicate sampling locations of [4], solid yellow squares indicate sampling locations of [5] and [6] and solid green squares indicate sampling locations of this study.

Table 1. The major oxides abundance (in %) in several igneous rock samples at Botorubuh Beach (this study) and some data from the previous studies [4], [5] and [6].

	This study				Previous study [4]				Previous study [5] in [6]				
	Andesitic lava				Andesitic lava		Andesitic intrusion		Andesitic lava		Dioritic intrusion		
	BR01	BR02	BR03	BR04	AH 8	AH43	AH15	AH16	SAT15	SAT41	SAT20	SAT49	SAT56
SiO_2	60.30	58.60	59.60	61.00	60.00	58.80	61.56	61.18	60.43	59.03	61.69	62.07	63.16
TiO_2	0.57	0.55	0.56	0.57	0.49	0.55	0.49	0.47	0.54	0.46	0.48	0.48	0.46
Al_2O_3	17.30	17.45	17.65	17.60	17.14	17.35	16.95	16.83	17.81	17.06	17.32	17.56	17.47
Fe_2O_3	6.87	6.33	6.32	5.96	5.37	5.26	5.29	4.97	5.82	5.93	5.6	6.29	5.26
MnO	0.12	0.06	0.06	0.05	0.11	0.1	0.13	0.11	0.08	0.12	0.09	0.12	0.12
MgO	2.96	2.89	2.64	2.02	2.87	1.91	2.28	2.27	2.37	2.27	1.9	2.09	1.89
CaO	6.49	6.73	6.90	5.92	5.3	7.57	6.44	6.24	6.55	5.77	6.56	6.15	5.96
Na_2O	3.34	3.09	3.08	3.24	3.8	3.39	3.53	3.42	3.62	3.41	2.97	3.89	3.61
K_2O	0.78	0.46	0.48	0.79	1.06	0.61	0.83	0.73	0.57	0.65	0.62	0.62	0.79
P_2O_5	0.14	0.13	0.14	0.14	0.13	0.15	0.01	0.13	0.13	0.14	0.12	0.12	0.13
SrO	0.03	0.03	0.03	0.03	0.04	0.04	0.07	0.06	0.04	0.04	0.04	0.04	0.05
BaO	0.01	0.01	0.02	0.02	0.01	0.01	0.02	0.02	0.01	0.02	0.02	0.02	0.04
LOI	1.29	3.13	3.98	3.56	3.67	4.25	2.39	3.56	2.05	5.28	2.46	0.46	1.06
Total	100.20	99.46	101.46	100.90	99.99	99.99	99.99	99.99	100.02	100.18	99.87	99.91	100.00

Table 2. Trace element concentration (in ppm) in several igneous rock samples at Botorubuh Beach (this study) and some data from the previous studies [4] and [5].

	This study				Previous study [4]				Previous study [5]				
	Andesitic lava				Andesitic lava		Andesitic intrusion		Andesitic lava		Andesitic intrusion		
	BR01	BR02	BR03	BR04	AH 8	AH43	AH15	AH16	SAT15	SAT41	SAT20	SAT49	SAT56
Ba	130.50	126.00	146.50	150.50	117.00	113.50	219.00	134.00	125.00	184.50	60.00	144.00	366.00
Ce	13.00	13.00	18.60	13.30	14.10	15.20	16.50	15.50	15.50	14.80	17.00	16.50	17.70
Cr	20.00	10.00	20.00	10.00	10.00	20.00	10.00	10.00	10.00	10.00	10.00	10.00	10.00
Cs	0.65	0.34	0.13	0.59	0.48	0.52	1.28	1.55	0.18	5.83	0.83	0.87	1.21
Dy	2.07	2.02	7.77	2.23	1.84	1.71	1.80	1.62	2.00	1.81	1.81	1.56	1.68
Er	1.20	1.18	4.45	1.26	0.94	0.91	1.05	0.94	1.13	1.10	1.18	1.05	0.98
Eu	0.68	0.72	2.99	0.79	1.30	3.40	1.60	1.50	0.74	0.72	0.76	0.78	0.78
Ga	17.10	18.80	18.50	16.60	17.40	19.00	19.00	18.40	18.20	18.30	19.20	19.20	19.50
Gd	2.50	2.01	11.05	2.86	1.99	2.16	2.25	1.90	2.13	2.01	2.01	1.80	2.10
Hf	1.80	1.80	1.80	2.10	1.70	1.80	1.70	1.80	1.90	2.00	2.00	1.80	2.10
Ho	0.43	0.41	1.51	0.43	0.34	0.32	0.38	0.33	0.38	0.39	0.34	0.70	0.33
La	5.70	6.00	55.40	6.40	6.20	6.70	7.80	6.90	6.90	6.80	8.20	8.00	8.80
Lu	0.17	0.18	0.81	0.18	0.15	0.13	0.17	0.14	0.15	0.17	0.16	0.14	0.16
Nb	1.20	1.10	2.70	1.30	1.30	3.40	1.60	1.50	1.30	1.50	1.60	1.60	1.50
Nd	8.50	8.00	57.00	9.20	8.30	9.10	9.60	8.90	9.20	9.00	10.00	10.00	10.40
Pr	1.81	1.79	13.05	2.05	1.96	2.06	2.31	2.05	2.27	2.07	2.43	2.20	2.47
Rb	21.30	3.90	5.10	19.60	16.00	14.40	22.90	17.30	9.60	18.10	18.60	18.20	25.10
Sm	2.18	2.26	11.00	2.41	1.91	2.31	2.38	2.05	1.94	1.81	1.08	2.03	1.91
Sr	308.00	327.00	330.00	305.00	327.00	395.00	633.00	491.00	393.00	364.00	391.00	402.00	424.00
Ta	0.40	0.20	0.20	0.30	0.10	0.10	0.10	0.10	0.20	0.20	0.20	0.20	0.20
Tb	0.36	0.33	1.35	0.37	0.30	0.31	0.34	0.29	0.31	0.26	0.33	0.26	0.28
Th	1.74	1.79	1.83	1.92	1.41	1.60	1.80	1.69	1.63	1.60	1.88	1.82	2.04
Tm	0.19	0.17	0.64	0.21	0.14	0.14	0.14	0.14	0.11	0.15	0.15	0.14	0.16
U	0.45	0.47	0.46	0.49	0.36	0.36	0.42	0.37	0.31	0.32	0.47	0.41	0.45
Y	10.90	11.40	41.00	10.60	9.40	9.50	10.70	9.40	10.70	10.00	10.50	9.60	10.40
Yb	1.16	1.17	4.79	1.16	0.93	0.85	0.96	0.86	0.95	0.97	1.40	0.91	1.01
Zr	59.00	58.00	58.00	63.00	59.00	63.00	61.00	63.00	64.00	70.00	66.00	66.00	70.00

4. Results

4.1. Geological Study Area

Based on its dominance, the lithology in the study area can be grouped into three rock units, which is andesite lava units, andesite breccia units, and limestone units. The geological map of the research area can be seen on Figure 3. These igneous rocks are expressed in form of an isolated hilly morphology where andesite lava unit is surrounded by andesite breccia unit. Columnar jointing structure and shear joints of andesite lava are well exposed on coastal cliffs of studied area (Figures 4 and 5, respectively). This research only focuses on the andesite lava units. Megascopically, the andesite is characterized by grayish black (least altered) and brownish gray (slightly weathered) with a massive structure, a porphyritic texture, holocrystalline, subhedral crystal form. Phenocrysts are composed of abundant plagioclase (0.2-4 mm in size) and hornblende, while the groundmass with crystal size <0.01mm is composed of plagioclase and mafic minerals. The samples from previous studies [4], [5] and [6] show that andesitic intrusions [4] and dioritic intrusions [6] are dominated by plagioclase (0.2–6 mm in size), and hornblende and opaque minerals which also can be identified as phenocryst, while the groundmass is dominated by plagioclase. Andesite lava is characterized by plagioclase (<0.01-2 mm in size) both as phenocryst and groundmass. Hornblende and opaque minerals are also common in the andesite lava.

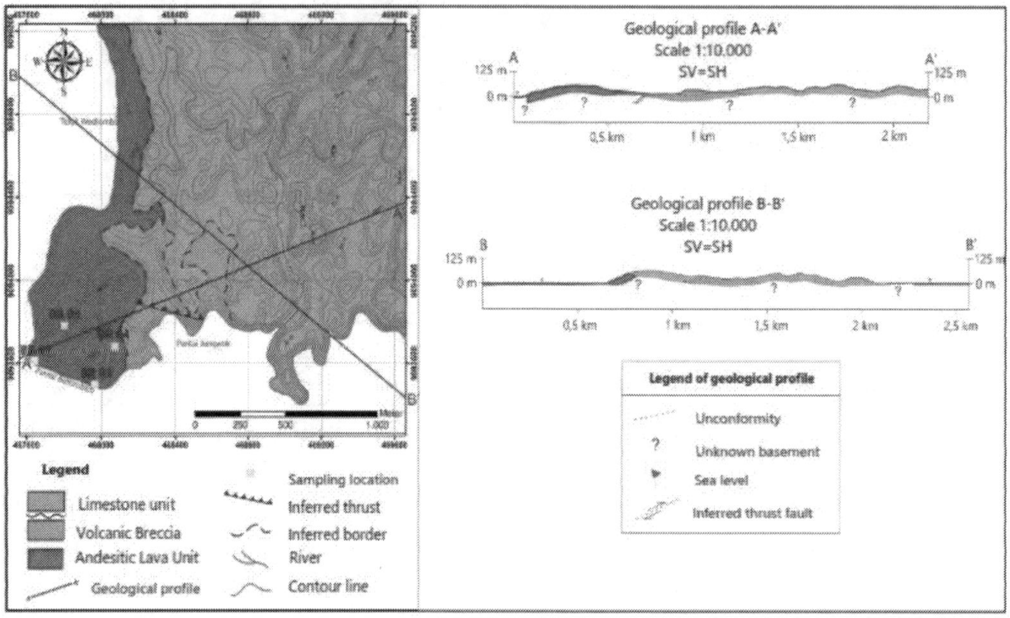

Figure 3. Geological map, geological profile and rock sampling locations of this study.

Figure 4. Andesite columnar jointing structure is well exposed on coastal cliffs at the studied area.

Figure 5. Shear joints of andesite lava at the studied area.

4.2. Petrography Characteristics

Petrographic observations show that the andesite is dominated by plagioclase, clinopyroxene, and hornblende minerals with a crystal size of 0.01 - 4 mm, with porphyritic texture, and typical textures of trachytic texture and oscillatory zoning (Figures 6A-H). The abundance of minerals is dominated by plagioclase either as phenocrysts or as microcrystal in the groundmass. Hornblende and clinopyroxene also present as phenocrysts.

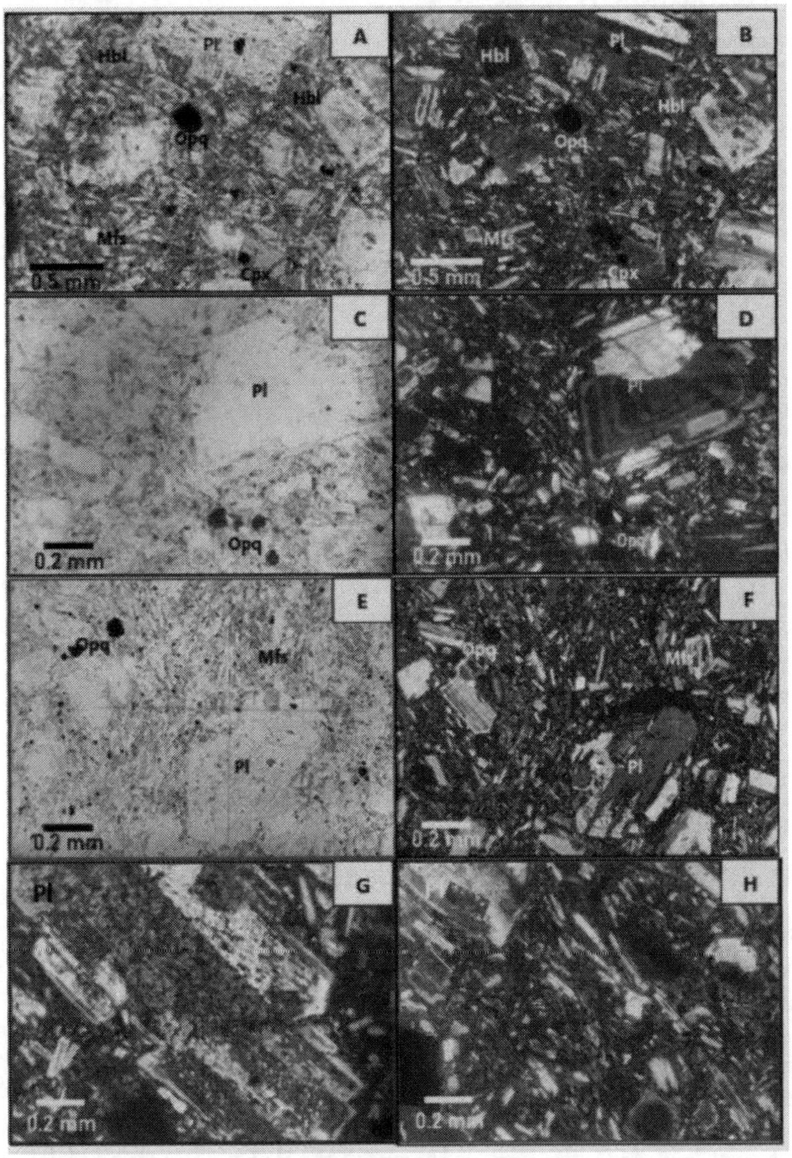

Figure 6. Photomicrographs of andesite samples in the studied area that show some special textures: porphyritic (A-B, E-F), oscillatory zoning (C-D), sieve (G) and trachytic texture (H).

Porphyritic textures were identified of the samples as shown on Figures 6A-B dan Figures 6E-F. The oscillatory zoning texture is recognized in phenocryst of plagioclase that suggests variations of plagioclase composition from the center of the crystal (core) to the edge of the crystal (rim), resulting in the appearance of a low-high extinction angle that changes concentrically (Figures 6C-D). Sieve textures (Figure 6G) are identified in almost all plagioclase phenocrysts together with the oscillatory zoning texture as a characteristic texture of the disequilibrium state. Trachytic texture is clearly recognizable on Figure 6H showing an orientation of plagioclase minerals. The texture of trachytic is a product of flow materials, so that the naming of the andesite lava is supported by evidence of petrographic analysis which indicates the presence of this trachytic texture.

4.3. Geochemical Characteristics

Igneous rocks on Botorubuh Beach including rock samples from the previous researches ranges from 58.6 to 63.16%, while the total alkali ($Na_2O + K_2O$) ranges from 3.55 to 4.86 wt% (Table 1). Based on geochemical data plotting on the diagram of Total Alkali vs. Silica (Figure 7A) and diagram of Nb/Y vs. Zr/Ti (Figure 7B), the rocks fall within andesite fields. A FeO*/MgO vs. SiO_2 diagram with assist line from [10] has been applied to observe the calc-alkaline characteristics of the rock samples (Figure 8A). The magma suites of the AFM ternary diagram (Irvine and Baragar, 1971 in [11]) of the studied area and previous studies samples is shown on Figure 8B. The magma suites can also be determined using the trace element ratio as the discriminant, namely Ta / Yb vs. Th / Yb [12] is shown on Figure 8C. On the diagram (Figure 8C), magma suites in the study area and previous studies are also classified as the calc-alkaline magma suites.

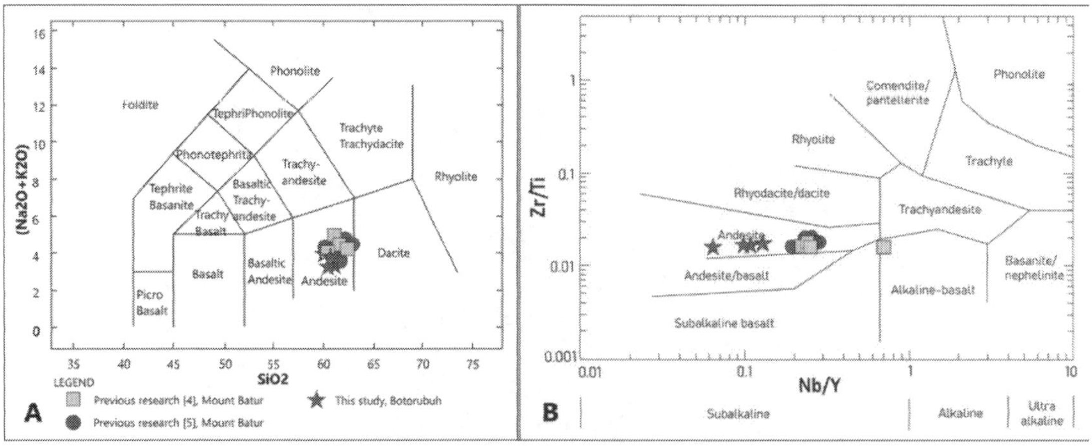

Figure 7. (A) Silica versus total alkali diagram for igneous rocks of studied area and its surrounding with Total Alkali Silica (TAS) classification (after Le Bas and Streckeisen, 1991 in [13]). (B) Igneous rock of Botorubuh area and its surrounding plotted on the Winchester and Floyd classification scheme [14] based on trace element content of Nb/Y vs. Zr/Ti, with most samples plotting in the andesite fields.

The N-MORB normalized [15] trace element spider diagram for the igneous rocks in study area show that trace elements of LIL (Large Ion Lithophile) which are Rb, Ba, Sr, K are enriched relatively to HFSE (High-Field-Strength Elements) such as Th, Nb, Zr, Ti. The negative anomaly of the trace elements is shown by Nb (Figure 9). The pattern of the HREE elements (Eu, Gd, Tb, Dy, Y, Ho, Er, Tm, Yb, Lu) in the diagram shows depletion in relative to the LREE pattern (La, Ce, Pr, Nd, Sm) (Figure 10). Moreover, the REE diagram show a slight negative anomaly of Eu for the andesitic samples of Botorubuh area and previous research of [5]. In contrast, Eu element displays a positive anomaly on previous research of [4] samples (Figure 10). It means that the Eu concentration in the mineral is enriched relative to the other rare-earth elements on those samples. Anomalies of the BR03 sample (Figures 9 – 10) indicate enrichments in the average of trace element concentration in comparison to other samples in the study area. A contamination of subducted sediment will show a negative anomaly of Ce element relative to other rare earth elements [16]. This is consistent with the rare earth elements diagram on the study area showing a negative anomaly of Ce element for the BR03 sample (Figure 10). Therefore, the anomalies of the BR03 sample are considered and assumed due to the condition of the sample which is as the result of incorporation of subducted sediment to the arc magma sources.

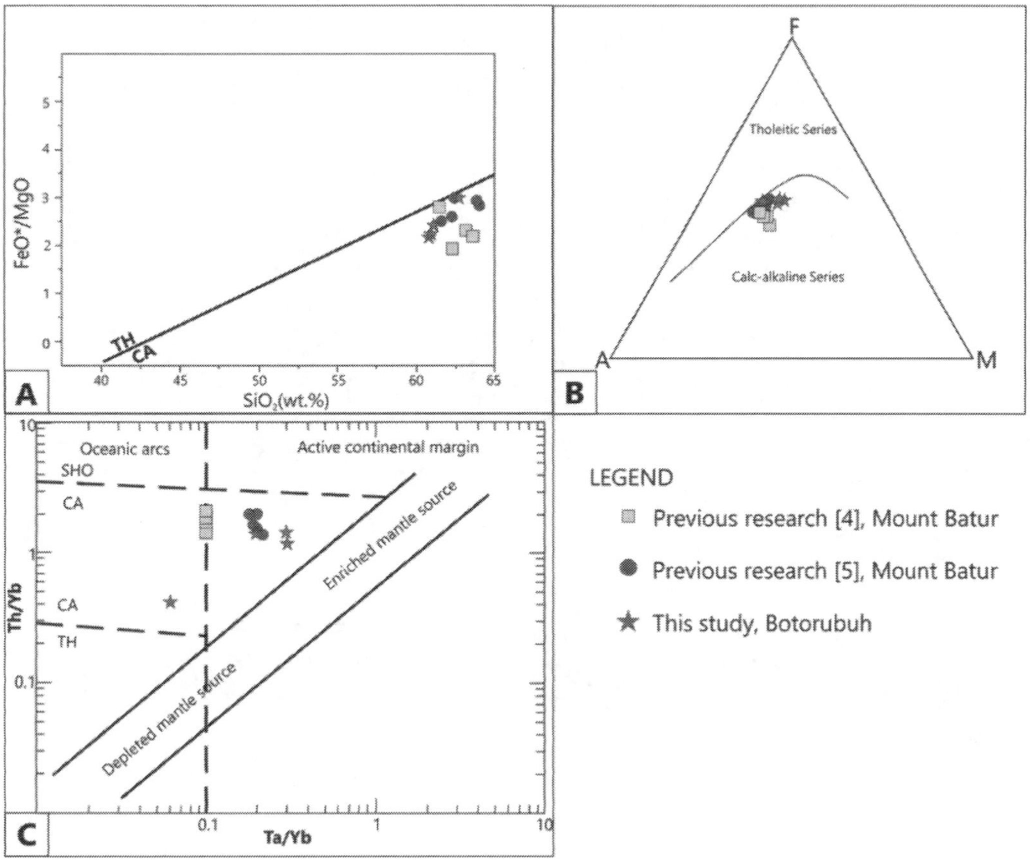

Figure 8. (A) Binary diagram of magma suites determination based on FeO*/MgO and SiO_2 content of Botorubuh samples and previous researches in Mount Batur (with assist line from [10]). (B) The AFM diagram of studied area samples shows the calc-alkaline magma suites (After [18]). (C) Magma series determination of studied area samples indicates a calc-alkaline magma suite (After [12]).

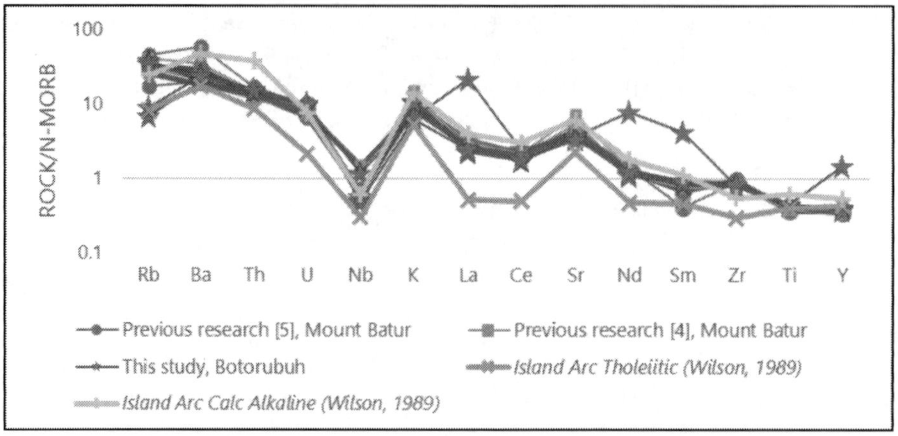

Figure 9. The N-MORB normalized patterns of trace element in some geological settings from [10] are compared with rock sample data of the studied area (Botorubuh) and its surroundings (Mount Batur).

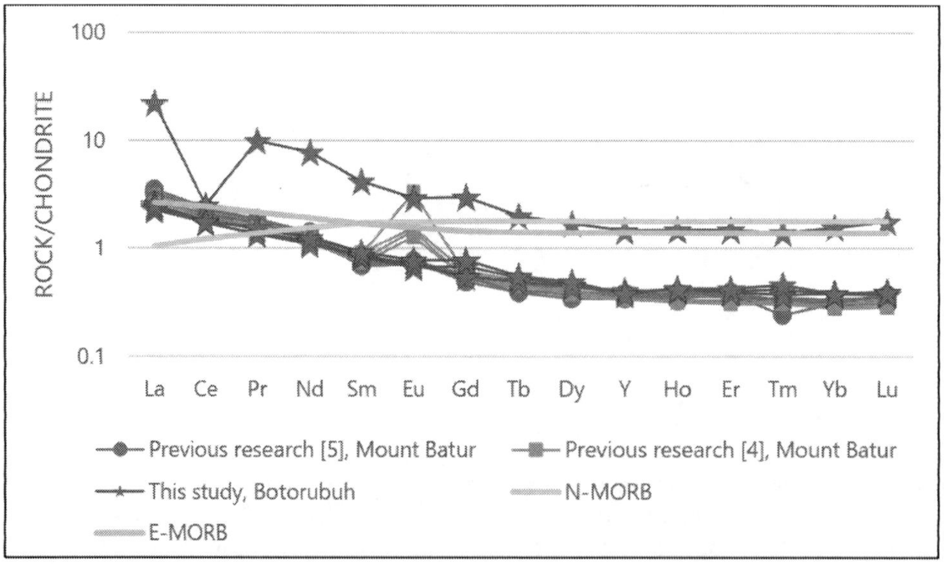

Figure 10. Chondrite normalized of REE spider diagram of studied area samples at Botorubuh and its surroundings (previous researches at Mount Batur).

5. Discussion

Petrographic analysis shows that the porphyritic texture is identified in all rock samples of the study area showing a different crystal size of plagioclase groundmass and plagioclase phenocryst. This difference in the size of the plagioclase crystals indicates the occurrence of two-stage cooling. Referring to the statement of [20], it can confirm that the indication of the two-stage cooling suggests a difference in cooling rate of the magma source of andesitic rock in the research area. It can also be an indication of occurrence of two phases of the crystallization process that is the earlier phase crystallization of phenocryst minerals and the later phase crystallization of groundmass minerals. Typical textures including trachytic texture, oscillatory zoning, and sieve texture are also identified in the rock samples. The trachytic texture in Figure 6H shows that the smaller plagioclase minerals show an alignment pattern around the phenocryst. This texture is common in a product of flow materials. Crystallization of flow materials causes in a tendency for the crystals to position themselves in the direction of the flow. Therefore, the presence of a trachytic texture indicates that andesite in the study area might be associated with a flow process. The oscillatory zoning texture (Figures 6C-D) shows that there are variations in plagioclase composition due to crystal responses to fluctuating external conditions, or due to repeated influx of new magma and mixing of magma with more evolved magma [20]. This zoning texture is able to be an indication of occurrence of pressure and temperature changing during magma movement and volatiles release during magmatic eruptions leading to differentiation of the plagioclase composition (Fenner, 1926 in [21]). The sieve texture (Figure 6G) shows the appearance of holed plagioclase like corrosion (corroded-like) and then the holes are filled with the later phase magma fluid.

Based on the results of plotting on the diagrams of binary FeO*/MgO vs SiO_2 (with assist line from [10]), the AFM magma series by [18] and Ta / Yb vs. Th / Yb by [12] which are shown on Figures 8A-C, respectively, andesite in the Botorubuh coast area is categorized as a calc-alkaline suite. The calc-alkaline suites are a characteristic magma suite on a convergent boundary of subduction zone or on an active continental margin of subduction zone [11]. The suggestion is supported by plotted result of Total Alkali Silica (TAS) diagram and diagram of Nb/Y vs. Zr/Ti (Figures 7A-B, respectively) showing the rock samples classified on andesite, indicate to magma type which formed the andesite rocks, is the intermediate magma. This magma is a typical volcanic rock on the subduction zone [11].

Trace elements of LILE (Large Ion Lithophile Elements) such as Rb, Ba, Sr, K which are enriched relatively to HFSE (High-Field-Strength Elements) such as Th, Nb, Zr, Hf, Ti for andesite of studied area (Figures 9 and 10) suggests a typical pattern of calc alkaline arc. By refers to [22] and is supported by reference work of [11], it can be suggested that the trace element patterns of the rock samples in the study area and in Mount Batur (previous studies) have similarities with the trace element patterns of the calc alkaline island arc tectonic setting (Figure 9). The REE pattern (Figure 10) shows a consistentcy with the REE pattern of E-MORB magma that the pattern of E-MORB was summarized from [19]. Additionally, the spider diagram of REE shows the enrichment LREE and HREE depletion. Typical subduction zone magma shows a similar phenomenon of that, which is an enrichment of LREE relative to MREE and HREE [23]. Therefore, this typical pattern of trace elements and REE indicates that andesite at studied area was performed in subduction zones and generated from same source. However, there is an exception of one sample (BR03 sample from studied area) that shows a negative anomaly of Ce element, which is considered and assumed due to as the result of incorporation of subducted sediment to the arc magma sources. Plagioclase fractionation is commonly assumed responsible for negative Eu anomalies in calc-alkaline volcanic arc suites (e.g., [24], [25]). The abundance of plagioclase phenocrysts in andesite of studied area (Figures 6A-H) provides supporting petrographical evidence for this interpretation. Consequently, the slight negative Eu anomaly of the Botorubuh andesite samples (Figure 10) suggests a plagioclase fractionation within the source magma. Plotted REE on spider diagram of Hendrawan's samples [4] which shows positive Eu anomalies (Figure 10) suggests that magma source accumulated plagioclase crystals before solidification and its consequent was most of the Eu would be incorporated into the plagioclase, causing a higher concentration of Eu in the mineral versus other REE in that mineral. That produced rock composition which would display a relatively positive Eu anomaly. It can be suggested that the Hendrawan's samples [4] at Mount Batur had experienced solidification at earlier magmatic phase in relative to the other samples from the studied area at Botorubuh. This interpretation is supported by [6] which considered that andesite lava of Mount Batur was formed in the initial magmatic phase (position of Hendrawan's samples), then followed by the diorite intrusion on the area which was formed in the next magmatic phase in Calc-Alkaline Arc.

Refers to the regional geological map [2], the andesitic rocks of Botorubuh studied area can be classified as the Wuni Formation of the Middle - Late Miocene Epoch. Additionally, according to [9] who have reconstructed Tertiary volcanic centers in Java Island and identified Mount Batur as one of the Lower Miocene volcanic center, and also refers to [8] who have explained the three phases of magmatic arcs on the Java Island, therefore it can be summaried that the andesitic rocks of Botorubuh in studied area related with a paleomagmatism of the Middle – Late Miocene Epoch, and in relation to the Late Miocene-Pliocene subduction zone.

6. Conclusion

Andesitic rocks at Botorubuh Beach are characterized by porphyritic, trachytic, oscillatory zoning, and sieve textures. The results of geochemical analysis (major and trace elements) show that the andesite is typified by calc alkaline suites. These rocks are enriched in LILEs (Rb, Ba, K) and depleted for HFSEs (Nb, Ti, Ce). In addition, REE shows a slight enrichment of light-REE and a slight negatively anomaly of Eu. The patterns of the trace elements including REE show a typical pattern of calc alkaline arc. The petrographical and geochemical characteristics of the andesite rocks in the research area suggests that the lithology was produced by magma differentiation process which was by a mechanism of crystallization fractionation. The andesite was formed in relation to a Middle – Late Miocene paleomagmatism and the Late Miocene - Pliocene subduction zone.

Acknowledgment

The authors would like to thank the Department of Geological Engineering, Faculty of Engineering, Universitas Gadjah Mada which has given funding for this publication.

References

[1] Surono et al 1992 *Peta Geologi Lembar Surakarta Giritontro skala 1:100.000* (Bandung: Pusat Penelitian dan Pengembangan Geologi)
[2] Hartono G dan Bronto S 2007 *Asal-usul Pembentukan Gunung Batur di daerah Wediombo, Gunungkidul, Yogyakarta*, Jurnal Geologi Indonesia **2(3)** 143-158
[3] Mustakim W Y, Idrus A, Setiadji L D and Warmada I W 2014 *Geologi dan Alterasi Hidrotermal di Gunung Batur, Wediombo, Kabupaten Gunung Kidul, Provinsi D.I. Yogyakarta*, Proc. Nat. seminar of Earth Science vol 7 (Yogyakarta: Geological Engineering Department, Faculty of Engineering, Universitas Gadjah Mada) p 657-664
[4] Hendrawan A 2017 *Genesa Batugamping Merah di Daerah Siung dan Sekitarnya, Kecamatan* Tepus *dan Girisubo, Kabupaten Gunungkidul, Yogyakarta* [thesis] (Yogyakarta: Geological Engineering Department, Faculty of Engineering, Universitas Gadjah Mada)
[5] Nugroho Y S 2019 *Karakteristik Mineralogi dan Geokimia Batuan Vulkanik Gunung Batur Daerah Pantai Wediombo dan Sekitarnya, Kecamatan Girisubo, Kabupaten Gunungkidul, Provinsi Daerah Istimewa Yogyakarta* [thesis] (Yogyakarta: Geological Engineering Department, Faculty of Engineering, Universitas Gadjah Mada)
[6] Hakim F, Nugroho Y S, Dana C D P, and Titisari A D 2019 *J. App. Geol.* **4(1)** 32-42
[7] Fillah A A, Widyatmaji B N, Sutikno A, and Hendratno A 2019 *Studi Kajian Geowisata Pantai Botorubuh dalam Aspek Geosite dan Geomorphosite pada Kawasan Subzona Gunung Sewu,* Proc. Nat. seminar of Earth Science vol 12 (Yogyakarta: Geological Engineering Department, Faculty of Engineering, Universitas Gadjah Mada) p 1327-1340
[8] Soeria-Atmadja R, Maury R C, Bellon H, Pringgoprawiro H, Polve M, and Priadi B 1994 *J. Southeast Asian Earth Sci.* **9(1/2)** 13-27
[9] Setijadji L D, Kajino S, Imai A, and Watanabe K 2006 *Resour. Geol.* 56(3) 267–92
[10] Miyashiro, A 1974 Am. J. Sci. **274** 321–355
[11] Wilson M 1989 *Igneous Petrogenesis* (Dordrecht: Springer)
[12] Pearce J A 1983 Role of the sub-continental lithosphere in magma genesis at active continental Margins *Continental basalts and mantle xenoliths* ed Hawkesworth C J and Norry M J (Cheshire: Shiva Publications Nantwich) pp 230-49
[13] LeMaitre R W, Streckeisen A, Zanettin B 2002 *Igneous Rock: A Classification and Glossary of Terms*, Cambridge University Press, New York
[14] Winchester J A and Floyd P A 1977 *Chem. Geol.* **20** 325-343

[15] Sun and McDonough W F 1989 *J Geol Soc London* **42** 313-45
[16] Shimizu H, Sangen K and Masuda A 1982 *Geochem. J.* **16** 107-17
[17] Rollinson H R 1993 *Using Geochemical Data: Evaluation, Presentation, Interpretation* (London: Longman Group UK) p 352[18] Irvine T N and Baragar W R A 1971 Can. J. Earth Sci **8** 523-548
[19] Tatsumi Y and Eggins S 1995 *Subduction Zone Magmatism* (Cambridge: Blackwell Science Inc)
[20] Gill R 2010 Igneous Rocks and Processes: A Practical Guide (Chichester: Wiley-Blackwell)
[21] Johannsen A 1939 *A Descriptive Petrography of the Igneous Rocks*, University of Chicago
[22] Pearce J A and Parkinson I J 1993 *Geol. Soc. Lond. Spec. Publ.* **76** 373-403
[23] Winter J D 2014 *An Intoduction to Igneous and Metamorphic Petrology* (New Jersey: Prentice-Hall Inc)
[24] Murphy J B 2007 *Geosci. Can.* **34** 7-35
[25] MacDonald R, Belkin H E, Fitton J G, Rogers N W, Nejbert K, Tindle A G and Marshall A S 2008 *J. Petrol.* **49** 1515-47

IOP Conf. Series: Earth and Environmental Science **851** (2021) 012030 doi:10.1088/1755-1315/851/1/012030

Textural and mineralogical characteristics of sand grains of Oyo River Deposit at Yogyakarta, Indonesia: implications on sedimentation and origin

I F Lubis[1], **A D Titisari**[1*] **and N S Irsani**[1]

[1]Geological Engineering Department, Faculty of Engineering, Universitas Gadjah Mada, Jalan Grafika Bulaksumur 2, Yogyakarta 55284, Indonesia
*E-mail: adewititisari@ugm.ac.id

Abstract. Oyo river deposit is one of the potential resources for small scale sand mining in Yogyakarta. The deposit extends along the upstream to downstream area of the river. A detailed study on textural and mineralogical characterization is aimed for better understanding of Oyo River sedimentation and provenance. Ten sediment samples were analyzed by using granulometry method, these include grain size, morphological and mineralogical analysis. The result shows that the average of grain size changes from coarse sand (0.88 mm) in the upstream to medium sand (0.37 mm) in the downstream. Sorting value is dominated by moderately sorted (0.71φ – 0.94φ), skewness is dominated by very fine skewed (0.3φ - 1.10φ), and kurtosis is dominated by extremely leptokurtic (2.8φ – 6.5φ). Meanwhile for the grain shape is dominated by oblate and bladed, sphericity changes from elongate to subequant, roundness is dominated by subangular and subequant. Mineralogical composition consists of magnetite, hematite, pyroxene, amphibole, quartz, feldspar, and lithic fragment. Plotting result on triangular QFL (Quartz Feldspar Lithic-fragment) diagram shows that the provenance of Oyo River sand deposit is coming from dissected arc tectonic setting.

1. Introduction

The Oyo River is one of the rivers where the upstream is located on the Gajah Mungkur Mountains, Wonogiri Regency, while the downstream is located at the Bantul Regency. The length of main river is 106.75 km and watershed area are 639 km^2. This watershed area is bordered by Baturagung Mountains in the north and the Wonosari Basin in the south. Sedimentation and provenance of Oyo River is interesting to be observed because the position of the upstream river is in the inactive volcanic area while most of the river bodies are in the karst area. It is impacted on the different dynamics compared to other rivers in Yogyakarta. Thus, the objective of this study is to determine the changes in sediment characteristics from upstream to downstream and the origin rocks of the sediments in the Oyo River based on the statistical values.

2. Regional Geology

Physiographically, the Oyo River is in the Southern Mountain Zone where the main physiography through which passed is the Wonosari Basin in the north and the Baturagung Mountains in the west [1]. The south is bordered with karst area and the east is bordered with Panggung and Plopoh Mountains. The geomorphic process which controls or works on the Oyo River are dominated by fluvial process.

The regional geological map of Yogyakarta and Surakarta – Giritontro, with a scale of 1:1000,000 by [2] and [3] which passed by Oyo River from the oldest to the youngest formations consists of 5 main formations such as Semilir Formation (Tuff, lapilli tuff, breccia and shale), Nglanggran Formation (Breccia, agglomerate, tuff and lava flow), Sambipitu Formation (sandstone, siltstone, and stone stone), Oyo Formation (tuff, tuff marl and limestone), and Wonosari Formation (limestone and tuff inserts and tuff marl). These formations are passed by the Oyo River as seen in Figure 1. Beside the formation which have mentioned, there are also some types of sediment along the Oyo River such as sediment of mount Merapi which are composed of breccia, lava, and tuff.

The geological structure which develops around the flow of Oyo River includes 3 geological structures [4]. The first is in the Nglipar area of Gunung Kidul Regency—in the form of a descending fault, the second is in the Patuk area of Gunung Kidul Regency which is sinistral shear fault, and the third is in Imogiri Regency—in the form of a dextral shear fault.

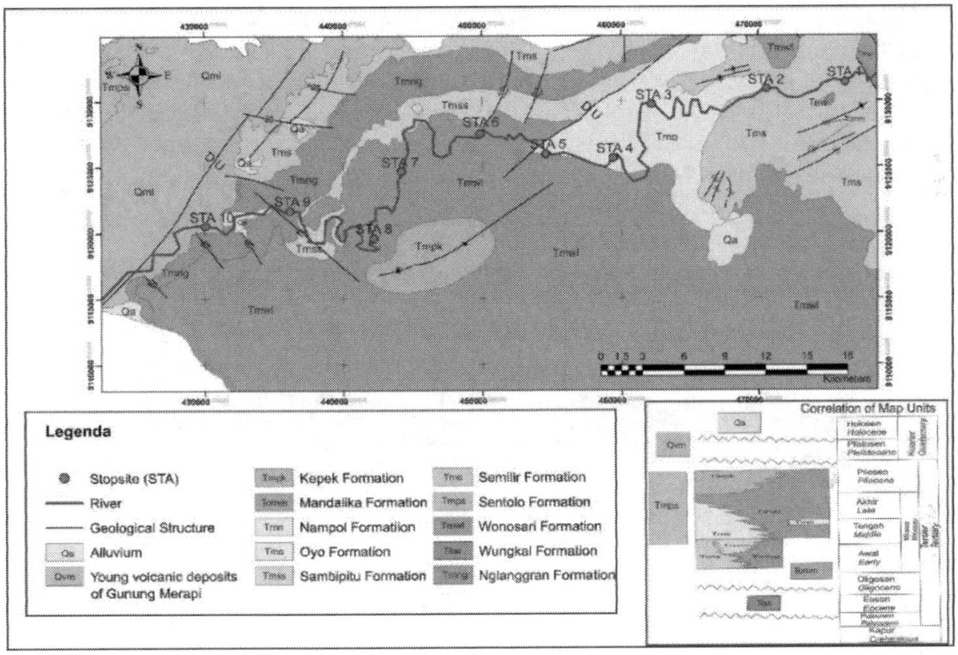

Figure 1. Regional geological map of the Surakarta-Giritontro sheet from east to west [3] and the southeastern part of Yogyakarta sheet [2] accompanied by a research path on the Oyo river.

3. Research Method

The research method used is laboratory analysis on the sand-sized sediment samples [5]. Sediment sample is taken from ten observation locations along the Oyo River, Central Java Province and D.I. Yogyakarta Province.

The laboratory analysis is carried out in the form of granulometric analysis and the composition of the sand sediment with each sample represent an observation location [6]. Granulometric analysis is carried out by visual comparison method [7]. The sieving process in this granulometric analysis uses mesh sizes 18 (very coarse sand), 35 (coarse sand), 60 (medium sand), 120 (fine sand), 270 (very fine sand) and> 270 (silt) [8].

The parameters used to interpret sedimentation mechanism are average grain size (mean), grain size distribution value (sorting), the symmetrical value of the frequency curve (skewness), the curve puncturing value (kurtosis) [9]. Meanwhile, determination of the provenance is carried out by separating heavy and light mineral content continued with petrographic observations using binocular microscope [10].

IOP Conf. Series: Earth and Environmental Science **851** (2021) 012030 doi:10.1088/1755-1315/851/1/012030

4. Results

4.1. Granulometric Data

4.1.1. *Mean.* Mean is the average value of grain size. Based on the mathematical calculation, the results are obtained from 5 samples of coarse sand and 5 samples of medium size sand as seen in Figure 2a. Based on the mean graph, it is obtained that the fluctuation of the change in the average grain size value from upstream to downstream. Generally, the curve shows smooth trend from stop site 1 to stop site 8 then it is continued to rough trend at stop site 9 and stop site 10.

4.1.2. *Sorting.* Sorting is the standard deviation value of grain sizes which shows the level of uniformity. Based on mathematical calculation, it is obtained the value that 9 samples have moderately sorted values, and 1 sample has moderately well sorted value as seen in Table 1 and Figure 2b. Based on the graph, it is found that the sorting value shows the relatively stable values in the moderately sorted range.

4.1.3. *Skewness.* Skewness shows the symmetric value of the frequency curve. Based on mathematical calculation, it is found that 9 samples have very fine skewed value, and 1 sample has near-symmetrical value. Based on the graph, it is known that the skewness value decreases which means that the further downstream is dominated by smoother grain size. However, at stop site 9 and stop site 10 increase to be the dominant of coarser grain.

4.1.4. *Kurtosis.* Kurtosis is the key value of the curve. Based on mathematical calculation, it is found the result that all samples have very leptokurtic value and extremely leptokurtic value. Based on the graph, the kurtosis value shows the increasing trend as it approaches the downstream. However, but the trend changes downward at stop site 9 and stop site 10 showing poor sorting dominance.

Table 1. Granulometric data consist of mean, sorting, skewness, and kurtosis.

Stopsite (STA)	Mean	Sorting		Skewness		Kurtosis	
		Value	Classification	Value	Classification	Value	Classification
1	0.18	0.79	Moderately Sorted	1.13	Very Fine Skewed	4.12	Extremely Leptokurtic
2	0.61	0.89	Moderately Sorted	0.49	Fine Skewed	2.76	Very Leptokurtic
3	0.89	0.81	Moderately Sorted	0.54	Fine Skewed	3.43	Extremely Leptokurtic
4	1.11	0.82	Moderately Sorted	0.52	Fine Skewed	3.17	Extremely Leptokurtic
5	1.27	0.81	Moderately Sorted	0.43	Fine Skewed	2.92	Very Leptokurtic
6	1.30	0.92	Moderately Sorted	0.03	Near Symmetrical	2.82	Very Leptokurtic
7	1.32	0.95	Moderately Sorted	0.30	Fine Skewed	2.89	Extremely Leptokurtic
8	1.43	0.78	Moderately Sorted	0.40	Fine Skewed	3.19	Extremely Leptokurtic
9	0.63	0.71	Moderately Sorted	0.49	Fine Skewed	3.63	Extremely Leptokurtic
10	0.36	0.67	Moderately well Sorted	1.02	Very Fine Skewed	6.57	Extremely Leptokurtic

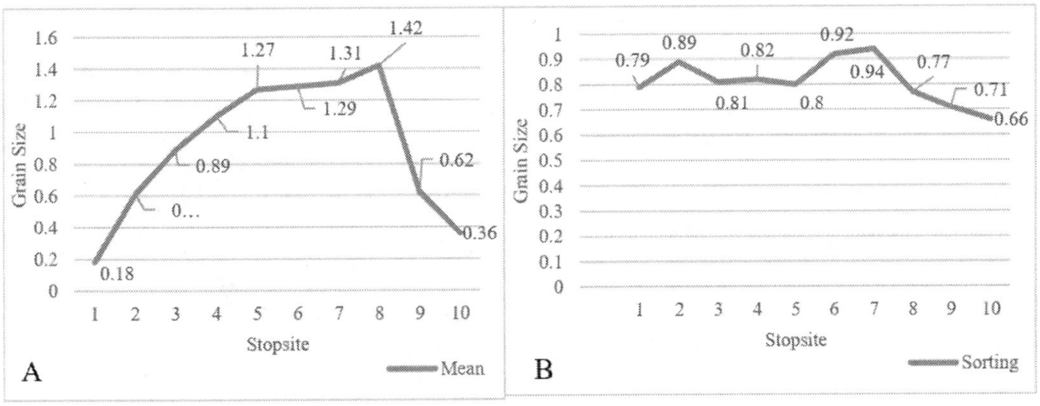

Figure 2 (a) The graphic of mean (b) The graphic of sorting from Oyo River granulometric data.

4.2 Sand Grain Morphology

4.2.1 *Grain Shape.* Based on the results of visual observations uses binocular microscope, it is found that the oblate and equant grains dominate good on the grain size of the sand as seen in Figure 3a. The shape of the oblate grain is characterized by the ratio of the longest and middle axes which are relatively similar. However, the ratio of them with the shortest axis is relatively different. While the equant grain shape is characterized by the ratio of the longest to the shortest axes which are relatively similar. The shape of the available oblate grain and equant grain can be formed by transport mechanism which tends to suspense or bedload. Besides, it can be caused by the initial shape of the grain before transporting has been oblate or equant.

4.2.2 *Sphericity.* Sphericity is a measure of how a grain size approaches spherical shape. Based on the determination of the sphericity of sand grains by comparing it with the visual image, it is conducted the result that sphericity result at the quartz mineral is dominated by sub-elongate, the sphericity result of the feldspar mineral is dominated by sub-equant, the sphericity result of the mineral lytic is dominated by sub-equant, the result of sphericity is obtained that in pyroxene mineral is dominated by very elongated as seen in Figure 3b. The value on the Oyo River indicates that the more downstream has tendency the form more roundness from stop site 1 to stop site 8 then the value decreases when in the stop site 9 and stop site 10. This can occur due to the erosion process and sedimentation which take place on the Oyo River. This process is also influenced by the level of resistance at the mineral which observed different. This can be seen on the difference in the dominance of different forms of sphericity.

4.2.3 *Roundness.* Roundness relates to the sharpness of the edges or corners of the sediment grains. Based on the determination of the roundness by comparing it with the visual, the roundness results of the quartz mineral are dominated by subangular, the roundness results of the mineral feldspar are dominated by rounded, the roundness results are obtained in the mineral lithic dominated by subrounded, the roundness results in pyroxene minerals are dominated by angular as seen in Figure 3c. The value that is on the Oyo River indicates that the downstream tends to be more rounded from stop site 1 to stop site 8 then this value decreases when it is at stop site 9 and stop site 10. This can occur due to the process of erosion and sedimentation that is taking place on the Oyo River. This process is also influenced by the level of resistance in minerals which is observed different, this can be seen the difference in the dominance of different degree of roundness.

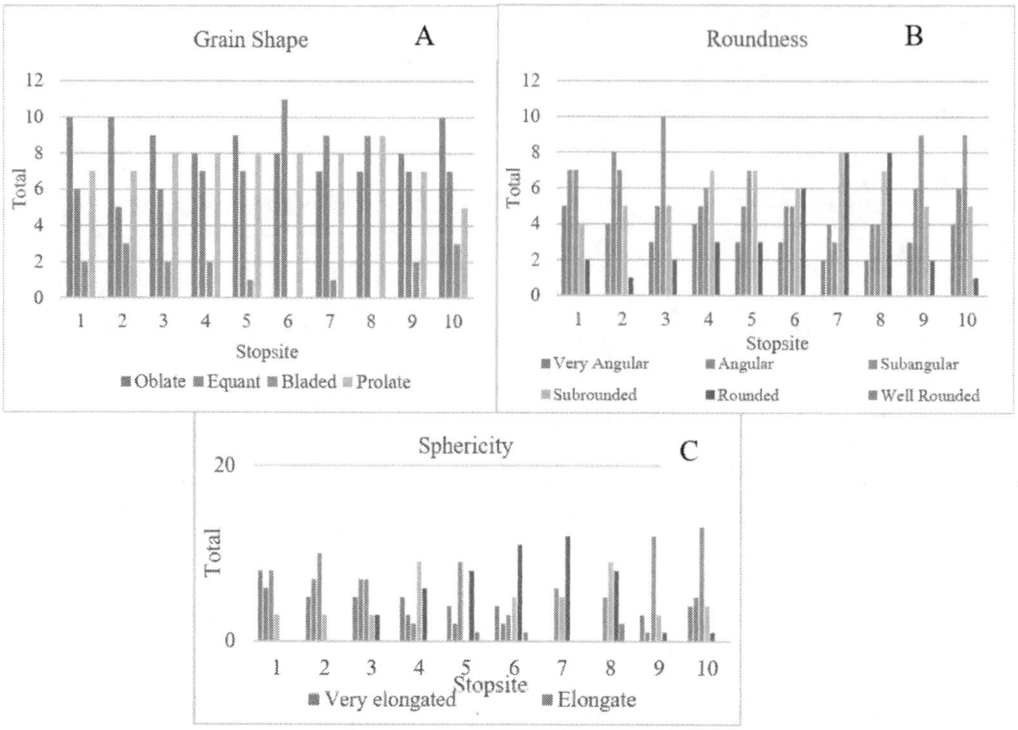

Figure 3. (a) The graphic of grain shape (b) The graphic of roundness (c) The Graphic of sphericity from Oyo River data.

4.3 Sand Grain Composition Data

The sand grain samples which are analyzed for the observation is 10 samples from all the observation locations. From the 10 samples, each sample are taken about 100 grams sample on a mesh size of 60. Based on the composition analysis of the grain sand, the results of light minerals are quartz, feldspar, and mineral lytic. The composition of light minerals changes with the addition of new sedimentary materials and the distance of the stop sites to light mineral resistance. At stop site 1 as seen in Figure 4, it is observed that the presence of feldspar minerals is quite a lot, then decreases as the addition of distance and time of the sedimentation to stop site 8, then at stop site 9 and stop site 10 the number increase again as seen in Table 2 and Table 3. Heavy mineral obtained are hematite, magnetite, amphibole, pyroxene, pyrite, rutile, ilmenite, and olivine. Heavy mineral is dominated by opaque mineral in the form of magnetite, hematite, stable meta minerals, amphibole, and pyroxene as seen in Figure 5.

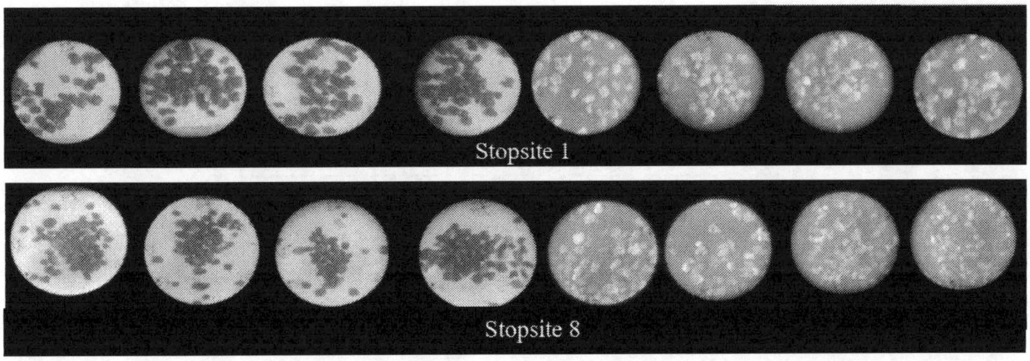

Figure 4. The mineralogy data which was observed under binocular microscope, both for heavy (left) and light (right) minerals.

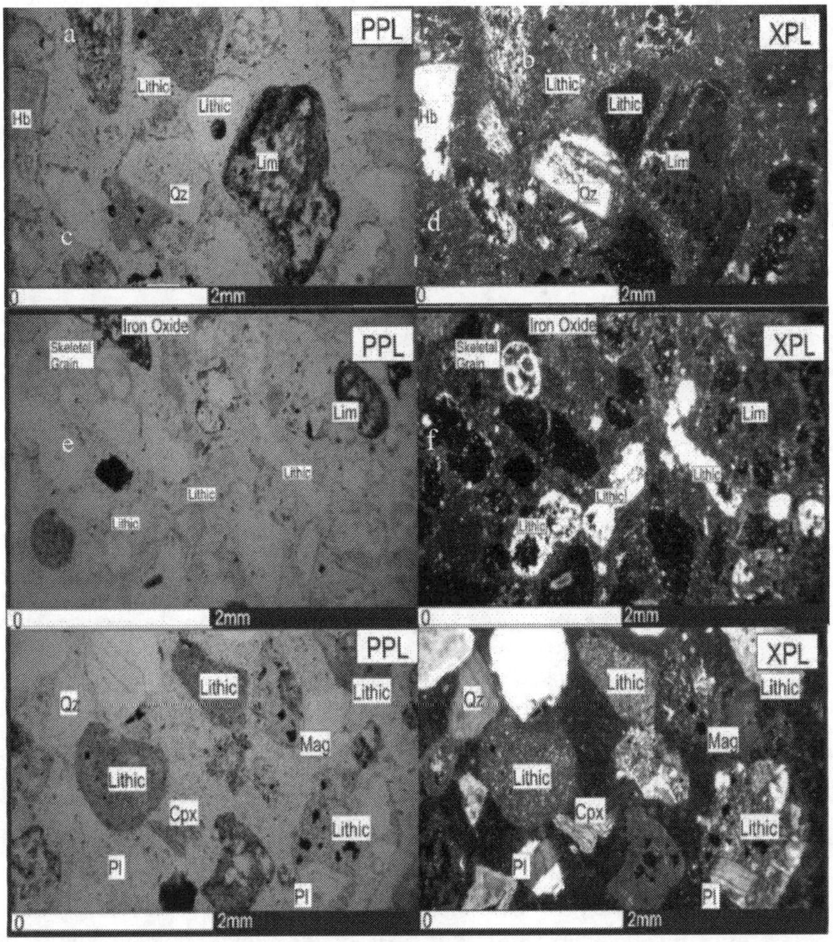

Figure 5. Petrography analysis PPL and XPL of the mineral to validate mineral composition (a-b) Stopsite 1 (c-d) Stopsite 8 (e-f) Stopsite 10.

Table 2. Heavy minerals counting from each stop site. Ilm = Ilmenite, Mt = Magnetite, Ol = Olivine, Px = Pyroxene, Hm = Hematite, Py = Pyrite, Amph = Amphibole.

Stop Site	Ilm	Mt	Ol	Px	Hm	Amph	Lithics
1	16	147	7	55	118	135	22
2	15	166	5	46	105	137	26
3	13	149	3	33	127	125	50
4	14	155	3	23	121	117	67
5	14	145	3	24	133	118	63
6	14	161	3	30	132	116	44
7	18	148	3	26	138	116	50
8	15	158	2	32	134	122	37
9	19	149	6	42	142	136	45
10	25	112	8	44	122	135	53

Table 3. Light minerals counting from each stop site. Qz = Quarts, Fs = Feldspar, Chl = Chlorite.

Stop Site	Qz	Fs	Lithic Fragment	Fossil
1	110	123	119	36
2	151	123	114	33
3	182	106	109	33
4	173	100	110	41
5	158	102	96	76
6	157	101	104	73
7	163	103	111	94
8	150	110	117	94
9	183	114	119	36
10	158	112	109	26

4.4 Origin of Rock

In the original rock type determination, it is used the QtFL triangle (Qt = total quartz, F = feldspar, L = lithic) according to [11]. The result of plotting was obtained that the original rock originated from a dissected arc can be interpreted as the result of volcanic material which lined up on the island arc which is still active and also from volcano which eroded at the continental boundary. Meanwhile, the paleoclimate determination is based on the QFL triangle diagram (Q = quartz, F = feldspar, L = lithic). The result of plotting was obtained that the original rock was formed on the humid climate and dry climate in plutonic rocks.

5. Discussion

5.1 Sedimentation

The interpretation of the sedimentation mechanism which occurs in the sand deposits along the Oyo River is based on the results of the analysis of grain size and grain morphology. The results of the analysis show that the deposited material through the transport mechanism by bedload is a combination of rolling, sliding, and saltation. However, there is also a transport mechanism in the form of suspension on a small portion of the sediment particles which are very smooth in size.

If it is seen from the grain morphology (grain shape) in the four types of minerals such as quartz, feldspar, lytic and pyroxene are gotten the dominant form of equant and oblate. This grain shape indicates that the transportation process which occur is more dominated by rolling and saltation, especially near the downstream part of the stop site 8, the value of minerals that have this shape increases well.

If it is seen from grain morphology (degree of sphericity) on the three types of minerals such as quartz, feldspar, pyroxene, and lithic. Their shape is dominated by very elongated to equant depends on the resistance value of the observed minerals. The tendency of sand grain is more swollen or the value of spericity is getting bigger suitable with the distance from the sediment source, stop site 8 as the downstream area has a greater value than the upstream area at stop site 1.

If it is seen from the sand morphology (degree of roundness) in the three types of minerals and lithic, it is found that the dominance of smooth forms starts from very angular to rounded depend on the resistance value of the observed minerals. The tendency of sand grains is observed to have smoother degree of roundness or more roundness, suitable with the distance to the sediment source, stop site 8 as the downstream area has bigger value or smoother value than upstream area at stop site 1.

5.2 Provenance (Origin)

The presence of several minerals which are less stable (meta-stable) such as olivine and pyroxene are found on the enough far distance from the source indicate that the origin area on the area where has high relief. These minerals can be found, it is possible because the supply of minerals from the surrounding rocks are still quite a lot although the level of weathering is high too [5]. Based on the heavy mineral association it can be estimated that the sand samples along Oyo River come from the Provenance type of intermediate igneous rock. If it is seen in **Figure 6** from the result of triangular diagram [11], shows the provenance type originates from the dissected arc area which represents the magmatic arc zone. In the plotting results of diagram from [12] it was known that the climate of deposition is available on the humid to arid climates. The presence of opaque minerals such as magnetite, hematite and ilmenite in the Oyo River area are interpreted from the rocks in the Semilir Formation and Nglanggran Formation which contains altered volcanic rocks.

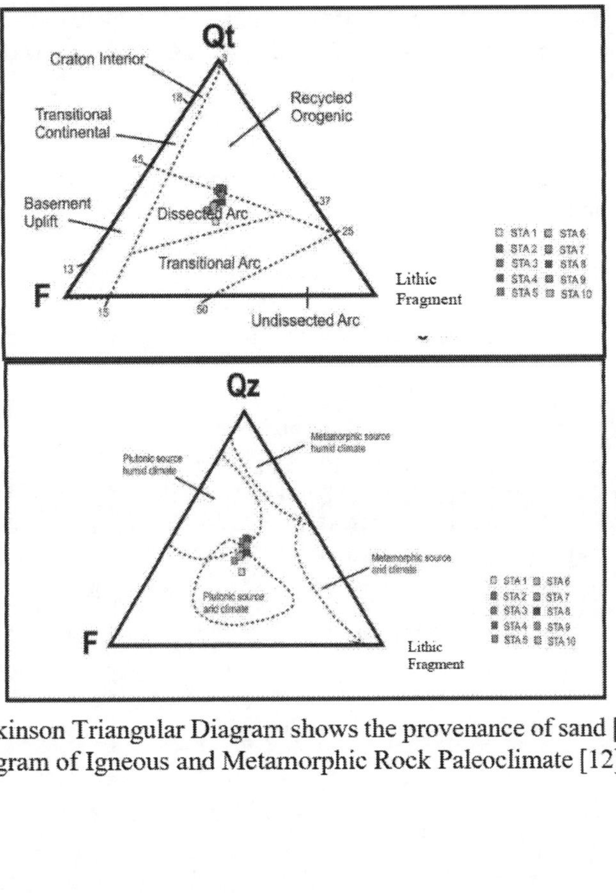

Figure 6. Dickinson Triangular Diagram shows the provenance of sand [11] (top). QFL Diagram of Igneous and Metamorphic Rock Paleoclimate [12] (Bottom).

6. Conclusion

1. Sand deposits along Oyo River have an average grain size value dominated by coarse sand to medium sand, with dominant sorting of moderately sorted, skewness dominated by very fine skewed, and extremely leptokurtic kurtosis.
2. The shape of the grains of sand is dominated by oblate and equant, the value of sphericity is dominated by very elongated to equant, and the rounded value is dominated by very angular up to rounded.
3. The rock type from sand deposit is interpreted to be intermediate igneous rock originating from the erosion of the Semilir or Nglanggran Formations.
4. The sedimentation process of sand deposits along the Oyo River is dominated by a bedload mechanism with a combination of rolling, sliding, saltation along with a slight suspension process.
5. The provenance of Oyo River sand deposit is coming from dissected arc tectonic setting.

Acknowledgment

The authors would like to thank the Department of Geological Engineering, Faculty of Engineering, Universitas Gadjah Mada which has given funding for this publication and people alongside Oyo River who are very nice and kind when this research was done.

References

[1] Husein, S. dan Srijono. 2010. Peta Geomorfologi Daerah Istimewa Yogyakarta. Symposium Geologi Yogyakarta: 1-6. (in Bahasa Indonesia)
[2] Raharjo, W., Sukandarrumidi, dan H. M. S. Rosidi. 1995. Peta Geologi Lembar Yogyakarta, skala 1:100.000. Pusat Penelitian dan Pengembangan Geologi, Bandung (in Bahasa Indonesia)
[3] Surono, dan Permana. A. 1992. Peta Geologi Lembar Surakarta-Giritontro, skala 1:100.000. Pusat Penelitian dan Pengembangan Geologi, Bandung. (in Bahasa Indonesia)
[4] Prasetyadi, S. Sudarno, I. Indranadi, V. B. dan Surono. 2011. Pola dan Genesa Struktur Geologi Pegunungan Selatan, Provinsi Daerah Istimewa Yogyakarta dan Provinsi Jawa Tengah. Jurnal Sumber Daya Geologi Vol 21 No. 2 (in Bahasa Indonesia)
[5] Surjono, S. S. dan D. H. Amijaya. 2017. Sedimentologi. Gadjah Mada University Press. Yogyakarta. (in Bahasa Indonesia)
[6] Pettijohn, F. J. 1975. *Sedimentary Rocks*. 3rd ed. Harper & Row Publishing Co. New York.
[7] Wadell, H. 1932. Volume, Shape and Roundness of Rocks Particles. The Journal of Geology **40**(5), 443-51
[8] Boggs, S. Jr. 1987. *Principles of Sedimentology amd Stratigraphy*. Merril Publishing Co. Columbus
[9] Wentworth, C. K. 1922. A Scale of Grade and Class Terms for Clastic Sediments. The Journal of Geology, **30**(5), 37
[10] Tucker, M. E. 1991. *Sedimentary Petrology: An Introduction to The Origin of Sedimentary Rocks*. 2nd ed. Blackwell Scientific Publications. London
[11] Dickinson, W. R. and Suczek, C.A.,1979. *Plate Tectonics and Sandstone Compositions.* The American of Petroleum Geologist Bulletin. V.63. No 12
[12] Suttner, L.J., Basu, A.S. and Mack, G.H. 1981. Climate and Origin of Quartz Arenites. Journal of Sedimentary Petrology **51**(4), 1235 – 1246

Preliminary study of the origin of mafic mineral megacrysts in volcanic rocks in the southwestern part of Kulon Progo Mountains, Indonesia

L Aryani[1], F Hakim[1]*, A D Titisari[1], G Nugraha[1] and C A Mahardika[1]

[1] Geological Engineering Department, Engineering Faculty, Universitas Gadjah Mada, Grafika Street No.2, Bulaksumur, Yogyakarta 55284, Indonesia
*E-mail: fahmihakim@ugm.ac.id

Abstract. The Bagelen district in the southwestern part of Kulon Progo Mountains has a unique aspect in the form of the occurrence of mafic mineral megacrysts in volcanic rocks. This research is conducted to determine the type, the distribution, and the origin of those megacrysts, which are abundant in pyroclastic rocks. Detailed geological mapping on an area of 4 x 4 km with a scale of 1: 25,000 and petrographic analysis were done to explain the geological aspects that led to the presence of mafic mineral megacrysts in the study area. Petrographic analysis shows that the types of mafic mineral megacryst are hornblende, clinopyroxene, and plagioclase. Their features under the microscope observation are oscillatory zoning, half oscillatory zoning, sieve, and breakdown texture. These minerals are found as single crystal fragments and also as clinopyroxene and hornblende cumulate fragments in pyroclastic breccia. Based on mineralogical composition and texture, these mafic mineral megacrysts were formed by fractional crystallization process in the form of crystal settling mechanism. These minerals were initially accumulated in the bottom of magma chamber, then carried upward by rapid magma rising in association with explosive eruption event of Mount Ijo. These minerals were then transported on the surface through the mechanism of pyroclastic flow and were deposited in the valley to form an alluvial fan morphology in the southwestern part of Kulon Progo Dome.
Keywords: megacrysts, hornblende, cumulate, pyroclastic flow, Kulon Progo

1. Introduction

Kulon Progo Mountains, which are located in the west of the city of Yogyakarta are mainly composed of Eocene to Pliocene rocks. Many of these rocks are volcanic rocks of Oligocene to Miocene age, which presents many interesting aspects to study. One of the interesting locations to study in Kulon Progo Mountains is located in the Bagelen District, Purworejo Regency, Central Java, which is specifically located in the southwestern part of Kulon Progo Mountains (Figure 1). This location has a unique aspect because some of giant mafic minerals are found in volcanic rocks which have been interpreted before as hornblende. Hornblende and other mafic minerals like pyroxene and olivine are certainly very common in igneous and volcanic rocks in the Kulon Progo area. However, mafic minerals with large crystal size (>1 cm) which considered as megacryst are quite rare. Occurrence and abundance of the megacrysts in the study area are very intriguing to study, especially their origin.

Megacryst is a large crystal (1 to 7 cm long xenocryst and phenocryst) with at least one crystal face [1] usually euhedral crystal set in fine grained groundmass or matrix. This term does not have genetic implications like phenocryst [2]. Megacryst also refers to monomineralic aggregates which have an uncertain relationship with the host rock [3].

Megacryst minerals in the study area are found as fragments in pyroclastic rocks. It is very difficult to determine whether the mineral is hornblende or pyroxene by hand specimen observation. Previous studies related to the mineralogy in the region are limited in terms of the origin of igneous rock [4] [5] [6] [7] and no research has yet discussed the presence of these megacrysts. These igneous rocks in Semono Area, Bagelen District are pyroxene andesite, hornblende andesite, and dacite [6]. Other studies suggest that the rock is composed of pyroxene basaltic andesite and hornblende basaltic andesite as lava flows and shallow intrusions [4] [5] [7]. This study shows the results of the analysis of megacryst minerals in pyroclastic rocks along with their mineral types to their origins from a textural and mineralogical perspective.

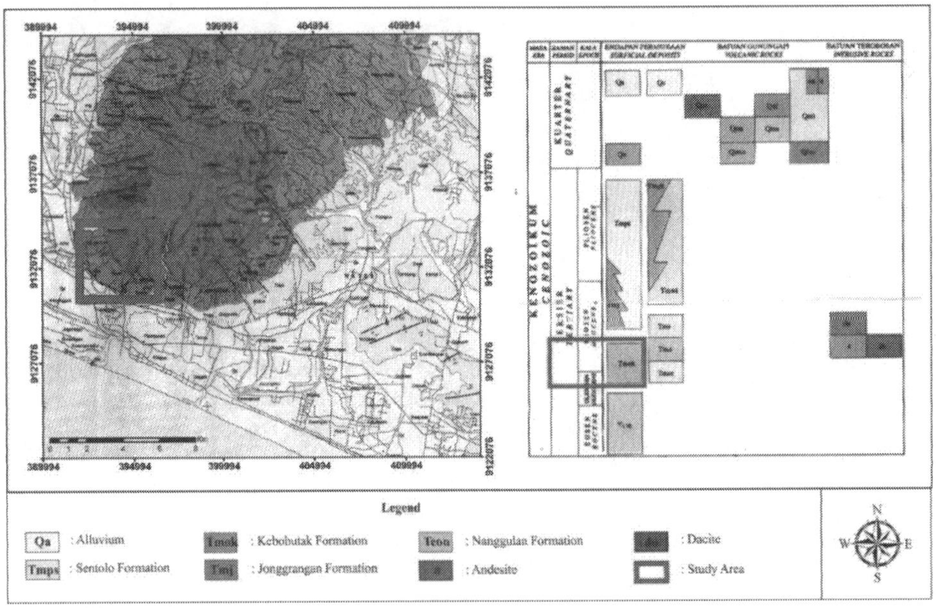

Figure 1. Regional geological map of the southwestern part of Kulon Progo Mountains from [9] and the location of studied area (red box).

2. Regional Geology

Kulon Progo Mountains is a part of the South Serayu Mountains Zone which is composed of Eocene to Miocene volcanic rocks and known as Kulon Progo Dome or "oblong dome" because of the unique morphological expression of an elongated dome [8]. The regional stratigraphy of Kulon Progo Dome is composed of four formations, from the old to the young in succession is Nanggulan Formation, Old Andesite Formation, Jonggrangan Formation, Sentolo Formation, and Alluvial Deposit [9].

The Old Andesite Formation [8] or Kebo Butak Formation [9] refers to all volcanic rocks in the Kulon Progo Mountains. Volcanism during the Tertiary period resulted in the formation of a volcanic complex in the Kulon Progo Mountains. Volcanic activity in Kulon Progo began with the formation of the Gajah Volcano formation during the Oligocene [10]. Then in the Middle Miocene, the southern side was intruded by the Ijo Volcano. Furthermore, Menoreh volcano emerges at Late Miocene in the northern part of Kulon Progo Mountains. The Oligocene - Miocene volcanism are controlled by the North - South compression stress and produced a sinistral fault with the north-northeast direction [11].

3. Research Methods

The detailed geological mapping covered an area of 4 x 4 km with scale of 1:25,000, consist of lithology, stratigraphy, geomorphology, and geological structures has been conducted to identify the

IOP Conf. Series: Earth and Environmental Science **851** (2021) 012031 doi:10.1088/1755-1315/851/1/012031

geological conditions, distribution of rock types, and the distribution of megacryst minerals. This research uses a smart mapping method and fresh samples collected only from locations that represent all lithological types in this study area.

Textural characterization of the rocks and minerals was conducted by megascopic and petrographic analysis. A total of fifteen (15) sections from twelve (12) rocks and one crystal sample were analysed with a polarization microscope. The petrographic observations were undertaken at the Get-In Cicero Laboratory of the Geological Engineering Department, Universitas Gadjah Mada. Topography map and Digital Elevation Model (DEM) also used to analyze morphological features including geological landscape and the distribution of intrusion bodies.

4. Result

4.1 Geology of the Study Area

The geomorphological unit in the study area is divided into pyroclastic flow ridge unit, intrusion hill unit in the north, and alluvial plain unit in the west. The pyroclastic flow ridge unit is characterized by ridges separated by several valleys that have an appearance of an alluvial fan and have a northeast-southwest trending morphological alignment. The intrusion hill unit has a rough morphological appearance and is formed from the erosion process of the tertiary intrusion body in Mount Ijo. The alluvial plain unit is a flat morphology that can be easily distinguished from other units because of its smooth appearance.

The research area consists of three types of lithological units, which are sorted from oldest to youngest, namely pyroclastic breccia units, andesite intrusions, and alluvial deposits. In several locations, can be found pyroclastic breccia were intruded by andesite (Figure 2A). Pyroclastic breccia in this study area is the main object of research which is composed of various volcanic fragments (Figure 2B, 3C) and found a lot of megacrysts as well as cumulate rocks (Figure 3E). Fragments are composed of hornblende, plagioclase, pyroxene, lithic tuff (Figure 3A), lapilli tuff (Figure 3B, 3F), and andesite (Figure 3D).

Figure 2. Pyroclastic breccia outcrop contact with andesite intrusion (A). Pyroclastic breccia composed of various volcanic fragments (B).

Figure 3. Rock samples from the study area which are most of them consist of mafic megacrysts. Lithic tuff (A), Lapilli tuff (B), Tuff breccia (C), Andesite (D), Contact of pyroxene cumulate with lapilli tuff (E), Lapilli tuff with hornblende megacryst fragments (F).

The geological map of the study area can be seen in Figure 4. The geological structure found in the form of an uncertain sinistral strike slip fault is thought to be trending northeast southwest through the andesite intrusion unit and the breccia pyroclastic unit.

4.2 Lithological Characteristics

The various types of pyroclastic breccia unit are lapilli tuff, tuff, and pyroclastic breccia. Lapilli tuff has a blackish gray color, massive structure, matrix supported, and strongly consolidated. It is composed of hornblende, pyroxene, and plagioclase minerals that appear as fragments, while the matrix is composed of plagioclase and volcanic ash. In some parts (Figure 5B), hornblende fragments are found with flow structure.

Megascopically, pyroclastic breccias are composed of fragments (1 - 30 mm) and a matrix (<1 mm) with angular roundness, poor sorting, matrix supported, and massive structures. Fragments are composed of hornblende, plagioclase, pyroxene, lapilli tuff, lithic tuff and andesite. Hornblende and pyroxene present as individual crystal fragments and as cumulates (Figure 5A; 5D). In some locations, pyroxene appears as a large crystal with cumulate structure and commonly in contact with lapilli tuff (Figure 5C). Both of them are fragments in pyroclastic breccia (Figure 5D).

The andesite intrusion unit is composed of pyroxene andesite, hornblende andesite, and basaltic andesite. Pyroxene andesite and hornblende andesite, both have porphyroaphanitic textures, but they are distinguished based on the type of phenocryst. Whereas basaltic andesite has an aphanitic texture.

This research is focused on pyroclastic breccia units which consist of mafic megacryst fragments. The pyroclastic breccia occupies an area around 7 km^2 with fresh to slightly weathered rock conditions. Characteristics of mafic mineral megacrysts (hornblende and pyroxene) are discovered in the lapilli tuff and pyroclastic breccia in the southwestern part of Mount Ijo, Kulon Progo Mountains.

IOP Conf. Series: Earth and Environmental Science **851** (2021) 012031 doi:10.1088/1755-1315/851/1/012031

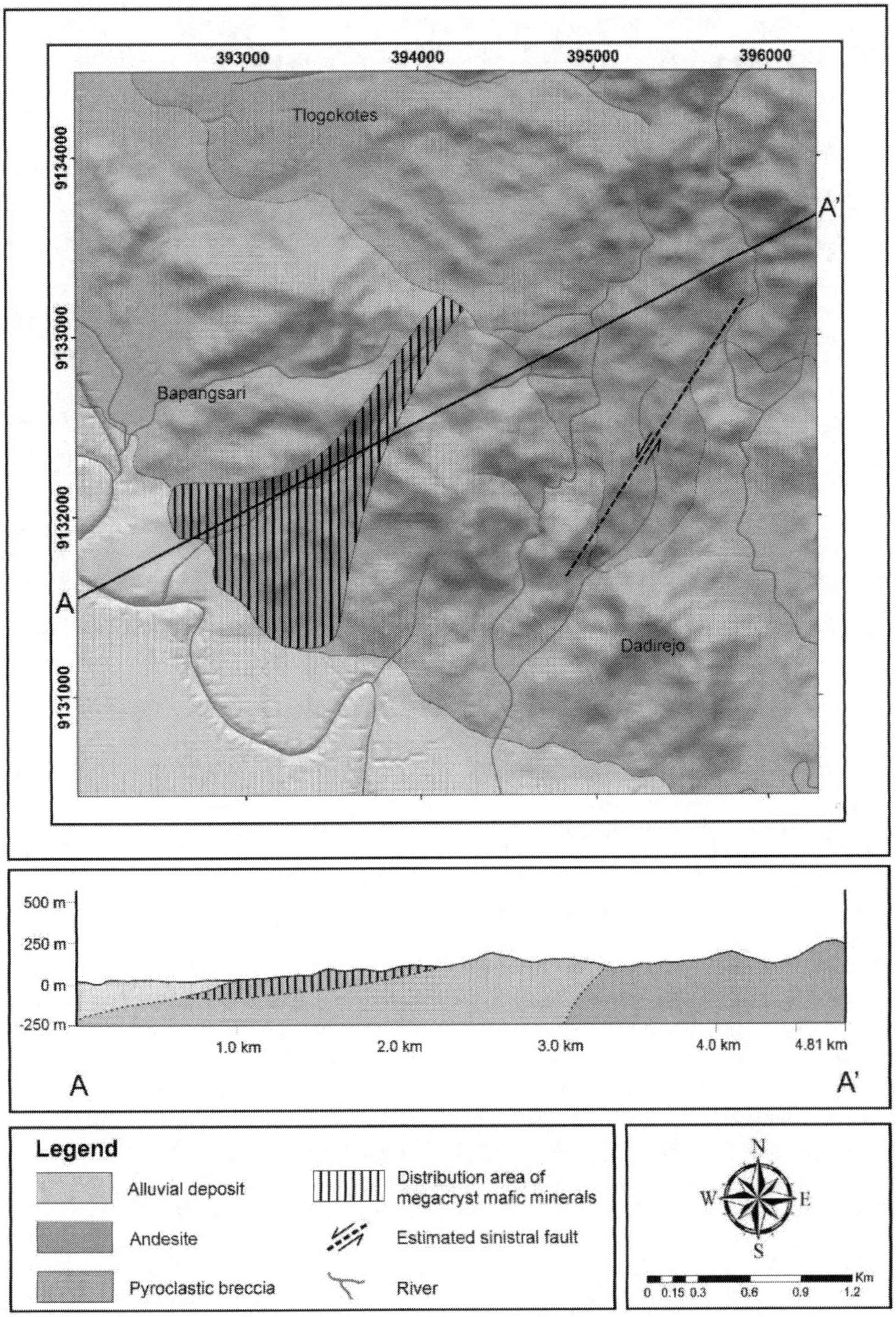

Figure 4. Geological map and geological profile of the study area.

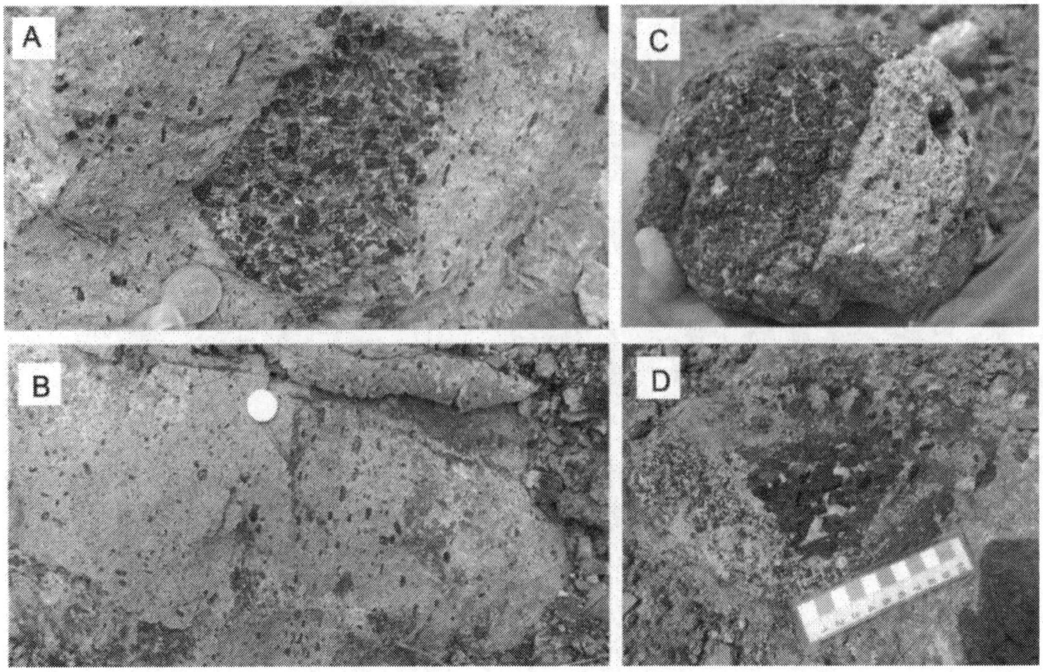

Figure 5. Hornblende cumulate as fragment in pyroclastic breccia (A), hornblende with flow structure (B), contact of pyroxene cumulate with lapilli tuff (C), pyroxene megacrysts as fragment in pyroclastic breccia (D).

4.3 Petrographic Characteristic

4.3.1. Types of Mafic Minerals.

Based on petrographic observations, lapilli tuff is composed of fragments in the form of hornblende, plagioclase, clinopyroxene, and opaque minerals while the matrix is composed of plagioclase minerals, volcanic glass, and mafic microliths. Typical mineral assemblages in the tuff consist of plagioclase, hornblende, clinopyroxene, volcanic glass, and opaque minerals. Opaque minerals are more common in this lithology (Figure 6a).

Pyroclastic breccia is composed of fragments in the form of hornblende, plagioclase, clinopyroxene, opaque minerals and andesite rocks as lithic. The matrix is composed of lithic volcanic, volcanic glass, plagioclase, and mafic microliths. Plagioclase with subhedral crystals is present both as fragment and matrix.

According to the optical properties, the clinopyroxene cumulate fragments (Figure 6b) are characterized by medium grain size (1 - 10 mm) and subhedral prismatic shaped. This clinopyroxene cumulate fragments are rarely composed of olivine and plagioclase crystals.

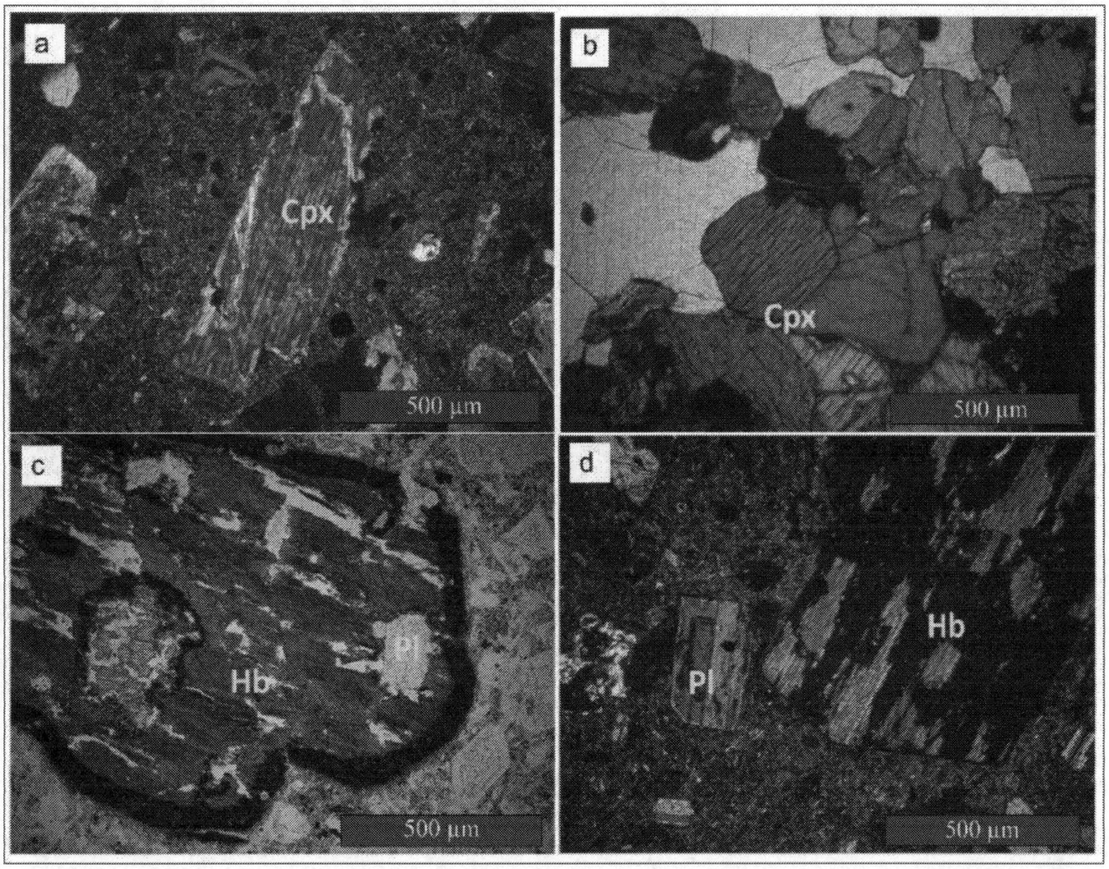

Figure 6. Pyroxene with oxide rim in lapilli tuff in cross polarized (a), clinopyroxene (Cpx) cumulate in cross polarized (b), oxy-hornblende (Hb) with remnant of plagioclase (Pl) in plane polarized (c), half oscillatory zoning in plagioclase with breakdown texture of hornblende in cross polarized (d).

4.3.2 Textural Characteristics

Hornblende minerals in lapilli tuff found in the form of oxy-hornblende which some of them have plagioclase and opaque mineral inclusions (Figure 6c). Oxy-hornblende fragments are the most common mafic minerals in this lithology. The oxy-hornblende has green-dark brown color, high relief with rim margin that is composed of oxide minerals. The form of oxy-hornblende crystals is mostly prismatic subhedral with high relief and green to brown pleochroism. Other samples show remnant of pyroxene and plagioclase inside hornblende crystals. In general, the condition of the hornblende shows different breakdown textures which represent as internal breakdown patches and rim margin (Figure 6c; 6d).

Pyroxenes in lapilli tuff are characterized by internal breakdown patches and iron oxide rim texture (Figure 6a). The pyroxene found in this study area have second order blue reference color, 0.5 mm in size, and inclined extinction angle which is typical of clinopyroxene.

Some of plagioclase show the patchy-zoned texture with patchy-spongy cellular core type (Figure 7a) [12] and also show sieve texture (Figure 7b). It is composed of oscillatory zoned rims with a core of irregular-shaped bright and dark patches. Fewer plagioclase crystals are anhedral and surrounded by pyroxene in cumulate rock fragments (Figure 7d).

Most of the plagioclase crystals show oscillatory zoning texture (Figure 7c). Half oscillatory zoning is also found in some plagioclase (Figure 6d). Those crystals show albite twinning while some of them was altered into clay minerals. Most of plagioclase shows various breakdown textures.

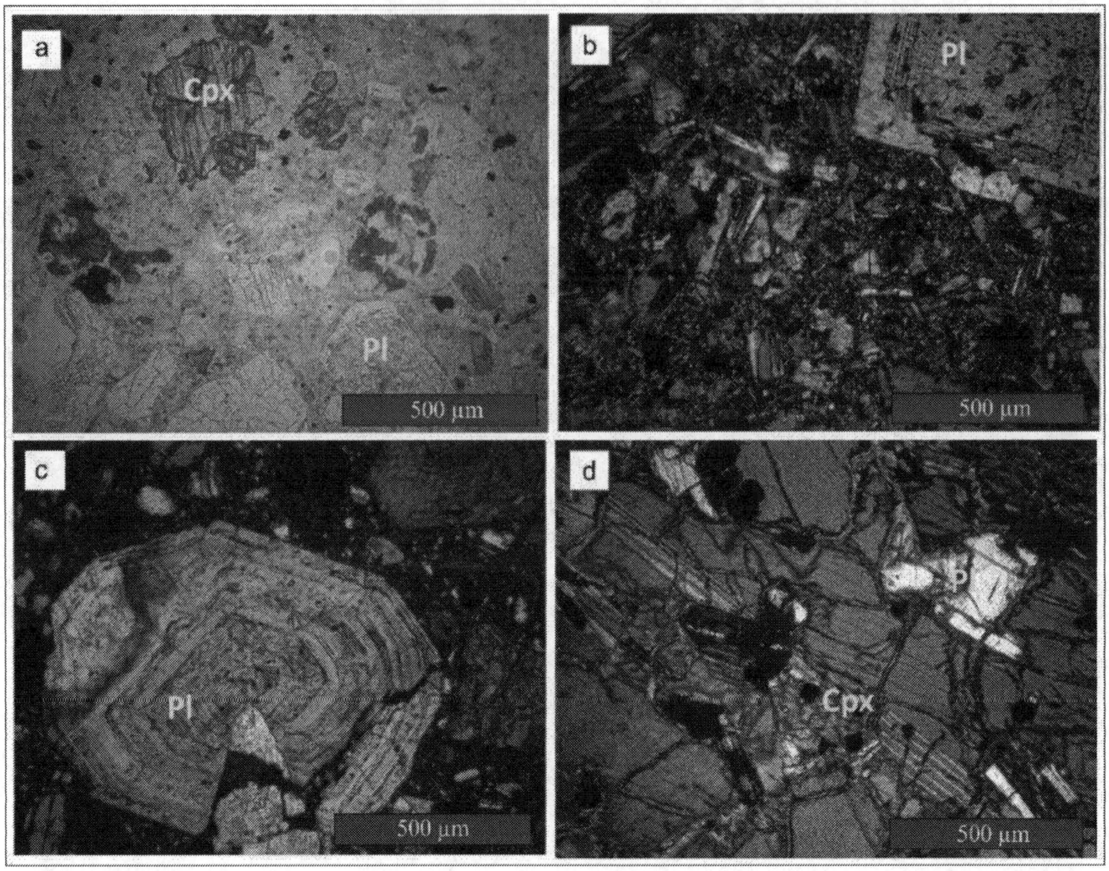

Figure 7. Plagioclase (Pl) shows the patchy-zoned texture with patchy-spongy cellular core type in plane polarized (a), sieve textured plagioclase in cross polarized (b), oscillatory zoning in plagioclase in cross polarized (c), anhedral plagioclase (Pl) surrounded by clinopyroxene (Cpx) with opaque minerals inclusion in cross polarized (d).

5. Discussion

5.1 Geological Control of the Distribution of Mafic Mineral Megacrysts

Field observations show that most of the pyroclastic rocks in the study area has large mafic minerals as fragments. Based on the distribution map of megacryst minerals and Digital Elevation Model (DEM) analysis, these minerals were carried by pyroclastic flows that passed through the valleys and were deposited in an alluvial fan (Figure 4) in the southwest distal part of an ancient volcano (Mount Ijo). The distribution of these minerals is mostly exposed at the distal area of the volcano as a result of an explosive eruption that produces pyroclastic flows towards the southwest of the volcano. These minerals were deposited in the valley to form an alluvial fan morphology in the southwestern part of Kulon Progo Dome.

The same orientation of mafic minerals i.e hornblende with the direction of the slopes of alluvial fan could indicates pyroclastic flow deposition mechanism. Other supporting data, such as rock texture which are poor sorting and matrix supported, can bolster this interpretation. The mafic minerals were carried away by a pyroclastic flow during the explosive eruption of Mount Ijo so that it was mixed with glass material and fine-grained plagioclase minerals. The products of the eruption were in the form of lapilli tuff and pyroclastic breccia.

IOP Conf. Series: Earth and Environmental Science **851** (2021) 012031 doi:10.1088/1755-1315/851/1/012031

5.2. Interpretation of Their Origin

The oscillatory zoning texture of plagioclase showing the magma differentiation process. Oscillatory zoning can be explained as a crystal's response to fluctuating external conditions. This texture is a reflection of composition changes of the magma by external causes [13]. This texture can be formed due to several things, as stated in [14]: a. movement of crystals in the magma and crystal may settle due to differentiation whether by gravitation; b. Movement of the magma as a whole into a region with different conditions of temperature and pressure; c. additional magma into the crystallizing liquid; d. loss of volatile constituent due to rapid crystallization. Meanwhile, based on [15] oscillatory zoning textures are due to kinetics of crystallization. The oscillatory zoning, breakdown and sieve texture in plagioclase minerals exhibit evidence of mineral-melt disequilibrium [16]

Furthermore, Patchy zoning of plagioclase is formed due to open-system processes or due to diffusional chemical re-equilibration of a crystal zoning [17]. Disequilibrium condition initiates a spongy-like dissolution of the plagioclase crystal and crystallization can occur with composition reflecting changes in conditions [17]. Pyroxene megacrysts contain opaque and plagioclase inclusions (Figure 7d) that may indicate magma mixing or disequilibrium crystallization during the volcanic evolution [16].

The oxy-hornblende found in pyroclastic rocks is a product of green hornblende alteration [18] and is possibly associated with hot gases at the late magmatic stage [19]. The breakdown texture in hornblende indicates the time spent by the crystal outside its field of stability [20] although this depends on the ambient conditions [21].

During the eruption period, crystal settling of mafic minerals (pyroxene, hornblende, as well as plagioclase) was occurring in the magma chamber (Figure 8). The pyroxene that found in the field was in the form of cumulate while hornblende was found as individual crystals, although in some locations, hornblende cumulates were also found. The mafic minerals occur in the surface as fragments in the pyroclastic breccia.

The difference in texture between hornblende and pyroxene indicates a magma differentiation through fractional crystallization process. Pyroxene which formed earlier than hornblende, was concentrated at the bottom part of magma chamber. When the pyroxene began to crystallize, plagioclase also began to crystallize. Pyroxene and plagioclase will settle at the bottom of the magma chamber due to density (gravity settling). This process causes the formation of a pyroxene layer at the bottom of the magma chamber. After that, hornblende started to crystallize. The zoning texture in the plagioclase becomes supporting data that there is a change in composition due to the differentiation process and the presence of crystal settling. [14].

For some reason, an explosive eruption occurred when the formation of hornblende was taking place so that the mafic minerals that had been formed were carried to the surface. Changes in crystallization conditions are characterized by the presence of a patchy-spongy texture [17].

When an explosive eruption occurs, the rate of magma cooling becomes faster, so that fine-sized plagioclase crystals (<1 mm) and volcanic glass are formed. This explosive eruption resulted in a product in the form of pyroclastic breccia with mafic mineral megacrysts as fragments, and plagioclase with volcanic glass as matrix. These minerals are carried by pyroclastic flows so that the mafic minerals condition is in the form of destruction. Most of the pyroxenes found as fragments have a cumulate texture, while most of hornblende are found as individual crystals. This indicates that the pyroxene has undergone crystallized or solidified before being carried out of the magma chamber.

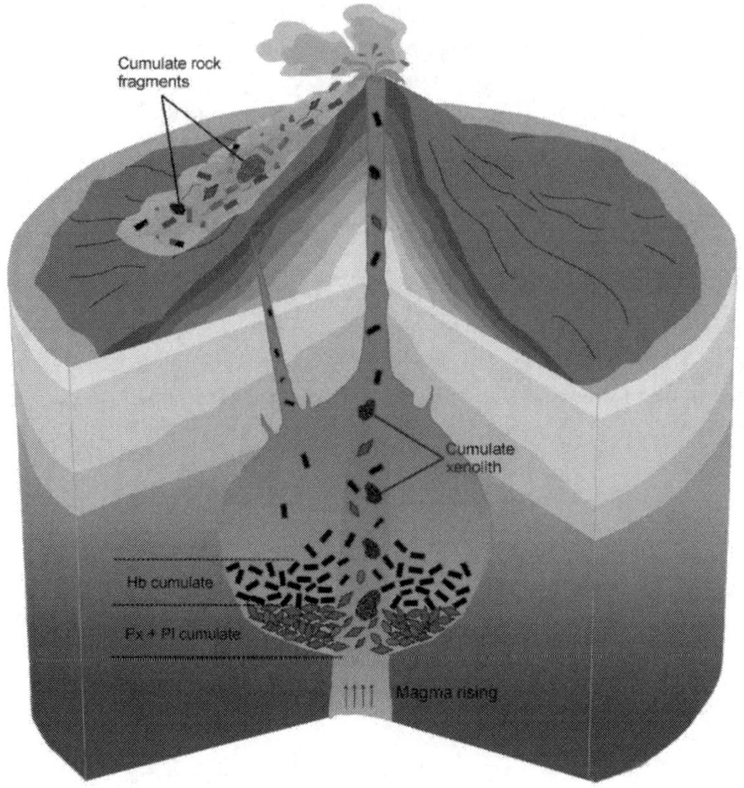

Figure 8. Petrogenetic model to illustrate the occurrence of mafic mineral megacryst fragments in pyroclastic breccia in the study area.

6. Conclusion

Mafic mineral megacrysts found in the study area consists of hornblende, pyroxene, and plagioclase. They are found as fragments in pyroclastic breccias. Most of the pyroxene that found in the field was in the form of cumulate while hornblende often found as individual crystals, although in some locations, hornblende cumulates were also found. Based on mineralogical composition and texture, mafic mineral megacrysts in the study area are interpreted to have formed as a result of fractional crystallization process through a crystal settling mechanism in the magma chamber. These minerals were carried out to the surface along with the rapid rise of magma and were transported as pyroclastic flows in an explosive eruption event of Mount Ijo. These pyroclastic flows passed through the valleys and were deposited in an alluvial fan morphology in the southwestern part of Kulon Progo Dome.

References

[1] King P L, Hervig R L, Holloway J R, Vennemann T W and Righter K 1999 Oxy-substitution and dehydrogenation in mantle-derived amphibole megacrysts *Geochim. Cosmochim. Acta* **63** 3635–51
[2] Michael A 2008 *A Dictionary of Earth Sciences* (New York: Oxford University Press)
[3] Pearson G, Canil D and Shirey S 2003 Mantle Samples Included in Volcanic Rocks: Xenoliths and Diamonds *Treatise on Geochemistry* **2** 171–275
[4] Pinasthi M and Hendratno A 2016 Studi Geologi dan Kualitas Andesit di Daerah Hargorojo, Kecamatan Bagelen, Kabupaten Purworejo sebagai Bahan Bangunan *Seminar Nasional*

Kebumian Ke-9 pp 485–96
[5] Ismail I and Hendratno A 2016 Studi Petrogenesis Andesit di Daerah Hargorojo dan sekitarnya, Kecamatan Bagelen, Kabupaten Purworejo, Provinsi Jawa Tengah *Seminar Nasional Kebumian Ke-9* pp 594–611
[6] Pratama I wahyu, Hanif I M, Hidayatullah and Pramumijoyo S 2017 Studi Petrogenesa Batuan Beku Di Daerah Semono Dan Sekitarnya, Kecamatan Kaligesing Dan Bagelen, Kabupaten Purworejo, Provinsi Jawa Tengah Dengan Metode Sayatan Tipis *Seminar Nasional Kebumian Ke-10 Peran Penelitian Ilmu Kebumian Dalam Pembangunan Infrastruktur Di Indonesia*
[7] Nurhaci D S 2018 Studi Petrografi Daerah Bagelen Kabupaten Purworejo Provinsi Jawa Tengah *J. Sains Dasar* **7** 5–11
[8] van Bemmelen 1949 The Geology of Indonesia *Journal. Pract.* **1 A**
[9] Rahardjo, W; Sukandarrumidi; Rosidi H 2012 Peta Geologi Lembar Yogyakarta *Pus. Surv. Geol. Energi dan Sumberd. Miner.*
[10] Barianto D H, Aboud E and Setijadji L D 2009 Structural analysis using landsat TM, gravity data, and paleontological data from tertiary rocks in yogyakarta, Indonesia *Mem. Fac. Eng. Kyushu Univ.* **69** 65–77
[11] Widagdo A, Pramumijoyo S, Harijoko A and Setiawan A 2016 Kajian Pendahuluan Kontrol Struktur Geologi terhadap Sebaran Batuan-batuan di Daerah Pengunungan Kulonprogo Yogyakarta *Proceeding Seminar Nasional Kebumian* pp 9–20
[12] Kiss B, Harangi S, Ntaflos T, Mason P R D and Pál-Molnár E 2014 Amphibole perspective to unravel pre-eruptive processes and conditions in volcanic plumbing systems beneath intermediate arc volcanoes: A case study from Ciomadul volcano (SE Carpathians) *Contrib. to Mineral. Petrol.* **167** 1–27
[13] Fenner C N 1926 The Katmai Magmatic Province *J. Geol.* **34** 673–772
[14] Phemister J 1934 Zoning in plagioclase feldspar *Mineral. Mag. J. Mineral. Soc.* **23** 541–55
[15] Shcherbakov V D, Plechov P Y, Izbekov P E and Shipman J S 2011 Plagioclase zoning as an indicator of magma processes at Bezymianny Volcano, Kamchatka *Contrib. to Mineral. Petrol.* **162** 83–99
[16] Temizel I and Arslan M 2008 Petrology and geochemistry of Tertiary volcanic rocks from the İkizce (Ordu) area, NE Turkey: Implications for the evolution of the eastern Pontide paleo-magmatic arc *J. Asian Earth Sci.* **31** 439–63
[17] Streck M J 2008 Mineral textures and zoning as evidence for open system processes *Rev. Mineral. Geochemistry* **69** 595–622
[18] Winchell A N 1951 *Elements of Optical Mineralogy Part II* (New York: John Wiley and Sons)
[19] Rogers A F 1940 Lamprobolite, a new name for basaltic hornblende *Am. Mineral.* **25** 826–8
[20] Rutherford M J and Hill P M 1993 Magma ascent rates from amphibole breakdown: an experimental study applied to the 1980-1986 Mount St. Helens eruptions *J. Geophys. Res.* **98**
[21] Rutherford M J and Devine J D 2003 Magmatic conditions and magma ascent as indicated by hornblende phase equilibria and reactions in the 1995-2002 Soufrière Hills magma *J. Petrol.* **44** 1433–54

IOP Conf. Series: Earth and Environmental Science **851** (2021) 012032 doi:10.1088/1755-1315/851/1/012032

Reconstruction of groundwater level at Kumamoto, Japan by means of deep learning to evaluate its increase by the 2016 earthquake

K Yokoo[1*], K Ishida[2,3], T Nagasato[4], Y Kawagoshi[2] and H Ito[2]

[1]Graduated School of Science and Technology, Kumamoto University, 2-39-1 Kurokami, Kumamoto 860-8555, Japan
[2] Center for Water Cycle, Marine Environment, and Disaster Management, Kumamoto University, 2-39-1 Kurokami, Kumamoto 860-8555, Japan
[3] International Research Organization for Advanced Science and Technology, Kumamoto University, 2-39-1 Kurokami, Kumamoto 860-8555, Japan
[4] Department of Civil and Environmental Engineering, Kumamoto University, 2-39-1 Kurokami, Kumamoto 860-8555, Japan
*E-mail: 207d2121@st.kumamoto-u.ac.jp

Abstract. Groundwater is a very important water resource at Kumamoto City. Kumamoto City is the capital city of Kumamoto Prefecture, which is located in the Kyushu region, Japan. All domestic water is obtained from groundwater in Kumamoto City. Modeling groundwater is a difficult issue. Conditions under the ground are complex, and difficult to be obtained. Even the delineation of a groundwater basin is frequently unknown. Nowadays, deep learning is a hot topic in many research fields including geoscience. A recurrent neural network (RNN) is a type of deep learning that is suitable for time series modeling. Then, it has been successfully applied for groundwater modeling. Therefore, this study utilized a new type of RNN, Long and Short-Term Memory (LSTM) network, to model groundwater level at a monitoring well within Kumamoto City. The results in this study showed good agreement with the observed groundwater. In addition, it is known that severe earthquakes in April 2016 affected the groundwater level around Kumamoto City. The groundwater level model by LSTM was also utilized to estimate the effects of the severe earthquakes on the groundwater level. The results indicated that the earthquakes may have increased the groundwater level at Kumamoto City by more than 3 m.

1. Introduction

Groundwater is a very important water resource at Kumamoto City. Kumamoto City is the capital city of Kumamoto Prefecture, which is located in the Kyushu region, Japan (Figure 1). The population of the Kumamoto Prefecture is approximately 1.74 million. The population of Kumamoto City is approximately 0.73 million. Approximately 42% of the population in Kumamoto Prefecture is in Kumamoto City. Meanwhile, Kumamoto City is the third-largest city in the Kyushu region. Then, approximately 80% of domestic water is obtained from groundwater in Kumamoto Prefecture. All domestic water is obtained from groundwater in Kumamoto City. Also, groundwater is utilized not only for domestic water but also for agricultural and industrial water. It is very important to investigate the conditions of groundwater around Kumamoto City.

Severe earthquakes hit Kumamoto Prefecture in 2016. The first huge earthquake hit at 21:26 (JST) on April 14. The magnitude of the earthquake was 6.5. The second huge earthquake hit at 1:25 (JST) on April 16. The magnitude of the second one was 7.3. Besides, many earthquakes occurred around Kumamoto Prefecture. 1,223 earthquakes were recorded on April 16, 2016. The series of earthquakes caused not only property losses but also human loss. According to the report of Kumamoto Prefecture, 1665,588 houses were damaged or lost, 93 people were killed, and 2,346 people were injured by the earthquakes. It is also known that these severe earthquakes affected the groundwater.

It is not easy to model groundwater. Groundwater can be modeled with physically-based models, conceptual models, or black-box models. A physically-based model such as ModFlow [1] is based on the governing equations of groundwater. It has several important features. It is literarily consistent with the physical processes of groundwater. It can also describe the behavior of groundwater flow. However, it is generally very computationally intensive. Meanwhile, it requires a lot of information under the ground. When it is difficult to obtain sufficient information under the ground, it is difficult to model groundwater properly. A conceptual model is between a physically-based model and a black-box model. A well-known conceptual model may be a tank model. A tank model was applied for modeling groundwater around Kumamoto City [2]. However, a conceptual model generally requires knowledge and experience to properly set up its structure, and it may also require some information of the study area although it will be much less than that required by a physically-based model.

When there is not sufficient information about the study area, it may be a good option to use a black-box model for groundwater modeling. Also, it may be possible to sufficiently supplement the data required for rainfall runoff modeling by using the reanalysis data. Nowadays, machine learning is a hot topic in many research fields, which is known as deep learning. Deep learning was originally developed mainly in the computer science field. However, it has recently been applied or developed in other research fields. There is a type of deep learning that is suitable for time series modeling. The type of deep learning is called Recurrent Neural Network (RNN). Especially, it is known that a new type of RNN, Long and Short-Term Memory (LSTM) network, has the potential to accurately model time series. Therefore, LSTM has been utilized for time series modeling in various research fields including natural science. For instance, LSTM has been utilized for river discharge modeling [3], river

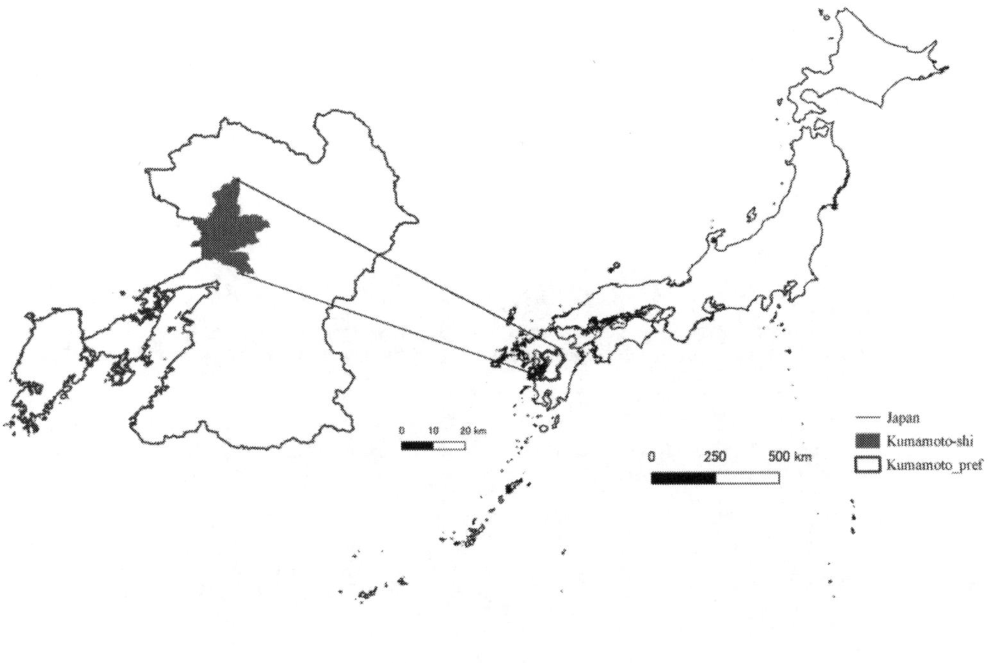

Figure 1. Study area.

discharge forecasting [4–9], precipitation estimation [10–12], reservoir operation modeling [13], lake water level forecasting [14,15], sea surface level modeling [16] and so on. Besides, it has been also employed for some applications of groundwater modeling or forecasting, and these studies were very successful [17–20].

In this context, this study tries LSTM for groundwater modeling at a monitoring well within Kumamoto City and then investigates its applicability by checking the model accuracy. Also, the developed model is utilized to reconstruct groundwater level after the earthquakes. After that, the reconstructed groundwater level is compared with the observation in order to try to quantify the effects of the earthquakes on the groundwater level at Kumamoto City.

2. Study area and Data

This study focuses on groundwater level at a monitoring well within Kumamoto City. The location of the gauging station is non-public because of the regulation. The groundwater level data are available at the daily temporal-scale from April 2011 to December 2019. This station was selected because of the data availability (longer period and fewer missing parts). The groundwater level has yearly fluctuation. Meanwhile, there is some gap in the groundwater level between the periods before and after April 2016. As aforementioned, severe earthquakes happen in the middle of April 2016. Just after the earthquakes, the groundwater around Kumamoto City clearly increased, which was caused by water release due to ruptures at the surrounding mountain aquifers accordint to Hosono et al. (2020) [21].

Daily precipitation and daily mean air temperature data are used as input for the groundwater level modeling in this study. It is expected that a deep learning method can reflect effects of air temperature on the groundwater level thorugh evapotranspiration. These data were obtained at the Kumamoto meteorological station from a website operated Japan Meteorological Agency (https://www.jma.go.jp/jma/). The Kumamoto meteorological station is also located in Kumamoto City. The meteorological data are available from 1891 to the present.

3. Methods

3.1. Long and Short-Term Memory network

The Long and Short-Term Memory (LSTM) network is a kind of recurrent neural network (RNN). RNN is a type of neural network, which is categorized into deep learning. RNN has a recurrent block or multiple recurrent blocks. A recurrent block contains learnable parameters such as weights and biases, similar to a traditional neural network.

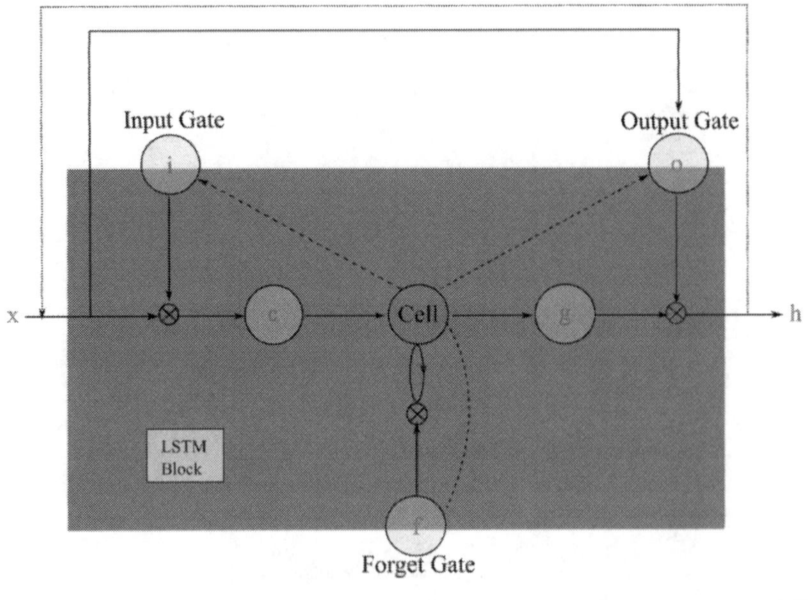

Figure 2. LSTM block.

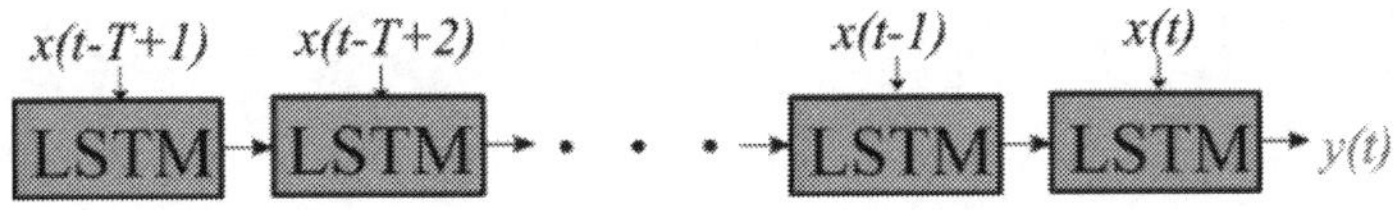

Figure 3. Relations between input time series and the output of LSTM.

Then the learnable parameters of the recurrent block(s) are trained with a set of input data and target data. When RNN receives time series data as input, the recurrent bock is literarily utilized recurrently. Because of this feature of RNN, RNN can work well with sequential data including time. However, a traditional RNN has a serious problem in its structure. Sometimes a traditional RNN cannot learn long-term dependencies between input data and target data. In general, a deep learning method including RNN is trained with a gradient descent method. During its training process with a gradient descent method, the learnable parameters cannot sometimes be updated properly.

To solve this issue, LSTM was developed by Hochreiter and Schmidhuber (1997)[22] and Gers et al. (2000)[23]. LSTM has several additional functions compared to a traditional RNN. A typical recurrent block of LSTM, which is referred to be as an LSTM block, consists of a cell state and four gates. A cell state works as an internal memory which is considered to keep information during a long-term period. The four gates are called the cell gate, the input gate, the output gate, and the forget gate. The structure of LSTM is illustrated in Figure2. When using time series data as input, an LSTM block receives input data at a time step. The input data are given to the four gates; the cell gate, the input gate, the output gate, and the forget gate. The information passing through the input gate and the cell gate are integrated. The cell state at the previous step passes through the forget gate. These two kinds of information are kept as the cell state (memory) at the current time step. Then, the information of the memory goes through the output gate. The information out from the output gate is called a hidden state. The hidden state at the current time step is utilized at the next time step together with the input data at the next step. Finally, the hidden state is transformed by the Affine transformation to the shape of the target data, which becomes the output of LSTM. A LSTM block is recurrently used along the time of input time series as shown in Figure3. After the learnable parameters of RNN are trained, the trained parameters are the same along time.

3.2. Application and model configuration

This study implemented a daily groundwater level model at a single point within Kumamoto City utilizing LSTM with daily precipitation data and daily mean air temperature data as inputs. Similar to other deep learning methods, LSTM has several model options, which are called hyper-parameters. This study set the input data length (IDL) as 365 (days). The hidden state length is set to be 50. The aforementioned LSTM bock can be multi-layered. This study tried one, two, and four LSTM layers. LSTM is trained with a mini-batch gradient descent method. The batch size is set to be 512. The shuffling batch selection method is utilized to extract batches from the dataset. As the loss function, the mean square error (MSE) is utilized. To adjust the learning rate of the gradient descent method, the Adam optimization algorithm [24] is employed. The activation function is the linear function in this study.

The dataset before the severe earthquakes is utilized to train and verify the model. The dataset before the severe earthquakes is divided into three parts: the training dataset from April 2011 to March 2014, the validation dataset from April 2014 to March 2015, and the test dataset from April 2015 to March 2016. The training dataset and the validation dataset are utilized for the training process. The learnable parameters are updated with the training dataset. The overfitting is checked with the validation dataset.

This training process is conducted with a single LSTM layer, two LSTM layers, and four LSTM layers, respectively. There is some randomness in the training process such as the initial conditions of the learnable parameters. In consideration of the randomness, the training process is conducted 100 times with each configuration. Then, the best result with respect to the loss for the validation dataset is selected as the final trained model for each configuration. Finally, the test dataset is utilized to check the accuracy of the final trained model.

After testing the model, the trained model is also run with the dataset for the period from April 2016 to March 2019, which is the period after the severe earthquakes. Then, the simulated groundwater level is compared with the corresponding observations.

4. Results and Discussions

Table 1 shows the goodness-of-fit test for the test dataset that was obtained by LSTM with different numbers of layers. The correlation coefficient (R), the Nash-Sutcliffe efficiency (NSE), and the room mean square error (RMSE) obtained by LSTM with a single layer are 0.737, 0.271, and 2.361 m, respectively. Those with the two layers are 0.949, 0.850, and 1.070 m, respectively. Then, these statistical values with the four layers are 0.948, 0.858, and 1.042 m, respectively. LSTM with the four layers showed the best accuracy with respect to NSE and RMSE. Contrarily, R is the largest with the two layers. A larger number of layers of LSTM increases the number of learnable parameters (weights and biases). This study set the hidden state length the same among the three configurations. The number of learnable parameters with two layers is double that with a single layer. An increase in the learnable parameters helps LSTM to properly learn the fluctuation of groundwater level.

Figure 4 shows the time series obtained by LSTM with a single layer. LSTM with a single layer clearly underestimates the groundwater level except for the validation period, compared to the corresponding observations. For instance, the maximum groundwater level during the training period (April 2011- March 2014) is 27.7 m in the observation while it is 22.9 m in the simulated results. Figures 5 and 6 show the time series of the groundwater level obtained by LSTM with two layers and with four layers. Both of the simulated results show good agreement with the corresponding observation for all three periods. These results indicate that LSTM has a high potential to model groundwater level at the daily temporal scale.

This study utilized daily precipitation and daily mean air temperature only at a single meteorological station as input to the model. The recharge areas of groundwater around Kumamoto City are much wider. Meanwhile, precipitation generally does not evenly fall. If using two-dimensional precipitation field data as input may improve the model accuracy. On the other hand, it is known that the groundwater at Kumamoto City is also affected by precipitation fall around the city. Precipitation distribution is heterogeneity. However, this study shows high accuracy even though precipitation was obtained from a single point. It may indicate that the flexibility of LSTM for time series modeling.

Table 1. R, NSE. RMSE for the test period obtained by LSTM with each number of layers.

Layers	R	NSE	RMSE
1	0.845	0.903	0.878
2	0.847	0.841	0.850
4	0.837	0.912	0.909

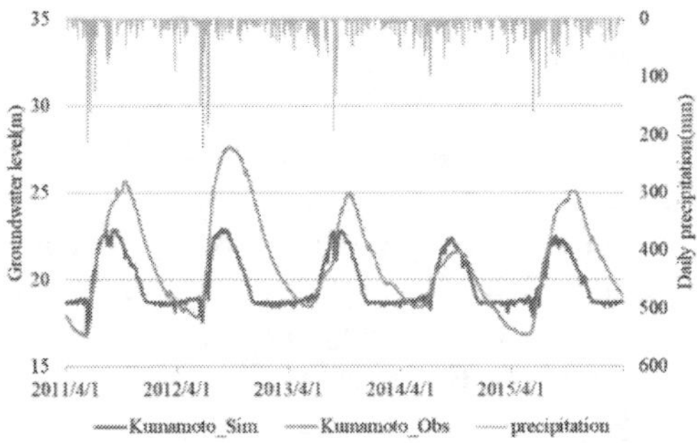

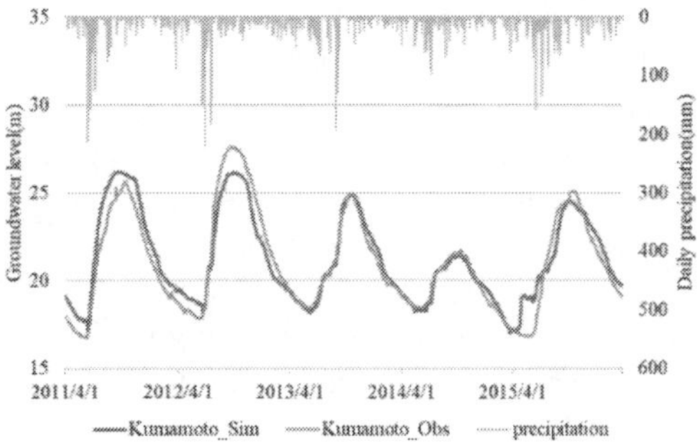

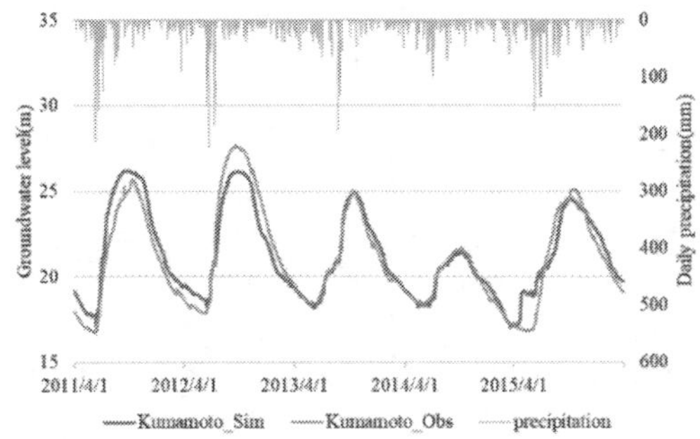

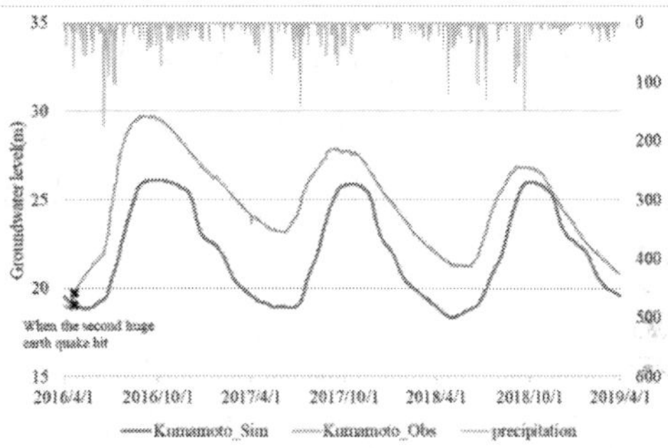

Figure 4. Results obtained by LSTM with a single layer from April 2011 to March 2016.

Figure 5. Results obtained by LSTM with two layers from April 2011 to March 2016.

Finally, the trained model with the four layers was run for the period from April 2016 to quantify the increase in the groundwater level after the severe earthquakes. The difference in the groundwater level between the simulated results and the corresponding observations is 0.206 m during the period from April 2015 to March 2016 on average. Contrarily, it suddenly becomes large just after April 2016 as shown in Figure 7. The difference is 3.29 m from April 2016 to March 2017, 3.36 m from April 2017 to March 2018, and 1.81 m from April 2018 to March 2019. The observed groundwater level is more than 3 m larger than simulated results for two years. Then the difference reduces in 2018. Because the trained model accurately models the groundwater level before the severe earthquakes, it may be able to be considered that the gap between the simulated results and the corresponding observations is caused by the severe earthquakes.

5. Conclusions

This study utilized a deep learning method, LSTM, to model groundwater level at a single point within Kumamoto City, Japan. The results show that LSTM is capable to model groundwater level in the study area. Meanwhile, the trained model is utilized to quantify the increase in the groundwater by the severe earthquakes that occurred in April 2016. The results in this study indicate that severe earthquakes increase the groundwater level at the study point by more than 3 m.

Acknowledgments

This study was supported by the Integrated Research Program for Advancing Climate Models (TOUGOU) Grant Number JPMXD0717935498 from the Ministry of Education, Culture, Sports, Science and Technology (MEXT), Japan. We also thank Kumamoto City Waterworks and Sewerage Bureau for providing groundwater level data.

References

[1] Anon 1988 *A modular three-dimensional finite-difference ground-water flow model*

[2] Hong N, Hama T, Suenaga Y, Aqili S W, Huang X, Wei Q and Kawagoshi Y 2016 Application of a modified conceptual rainfall–runoff model to simulation of groundwater level in an undefined watershed *Sci. Total Environ.* **541** 383–90

[3] Kratzert F, Klotz D, Brenner C, Schulz K and Herrnegger M 2018 Rainfall-runoff modelling using Long Short-Term Memory (LSTM) networks *Hydrol. Earth Syst. Sci.* **22** 6005–22

[4] Tian Y, Xu Y-P, Yang Z, Wang G and Zhu Q 2018 Integration of a Parsimonious Hydrological Model with Recurrent Neural Networks for Improved Streamflow Forecasting *Water* **10** 1655

[5] Song T, Ding W, Wu J, Liu H, Zhou H and Chu J 2019 Flash Flood Forecasting Based on Long Short-Term Memory Networks *Water* **12** 109

[6] Li W, Kiaghadi A and Dawson C N 2020 High Temporal Resolution Rainfall Runoff Modelling Using Long-Short-Term-Memory (LSTM) Networks

[7] Zhu S, Luo X, Yuan X and Xu Z 2020 An improved long short-term memory network for streamflow forecasting in the upper Yangtze River *Stoch. Environ. Res. Risk Assess.* 1–17

[8] Kao I-F, Zhou Y, Chang L-C and Chang F-J 2020 Exploring a Long Short-Term Memory based Encoder-Decoder Framework for Multi-Step-Ahead Flood Forecasting *J. Hydrol.* 124631

[9] Xiang Z, Yan J and Demir I 2020 A Rainfall-Runoff Model With LSTM-Based Sequence-to-Sequence Learning *Water Resour. Res.* **56**

[10] Miao Q, Pan B, Wang H, Hsu K and Sorooshian S 2019 Improving monsoon precipitation prediction using combined convolutional and long short term memory neural network *Water* **11** 977

[11] Misra S, Sarkar S and Mitra P 2018 Statistical downscaling of precipitation using long short-term memory recurrent neural networks *Theor. Appl. Climatol.* **134** 1179–96

[12] Tran Anh D, Van S P, Dang T D and Hoang L P 2019 Downscaling rainfall using deep learning long short-term memory and feedforward neural network *Int. J. Climatol.* **39** 4170–88

[13] Zhang D, Lin J, Peng Q, Wang D, Yang T, Sorooshian S, Liu X and Zhuang J 2018 Modeling and simulating of reservoir operation using the artificial neural network, support vector regression, deep learning algorithm *J. Hydrol.* **565** 720–36

[14] Hrnjica B and Bonacci O 2019 Lake Level Prediction using Feed Forward and Recurrent Neural Networks *Water Resour. Manag.* **33** 2471–84

[15] Liang C, Li H, Lei M and Du Q 2018 Dongting Lake water level forecast and its relationship with the Three Gorges Dam based on a long short-term memory network *Water* **10** 1389

[16] Ishida K, Tsujimoto G, Ercan A, Tu T, Kiyama M and Amagasaki M 2020 Hourly-scale coastal sea level modeling in a changing climate using long short-term memory neural network *Sci. Total Environ.* **720** 137613

[17] Bowes B D, Sadler J M, Morsy M M, Behl M and Goodall J L 2019 Forecasting groundwater table in a flood prone coastal city with long short-term memory and recurrent neural networks *Water* **11** 1098

[18] Jeong J and Park E 2019 Comparative applications of data-driven models representing water table fluctuations *J. Hydrol.* **572** 261–73

[19] Jeong J, Park E, Chen H, Kim K Y, Shik Han W and Suk H 2020 Estimation of groundwater level based on the robust training of recurrent neural networks using corrupted data *J. Hydrol.* **582** 124512

[20] Zhang J, Zhu Y, Zhang X, Ye M and Yang J 2018 Developing a Long Short-Term Memory (LSTM) based model for predicting water table depth in agricultural areas *J. Hydrol.* **561** 918–29

[21] Hosono T, Yamada C, Manga M, Wang C-Y and Tanimizu M 2020 Stable isotopes show that

earthquakes enhance permeability and release water from mountains *Nat. Commun.* **11** 2776
[22] Hochreiter S and Schmidhuber J 1997 Long Short-Term Memory *Neural Comput.* **9** 1735–80
[23] Gers F A, Schmidhuber J and Cummins F 2000 Learning to Forget: Continual Prediction with LSTM *Neural Comput.* **12** 2451–71
[24] Kingma D P and Ba J 2014 Adam: A Method for Stochastic Optimization

Effects of the spatial and temporal resolution of meteorological data on the accuracy of precipitation estimation by means of CNN

T Nagasato[1], K Ishida[2,3*], K Yokoo[4], M Kiyama[5] and M Amagasaki[5]

[1]Department of Civil and Environmental Engineering, Kumamoto University, 2-39-1 Kurokami, Kumamoto 860-8555, Japan
[2]Center for Water Cycle, Marine Environment, and Disaster Management, Kumamoto University, 2-39-1 Kurokami, Kumamoto 860-8555, Japan
[3]International Research Organization for Advanced Science and Technology, Kumamoto University, 2-39-1 Kurokami, Kumamoto 860-8555, Japan
[4]Graduated School of Science and Technology, Kumamoto University, 2-39-1 Kurokami, Kumamoto 860-8555, Japan
[5]Faculty of Advanced Science and Technology, Kumamoto University, 2-39-1 Kurokami, Kumamoto 860-8555, Japan
*E-mail: keiishida@kumamoto-u.ac.jp

Abstract. Future climate projections are valuable datasets to investigate the impacts of future climate changes on natural disasters such as intense precipitation, severe flood, and drought. However, they are too coarse depending on the purpose, and downscaling is required in such a case. There is nowadays a downscaling technique using deep learning such as CNN. Atmospheric information can be used as an input for precipitation downscaling by means of Convolutional Neural Network (CNN). For such precipitation downscaling, the spatial and temporal resolution of the atmospheric information may be important. This study obtained atmospheric information from a coarser reanalysis dataset and a finer reanalysis dataset as input for precipitation downscaling. As a coarser reanalysis dataset, ERA-Interim was selected. As a finer reanalysis dataset, ERA5 was utilized. Then, this study investigated the effect of spatial and temporal resolution of input data on the estimation accuracy of precipitation downscaling by CNN. For simplification, daily average precipitation at a watershed was used as the target data. The results show advantage of the use of a higher resolution as input can improve the model accuracy.

1. Introduction

In recent years, meteorological disasters have become more frequent and severe in the world. Global warming is considered to be a cause. There is a possibility that the magnitude and frequency of meteorological disasters will increase due to global warming in the future. Under such circumstances, it is very important to predict future precipitation in consideration of global warming to reduce damage caused by meteorological disasters such as floods and droughts.

Future climate projections are often used to predict future precipitation. Basically, future climate projections are obtained by numerical simulation using a general circulation model based on future prediction scenarios. Hence, future climate projections are valuable to assess the impacts of climate change.

IOP Conf. Series: Earth and Environmental Science **851** (2021) 012033 doi:10.1088/1755-1315/851/1/012033

However, resolutions of future climate projections data are generally too low for regional or watershed scale assessments. Downscaling is necessary for attaining high-resolution data from future climate projections. Dynamic downscaling and statistical downscaling have mostly been utilized as downscaling methods. In addition to these methods, nowadays, deep learning is used for downscaling. In the field of hydrology, many studies have been conducted on the applicability of various types of deep learning methods. There are various types of deep learning methods, and in recent years Long Short-Term Memory (LSTM) network, Convolutional Neural Network (CNN), and Convolutional LSTM Network (ConvLSTM) have been mainly applied in the field of hydrology. PAN et al. (2019) [1] estimated precipitation by utilizing CNN. The GPH and PW field data are used as inputs. The results showed CNN has applicable to precipitation estimation. Sadeghi et al. (2019) [2] proposed Precipitation Estimation from Remotely Sensed Information Using Artificial Neural Networks CNN (PERSIANN-CNN). In the study, CNN was trained with bispectral satellite imagery (infrared and water vapor) as input data and precipitation rate as target data. The study showed that CNN outperforms the baseline model. Kratzert et al. (2018) [3] successfully utilized LSTM for rain-runoff modeling. SHI et al (2015) [4] proposed the ConvLSTM for precipitation nowcasting. The ConvLSTM used the radar echo data (it recorded every 6 minutes) used as input data and predicted 15 frames ahead. All in all, the high applicability of deep learning methods to the field of hydrology has been shown. Compared to these deep learning methods, ConvLSTM requires a huge calculation cost. As for LSTM, it is good at analyzing time-series data. However, LSTM cannot reflect spatially information in input data. Hence, LSTM is not very suitable for downscaling where the input data is two-dimensional. Then, CNNs are thought to be able to extract spatial features efficiently. Therefore, CNNs are would be more suitable for downscaling among deep learning methods.

The applicability of CNN to downscaling precipitation was investigated in many previous studies. In precipitation downscaling by means of CNN, in general, precipitation distribution is used as input and spatial information is refined, or precipitation is estimated from atmospheric variables. Regarding the latter, Miao et al. (2019) [5], implemented 2D CNN for precipitation downscaling and showed the high possibility of application because of its high estimation accuracy. Nagasato et al. (2020) [6] investigated the characteristics of precipitation downscaling using 2D CNN. Additionally, they revealed that CNN can estimate precipitation at basin levels with high accuracy. These results of the studies show that CNNs have a large potential to estimate time-series precipitation with high accuracy. Nagasato et al. (2020) suggest that the estimation accuracy may change when the horizontal resolution of the input data changes. In recent years, ERA5, which has a higher spatiotemporal resolution are available although the previous studies used ERA-interim. The temporal resolution and spatial resolution of ERAI are 6 hours and 0.75 degrees, respectively. On the other hand, ERA5 has 6 times higher temporal resolution and 3 times higher spatial resolution than ERAI. It may be possible to improve the results of precipitation estimation by using ERA5 as input.

Therefore, this study investigated the effect of spatiotemporal resolution of input data on estimation accuracy in precipitation downscaling by CNN. CNN was implemented with the atmospheric variables extracted from ERA5 and ERA-Interim. As the target data, the daily average precipitation at the basin level was used. Similarly, this study implemented a CNN that uses atmospheric variables extracted from ERA-interim as input and made a detailed comparison between the two. Regarding the comparison between the two, detailed comparisons and examinations were conducted under various structural and hyperparameter setting conditions.

2. Methodology

2.1. Convolutional Neural Network

Convolutional Neural Network (CNN) which is a kind of deep Neural Network has been successfully applied to image recognition. CNN is composed of several types of layers such as convolutional layers, pooling layers, and fully connected layers. Generally, CNN has a structure in which convolution layers and pooling layers are connected alternately. The most characteristic algorithm of

CNN is the processing in the convolutional layer. In the convolution layer, the filter is slid into the input data and the feature is extracted by performing the product-sum operation (convolution operation). In the pooling layer, a filter is applied to the data output from the convolution layer to reduce the spatial size of the input data. A pooling layer is also used to accommodate changes in the position of features in the feature map. Hereafter, the filters used for input/output data and feature extraction in the convolution layer are called feature maps and kernels, respectively. In the fully connected layer, feature maps are formed into output shape. The kernel at the convolutional layer, bias, and weights at the fully connected layer are trained with the gradient descent procedure.

3. Study area

As the study area, Shira River Basin (SRB) was selected as the study area (Figure 1). SRB is located in the northern part of the Kyushu region of Japan. The length of the Shira River is 74 km. The area size of the SRB is 480 km^2. The upstream of Shira River is located in Aso caldera. In addition, there is a raised riverbed downstream of Shira River.

4. Dataset

This study uses atmospheric reanalysis data as input, and uses observed precipitation data as target data. The daily watershed-scale precipitation was calculated from observed daily precipitation data. The observation precipitation data were obtained from the Asian Precipitation-Highly-Resolved Observational Data Integration Towards Evaluation Data of Water Resources (APHRODITE) [7]. APHRODITE is continental-scale daily precipitation grid data based on rain gauge observation data with a resolution of 0.05-degree x 0.05-degree (about 5km). It covers the entire Japan spatially, and the 112 years from 1900 to 2015 temporally. In this study, the daily average precipitation in the target basin was calculated from APHRODITE.

Atmospheric variables were obtained from ERA-interim (ERAI) and ERA5 atmospheric reanalysis data, from the European Centre for Medium-Range Weather Forecasts (ECMWF). ERA-interim and ERA5 are atmospheric reanalysis dataset. ERAI covers the entire world and from 1 January 1979 to 31 August 2019. ERA5 covers the entire world and from 1 January 1979 to the present. The spatial resolution of ERAI is 0.75-degree x 0.75-degree (about 80km) and the temporal resolution is 6 hourly. On the one hand, the spatial resolutions and temporal resolution of ERA5 are 0.25-degree x 0.25-degree and hourly. ERA-Interim and ERA5 contain three-dimensional atmospheric variables on 37 pressure levels from 1 hPa to 1000 hPa. From ERA-interim and ERA5, this study obtained various atmospheric variables at 300/500/700/800/850/925/1000 hPa (Table.1). Regarding variables, U component of wind, V component of wind, W component of wind, specific humidity were selected based on the research results of Nagasato et al. (2020).

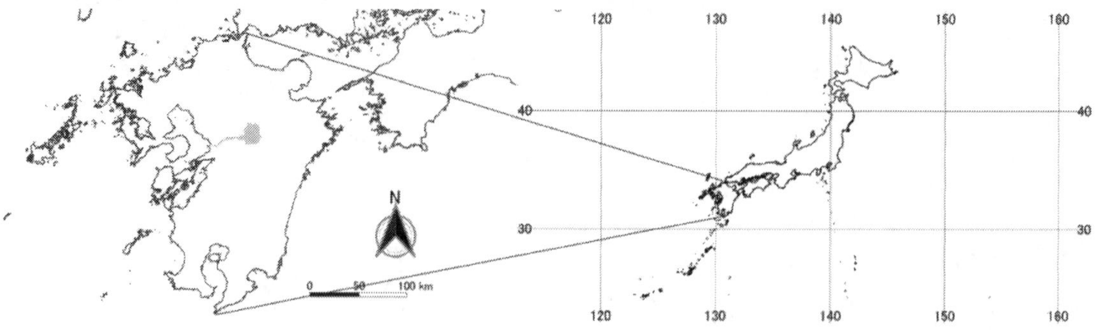

Figure 1. Shira River Basin, Kyushu region, Japan.

5. Model Implementation

This study employed PyTorch which is a framework of deep learning in Python. As the loss function, the Mean Square Error (MSE) is selected. Adam [8] is used as the optimizer. Epoch is set to 200. As a pooling method, Max-pooling was selected for pooling layers to emphasize the edges in the input data. The softplus function [9] is set as the activation function so that precipitation is not estimated to be a negative value. The mini-batch gradient descent method was used to save cost in the calculation. The number of data contained in the subset is called the batch size and was set to 512 in this study. The basic architecture of CNN is illustrated in Figure 2.

As aforementioned, the horizontal resolution is different between ERAI and ERA5. The horizontal resolution of ERA5 is three times finer than that of ERAI. This study obtained variables from two horizontal ranges: 6x6 and 8x8 grids for ERAI, and 16x16 and 22x22 for ERA5. These smaller and larger grid areas of each reanalysis dataset cover the same areas, respectively. The temporal resolution is also different between ERAI and ERA5. The temporal resolution of ERAI is 6 hourly. there are four time steps of 3:00, 9:00, 15:00, and 21:00 within a day in Japan Standard Time. The temporal resolution of ERA5 is hourly. There are 24 time steps within a day. To estimate daily basin-average precipitation at the study watershed, this study uses six time steps of ERAI. In addition to the above four time steps within the day, two time steps (21:00 on the previous day and 3:00 on the next day) are utilized.

With these input data, the model accuracy is compared with ERAI and ERA5. For the comparisons, several configurations are tried to increase the reliability of the comparison results. First, two and three sets of the layers (the convolutional layer, the batch normalization layer, and the max pooling layer) are tried, which are referred to be as 2L and 3L. In addition, the number of the output channel of the first convolutional layer is set to be 16, 32, or 64. As described above, two different sizes of the grid area are utilized. Then, to investigate the effects of the spatial and temporal resolution separately, this study utilizes not only 24 steps within a day from ERA5, but also extracted the same six time steps from ERA5 as ERAI. The configurations used for the comparisons are shown in Table 1-2.

With each configuration, CNN is trained with the training dataset during the period 1980-2005, is validated with the validation dataset during 2006-2010, and is tested with the test dataset during 2011-2015. During the validation period, it is confirmed whether overfitting has occurred, and during the test period, the estimation accuracy of the model is evaluated. The learning process of CNN has randomness depending on the initial conditions and batch selection. Hence, in this study, learning is performed 50 times with each configuration. Among the 50 trained results, the best model is selected, and then it is tested with the test dataset. To evaluate model estimation accuracy, this study utilizes the root mean square error (RMSE), the correlation coefficient (R), and the Nash-Sutcliffe efficiency (NSE).

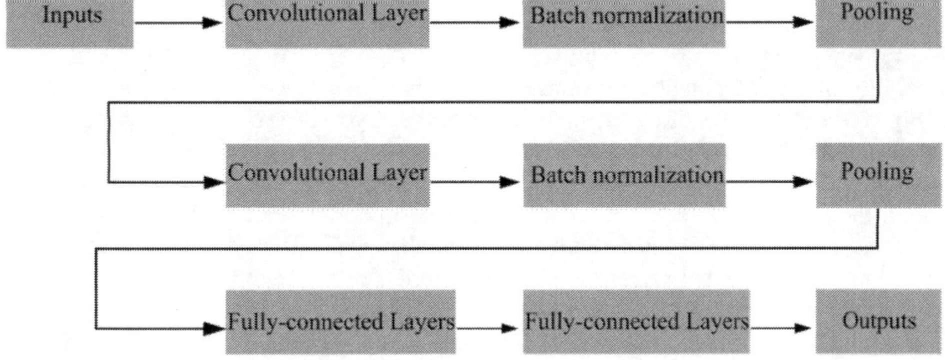

Figure 2. Basic architecture of CNN.

Table 1. The configurations used for the comparisons of spatial resolution.

Case		data	Channel	Grid size	Structure
S1	S1-1	ERA-interim	16	6 x 6	2L
	S1-2	ERA5		16 x 16	
S2	S2-3	ERA-interim	32	6 x 6	
	S2-4	ERA5		16 x 16	
S3	S3-5	ERA-interim	64	6 x 6	
	S3-6	ERA-interim		8 x 8	
	S3-7	ERA5		16 x 16	
	S3-8	ERA5		22 x 22	
S4	S4-9	ERA-interim	16	6 x 6	3L
	S4-10	ERA5		16 x 16	
S5	S5-11	ERA-interim	32	6 x 6	
	S5-12	ERA5		16 x 16	
S6	S6-13	ERA-interim	64	6 x 6	
	S6-14	ERA-interim		8 x 8	
	S6-15	ERA5		16 x 16	
	S6-16	ERA5		22 x 22	

Table 2. The configurations used for the comparisons of temporal resolution.

Case		Time range	Structure	Grid size
T1	T-1	6	2L	16 x 16
	T-2	1		
T2	T-3	6	3L	
	T-4	1		

6. Results and Discussion

First, ERAI and ERA5 were compared with the three numbers of output channels and with 2L and 3L. In these cases, only six time steps were utilized from ERA5 in order to investigate effects of the spatial resolution of the model. Meanwhile, 6x6 and 16x16 of the grid size were utilized for ERAI and ERA5, respectively. Table 3-4 shows the three statistics (RMSE, R, RMSE) for the test period. Except in S5, ERA5 show better statistical values compared to ERAI. For instance, in S2, the use of ERA5 improved the NSE value by 0.011-0.073, the RMSE value by 0.21-1.098 mm, and the R value by 0.002-0.044. Figure 3 shows the NSE values for the two periods: the training period, the validation period. Because there are 100 trained results with each reanalysis dataset with each configuration, the results are illustrated by boxplots. These boxplots more clearly show the advantage of ERA5. For instance, the median values of NSE obtained with ERA5 are higher than those with ERAI in all the cases. The median values were improved by 0.087-0.132 for the training period and 0.001-0.012 for the validation period by using ERA5.

The same comparisons are conducted with 8x8 and 16x16 of grid size for ERAI and ERA5, respectively. These results (S3-6 and S3-8, and S6-14 and S6-16) are also shown in Table 3 and Figure

3. Basically, the increase in the grid size does not clearly increase the model accuracy with ERAI and ERA5. For the training period, the three statistical values with 2L were worsen by the larger grid size for both of the reanalysis dataset. They were improved with 3L. However, the NSE values were generally worsen for the validation period (Figure 3). Consequently, the use of ERA5 has still advantage over ERAI in these cases, too.

Table 3. NSE values for the test period (Effects of the spatial resolution of input data on the accuracy).

		Training		Validation		Test		
Case		NSE_max	NSE_median	NSE_max	NSE_median	NSE	RMSE	R
S1	S1-1	0.948	0.743	0.742	0.718	0.675	11.8	0.827
	S1-2	0.956	0.851	0.764	0.719	0.686	11.6	0.829
S2	S2-3	0.934	0.757	0.745	0.719	0.642	12.4	0.805
	S2-4	0.941	0.861	0.766	0.727	0.715	11.0	0.849
S3	S3-5	0.944	0.753	0.755	0.721	0.675	11.8	0.827
	S3-6	0.955	0.765	0.755	0.717	0.644	12.3	0.804
	S3-7	0.947	0.840	0.753	0.727	0.722	10.9	0.854
	S3-8	0.954	0.855	0.752	0.723	0.669	11.9	0.820
S4	S4-9	0.924	0.749	0.748	0.716	0.659	12.1	0.815
	S4-10	0.952	0.880	0.756	0.727	0.703	11.3	0.841
S5	S5-11	0.949	0.742	0.745	0.721	0.703	11.3	0.848
	S5-12	0.958	0.870	0.768	0.732	0.684	11.6	0.833
S6	S6-13	0.935	0.738	0.744	0.720	0.631	12.6	0.807
	S6-14	0.940	0.740	0.744	0.715	0.693	11.5	0.836
	S6-15	0.946	0.854	0.764	0.732	0.693	11.5	0.835
	S6-16	0.965	0.852	0.752	0.727	0.715	11.0	0.856

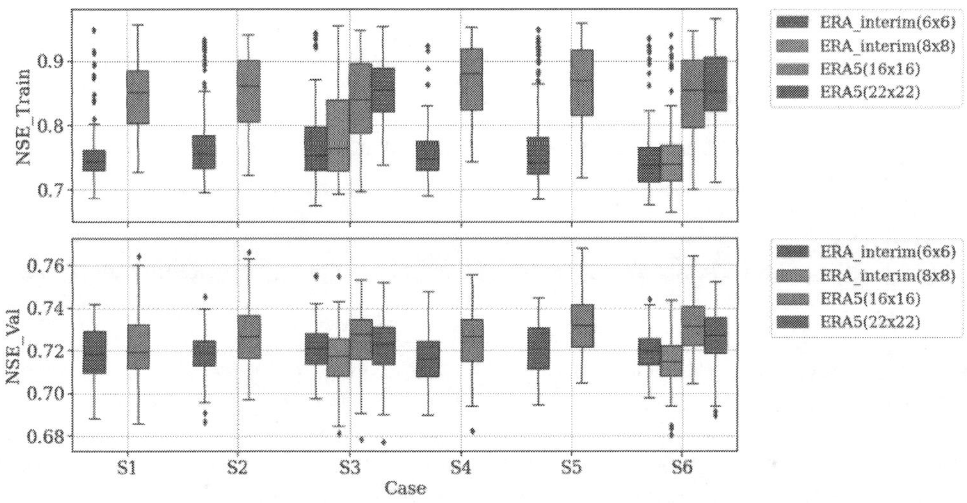

Figure 3. NSE values for the training and validation periods (Effects of the spatial resolution of input data on the accuracy).

The effects of the used time steps of ERA5 were also investigated as shown in Table 4 and Figure 4. The use of the hourly time steps (24 hours within a day) improved the model accuracy compared to the use of the six-hourly time steps. For the test period, the NSE, RMSE, and R values were improved by 0.041, 0.84 mm, and 0.021, respectively, with 2L. They were improved by 0.019, 0.36 mm, and 0.010, respectively, with 3L. There are some reductions in the NSE values for the training period by using the hourly time steps. However, generally NSE value for the training period is clearly higher than for the other two periods. The results may indicate that the use of the hourly time steps mitigates overfitting to the training dataset and improve the results for the other periods.

Table 4. NSE values for the test period (Effects of the temporal resolution of input data on the accuracy).

Case		Training		Validation		Test		
		NSE_max	NSE_median	NSE_max	NSE_median	NSE	RMSE	R
T1	T-1	0.947	0.840	0.753	0.727	0.722	10.9	0.854
	T-2	0.928	0.775	0.771	0.742	0.763	10.1	0.875
T2	T-3	0.946	0.854	0.764	0.732	0.693	11.5	0.835
	T-4	0.948	0.785	0.772	0.745	0.712	11.1	0.845

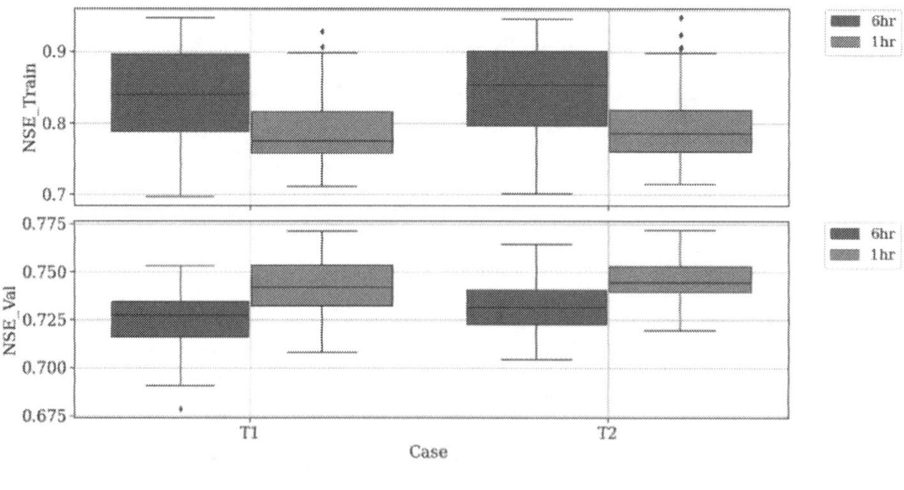

Figure 4. NSE values for the training and validation periods (Effects of the temporal resolution of input data on the accuracy).

7. Conclusion

This study investigated effects of the spatial and temporal resolution of meteorological data on the accuracy of precipitation estimation by means of CNN. For the purpose, the two reanalysis dataset with different spatial and temporal resolutions, ERAI and ERA5, were utilized. The results show advantage of the use of a finer resolution as input can improve the model accuracy. Not only the spatial resolution of the meteorological data, but also their temporal resolution affects the model accuracy. The goal of the series of this study is to develop the methodology to accurately downscale precipitation from coarse resolution data because future climate projections are generally coarse. Increase in resolution of future climate projections them may be valuable.

References

[1] Pan B, Hsu K, AghaKouchak A and Sorooshian S 2019 Improving Precipitation Estimation Using Convolutional Neural Network *Water Resour. Res.* **55** 2301–21

[2] Sadeghi M, Asanjan A A, Faridzad M, Nguyen P H U, Hsu K, Sorooshian S and Braithwaite D A N 2019 PERSIANN-CNN: Precipitation estimation from remotely sensed information using artificial neural networks–convolutional neural networks *J. Hydrometeorol.* **20** 2273–89

[3] Kratzert F, Klotz D, Brenner C, Schulz K and Herrnegger M 2018 Rainfall-runoff modelling using Long Short-Term Memory (LSTM) networks *Hydrol. Earth Syst. Sci* **22** 6005–22

[4] Shi X, Chen Z, Wang H, Yeung D-Y, Wong W-K, Woo W-C and Kong Observatory H 2015 *Convolutional LSTM Network: A Machine Learning Approach for Precipitation Nowcasting* vol 28

[5] Miao Q, Pan B, Wang H, Hsu K and Sorooshian S 2019 Improving monsoon precipitation prediction using combined convolutional and long short term memory neural network *Water (Switzerland)* **11**

[6] Takeyoshi Nagasato, Kei Ishida, Makoto Ueda, Kazuki Yokoo, Masato Kiyama M A 2020 CHARACTERISTICS OF PRECIPITATION DOWNSCALING BY MEANS OF DEEP LEARNING METHOD *J. Japan Soc. Civ. Eng. Ser. B1* **76** 373–8 (in Japanese)

[7] Kamiguchi K, Arakawa O, Kitoh A, Yatagai A, Hamada A and Yasutomi N 2010 Development of APHRO_JP, the first Japanese high-resolution daily precipitation product for more than 100 years *Hydrol. Res. Lett.* **4** 60–4

[8] Kingma D P and Ba J L 2015 Adam: A method for stochastic optimization *3rd International Conference on Learning Representations, ICLR 2015 - Conference Track Proceedings* (International Conference on Learning Representations, ICLR)

[9] Dugas C, Bengio Y, Belisle F, Nadeau C and Garcia R *Incorporating Second-Order Functional Knowledge for Better Option Pricing*

Identification of Paleotsunami Deposits Using XRF and Artificial Intelligence Methods on the Southern Coast of Lebak, Banten

U Rikhasanah[1], **M R Septyandy**[1*] **and Supriyanto**[2]

[1] Program Study Geology, Universitas Indonesia, Indonesia
[2] Program Study Geophysics, Universitas Indonesia, Indonesia
*E-mail: m.rizqy@sci.ui.ac.id

Abstract. The southern coastal area of Lebak, Banten is the southern region of Java which is prone to tsunami disasters. There is evidence of past tsunami events in the southern region of Java. However, not all tsunami deposits have identifiable sedimentological and micropaleontological traces. Geochemical proxies and artificial intelligence with machine learning methods can be used to identify paleotsunami deposits. Machine learning methods that can be used to cluster paleotsunami deposits are Agglomerative Hierarchical Clustering (AHC) and Support Vector Machine (SVM) with validation of model accuracy using the Receiver Operating Characteristic (ROC) and Area Under Curve (AUC) methods. Input data are XRF analysis data and macroscopic core sample description data. The output from data processing is in the form of prediction of tsunami and non-tsunami deposits at each depth of core data sample. The data is correlated and interpreted to identify tsunami events that occurred in the past. The identification results show that the tsunami deposition in the research area, namely the area around the Bagedur coastal coast and Bolang village had tsunami deposit characteristics based on macroscopic description, XRF analysis, and artificial intelligence clusterization. It is also thought to be correlative with the tsunami deposition of previous research in the area around the Binuangeun coastal coast.

1. Introduction

One of the areas in southern Java that is prone to tsunami disasters is the area around southern coast of Lebak, Banten. Previous research found a correlation between paleotsunami deposits in the Lebak and Pangandaran areas with the interpretation that these deposits were deposited from the same tsunami event because they are estimated to have an extension of 300 km [1]. The proxies most often used in paleotsunami sediment research are the proxies for sedimentology and micropaleontology. However, one disadvantage of these proxies is that not all tsunami deposits have observable sedimentological and micropaleontological traces, so other proxies are proposed, such as geochemical proxies. The use of geochemical proxies to help identify paleotsunami deposits was first made about 30 years ago, namely in the 1990s [2].

Technological developments greatly helped in discovering methods for clustering paleotsunami deposits by studying their patterns. One method that can be used to predict paleotsunami deposition patterns is the use of artificial intelligence with machine learning methods. One of the applications of geochemical proxies using machine learning in determining paleotsunami deposits has been carried out [3]. This method was applied to the 2011 Tohoku tsunami deposit in Japan. Based on the research results, machine learning can display prediction of tsunami and non-tsunami deposits after going through several series of experiments to obtain the maximum predictive accuracy model by recognizing the pattern of the chemical content of the studied sediment like increases compounds in TiO_2, Fe_2O_3, K_2O, CaO, and MnO. Data processing from geochemical proxies using machine learning can be a method that is expected to predict tsunami event patterns and identify the process so that it is useful in mitigating future tsunami disasters.

2. The Potential of Tsunami Disaster in Lebak Area, Banten

The southern coastal area of Lebak is located in Lebak Regency, Banten Province. The area around the southern coast of Lebak, Banten is one of the areas on the South Coast of Java Island which is close to the southern subduction zone of Java trench which has the potential to create earthquakes that can cause tsunamis (figure 1) [4]. This tsunami determined to generate earthquake based on the scenario of historical large earthquakes that have occurred in the subduction zone of the southern region of Java. The value of fault parameters such as magnitude, length, width and slip are determined based on the tectonic arrangement and the scaling law of earthquakes (table 1). Lebak Regency has a coastline directly facing the Indian Ocean, which is geologically an active continental margin because it is a meeting point between the Indo-Australian oceanic plate and the Eurasian continental plate which is characterized by high seismic activity (the red round show on the map in the right side of figure 1) [4].

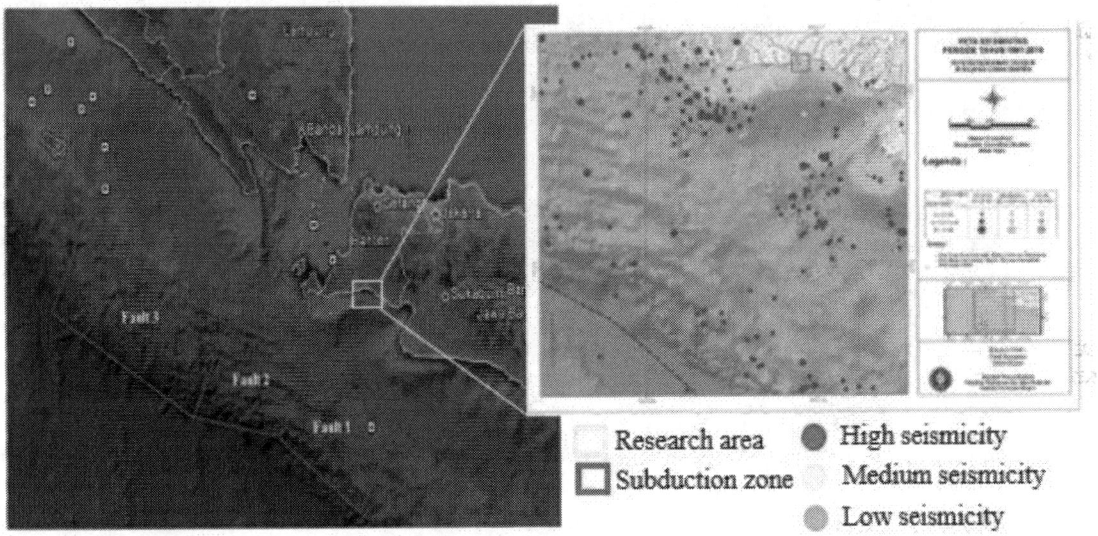

Figure 1. Subduction zone which has the potential to create earthquakes that can cause tsunamis on the south of Java Island (left) & seismicity map of south coast Lebak, Banten (right) [4].

Table 1. The scenario of historical large earthquakes that have occurred in the subduction zone of the southern region of Java based on the tectonic arrangement and the scaling law of earthquakes [4].

Fault	Length (km)	Width (km)	Loc (°S)	Loc (°E)	Mag (Mw)	Strike (°)	Dip (°)	Rake (°)	Slip (m)	Depth (km)
Fault 1	128	100	8,73	105,83	8,7	315	10	89	20	10
Fault 2	115	100	8,05	104,87	8,7	281	10	100	20	10
Fault 3	215	100	7,83	103,82	8,7	300	10	114	20	10

3. Geochemical Proxies for Paleotsunami Deposits

Geochemical proxies of the geology of the study area, and a good understanding of depositional and chemical processes can help differentiate and understand chemical signatures in tsunami deposits. Geochemical proxies for current and past tsunami deposits are considered useful proxies and can be traced quantitatively, especially when other proxies cannot be used. Geochemical proxies describe sediment sources (based on mineralogy), co-components (biomaterials such as shells / organic matter from land or sea), and transport media [2]. Chemical analysis of Si, Al, Mg, Fe, K, Mn, Ti, P and Ca was performed by means of the X-ray fluorescence (XRF) spectrometer of analytical SiO_2 content slightly increases (4–5%) in the tsunami deposit while other elements occur in much lower concentrations [2]. Depletion in Fe and Ti, essentially found as rutile (TiO_2) and titanomagnetite ($Fe_{2x}Ti_xO_3$), matches the low values of magnetic susceptibility observed in the tsunami deposit [5]. The main geochemical signatures of the tsunami deposit are originated from inherited phyllosilicates contained in the adjacent clayish units and diluted within a huge volume of non-magnetic minerals (i.e., calcite and silicon). Quantitative data normalized by Al_2O_3 confirm the connection between the signature of the tsunami deposit and those of the mudflat. These results suggest an external source of SiO_2 and CaO (shell fragment and detrital calcite & silicates). The analysis displays an excellent positive correlation between Al_2O_3 and major elements commonly associated with inherited minerals (TiO_2, K_2O, Na_2O, Fe_2O_3). Conversely, SiO_2 and CaO display a negative correlation with Al_2O_3 reaching a correlation coefficient of 0.95. In addition to inherited phyllosilicates from the clays and biogenic/detrital carbonates from the beach, a slight but notable increase in other elements (K_2O, Na_2O and MgO) is observed (Figure 2). MgO is mostly associated to dolomite while Na_2O and K_2O can be related to micaceous minerals and feldspars [5].

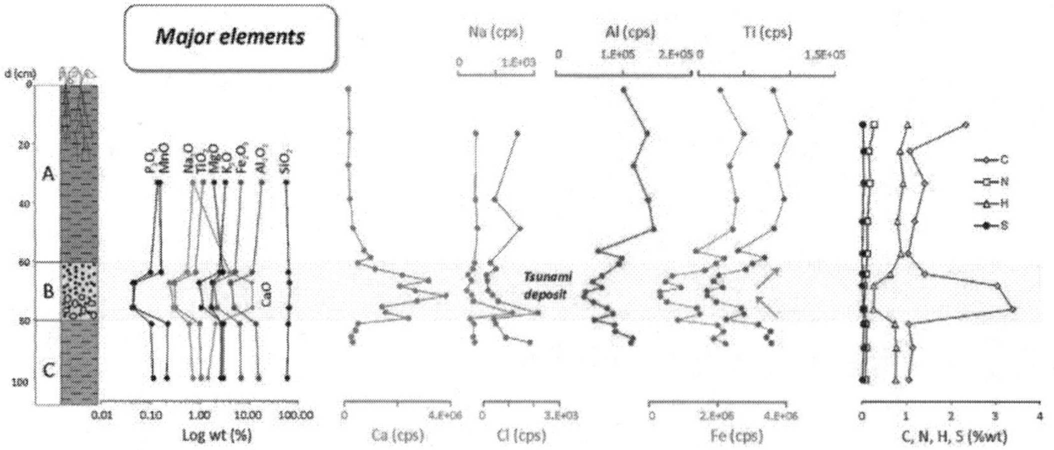

Figure 2. Geochemical data of major elements on Lisbon tsunami deposit at Boca do Rio Estuary: quantitative data (wt.%) and trends in marine seawater (Ca, Na, Cl) and terrestrial/high-energy (Al, Fe, Ti, Zr) indicators (in cps). Elemental analysis (C, N, H, S) is also shown [5].

4. The Concept of Artificial Intelligence in Paleotsunami Deposits Research

The application of artificial intelligence is very useful in identifying paleotsunami deposits and practical method with quantitative models. The use of artificial intelligence methods is of great interest to help identify paleotsunami deposits. This study utilizes XRF analysis data to identify the geochemical proxies in paleotsunami deposits and uses machine learning techniques to process the data. The machine learning technique used in this research is unsupervised clustering method of the XRF analysis processed using Agglomerative Hierarchical Clustering (AHC) method to cluster paleotsunami and non-paleotsunami deposits at each depth [6]. Machine learning techniques with the AHC method will classify according to the character of each element at each depth. It is necessary to try several suitable affinity and linkage methods to the prediction of the classification of paleotsunami and non-paleotsunami

IOP Conf. Series: Earth and Environmental Science **851** (2021) 012034 doi:10.1088/1755-1315/851/1/012034

deposits is in accordance with the manual prediction at the beginning by taking into account the peculiarities of the paleotsunami geochemical deposits. Then the results of the AHC clustering data are used as a reference for the descriptive data to classify paleotsunami deposits at each depth. Data result from AHC processed to Support Vector Machine (SVM) method. The application of the SVM method in this study is to provide prediction results in the form of prediction of paleotsunami and non-paleotsunami deposits on data whose patterns have been known from the processing of the AHC method. The classification results of the SVM method are checked for the accuracy of the prediction results by performing validation to test the accuracy or success of an AHC and SVM model in producing paleotsunami sediment predictions. The validation method used is the Receiver Operating Characteristic (ROC) Curve and Area Under Curve (AUC). Receiver Operating Characteristic (ROC) Curve and Area Under Curve (AUC) are validation techniques commonly used in comparing the results of artificial intelligence prediction models [7]. The ROC graph has a range of AUC values from 0 - 1. A good and ideal model has an AUC value close to 1 indicating that the model has very good accuracy, while an AUC value close to 0.5 indicates that the model is not accurate.

5. Research methods

The data in this study area were obtained from sampling in the form of coring data (wells) with a depth ranging from 1-2 m in a predetermined area which is thought to have paleotsunami deposits. Sampling was carried out at Bagedur Beach and Bolang Village, Malingping, Lebak Regency, Banten (figure 3). At Bagedur Beach one core data sample is taken and in Bolang Village three core data samples are taken, with the location coordinates as follows:

- Bagedur Beach (BGD-04): (6°42'55.6"S, 105°59'32.5"E)
- Bolang village 1 (BLG-01): (6°47'39.7"S, 105°58'18.0"E)
- Bolang village 2 (BLG-02): (6°47'41.0"S, 105°58'20.5"E)
- Bolang village 3 (BLG-03): (6°47'42.0"S, 105°58'21.6"E)

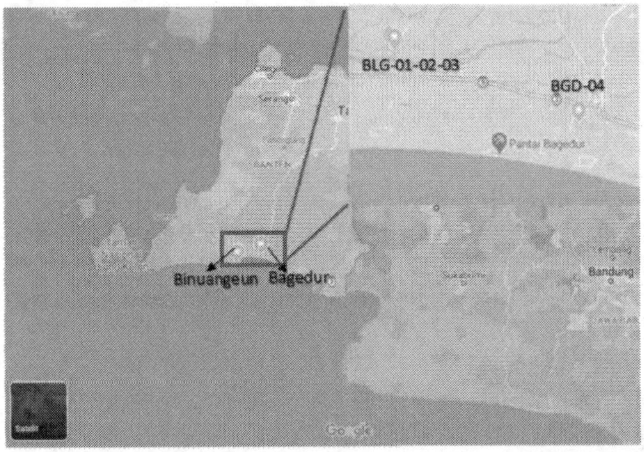

Figure 3. Sampling location on the study are (modified from google maps).

The results of the well data are then described macroscopically. After that the data is processed by doing XRF analysis and artificial intelligence with the following stages:

- XRF analysis is carried out for descriptive data that are suspected of having the characteristics of paleotsunami deposits to determine the content of the compounds contained at each depth.
- The content of compounds that are known from each depth in the XRF analysis, a prediction is made at the depth where tsunami deposits are suspected.
- Prediction is used as a reference in determining the most appropriate clustering method in the Agglomerative Hierarchical Clustering method whose prediction results will be used to classify paleotsunami and non-paleotsunami deposits in the description data.

- Predicted data on the Agglomerative Hierarchical Clustering method are used as training and validating data which will be processed using the Support Vector Machine (SVM) method with a percentage of 80% training data and 20% validating data.
- Validation is done to test the accuracy or success of the model. The validation method used is the receiver operating characteristic (ROC) curve and area under curve (AUC).
- After obtaining a good accuracy value, testing is carried out on the new description data to predict both paleotsunami and non-paleotsunami deposits in the description data.
- The prediction results will be used as a reference in the depth correlation process at sample points where paleotsunami deposits are suspected.

6. Result & Discussion

6.1. Results of data processing

The process of clustering paleotsunami and non-paleotsunami deposits using artificial intelligence methods by processing input data in the form of XRF analysis data of 25 samples from data core of BGD-04 (0-100). Before being processed using machine learning, manual predictions of tsunami deposition predictions were carried out on sample data based on the results of research on the characteristics of paleotsunami geochemical deposits from several previous studies. The results of manual prediction resulted in the conclusion that at the predicted depth the tsunami deposit had a pattern in the form of high Fe_2O_3 and TiO_2 content and low SiO_2 and Al_2O_3 content. In general, based on previous research, the geochemical characteristics of tsunami deposits are an increase in TiO_2 compounds and elements of Fe, Ti, Ca and K and a decrease in phylosilicate minerals that contain Si & Al, which are the most common soil mineral classes. The graph of each compound's content also shows a unique pattern in paleotsunami deposits. The following is a table of the results of manually predicting the estimated depth of tsunami deposits based on the chemical compound content and the graph of the compound content in the XRF analysis results. The graph shows the compound with the blue line graph has increased at several depths and the graph with the red line has decreased at several depths.

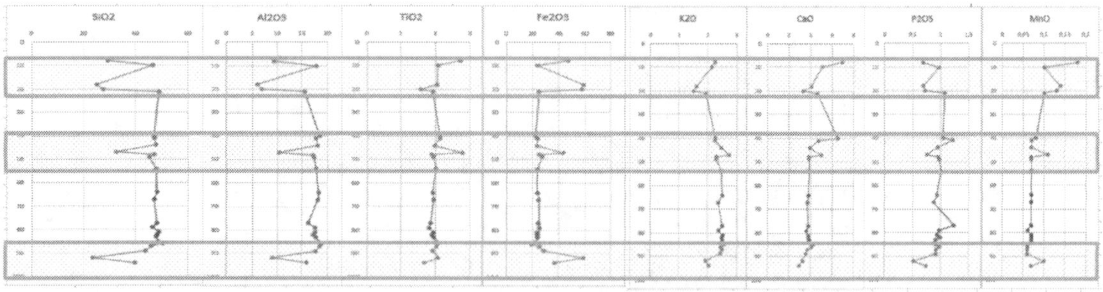

Figure 4. The graph shows pattern in the suspected tsunami deposition layer on several depth (cm). Increases compounds were seen in TiO2, Fe2O3, K2O, CaO, and MnO at depths of 8, 18, 20, 47, and 92 cm. Meanwhile, the SiO2, Al2O3, and P2O5 compounds also decreased at depths of 8, 18, 20, 47, and 92 cm.

The results of manual prediction are used as a reference in finding the best combination of methods in machine learning with the Agglomerative Hierarchical Clustering (AHC) method to predict XRF analysis data. The sample data used in this study were relatively small (25 samples). It is hoped that by obtaining the appropriate combination of methods, it can be helped to process data in the future with a larger amount of data because we know the best combination of methods in predicting the pattern of chemical compound content of paleotsunami deposits. After going through several experiments with the combination of the affinity and linkage methods used in the AHC method, the best method for predicting paleotsunami deposits based on geochemical characteristics is the affinity "cosine" and "average" linkage type methods. The experimental results of several method combinations are attached in the appendix. The prediction is exactly same as the manual prediction that has been done previously based on the geochemical characteristics of the tsunami deposit from previous studies. By finding an accurate model, it is hoped that more research data in the future will be more effective in processing and no need to use manual predictions.

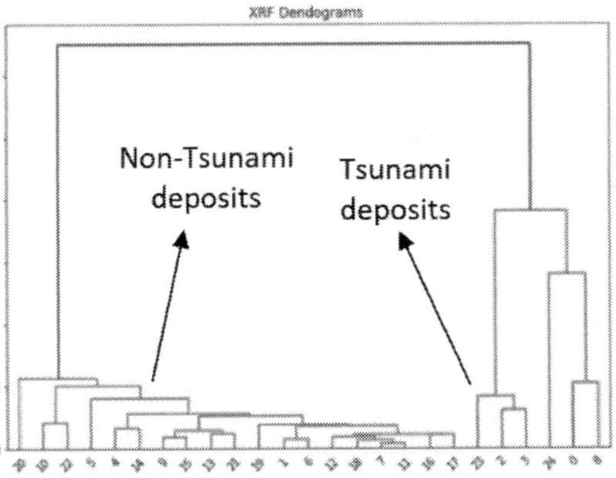

Depth	SiO_2	Al_2O_3	TiO_2	Fe_2O_3	K_2O	CaO	P_2O_5	MnO	Prediction	AHC Result
8	29,49	9,27	2,72	46,86	2,26	6,96	0,69	0,18	1	1
10	46,82	17,9	2,09	23,85	2,16	5,05	0,97	0,1	0	0
18	25,08	6,07	2,07	58,62	1,6	4,09	0,69	0,14	1	1
20	27,54	6,82	1,58	57,38	1,49	3,25	0,71	0,13	1	1
21	48,89	15,53	1,92	25,35	1,93	4,65	1,07	0,1	0	0
40	46,87	18,48	2,14	21,85	2,26	6,54	1,05	0,08	0	0
41	47,18	17,69	2,13	23,46	2,24	4,74	1,21	0,07	0	0
44	47,86	17,96	1,99	23,91	2,47	3,91	0,95	0,07	0	0
47	32,74	10,46	2,8	43,65	2,74	5,01	0,76	0,11	1	1
48	46,99	17,05	1,9	25,84	2,31	3,82	0,96	0,07	0	0
49	45,11	17,33	1,95	27,28	2,29	3,83	0,97	0,07	0	0
54	47,8	17,61	2	24,02	2,48	3,88	1	0,07	0	0
64	48,14	18,22	1,93	23,47	2,51	3,78	0,93	0,07	0	0
67	46,88	18,03	1,94	25,15	2,36	3,71	0,88	0,07	0	0
77	47,98	16,22	1,88	25,01	2,49	3,8	1,24	0,07	0	0
79	46,4	17,55	1,83	26,21	2,36	3,57	0,97	0,06	0	0
81	48,75	17,47	1,94	23,65	2,53	3,75	0,94	0,07	0	0

Figure 5. The left side is AHC clustering result from sample BGD-04 (0-100). The right side show the table of tsunami deposit prediction from XRF analysis and machine learning clustering (blue highlight show the tsunami deposits). The result shows the same prediction both XRF analysis & machine learning.

The prediction results of tsunami suspicion deposits in the XRF analysis data are transferred to the description data according to the predicted depths obtained because of the limitations of XRF analysis data that cannot be read at each depth of the data described. Descriptive data processing in sample BGD-04 (0-100) is used as training data and validating as core data for machine learning to recognize patterns from tsunami deposits based on the data provided. The results of the prediction data are then processed using the Support Vector Machine (SVM) method. As many as 80% (57 samples) of data were used as training data and 20% (15 data) of data were used as validating data. Before being processed using the SVM method, the data must be encoded first to change the information data so that machine learning can be understood. The following are the results of the encoding on the sample description data BGD-04 (0-100). The encoded data is processed using the SVM method. In the training data, the prediction results were obtained in the form of a tsunami deposition prediction of 7 points from 57 data. Meanwhile, for the validating data, the prediction results obtained from the predicted tsunami deposits are 3 points from 13 data points. Following are the results of the prediction of training data and validating of tsunami suspected deposits from the SVM processing method.

The SVM model that has been created in ppython programming produces a ROC-AUC value of training data and validation of 100%. This result is very satisfying, considering that it is very rare to find 100% accuracy for both. However, it also needs to be evaluated that the input data on which machine learning is based is still small and not too complex, so the resulting accuracy is quite good because the input data is relatively simple. It is hoped that this model can be the basis for model development in future studies with more complex parameters. In addition, the results of this model need to be compared with other proxies that support evidence of suspected tsunami deposits or can be completed from previous research results so that it can be proven that there is a correlation between current research and previous research that has been carried out in relation to the existence of tsunami deposits in the past. Following are the results of the machine learning accuracy assessment using the ROC-AUC method on training and validating data.

After obtaining good accuracy values in the training and validating data, then testing the other descriptive data on samples BLG-01, BLG-02, and BLG-03. Before testing, all core data are put together in one data file and encoding is carried out as in the previous data training and validating stages. The output of this testing data is the prediction of the presence of tsunami suspicion deposits in each sample. Then the prediction data obtained from each core data is correlated to determine the temporal variation of the tsunami event, the characteristics of the sediment, and the depositional environment of each correlated data.

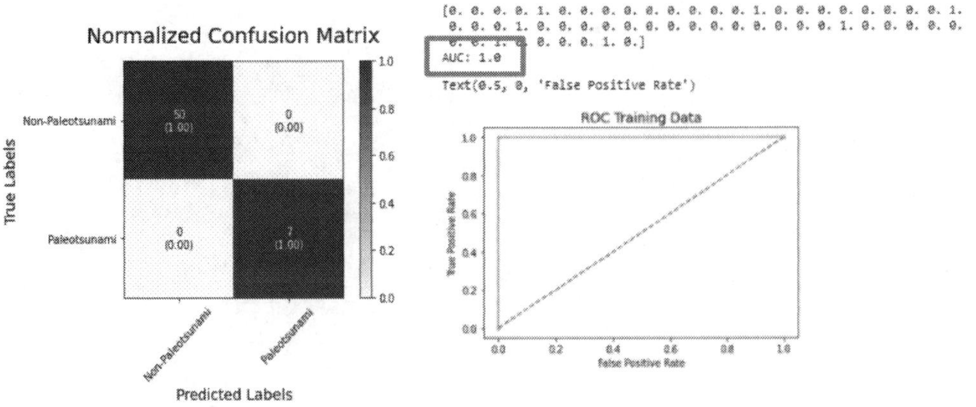

Figure 6. The confussion matrix of training data show prediction of tsunami deposits in 7 data and non-tsunami 50 data (left). The accuration value of model based on ROC-AUC is 100% (right).

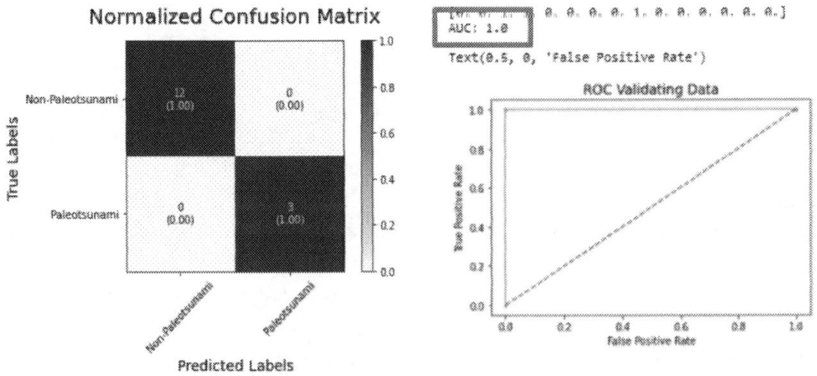

Figure 7. The confussion matrix of validating data show prediction of tsunami deposits in 3 data and non-tsunami 12 data (left). The accuration value of model based on ROC-AUC is 100% (right).

6.2. Characteristics of paleotsunami deposits in the research area

The characteristics of paleotsunami deposits on the research area based on XRF analysis data, macroscopic descriptive data, and artificial intelligence methods have their own patterns in each of the analyzed data. The XRF analysis results show that the suspected tsunami deposits in the study area tend to increase in TiO_2 and Fe_2O_3 compounds and decrease in SiO_2 and Al_2O_3. This is possible because Fe and Ti are heavy minerals which are indicators of high energy deposition, while Si and Al are characteristics of minerals from land. The following table compares the content of compounds in paleotsunami deposits in the tsunami-suspected sediment layer.

Table 2. Summary of the characteristic description component on tsunami deposits.

Depth	Sediment	white flakes	Root fibers	Pieces of logs	Quartz mineral	Shell fragments	Prediction
BGD-04 (0-100)							
8	Clay	1	0	0	1	1	Tsunami deposit
9	Clay	1	0	0	1	1	Tsunami deposit
19	Clay	1	0	0	1	1	Tsunami deposit
20	Sand	1	0	0	1	1	Tsunami deposit
47	Clay	1	0	0	1	1	Tsunami deposit
90	Clay	1	0	0	0	1	Tsunami deposit
91	Clay	1	0	0	0	1	Tsunami deposit
92	Clay	1	0	0	0	1	Tsunami deposit
93	Clay	1	0	0	0	1	Tsunami deposit
BLG-01 (0-90)							
22	Clay	0	0	0	0	1	Tsunami deposit
39	Clay	1	1	0	0	1	Tsunami deposit
BLG-02 (0-50)							
1	Sand	1	1	0	1	1	Tsunami deposit
19	Sand	1	0	0	1	1	Tsunami deposit
48	Sand	1	0	0	1	1	Tsunami deposit
BLG-03 (0-90)							
19	Sand	0	0	0	0	1	Tsunami deposit
28	Sand	1	0	0	0	1	Tsunami deposit
29	Sand	1	1	0	0	1	Tsunami deposit
30	Sand	1	1	0	0	1	Tsunami deposit
40	Clay	0	1	0	0	1	Tsunami deposit
41	Clay	0	1	0	0	1	Tsunami deposit
48	Clay	0	0	0	0	1	Tsunami deposit

Based on descriptive data and artificial intelligence, it is known that there are distinctive characteristics of tsunami-suspected deposits in the descriptive data recognized by machine learning. The presence of shell chips is a key factor in determining whether sediment is tsunami deposition or not. The presence of shell fragments indicates a source of material originating from the sea which is thought to have originated from the tsunami phenomenon. In addition, the description data that has a shell fragment component is also correlative with the results of XRF analysis on the presumed paleotsunami deposit layer. White flakes in tsunami deposits usually indicate the presence of foraminifera microfauna on tsunami deposit. the presence of root fibers and pieces of wood indicates the depositional environment on land so that it is identified as not tsunami deposits.

6.3. Identification of past tsunami events

Based on the results of the comparison of the log correlation between the study area and the previous research around the Pesisir Selatan area of Lebak, Banten, it shows that there is a correlation between the thickness of the tsunami deposits at two layers of depth. The log data in figure 8a is the log from the research area located around Bagedur Beach, while the log in figure 8b is the log from the Binuangeun Coastal research area, Wanasalam Area, Banten by [10]. The correlation results show that a depth of 45-80 cm in the coastal logs of Binuangeun Beach contains tsunami deposits. It is estimated that the thickness of the tsunami deposit in the Binuangeun Coastal area is around 35 cm. These deposits are thought to be correlated with tsunami deposits on the Bagedur Coast which are found at a depth range of 18-49 cm which means that they have a thickness of about 31 cm so they are similar to tsunami deposits on the Binuangeun Coast. However, tsunami deposits on the Bagedur Coast are not continuous as in the Binuangeun Coastal sediments, but rather occur between layers of clay and sand. This shows the possibility of differences in the depositional environment in the study area with the previous research area located around the South Coast of Lebak Beach, Banten.

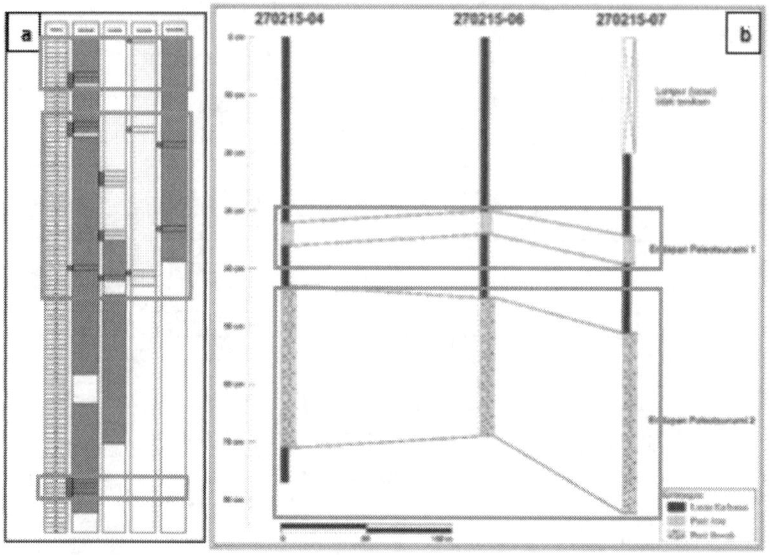

Figure 8. Correlation between research area in Bagedur Coast & Bolang Village (left) and previous research in Binuangeun Coastal Coast (right) [4]. Green colour is sedimentation layer of clay, yellow colour is sedimentation layer of sand, the blue square shows the prediction of tsunami deposits.

Based on description of data sample and compound's content of XRF analysis we conclude that the research area sedimentation environment is lacustrine. The lacustrine environment is an intermediate area between land and sea. This is concluded from the depositional pattern which indicates an alternating sediment of clay and sand. Sand sediment is found among the dominant clay sediment and the presence of mollusk shells on the edge of the lake and their shells forming clusters several centimeters in thickness [8]. The chemical compound showed the similar pattern with geochemical data of major elements on Lisbon tsunami deposit at Boca do Rio Estuary [2].

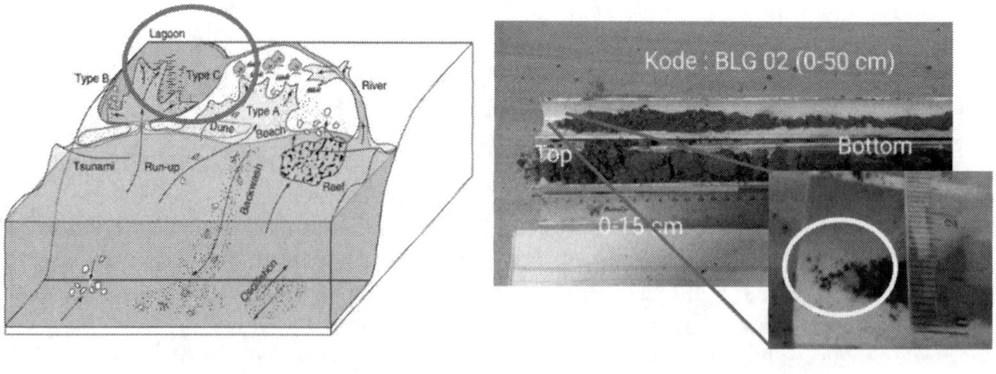

Figure 9. Interpretation of tsunami sedimentation types (left) [9] and the presence of molusk shell fragment in research area show by yellow circle (right).

The tsunami depositional mechanism in the study area was identified from the features and sediment patterns recorded in the tsunami deposition candidate layer. Based on the features and depositional patterns, the authors classify the tsunami deposits into the coastal lacustrine basin deposit mechanism (coastal lacustrine basin). The existence of a barrier topography prevents the invasion of tsunami wave deposits, so that tsunami deposits in the lacustrine environment are generally found as a layer of sand interspersed with mud sediments. The tsunami deposition mechanism of this type indicates a tsunami inundation carrying coastal sediments with a fast deposition process [9]. The following is the sedimentological profile and the mechanism of deposition in the lacustrine environment.

7. Conclusion

Based on the results and analysis of research conducted by the author. Then it can be concluded as follows:

- The graph of XRF analysis show unic pattern with increasing/decreasing of some compound on several depth of paleotsunami deposit.
- The machine learning prediction show the same prediction with XRF analysis of paleotsunami deposit.
- The result of prediction data using XRF analysis & machine learning can be used to predict other sample data without XRF analysis.
- The macroscopic description data sample of paleotsunami show that the presence of shell fragment and white flakes are the key to determine tsunami deposit that indicates a source of material originating from the sea which is thought to have originated from the tsunami phenomenon.
- Based on description of data sample and compound's content of XRF analysis, also based on study literature of sedimentation environment of tsunami deposit from any previous research we conclude that the research area sedimentation environment is lacustrine

Acknowledgments

The Authors would like to thank Program Study Geology and Geophysics Universitas Indonesia for Supporting this research. This Research was funded by Basic Research for Higher Education Grant, Ministry of Research and Technology.

References

[1] Putra P S 2015 *Studi Paleotsunami di Selatan Jawa* Pemaparan Hasil Penelitian Geoteknologi 2015 ISBN 978- 979-8636-30-1

[2] Chagué-Goff. C, Szczuci ński W, Shinozaki T 2016 *Applications of geochemistry in tsunami research: A review* Earth-Science Reviews. Elsevier

[3] Kuwatani T, Nagata K, Okada M, Watanabe T, Ogawa Y, Komai T, Tsuchiya N 2014 Machine-learning techniques for geochemical discrimination of 2011 Tohoku tsunami deposits *Scientific Reports* DOI: 10.1038/srep07077

[4] Sugianto D, Nurjaya, I W, Natih N M and Pandoe W W 2017 Potensi Rendaman Tsunami di Wilayah Lebak Banten Pusat Gempa Bumi dan Tsunami BMKG DOI: 10.15578/jkn.v12i1

[5] Font E, Veiga-Pires C, Pozo M, Nave S, Costas S, Ruiz M F, Abad M, Simões N, Duarte S and Rodríguez-Vidal J 2013 Benchmarks and sediment source(s) of the 1755 Lisbon tsunami deposit at Boca do Rio Estuary *Mar Geol* **343**, 1–14

[6] Nugroho A S, Witarto A B, Handoko D 2003 Support Vector Machine–Teori dan Aplikasinya dalam Bioinformatika *Proceeding of Indonesian Scientific Meeting in Central Japan*. Gifu-Japan

[7] Darabi H, Choubin B, Rahmati O, Haghighi A T, Pradhan B and Klove B 2018 Urban flood risk mapping using the GARP and QUEST models: A comparative study of machine learning techniques *Journal of Hydrology* **569** (2019) 142–154 Elsevier Ltd

[8] Sugawara D, Minoura K, and Imamura F 2008 *Tsunamis and tsunami sedimentology*. In ''Tsunamiites—Features and Implications''

[9] Shiki T, Tsuji Y, Yamazaki T and Minoura K 2008 *Tsunamiites: Features and Implications* Elsevier Radarweg 29, PO Box 211, 1000 AE Amsterdam, The Netherlands Linacre House, Jordon Hill, Oxford OX2 8DP, UK

[10] Zulkarnain D A, Amijaya H, Yulianto E 2017 *Karakteristik Endapan Paleotsunami di Pesisir Binuangeun, Daerah Wanasalam, Banten* Proceeding, Seminar Nasional Kebumian ke-10

Artificial intelligence approach to depositional facies characterization based on electrical log data

G E Putri[1], A Haris[2*] and M R Septyandy[3]

[1]Reservoir Geophysics, Department of Physics, Universitas Indonesia
[2]Geophysics Study Program, Universitas Indonesia
[3]Geology Study Program, Universitas Indonesia
*E-mail: aharis@sci.ui.ac.id, m.rizqy@sci.ui.ac.id

Abstract. This research aims to determine the depositional facies from electrical log data using the gradient boosting classifier method, which comprises a powerful algorithm. The electrical logs used are gamma-ray (GR), resistivity (ILD), neutron porosity(NPHI), and density (RHOB), while the output is in the form of images. The training data consists of 4 wells in Jambi sub-Basin, South Sumatera Basin, while the testing data comprises 5 wells with gamma-ray, resistivity, NPHI, and RHOB as input. Several scenarios are used to predict the facies model, namely training and validation dataset by using and isolating facies in well combination input, and with or without feature augmentation. Furthermore, the values collected were validated using F1 score. The result showed that 85.5% and 84.7% of F1 scores were allocated to training and validation to increase accuracy in scenarios without facies isolation and with feature augmentation. Therefore, the gradient boosting classifier method is reliable enough to characterize depositional facies in the associated area of interest.

1. Introduction

It is imperative to correlate lithofacies with well log data and other geological or geophysical information described from core samples for proper connection [1]. However, such information is not always provided to the entire length of the wells, therefore, pattern recognition needs to be applied for adequate prediction. Data can be separated into several subsets, with each element consisting of similar attributes, irrespective of its type. Furthermore, it is possible to classify elements based on some common characteristics obtained from the dataset because it is the basic principle of automatic classification of a new point [2]. The numerous approaches to automatic classification of data are divided into supervised and unsupervised methods [3].

The supervised methods are based on classification labels (targets) knowledge, while the unsupervised cluster the samples into subsets through statistical and other mathematical approaches [2]. This research used the merge interpretation method with well logs and core data in Jambi Sub-Basin, South SumateraBasin area, to classify the depositional facies.

2. Problem Formulation

Well log data contains most information capable of determining depositional facies. Therefore, through logging, it is possible to obtain a detailed description of rock formations at different depth levels by measuring a wide variety of its properties [4]. However, due to the high cost of coring, it is important to find another method capable of determining the depositional facies. One of such methods is the use of machine learning to predict facies.

The application of supervised learning algorithms in the machine enables it to recognize the complicated correlations between various data sets. It also allows for the extraction of essential information to improve the precision of reservoir quality assessment, minimize exploration risks, and identify break-even points [5].

3. Facies Classification Using Gradient Boosting Classifier

In this research used 4 electrical logs as input, namely Gamma Ray (GR), Induction Resistivity (ILD), Neutron Porosity (NPHI), and Density (RHOB) for radioactivity formation, electrical resistivity, porosity of facies, and density of lithology in the subsurface, respectively.

After a brief determination of the lithology from the interpreted depth, the facies were classified from the basic well log form that characterizes a depositional environment such as a cylindrical, irregular, bell, funnel, symmetrical, and asymmetrical [6]. However, this research only used 3 forms in this area, namely cylindrical, symmetrical, and serrated. The cylindrical is associated with facies accumulation which is heterogenic to the shallow water environment. The symmetrical occur where there is a conformity combination of bell-funnel form and a combination of coarsening-fining upward. It forms as a result of reworked offshore build-up, from regressive to transgressive shoreface. The serrated form is associated with a storm-dominated shelf and distal deep marine slope, indicating thin sand interbedded with shale.

Based on well log form, 3 lithofacies characterizing the layer at depth d of the well is proposed as follows.

1. Distributary Channel Mudstone Sandstone (DCMdSt)
2. Shoreface Delta Mudstone Sandstone (SDMdSt)
3. Deep Marine Slope Interbedded Mudstone Sandstone (DMSMdSt)

The available attributes are concatenated from the machine learning process and associate with each depth and a feature vector as defined in equation 1.

$$\boldsymbol{f}_{d,w} = [\, f^{GR}_{d,w}, f^{Res}_{d,w}, f^{Por}_{d,w}, f^{Den}_{d,w} \,] \tag{1}$$

In addition, each depth is associated with a class label $\boldsymbol{cd}$, $\in$ {DCMdSt, SDMdSt, DMSMdSt} which indicates the facies associated with the layer.

Gradient boosting classifier works by splitting training data observations into different subsets [8]. A limited number of features are selected from each subset to train a separate decision tree, thereby solving a gradient-based minimalization problem. A new feature vector f is classified by testing each tree in the forest to provide a candidate label. After that, the results from all trees are merged into a single decision [4].

4. Research Methodology

The gradient-boosting classifier (GBC) attempts to decrease error by resampling and varying the weights for individual weak learners to increase classification accuracy [7]. This method provides a metric to assess the relative influence of each parameter included in the classifier [8]. GBC also provides the output in images, symbols, and graphics with its various characteristics.

This research uses the general supervised machine learning flow divided into training, validation, and testing phases. During training, a set of labeled observations from used wells in Jambi Sub-Basin, South Sumatera Basin, is used to determine a function that maps features to class labels. After training, the model formed must be validated with input data which is not used as a training data set with an accuracy value close to the obtained result. When the validation output is far from the training result, it is called underfitting or overfitting. In the testing phase, the classifier is used to estimate class labels from any unlabelled feature vector f obtained from new well logs [4].

The electrical logs used to characterize the depositional facies are gamma-ray (GR), resistivity (ILD), neutron porosity (NPHI), and density (RHOB). These parameters are used to train, validate and test data. This study utilized 4 wells as the training and validation data, namely E, I, K, and L, and 5 wells as the testing data, including C, G, J, M, and N. The workflow of this research is shown in Figure 1.

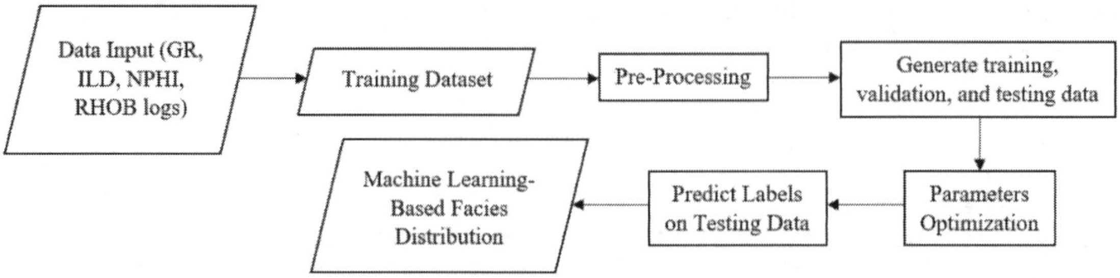

Figure 1. Depositional facies determination with artificial intelligence workflow.

5. Results and Discussion

The input data used in this research are 3 facies with GR, ILD, NPHI, and RHOB logs. The training and validation data were obtained from E, I, K, and L wells, while the testing data were from C, G, J, M, and N wells. Correlations between parameters are shown in Figure 2. The result shows that correlations of the logs well illustrate the facies. This is seen through facies log plots according to the depth of each well, which is adequately described as shown in Figure 3. Therefore, to ensure the data is useful, feature distribution was carried out as shown in Figure 4.

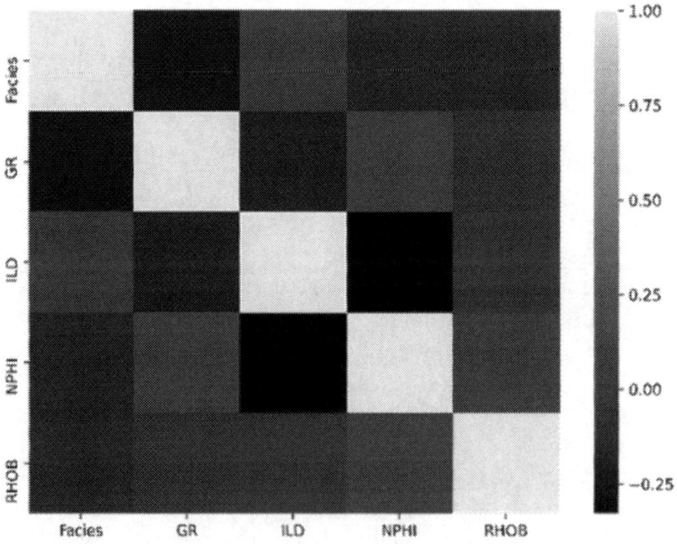

Figure 2. Correlation between parameters

IOP Conf. Series: Earth and Environmental Science **851** (2021) 012035 doi:10.1088/1755-1315/851/1/012035

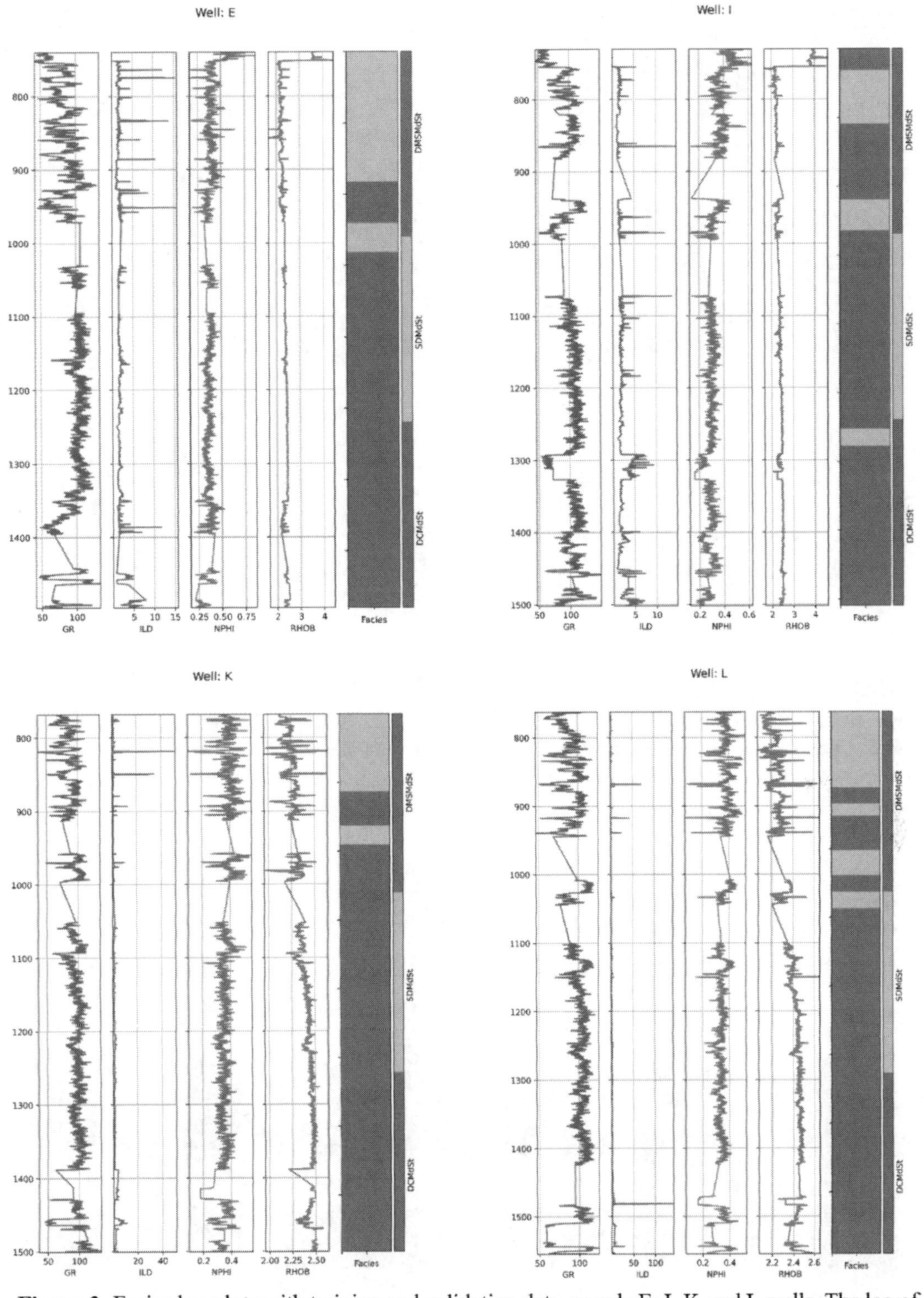

Figure 3. Facies log plots with training and validation data, namely E, I, K, and L wells. The log of each well illustrates its facies such as distributary channel mudstone sandstone (DCMdSt) coloured in dark blue, shoreface delta mudstone sandstone (SDMdSt) coloured in light blue, and deep marine slope interbedded mudstone sandstone (DMSMdSt) coloured in purple. Lithofacies with low GR, low ILD, high NPHI and low RHOB represents low radioactivity, high resistivity, high porosity, and low density.

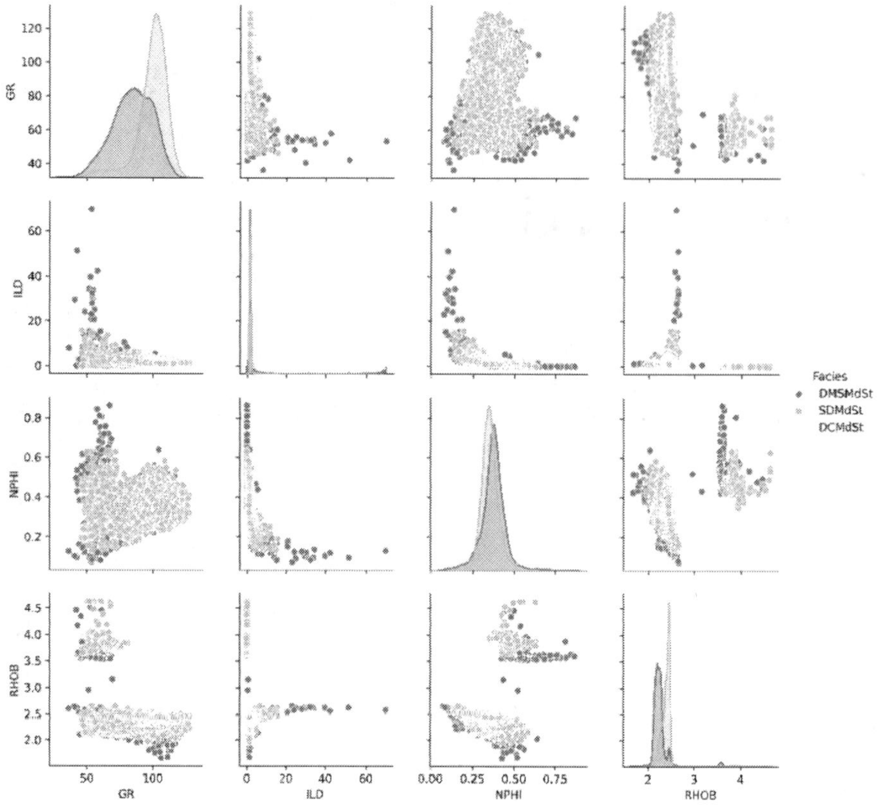

Figure 4. Feature Distribution is between fellow logs such as NPHI with NPHI have good facies distributions which are showed by the curves formed. Meanwhile, the facies distributions between different logs such as ILD with GR are not good enough.

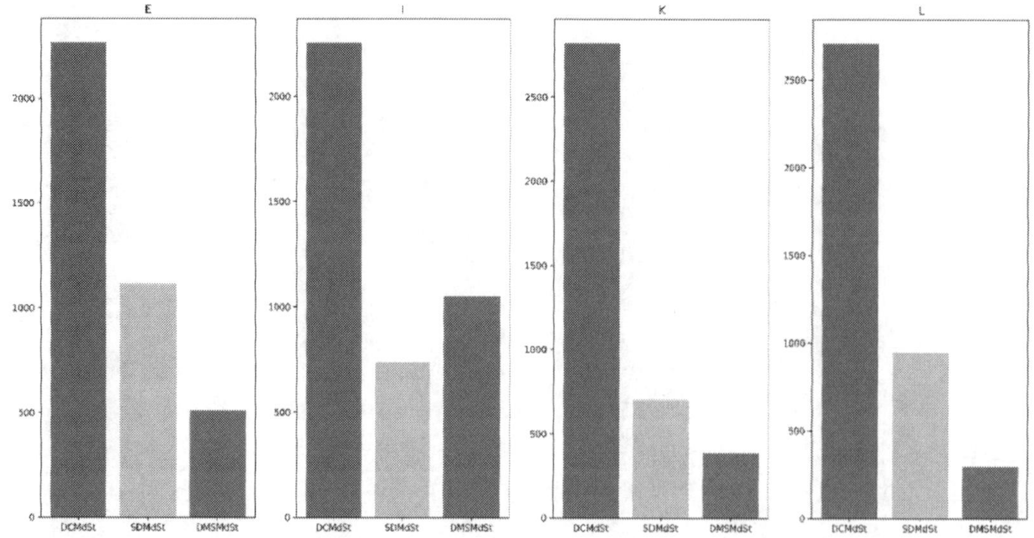

Figure 5. Facies Distribution is the most common facies in each well with distributary channel mudstone sandstone (DCMdSt), followed by shoreface delta mudstone sandstone (SDMdSt), and deep marine slope interbedded mudstone sandstone (DMSMdSt).

IOP Conf. Series: Earth and Environmental Science **851** (2021) 012035 doi:10.1088/1755-1315/851/1/012035

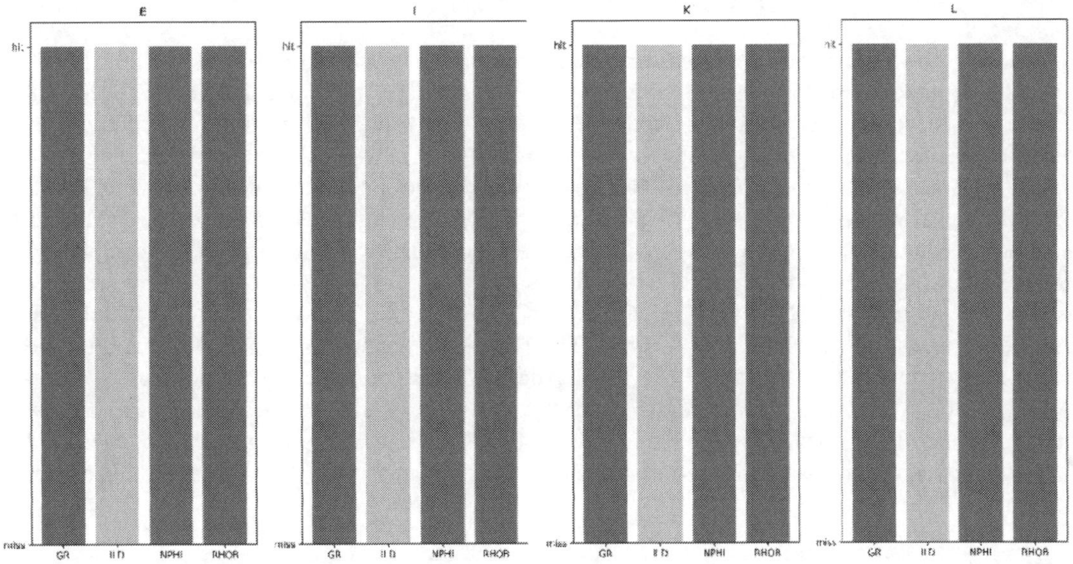

Figure 6. Features per well for all logs with the parameters consisting of the same amount of high accuracy value.

The distributary channel mudstone sandstone facies (DCMdSt) are the most commonly found in eachwell, as shown in Figure 5. Meanwhile, Figure 6 shows that all input needs to be balanced to geta high accuracy value with a similar amount of data.

In this research, the authors conducted 4 scenarios to obtain the best results needed to predict lithofacies in this area. The first is training and validation dataset using isolated facies in well combination input with feature augmentation. The second is similar to the first and without feature augmentation. The third is training and validation dataset without isolated facies and with feature augmentation. Meanwhile, the last is similar to the third without feature augmentation. Furthermore, the average values of F1 and confusion matrix scores were used to strengthen the accuracy results, while the standard scalerfunction was used to normalize the functions. From the 4 scenarios model, the best average F1 score and confusion matrix score from the third scenario for both training and validation with the accuracy of the F1 score average are 0.855 (training) and 0.847 (validation), as shown in Figure 7.

A

Pred / True	DCMdSt	SDMdSt	DMSMdSt	Total
DCMdSt	4720	300	510	5530
SDMdSt	338	1306	3	1647
DMSMdSt	190	5	486	681

Precision	0.90	0.81	0.49	0.85
Recall	0.85	0.79	0.71	0.83
F1	0.88	0.80	0.58	0.83

B

Pred / True	DCMdSt	SDMdSt	DMSMdSt	Total
DCMdSt	4680	316	534	5530
SDMdSt	372	1248	27	1647
DMSMdSt	221	7	453	681

Precision	0.89	0.79	0.45	0.83
Recall	0.85	0.76	0.67	0.81
F1	0.87	0.78	0.53	0.82

C

Pred / True	DCMdSt	SDMdSt	DMSMdSt	Total
DCMdSt	5535	419	85	6039
SDMdSt	422	1654	22	2098
DMSMdSt	422	3	916	1341

Precision	0.87	0.80	0.90	0.86
Recall	0.92	0.79	0.68	0.86
F1	0.89	0.79	0.77	0.85

D

Pred / True	DCMdSt	SDMdSt	DMSMdSt	Total
DCMdSt	5527	406	106	6039
SDMdSt	441	1603	54	2098
DMSMdSt	462	1	878	1341

Precision	0.86	0.80	0.85	0.84
Recall	0.92	0.76	0.65	0.84
F1	0.89	0.78	0.74	0.84

Figure 7. Interaction of facies in the confusion matrix in scenarios 1 (A), 2 (B), 3 (C), and4 (D). The best model is scenario 3 with F1-Score 0.855.

IOP Conf. Series: Earth and Environmental Science **851** (2021) 012035 doi:10.1088/1755-1315/851/1/012035

Figure 8 shows the normalized confusion matrix consisting of predicted and true labels. However, it is majorly associated with determining the value of false positives and negatives. Therefore, to acquire good accuracy, both amounts need to be smaller than true positives and true negatives. In the confusion matrix, good results are shown by the interaction between facies in producing the most significant number when compared with others. From scenario 3, the relationship between DCMdSt, SDMdSt, and DMSMdSt is obtained with values of 0.92, 0.79, and 0.68. The accuracy value of the detector is obtained by averaging diagonal values of the confusion matrix. A perfect detector tends to produce a diagonal confusion matrix [4] which found that the best lithofacies predicted by the model is distributary channel mudstone and sandstone (DCMdSt), while the worst is deep marine slope interbedded mudstone sandstone (DMSMdSt). This is possible because the DCMdSt facies is the largest in this study while the DMSMdSt is the least. Another interesting fact is the number of facies of shoreface delta mudstone sandstone (SDMdSt) and deep marine slope interbedded mudstone sandstone (DMSMdSt) which are predicted to be distributary channel mudstone sandstone (DCMdSt). This is the basis for increasing the amount of data on the two facies to enable them to produce accurate predictions.

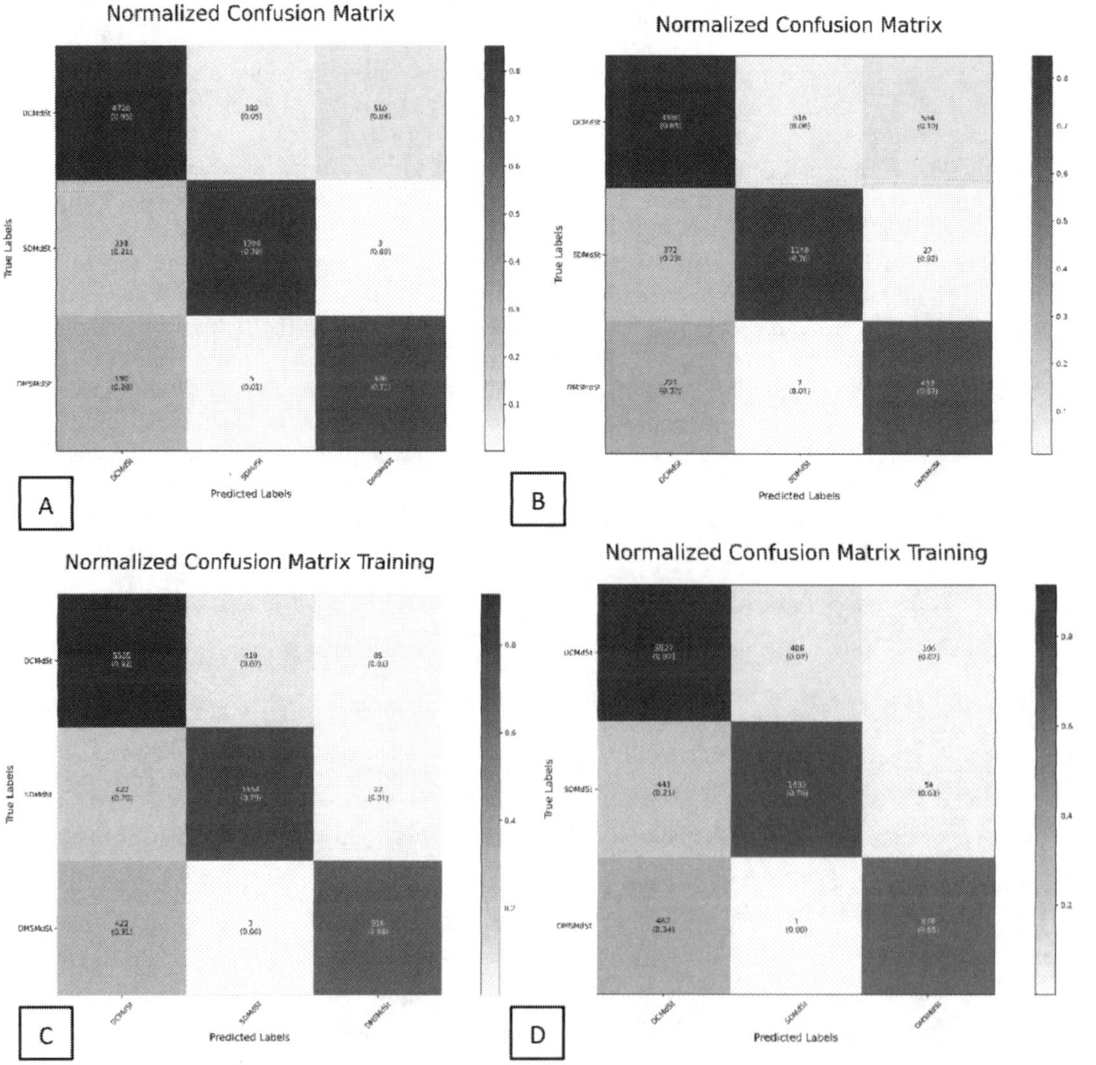

Figure 8. Normalized Confusion Matrix. Dark blue part shows good value, which is close to 1. This matrix shows relationships between fellow facies in scenarios 1 (A), 2 (B), 3 (C), and 4 (D).

IOP Conf. Series: Earth and Environmental Science **851** (2021) 012035 doi:10.1088/1755-1315/851/1/012035

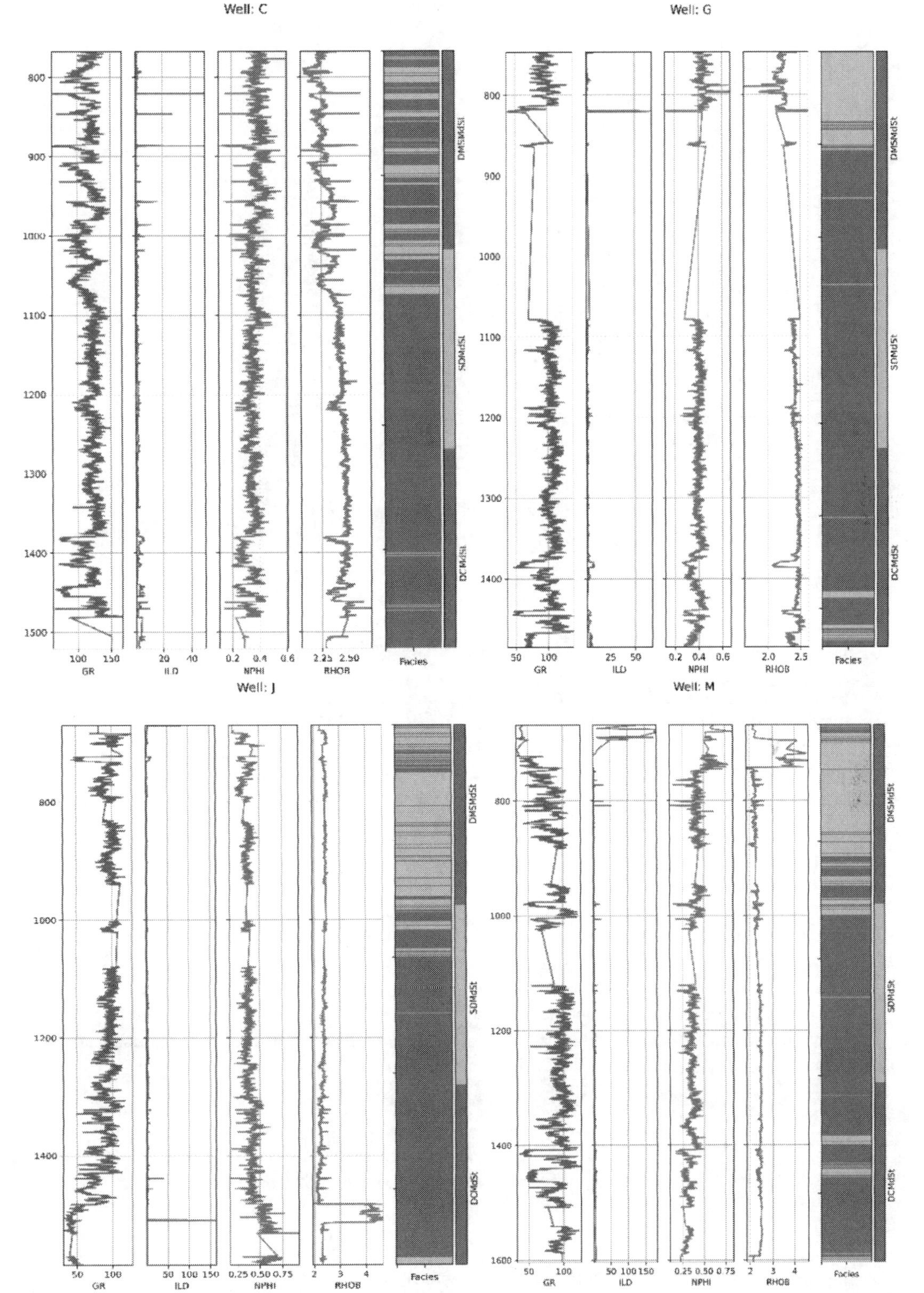

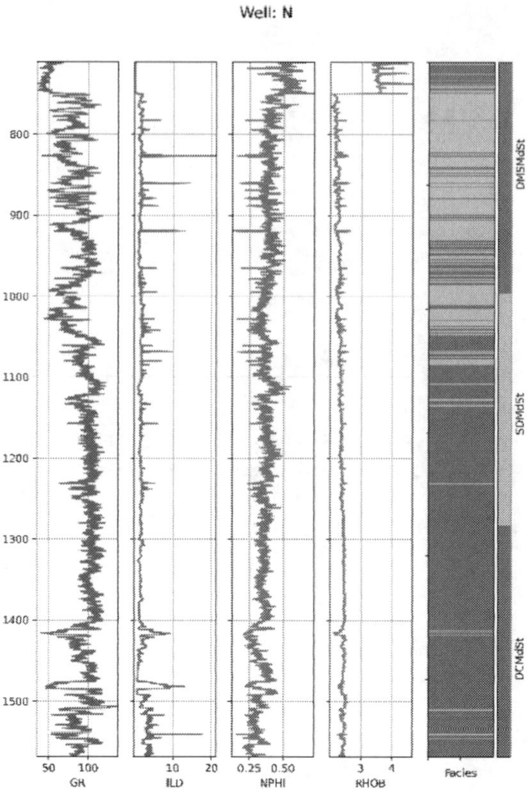

Figure 9. Predicting Labels to the Testing Data (C, G, J, M, N wells).

At the end of this research, classification techniques were used to the test data collected. The selection of test data is based on wells that need to have the same input data as the training and validation such as GR, ILD, NPHI, and RHOB. The wells used as the testing data are C, G, J, M, and N. The disadvantage of using this method is that it consists of numerous facies, therefore it tends to affect the accuracy. Therefore, the alternative solution is to reduce the amount of facies to obtain accurate values.

6. Conclusion

In conclusion, the gradient boosting classifier method help in determining depositional facies in any area of interest with limited core data. The input data used to determine it are GR, ILD, NPHI, and RHOB logs, while F1 score and confusion matrix were used as the accuracy techniques. The training and validation dataset without isolated faciesin scenario 3 showed an F1 score of 85.5 % on training and 84.7% on validation. This is in accordance with the result of the predicted labels to the test data, which shows the classification of depositional facies. The accuracy value is high, indicating that the gradient boosting classifier method effectively determines the process.

Acknowledgments

The authors are grateful to the Reservoir Geophysics, Department of Physics, Geophysics and GeologyStudy Program Universitas Indonesia for supporting this research.

References

[1] Sanchetta A C, Leite E P and Honório B C Z 2013 Facies recognition using a smoothing process through fast independent component analysis and discrete cosine transform. *Computers & Geosciences* **57**, 175-182

[2] MacQueen J 1967 Some methods for classification and analysis of multivariate observations. In *Proceedings of the fifth Berkeley symposium on mathematical statistics and probability* Vol. **1**, No. 14, pp. 281-297.
[3] Duda R O and Hart P E 1973 *Pattern classification and scene analysis* Vol. 3, pp. 731-739. New York: Wiley.
[4] Bestagini P, Lipari V and Tubaro S 2017 A machine learning approach to facies classification using well logs. In *Seg technical program expanded abstracts 2017* (pp. 2137-2142). Society of Exploration Geophysicists.
[5] Dev V A and Eden M R 2019 Formation lithology classification using scalable gradient boosted decision trees. *Computers & Chemical Engineering* **128**, 392-404.
[6] Catuneanu O, Galloway W E, Kendall C G S C, Miall A D, Posamentier H W, Strasser A and Tucker M E 2011 Sequence stratigraphy: methodology and nomenclature. *Newsletters on stratigraphy* **44**(3), 173-245.
[7] Schapire R E, Freund Y, Bartlett P and Lee W S 1998 Boosting the margin: A new explanation for the effectiveness of voting methods. *The annals of statistics* **26**(5), 1651-1686.
[8] Friedman J H 2001 Greedy function approximation: a gradient boosting machine. *Annals of Statistics*, 1189-1232.

IOP Conf. Series: Earth and Environmental Science **851** (2021) 012036 doi:10.1088/1755-1315/851/1/012036

Gas geochemistry analysis in Candradimuka Crater, Mount Lawu, Central Java, Indonesia

M F Bagaskara[1], A D Oktaviani[1], R Setyadi[1], D A Yoga[1], D N Sahdarani[1*], D T Kurniadi[1]

[1]Geology Study Program, Faculty of Mathematics and Natural Science, Universitas Indonesia, Kampus UI Depok, Depok 16424, West Java, Indonesia
*E-mail: dyahnindita@ui.ac.id

Abstract. Mount Lawu is a stratovolcano in Central Java that holds a large geothermal energy potential. Within Mount Lawu Geothermal Area, several thermal manifestations could be located one of them being Candradimuka Crater located at the proximity of Mount Lawu peak. This study aims to assess the gas geochemistry in Candradimuka Crater by obtaining gas sample within the crater area. Gas geochemistry analysis uses geothermometer, geoindicator, and N_2-He-Ar ternary diagram to analyse the subsurface condition. The analysis reveals that the fluid circulating in the crater originated from meteoric water and that Mount Lawu reservoir temperature ranged between 250°C - 289°C. Based on geological observation and gas geochemistry, Candradimuka Crater is located within the upflow zone of Mount Lawu.

1. Introduction

Indonesia holds a large amount of geothermal energy potential approximately around 28 GW energy or almost equal to 40% of the world's total geothermal energy potential [1]. The large sum of geothermal energy can be contributed to Indonesia's proximity to in the convergence boundary of several major tectonic plates, such as Eurasia, Indo-Australia, Philippine, and Pacific plate or also known as the Ring of Fire [2]. The convergence plate boundary resulted in active volcanic activity in the region, especially in Java Island of Indonesia. Volcanic formation stretches along the archipelago of Indonesia resulting in around 252 geothermal fields where 71 of the fields are located in Java Island [3]. Mount Lawu is an established Geothermal Working Area located in the border of Central Java and East Java (Figure 1). Mount Lawu contains 137 MW hypothetical energy and has been set as a Geothermal Working Area by Indonesia Ministry of Energy and Mineral Resources in 2012 [1].

Geological and geochemical investigations are used to evaluate a geothermal system. Geological investigations focus on determining lithology variation, stratigraphy, and structural geology in the area. Geochemical investigations focus on the chemical characteristics of the thermal manifestations that can provide an overview of the subsurface condition. These data are combined to get a better understanding of the geothermal system in the area [4].

Assessing the geothermal system of Mount Lawu using gas geochemistry investigation is the objective of this research. The gas sample is obtained from Candradimuka Crater fumaroles. Gas sample analysis is considered the most suitable method for this type of thermal manifestation (fumaroles) because fumaroles release an intense geothermal gas to the surface.

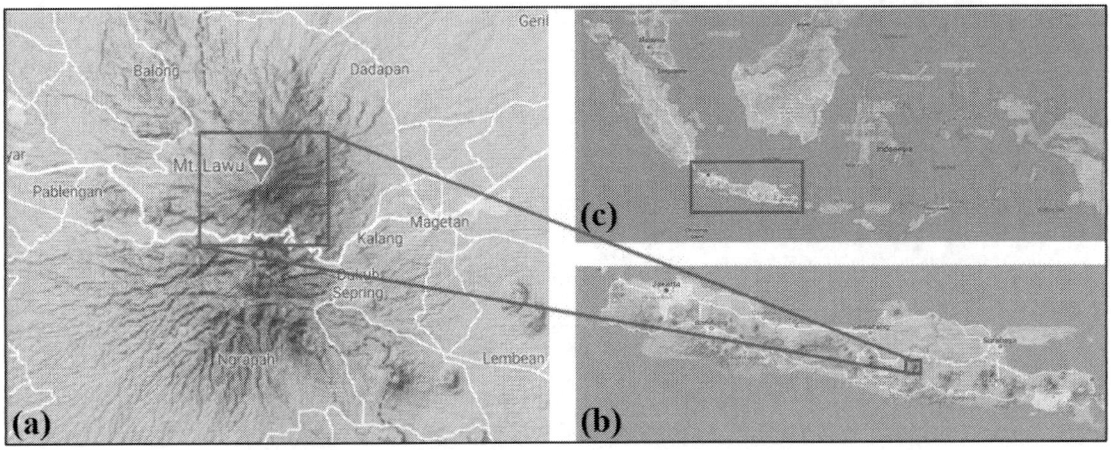

Figure 1. (a) Location of Candradimuka Crater, Mount Lawu, Java, (b) Location of Mount Lawu in Java, (c) Location of Java in Indonesia Archipelago.

2. Geological overview

Mount Lawu is an active volcano located in the border between Central Java and East Java. It still has an active geothermal system that is marked by some active thermal manifestations, such as fumaroles, hot pools, and hot springs. Some of the rock within the area has been altered [4]. The research area is primarily composed of 2 rock units, which are Candradimuka Lava and Lawu Volcanic Rocks. Candradimuka Lava consists of andesitic lava, while Lawu Volcanic composed of tuff and volcanic breccia intercalating with andesitic lava. Both of these rock units formed during the Holocene period [4,5].

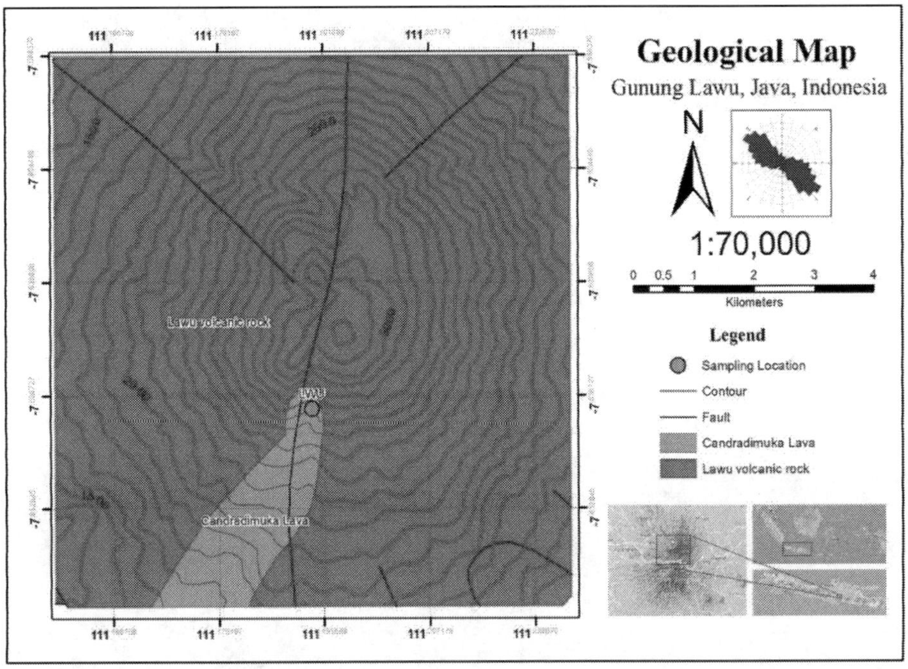

Figure 2. Geological Map of Mount Lawu (modified from [5] and Hermawan & Permana, 2018).

Based on regional geological map by [5], there are several structural features that could be observed. The major fault found in the area has NW-SE orientation as its dominant orientation (Figure 2). Other minor structure consists of faults trending N-S, E-W, and NE-SW. The appearance of thermal manifestation around Mount Lawu are presumably controlled by faults with most of the manifestations are located 200-700 meters not far from the faults [5,6].

Based on the water geochemistry analysis conducted by Sahdarani *et al.* (2020), thermal manifestations around Mount Lawu comprised of all different type of geothermal water such as acidic sulphate water and diluted chloride carbonate water. Some of the manifestation located on the western flank contains a high amount of Fe and As indicating an intense rock-water interactions. Mount Lawu has only one reservoir underlaying in the subsurface based on the Cl/B ratio comparison from different manifestations. The estimated reservoir temperature is around 160°C - 170°C based on the Na/K geothermometer calculation, therefore Mount Lawu is considered as an intermediate-high enthalpy system [4].

3. Research Methodology

The Candradimuka Crater fumarole gasses were sampled. The number of samples obtained from the crater is one sample with LWU as its sampling code. The sampling location is marked by the green plot on Figure 2. Before conducting gas sampling, the gas sample bottle needs to be prepared by vacuuming the bottle with 100 mL NaOH solution inside the flask. A stainless-steel funnel is needed to capture the gas from the crater. The funnel needs to be connected to the gas sample bottle by using heat proof silicone tube. The bottle must be placed inside a water filled bowl during the sampling process. The gas will start flowing to the bottle when the bottle's valve is opened.

The chemical components of the gas samples were analysed in Central Volcanology and Geological Hazard Mitigation Laboratory. The chemical components obtained from the analysis are CO_2, H_2S, He, H_2, N_2, Ar, and CH_4. These components are useful to evaluate the geothermal system.

Based on gas chemical analysis, several calculations on geothermometer and geoindicator were performed. The interpretations include N_2-He-Ar ternary diagram analysis. Geoindicator is used to indicate the flow zone of the manifestation to determine the upflow or outflow zone. Meanwhile, N_2-He-Ar ternary diagram is used to determine the fluid origin. Subsequently, we conducted geothermometer most appropriate to the chemical analyses result to calculate the reservoir temperature [7].

The most appropriate geothermometer analysis are H_2S, H_2, FT-CO_2, and CAR-HAR. CO_2 and CH_2/CO_2 cannot be used because the predicted reservoir temperature from the analysis is beyond the possible temperature range (Table 1). Moreover, CO_2/H_2 and H_2S/H_2 is considered unsuitable because these geothermometers are only producing valid number when the reservoir temperature is between 200°C - 300°C with chloride concentration above 500 ppm [8]. Chloride concentration in Candradimuka Crater is relatively low based on the Cl/B ratio [4], therefore CO_2/H_2 and H_2S/H_2 value are not valid.

4. Results and Discussion

Candradimuka Crater main fumaroles covering an area of approximately 10 m x 8 m. The fumarole temperature is around 90°C. The acidity level of the gas is around 3. Based on field observation, there were abundant of sulphur deposits that surrounded the gas outlet.

Figure 3. Candradimuka Crater fumaroles.

IOP Conf. Series: Earth and Environmental Science **851** (2021) 012036 doi:10.1088/1755-1315/851/1/012036

Fluid origin interpretated by using N_2-He-Ar diagram (Figure 4, red plot), revealing that Candradimuka Crater fluid originated from meteoric water. The geothermal fluid has N_2/Ar value of 54.67 where this value is ranged between the Air ratio composition ($N_2/Ar = 84$) and ASW ratio composition ($N_2/Ar = 38$).

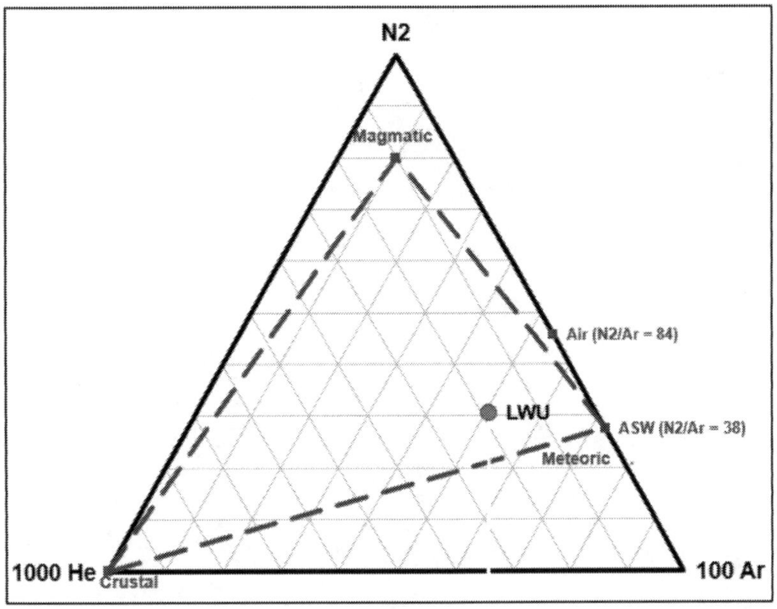

Figure 4. Fluid origin of Candradimuka Crater based on N_2-He-Ar ternary diagram.

Based on FT-CO_2, CAR-HAR, H_2S, and H_2 geothermometer analysis, Mount Lawu geothermal field reservoir temperature is around 250°C (Figure 5, red plot), 260°C (Figure 6, red plot), 289°C, and 264°C, respectively. It is considered as a high enthalpy system [9]. Besides reservoir temperature, the state of the fluid inside the reservoir can also be determined. Based on CAR-HAR geothermometer (Figure 6, red plot), the plotting result lies just below the equilibrated liquid line. It indicates that the reservoir is a liquid-dominated reservoir.

Table 1. Gas geothermometer analysis result.

Gas Geothermometer (in °C)							
CO_2	H_2S	H_2	CO_2/H_2	H_2S/H_2	CH_2/CO_2	FT-CO_2	CAR-HAR
340	289	264	270	241	530	250	260

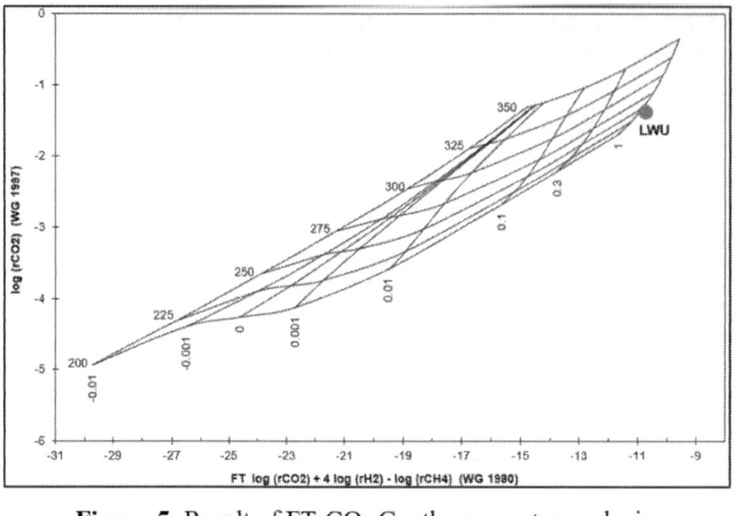

Figure 5. Result of FT-CO_2 Geothermometer analysis.

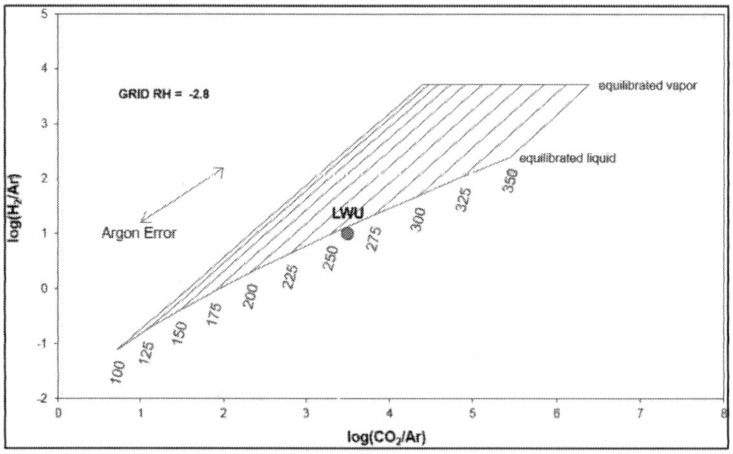

Figure 6. Result of CAR-HAR Geothermometer analysis.

Candradimuka Crater is located in the upflow zone based on the manifestation type, temperature, and gas geochemical analysis (CO_2/H_2S ratio). The fumaroles emits distinguishable sulphuric gas at 90°C which signified an upflow zone [4,7]. The findings of CO_2/H_2S ratio in Candradimuka Crater is calculated at 8.32, which is considered as a relatively low value. The lower CO_2/H_2S ratio indicates the less interaction between the geothermal gas and the host rock. This minimum interaction shows that the migration of the gas is not far away from the magmatic source [7], so it concluded that Candradimuka Crater is in the upflow zone.

As it is in the upflow zone, Candradimuka Crater produces numerous amounts of gas. The gas is formed when there is a reduction of the confining pressure in the geothermal fluid which makes the fluid boiled. The boiling fluid then separated into two phases, i.e., liquid and vapour phase. Then, the vapour tends to migrate vertically along the upflow zone, while the liquid phase tends to migrate horizontally along the outflow zone [7]. This characteristic makes Candradimuka Crater able to produce a massive amount of gas with the help of fractures as its pathway.

In the case of Candradimuka Crater, gas analysis is more reliable than water analysis. The lightweight properties of gas drives gas migration along the upflow path and will be less likely to be mixed with other substances, such as groundwater. Therefore, gas chemical characteristic has more retention and can reflect the reservoir characteristics. However, the reservoir type also has a role in determining which analysis should be used. In liquid dominated reservoirs, water analysis is more reliable compared to gas analysis, because the thermal manifestation found in geothermal area tends to be a hot spring or hot pool

IOP Conf. Series: Earth and Environmental Science **851** (2021) 012036 doi:10.1088/1755-1315/851/1/012036

type. Furthermore, based on statistic reasoning, the small number of manifestations appearance makes the data less reliable [10]. The existence of water-type manifestation has a greater number than the gas-type manifestation in Mount Lawu. In conclusion, gas analysis is less reliable than water analysis in Mount Lawu because of the liquid-dominated reservoir and the appearance of only one fumarolic manifestation area which located in Candradimuka Crater.

5. Conclusion

Mount Lawu fluid originated from meteoric water. Based on the geothermometer analysis, the reservoir temperature of Mount Lawu is approximately at 250°C - 289°C and can be classified as a high enthalpy system. The reservoir of the system is a liquid-dominated reservoir. Candradimuka Crater is within the upflow zone of the Mount Lawu geothermal system. Candradimuka Crater presence within the upflow zone makes it emits gas in a large amount. However, to have a better understanding of Mount Lawu geothermal system, water geochemistry analysis is more reliable than gas geochemistry analysis due to the phase of the reservoir and type of manifestation.

Acknowledgments

This research is supported by the funding from Dana Hibah Publikasi Terindeks Internasional (PUTI) Prosiding 2020 Nomor: NKB-1032/UN2.RST/HKP.05.00/2020.

References

[1] Ministry of Energy and Mineral Resources. Indonesia Geothermal Potential Book 1. Jakarta: Ditjen EBTKE; 2017.
[2] Darman H, Sidi FH. An Outline of The Geology of Indonesia. Jakarta: Ikatan Ahli Geologi Indonesia; 2000.
[3] Suharmanto P, Fitria AN, Ghaliyah S. Indonesian Geothermal Energy Potential as Source of Alternative Energy Power Plant. KnE Energy. 2015;1(1):119–24.
[4] Sahdarani DN, Ponka MA, Yoga DA, Kurniadi DT. Hydrogeochemical Characteristics of Geothermal Water in Mount Lawu, Indonesia. In: *Proceedings 42nd New Zealand Geothermal Workshop*; 2020; Waitangi. New Zealand: University of Auckland.
[5] Sampurno, Samodra H. Geological Map of The Ponorogo Quadrangle, Java. Bandung: Geological Research and Development Centre; 1997.
[6] Helmi H, Kurniawan H, Adam W. Kontrol geologi terhadap permunculan manifestasi panas bumi di kawasan gunung lawu. KURVATEK. 2020;5(1):1–9.
[7] Nicholson K. *Geothermal Fluids*. Berlin, Heidelberg: Springer Berlin Heidelberg; 1993.
[8] Ping Z. Gas geothermometry and chemical equillbria of fluids from selected geothermal fieids. 1991;1–46.
[9] Moeck IS. Catalog of geothermal play types based on geologic controls. Renew Sustain Energy Rev. 2014;37:867–82.
[10] Marini L. Geochemical techniques for the exploration and exploitation of geothermal energy. Genova: University of Genova; 2000.

IOP Conf. Series: Earth and Environmental Science **851** (2021) 012037 doi:10.1088/1755-1315/851/1/012037

Hummock size and alignment in Gadung debris avalanche deposit, Raung Volcanic Complex, East Java, Indonesia

M L A Moktikanana[1], H E Wibowo[1*], E Rahayu[1], and A Harijoko[1,2]

[1]Department of Geological Engineering, Faculty of Engineering, Universitas Gadjah Mada. Jalan Grafika no 2, Yogyakarta 55281, Indonesia

[2]Center of Disaster Studies, Universitas Gadjah Mada. Jalan Mahoni C 16, Yogyakarta 55281, Indonesia

*E-mail: haryo.edi.w@ugm.ac.id

Abstract. Debris avalanche deposit is formed by the failure of volcanic edifice and has a high potential to directly impact human civilization. The famous characteristic landform of debris avalanche deposit is hummocky hills. To understand the dynamics of avalanche flow, we investigate the morphometric characters of debris avalanche hummocks from Gunung Gadung in Raung Volcanic Complex. The collapse of Gunung Gadung follows two main flow direction with two different kinds of debris avalanche, i.e., freely spreading and valley filling. Our study recorded that there is no significant correlation between hummock size and distance from the source. The distribution of Gadung hummock size is mainly controlled by pre-existing morphologic feature such as Iyang-Argapura Volcanic Complex and Meru Betiri Mountains. We identify four domains area of Gadung debris avalanche flow based on hummock size distribution, which are Toreva domain, Hummock domain, Collision domain, and Oblique collision domain. Hummock orientation and displacement angle of Gadung debris avalanche deposit depend on structural regime existed in certain area. Compressional regime may occur because of collision between Gadung avalanche flow and Iyang-Argapura Volcanic Complex, or shifting of main flow direction. Meanwhile, extensional regime may happen due to decreasing in slope and spreading of avalanche flow.

1. Introduction

Sector collapse is a large-scale landslide of a volcanic edifice that has a high potential to damage human society directly through a debris avalanche or a secondary catastrophe such as lahars and tsunami [1, 2, 3, 4, 5, 6]. Volcanic debris avalanche, as the primary product of sector collapse, could bring few thousand to several million cubic rocks transported in >100 m/s velocity with run outs up to 70 km [7, 8, 9]. Debris avalanche can also produce debris flow due to increasing water content from water incorporation of snow or ice melt [10, 11, 12, 8] or as a result of the breaching of lake-induced debris avalanche deposit [13, 8]. Sector collapse may have more devastating impact if associated with water body such as lake or sea, as occurred at Anak Krakatau Volcano, Indonesia. A small-scale sector collapse of Anak Krakatau in December 2018 has triggered a tsunami that killed almost 400 people [14].

Hummocks are the most prominent characteristic landform of debris avalanche deposit. These conical shaped hills usually associated with the volcanic collapse scars [15, 16, 17, 18, 6]. Hummocks were formed during the failure of volcanic body. The landslide body of volcanic masses are separated from each other and forms hundred to thousand conical hills depending on the avalanche volume [18]. Hummocks may be characterized based on the basal shape, surface morphology, and orientation with respect to flow and slide direction. Hummock elongation can be parallel, perpendicular or randomly oriented according to the deformation resulting from the flow direction [19]. The geomorphological characters of hummocks might be a key to understanding the formation and of debris avalanche and the dynamics of the avalanche flow.

The forementioned conical morphology was found around western part of Raung Volcanic Complex, together with the collapse scars that form an elongated horseshoe-shaped caldera opening to the west. The collapse scars have 14 km long and 3 km wide. The hilly area of hummocks located west-southwest of the horseshoe-shaped caldera and covered approximately more than 1000 km^2 area in the west of Gunung Gadung [7]. This represents a largest caldera collapse that has ever happened in Indonesia and could turn into a potentially disastrous volcanic catastrophe. Therefore, the study of volcanic sector collapse especially in Raung Volcanic Complex is really important to understand the emplacement mechanism as an attempt of disaster risk reduction.

In this study, we examine the geomorphological significance of debris avalanche hummocks of Gunung Gadung in Raung Volcanic Complex to understand the dynamics of avalanche flow through quantitative analysis. Similar studies had been done for some volcano in Japan by [5, 6, 20, 21, 22]. Previous studies on debris avalanche hummocks by [7, 16, 23, 24, 25] mainly focused to hummock morphology as evidence of sector collapse.

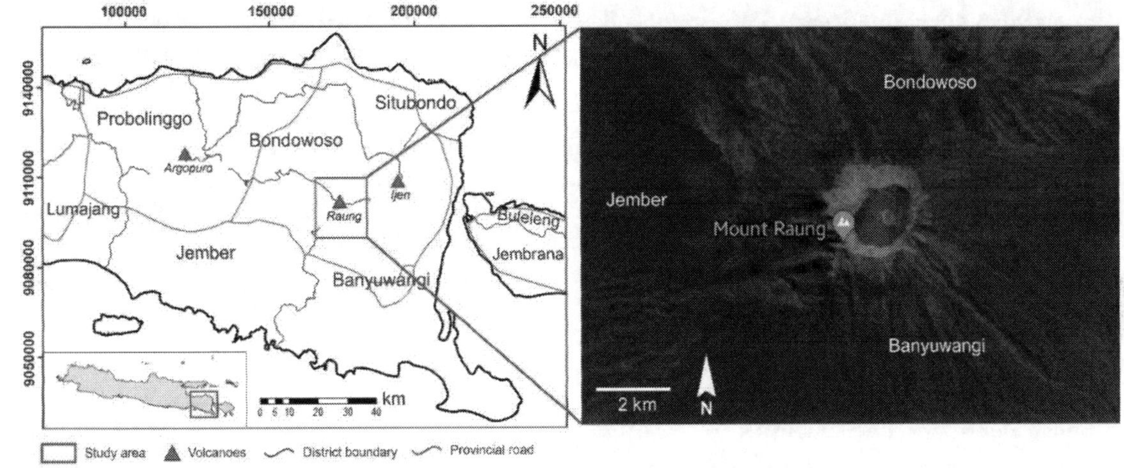

Figure 1. Location of study area.

2. Geology of Raung Volcanic Complex

Raung volcano (3332 masl) is one of the most active Quaternary volcanoes located on the border of 3 regencies in East Java, Indonesia, which are Jember, Banyuwangi, and Bondowoso (Figure 1). The morphology of Raung volcano is represented by a 1.75 x 2.25 km wide and 400-550 m deep caldera, indicating the enormity of highly explosive eruption that has been occured in Raung history. Raung volcanic complex consists of Gunung Raung (3332 masl), Gunung Suket (2750 masl) on the northern part, Gunung Lempeh (2932 masl) on the northeastern part, Gunung Jampit (2338 masl) on the eastern part, Gunung Wates (2796 masl) on the southern part, Gunung Gadung (2290 masl) and Gunung Pajungan (2352 masl) on the western part.

The geology of Raung volcano has been reported by [26] on a 1: 100,000 scale geological map. Raung volcanic activity consists of 3 periods, from the oldest to youngest there are Old Raung Volcanism, Gadung Volcanism, and Young Raung Volcanism. Raung volcanic products generally composed of lava flows, pyroclastic flow deposits, pyroclastic fall deposits, and lahar with cinder cones present on the western flank and peak. Historical eruption records show that Raung volcano is characterized by the explosive eruption mainly producing volcanic ash and pyroclastic flow. The first activity of Raung volcano has been recorded in the human history since the 16th century (1586), and the last activity of Raung has also been recorded since January 2021. Some paroxysmal eruption documented in 1586, 1597, 1638, 1890, 1953, and 1956 producing an eruption column up to 12 km high, with ashfall distribution reaching a radius of up to 200 km, covering Bali and Surabaya [27]. Eruption record since 1902 shows that the shortest hiatus of Raung volcano activity is 1 year, while the longest hiatus is 15 years [28]

[7] and [24] studied the debris avalanche-sector collapse deposit on Raung Volcanic complex and mentioned that it was the largest known debris avalanche in Indonesia. Based on their research, the debris avalanche deposit extended 60 km from its source on Gunung Gadung, 4 km to the west from the Raung center caldera rim. The latest study has traced the deposit to 79 km from the source [28]. Referring to Java physiographic zone by [29], Raung debris avalanche deposit located on the central depression zone of Java, namely Solo Depression Zone. The debris avalanche deposit is bordered by Iyang-Argapura Volcanic Complex on the western part, Meru Betiri Mountains on the southern part, and Kendeng Anticlinorium Zone on the northern part. The deposits filling a broad valley between Iyang-Argapura Volcanic Complex and Meru Betiri Mountains.

The Gadung debris avalanche deposit has a proximal area extending up to 30 km from the source towards the Jember, consists of abundant hummocks to proximately 75 m in height. On the distal area, the height of each hummock decreased to ±20 m, fewer in size, and more widely scattered [7]. There are no new magmatic material present in the avalanche deposit which indicates that the deposit was produced by non-eruption associated sector collapse [7]. In some areas, a thin unit at the top of hummock resembles a lateral-blast deposit [7] but was not discovered in the interhummock area or outside the avalanche deposit. Post collapse eruptions had construct Pajungan volcano, a new conical-shaped stratovolcano within the horseshoe-shaped caldera [7].

At the proximal area, the hummock deposits are consisted of various segments of lava flows in wide stage of fragmentation and thick sequences of half-litified rock, covered by tephra layers up to tens centimeter thick. Mixed facies are commonly located in distal portion of the deposit that deflected to the south by Iyang-Argapura Volcanic Complex. These facies usually contain subangular clast in fine matrix deposits. At the distal area, hummocks are mainly consist of stratified tephra layers as found at proximal area [7].

3. Methods

We conducted morphological analysis on the Gunung Gadung debris avalanche deposit using DEMNAS digital elevation model published by The Indonesian Geospatial Information Agency. The digital elevation models were processed using multi-directional hillshade visual effect to produce terrain map. Multi-directional hillshade was created using 8 light azimuths (0 °, 45 °, 90 °, 135 °, 180 °, 225 °, 270 °, and 315 °) with 30 ° altitude. We trace the outline of each hummock on plan view from the terrain map, referring to DEMNAS, World Hillshade by ESRI, and Aerial Photographs on Google Maps. DEMNAS data has a resolution of 8.5 m with *Root Mean Square Error* of 2.79 m and -0.13 m bias error, while the ESRI World Hillshade has a resolution up to 25 m in Indonesia, specifically. The parameter of hummock morphometry are examined, such as size of each hummock, major axis, minor axis, elongation ratio, distance of individual hummocks from the sector collapse source, flow direction, and displacement angle. We refer for each hummock parameter to the previous study by [5, 6, 20, 22, and 30].

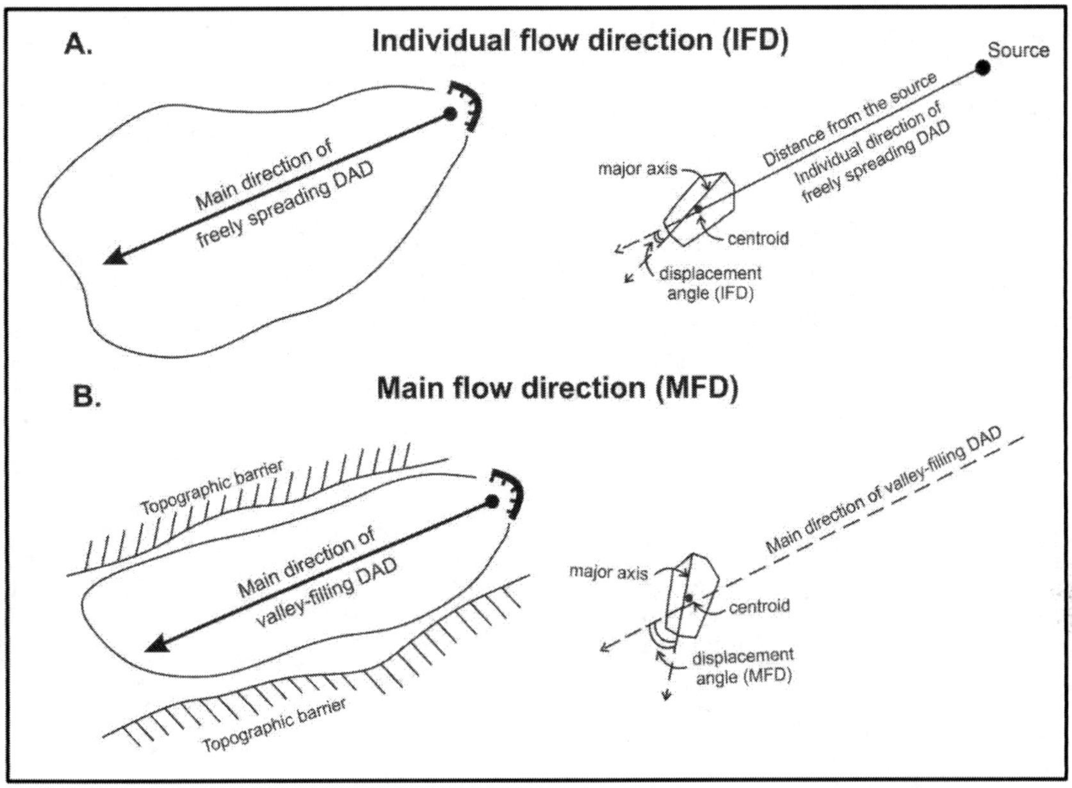

Figure 2. Morphometric parameter of debris avalanche hummock for (a) freely spreading and (b) valley-filling debris avalanche deposit, adapted from [20]. DAD refers to "debris avalanche deposit".

We define the hummock size by the boundaries formed by the hummock outline. Major axis is the longest segment that connecting two vertices of hummock polygon while minor axis is the shortest segment (Figure 2). Both major and minor axis were calculated to get the elongation ratio, which is the ratio between major axis to the minor axis [22] Distance from the source is the length of a horizontal line between the source and the center of each hummock (Figure 2).

The collapse of Gadung edifice follows two main flow direction. The closest flow from the source is MFD 1, categorized as "freely spreading debris avalanche" with the main direction of N273°E. 20 km from the source, the main flow direction shifted to the southwest in the direction of N241°E due to the presence of topographical barrier, which are Iyang-Argapura Volcanic Complex to the west and Meru Betiri Mountains to the south. It is called as "valley filling debris avalanche", because the flow had been nearly enclosed by surrounding topography. The valley filling debris avalanche is then defined as MFD 2 (Figure 3).

We divide the research area into 3 zones based on the main direction of debris avalanche flow (Figure 3). The area with freely spreading debris avalanche affected by MFD 1 is called MFD 1 zone, while the one affected by MFD 2 with valley filling type is referred as MFD 2 zone. Right in the location where the flow direction change, there is a zone called MFD transition zone. This zone has not experienced direct contact with the topographic barriers; therefore, it is still considered as freely spreading debris avalanche and affected mostly by MFD 1.

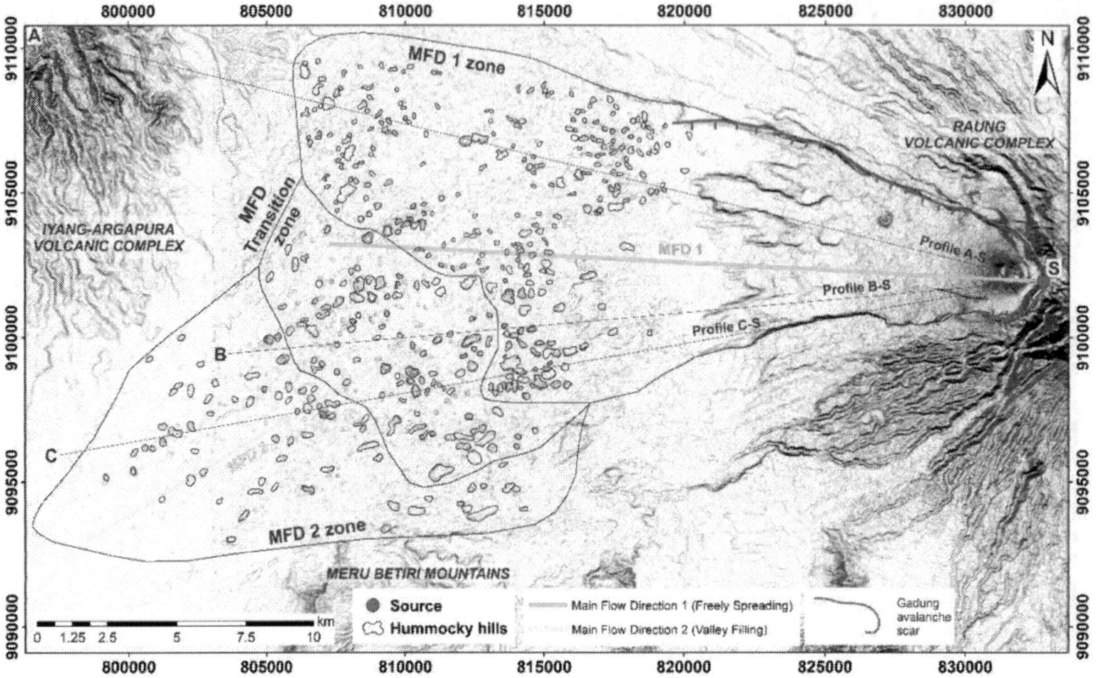

Figure 3. Detailed map of Gadung debris avalanche deposit showing main flow directions, hummocks delineation, and MFD zones division.

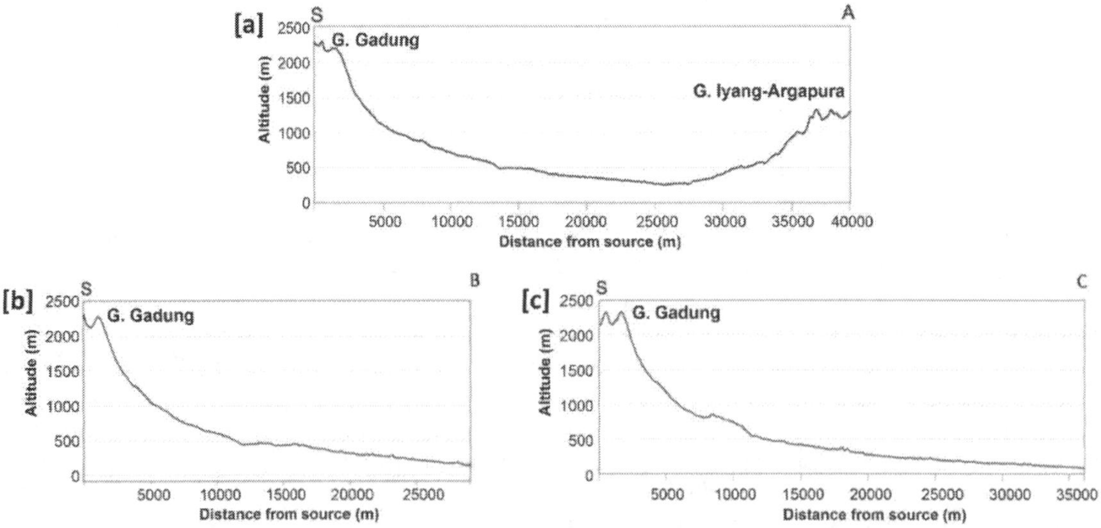

Figure 4. Topographic profile of Gadung edifice along the (a) MFD 1 Zone, (b) MFD Transition Zone, and (c) MFD 2 zone.

The determination of the displacement angle is carried out in different ways for each different type of debris avalanche. For freely spreading debris avalanche such as in MFD 1 zone and MFD Transition zone, the displacement angle is calculated using "individual flow direction" (IFD) or the orientation connecting the center of each hummock to the center of the source (Figure 2). The displacement angle for the freely spreading debris avalanche is defined as the angle between the major axis of each hummocks and the IFD. For valley filling debris avalanche such as in MFD 2 zone, displacement angle is calculated using the angle between the major axis of each hummocks and the main flow direction (Figure 2). The extraction of morphometric data was performed on a total of 428 hummocks: 266 hummocks from MFD 1 zone, 92 hummocks from MFD transition zone, and 70 hummocks from MFD 2 zones. A total of 3 topographic profiles were made from the avalanche source, each passing through the MFD 1, MFD Transition, and MFD 2 zones in order to provide an overview of the slope conditions and its effect on the sector collapse mechanism (Figure 4). The processes mentioned above are performed using Software ArcGis 10.4, ArcGis Pro 2.0, and Global Mapper 19.

4. Results

4. 1. Hummock size and distribution

We selected the most representative data for hummock morphometry based on the elongation ratio. Elongation ratio is the comparative value between hummock's major axis and the minor axis. Figure 5 shows the positive correlation between major axis and elongation ratio. Larger major axis and elongation ratio represents more elongated hummock shape, while lower major axis and elongation ratio indicates more equant hummock shape. Data selection was conducted in order to eliminate hummocks with relatively equant shape because of their uncertainty in orientation and displacement angle. We picked the more elongated hummocks with elongation ratio greater than 1, so that the data really represents the accurate hummock size and orientation.

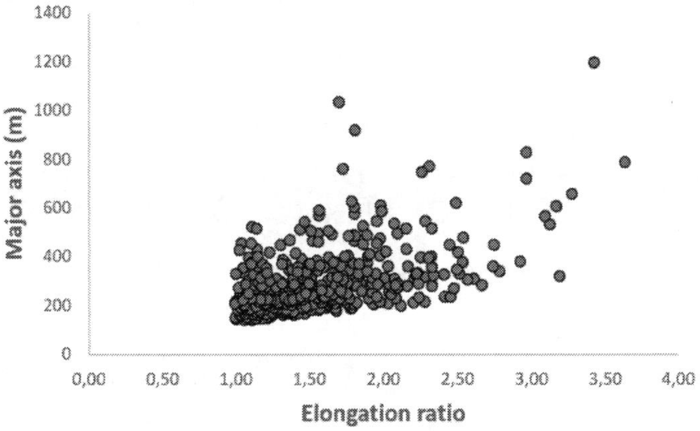

Figure 5. Relation between major axis and elongation ratio of Gadung hummocks shows positive correlation.

The distribution of hummock size in relation to distance from the source has a fairly wide range. The hummock area on MFD 1 zone, located between 12000 to 27000 m from the source, ranges between 14000 to 340000 m^2. On the MFD Transition zone, which is 18000 to 27000 m from the source, the hummock area ranges from 26000 to 310000 m^2. Meanwhile, on the farthest area or MFD 2 zone, the hummock area ranges between 27000 to 302000 m^2 at a distance of 20000 to 35000 m from the source.

Bivariate plot of hummock area and distance from the source not showing any downsizing or enlargement in hummock size, or the positive or negative trend of hummock size with increasing distance from the source (Figure 6). This may be due to the rapid depositional processes of high-flow velocity and dry debris avalanche, so that no significant sorting actually happens. However, it should be noted that the debris avalanche flow velocity does not depend only on the distance from the source. The conditions of pre-existing morphology such as topographic barrier or pre-depositional slope also plays a role on flow velocity. The effect of the pre-existing morphology on flow velocity cannot be pictured only from simple bivariate plot, therefore, we created a hummock size distribution map to evaluate the effect of pre-existing morphology on hummock size.

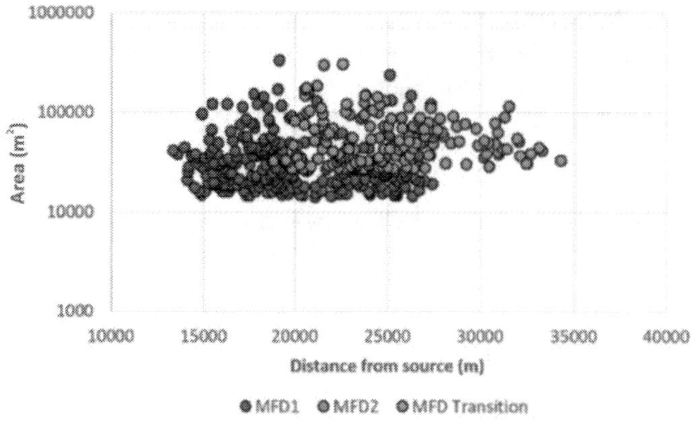

Figure 6. Bivariate plot of hummock area and distance from the source is not showing any significant relation between hummock size and increasing distance.

The hummock size is classified into medium and big based on cumulative probability plot for each zone (Figure 7). Hummocks are considered to be big size if the hummock area exceeds >52000 m^2 in MFD 1 zone, >62000 m^2 in MFD Transition zone, and >100000 m^2 in MFD 2 zone. Hummocks with area below those values are considered as medium size. The hummock size data is then interpolated with Natural Neighbor method to generate the hummock size distribution map (Figure 8).

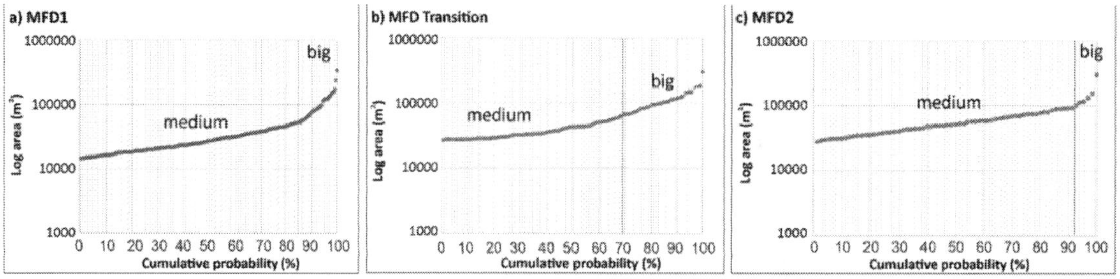

Figure 7. Cumulative probability plot of hummock area to determine the hummock size class for each MFD zone.

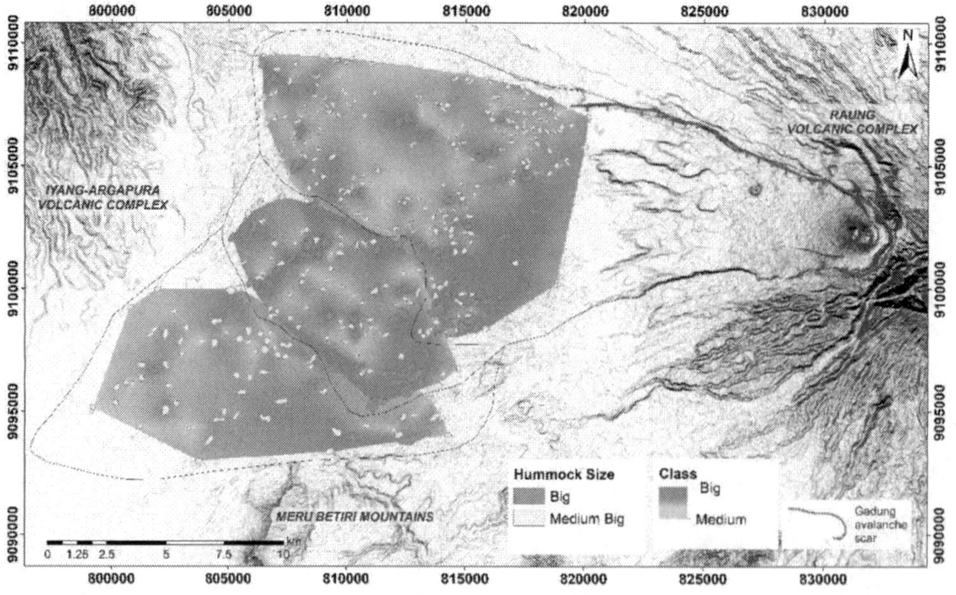

Figure 8. Hummock size distribution maps of Gadung debris avalanche deposit.

4. 2. Hummock alignment and displacement angle

The rose diagram of MFD 1 zone hummock alignment shows that the dominant hummock orientation is in the same direction as the main flow direction (Figure 9a). Meanwhile on the MFD Transition zone and MFD 2 zone, the dominant orientation of hummock is not the same as the main flow direction, even some are aligned perpendicular to the main flow direction (Figure 9b-c). Thus, it can be concluded that hummock alignment is not always exactly in the same direction with the main flow direction.

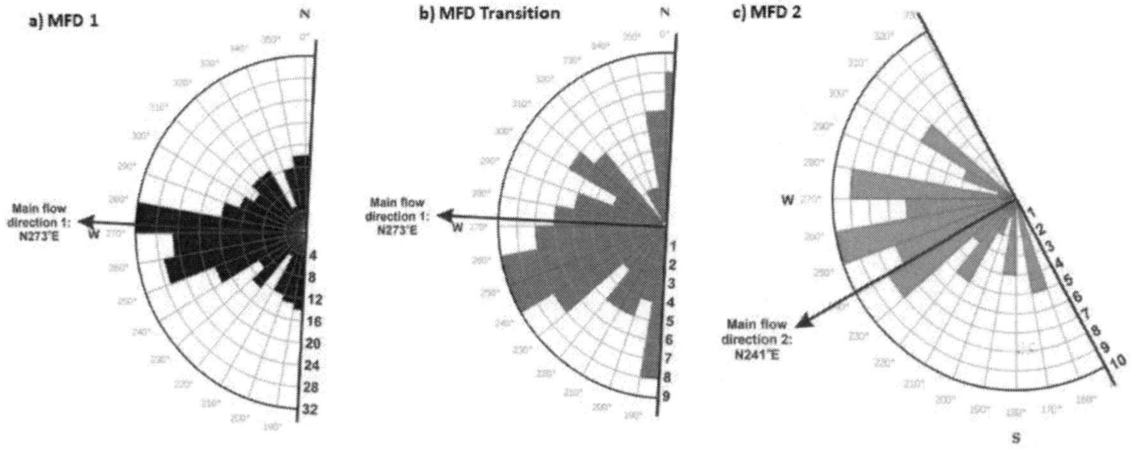

Figure 9. Hummock alignment rose diagram from (a) MFD 1 zone, (b) MFD transition zone, and (c) MFD 2 zone of Gadung debris avalanche deposit.

[20] and [30] studied the change in hummock alignment in relation to debris avalanche path in Japan by using displacement angle parameter. The displacement angle lower than 45° is considered as parallel to MFD, while the one with value greater than 45° is defined as perpendicular to MFD. Bivariate plot of the hummock displacement angle versus distance from the source shows a broad dispersal of data, suggesting that the displacement angle varies from parallel to perpendicular either at the close or far distance (Figure 10).

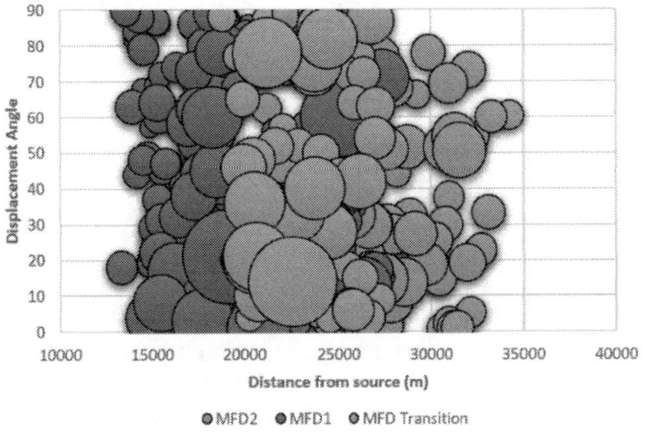

Figure 10. Hummock alignment rose diagram from (a) MFD 1 zone, (b) MFD transition zone, and (c) MFD 2 zone of Gadung debris avalanche deposit.

To examine the correlation between distance from the source, debris avalanche type, and displacement angle more clearly, we created a bivariate plot of the average displacement angle versus distance from the source for each MFD zone (Figure 11). The average displacement angle was calculated for 2000 m spacing.

IOP Conf. Series: Earth and Environmental Science **851** (2021) 012037 doi:10.1088/1755-1315/851/1/012037

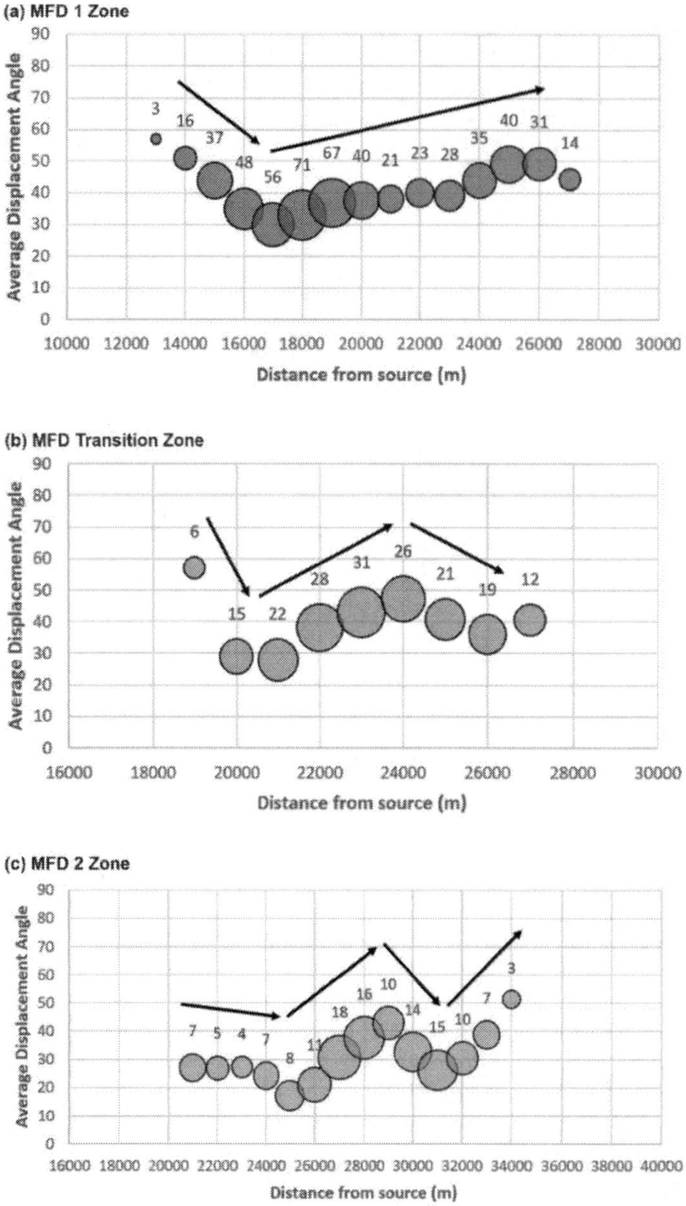

Figure 11. Bivariate plot of hummock average displacement angle and distance from the source of (a) MFD 1 zone, (b) MFD Transition zone, and (c) MFD 2 zone.

Further explanation about the changes in the average displacement angle for each zone is described as follows.

4. 2. 1. MFD 1 zone. It covers the area of freely spreading debris avalanche which is bounded to the west by Iyang-Argapura Volcanic Complex. Change in displacement angle in this zone coincides with the change in slope. At 13000-17000 m from source, the hummock alignments are perpendicular to parallel with MFD, slope constant to decreasing. At 17000-25000 m from source, hummock alignments shift from parallel to perpendicular with MFD along with the change of slope from decreasing to increasing.

4. 2. 2. MFD Transition zone. Within this zone, there is a shifting in main flow direction to the southwest. The slope is constantly decreasing but the displacement angle undergone several changes.

At 19000-20000 m from source, the hummock displacement angle is perpendicular to parallel. At 20000-24000 m from source, hummock displacement angle changes from parallel to perpendicular, and from 24000-26000 m the displacement angle returns to parallel.

4. 2. 3. MFD 2 zone. It covers the areas of valley filling debris avalanche, bordered by the topographic barrier of Iyang-Argapura to the north and Meru Betiri to the south. The variation in average displacement angle in respect of distance from source shows an irregular pattern with constantly decreasing slope. At 19000-24000 m from source, the displacement angle is constantly parallel, then begins to decrease at 24000-25000 m. The displacement angle changes from parallel to perpendicular at a distance of 25000-29000 m. At 29000-31000 m, displacement angle shifts again from perpendicular to parallel, then returns to parallel to perpendicular at 31000-34000 m.

Similar to hummock size, the hummock orientation also depends on the pre-existing morphology which cannot be depicted using only a simple bivariate plot. Therefore, we made a displacement angle distribution map to determine the impact of pre-existing morphological conditions to hummock alignment (Figure 12).

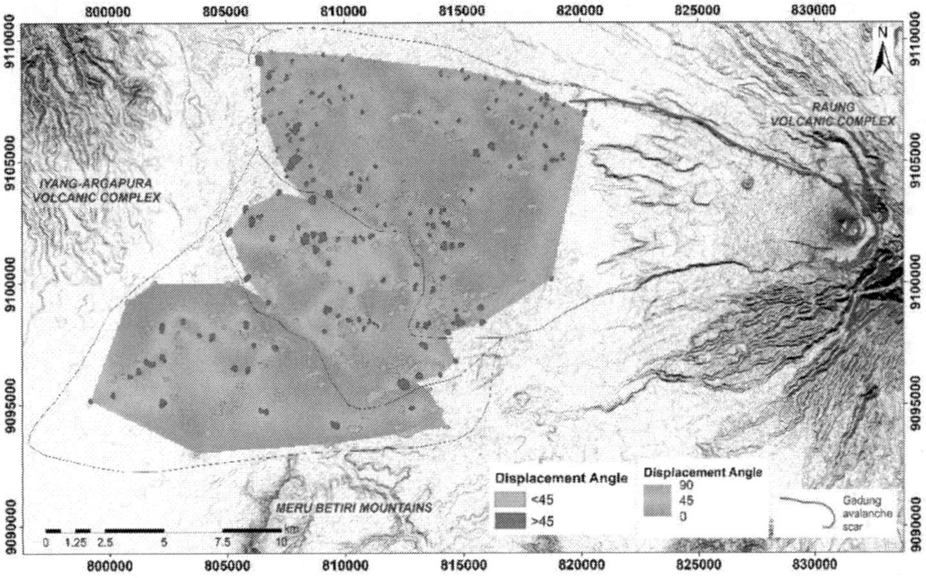

Figure 12. Hummock displacement angle distribution maps of Gadung debris avalanche deposit.

5. Discussion

5. 1. Relation between hummock size and pre-existing morphology

The volcanic debris avalanche analog model proposed by [31] and [19] divides the failure zone into 2 domains. The upper sliding flank is called Toreva domain, which is the initial part of flank collapse where normal faults are formed and produce huge blocks. On the lower sliding flank, fragmentation of Toreva blocks occur due to the avalanche continuing to spread, forming more abundant normal and transtensional faults so that the hummock is broken into smaller size. These two zones are also observed in Gadung debris avalanche.

We identify the presence of four domains area of Gadung debris avalanche deposit based on the hummock size distribution map (Figure 13). The closest area from source is "**Toreva domain**" where relatively larger hummocks are clustered in the proximal zone. The large hummocks of Toreva domain were created from Toreva blocks that spread outwards, experience extension force, and teared apart into smaller blocks. Behind Toreva domain we have "**Hummock domain**" extending from north to the south, which is a region where fragmentation happens more intensively, hence generating smaller hummocks than those of the Toreva domain. In the area bordered directly by Iyang-Argapura Volcanic Complex, there is a group of large hummocks namely **"Collision domain"**. This domain is formed by the collision of Gadung debris avalanche flow with Iyang-Argapura Volcanic Complex in the west. The debris avalanche flow velocity drops dramatically due to collision, causing hummocks to merge around each other. Outside the border of the Gadung Amphiteather in the south, there is one region where large hummocks also accumulate. This area is called "**Oblique collision domain**", which is influenced by Meru Betiri Mountains. The presence of Meru Betiri Mountains in the south causes an oblique collision with the flow of Gadung debris avalanche, creates friction force, allowing the debris avalanche flow velocity to slow down so that fragmentation did not occur.

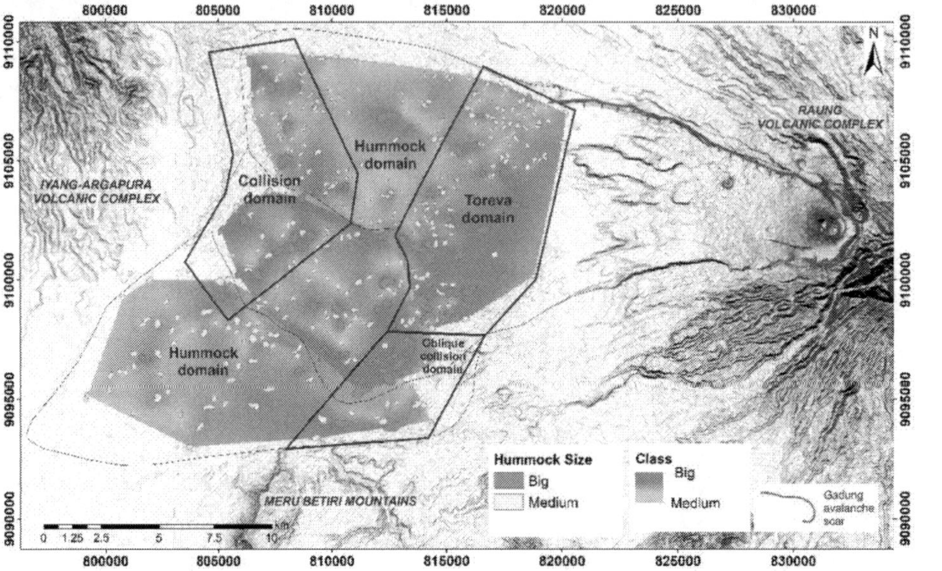

Figure 13. The four domains area based on hummock size distribution maps of Gadung debris avalanche deposit consists of (1) Toreva domain, (2) Hummock domain, (3) Collision domain, and (4) Oblique collision domain.

5. 2. Relation between hummock alignment and pre-existing morphology

The Hummocks formation model in volcanic setting proposed by [19] elaborates that when an avalanche starts to transport, hummocks initially form through extensional faulting. The avalanche materials are forced to spread further as they slide down from volcano edifice. Depending on the condition of topography and morphology, the traveling down of avalanche material may result in compressional or extensional structures. Extensional regime may be generated in an open depositional environment, while compressional regime may be established if there are any topographic barriers either parallel or perpendicular to the flow direction

Hummock orientation and displacement angle depend on whether an area is encountered a compressional or extensional regime. Shifts in displacement angle from parallel to perpendicular represents a change in structural regime from extensional to compressional. On the other hand, displacement angle changing from parallel to perpendicular indicates that the structural regime shifts from compressional to extensional [20]

The bivariate plot between average displacement angle and the distance from the source is correlated with the displacement angle distribution map in order to observe the relationship between hummock alignment and morphology. We can divide the displacement angle distribution map into several zones (Figure 14). Red colored zone indicates the dominance of the perpendicular displacement angle with the compression regime. Green colored zone shows the dominance of parallel displacement angle with extensional regime.

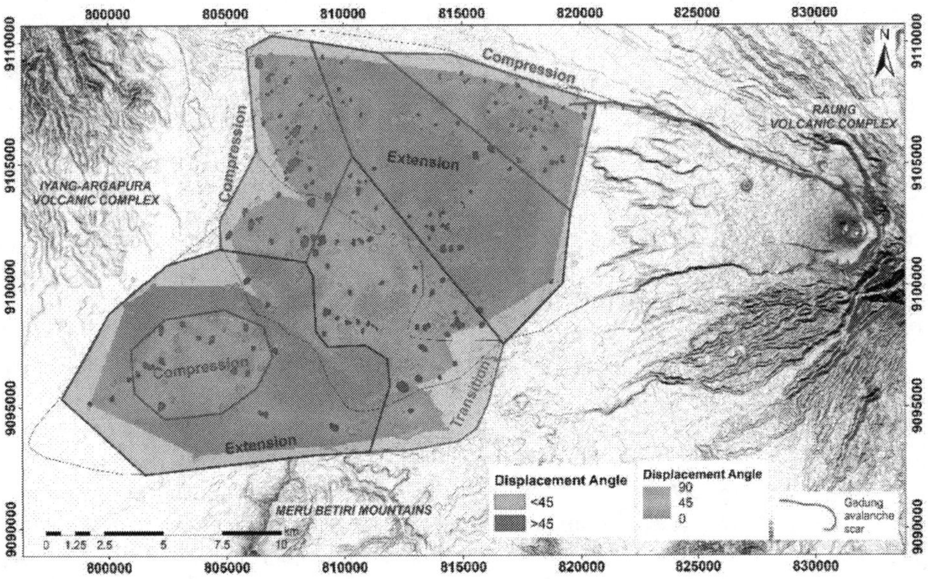

Figure 14. The distribution of compression and extension regime from Gadung debris avalanche deposit based on the displacement angle distribution map.

5. 2. 1. Hummock alignment in MFD 1 zone. On the displacement angle distribution map, this zone is shown by a change from compression to extension regime, and then the structural regime changes again to compression on the distal part. This condition is also correlated with the plotting of the average displacement angle versus distance from the source. At a distance of 13000-17000 m from the source, the hummock displacement angle changes from perpendicular to parallel. It shows a change in structural regime from compression to extension. Extensional regime dominates this section due to constant to decreasing slope and the avalanche flow spreads both downslope and laterally. The displacement angle begins to change from parallel to perpendicular at 17000-25000 m. Change in structural regime also occurs from extension to compression. Compression may happen as the avalanche flow is confined by the Iyang-Argapura in the west and causes a significant drop in flow velocity due to the change in slope from decreasing to increasing.

5. 2. 2. Hummock alignment in MFD Transition zone. There is a shift of main flow direction in this transition zone. No dominance of compression or extension regimes because the distribution is relatively similar. Therefore, this zone is also defined as "transition zone" by the character of hummocks displacement angle. At a distance of 19000–20000 m from the source, the displacement angle shows perpendicular to parallel. The structural regime changes from extension to compression at 20000-24000 m, as opposed to the decreasing the slope. This implies that compression possibly occurs because of a factor other than the slope condition, which is most likely the change in main flow direction due to topographic barrier. At 24000-26000 m, the displacement angle changes again from perpendicular to parallel because the main flow direction has shifted to the southwest, so that the extension force is more influenced by the lowering of topography.

5. 2. 3. Hummock alignment in MFD 2 zone. The variation in average displacement angle with respect to the distance from source in MFD 2 zone is very random and irregular, even though the slope is constantly decreasing. This case is similar to Kannongawa valley-filling debris avalanche from [20]. [20] states that the variation in displacement angle in Kannongawa debris avalanche may be attributed to the existence of lateral topographical barriers. In the case of Gadung debris avalanche on MFD 2 zone, which is also a valley-filling debris avalanche, the most influential lateral topographic barrier is the southeast slope of Iyang-Argapura which is directly adjacent to MFD 2 zone. The pre-depositional slope of the Iyang-Argapura southeast flank is aligned southward, relatively perpendicular to the Gadung MFD 2. The existence of this topographic barrier causes hummocks to be randomly oriented because of the friction and compression force, and the orientation of the hummock follows the topography of the Iyang-Argapura slope as well.

6. Conclusion

We classify the area covered by Gadung debris avalanche deposit into 3 zones depending on the main direction of the avalanche flow. The freely spreading debris avalanche zone is referred as the MFD 1 zone, while the valley filling type area is referred to as the MFD 2 zone. Between MFD 1 and MFD 2 zone, right in the location where the flow direction change, there is a zone called MFD transition zone. According to our research, it can be concluded that there is no significant correlation between hummock size and distance from the source of Gadung debris avalanche deposit. The Gadung hummock size is mainly affected by debris avalanche flow velocity, which is controlled by pre-existing morphologic feature such as Iyang-Argapura Volcanic Complex and Meru Betiri Mountains. Based on hummock size distribution map, we could identify four domains area of Gadung debris avalanche flow mechanism, which are Toreva domain, Hummock domain, Collision domain, and Oblique collision domain.

The hummock orientation and displacement angle of Gadung debris avalanche deposit depends on the structural regime existed in certain area. The proximal area of MFD 1 zone is dominated by extensional regime due to constant to decreasing slope and the spreading of avalanche flow downslope and laterally. Change of structural regime from extensional to compressional occurs in the distal part of MFD 1 zone because of the collision between Gadung avalanche flow and Iyang-Argapura Volcanic Complex. In the MFD Transition zone, the change of structural regime from extension to compression is opposed to the decrease in the slope, possibly because of the change in main flow direction to the southwest. As for MFD 2 zone, the irregular pattern of displacement angle with respect to the continuously decreasing slope may happen due to the existence of lateral topographical barriers such as the southeast flank of Iyang-Argapura Volcanic Complex.

Acknowledgments

This work is fully funded by the Ministry of Research and Technology of Indonesia through *Penelitian Terapan Unggulan Perguruan Tinggi (PTUPT)* funding program for year 2021, entitled "*Pengurangan Risiko Bencana Gunung Api Berbasis Integrasi Ilmu Kebumian: Studi Kasus Gunung Raung, Jawa Timur*".

References

[1] Siebert L 1992 Threats from debris avalanches *Nature.* **356** 658–659.
[2] Capra L, Macı´as J L 2002 The cohesive Naranjo Debris-Flow Deposit (10km^3): a dam breakout flow derived from the Pleistocene Debris-Avalanche Deposit of Nevado de Colima Volcano (Mexico) *J. Volcanol. Geotherm. Res.* **117** 213-35
[3] Thouret J -C 2005 The stratigraphy, depositional processes, and environment of the late Pleistocene Polallie-period deposits at Mount Hood volcano, Oregon, USA *Geomorphology* **70** 12–32.
[4] Carrasco-Núñez G, Díaz-Castellón R, Siebert L, Hubbard B, Sheridan M F, Rodríguez S R, 2006 Multiple edifice-collapse events in the Eastern Mexican Volcanic Belt: the role of sloping substrate and implications for hazard assessment *J. Volcanol. Geotherm. Res.* **158** 151–76.
[5] Yoshida H, Sugai T, Ohmori H 2010 Longitudinal downsizing of hummocks by the freely spreading volcanic debris avalanches in Japan *The Quaternary Research (Daiyonki-Kenkyu),* **49 (2)** 55–67.
[6] Yoshida H, Sugai T, Ohmori H 2012 Size–distance relationships for hummocks on volcanic rockslide-debris avalanche deposits in Japan. *Geomorphology* **136 (1)** 76–87

[7] Siebert L 2002 Landslides resulting from structural failure of volcanoes in Evans S G, De Graff J V (ed) Catastrophic landslides: effects, occurrence, and mechanisms. *Geol. Soc. Amer. Rev. Eng. Geol.* **15** 209–35
[8] Bernard B, van Wyk de Vries B, Leyrit H 2009 Distinguishing volcanic debris avalanche deposits from their reworked products: The Perrier Sequence (French Massif Central) *Bull. Volcanol.* **71 (9)** 1041–56.
[9] van Wyk de Vries B and Davies T 2015 Landslides, debris avalanches, and volcanic gravitational deformation in Sigurdsson H, Houghton B F, McNutt S R, Rymer H, Stix J (eds) *Encyclopedia of Volcanoes* 665–8 (San Diego: Academic Press)
[10] Palmer A B, Alloway B V, Neall V E 1991 Volcanic debris avalanche deposits in New Zealand lithofacies organization in unconfined wet-avalanche flows in Fisher R V, Smith G A (eds) Sedimentation in volcanic settings. *SPEM Special Publication* **45** 89–98
[11] Glicken H 1996 Rockslide-debris avalanche of May 18, 1980, Mount St. Helens Volcano, Washington. *US Geol. Surv. Open-File Rep.* **96–0677** 1–90
[12] Kerle N and van Wyk de Vries N 2001 The 1998 debris avalanche at Casita volcano, Nicaragua: investigation of structural deformation as the cause of slope instability using remote sensing. *J. Volcanol. Geotherm. Res.* **105 (1)** 49-63
[13] Costa J E and Schuster R L 1988 The formation and failure of natural dams. *Geol. Soc. Am. Bull.* **100** 1054-1068.
[14] Walter T R, Haghshenas Haghighi M, Schneider, F M *et al* 2019 Complex hazard cascade culminating in the Anak Krakatau sector collapse *Nat. Commun.* **10** 4339
[15] Crandell D R, Miller C D, Glicken H X, Christiansen R L, Newhall C G 1984 Catastrophic debris avalanche from ancestral Mount Shasta volcano, California *Geology* **12 (3)** 143–46.
[16] Siebert L 1984 Large volcanic debris avalanches: characteristics of source areas, deposits, and associated eruptions. *J. Volcanol. Geotherm. Res.* **22** 163–97
[17] Ui T, Takarada S, Yoshimoto M 2000 Debris avalanches in Sigurdsson H, Houghton B F, McNutt S R, Rymer H, Stix J (eds) *Encyclopedia of Volcanoes* 617–26 (San Diego: Academic Press)
[18] Shea T, van Wyk de Vries B 2008 Structural analysis and analogue modeling of the kinematics and dynamics of rockslide avalanches *Geosphere* **4** 657–86.
[19] Paguican E M R, van Wyk de Vries B, Lagmay A M F 2014 Hummocks: how they form and how they evolve in rockslide-debris avalanches *Landslides* **11** 67–80.
[20] Yoshida H 2014 Hummock alignment in Japanese volcanic debris avalanches controlled by pre avalanche slope of depositional area *Geomorphology* **223** 67–80.
[21] Yoshida H 2016 Magnitude-frequency distribution of hummocks on rockslide-debris avalanche deposits and its geomorphological significance *Geosciences* **6** 5
[22] Hayakawa Y S, Yoshida H, Obanawa H, Naruhashi R, Okumura K, Zaiki M, and Kontani R 2018 Characteristics of debris avalanche deposits inferred from source volume estimate and hummock morphology around Mt. Erciyes, Central Turkey *Nat. Hazards Earth Syst. Sci.* **18** 429–44
[23] Escher B G 1925 Leboulement prehistorique de Tasikmalaja et le Vulcan Galoungoung (Java) *Leids. Geol. Mededel.* **1** 8-21
[24] Neumann van Padang M 1939 Uber die vielen tausend Hiigel in westlichen Vorlande des Raoeng-Vulkans (Ostjava) *Ing. Ned. Indies* **6 (4)** 35-4
[25] Bronto S 1989 *Volcanic Geology of Galunggung, West Java, Indonesia* (PhD dissertation) (Cristchurch: University of Canterbury)
[26] Sutawidjaja I S, Suparman, Sitorus K 1996 *Geological Map of Raung Volcano, East Java* (Bandung: Volcanological Survey of Indonesia)
[27] Center of Volcanology and Geological Hazard Mitigation (CVGHM) 2014 *Data Dasar Gunung Api Raung* accessed 16 August 2019 from http://www.vsi.esdm.go.id/index.php/gunungapi/data-dasar-gunungapi/526-g-raung

[28] Global Volcanism Program 2013 Raung (263340) in Volcanoes of the World, v. 4.9.3 (01 Feb 2021) Venzke E (ed.) (Washington D C: Smithsonian Institution)
[29] van Bemmelen R W 1949 *General Geology of Indonesia and Adjacent Archipelagoes* (The Hague: Government Printing Office)
[30] Yoshida H and Sugai T 2010. Quantitative examination of hummock alignment in the debris avalanche deposits: Zenkoji debris avalanche, Usu volcano, Japan. *Geographical Review of Japan Series B* **83** 64–72.
[31] Andrade S and van Wyk de Vries B 2010 Structural analysis of the early stages ofcatastrophic stratovolcano flank-collapse using analogue models *Bull. Volcanol.* **72** 771–789.

Re-interpretation of distribution of Lautan Pasir caldera-forming eruption products, Bromo-Tengger Caldera Complex, East Java

R M P P Gunawan[1], G Ikhwanushova[1], A Harijoko[2*], H E Wibowo[1] and A Setianto[1]

[1]Geological Engineering Department, Faculty of Engineering, Universitas Gadjah Mada, Yogyakarta, Indonesia
[2]Central Study of Disaster, Universitas Gadjah Mada, Yogyakarta, Indonesia
*E-mail: aharijoko@ugm.ac.id

Abstract. We present the result of morphostratigraphic analysis and geological field observation to identify the distribution of scoria flow deposit related to the caldera-forming eruption of Lautan Pasir caldera as the youngest caldera in Bromo-Tengger caldera complex. Morphostratigraphic analysis derived from digital elevation model and volcanic stratigraphy shows that products of Lautan Pasir caldera-forming eruption(s) mainly distributed to the north filled up the Sapikerep valley, and further to the north formed Sukapura distal fan. While to the south, it filled the valley between Ijo and Old Semeru. Field observation in the deeply dissected part of Sapikerep valley found two massive pyroclastic flow deposits separated by lava flow. Above the lava flow is scoria rich pyroclastic flow deposit typical of Lautan Pasir caldera-forming eruptions. Below the lava flow is scoria poor pyroclastic deposit of older products, most likely, of Ngadisari caldera. The distribution of Ngadisari caldera-forming eruption products is very limited considering the following massive volcanism of Lautan Pasir caldera. This outcrop might provide the key information to further reconstruct the volcanism stage of Bromo-Tengger caldera complex

1. Introduction

The difficulty in identifying caldera edifices in the caldera complex makes caldera-forming study very challenging. Since a caldera complex morphology represents overlapping calderas of different ages [1], it caused difficulty in determining the volcanic units on the field and eventually made the fieldwork less effective. The limited access to the outcrop and large deposit distribution also gives more challenge to the volcanologist who works in this area. Many studies used remote sensing analysis to make interpretations of the volcanic units (e.g., [2], [3]). This method helps reconstruct caldera edifices and determine observation points for more efficient fieldwork.

The order of the formation in the volcanic area can be represented in a morphostratigraphic map. A morphostratigraphic unit represents lithology on a particular landform and its relative age with the other units [4]. The early studies of morphometric analysis were mostly applied on monogenetic volcanoes (e.g., [5], [6], [7], [8]) because their formation process is thought to be less complicated than that on polygenetic volcanoes. There are also morphostratigraphy studies on stratovolcanoes and caldera (e.g. [2], [3]) that used this method to determine morphostratigraphic units and proposed morphometric parameters in characterizing volcanic edifices using Digital Elevation Model (DEM). This method is necessary to understand the historical of volcanism of a volcanic complex which is important to the hazard assessment [11]. We applied this method to a caldera complex morphology such as the Bromo-Tengger in East Java, Indonesia. Here, we tried to combine several morphological aspects based on DEM data's derivation to determine the caldera edifices in the Bromo-Tengger caldera complex.

We here combine the DEM-derived topographic map, geological map, and fieldwork observation to make a morphostratigraphic map of the Bromo-Tengger caldera complex. We propose an alternative interpretation of the distribution of the Lautan Pasir caldera-forming eruption products and supported by field evidences.

2. Geological Background

Bromo-Tengger caldera complex shows a complicated formation process represented by its complex morphology (landform) and various lithologies. This caldera complex is part of Quaternary volcanic arc in the east side of Java Island [12]. The volcanic products are divided generally into two groups, old (before Lautan Pasir caldera-forming) and young (after Lautan Pasir caldera-forming) Quaternary volcanic deposits [9]. Based on a geological map with a scale of 1:50,000, Zaennudin et al. [10] proposed that the Bromo-Tengger caldera complex is formed of three caldera-forming activities. These activities were estimated to come from Ijo, Ngadisari, and Lautan Pasir caldera-forming processes.

Based on the study from Zaennudin et al. [10], Old Tengger volcanism is commonly produced andesitic lava and formed another volcanic cone called Sedaeng volcanic cone. Ngadisari caldera-forming eruption generated lava and ignimbrite that filled the Sapikerep Valley and distributed widely to Madura Strait. On the other hand, the Lautan Pasir deposit consists of pyroclastic fall, pyroclastic flow, basaltic lava, and andesitic lava. Post-caldera volcanisms are remarked by several monogenetic volcanoes inside the Lautan Pasir caldera namely Mt. Widodaren, Mt. Kursi, Mt. Segara Lor, Mount Segara Kidul, Mt. Bromo, and Mt. Batok [9]. The most recent volcanic activity in the Bromo-Tengger caldera complex, by the writing of this report, came from Mt. Bromo on December 23rd – 29th 2020 [13].

3. Methods

Essential aspects of morphostratigraphy study are landform morphology, lithology, and the relative order of age. Therefore, the morphostratigraphic units defined here considers the volcanic sources, deposit distribution, and erosional process displayed on the surface shape variation. This study used DEM with a maximum spatial resolution of 8.5 m from open source DEMNAS (tides.big.go.id/DEMNAS) and a geological map of Bromo-Tengger with a scale of 1:50,000 [10]. Spatial data were obtained using software like ArcGIS 10.5, Global Mapper, PCI Geomatics, and Google Earth Pro. These data were processed to produce several maps such as hillshade, slope, ridge alignment, and drainage pattern. Based on these maps, the morphostratigraphic units were determined.

Volcanic sources and deposit distribution in the Bromo-Tengger caldera complex were determined by hillshade, slope, and ridge alignment, as well as lithology types from the geological map and fieldwork observation. Hillshade and slope maps are generated in ArcGIS by *hillshade* and *slope* tool, respectively. Hillshade map was made from the sum of 4 data with 270°, 315°, 45°, and 90° light directions by *weightedSUM* tool. It represents a 3D relief of the area. Meanwhile, the slope map shows a 2D relief variation of the area. Also, the hillshade map derived from DEM was processed using the PCI Geomatics software algorithm to show the volcanic product direction presented in the rose diagram. This result needs to be corrected manually by the lineament of the ridge. Combining the distinctive features on data of 3D relief, 2D relief, and volcanic direction, we delineated the geomorphologic units. Furthermore, these geomorphologic boundaries need to be compared with the geological map and volcanic deposits observed on the field.

The erosional process is one of the causes of the landform variation in the volcanic area. This process affected the valley depth and eventually the drainage pattern on a river basin [14]. The determination of the drainage pattern used here is derived from the drainage map. This map was made by *hydrology* tool in ArcGIS. The river network from this map was compared with the pattern on the drainage model [15].

Another important parameter to determine the drainage pattern is drainage density (Dd) which calculated by the equation (1) from [16]:

$$Dd = \frac{\sum L}{A} \tag{1}$$

where Dd is drainage density (km/km^2); ΣL is total river length on a basin (km); and A is basin area (km^2). This equation was commonly used in river basin studies [14], [17] but still can be useful in this study to interpret the resistance of rocks to erosional processes in the volcanic area [18]. In addition, the variation of the valley depth due to different lithology and erosional process need to be considered. In this study, the measurement of the highest and the lowest points on a valley morphology was done by Global Mapper software.

Another important criterion of morphostratigraphy is the order of formation which is reflected in the cross-cutting relationship of each unit. Thus, we determined the morphostratigraphic maps based on 1) eruption sources; 2) cross-cutting relationship; 3) drainage density; 4) lithology types; 5) degree of erosion on each unit. Then we validated the unit by making sections to ensure that every unit has distinctive characteristics. Finally, the morphostratigraphy of the Bromo-Tengger caldera complex is presented on a map with a scale of 1:50,000, and the unit's name used here is principally similar to Brahmantyo and Bandono [19].

4. Results and Discussion

The identification of the morphostratigraphic units of the caldera edifices in the Bromo-Tengger caldera complex helps to infer the stage of caldera-forming events. The combination of DEM-derived maps and secondary data from previous study used to determine the geomorphologic units of the area. Then, we identified the lithology and determined prospective location based on these geomorphologic units. As a result, the Bromo-Tengger caldera-forming order could be interpreted based on the characteristics of each unit.

4.1. Volcanic edifice determination

Caldera complex morphology consists of several caldera edifices. Each edifice represents a phase of formation. During the phases, the older volcanic products could be covered by the younger deposits. This condition makes volcanic edifice determination is difficult [20]. Meanwhile, each formation phase in caldera complex commonly consisted of different type of rocks and deposited on different period of time. Hence, the morphostratigraphic map will show the relation of each volcanic edifice.

Based on eruption sources, cross-cutting relationship, drainage density, lithology types, and degree of erosion, there are five caldera edifices comprising the Bromo-Tengger caldera complex (Figure 1). Each unit has distinctive characteristics of morphostratigraphic parameter shows in Table 1. Higher drainage density in an area indicates weaker and impermeable subsurface material, sparse vegetation, and higher relief [14]. High Dd values in the Bromo-Tengger caldera complex is shown in the edifices of Ngadisari caldera volcanism, Ijo caldera volcanism, post-Lautan Pasir caldera volcanism, and Lautan Pasir cladera volcanism which dominantly consisted of pyroclastic deposits. Meanwhile, lower Dd value is shown in the Old Tengger edifice which dominantly consisted of lavas that is higher resistant and low moderate relief. The exceptionally high value of Dd (7.59) is shown in Sapikerep valley. This can be caused by the confined morphology that intensify erosional process as represented on its deepest valley depth (67 m) and highest Dd value in compare to the other units.

IOP Conf. Series: Earth and Environmental Science **851** (2021) 012038 doi:10.1088/1755-1315/851/1/012038

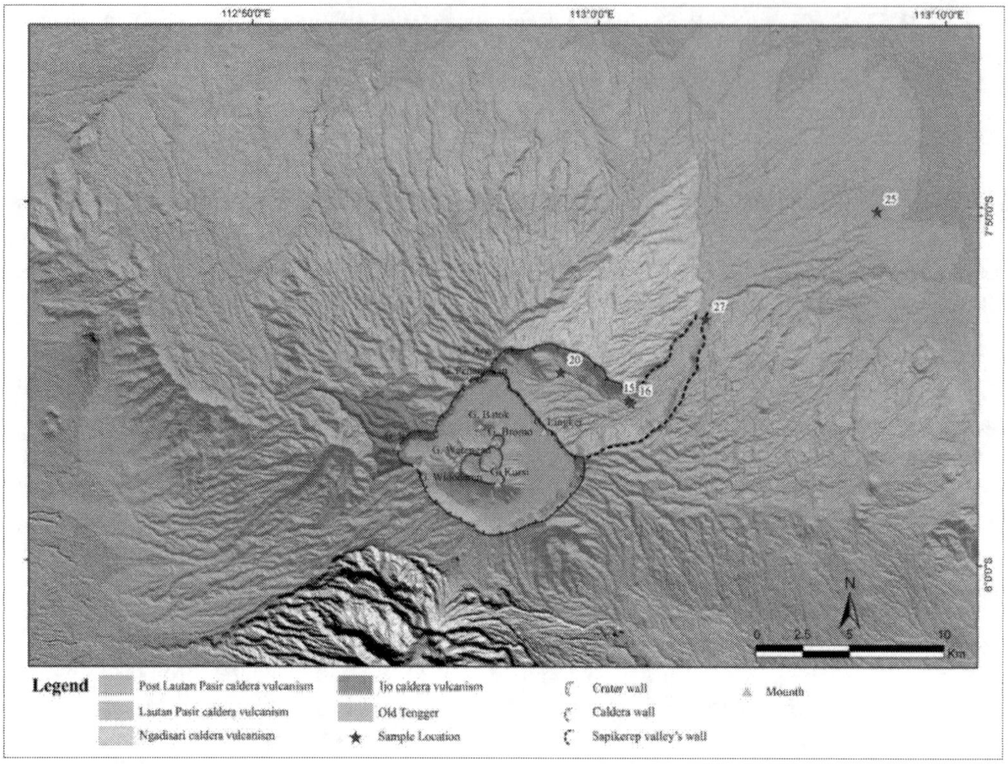

Figure 1. Edifice map of Bromo-Tengger Caldera Complex showing caldera and crater wall. The biggest caldera is Lautan Pasir, the eastern part of Lautan Pasir caldera is Ngadisari caldera and the western part of the Lautan Pasir caldera is Ijo crater.

Table 1. Morphostratigraphic parameter of caldera edifices in the Bromo-Tengger caldera complex.

Unit area (km^2)	River length (km)	Drainage density (km/km^2)	Slope (°)	Valley depth (m)	Lithology	Edifice
1018	3721.24	3.65	17	58	Tengger lava	Old Tengger volcanism
11	69.68	6.45	12	31	Ijo pyroclastic deposit	Ijo caldera volcanism
56	307.21	5.51	17	31	Ngadisari pyroclastic deposit	Ngadisari caldera volcanism
14	105.22	7.59	20	67	Lautan pasir lava and pyroclastic deposit	Sapikerep valley
298	1309.73	4.39	17	30	Lautan Pasir pyroclastic deposit	Lautan Pasir caldera volcanism
18	93.39	5.26	7	12	Intra-cone pyroclastic deposit	Post-Lautan Pasir caldera volcanism

4.2. Fieldwork and lithology identification

Checking the lithology variation in the field on each interpreted morphostratigraphic unit is essential. Several observation points in deeply dissected valley which preserved various lithologies from different eruption sources [9] was chosen to be representative of this method. The location, stratigraphy data, and outcrop description refer to Alghifari [21]. The outcrop varied of alternating pyroclastic fall and flow deposit (Figure 2a, 2b) which sometimes lied on brown older pyroclastic flow deposit (Figure 2c). In between them, there were pyroclastic surge deposit layers. At another location, older pyroclastic flow deposit found bellow basaltic lava (Figure 2d).

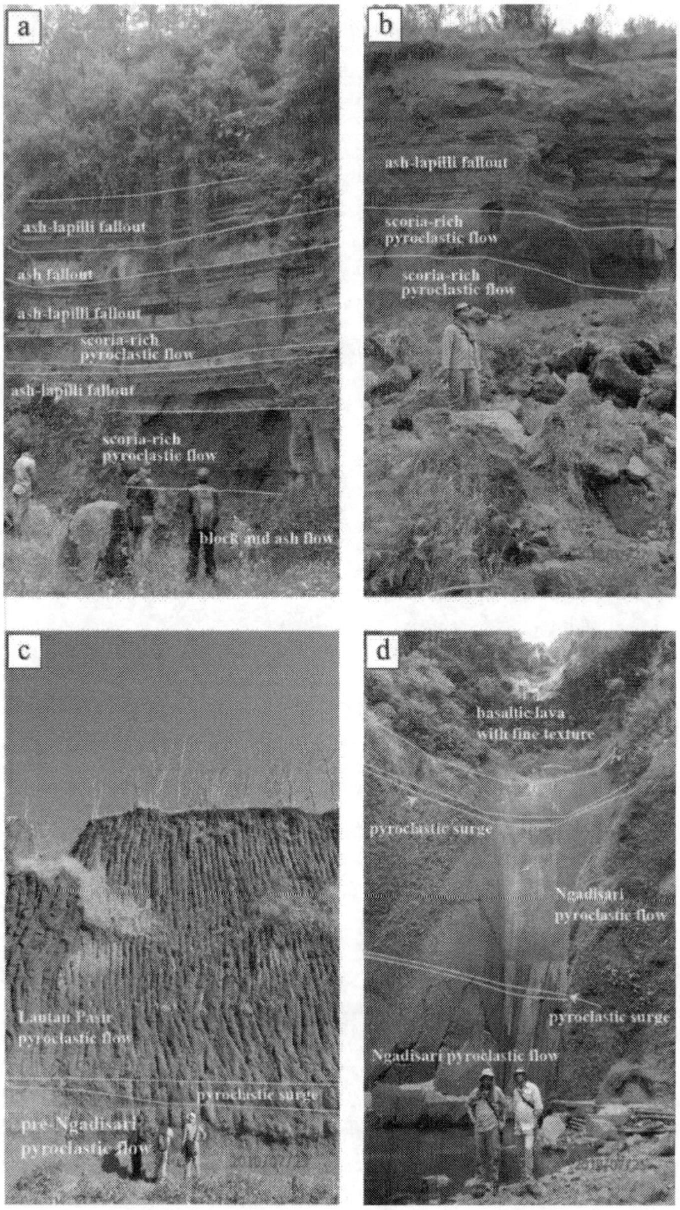

Figure 2. Volcanic deposit variation around Bromo-Tengger caldera complex. (a) and (b) pyroclastic flow and pyroclastic fall layers of Lautan Pasir caldera volcanism at STA 20 and STA 15, respectively; (c) contact of pra-Ngadisari caldera deposit and Lautan Pasir caldera deposit at STA 25 in Sukapura distal fan; (d) Ngadisari caldera pyroclastic deposits at STA 27.

The deposit sequence observed from the distal (STA 25, STA 27) to the proximal area (STA 15, STA 16). The variation of the stratigraphy in the valley shows distinctive characteristics of each unit (Figure 3). We divided the lithology units in this area into three groups, they are (1) Lautan Pasir caldera-forming deposit; (2) Pre-Lautan Pasir deposit; and (3) Ngadisari caldera-forming deposit. The first unit is composed of grey, scoria rich pyroclastic flow which is called as Sukapura breccia, and ash to lapilli fall deposit sometimes occurred in between the flow deposits. The second unit consisted of andesitic lava with coarse texture, lahar deposit, a basaltic lava with medium texture, brownish pyroclastic flow, and alternating fall deposit with surge deposit from lower to upper position, respectively. The last unit is comprised by grey pyroclastic flow with less scoria content than that in the second unit. The name of the unit is assumed to be the process contributed to the deposition of each sequence.

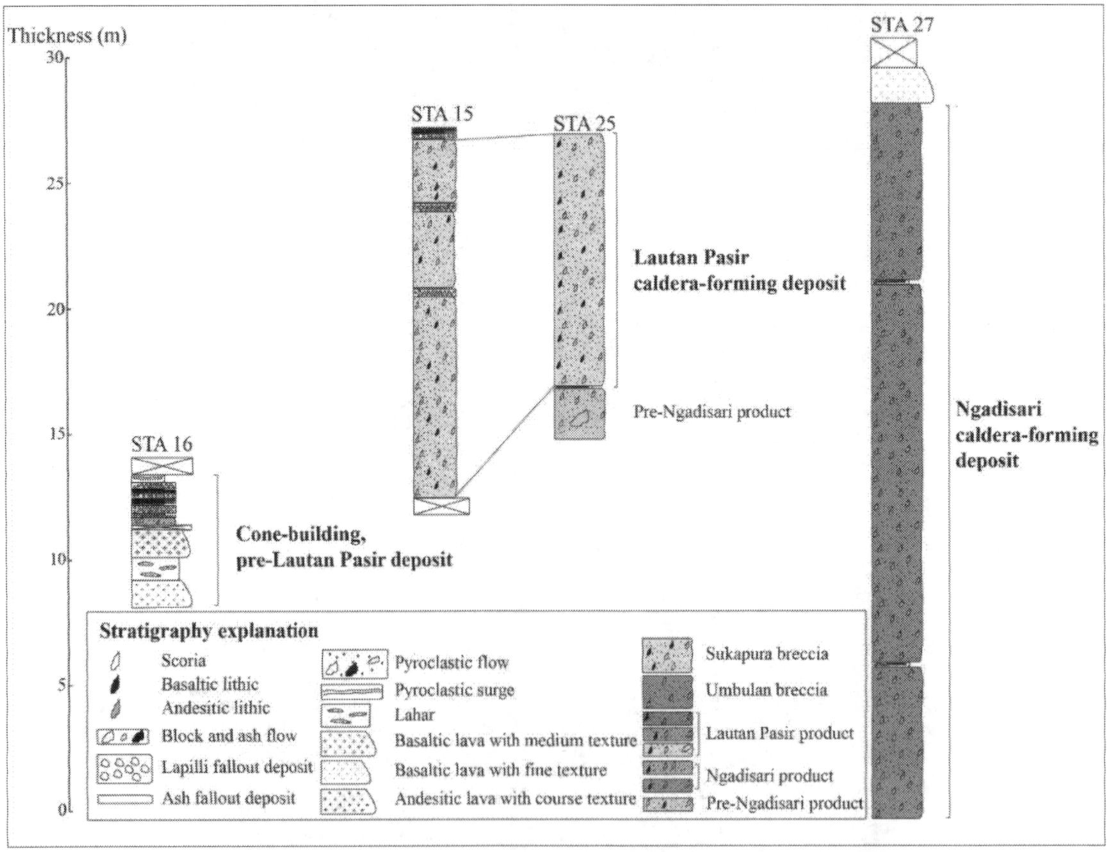

Figure 3. Stratigraphy sections of the Bromo-Tengger caldera deposit, showing the variation of Lautan Pasir and Ngadisari caldera volcanism, which make up the deeply dissected valley.

4.3. Interpretation of the distribution of caldera-forming eruption products of Lautan Pasir Caldera

The identification of the caldera edifices in this study is based on the morphostratigraphic parameters and the lithology checking on the designated locations. The results of this study showed that there are three calderas in the Bromo-Tengger caldera complex. These are Ijo caldera, Ngadisari caldera, and Lautan Pasir caldera. Product of Ijo caldera is distributed on the southeastern part of the Bromo-Tengger caldera complex. Products of Ngadisari caldera is likely distributed inside the Sapikerep valley and mostly are covered by following products of Lautan Pasir caldera.

This interpretation is built from field evidence of the two massive pyroclastic flow deposits separated by lavas. The scoria rich pyroclastic deposits above the lava represents the youngest caldera-forming eruption event of Lautan Pasir caldera. The lava itself might resembles the construction phase of Lautan Pasir volcano. While the scoria poor pyroclastic flow deposit below the lava might represents the older caldera-forming eruption event of Ngadisari caldera. However, the outcrop of this deposit is very limited. These chronological events are supported by morphostratigraphic analysis showing higher drainage density of Ngadisari edifice than that of Lautan Pasir.

5. Conclusion

There are five volcanism phases resulting the Bromo-Tengger caldera complex morphology. These phases based on morphostratigraphy analysis are Old Tengger, Ijo, Ngadisari, Lautan Pasir, and Post-Lautan Pasir volcanism. The determination of each morphostratigraphic unit is deducted from DEM-derived maps in combination with geological maps and field observation to interpret eruption sources, cross-cutting relationship, drainage density, lithology types, and degree of erosion. We determined the several observation points that can show the variation in lithology as one of criteria in defining the morphostratigraphic unit. Furthermore, Lautan Pasir deposit is characterized by a scoria rich pyroclastic deposit above the lava flow while Ngadisari deposit is characterized by a scoria poor of older products bellow the lava flow in deeply dissected valley. To sum up, most of the Lautan Pasir deposits filled up Sapikerep valley to the north and further formed the Sukapura distal fan, also it filled the valley between Ijo and Old Semeru. In contrast, Ngadisari deposit is limited to be found considering the following massive volume of Lautan Pasir products.

Acknowledgments

This research is conducted by the financial support of PDUPT number 2566/UN1.DIT-LIT/DIT-LIT/LT/2019 by Ministry of Education and Cultural of Indonesia.

References

[1] Cole J W, Milner D M and Spinks K D 2005 Calderas and caldera structures: a review *Earth Sci. Rev.* **69** 1-26
[2] Espinasa-Pareña R and Martín-Del Pozzo A L 2006 Morphostratigraphic evolution of Popocatépetl volcano, México *Conf. Paper in Geol. Soc. Am.* Special Paper 402 pp 101-123
[3] Grosse P, van Wyk de Vries B, Euillades P A, Kervyn M and Petrinovic I A 2012 Review: systematic morphometric characterization of volcanic edifices using digital elevation models *Geomorphology* **136** 114-131
[4] Lowe J J and Walker M C J 1984 *Reconstructing Quaternary environments* (New York: John Wiley & Sons)
[5] Perter 1972 Distribution, morphology, and size frequency of cinder cones on Mauna Kea volcano, Hawaii *Geol. Soc. Am. Bull.* **83** 3607-12
[6] Wood C A 1980a Morphometric evolution of cinder cones *J. of Volcanology and Geothermal Research* **7** 387-413
[7] Wood C A 1980b Morphometric analysis of cinder cone degradation *J. of Volcanology and Geothermal Research* **8** 137-160
[8] Settle M 1979 The structure and emplacement of cinder cone fields *Am. J. Sci.* **278** 1089-1107
[9] Zaennudin A 1990 *Stratigraphy and nature of the stratocone Mt. Cemara Lawang in Bromo-Tengger Caldera, East Java, Indonesia* Master thesis (New Zealand: Victoria University of Wellington)
[10] Zaennudin A, Hadisantono R D, Erfan R D and Mulyana A R 1994 *Peta Geologi Gunungapi Bromo-Tengger, Jawa Timur* Geological map (Bandung: Pusat Penelitian dan Pengembangan Geologi)
[11] Freski Y R 2017 *Morphostratigraphy of Young Lawu Volcano, Central Java and East Java Province, Indonesia in the hazard assessment of future eruption* Master thesis (Yogyakarta: Universitas Gadjah Mada)

[12] Van Bemmelen R W 1949 *The geology of Indonesia* Vol. 1A (The Hague: Government Printing Office)
[13] Global Volcanism Program 2020 *Report on Tengger caldera (Indonesia) In: Sennert, S K (ed), weekly volcanic activity report, 23 December-29 December 2020* Smithsonian Institution and US Geological Survey
[14] Rai P K, Mohan K, Mishra S, Ahmad A and Mishra V N 2014 A GIS-based approach in drainage morphometric analysis of Kanhar River Basin, India *Appl. Water Sci.* DOI 10.1007/s13201-014-0283-y
[15] Zhang L and Guilbert E 2012 A study of variables characterizing drainage patterns in river networks *Intl. Archives of the Photogrammetry, Remote Sensing and Spatial Info. Sci.* **XXXIX-B2** 29-34
[16] Horton R E 1945 Erosional development of streams and their drainage basins; hydrophysical approach to quantitative morphology *Bull. Geol. Soc. Am.* **56** 275-370
[17] Albaroot M, Al-Areeq M, Aldharab H S, Alshayef M and Ghareb S A 2018 Quantification of morphometric analysis using remote sensing and GIS techniques in the Qa'Jaharan Basin, Thama Province, Yamen *Intl. J. of New Tech. and Research* **4** Issue 8 12-22
[18] Sipahutar S R, Sukiyah E and Sulaksana N 2018 Morphometry and morphotectonic of Cianten and Cisaat watershed on Quaternary volcanic terrain, Garut, West Java *J. of Geological Sci. and Applied Geol.* **2**
[19] Brahmantyo B and Bandono 2006 Landform classification for geomorphology maping on a scale of 1:25.000 and application for spatial planning (Indonesian Language) *Jurnal Geoplika* **1** 1-7
[20] Thouret J C 1999 Volcanic geomorphology-an overview *Earth-Sci. Rev.* **47** 95-131
[21] Alghifari R 2019 *Sapikerep valley pyroclastic deposits of Bromo-Tengger caldera complex, Sukapura, Lumbang, Kuripan, and Wonomerto, Probolinggo Regency, East Java* Undergraduate final report (Yogyakarta: Universitas Gadjah Mada)

Effect of power plant operating conditions on fly ash and bottom ash composition: a case study from power plant in Lampung

D A A Besari[1], F Anggara[2,3*], H T B M Petrus[3,4], W. Astuti[5] and W.A. Husnah[6]

[1]Graduate School of Geological Engineering, Department of Geological Engineering, Gadjah Mada University, Yogyakarta, Indonesia
[2]Department of Geological Engineering, Gadjah Mada University, Yogyakarta, Indonesia
[3]Unconventional Geo-resources Research Group, Faculty of Engineering, Universitas Gadjah Mada, Yogyakarta, Indonesia
[4]Department of Chemical Engineering (Sustainable Mineral Processing Group), Gadjah Mada University, Yogyakarta, Indonesia
[5]Research Unit for Mineral Technology, Indonesian Institute of Sciences (LIPI), Jl. Ir. Sutami Km. 15, Tanjung Bintang, Lampung, Indonesia
[6]PT. Bukit Asam, Tbk, Palembang, Indonesia
*E-mail: ferian@ugm.ac.id

Abstract. Fly ash and Bottom ash (FA-BA) are solid waste produced from the coal-generated power plant. FA consists of fine powder particles with diameters varying from 0.5-100 μm, while BA consists of coarser particles with a diameter of 0.125-2 mm. FABA and feed coal samples were collected from the power plant in Lampung to distinguish the effect of power plant operating conditions on FABA composition. Feed coal characteristics were examined by proximate and petrography analysis. Organic and inorganic components of FABA were identified using petrography analysis, while the detailed mineralogical composition of FABA was determined using X-Ray diffraction analysis. Moreover, major oxides of the feed coal and FABA were determined using inductive ICP-AES. From the analysis, it was found that FA composition is mainly organic material in the form of unburned carbon (UC). The abundance of UC content (up to 74.68%) indicates the inefficiency of the coal combustion process. In terms of inorganic, quartz is the most abundant mineral found in the FABA sample, up to 50.15%, in contrast to the glass content that was only 17.41%. A minor amount of spinel (0.03-2.00%) was found associated with Fe-oxide mineral. Kaolinite was only found in the BA sample from a landfill source and interpreted as external input from the dumping ground. The high content of UC and quartz will decrease the compressive strength of the material construction product providing a negative impact on the utilization of FABA.

1. Introduction

Fly ash is fine solid powders that remain after the coal combustion in a power plant recovered from an electrostatic dust filter [1]. FA particles generally have a spherical shape with a diameter ranging from 0.5-100 μm. BA consists of coarser particles (0.125-2 mm) than FA concentrated at the bottom of the furnace after combustion [1]. Based on Indonesia government regulation No. 11 Year 2014, FABA is classified as hazardous waste and toxic material, which has many mandatories of treatment and permits that must be completed [2]. Indonesia's power generation produced 4.38 million tons of FABA in 2015 and 166.2 million tons in 2019 [2]. Aside from the abundance of FABA, the characteristic will be undoubtedly essential to determine the potential utilization. FABA characteristics are affected by several factors. One of them is the coal-burning technology applied in the power plant. The technologies are Fluidized Bed Combustion (FBC) and Pulverized Coal Combustion (PCC). The main difference between those technologies is in the feed coal particle size in which powder is applied in the PCC technology. The characteristics of FBC-FABA are different than PCC-FABA, generally because of the wide variety of fuel mixtures, additive possibilities, combustion temperatures, boiler technology, and ash collection technology [3]. Previous studies on FABA compositions were limited to a power plant in Java, Indonesia [4]. This study will provide information and comprehensive discussion about the effect of operating conditions of a power plant on FABA characteristics.

2. Materials and Method

FABA and feed coal samples were collected in 2019 from Power Plant located in Lampung, Indonesia (see Figure 1). Sixteen samples of FA, six samples of BA, and two samples of feed coal were examined in this study (Table 1). The FA samples were collected directly from the silo, electrostatic precipitator (ESP), and landfill source, while BA samples were collected directly from the coal ash handling equipment and landfill. All samples were sealed in plastic bags and tightly closed to avoid contamination.

Table 1. List of samples collected from the study area.

Sample	Type	Source
FA-1	Fly Ash	Silo
FA-2		
FA-3		
FA-4		
FA-5	Fly Ash	Electristatic Precipitator (ESP)
FA-6		
FA-7		
FA-8		
FA-9		
FA-10		
FA-11		
FA-12		
FA-13		
FA-14	Fly Ash	Landfill
FA-15		
FA-16		
BA-1	Bottom Ash	Fresh
BA-2		
BA-3		
BA-4	Bottom Ash	Landfill
BA-5		
BA-6		
C-1	Feed Coal	Coal Feeder
C-2		

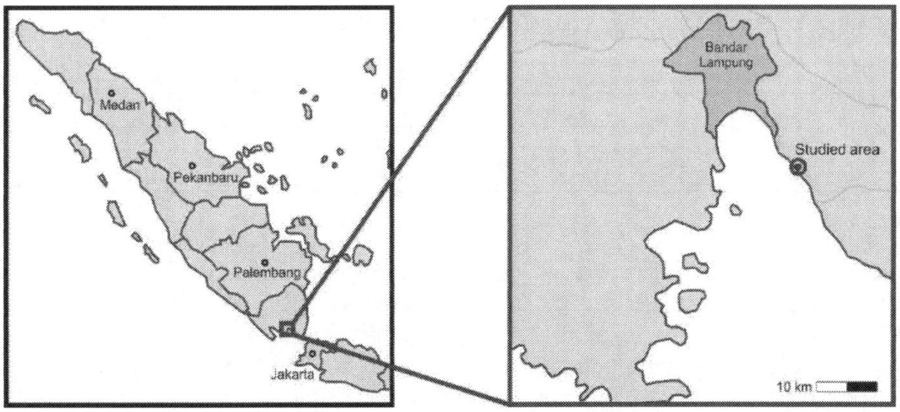

Figure 1. Maps showing the location of the Power Plant, Lampung, Indonesia.

Several analyses were conducted on this study to examine the effect of power plant operating conditions on FABA compositions as follows:

2.1. Proximate Analysis
Proximate analysis was conducted to evaluate ash yield, volatile matter, moisture content, and fixed carbon based on ASTM [5] in the laboratory of the Center for Mineral, Coal, and Geothermal Resources, The Ministry of Energy and Mineral Resources, Indonesia. The calorific value of feed coal was evaluated based on ASTM [6].

2.2. Geochemistry Analysis
Major oxides of FABA and feed coal samples were examined using inductively coupled plasma – atomic emission spectroscopy (ICP-AES) by ALS Geochemistry Laboratory, Kamloops, Canada.

2.3. Petrography Analysis
Polished blocks of feed coal and FABA samples were made using SpeciFix-20 Kit and Struers Labo System tools. Coal petrography was conducted based on the Standard Test Method for microscopical identification of maceral compositions [7]. Maceral classification followed the ICCP System 1994 [8-10](ICCP, 2001; Sýkorová et al., 2005; Pickel et al., 2017). Organic and inorganic components of FABA were identified using genetic classification [11-12]. All petrography analysis were conducted in the Department of Geological Engineering, UGM.

2.4. Mineralogy Analysis
X-Ray Diffraction (XRD), Panalytical E'xpert 3 Powder, analysis was conducted for the detailed mineralogical composition of FABA. Bulk XRD analysis was performed with 2θ from 2° to 60° and a scanning step of 0.02 in 2θ (degree) in the analytical laboratory of the Indonesian Institute of Sciences (LIPI), Lampung, Indonesia. XRD analysis could reveal the mineralogical transformation due to coal combustion in the power plant.

3. Results

3.1. Feed Coal Properties
Feed coal from the Power Plant has ash (A) yield in the range of 9.80-9.89 wt.%, moisture (M) contents of 24.40-24.50 wt.%, volatile matter (VM) of 32.69-32.96 wt.%, fixed carbon (FC) content of 32.65-33.11 wt.% while the calorific value (CV) is in the range of 4616-4630 cal/gr (Table 2).

Table 2. Result of proximate analyses (wt.%) and calorific values (cal/gr) for feed coals from the study area.

Sample	M_{adb}	A_{adb}	VM_{adb}	FC_{adb}	$CV_{, adb}$ (Cal/gr)
C-1	24.40	9.80	32.69	33.11	4630
C-2	24.50	9.89	32.96	32.65	4616

3.2. Organic and Inorganic Composition of Feed Coal and FABA

From the petrography, it can be seen that the feed coal samples are rich in vitrinite (66.00-70.30%, Table 3), while liptinite content is in the range of 15.65-21.40%. Inertinite is about 10.60-10.75%, and minerals matter from 2.00-3.30%. The main composition of mineral matter of feed coal is clay mineral, pyrite and quartz are also observed but in lower quantity (Figure 2).

Table 3. Feed coal compositions based on petrography analysis (%, vol).

Sample	Vitrinite	Liptinite	Inertinite	Mineral Matter
C-1	66.00	21.40	10.60	2.00
C-2	70.30	15.65	10.75	3.30

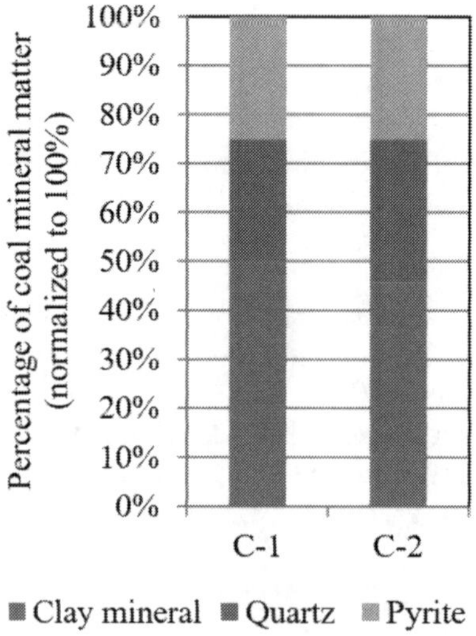

Figure 2. Percentages of coal mineral matter normalized to 100%.

Petrography analysis indicated that FABA samples consist of organic and inorganic components (Figure 3). Organic material presents in the form of unburned carbon (UC) with dark grey or black color. Inorganic components of FABA were identified as crystalline phase minerals and amorphous material. Amorphous material occurred as Glass. Glass based on their morphology variations can be identified as cenosphere and pleiosphere. Glass characterized by perfectly or not-perfect spherical form with light grey-dark grey color under reflected light examination. The crystalline phase mineral of FABA consists of quartz, fe-oxide minerals, and spinel. Quartz is identified as an angular-subrounded form with colorless or white color. The fe-oxide mineral is characterized by orange-red color with an angular-subangular shape. At the same time, spinel generally occurs as angular crystals with a dendritic texture associated with Fe-oxide minerals.

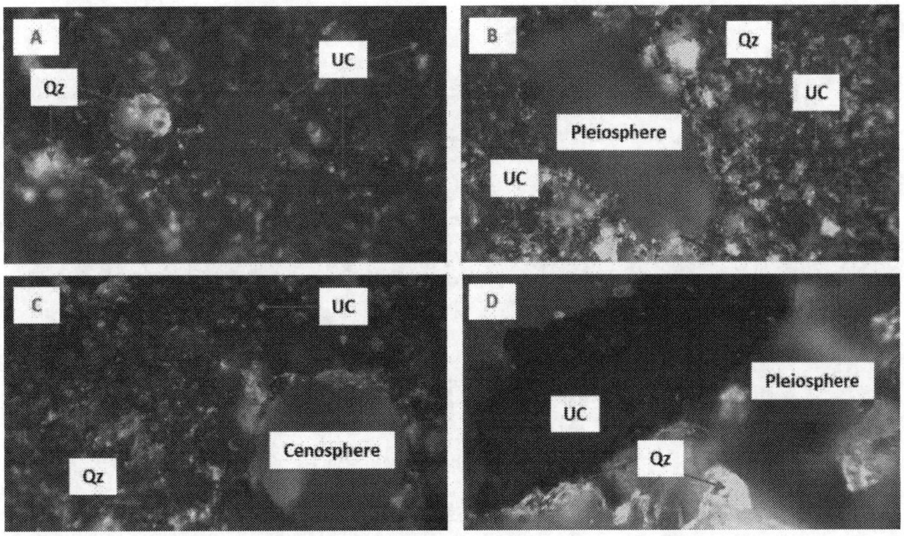

Figure 3. Reflected light photomicrographs of different FABA samples: A.) FA-Silo; B.) FA-Landfill, C.) FA-ESP, D.) BA-Landfill. Qz: Quartz; UC: unburned carbon.

UC dominates FA composition with abundance in the range of 40.00-74.68%. The other components occur from 16.98-43.94% for quartz, 4.85-15.45% for Glass, 0.38-2.00 for fe-oxide minerals, and 0.04-0.08% for spinel. Different compositions trends showed in BA samples (see Figure 4). BA compositions are dominated by quartz with abundances in the range 49.03-50.15%. The other compositions occur varies from 32.34-33.45% for UC, 16.85-17.41% for glass, 0.34-0.49% for Fe-oxide minerals, and 0.03-0.05% for spinel.

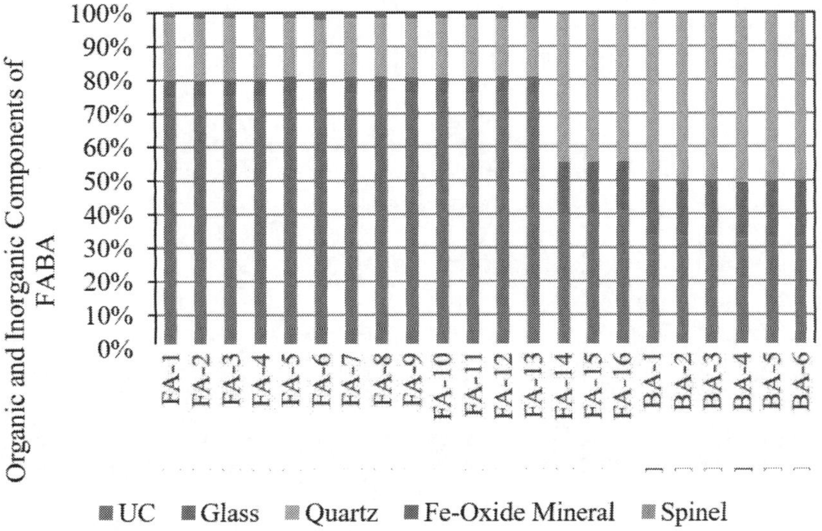

Figure 4. Percentage of organic and inorganic components on FABA. UC=unburned carbon.

3.3. Mineralogy Composition of FABA

Bulk XRD analysis was only conducted for six FABA samples (Figure 5). The diffractogram shows that FABA samples have a major peak of quartz minerals (Figure 6). Mineralogy variations based on XRD analysis were identified as quartz, hematite, pyrite, calcite, plagioclase, and kaolinite. Kaolinite only occurs in BA samples from landfill sources. Results of XRD analysis indicated that quartz content in

IOP Conf. Series: Earth and Environmental Science **851** (2021) 012039 doi:10.1088/1755-1315/851/1/012039

FA samples is higher than in BA, while plagioclase content in FA was lower than BA samples (Figure 5).

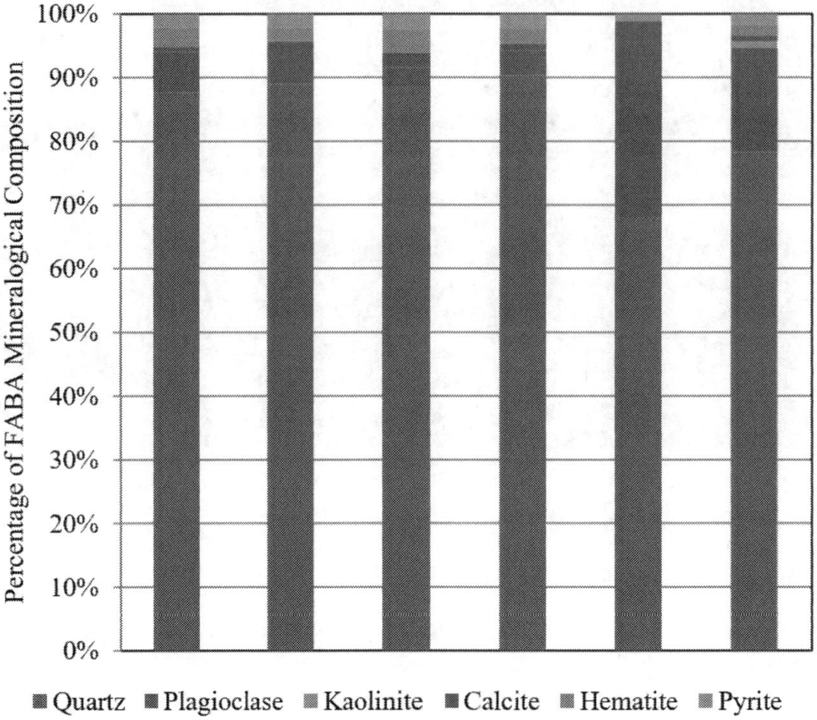

Figure 5. Mineralogical compositions of FABA.

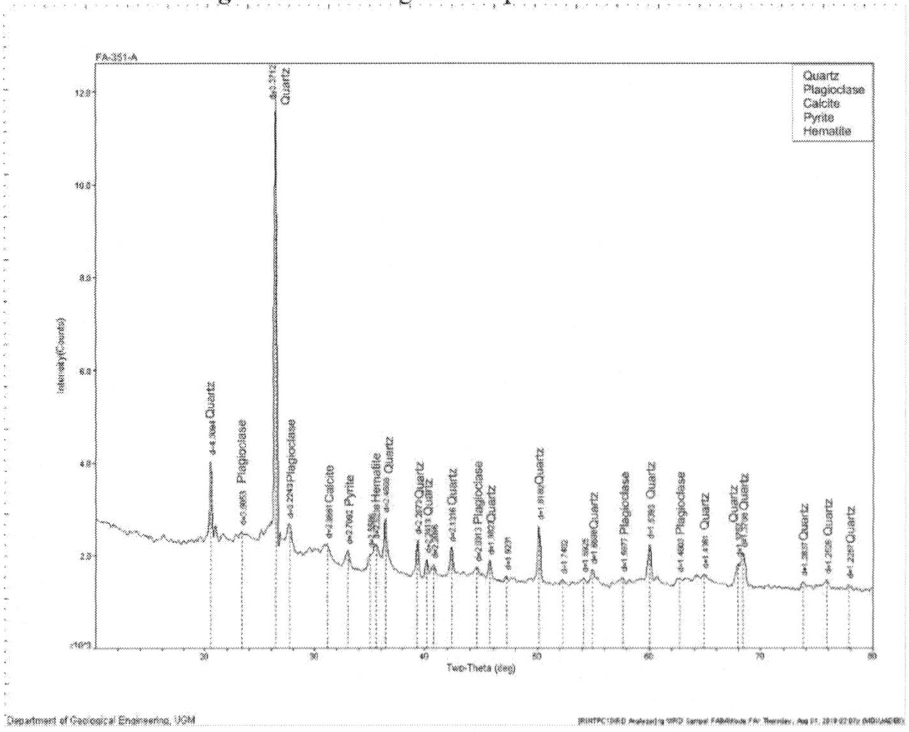

Figure 6. Detail mineralogical identification of sample FA-5 based on XRD analysis.

3.4. Feed Coal and FABA Geochemistry

The main major oxides of feed coal and FABA samples are SiO_2, Al_2O_3, Fe_2O_3, and CaO (see Figure 7). The most dominant major oxide both in feed coal and FABA is SiO_2 (≥55%). The dominant major oxides content (SiO_2, Al_2O_3, Fe_2O_3, and CaO) in the coal is generally related to clay mineral, quartz, and pyrite minerals abundances. The abundances of inorganic components found in FABA, including quartz, Glass, hematite, pyrite, spinel, plagioclase, calcite, and kaolinite, also correlate with dominant major oxides' content the FABA samples (SiO_2, Al_2O_3, Fe_2O_3, and CaO).

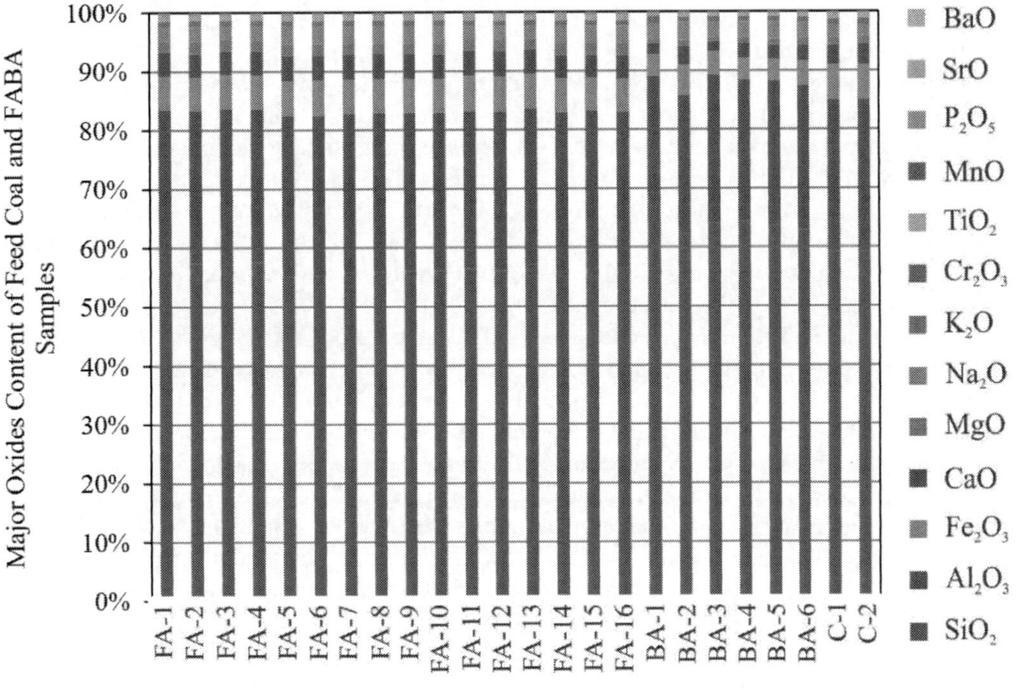

Figure 7. Major oxides content of feed coal and FABA samples.

3.5. Power Plant Operating Condition

The Power Plant of the study area uses circulating fluidized bed combustion (CFB) technology and has a capacity of 2×8 MW (MegaWatt), which uses limestone to desulfurize while operating and uses quartz sand particles (<1.49 mm) as bed material. It consumes about 50 tons of coal per hour. FABA samples were collected when the power plants were operating under normal conditions. The operating temperature was set to 850°C, the steam temperature at 470°C, and the steam pressure at 4.9 Mpa. The type of coal feeder system was used in this plant is a screw feeder system with the feed coal size specification ranged between 2-5 mm.

4. Discussion

The characteristics of FABA components are generally controlled by the feed coal characteristic, boiler, and operating conditions [13]. FA composition from the study area was dominated by organic material in the form of UC. The abundances of UC content (up to 74.68%) indicate inefficiency in the coal combustion process. UC in coal ash was produced from incomplete combustion of origin coal.

The occurrence of UC is caused by the rank and maceral composition of the coal and other factors such as combustion technology, off-spec feed coal particle size, and the use of unsuitable feed coal composition [12]. Feed coal plays a significant role in controlling the amount of carbon burnout, with increased carbon burnout being important for increased boiler efficiency [12]. A High proportion of >150μm feed coal particle decreased carbon burnout and negatively impacted efficiency [14]. According to Ganguli and Bandopadhyay [15], the higher the coal particle size < 76 μm, the higher the power plant efficiency. Referring to the specification of pulverized feed coal used in the Power Plant in the size of 2-5 mm (coarse particle), the inefficiency of coal-burning will most likely occur, leading to decrease the plant efficiency, and caused the abundance of UC formation in FA sample. That is why in the FCC technology, the burning efficiency is lower than that of the PCC due to the bigger particle size distribution of the feed coal. In addition, affecting the character of the FA resulted in higher unburned carbon content in the FCC.

Inorganic components of FABA from the study area were mainly composed of crystalline phase minerals, in contrast to pulverized coal combustion (PCC) FABA from Java, mainly consisting of spherical amorphous materials or Glass [4]. This difference is due to the lower operating temperature of CFB technology (700-900°C) compared to PCC technology at 1300-1700°C; this condition caused FBC boiler temperature is not enough to melt all inorganic particles [3]. The lower operating temperature condition in the study area resulted in the abundance of quartz up to 50.15%, in contrast to Glass produced during combustion is only 17.41%. Amorphous material or Glass was found less abundant in ashes from FBC than from pulverized-coal plants, mainly due to the lower operating temperature [12]. Quartz is the most abundant mineral found in the FABA sample from the study area. Quartz is the most common mineral pass through the coal combustion because it does not melt when heating at a temperature below 1400-1500°C [11-12]. Inside the CFB boiler, sand beds were floating together with fuel on a forced high-velocity airflow [3]. The function of bed material is to improve heat transfer and reduce temperature gradients [3]. Application of bed material using sand that was not suitable for the boiler affected the abundant quartz and resulted in less plagioclase on FABA components.

Fe-oxide mineral in FABA samples was identified as hematite. Fe-oxide minerals in FABA samples are produced from coal combustion, rich in Fe [16], while Fe-bearing mineral in origin feed coal sample presents as pyrite. Pyrite starts to decompose at around 400-500°C and produces the main product of iron oxide minerals such as hematite and magnetite [17]. A minor amount of pyrite found in the FABA sample varies from 1.76-3.64%. A minor amount of spinel (0.03-2.00%) was found associated with Fe-oxide mineral. Spinel is a secondary iron mineral produced from the decomposition of Fe sulfides and other Fe-bearing minerals in the coal [11]. Mineralogy of CFB FA is often complex due to limestone or dolomite as an in situ desulfurizing agent [12]. Calcite in the FABA sample was interpreted as the natural origin from coal mineral matter or as a secondary source related to limestone input during the combustion process. Kaolinite is only found in the BA sample from landfill sources and interpreted as external input from the dumping grounds.

5. Conclusions

UC dominates the composition of FABA from the study area up to 74.68%. In contrast, the maximum Glass produced is only 17.41%, while quartz content is up to 50.15%. It indicates inefficiency in the coal combustion process that may be due to applying the bed material that is not suitable for the boiler and/or the coal feed size. The higher UC and quartz and lower glass content will reduce the potential utilization of FABA as the raw material of construction.

Acknowledgments

The author would like to convey high appreciation to Universitas Gadjah Mada under the RTA scheme No. 3143/UN1.P.III/DIT-LIT/PT/2021, for partially supporting this research. Samples were provided by PT. Bukit Asam Tbk.

References

[1] Thenepalli T, Ngoc N T M, Tuan L Q, Son T H, Hieu H H, Thuy D T N, Thao N T T, Tam D T T, Huyen D T N, Van T T and Chilakala R 2018 Technological solutions for recycling ash s lag from the Cao Ngan coal power plant in Vietnam *Energies* **11** 8
[2] Kementerian Energi dan Sumberdaya Mineral 2016 Indonesia electricity development plan and Indonesia coal-ash management implementation *International Coal Based Power Conferences 2016* New Delhi India 17th March 2016
[3] Ohenoja K, Pesonen J, Yliniemi J and Illikainen M 2020 Utilization of fly ashes from fluidized bed combustion: A review *Sustainability* **12** 7 2988
[4] Besari D A A 2018 Komponen penyusun dan model keterdapatan rare earth elements and yttrium (REY) pada fly ash dan bottom ash *Bachelor thesis Gadjah Mada University*
[5] ASTM ASTM D7582-15 2015 Standard test methods for proximate analysis of coal and coke by macro thermogravimetric analysis
[6] ASTM D5865-13 2015 Standard test method for gross calorific value of coal and coke
[7] ASTM D2799-05a 2005 Standard test method for microscopical determination of the maceral composition of coal
[8] ICCP 2001 New inertinite classification (ICCP System 1994) *Fuel* **80** 459–71
[9] Sýkorová I, Pickel W, Christanis K, Wolf M, Taylor G H and Flores D 2005 Classification of huminite—ICCP System 1994 *Int. J. Coal. Geol.* **62** 1-2 85-106
[10] Pickel W, Kus J, Flores D, Kalaitzidis S, Christanis K, Cardott B J, Misz-Kennan M, Rodrigues S, Hentschel A, Hamor-Vido M and Crosdale P 2017 Classification of liptinite–ICCP System 1994 *Int. J. Coal. Geol.* **169** 40-61
[11] Hower J C 2012 Petrographic examination of coal-combustion fly ash *Int. J. Coal. Geol.* **92** 90-7
[12] Hower J C, Henke K R, Dai S, Ward C R, French D, Liu S and Graham UM (2017) Generation and nature of coal fly ash and bottom ash *Coal Combustion Products (CCP's)* Woodhead Publishing 21-65
[13] Kutchko B G and Kim A G 2006 Fly ash characterization by SEM–EDS *Fuel* **85** 17-18 2537-44
[14] Maier H, Spliethoff H, Kicherer A, Fingerle A and Hein K R G 1994 Effect of coal blending and particle size on NOx emission and burnout *Fuel* **73** 9 1447-52
[15] Ganguli R. and Bandopadhyay S., 2012, Relationship between Particle Size Distribution of Low-Rank Pulverized Coal and Power Plant Performance, Journal of Combustion, Article ID 786920
[16] Dai S, Zhao L, Peng S, Chou C L, Wang X, Zhang Y, Li D and Sun Y 2010 Abundances and distribution of minerals and elements in high-alumina coal fly ash from the Jungar Power Plant, Inner Mongolia, China *Int. J. Coal. Geol.* **81** 4 320-32
[17] Tomeczek J and Palugniok H 2002 Kinetics of mineral matter transformation during coal combustion *Fuel* **81** 10 1251-58

The potential of lithium enrichment in Lapindo Brantas, Mount Anyar, and Buncitan Mud Volcanoes, Sidoarjo District, East Java Province

N A Sari[1], I W Warmada[1]* and F Anggara[1]

[1] Department of Geological Engineering, Faculty of Engineering, Gadjah Mada University. Jl. Grafika No. 2 – Bulaksumur, Yogyakarta 55281 Indonesia.
*E-mail: warmada@ugm.ac.id

Abstract. Lithium is an alkaline, metallic element found primarily in the earth's crust and seawater. Recently, the industry's need, especially lithium batteries, has been increasing. Lithium exploration is directed to various geological environments that have the potential to contain lithium. Based on previous research, mud products from mud volcanoes can contain lithium. We conducted this research to determine the potential for lithium enrichment found in mud produced by mud volcanoes in Lapindo, Mount Anyar, and Buncitan. Samples were prepared to analyze petrography, X-ray diffraction (XRD), ICP-MS, and ICP-AES. Mineralogically, the mud sample comprises smectite, kaolinite, quartz, plagioclase, clinopyroxene, calcite, pyrite, and dolomite. Lithium in the mud sample has an average concentration of 92.5 ppm with the highest concentration of 130 ppm from the Lapindo Mud volcano. The lowest concentration of 70 ppm comes from the Buncitan Mud Volcano sample. Based on its mineralogy and structure, the dominant of smectite and kaolinite clay minerals can bind the lithium. Lithium in the study area is thought to come from altered rocks below the surface, which migrate and are bound to clay minerals predominantly found in subsurface mud.
Keywords: lithium enrichment, mud volcano, Lapindo Brantas, Mount Anyar, Buncitan.

1. Introduction

Elemental lithium is the lightest metal element and the densest of all the non-gaseous elements at 20°C and can float in water [1]. This element is not found freely in nature but is found in compounds with other minerals. Lithium is found in igneous rocks such as pegmatite, brine water, and clay minerals from hydrothermal alteration. The use of lithium in the industry is quite diverse. Lithium is an essential component in digital device and electric car batteries. Apart from the battery industry, lithium is also used in the metallurgical, ceramic, and glass industries, polymers, and health [2].

The potential for lithium sources in Indonesia is still open. One of the previous studies was carried out at BledugKuwu Mud Volcano by Rohmah *et al.* [3] by testing the mud's lithium content. Lithium content is found in montmorillonite clay minerals with a concentration of 29 ppm in wet conditions. Based on the research of Rohmah *et al.* [3], lithium has the potential to be found in mud from the natural phenomenon of mud volcanoes. Mud volcanoes are the appearances of a cone-shaped topography that occurs naturally from the materials below the earth's surface and consists of fine mud, liquid, and gas [4]. Lapindo mud is one of the largest mud sources that emerged in 2006 in the Kendeng Zone and the North Coast Alluvial Plains Zone [5]. Apart from Lapindo mud volcano, several other mud volcanoes are found, such as Mount Buncitan and Mount Anyar. This phenomenon produces mud products that can be utilized. The waste materials in the form of mud from mud volcanoes in the future can be a potential source of lithium.

Based on Tanikawa study [6], he concluded that Lapindo mud can contain lithium. The type of lithium-bearing clay minerals, the lithium concentration, and lithium enrichment in the mud have not been studied yet. Therefore, this research will test the lithium content in the mud in 3 locations that have never been carried out in previous research, namely in Lapindo Mud, Mount Anyar, and Buncitan, Sidoarjo, East Java. By knowing the lithium concentration and the mud mineralogy, we will know the possible factors controlling the lithium enrichment in the Lapindo mud volcano. With this research, it is hoped that lithium in volcanic mud in the Sidoarjo region can be identified and used in industry.

2. Geological setting

According to van Bemmelen [5], the research area belongs to the eastern Kendeng Zone and the North Coast Alluvial Plain Zone based on the physiographic division of eastern Java. The Kendeng Zone is characterized by a series of low hills with undulating morphology, with an altitude of 50–200 meters. The North Coast Alluvial Plain Zone is straight with 0–100 m with a relatively flat slope.

The research area's stratigraphy is based on the Regional Geological Map of the eastern part of the Surabaya and Sapulu sheets [7] and the northern part of the Malang sheet [8] (Figure 1). It is composed of the Lidah Formation from the Late Pliocene, which extends to the Pucangan Formation area. The Kabuh Formation is found in the north of the research area with sea facies and gradually changing to land facies, which is thought to be in conformity with the Jombang formation, and an unconformity with Quaternary volcanic rocks. Volcanic activity is located on the southern side of the study area. It is known to originate from the Lower Quaternary Volcano Formation, namely from Gendis Volcano rocks, Jombang Volcano rocks, and Anjasmara Muda volcanic rocks from the Middle Pleistocene. The Tuf Rabano Formation, namely from the rocks of Tengger Volcano, Arjuna Welirang Volcano, and Ringgit Volcano come from the Late Pleistocene-Holocene Period. The Upper Quaternary Volcano Formation, namely from the Penanggungan Volcano and Panderman Volcano, come from the Late Pleistocene-Holocene Period.

The research area is belong to the Java Island Central Depression Zone, which was formed due to the Eurasian plate's collision with the Indo-Australian plate. According to van Bemmelen [5], geological data show that the stratigraphy and tectonics of the eastern Kendeng Zone are still in an evolving state (tectonic processes are still ongoing). Duyfjes (1938, in Waluyo *et al.* [9]) concluded that the Gujangan anticline in the Porong area was intersected by a transversion fault with a descending eastern part. The Gujangan Fault is a sign of a transition between the tip of the Kendeng Zone, which falls in the Porong Delta, and the Madura Strait, which is still decreasing and filled with sediment, has not been folded. Along the central depression zone, the mud volcano phenomenon can be found in the Sidoarjo area, namely Lapindo Mud, Mount Anyar, and Buncitan mud volcanoes. Based on Putrohadi's research in Setiadi *et al.* [10], Sidoarjo mud originates from the Kujung Formation, the oldest formation in the Rembang mandala. The mud from the Kujung Formation under high pressure flowed upward by grinding the Pucangan Formation's claystone. This mudflow then penetrates the Watukosek fault zone, which passes through the Sidoarjo area.

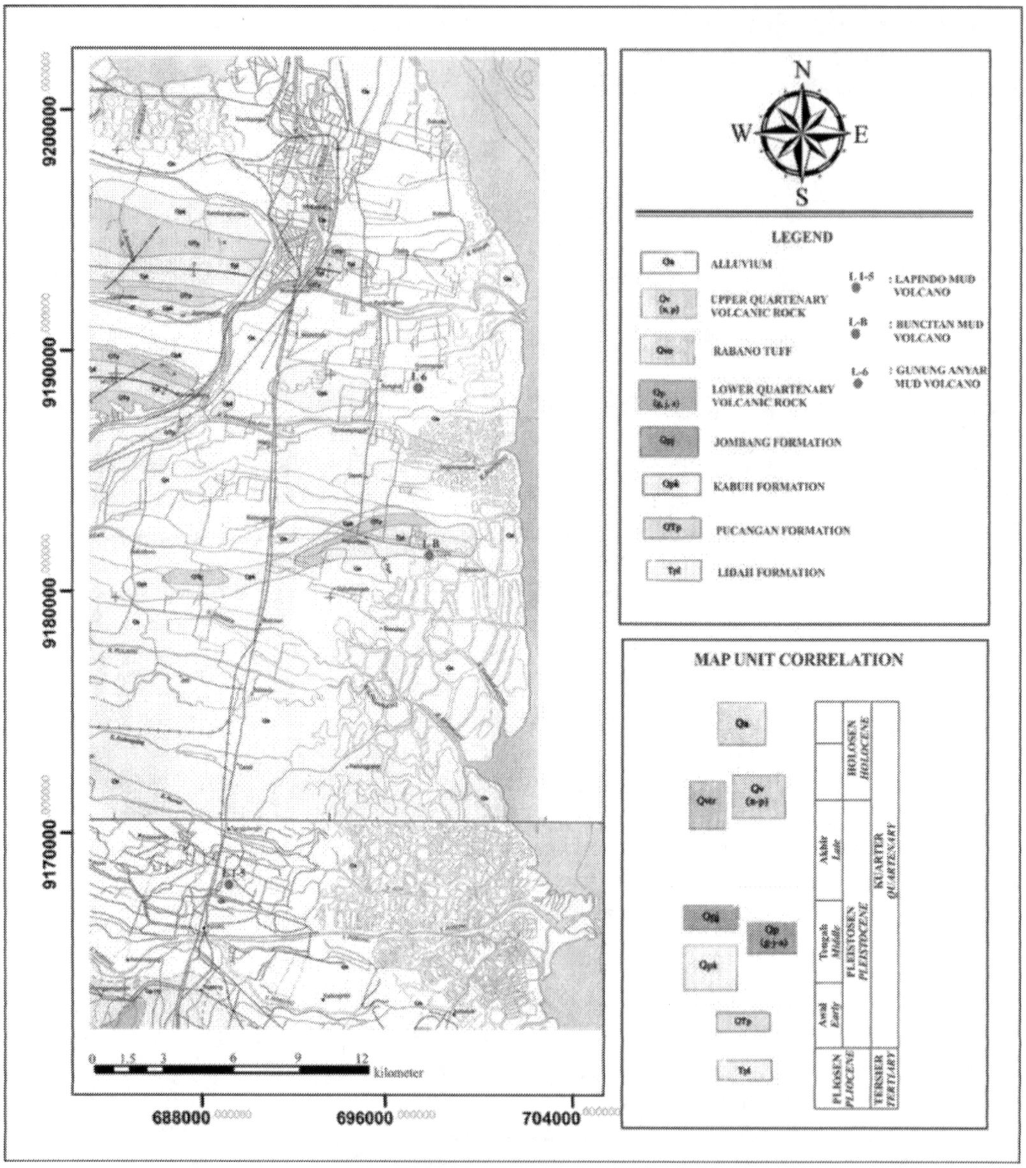

Figure 1. The regional geologic map of the eastern part of the Surabaya and Sapulu [6] and the northern part of the Malang [7], and the stratigraphic correlation of the study area. The locations of the research area are marked with a red symbol (•).

3. Methods

The study consisted of 8 samples of mud, with 5 mud samples collected from the Lapindo Mud Field, 1 mud sample from the Mount Anyar Field, and 2 mud samples from the mud volcano field in Buncitan Village. The analysis carried out in the study consisted of mineralogical analysis in the form of petrography and XRD and geochemical analysis in the form of ICP-MS, and ICP-AES. The petrographic analysis was carried out on 4 samples, divided into 2 fractions, namely the fraction <63 μm with a centrifuge and the fraction 63–125 μm. XRD, ICP-MS, and ICP-AES analyzes were performed on all samples. XRD analysis was carried out at the Central Laboratory, Geological Engineering Department UGM, while ICP-MS and ICP-AES were carried out at ALS Chemex, Canada. The analytical results were used to assess the mud's mineralogical and chemical characteristics, the concentration of lithium in the mud, and the factors causing the enrichment of lithium in the mud at Lapindo Mud Field, Mount Anyar, and Buncitan Village.

4. Results and Discussion

4.1. Mineralogical characteristics

Based on petrographic observations of thin slices of fraction <63 μm, it was found that the mineralogical composition of the mud samples included clay minerals, carbonate materials, plagioclase, quartz, opaque minerals, and pyrite. In the 63–125 μm fraction, the mineralogical composition was found, including clay minerals, carbonate materials, quartz, opaque minerals, plagioclase, and clinopyroxene (Table 1).

Table 1. Mineralogical composition of 4 samples from thin section petrography.

Sample code	Fraction <63 μm	Fraction 63–125 μm
L-3B	Clay minerals, carbonate material, quartz, opaque mineral	Clay minerals, clinopyroxene, quartz, plagioclase, opaque mineral, carbonate material
L-5B	Clay minerals, plagioclase, carbonate material, opaque mineral	Clay minerals, carbonate material, quartz, opaque mineral
L-6B	Clay minerals, pyrite, plagioclase, carbonate material, opaque mineral.	Clay minerals, carbonate material, quartz, opaque mineral.
L-BA	Clay minerals, carbonate material, plagioclase, opaque mineral.	Clay minerals, carbonate material, quartz, opaque mineral.

In the sample fraction <63 μm, the dominant clay minerals compose with a percentage of 57–75%, followed by 7–13% carbonate material, 7–10% opaque minerals, and 7–20% plagioclase. In one sample, quartz minerals with a percentage of 8.33% and pyrite with a percentage of 22% were found. In the 63–125 μm fraction of the sample, clay minerals continued to dominate with a percentage of 40–50%, followed by 11–14% quartz, and 9–14% opaque minerals. Also, in one sample, plagioclase minerals were found with a percentage of 11% and pyroxene 23%. Carbonate material in the form of foraminifera fossils were found with a percentage of 6–29%.

Based on XRD analysis, the mud sample's mineralogy consists of smectite, kaolinite, quartz, plagioclase, calcite, dolomite, and pyrite minerals (Figure 2). Smectite mineral is a clay mineral that dominates the sample, followed by kaolinite. The smectite clay minerals in the study area were formed by the feldspar diagenesis [11]. Kaolinite minerals were formed by weathering or hydrothermal alteration of aluminosilicate minerals [12].

The carbonate material, namely calcite, dolomite, and foraminifera fossils, is thought to have originated from the lithology of carbonate sedimentary rocks around the study area (Putrohadi in Setiadi *et al.* [10]). Meanwhile, the research sample's volcanic material is thought to come from volcanic activity around the research area [8].

4.2. Geochemical characteristics

Based on the results of the analysis using the ICP-MS and ICP-AES methods, the dominant mud samples were enriched by the main oxide compounds SiO_2, Al_2O_3, and Fe_2O_3, with a concentration range of SiO_2 47.1–60.8 wt.%, Al_2O_3 13.35–20 wt.%, and Fe_2O_3 6.63–12.9 wt.%. CaO were found with a concentration range of 1.64–3.12 wt.%, MgO 1.55–2.33 wt.%, Na_2O 1.92–2.85 wt.%, K_2O 1.06–1.6%, Cr_2O_3 0.006–0.014 wt.%, TiO_2 0.7–1.22 wt.%, MnO 0.09–0.15 wt.%, P_2O_5 0.09–0.16 wt.%, SrO 0.03–0.12 wt.%, and BaO 0.02–0.05 wt.%.

The dominant content of SiO_2 in the mud samples comes from quartz, feldspar, and clay minerals [13]. The high levels of Al_2O_3, Fe_2O_3, and K_2O are thought to be related to the abundance of clay minerals present in the sample. Besides, the Na_2O content is principally related to the presence of plagioclase feldspar. The opaque minerals present in the sample are related to TiO_2, dolomite minerals are related to the abundance of MgO, and calcite minerals are related to the abundance of CaO in the sample [13]. The highest concentration of trace elements in the sample, namely Sr reached 1065 ppm, element Ba reached 425 ppm, element V reached 360 ppm, and Zr reached 248 ppm.

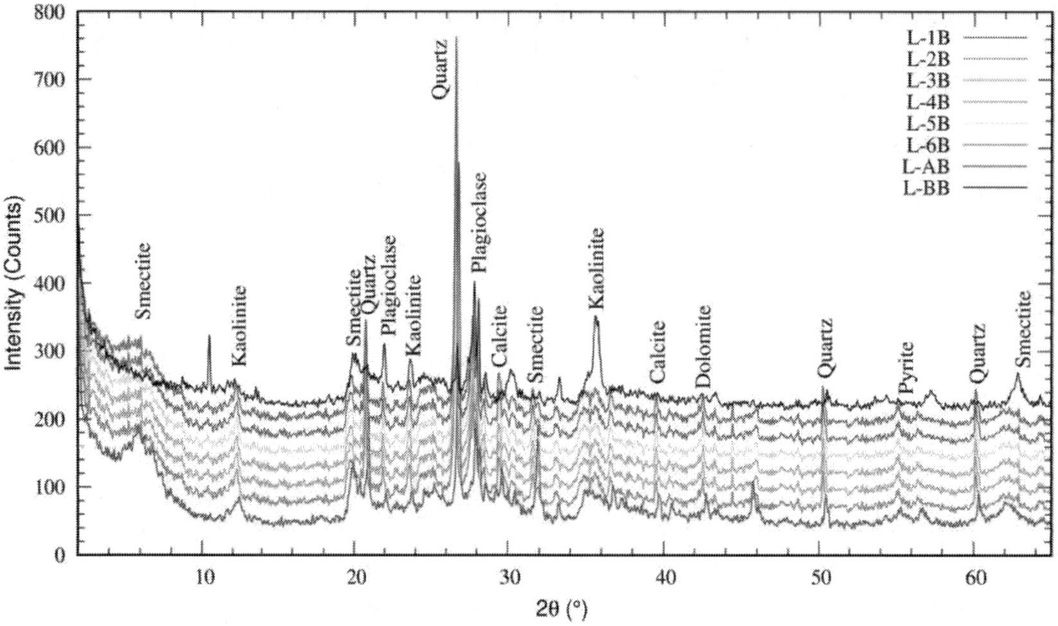

Figure 2. Mineral identification from XRD results for all samples based on random oriented measurement.

4.3. Lithium concentration

The lithium concentration in the mud sample can be seen in Table 2. The average concentration of lithium in the study area sample reached 92.5 ppm. The average concentration of lithium in the sample is above the average concentration of lithium in the earth's crust, which is 17–20 ppm [14].

Table 2. Lithium concentration in the samples.

Sample code	Li (ppm)
L-1B	90
L-2B	80
L-3B	130
L-4B	90
L-5B	100
L-6B	90
L-BA	90
L-BB	70
Average	92.5

The average concentration of lithium in the study sample, compared to the lithium concentration from several world brine water sources (Figure 2), is still classified as low. The economical range of lithium-containing brine sources in the world is 200–4000 mg/l [15].

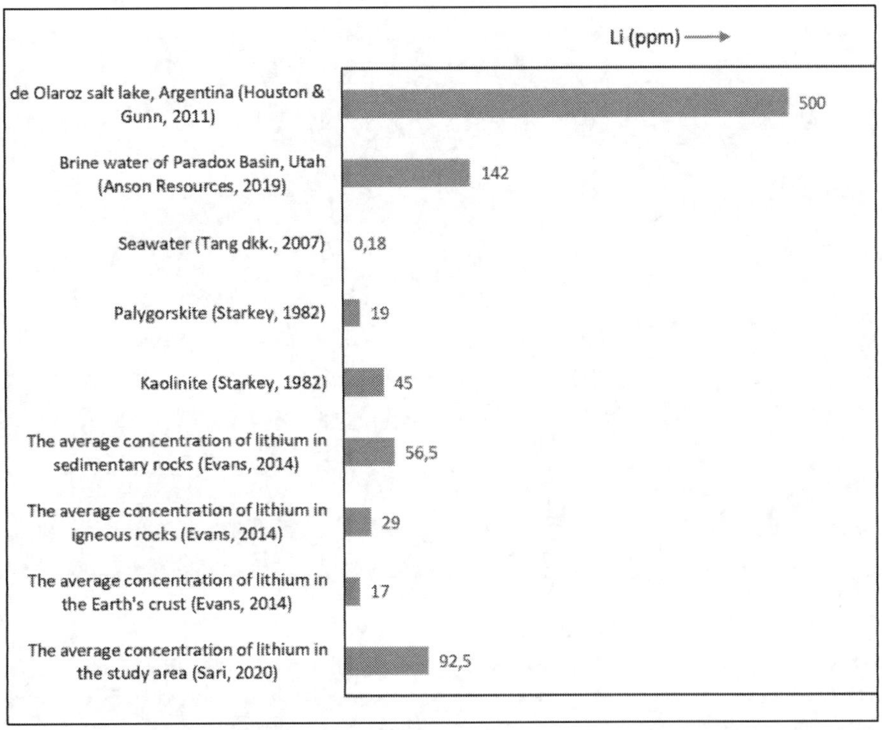

Figure 3. Bar diagram of comparison of lithium concentrations from mud samples in the study area with several sources of lithium in the world (Houston and Gunn [16]; Anson Resources [17]; Tang *et al.* [18]; Starkey [19]; Evans [14]; Sari *et al.* [this study]).

Based on its distribution, the highest lithium concentrations in the Lapindo mud are found on the eastern and northern sides of the mudflow center, while those on the south side tend to be further from the center of the eruption, have lower concentrations (Figure 4). In the mud samples from Mount Anyar and Buncitan, the lithium concentrations were found to be the same (Figure 5).

4.4. Lithium enrichment

Lithium is not found freely in nature but is found in compounds with other minerals. Weathering causes lithium to be released from its binding minerals. Lithium results from the weathering process can form sediment or bond in clay minerals, such as smectite, kaolinite, illite, and fibrous clay minerals. Lithium in clay minerals appears as impurities, inclusion materials, trapped in lattice structures, adheres to the surface of clay minerals, or become the isomorphic substitution of an element [20].

In research samples, lithium can be found in all mud samples containing kaolinite and smectite clay minerals. The dominant clay minerals compose the mud with a percentage based on petrography and XRD. Sample L-3B is a mud with the highest lithium concentration at 130 ppm. This sample is dominated by clay minerals smectite and kaolinite. The smectite mineral in the L-3B sample based on XRD has a relative percentage based on the semi-quantitative calculation of 27.3%, and kaolinite is 9.1%. The L-BB sample was a mud sample with the lowest lithium concentration at 70 ppm. The mineral content of smectite and kaolinite in the L-BB sample has a relative percentage based on semi-quantitative calculations, which tend to be lower than the other samples, namely 20% smectite 13.3% kaolinite.

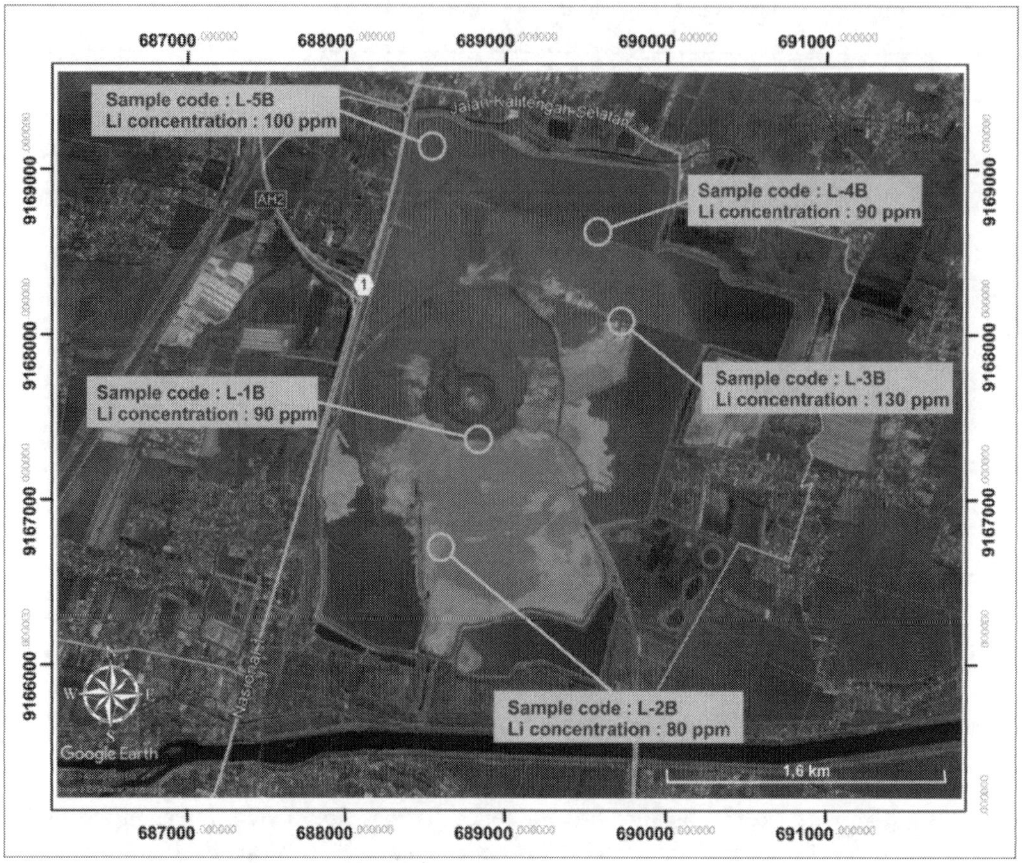

Figure 4. Map of the distribution of lithium concentrations in mud samples at Lapindo Mud Volcano. The sampling location is marked with a yellow circle symbol (○).

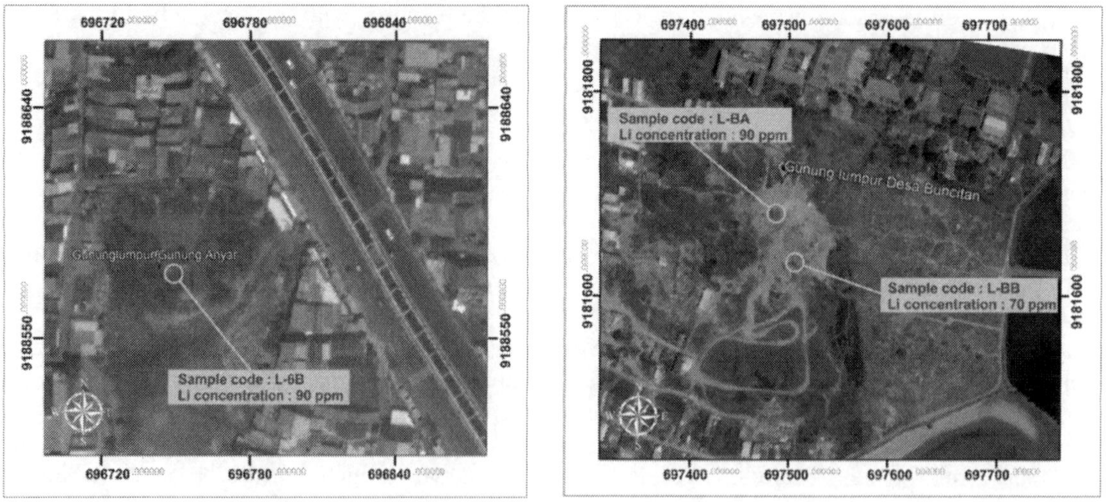

Figure 5. Lithium concentration distribution map in mud samples in a) Mount Anyar Mud Volcano and b) Buncitan Mud Volcano. The sampling location is marked with a yellow circle symbol (○).

IOP Conf. Series: Earth and Environmental Science **851** (2021) 012040 doi:10.1088/1755-1315/851/1/012040

According to Güven [20], smectite with lithium content is formed due to ion substitution. Ion substitution can occur between lithium and Al, Mg, and Fe. Lithium content is also found in the form of oxide compounds bound to the surface of kaolinite through an adsorption process [21]. Based on plotting between lithium and oxides of SiO_2, Al_2O_3, Fe_2O_3, Na_2O, CaO, MgO, K_2O, and TiO_2 from all mud samples, it is known that the correlation between lithium concentration and the percentage of oxide in the sample is known.

4.5. Model of lithium enrichment

The source of lithium in the study area is thought to come from rocks altered beneath the surface, which cause the release of lithium from its binding minerals. Lithium released from bound rock undergoes migration and is bound to clay minerals predominantly found in subsurface diapir. In the research area, which is included in the depression zone of East Java, mud diapir is easily found beneath the surface due to an overpressure process that causes the rock layers to bury and form mud.

Subsurface mud migrates to the surface of the ground through fault zones. The mud that has appeared on the surface over time gets a new supply of mud from within the surface and causes lithium to become increasingly enriched in the minerals in the mud, one of which is bound to or undergoes ion substitution with clay minerals found in the sample. The study area's lithium enrichment model is illustrated and shown in Figure 6.

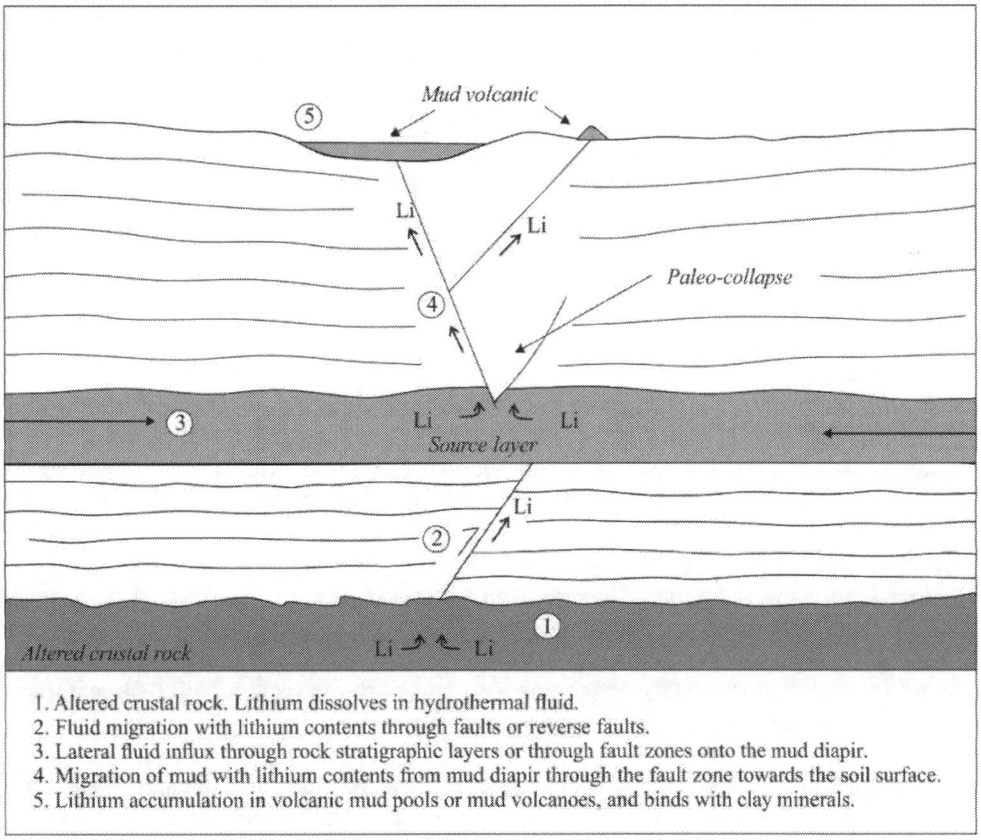

Figure 6. Lithium enrichment model for mud volcanoes in the study area.

5. Conclusions

Mineralogical characteristics of the mud consist of smectite, kaolinite, quartz, opaque minerals, calcite, plagioclase, dolomite, pyroxene, and pyrite. Chemically, the mud comprises dominant oxide compounds SiO_2 59.47%, Al_2O_3 20.24%, and Fe_2O_3 9.32%. The highest concentration of trace elements was found in Sr 1065 ppm, Ba 425 ppm, V 360 ppm, and Zr 248 ppm. Morphologically, the mud sample has an irregular-agglomeration appearance with few pores, cubic, and flaky.

The lithium concentration in the mud samples of the study area reached an average of 92.5 ppm with the highest concentration in the Lapindo mud sample, namely 130 ppm, and the lowest in the Buncitan mud sample, which was 70 ppm. The dominant smectite and kaolinite clay minerals in the mud act to bind lithium. Lithium in the study area is thought to come from altered rocks below the surface, which migrate and are bound to clay minerals predominantly found in subsurface mud.

Acknowledgments

The authors would like to thank the Department of Geological Engineering, Faculty of Engineering UGM, for providing grants to finance this research.

References

[1] Brown T, Walters A, Iodine N, Gunn G, Shawn RA, and Rayner D 2016 Mineral Profile Lithium. *Commodity Profiles*, 1-39

[2] Jaskula B 2016 *Lithium, in Mineral Commodity Summaries. Geological Survey (Ed.).* Government Printing Office

[3] Rohmah M, Lalasari L H, Wahyuadi J and Natasha N C 2018 Lithium recovery from BledugKuwu Mud volcano using water leaching method. In *2018 IEEE International Conference on Innovative Research and Development (ICIRD)* (pp. 1-5)

[4] Istadi B P, Pramono G H, Sumintadireja P and Alam S 2009 Modeling study of growth and potential geohazard for LUSI mud volcano: East Java, Indonesia. *Marine Petrol Geol* **26** 1724-1739

[5] van Bemmelen R W 1970. The geology of Indonesia (Vol. 1). Martinus Nijhoff

[6] Tanikawa W 2011 The Mechanism of Overpressure Generation in the LUSI Mud Volcano. In *E-Proceeding Symposium on Future Lusi*, BPLS, Surabaya, Indonesia

[7] Sukardi 1992 *Geologi Lembar Surabaya dan Sapulu, Jawa*. Pusat Penelitian dan Pengembangan Geologi

[8] Santoso S and Suwarti T 1992 *Peta Geologi Lembar Malang, Jawa* (Geological Map of the Malang Quadrangle, Jawa). Pusat Penelitian dan Pengembangan Geologi, Bandung.

[9] Waluyo G P and Utama W 2007. Deteksi pola patahan di desa Renokenongo, Porong Sidoarjo dengan metode geolistrik konfigurasi wenner. Laboratorium Geofisika FMIPA ITS Surabaya.

[10] Setiadi I, Darmawan A and Marjiyono M 2016 Pendugaan struktur geologi bawah permukaan daerah terdampak lumpur Sidoarjo (Lusi) berdasarkan analisis data geomagnet. *Jurnal Lingkungan dan Bencana Geologi* **7** 125-134.

[11] Morad S, Al-Ramadan K, Ketzer J M and De Ros L F 2010 The impact of diagenesis on the heterogeneity of sandstone reservoirs: A review of the role of depositional facies and sequence stratigraphy. *AAPG Bull* **94** 1267-1309.

[12] Nelson S A and Stephen A 2014 Weathering and clay minerals. *EENS* **2110** 1-7.

[13] Ali A J 2007 Petrography and major element geochemistry of Late Triassic Carpathian Keuper sandstones: Implications for provenance. *Bulletin de l'InstitutScientifique*, Rabat **29**, 1-14.

[14] Evans R K 2014 *Lithium. Critical Metals Handbook*, Ch.10. John Wiley & Sons Ltd. ISBN 978-0-470-67171-9.

[15] Kesler S E, Gruber P W, Medina P A, Keoleian G A, Everson M P and Wallington T J 2012 Global lithium resources: Relative importance of pegmatite, brine, and other deposits. *Ore Geol Rev* **48**, 55-69.

[16] Houston J and Gunn M 2011 Technical report on the Salar de Olaroz Lithium-Potash Project. NI 43Ā101 Report Prepared for Orocobre Ltd.

[17] Anson Resources 2019. *Due Diligence and Valuation Report*. Perth, Australia: Anson Resources Ltd. Retrieved September 18, 2020, from http://www.abid.co/sites/default/files/Anson %20Resources%20-%20ABID%20Report% 20-%2011%20March%202019.pdf

[18] Tang Y J, Zhang H F and Ying J F 2007. Review of the lithium Isotope System as a Geochemical Tracer. *Intl Geol Rev* **49**, p. 374–388

[19] Starkey HC 1982 *The role of clays in fixing lithium* (No. 1278). US Government Printing Office.

[20] Güven, N. 1991 On a definition of illite/smectite mixed -layer. *Clays and Clay Minerals* **39**(6), 661-662.

[21] Fairén AG, Losa-Adams E, Gil-Lozano C, Gago-Duport L, Uceda ER, Squyres SW and McKay CP 2015 Tracking the weathering of basalts on Mars using lithium isotope fractionation models. *Geochem, Geophys, Geosystems* **16**(4), 1172-1197.

IOP Conf. Series: Earth and Environmental Science **851** (2021) 012041 doi:10.1088/1755-1315/851/1/012041

Assessing the undiscovered resources of uranium and thorium in Mamuju, West Sulawesi

W Cakrabuana[1*], R C Ciputra[1] and H Syaeful[1]
[1]Center for Nuclear Minerals Technology (CNMT), Badan Tenaga Nuklir Nasional (BATAN), Jl. Lebak Bulus Raya No. 9, Jakarta, Indonesia
*E-mail: wira.cakrabuana@batan.go.id

Abstract. The exploration of nuclear minerals in Mamuju, West Sulawesi has been started in 2013 by the Center for Nuclear Minerals Technology (CMNT) – Badan Tenaga Nuklir Nasional (BATAN). There is a strong indication of uranium (U) and thorium (Th) minerals occurrences in the location. To follow the exploration result up, a resources assessment is needed. Due to the limited early exploration data, the concept of undiscovered resources can be used. Of all resource assessment methods suggested by International Atomic Energy Agency (IAEA), Energy Research and Development Administration (ERDA) method is the most appropriate method to be applied in the Mamuju case. In this research, Pocos de Caldas, Brazil (Osamu Utsumi Mine for U reference and Morro do Ferro Area for Th reference) was selected as control area due to its similarity on the geological condition and its more-advanced exploration stage compared to Mamuju. All the parameters for the assessment can be determined with the ERDA method based on the available information of the control area.

1. Introduction

CNMT-BATAN has started the exploration of nuclear minerals in Mamuju, West Sulawesi in 2013. After some passing years, enough evidence of the occurrence of U and Th mineralization in the location is indicated. To level the exploration stage up, an assessment of the number of resources is required.

The Mamuju area consists of hills and mountains with steep slopes. Hills in the study area are cone-formed and crater exist in some locations [1–3]. The exploration in Mamuju is still in the early stage so that the obtained data are still regional-scaled and quite simple. Those data would produce the value with a low certainty level if they were used to assess the resources. One of the resource types which can be assessed by using early-stage exploration data is the undiscovered resource. This resource is not quite usual in the mineral industry but more often applied for governmental affairs and inter-agency communication in both policymaking and commercial matters. By generating undiscovered resources of U and Th, the exploration in Mamuju can be leveled up to a more detailed exploration and the resource assessment result can be used as the early basis of exploration policymaking. Hence, this paper has the objective to assess the parameters for U and Th undiscovered resources in Mamuju using the most suitable method based on the available data.

2. The Geology of Study Area

Based on the results of geological mapping and petrographical observation, geology of Mamuju area consists of, from the oldest to the youngest, plutonic rocks, older sedimentary rocks, volcanic rocks, and younger sedimentary rocks [1–4]. The plutonic rocks were observed at Ampalas River in Ampalas Village, Kalukku District. The older sedimentary rocks exposed in this area are volcano-sedimentary in the form of claystones and sandstones. The volcanic rocks vary greatly in their either type or composition. According to regional geological data, volcanic rocks that existed in the study area consist of Talaya Volcanic Rocks and Adang Volcanic Rocks. In this study, Adang Volcanic Rocks are the only ones to discuss. Based on the results of satellite image interpretation and field geological mapping, Adang Volcanic Rocks can be classified into seven rock units, they are Tapalang, Ampalas, Adang, Malunda, Karampuang, Sumare, and Labuan Rano complexes. The younger sedimentary rocks are dominated by volcaniclastic products and carbonate rocks [1–5]. Figure 1 displays the geological map of Mamuju [4].

Generally, radioactive minerals are associated with felsic plutonic and volcanic rocks. Usually, U exists in the facies which is rich in vesicular and trachytic structures as well as volcanic breccia and spherulite glass. The enrichment of radioactive elements could also occur in mafic rocks. Leucite-basalt-composed volcanic rocks are formed in a potassic environment by magma with continental composition [6]. Mamuju Regency is the area with the highest radioactivity in Indonesia. The radioactivity there could reach the value of 2800 nSv/h. That high radioactivity value was mostly found in the area composed of volcanic rocks, especially Adang Volcanic Rocks. It was inferred that the high radioactivity was the result of the natural decay of radioactive minerals (containing U and Th). Because of their high abundances in nature, gamma ray intensity produced by the decay of U and Th can be easily detected by gamma-ray spectrometer. The calor produced by the fision reaction of isotope U-235 is the prime source of energy in nuclear power plant. [2,7].

According to the newest literature from IAEA which refers to the occurrences of the world's U and Th deposits, the exploration concept in Mamuju can be directed to volcanic and surficial deposit types. By using the concept, U could be accumulated on structural planes (structure-bound), bedding planes (strata-bound), or as volcano-sedimentary rocks [8,9]. The forming of U in volcanic rocks depends on the U extraction that occurred in felsic volcanic rocks during the transport of hydrothermal fluids and accumulation. The concentration of U in volcanic rocks increases in proportion to the increase in felsic content of the rocks [10]. The occurrences of radioactive minerals in Mamuju are related to the existence of intermediate-mafic Adang Volcanic Rocks. That thing might happen because of the advanced magma differentiation of the volcanic rocks [1,2,4].

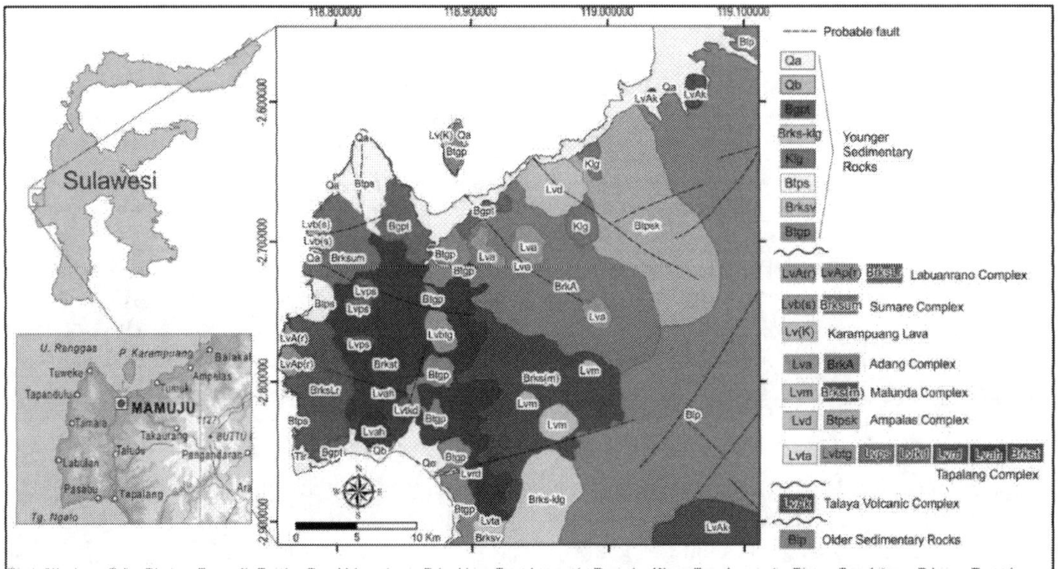

Qa = Alluvium, Qd = Stream Deposit, Bgpt = Coral Limestone, Brks-klg = Conglomerate Breccia, Klg = Conglomerate, Btps = Sandstone, Brksv = Breccia, Btgp = Limestone, LvA(r) = Labuanrano Andesite Lava, LvAp (r) Labuanrano Porphyritic Andesite Lava, BrksLr = Labuanrano Breccia, Lvb(s) = Sumare Breccia Lava, Brksum = Sumare Breccia, Lv(K) = Karampuang Lava, Lva = Adang Lava, BrkA = Adang Breccia, Lvm = Malunda Lava, Brks(m) Malunda Breccia, Lvd = Laba dome, Btpsk = Ampalas Breccia, Lvta = Taan Lava, Lvbtg = Botteng Lava, Lvps = Pangasaan Lava, Lvrd = Rante Doda Lava, Lvtkd = Takandeang Lava, Lvah = Ahu Lava, Brkst = Tapalang Breccia, LvAk = Kalukku Andesitic Lava, Blp = Claystone

Figure 1. The geological map of Mamuju (modified from [4]).

Some results obtained from geological and radiometric surveys of Mamuju are:

- U and Th minerals-bearing formation is Adang Volcanic Rocks;
- mostly consists of phonolite lava, autoclastic breccias (composed of phonolite), and limestones (clastic and reefal);
- regional structures are NE-SW and N-S-directed, it is inferred that the structures control the U and Th mineralizations;
- U and Th minerals' host rocks are phonolite-composed lava and autoclastic breccias;
- mean U and Th equivalent concentrations are 123 ppm and 416 ppm, consecutively; and
- there are three possibilities of mineralization mechanism: hydrothermal alteration (Hulu Mamuju, Taan, Pengasaan, Hulu Ampalas), dissolution and sedimentation by meteoric water (Botteng), and supergene enrichment (Ahu, Takandeang).

3. Basic Theory

Generally, a resource (mineral deposit, in this case) is a concentration of naturally occurring material in such form and amount that economic extraction is currently or potentially feasible [11]. There are some organizations which made the classification of resource. Table 1 attaches some resource classifications, especially which are related to nuclear minerals [12]. The undiscovered resource is an undiscovered body of mineral-bearing material whose existence is surmised from broad (regional) knowledge and theory [13]. Undiscovered resource intersects with some other terminologies, like potential resources, hypothetical resources, and speculative resources [12]. Assessment of undiscovered resources is significant to conduct for advanced exploration planning, economic evaluation, and land manage ment since the assessment provides information of the indication of mineral resource occurrences in the prospect area [11,14–16].

Table 1. The correlation of various terms in resource classifications [12,17].

Agency	Resource Classes				
DOE (1974-1982)	Reserves		Probable Potential Resources		Possible and Speculative Potential Resources
DOE (1983-current)	Reasonably Assured Resources (RAR)		Estimated Additional Resources (EAR)		Speculative Resources
DOI	Identified			Undiscovered	
	Measured Reserves	Indicated Reserves	Inferred Reserves	Hypothetical	Speculative
NEA/IAEA	Reasonably Assured Resources (RAR)		Estimated Additional Resources I (EAR=I)	Estimated Additional Resources II (EAR=II)	Speculative Resources (SR)
	Decreasing confidence in estimates ----->				

There are many methods to assess the undiscovered resource. Based on the data processing technique, there are qualitative methods (involving the delineation and ranking of prospective areas based on the potential level of mineral occurrence) and quantitative methods (estimating the tonnage of mineral to determine the geologically permissive area). Based on the data connecting technique, there are deterministic methods (premised on the causality between the tonnage of commodity and one or more geological features affecting to commodity) and probabilistic methods (connecting the tonnage of the commodity to geological features probabilistically). Based on the reference, there are data-driven method (premised on the geoscientific data in the study area) and knowledge-driven method (using conceptual geological models comprising geological features which support the mineralization) [11,14,15,18–21]. The probabilistic method is more prevalent to use in undiscovered resource assessment of various metal commodities [22], such as Platinum Group Elements (PGE) [23], Volcanic Massive Sulfides (VMS) [24], porphyry copper [25], and graphite [26]. Table 2 attaches the list of undiscovered resource assessment methods recommended by IAEA therewith the characteristics. [15].

Table 2. The list of undiscovered resource assessment methods recommended by IAEA [15].

Methods	Quantitative / Qualitative	Deterministic / Probabilistic	Objective / Subjective	Data-driven / Knowledge-driven	Brown Field / Green Field	Spatial / Non-Spatial
Boolean logic	Qualitative	Probabilistic	Subjective	Both	Both	Spatial
Index overlay	Qualitative	Probabilistic	Subjective	Both	Both	Spatial
Fuzzy logic	Qualitative	Probabilistic	Subjective	Both	Both	Spatial
Dempster-Shafer	Qualitative	Probabilistic	Subjective	Both	Both	Spatial
PROSPECTOR	Qualitative	Probabilistic	Subjective	Both	Both	Spatial
WofE	Qualitative	Probabilistic	Objective	Both	Brown	Spatial
Logistic regression	Qualitative	Probabilistic	Objective	Both	Brown	Spatial
Neural network	Qualitative	Probabilistic	Objective	Both	Brown	Spatial
Zipf's Law	Quantitative	Deterministic	Objective	Data	Brown	Non-spatial
Allais Model (1957)	Quantitative	Probabilistic	Objective	Data	Green	Non-spatial
ERDA	Quantitative	Probabilistic	Subjective	Both	Both	Spatial
NURE	Quantitative	Probabilistic	Subjective	Both	Both	Spatial
DSF	Quantitative	Probabilistic	Subjective	Both	Both	Spatial
3 part USGS	Quantitative	Probabilistic	Both	Both	Both	Spatial
One-level prediction	Quantitative	Probabilistic	Both	Both	Brown	Spatial
MPA + 3 part USGS	Quantitative	Probabilistic	Both	Both	Both	Spatial
MPA + one-level prediction	Quantitative	Probabilistic	Both	Both	Brown	Spatial

4. Methodology

Based on the exploration condition and data availability, the criteria of resource assessment method required are:

- quantitative: to produce the exact value of undiscovered resources as the reference for the next-stage exploration;
- probabilistic: due to the absence of adequate grade and tonnage data to be assessed deterministically;
- subjective: due to the inadequate data to generate objective model, geological and radiometric data are the only available data;
- data- and knowledge-driven: to prove the knowledge of the experts by geological and radiometric data obtained in the field;
- appropriate for greenfield: so that the method can accommodate the area with limited data; and
- spatial: because the only available data are maps and the expected results are maps too.

Of all the methods attached in Table 2, three methods comply with all the criteria required; they are Energy Research and Development Administration (ERDA) method, National Uranium Resources Evaluation (NURE) method, and the Deposit-Size Frequency (DSF) method [13,14]. ERDA method was selected to assess the undiscovered resources in Mamuju due to the consideration that the method is the most simple and can be adjusted to the available data. NURE and DSF methods are not appropriate enough to be applied since there are some unavailable parameters in both methods' formulas.

Most of the methods need the control area as the comparator of conditions to the study area being researched [18,20,27,28]. The control area is a geographical location where the mineralization, geological characteristics and parameters of the deposits, notably resources and average grades, have been well studied [15].

In the ERDA method, the undiscovered resource is termed a Potential Resource (PR). ERDA method has the following formula:

$$PR = N.F.U.T \tag{1}$$

Parameter N is the extent of the favorable area. A favorable area is a geographical location according to research data has geological features which bring an advantage to mineralization [13]. In this study, the favorable area was defined as the rocks unit according to field observation and laboratory test is proven to contain U and/or Th minerals. Parameter N was given in the m^2 unit.

Parameter F is the fraction of area favorability. This parameter represents the level of similarity between the study area and the control area. F was obtained by conducting the comparative study of geological conditions in both areas. A more similar condition means higher F. Parameter F was stated in the form of percentage fraction.

Parameter U is the fraction of unexplored area. This parameter implies the ratio of exploration stage between the study area and control area. U was acquired by undertaking the comparative study of the exploration stage in both areas. As the study area approaches the exploration stage of the control area, the higher U is. Parameter U was given in the form of a percentage fraction.

Parameter T is the density of mineralization. This parameter reflects how many mineral tonnages in a certain extent of the area. To determine this parameter, a literature study respecting mining activity in the control area was conducted. Parameter T was stated in the ton/m^2 unit.

After all the parameters were attained, PR could be assessed. It should be made sure that the units of all parameters were fit to the convention in the paragraphs above. By using Equation 1, PR was obtained in tons.

5. The Selection of Control Area

In this research, Pocos de Caldas, Brazil, was selected as the control area to the study area in Mamuju, Indonesia. Pocos de Caldas was selected due to its similarity in geological conditions and its more advanced exploration stage compared to Mamuju. Actually, among other alkaline volcanic complexes, only Pocos de Caldas associated with volcanogenic U deposit. Respecting to the exploration stage, two locations have been researched deeply in Pocos de Caldas, are Osamu Utsumi Mine (U potential site) and Morro do Ferro Area (Th potential site).

The alkaline complex of Pocos de Caldas is situated in the southern part of the State of Minas Gerais close to the boundary with the State of Sao Paulo. The complex is roughly circular and covers an area of about 1000 km^2. The complex is situated on a plateau and is essentially saucer-shaped in a cross-section reflecting the configuration of the collapsed caldera. The elevations vary from 1300 m in the central part of the depression to about 1600 m around the periphery [29].

The radioactive material occurs in rocks known as caldasite which is a mixture of zirconium oxide (baddeleyite) and zirconium silicate (zircon). For many years, all U exploration was focused on the caldasite deposits since it was thought that U could be extracted economically from refractory ores [29]. About 85 alkaline intrusive bodies are known in Brazil. These vary considerably concerning their age, size, and geographical distribution [30].

The complex is characterized by a wide variety of rock types which has given rise to the hypothesis of varied injections of magma accompanied by differentiation phenomena. The most common rock types belong to the family of the nepheline syenites, and in order of decreasing importance, these include tinguaites, phonolites, nepheline syenites, foyaites, lujaurites, and chibinites. The complex has a strong agpaitic tendency. Besides the above are found rocks of volcanic origin such as tuffs, agglomerates, ashes, and breccias. In the central part of the complex, the rocks have been highly altered by intense potassic metasomatism [29].

6. Parameters Assessment for Undiscovered Resources

Parameter N (the extent of the favorable area) was determined by delineating the areas which are proven to contain U and/or Th minerals (Adang Volcanic Rocks) based on the geological observation and have > 100 ppm equivalent U concentration based on radiometric measurement (100 ppm is a boundary value proposed by the experts in many best practices). From delineation, as many as three areas were obtained to be favorable. Area 1 consisted of Ahu, Pengasaan, Taan, Takandeang, Orobatu, and Botteng sectors. Area 2 consisted of Bebanga and Ampalas sectors. Area 3 consisted of the Hulu Mamuju sector. The extents of Area 1, Area 2, and Area 3 were 104,560,000 m^2; 165,720,000 m^2; and 4,580,000 m^2; respectively. Favorable area maps of Mamuju and their extents in m^2 (parameter N) are displayed in Figure 2.

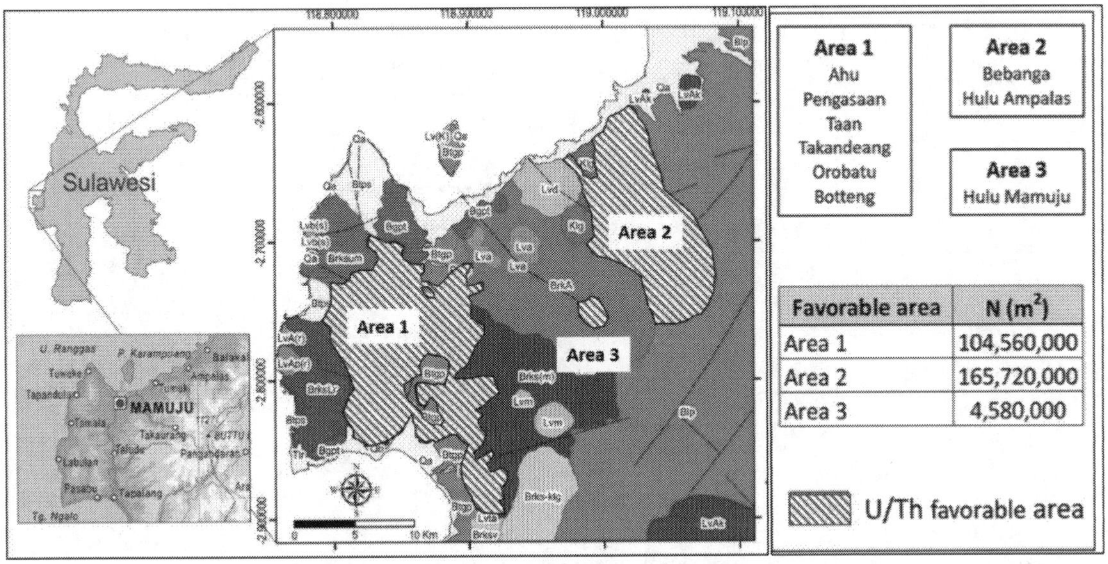

Figure 2. The determination of parameter N.

After parameter N was obtained, a comparative study of geological conditions between the study area (all exploration sectors of Mamuju) and the control area (the whole Pocos de Caldas) was conducted. Parameter F (the fraction of area favorability) was obtained by scoring the similarity of geological conditions in both areas. A more similar condition means a higher score. The score was given in the form of a percentage fraction. The comparative study results in an F value of 0.66. Table 3 attaches the result of parameter F determination.

As parameter F was obtained, a comparative study of the exploration stage between study area (all exploration sectors of Mamuju) and control area (Osamu Utsumi Mine for U reference and Morro do Ferro Area for Th reference) was undertaken. Parameter U (the fraction of unexplored area) was obtained by scoring the percentage of the study area's exploration stage to the control area's exploration stage. Parameter U was given in the form of percentage fraction as well as parameter F. The comparative study results in the U value of 0.34 for U and 0.41 for Th. Table 4 attaches the result of parameter U determination.

The last parameter required in the ERDA method is T (the density of mineralization). T is the ratio of the control area's deposit tonnage (in ton) to the extent of the control area (in m^2). The tonnage and the extent of the control area were obtained through the literature study respecting the mining activity in Osamu Utsumi Mine and Morro do Ferro Area. The tonnage of U in Osamu Utsumi Mine was as many as 21,800 tons (in form of U_3O_8) in an area extent of 20,000,000 m^2. The tonnage of Th in the Morro do Ferro Area was as many as 17,500 tons (in form of ThO_2) in an area extent of 3,140,000 m^2. Parameter T was determined as many as 0.0011 for U and 0.0056 for Th. Table 5 attaches the result of parameter T determination.

Table 3. The determination of parameter F (summarized from [1–5,29,31,32]).

Parameter	Mamuju, Indonesia	Pocos de Caldas, Brazil	F
Regional Parameter			
Tectonic setting	situated on the edge of Eurasia continental plate.	situated on the center of South America continental plate.	0.5
Stratigraphy	from older to younger, Mamuju is composed by Latimojong Fm (metamorphic rocks), Talaya Volcanic Rocks (autoclastic breccia, tuff, lava), Adang Volcanic Rocks (tuff, lava, autoclastic breccia), intrusive rocks (granite, diorite), Mamuju Fm (limestone), Tapalang Mb, aluvium deposits.	from older to younger, Pocos de Caldas is composed by lujaurite/cibinite, tuff and pyroclastic rocks, foyaite, tinguaite, phonolite, and potassic rocks.	0.6
Structure	reverse faults, fault/joint lineaments, NE-SW dominant strike	ring fault system composing caldera, NW-SE and NE-SW dominant strikes	0.5
Geomorphology	mountains and hills	caldera	0.4
Climate	tropical	tropical	0.9
Local Parameter			
Mineralization type	volcanic, surficial	volcanic, surficial	0.9
Source rocks	Middle Miocene - Upper Miocene phonolites	Upper Cretaceous caldasites	0.4
Mineral transportation	hydrothermal fluids flow, meteoric water flow	hydrothermal fluids flow, meteoric water flow	0.9
Host rocks	phonolite	phonolite and nepheline syenite	0.6
Mineral trap	structural	structural	0.9
Mineral/element products	primary (thorite, thorianite, davidite), secondary (gummite, autunite)	primary (pitchblende, U black oxides, thorite), secondary (pitchblende, U black oxides, thorite)	0.4
Preservation of mineral deposits	chemical weathering	chemical weathering	0.9
		Mean F (0-1)	0.66

Table 4. The determination of parameter U (summarized from [1–5,29,31,32]).

Exploration activity	Mamuju (all sectors)	Osamu Utsumi Mine (OUM)	Morro do Ferro Area (MFA)	Mamuju to OUM percentage	Mamuju to MFA percentage
A. Geological survey	1:25000 scale	1:1000 scale	1:5000 scale	0.5	0.6
B. Geochemical survey	1:25000 scale	1:1000 scale	1:5000 scale	0.5	0.6
C. Geophysical survey					
1. Radiometric mapping	200 m grid	25 m grid	50 m grid	0.25	0.3
2. Geomagnetic mapping	2 km grid	250 m grid	500 m grid	0.25	0.3
3. Geophysical logging	spectral gamma ray logging, 80 m average depth	spectral gamma ray logging, 150 m average depth	spectral gamma ray logging, 120 m average depth	0.5	0.6
E. Technological survey					
1. Exploration boring	5 spots / 100 km^2, 80 m average depth	5 spots / km^2, 150 m average depth	5 spots / 10 km^2, 120 m average depth	0.5	0.6
F. Topographic mapping	1:25000 scale	1:1000 scale	1:5000 scale	0.25	0.3
G. Laboratory analysis					
1. XRF	1:25000 scale	1:1000 scale	1:5000 scale	0.25	0.3
2. Fluorometry	1:25000 scale	1:1000 scale	1:5000 scale	0.25	0.3
3. Petrography	1:25000 scale	1:1000 scale	1:5000 scale	0.25	0.3
4. Mineragraphy	1:25000 scale	1:1000 scale	1:5000 scale	0.25	0.3
		Mean U (0-1)		0.34	0.41

Table 5. The determination of parameter T (summarized from [29,31–33]).

Control area	U_3O_8 tonnage (ton)	ThO_2 tonnage (ton)	The extent of control area (m^2)	T U (ton/m^2)	T Th (ton/m^2)
Osamu Utsumi Mine	21,800		20,000,000	0.0011	
Morro do Ferro Area		17,500	3,140,000		0.0056

From the analysis above, it is known that the most suitable undiscovered resource assessment method is the ERDA method and the parameters for the assessment can be determined by selecting the appropriate control area. Furthermore, the potential resource can be obtained from the parameters. The selected control area, Pocos de Caldas in Brazil has a similar geological condition as Mamuju Area. Pocos de Caldas is even more suitable for being a control area as it has areas for both U and Th mineralizations, which are the exploration target of CNMT BATAN in Mamuju Area. All the parameters in the ERDA method, except N, are largely affected by the exploration of the control area. Thus, the potential resource of U and Th in Mamuju Area obtained from the assessment can be a basis for the continuation decision of exploration in the area, but for a more confident resource to assess the economic potential of the deposit, more advanced exploration method for identified resource estimation should be done.

7. Conclusion

The required criteria of the method to assess the undiscovered resources of Mamuju are quantitative, probabilistic, subjective, data- and knowledge-driven, appropriate for greenfield, and spatial. The method which meets all those criteria is the ERDA method. The favorable target to assess in Mamuju is Adang Volcanic Rocks which is proven to contain U and/or Th minerals and the area on the formation which has > 100 ppm U equivalent concentration. Pocos de Caldas in Brazil was selected as the control area to Mamuju due to its similarity in the geological conditions and its more advanced exploration stage. All the parameters for the assessment can be determined with the ERDA method based on the available information of the control area.

8. Recommendation

To increase the certainty level of the U and Th resources assessed in Mamuju, it is necessary to upgrade the exploration stage by expanding and detailing all kinds of exploration activity.

Acknowledgments

The authors express a lot of gratitude to Mr. Yarianto Sugeng Budi Susilo as the Head of CNMT BATAN who has facilitated the entire research. The authors express the gratitude as well to all personnel of the Exploration Division, CNMT BATAN who have devoted support, help, criticism, and suggestion during the process of the research.

References

[1] Sukadana I G, Syaeful H, Suharji, Haryanto D dan Widodo 2018 Radioactive Elements Resources and Rare Earth Elements Potency in Taan Sector, Mamuju, West Sulawesi *MGEI Southwest Pacific Resources 2018* (Makassar: Masyarakat Geologi Ekonomi Indonesia) 103–7

[2] Syaeful H, Sukadana I G dan Sumaryanto A 2014 Radiometric mapping for Naturally Occurring Radioactive Materials (NORM) assessment in Mamuju, West Sulawesi *Atom Indones.* **40** 33–9

[3] Sukadana I G dan Syaeful H 2016 Uranium Exploration in Sulawesi *MGEI 8th Annual Convention 2016* (Bandu: Masyarakat Geologi Ekonomi Indonesia) hal 112–5

[4] Sukadana I G, Harijoko A dan Setijadji L D 2015 Tataan Tektonika Batuan Gunung Api di Komplek Adang, Kabupaten Mamuju, Provinsi Sulawesi Barat *Eksplorium* **36** 31–44

[5] Indrastomo F D, Sukadana I G dan Suharji 2017 Identifikasi Pola Struktur Geologi sebagai Pengontrol Sebaran Mineral Radioaktif berdasarkan Kelurusan pada Citra Landsat-8 di Mamuju , Sulawesi Barat *Eksplorium* **38** 71–80

[6] Burwash R A dan Cavel P A 1978 Uranium-Thorium Enrichment in Alkali Olivine Basalt Magma - Simpson Island Dyke, Northwest Territories Canada *Contributions to Mineralogy and Petrology* (Springer-Verlag)

[7] Iskandar D, Syarbaini dan Kusdiana 2007 *Map of Environmental Gamma Dose Rate of Indonesia* (Jakarta)

[8] Dahlkamp F J 1993 *Uranium Ore Deposits* (Berlin: Springer-Verlag)

[9] Dahlkamp F J 2009 *Uranium Deposits of the World: Asia* (Berlin, Heidelberg: Springer-Verlag)

[10] Breit G N dan Hall S M 2011 Deposit Model for Volcanogenic Uranium Deposits *U.S. Geol. Surv. Open-File Rep.* **1255**

[11] Singer D A dan Menzie W D 2010 *Quantitative Mineral Resource Assessments - An Integrated Approach* (New York: Oxford University Press)

[12] International Atomic Energy Agency 1998 Classification of uranium reserves/resources *IAEA TECDOC* **1035** 93

[13] Finch W I dan McCammon R B 1987 Uranium Resource Assessment by the Geological Survey: Methodology and Plan to Update the National Resource Base *U.S. Geol. Surv. Circ.* **994** 31
[14] International Atomic Energy Agency 1992 Methods for the estimation and economic of undiscovered uranium endowment and resources: An instruction manual *IAEA-TECDOC* **344** 1992
[15] International Atomic Energy Agency 2018 Quantitative and Spatial Evaluations of Undiscovered Uranium Resources *IAEA-TECDOC* **1861** 627
[16] Barnes R G, Jaireth S, Miezitis Y dan Suppel D W 1999 Regional mineral potential assessment for land use planning: GIS-based examples from Eastern Australia *Australas. Inst. Min. Metall.* **4** 601–5
[17] McKelvey V E 1967 Mineral resource estimates and public policy *Am. Sci.* **60** 3240
[18] Harris D P dan Agterberg F P 1981 The appraisal of mineral resources *Econ. Geol.* **75th Anniv** 897–938
[19] Porwal A K dan Kreuzer O P 2010 Introduction to the Special Issue: Mineral prospectivity analysis and quantitative resource estimation *Ore Geol. Rev.* **38** 121–7
[20] Singer D A 2010 Progress in integrated quantitative mineral resource assessments *Ore Geol. Rev.* **38** 242–50
[21] Taylor R B dan Steven T A 1983 Definitions of mineral resource potential *Econ. Geol.* **78** 1268 70
[22] Rasilainen K, Eilu P, Halkoaho T, Heino T, Huovinen I, Iljina M, Juopperi H, Karinen T, Kärkkäinen N, Karvinen A, Kontinen A, Kontoniemi O, Kousa J, Lauri L S, Lepistö K, Luukas J, Makkonen H, Manninen T, Niiranen T, Nikander J, Pictikäinen K, Räsänen J, Sipilä P, Sorjonen-Ward P, Tiainen M, Tontti M, Törmänen T dan Västi K 2017 Assessment of undiscovered metal resources in Finland *Ore Geol. Rev.* **86** 896–923
[23] Rasilainen K, Eilu P, Halkoaho T, Iljina M dan Karinen T 2010 Quantitative mineral resource assessment of undiscovered PGE resources in Finland *Ore Geol. Rev.* **38** 270–87
[24] Juliani C dan Ellefmo S L 2018 Resource assessment of undiscovered seafloor massive sulfide deposits on an Arctic mid-ocean ridge: Application of grade and tonnage models *Ore Geol. Rev.* **102** 818–28
[25] Hammarstrom J M, Mihalasky M J, Ludington S, Phillips J D, Berger B R, Denning P D, Dicken C L, Mars J C, Zientek M L, Herrington R J dan Seltmann R 2017 Undiscovered porphyry copper resources in the Urals—A probabilistic mineral resource assessment *Ore Geol. Rev.* **85** 181–203
[26] Cui N, Sun L, Bagas L, Xiao K dan Xia J 2017 Geological characteristics and analysis of known and undiscovered graphite resources of China *Ore Geol. Rev.* **91** 1119–29
[27] Drew L J 1986 Quantitative estimation of undiscovered mineral resources - a case study of US Forest Service Wilderness tracts in the Pacific Mountain system. *Econ. Geol.* **81** 80–8
[28] Singer D A dan Mosier D L 1981 Review of regional mineral resource assessment methods *Econ. Geol.* **76** 1006–15
[29] Loureiro F E D L dan Santos R C D 1988 The Intra-Intrusive Uranium Deposits of Pocos de Caldas, Brazil *Ore Geol. Rev.* **3** 227–40
[30] Ulbrich H H dan Gomes C B 1981 Alkaline rocks from continental Brazil: a review *Earth Sci.* **17** 135–54
[31] Santos R C D 1978 Geologia e desenvolvimento mineiro: C-09 - Mina Osamu Utsumi *Reunion del Grupo Asesor sobre Geologia del Uranium para America Latina* (Lima: International Atomic Energy Agency)
[32] Waber N 1992 The supergene thorium and rare-earth element deposit at Morro do Ferro, Poços de Caldas, Minas Gerais, Brazil *J. Geochemical Explor.* **45** 113–57
[33] International Atomic Energy Agency 2019 Thorium Resources as Co- and By-products of Rare Earth Deposits *IAEA-TECDOC* **1892** 69

IOP Conf. Series: Earth and Environmental Science **851** (2021) 012042 doi:10.1088/1755-1315/851/1/012042

Characteristics and utilization of clay minerals in Kuwu Mud Volcano, Kradenan District, Grobogan Regency, Central Java Province

I G A A S Sari[1] and I W Warmada[1*]

[1]Department of Geological Engineering, Faculty of Engineering, Gadjah Mada University. Jl. Grafika No. 2 – Bulaksumur, Yogyakarta 55281 Indonesia.
*E-mail: warmada@ugm.ac.id

Abstract. The benefits of clay minerals are currently numerous. Therefore, this study aims to determine clay minerals' characteristics in the mud material recommended for use, especially in the ceramic and cosmetic industry. This research is located in Bledug Kuwu, in Kuwu Village, Kradenan District, Grobogan Regency, Central Java Province. Bledug Kuwu produces mudflows composed of gravel, sand, silt, plastic clay, and water. Field observations and sampling were carried out on any differences in the morphology of the existing mud volcanoes, namely inactive craters and active griffins, with an average area of about 0.8–5 meters of mud material distribution. Samples were prepared to analyze X-ray diffraction (XRD), cation exchange capacity (CEC), pH, grain size distribution, plasticity test, swelling test, shrinkage limit test, and specific gravity. The results of the analysis show that the dominant clay minerals were smectite, followed by kaolinite and illite minerals, with dominant clay to silt grain sizes, plasticity levels of 24.78% to 55.48%, swelling percentage of 7.8%. up to 12.5%, with shrinkage limits of 9.36% to 14.83%, specific gravity values of 2.50–2.84, cation exchange capacity (CEC) values of 9.96–20.47 meq/100g, and the pH value shows the numbers 8.01–8.40. These characteristics can be recommended to use clay minerals as raw material for cosmetics and ceramics.

1. Introduction

Mud volcanism is a unique phenomenon. The periodic eruptions of large quantities of gases (mainly hydrocarbons) included rock fragments ranging in size from small to several meter blocks. The eruptive products accumulate near volcanic cone vents and craters. The size of a volcanic cone depends on the frequency of eruptions and the eruptive material's character [1]. Mud volcanoes have long been known in Indonesia. There are several active mud volcanoes, including Bledug Kuwu, Bledug Medang, Bledug Kesongo in Central Java Province [2]. Bledug Kuwu is a mud volcano phenomenon located in Kuwu Village, Kradenan District, Grobogan Regency (Figure 1). This 45-hectare area is unique and is one of the mainstay tourist attractions in Grobogan Regency. The mud is composed of gravel, sand, silt, plastic clay, and water. The material originates from the clay facies of the Pucangan Formation or the Pleistocene Lidah Formation from a depth of 750 to 1900 meters [3]. According to Ibrahim *et al.* [4], the mud volcano formed millions of years ago (5 million years) can upward to the surface until now due to tectonic pressure. Bledug Kuwu Mud Volcano lies on the active mud volcano that stretches from Purwodadi, Cepu, Bojonegoro to Porong.

Research on mud volcanoes in Indonesia has been widely carried out by researchers such as [5,6,7,8,9]. The research includes a discussion of the dynamics of mud volcano movement, the characteristics of swelling and shrinkage in clay minerals, physicochemical properties of clay minerals in mud volcanoes, identification of mud volcano properties and their utilization, geological conditions, and characteristics of clay minerals in mud volcanoes. These studies took data from several fields, i.e., active mud volcano complexes such as the Kradenan Mud Volcano Complex, Cengklik mud volcano, and Sidoarjo mud volcano. In 2016, Utama researched the implications of the constituent materials' characteristics and the mud's viscosity value on the morphological formation of the Kesongo Mud Volcano in Blora Regency, Central Java Province [10]. This study aims to determine the mud material's mineralogical and physical characteristics in the Bledug Kuwu mud volcano (Figure 1) and determine the possible utilization of clay minerals that have economic value in various industrial fields, especially cosmetics and ceramics.

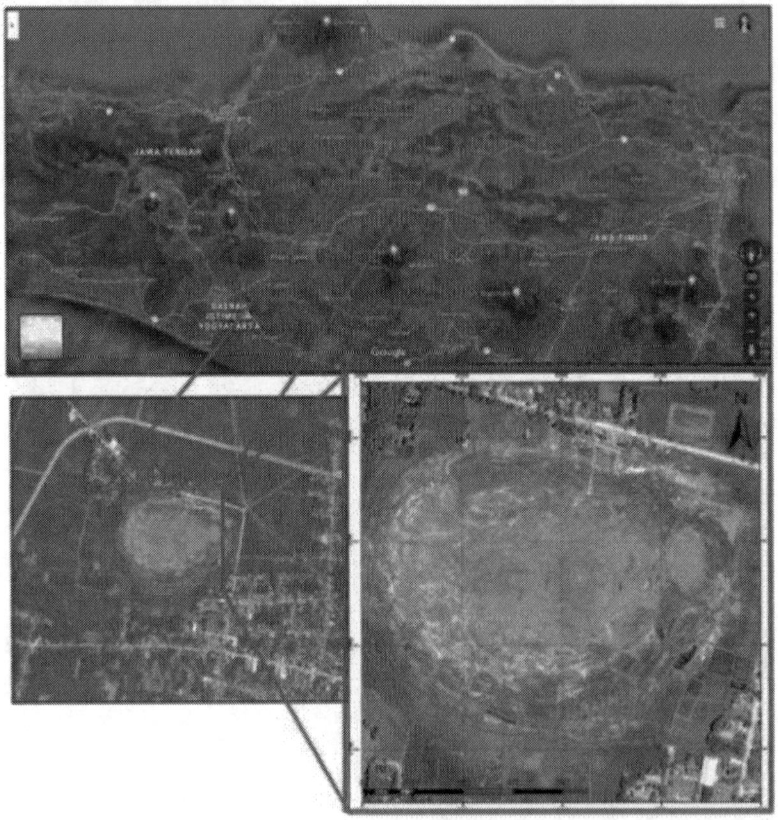

Figure 1. The research location is Bledug Kuwu mud volcano, Kradenan District, Grobogan Regency, Central Java Province. Map of Java Island and map of mud volcano locations (adapted from Google Maps).

2. Geological setting

The research area is located in the central part of Java Island. It is included in the Randublatung Zone, elongated depression or valley between the Kendeng Hills and Rembang Hills. According to Van Bemmelen, the Randublatung Depression was formed as a subsidence area, part of a regional isostasy equilibrium when the Rembang Hills and Kendeng Hills experienced tectonic uplift at the end of the Tertiary [2]. Stratigraphically, the Randublatung Zone has rock sequences such as those that develop in the Rembang Zone and Kendeng Zone. The stratigraphic order of the Randublatung Zone, according to Datun *et al.* [11], is the Tawun Formation, Ngrayong Formation, Wonocolo Formation, Ledok Formation, Mundu Formation, Selorejo Formation, and Tambakromo Formation. The anticline folding pattern in the Randublatung Zone still follows the Kendeng Zone fold pattern. This shows that the negative isostasy process is not the main factor in forming the Randublatung Zone. There are also compressive tectonic factors in the formation of the zone, as happened in the Kendeng Zone [12].

3. Methods

The research location is in Kuwu Mud Volcano, Kuwu Village, Kradenan District, Grobogan Regency, Central Java Province, the Kradenan Mud Volcano Complex (KMVK) in the Randublatung Depression Zone. This research uses grain size distribution analysis methods, X-Ray Diffraction (XRD) analysis, plasticity test, swelling test, shrinkage test, specific gravity analysis, cation exchange capacity (CEC) analysis and pH analysis. The interpretation of X-ray diffraction data was based on [13,14].

4. Results

4.1. Field data

Based on the field observations, there are two morphological types of mud volcano in the study area, i.e., active craters and active gryphons (Figure 2). The data of each morphology can be seen in the Table 1.

Table 1. Field Data of Kuwu Mud Volcano measurement.

Sample code	Coordinate		Morphology Type	Length (m)	Width (m)	Height (m)	Spread of Mud Materials	
	X	Y					Length (m)	Width (m)
STA 1	513317	9213218	Active crater	3	2	-	4	4
STA 2	513425	9213223	Active crater	5	5	-	5	5
STA 3	513509	9213247	Active gryphon	1.5	0.8	0.2	0.3	0.1
STA 4	513572	9213291	Active crater	5	2	1	6	3
STA 5	513546	9213178	Active crater	0.5	0.7	0.5	0.8	0.9
STA 6	513496	9213313	Active gryphon	1	1.5	0.5	2	2.5

4.2. Laboratory analysis results

4.2.1. *Physical properties of mud material*

The dominant grain size distribution of mud material ranges from clay to silt size, with a plasticity level of 24.78% to 55.48%, a swelling percentage of 7.8% to 12.5%, with a shrinkage limit of 9.36% to 14.83 %, and the value of specific gravity are 2.50–2.84. The chemical characteristics of clay minerals produce a Cation Exchange Capacity (CEC) value of 9.96–20.47 meq/100g and a pH value of 8.01–8.40.

4.2.2. *Mineralogical characteristics of mud material*

Clay minerals characteristics were based on the results of X-ray Diffraction (XRD) analysis (Figures 3–4). XRD analysis using both random powder and clay oriented was carried out on 4 selected samples, namely KW-01, KW-02, KW-03, and KW-04. The results of XRD analysis showed that the mud samples at Kuwu Mud Volcano consisted of clay minerals and non-clay minerals. Clay minerals have a greater abundance than that non-clay minerals. The clay mineral with the greatest abundance is smectite, followed by kaolinite and illite minerals. The non-clay minerals that appear are halite, calcite, quartz, and plagioclase minerals (Figure 3).

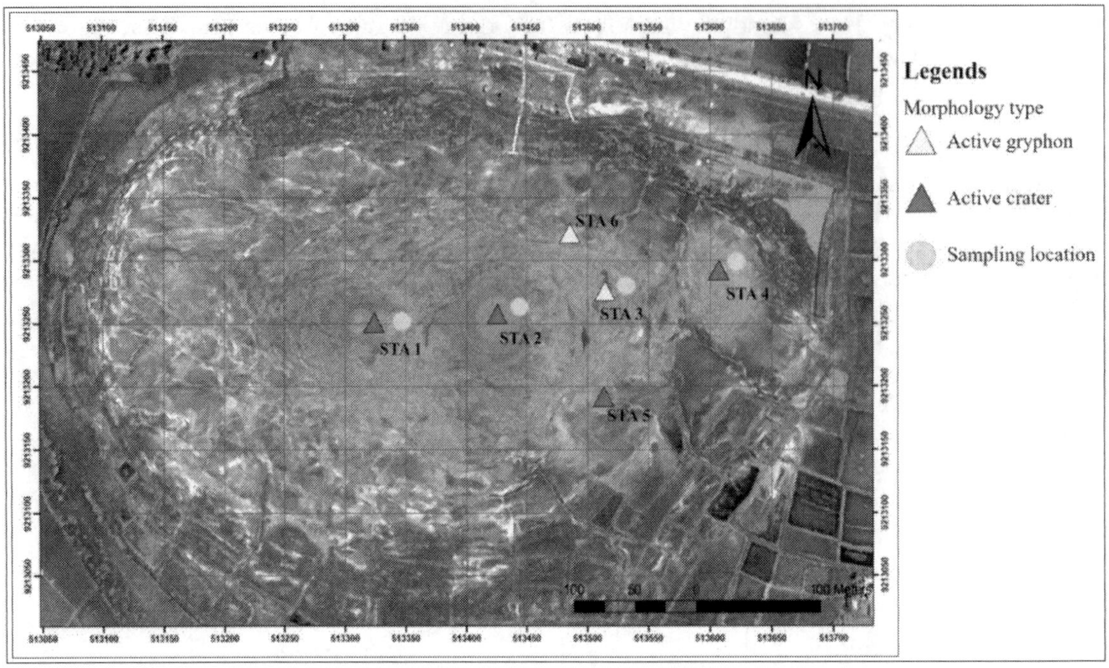

Figure 2. Map of the morphological distribution in the Kuwu Mud Volcano.

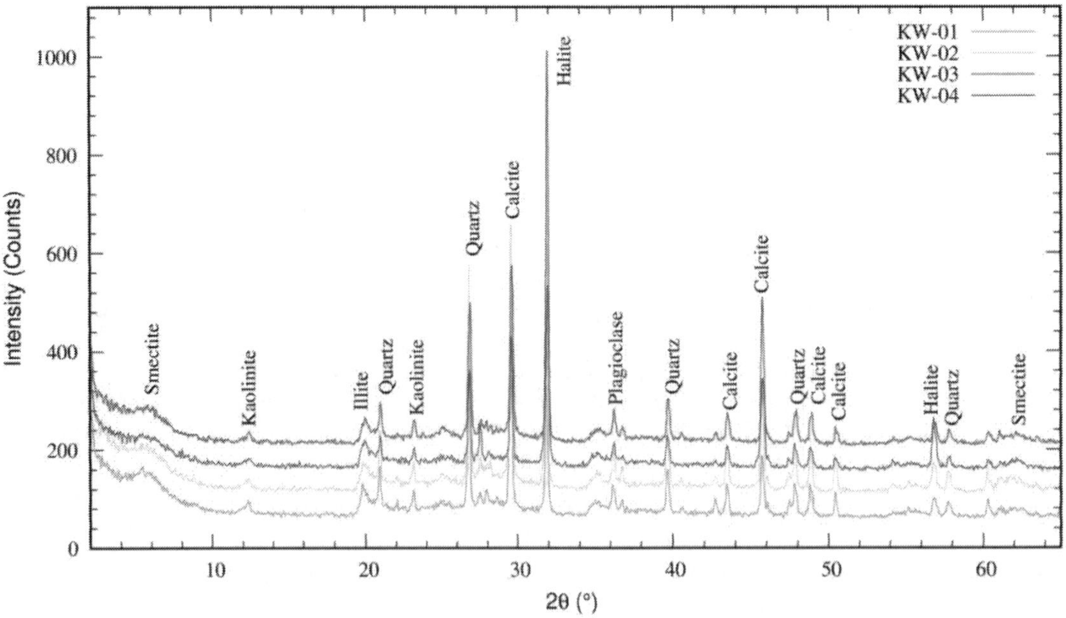

Figure 3. Mineral identification from XRD results for all four samples based on random oriented measurement.

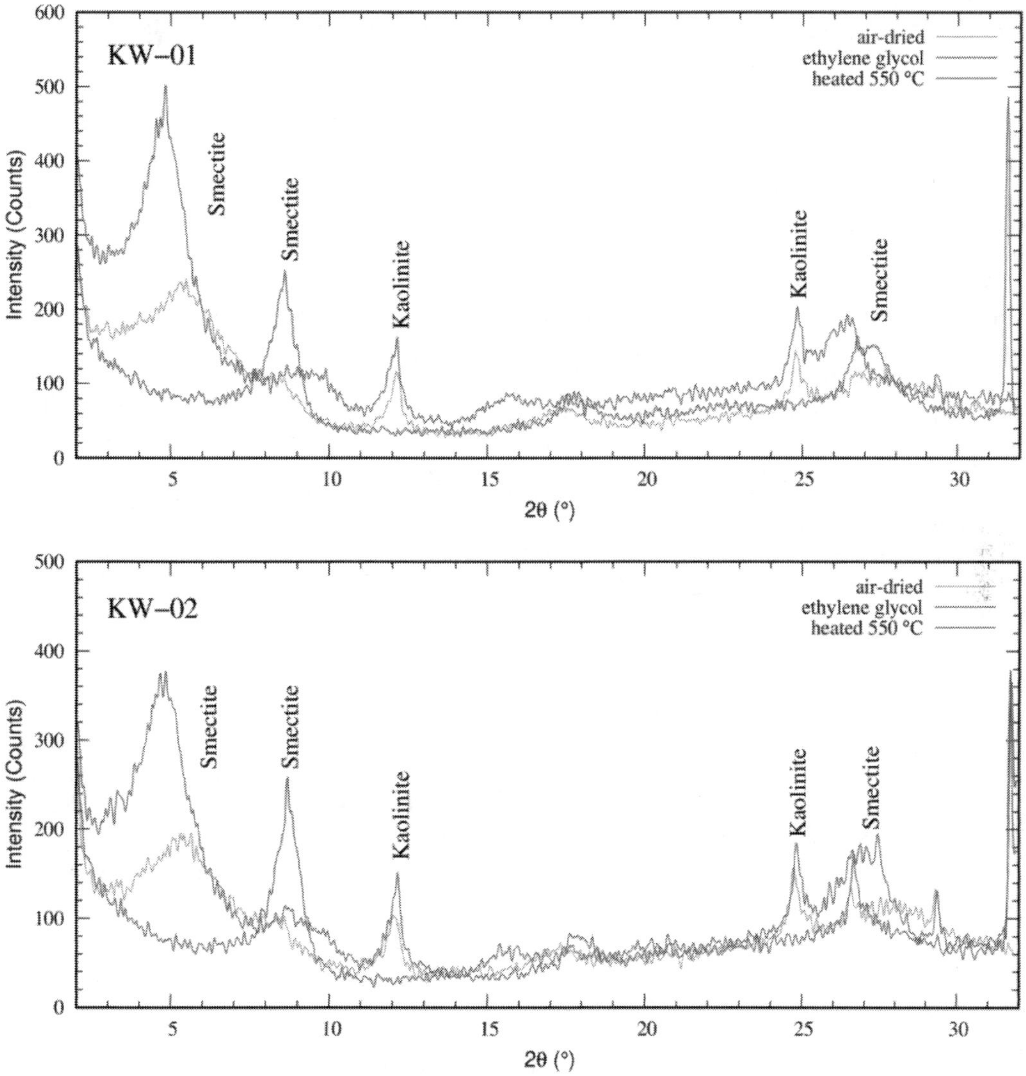

Figure 4. X-ray diffractogram of selected mud samples from the Kuwu Mud Volcano measured by clay treatment (air-dried, ethylene glycol, and heated 550°C).

Smectite is a clay mineral that is dominant in all mud samples. In the bulk analysis, smectite minerals will appear at peak *d* 15.05–16.70 Å (2θ = 5.28–5.78°). In the clay treatment by air-dried, ethylene glycol and heated 550°C, smectite appeared at peak *d* 17.72–18.79 Å (2θ = 4.99–4.60°); *d* 2.81–2.82 Å (2θ = 31.60–30.30°); *d* 3.03–5.03 Å (2θ = 29.3–17.5°). Besides, smectite has characteristics that when treated with ethylene glycol, the peak will shift from the air-dried treatment that has been done before. When it is heated to 550°C, the smectite peak will shift, or possibly even the peak will disappear (Figure 4).

Kaolinite is a clay mineral with the second largest percentage of all mud samples. In the bulk analysis, kaolinite minerals will appear at the *d* peak of 7.14–7.17 Å (2θ = 12.38–12.31°). In the clay treatment by air-dried, ethylene glycol and heated 550°C, kaolinite appeared at peak *d* 7.26–7.34 Å (2θ = 12.15–12.02°); *d* 3.58–3.60 Å (2θ = 24.90–24.80°). Also, kaolinite has characteristics that when treated with ethylene glycol, the peak will not change from the air-dried treatment that has been done before. When it is heated at 550°C, the peak kaolinite will disappear completely.

Ilite is a clay mineral with the lowest percentage in the entire mud sample. In the bulk analysis, illite mineral will appear at peak *d* 9.04–10.99 Å (2θ = 9.76–8.03°). In the clay treatment by air-dried, ethylene glycol and heated 550°C, illite appeared at peak *d* 10.15–10.78 Å (2θ = 8.85–8.20°); *d* 5.02–5.11 Å (2θ = 17.50–17.02°); *d* 3.25–3.36 Å (2θ = 27.3–26.5°). Also, illite has characteristics when treated with ethylene glycol, so the peak will not change from the air-dried treatment that has been done before; even when given heated 550°C, the peak mineral illite remains the same.

5. Discussion

5.1. Morfological distribution on the Kuwu Mud Volcano

The main morphology of the Kuwu Mud Volcano area is in the form of active gryphons and active crater holes with varying dimensions spread widely and unevenly. The large gryphons and main crater holes are located in the central and western parts of the Kuwu Mud Volcano manifestation area. Based on field observations, it can be seen that there are two types of mud volcano morphology in this area, namely in the form of active craters and active gryphons. The active eruption has resulted in a more gentle or flat morphological distribution in the west of the Bleduk Kuwu area.

5.2. Clay mineral utilization recommendations

Based on the analysis results, it can be seen that the mineralogical characteristics, physical properties, and chemical properties of clay minerals contained in the mud samples in the Kuwu Mud Volcano area, so that clay minerals can be recommended as raw materials in making cosmetics according to the specifications.

The comparison of physical and chemical characterization shown in the table shows that the mud samples from the Kuwu Mud Volcano contain predominantly clay minerals of the kaolinite type, which can be recommended as raw materials in making cosmetics, especially powder cosmetics (Table 2). However, there is a significant comparison between CEC values (Table 3). According to standard specifications, in the clay samples from Kuwu Mud Volcano with CEC values, the use of clay minerals for cosmetics is extensive. The research must be specific to only one type of mineral. Apart from the cosmetic industry, it turns out that the clay minerals in the mud samples in the Kuwu Mud Volcano area can also be recommended in the manufacture of ceramic raw materials according to the right specifications (Table 4 & 5).

Table 2. Comparison of physical characteristics of clay in Kuwu Mud Volcano with standard specifications for making cosmetics in the form of powder.

	Physical Properties	Remarks
Kuwu Mud Volcano samples	Grain size distribution	Clay–silt
	Plasticity Index (PI)	24.78 – 55.48%
	Swelling	7.8 – 12.5%
	Shrinkage Limit	9.36 – 14.83%
	Specific gravity	2.50 – 2.84
Standard Specification (Source: ClearOFF™ Minerals – Natural Mineral Solutions [15])	Grain size distribution	Clay
	Plasticity Index (PI)	-
	Swelling	10 – 40%
	Shrinkage Limit	-
	Specific gravity	0

The comparison of physical and chemical characterizations shown in the table shows that the mud samples at the Kuwu Mud Volcano contain clay minerals of smectite, illite, and kaolinite types, which can be recommended as raw materials for making ceramics. However, there is a significant comparison between the CEC value in the clay sample of Kuwu Mud Volcano and the CEC value according to the specification standard. The use of clay minerals for this ceramic is extensive, and the research must be specific to only one type of mineral. Therefore, the CEC and pH values will result in reactions between the bonds between the clay mineral structures, which will later affect the ceramic formation process's strength.

Table 3. Comparison of chemical characteristics of clay in Kuwu Mud Volcano with standard specifications for making cosmetics in the form of powder.

	Chemical Properties	**Remarks**
Kuwu Mud Volcano samples	Cation Exchange Capacity (CEC)	9.96 – 20.47 meq/100g
	pH	8.01 – 8.40
Standar Specification (Source: ClearOFF™ Minerals – Natural Mineral Solutions [15])	Cation Exchange Capacity (CEC)	74 – 90 meq/100g
	pH	6 – 10.8

Table 4. Comparison of physical characteristics of clay in Kuwu Mud Volcano with standard specifications for making ceramics.

	Physical Properties	**Remarks**
Kuwu Mud Volcano samples	Grain size distribution	Clay–silt
	Plasticity Index (PI)	24.78 – 55.48%
	Swelling	7.8 – 12.5%
	Shrinkage Limit	9.36 – 14.83%
	Specific gravity	2.50 – 2.84
Standard Specification [16]	Grain size distribution	Clay–silt
	Plasticity Index (PI)	35 – 45%
	Swelling	10 – 30%
	Shrinkage Limit	3.36 – 10%
	Specific gravity	1.75 – 2.95

Table 5. Comparison of chemical characteristics of clay in Kuwu Mud Volcano with standard specifications for making ceramics.

	Chemical Properties	**Remarks**
Kuwu Mud Volcano samples	Cation Exchange Capacity (CEC)	9.96 – 20.47 meq/100g
	pH	8.01 – 8.40
Standard Specification [16]	Cation Exchange Capacity (CEC)	3.73 – 5.97 meq/100g
	pH	9 – 9.1

6. Conclusions

Based on the results of the data exposure and discussion described in the previous sections, the conclusions that can be drawn for this research are as follows:

1. The characteristics of clay mineralogy in the Kuwu Mud Volcano area are predominantly smectite-type clay minerals followed by kaolinite and illite minerals. For physical characteristics, the dominant grain size distribution from clay to silt size, with a plasticity level of 24.78% to 55.48%, a swelling percentage of 7.8% to 12.5%, with a shrinkage limit of 9.36% to 14.83. % and the value of specific gravity are 2.50–2.84. The chemical characteristics of clay minerals produce a Cation Exchange Capacity (CEC) value of 9.96–20.47 meq/100g and a pH value of 8.01–8.40.
2. Through the analysis and processing of data on mineralogical characteristics, physical and chemical properties of the previous clay minerals, it is recommended that the use of clay minerals

in the Kuwu Mud Volcano area in the industrial sector, namely as a raw material for making cosmetics, especially powder and for making pottery-type ceramics.

Acknowledgments
The author would like to thank the Department of Geological Engineering, Faculty of Engineering UGM, for providing grants to finance this research. Also, Rizka Dwi Desiana has collaborated during sampling in the field and analysis in the laboratory.

References
[1] Guliyev I S, Feizullayev A A and Belov I S 2000 *All about mud volcanoes*. Geology Institute, Azerbaijan Academy of Sciences, Baku, Azerbaijan
[2] Van Bemmelen R W 1949 *The Geology of Indonesia: General Geology of Indonesia and Adjacent Archipelagoes*. The Hague, Batavia: Government Printing: Batavia
[3] Sudarsono U and Sujarwo I B 2008 Aspek Geologi Teknik Lumpur Sidoarjo Jawa Timur. *Buletin Badan Geologi* **18** pp. 1–14
[4] Ibrahim G, Subardjo, Sendjaja P 2010 Tektonik dan Mineral di Indonesia. Pusat Penelitian dan Pengembangan, Jakarta: Badan Meteorologi Klimatologi dan Geofisika
[5] Bronto S, Asmoro P and Efendi M 2017 Gunung Api Lumpur di Daerah Cengklik dan Sekitarnya, Kabupaten Boyolali Provinsi Jawa Tengah. Pusat Survei Geologi: *Jurnal Geologi dan Sumberdaya Mineral*
[6] Siregar S 2016 Analisis dan Pemanfaatan Unsur Belerang dan Salinitas Lumpur Bledug Kuwu di Desa Kuwu, Kecamatan Kradenan, Kabupaten Grobongan, Jawa Tengah. *POSITRON* **6** pp. 40–42
[7] Salam A 2017 Identifikasi Sifat Lumpur Sidoarjo dan Potensi Pemanfaatannya. Departemen Ilmu Tanah Dan Sumberdaya Lahan Fakultas Pertanian Institut Pertanian Bogor
[8] Rakhman 2014 Karakteristik Sifat Fisik Kembang Susut Lumpur Sidoarjo dan Implikasinya Terhadap Potensi Bencana Amblesan Berikut Mitigasinya di Kecamatan Porong dan Sekitarnya, Kabupaten Sidoarjo, Propinsi Jawa Timur. Prosiding Seminar Nasional Aplikasi Sains & Teknologi (SNAST), Yogyakarta
[9] Novian M I, Utama P P and Husein S 2013 Penentuan Batuan Sumber Gunung Lumpur di Sekitar Purwodadi Berdasarkan Kandungan Fosil Foraminifera. Prosiding Seminar Nasional Kebumian ke-6, Jurusan Teknik Geologi FT UGM, Yogyakarta, pp. 519-534
[10] Utama P P 2016 Implikasi Karakteristik Material Penyusun dan Nilai Viskositas Lumpur pada Pembentukan Morfologi Gunung Lumpur Kesongo di Kabupaten Blora, Provinsi Jawa Tengah. Skripsi Departemen Teknik Geologi, Universitas Gadjah Mada: Yogyakarta
[11] Datun M, Sukandarrumidi, Hermanto B and Suwarna N 1996 Peta Geologi Lembar Ngawi, Jawa (Bandung: Pusat Penelitian dan Pengembangan Geologi)
[12] Husein S, Titisari A D, Freski Y R, Utama P P 2016 Panduan Ekskursi Geologi Regional 2016, Departemen Teknik Geologi Fakultas Teknik Universitas Gadjah Mada
[13] Chen F H 1977 Table of Key Lines in X-Ray Powder Diffraction Patterns of Minerals in Clays and Associated Rocks. Department of Natural Resources Geological Survey Occasional Paper 21. The State of Indiana Bloomington, Indiana
[14] More D M and Reynolds R C Jr 1997 X-Ray Diffraction and the Identification and Analysis of Clay Minerals. 2nd Ed. (Oxford: Oxford University Press)
[15] ClearOFF™ Minerals 2020 ClearOFF™ Minerals - Natural Mineral Solutions: Supplying Trade & Public Since 2010. URL: https://www.clearoffminerals.com/shop (Accessed on June 22, 2020)
[16] Agayev T and Kucukuysal C 2017 Ceramic properties of Uşak Clay in comparison with Ukrainian Clay. Mugla Sıtkı Kocman University, Faculty of Engineering, Department of Geological Engineering, Kotekli

IOP Conf. Series: Earth and Environmental Science **851** (2021) 012043 doi:10.1088/1755-1315/851/1/012043

Ore characterization of LSE gold deposit in "X" Pit, Toka Tindung Project, North Sulawesi

N A Raharjanti[1], I W Warmada[1*] and H T B M Petrus[2].

[1]Departement of Geological Engineering, Faculty of Engineering, Universitas Gadjah Mada, Yogyakarta, Indonesia.
[2]Departement of Chemical Engineering, Faculty of Engineering, Universitas Gadjah Mada, Yogyakarta, Indonesia.
*E-mail: warmada@ugm.ac.id

Abstract. Toka Tindung Project is a low sulphidation epithermal gold deposit, which is located in North Sulawesi, Indonesia. To be able to maximize the production yield of the Toka Tindung Project, it is important to understand the characteristics of the ore along with its mineral associations. This paper discusses the ore characterization and its mineral associations in "X" pit using thin section analysis to define the rock characteristics and hydrothermal alteration, X-Ray Diffraction (XRD) to identify the hydrothermal alteration minerals, and ore microscopy to identify the type of ore minerals. The study area is occupied by three alteration zones, namely silicification, advanced argillic and intermediate argillic alteration. The silicification is characterized by quartz with accessory minerals such as kaolinite, smectite, and celadonite. The advanced argillic alteration is typified by kaolinite, with minor quartz. The intermediate argillic alteration is characterized by clay minerals such as illite, mixed illite/smectite, smectite, and minor quartz minerals. Mineralization in "X" pit is in the form of sheeted vein type quartz with colloform texture, and the ore minerals are found as sulfides such as pyrite and sphalerite.

1. Introduction

Each type of deposit has different characteristics with a wide variety of ore and associated minerals. Characterization of ore minerals, ore mineral texture, and its associated minerals are several things that need to be known so that the economic minerals of a deposit can be utilized properly and optimally. The mineralogical characteristics of ore and its mineral association in an ore deposit greatly influence the extraction process and the recovery value obtained [1]. Thus, knowing deposit's ore characteristics and mineralogy and its variation is a part of exploration activities, which has a high risk in investment perspective. This variation of ore has a great impact on technical risks associated with project evaluation, mine plan, and metallurgical unit operations [1–3]. For this reason, it is necessary to conduct a study on the characterization of ore minerals and their associated minerals at the research location so that it can support ore mineral mining activities in the future.

The research area is one of the pits in the Toka Tindung Project gold mine, namely the "X" pit with area of about 500 m long and 400 m wide. Gold deposit of the Toka Tindung Project is included in the low-sulfidation epithermal gold deposit with a gold metal resource of 1.75Moz AuEq [4,5] with gold and silver as its main ore product and is deposited in the form of quartz-adularia vein system,with host rocks consisting of Pliocene-Pleistocene volcanic rocks [5–7]. The characteristics of ore are important factors of the gold recovery that can be obtained through a specific processing method [8]. Therefore,

IOP Conf. Series: Earth and Environmental Science **851** (2021) 012043 doi:10.1088/1755-1315/851/1/012043

ore characterization in the study area is important to be done as the preliminary study for the further research.

Literature studies, field observations and laboratory analysis were then carried out to identify the texture characteristics of the ore and gold-bearing ore mineralogy at the "X" pit in Toka Tindung Project. Identification of the texture and mineralogical characteristics of the ore in the study area was obtained through several analytical methods including thin section analysis, ore microscopic analysis and X-Ray Diffraction (XRD) analysis. It is hoped that this research can produce the overall characteristics of ore minerals including ore mineral types, ore mineral textures, alteration types and associated minerals, as well as mineralization characteristics in the study area.

2. Regional Geology

The research area is located in north part of Sulawesi Island, Indonesia[7]. The processes of tectonism and volcanism in northern Sulawesi are influenced by a double subduction process in the Maluku sea plate [9]which was formed due to collision between the Australian and Pacific plates in the Oligocene (25Ma). This process also takes effects on hydrothermal magmatic mineralization in north arm of Sulawesi, including the Toka Tindung gold project.

The Toka Tindung gold project is a low sulfidation epithermal deposit that is associated with volcanic rock. According to Effendi and Bawono [10], Toka Tindung Project is included in young volcanic rocks formation that are comprised of lava, lapilli, and volcanic ash formed by several young stratovolcanic in north Sulawesi. Van Leeuwen and Pieters [7]states that the lithologies of the Toka Tindung Gold Project consists of Pliocene-Pleistocene volcanic rock which is mainly composed of basaltic andesite flow and volcanic rock layers with varying compositions from basaltic andesite to rhyodacite, and in some places is associated with hydrothermal breccias and carbonaceous tuffaceous sediments. The thickest limestone beds are found in the northern part of the Toka Tindung area. Theselithology are then intruded by domes and rhyodacite dykes (Figure 1).

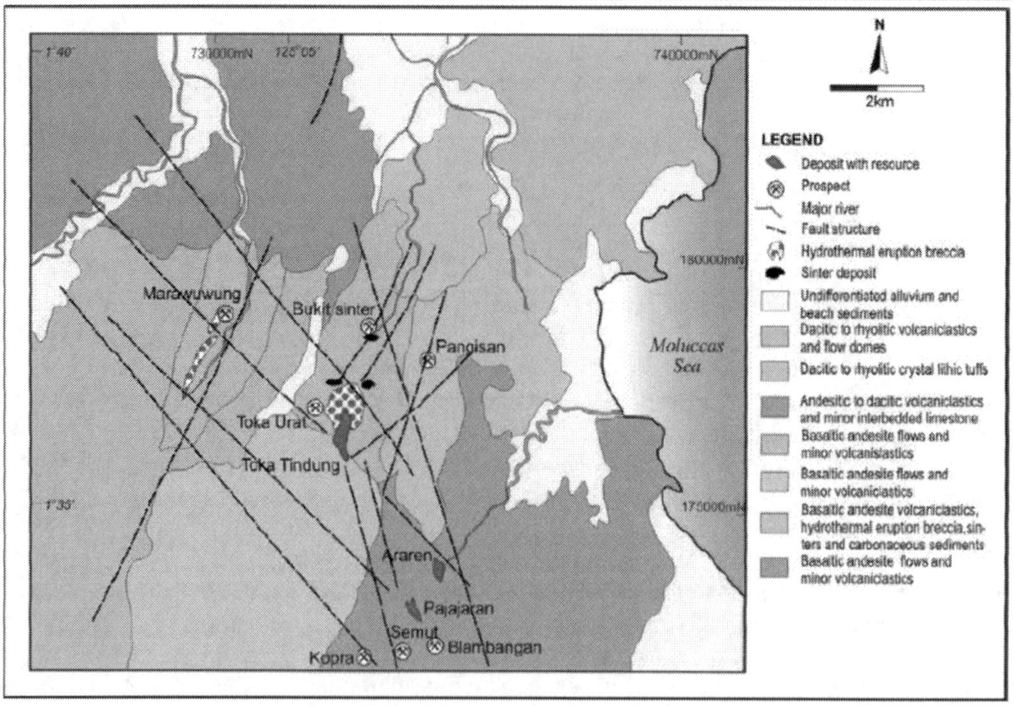

Figure 1. Geological map of the Toka Tindung Project, dominated by volcanic rocks (modified from Angeles 2001 in [7]).

Mineralization in Toka Tindung is controlled dominantly by North North West (NNW) trending faults. Other major geological structures that were formed in plio-pleistocene rocks have the East North East (ENE), North North East (NNE), and North West (NW) trends. Gold deposits in Toka Tindung were mainly controlled by NNW and ENE trending faults and manifested in the form of veins, stockwork, and hydrothermal breccias [7].

3. Methodology

The data in this study are divided into two based on how they were collected: primary and secondary data. The primary data are the data that author collected in the field such as samples, its further analysis, while the secondary data are the data given by the company of Toka Tindung Project, such as the pit map. The primary data were taken from field observations and laboratory analysis of the surface samples of "X" pit, which weredivided into two types of samples, namely vein and wallrock. Several analyses were then conducted to identify the type of ore characteristics and mineralogy in "X" pit and its association minerals, which is thin section analysis, ore microscopy, and X-Ray Diffraction (XRD).

3.1. Thin section analysis

Thin section (petrography) analysis was done to observe the mineral association and the abundance of the types of minerals. This analysis was also used to describe the host rock type, textures, alteration, and mineralization in the research area. The thin section analysis was done using polarized-light microscope to observe 12 thin section samples that were taken from "X" pit. This observation was carried out in Optical Geology Laboratory, Geological Engineering Department, Faculty of Engineering, Universitas Gadjah Mada.

3.2. Ore microscopy analysis

Ore microscopy analysis was done to 11 samples taken from the "X" pit using the reflected-light microscope for identify the opaque mineral in the research area. These ore minerals can be in form of native element, oxide, or sulfide minerals. This analysis was also used to observe the ore texture and other characteristics of the ore that can be used to understand the mineralization in the research area, particularly the ore minerals in the "X" pit. This observation was carried out in *GetIn Cicero* Laboratory, Geological Engineering Department, Faculty of Engineering, Universitas Gadjah Mada.

3.3. X-Ray Diffraction

XRD analysis is used to identify the type of minerals in rocks. The data collected in the XRD analysis was used to identify minerals in the rock based on a diffraction pattern called a diffractogram and written as X-ray vs 2θ (theta). This analysis was done using the qualitative method and was divided into two types of analysis, namely bulk and clay analysisincluding air dried, ethylene glycol immersed, and heated in 550 °C. The XRD method aims to find out mineral association in the research area, be it in the form of gangue minerals or ore minerals, mainly those which was not observed using microscopy methods. There were 4 samples used for this analysis, and the analysis was carried out using XRD machine type *Rigaku Multiflex* in Main Laboratory of Geological Engineering Department, Faculty of Engineering, Universitas Gadjah Mada.

4. Results

4.1. Hydrothermal alteration

Field observation shows that the research area is included in near surface low sulfidation epithermal mineralization. This causes the characteristics of hydrothermal alteration minerals in the study area in the form of minerals formed in the surface zone such as kaolinite mineral [11]. The research area is divided into three types of hydrothermal alteration based on their mineral association, namely silicification, advanced argillic, and intermediate argillic (Figure 2). The silicification alteration has

IOP Conf. Series: Earth and Environmental Science **851** (2021) 012043 doi:10.1088/1755-1315/851/1/012043

the widest distribution in "X" pit, covering the center part and the east side of the pit, while the advanced argillic alteration type is scattered to a limited extent in the graben zone and the intermediate argillic alteration type is scattered on the western edge of the "X" Pit.

The determination of the alteration type was carried out based on company data, field observations, thin section analysis, and XRD analysis. In general, it is easy to distinguish silicified alteration using thin incision analysis, given the relatively high abundance of quartz in the sample (Figure 3). However, to distinguish the types of clay minerals present in advanced argillic (Figure 4) and intermediate argillic (Figure 5) alterations, thin slice analysis needs to be supported by XRD analysis. With XRD analysis, not only clay minerals in advanced argillic alteration and intermediate argillic alterations can be identified, but also some accessory minerals in silicification alteration and opaque minerals present in the sample (Figure 6).

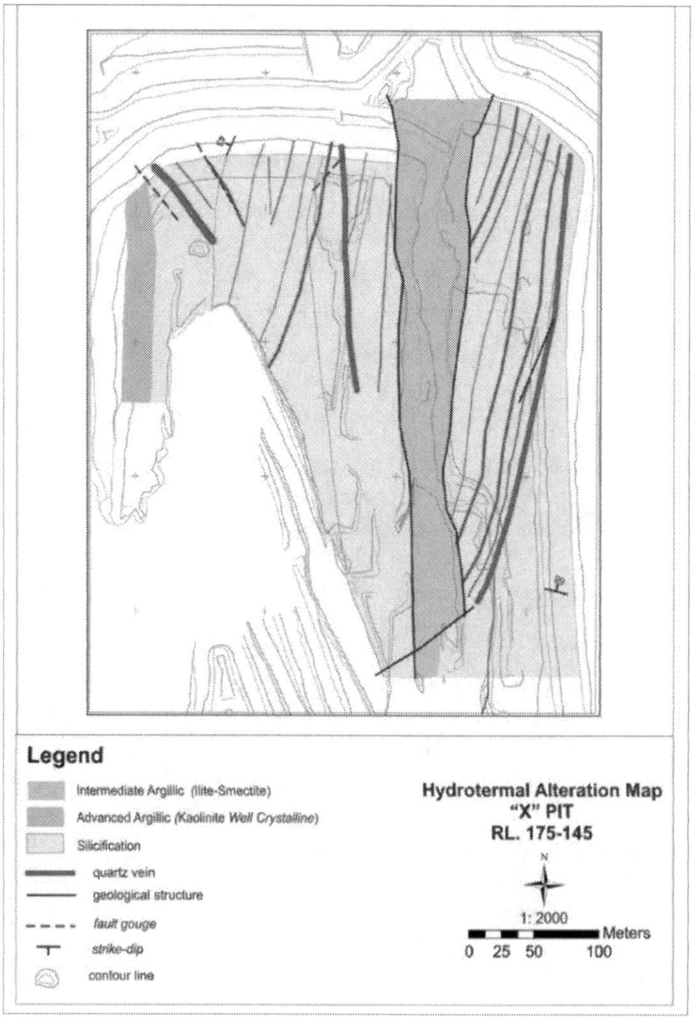

Figure 2. Hydrothermal alteration map of the study area ("X" pit) which shows three hydrothermal alteration zones, namely silicification, advanced argillic, and intermediate argillic.

Based on the thin section analysis, it can be observed that the silicified alteration has the same texture and structure as the original rock but has a higher level of hardness due to the intensive addition of silica in the form of secondary quartz (Figure 3). Advanced argillic alteration in the thin sections can be observed from the abundance of clay minerals which are quite high and the texture of the original rock that cannot be observed. Similar to advanced argillic alteration, intermediate argillic alteration also consists of a high abundance of clay minerals and an almost unidentified original rock texture. To differentiate between the two, XRD analysis needs to be carried out especially to identify kaolinite minerals which are the main features of advanced argillic alteration.

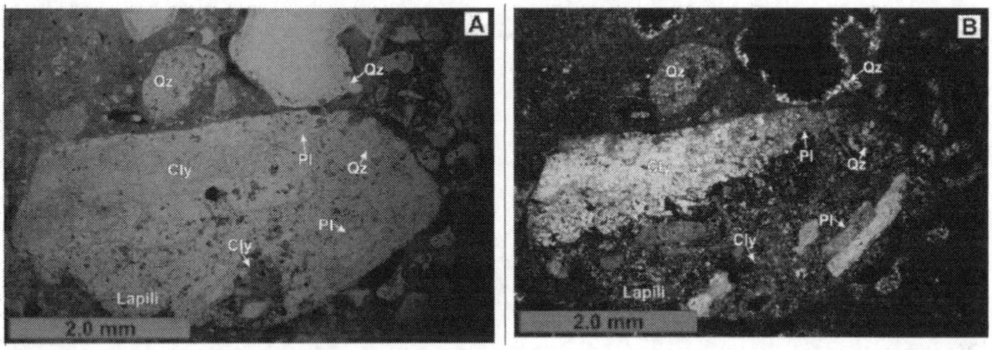

Figure 3. Thin section analysis of sample ANK_19 W's lapilli tuff which has undergone silicified alteration in parallel-polarised (A) and cross-polarised (B). Mineral abbreviation: Qz=Quartz, Pl=Plagioclase, Cly=Clay.

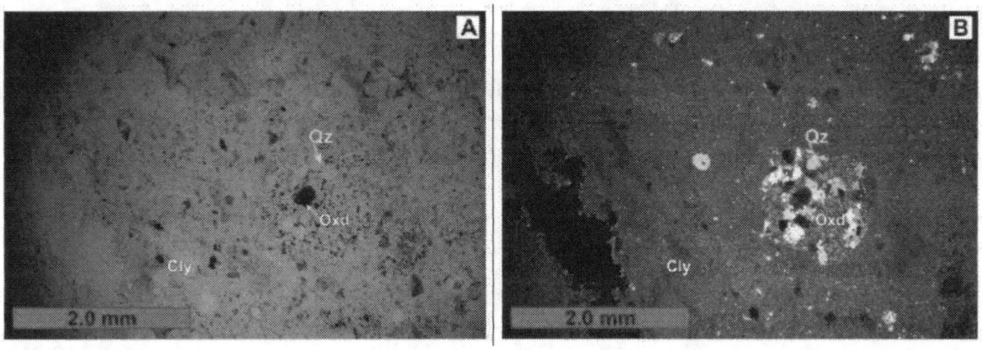

Figure 4. Thin section analysis of sample ANK_FS 11's tuff which has undergone advanced argillic alteration in parallel-polarised (A) and cross-polarised (B). Mineral abbreviation: Qz=Quartz, Cly=Clay, Oxd=Oxide.

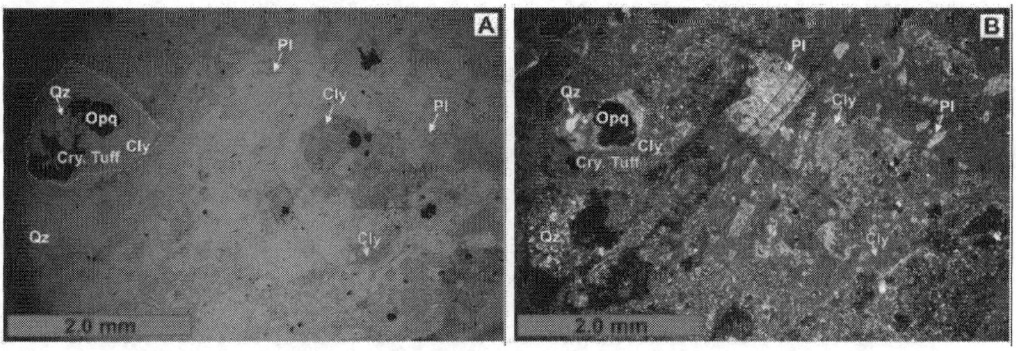

Figure 5. Thin section analysis of sample ANK_01's crystal tuff which has undergone intermediate argillic alteration in parallel-polarised (**A**) and cross-polarised (**B**). Mineral abbreviation: Qz=Quartz, Pl=Plagioclase, Cly=Clay, Opq=Opaque minerals.

IOP Conf. Series: Earth and Environmental Science **851** (2021) 012043 doi:10.1088/1755-1315/851/1/012043

After conducting the XRD analysis, it is known that the silicification alteration in "X" pit is characterized by quartz minerals with accessory minerals such as kaolinite, smectite, and celadonite found at several points (Figure 6A). Other than that, it is also known that the advanced argillic alteration in "X" pit is characterized by kaolinite clay mineral, with quartz present as a minor mineral (Figure 6B). Another type of alteration is the intermediate argillic alteration type which is characterized by clay minerals such as illite, mixed illite/smectite, smectite, and minor quartz and pyrite minerals (Figure 6C).

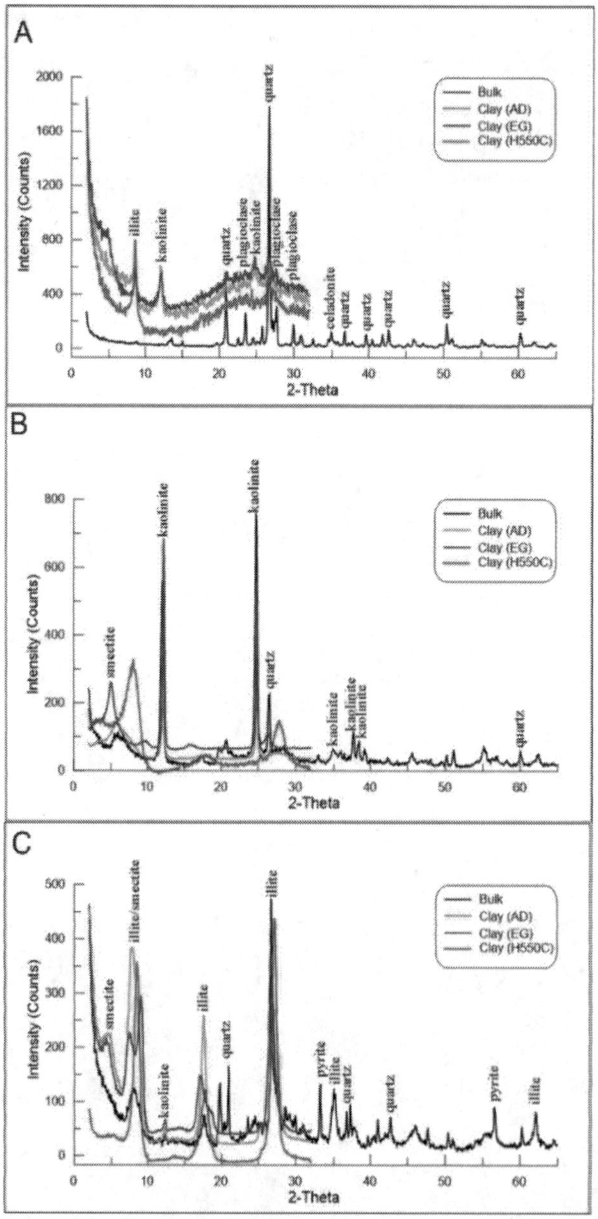

Figure 6. Hydrothermal alteration mineral association characteristics of each alteration zone: (**A**) silicification zone, (**B**) advanced argillic zone, (**C**) intermediate argillic zone. The abbreviations in clay analysis mean AD: Air Dried, EG: Ethylene Glycol, H550C: Heated in 550 °C.

4.2. Ore mineralogy and mineralization

The mineralization system formed in the study area is a low-sulfidation epithermal deposit in the form of veins with a sheeted vein structure that extends north-south following the geological fault structure found in "X" pit (Figure 2). Ore mineralization in the study area is deposited on quartz veins with the minor presence of chalcedony, adularia, and clay minerals as gangue minerals in the vein. Other minerals associated with the quartz vein are oxide minerals such as goethite, hematite, and manganese oxide. The vein texture formed in the study area is mainly in the form of banded texture with colloform banded texture as its main texture (Figure 7A, 7B, 7C), crustiform (Figure 7D), or mixed both (Figure 7C). Other minor textures in the form of open space filling textures are also present, namely as drusy cavities, cockade, and comb texture (Figure 7A, 7B, 7C).

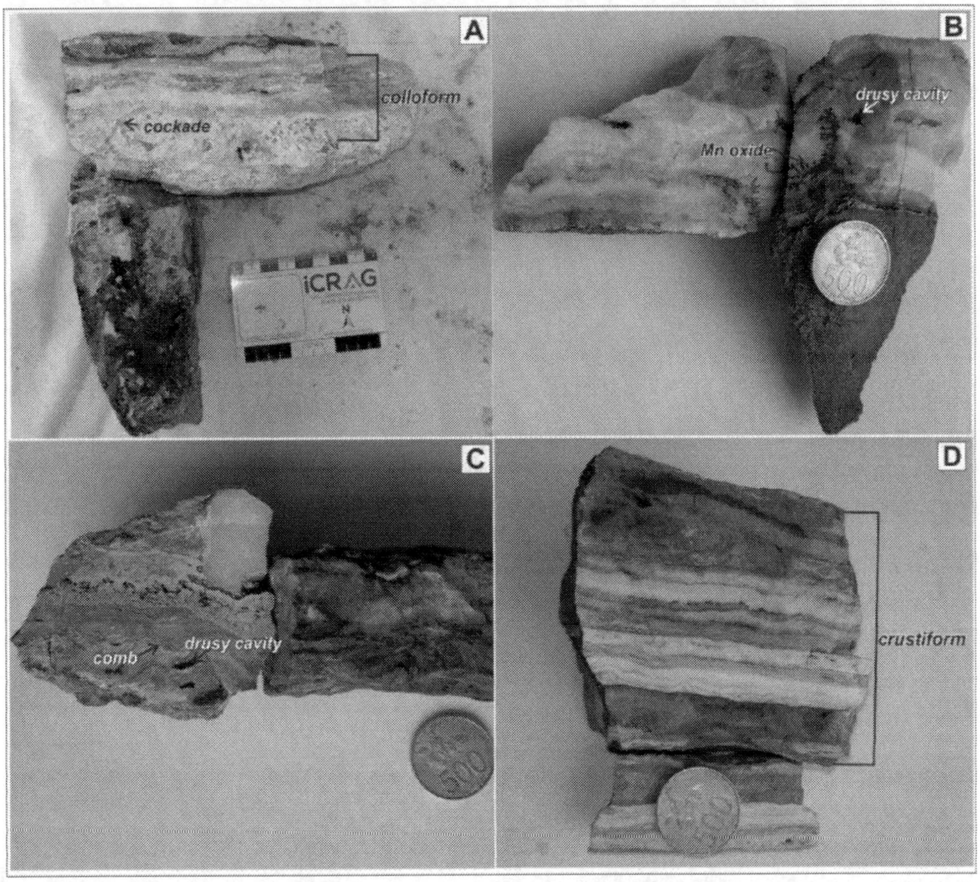

Figure 7. Vein textures on "X" pit: (**A**) quartz vein with colloform and cockade textures with yellow-black iron oxide on the surface, (**B**) Colloform and drusy cavity textures with manganese oxide on the surface, (**C**) veins with a colloform-crustiform texture, comb, and drusy cavities, (**D**) quartz veins with a crustiform texture.

There are very few ore minerals found in the veins, and the minerals that can be detected using the ore microscopy method in the vein are mainly in the form of sulfide and oxide minerals such as pyrite, sphalerite, and hematite (Figure 8), with a manganese oxide found at several points formed on the surface of the veins. After observing 11 samples using the ore microscopy method, it can be said that the gold minerals in the study area cannot be observed using the ore microscopy method.

The ore microscopy samples of the "X" pit are then divided based on the texture of the ore and ore minerals contained in each of these textures and observed in Table 1. In the table we can see that pyrite is present in all of the samples and hematite almost present in all of the samples, while sphalerite and goethite are present only in ANK_21V with colloform-cockade texture, and manganese oxide only present in ANK_FS 04V with colloform-drusy cavities texture.

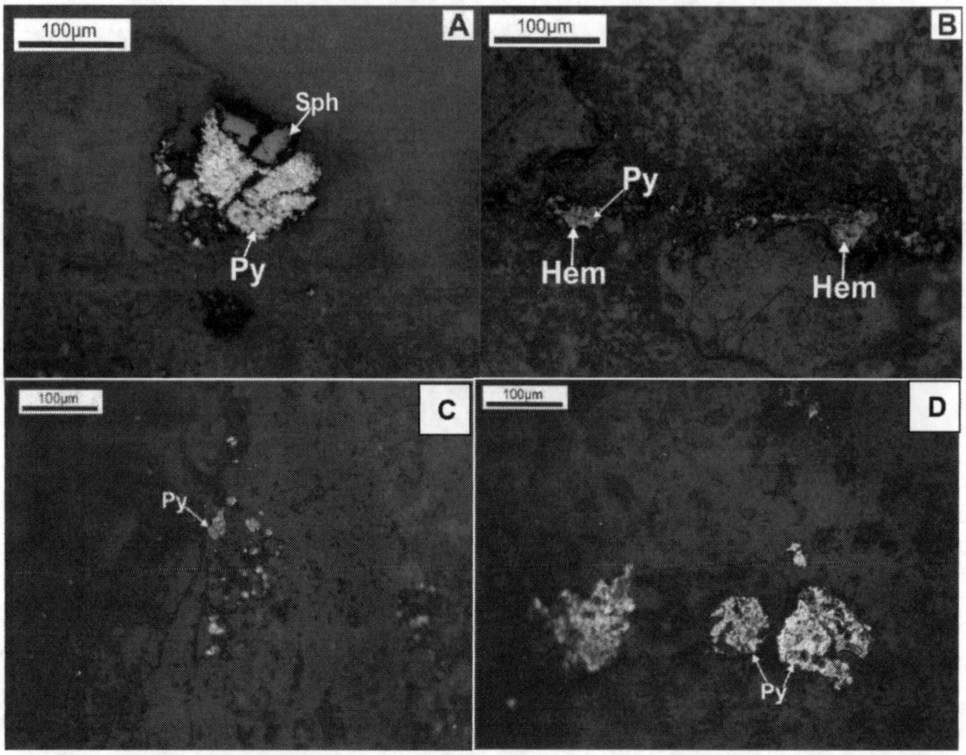

Figure 8. Ore minerals formed in "X" pit quartz veins, consist of (**A**) pyrite (Py) and sphalerite (Sph) in sample ANK_21V, (**B**) hematite (Hem) with pyrite inclusions (Py) in sample ANK_FS 05V, (**C**) pyrite in sample ANK_25V, and (**D**) pyrite in sample ANK_FS 13V.

Table 1. Summary of vein texture and ore minerals contained in pit "X".

Sample Number	Vein texture	Ore minerals
ANK_10V	Colloform	Pyrite
ANK_12V	Colloform, drusy cavities	Pyrite, hematite
ANK_13V	Crustiform	Pyrite, hematite
ANK_18V	Colloform-crustiform, comb	Pyrite, hematite
ANK_21V	Colloform, cockade	Pyrite, sphalerite, goethite
ANK_25V	Colloform, drusy cavities	Pyrite, hematite
ANK_FS 03V	Colloform	Pyrite, hematite
ANK_FS 04V	Colloform, drusy cavities	Pyrite, hematite, manganese oxide
ANK_FS 05V	Colloform	Pyrite, hematite
ANK_FS 08V	Colloform, cockade	Pyrite, hematite
ANK_FS 13V	Colloform	Pyrite

5. Discussion

The research area namely "X" pit is a part of the Toka Tindung Project which is characterized as a low - sulfidation epithermal gold deposit [4–6]. The "X" Pit is divided into three types of hydrothermal alteration based on their mineral association, namely silicification, advanced argillic, and intermediate argillic (Figure 2). The silicification alteration has the widest distribution covering the center part and the east side of the research area dominantly composed of quartz, with some clay minerals such as kaolinite, smectite, and celadonite formed as accessory minerals. The silicified wallrock is associated with banded quartz vein mineralization (Figure 7) showing that the low sulfidation epithermal deposit in the study area formed in a shallow environment (approximately 0-300 m) [11–13]. The second hydrothermal alteration formed limited on the western edge of the "X" Pit is intermediate argillic alteration characterized by clay minerals including illite, mixed illite/smectite, and smectite, and minor quartz, and pyrite minerals. The mineral association in the intermediate argillic alteration indicates that the alteration zone is located on the side of a low sulfidation epithermal system [11,14]. The third hydrothermal alteration type is advanced argillic located limited in the graben zone consisting of dominantly kaolinite, with quartz present as minor minerals. According to Hedenquist [11], advanced argillic alteration formed on the top of the low sulfidation epithermal system, which is defined as steam-heated alteration characterized by kaolinite blanket along with cristobalite, smectite, and locally alunite and native sulfur. Mineralization in the study area formed in banded textured quartz vein mainly colloform as its main texture (Figure 7A, 7B, and 7C), crustiform (Figure 7D), or both (Figure 7C), which was identified as typical low sulfidation epithermal deposit formed in a shallow environment (approximately 0-300 m) [4–6]. The sulfide minerals that can be observed through ore microscopy are pyrite (Figure 8) and sphalerite (Figure 8A) and the abundances of these minerals are very low. The gold minerals in the study area cannot be identified in ore microscopy method. Thus, this indicates that the gold in the study area falls in the invisible gold category [1].

6. Conclusion & Recommendation

Based on the results of the integration between the data analyses that have been carried out from samples taken from pit "X", the following conclusions can be drawn:

- Mineralization in pit "X" is a low sulfidation epithermal deposit in the form of sheeted vein and characterized by the presence of quartz, illite, smectite, and chalcedonyas gangue minerals, and pyrite, sphalerite, hematite, goethite, and manganese oxide as its ore minerals.
- Some of the characteristic textures of low sulfidation epithermal deposits are also present, such as banded textures (colloform and crustiform), open space filling textures (drussy cavities and comb), and cockade.
- Three types of alteration are recognized including silicification, advanced argillic, and intermediate argillic. The silicification alteration has the widest distribution in "X" pit, covering the center part and the east side of the pit with the abundance of secondary quartz as its key mineral and kaolinite, smectite, and celadonite as its minor minerals. The advanced argillic alteration type is scattered to a limited extent in the graben zone with kaolinite clay mineral, with quartz present as a minor mineral. The intermediate argillic alteration type is scattered on the western edge of the "X" pit with clay minerals such as illite, mixed illite/smectite, and smectite, and minor quartz minerals.
- Given that there's no gold observed in ore microscopy, it can be concluded thatgold in the research area is in the form of invisible gold.

Recommendation for further study:

- Since gold at the research location cannot be identified by the ore microscopy method, it is recommended that geochemical analysis and mineral mapping be carried out to determine with certainty the characteristics of the gold mineralogy in the study area.

Acknowledgments
This research was conducted in the project area of PT. Meares Soputan Mining (PT. MSM). The author would like to thank the management of PT. MSM who facilitated this research and permitted the use of the company's data. The author would also like to thank the GetIn Cicero Laboratory, Department of Geological Engineering, Faculty of Engineering, Universitas Gadjah Mada for facilitating data processing for this research.

References
[1] Zhou J and Gu Y 2016 Geometallurgical Characterization and Automated Mineralogy of Gold Ores *Gold Ore Processing: Project Development and Operations* ed M D Adams (Elsevier) pp 95–111
[2] Moon C J, Whateley M and Evans A M 2006 *Introduction to mineral exploration* (Oxford: Blackwell Publishing)
[3] Lipton I 2017 Using Geometallurgy to Improve Mine Profitability *MGEI Geometallurgy Convention 2017* vol 9th, ed Basuki N.I and Idrus A. (Yogyakarta: Masyarakat Geologi Ekonomi Indonesia) pp 5–7
[4] Williamson A 2011 Discovery and Development of Toka Tindung Low Sulphidation Epithermal Gold Project *The Sulawesi Mineral Recources 2011 Seminar MGEI-IAGI* ed N . Basuki (Manado: Masyarakat Geologi Ekonomi Indonesia) pp 259–66
[5] Moyle A J, Wake B A, Tuckey S H and Ariti J 1997 The Toka Tindung Gold Project, Northern Sulawesi, Indonesia *World Gold `97* (Singapore) pp 27–34
[6] Wake B A, Sinugroho I A and Kuswandi M D 1997 Epithermal Gold-Silver Mineralisation in a Fosil Hot Spring System, Toka Tindung, North Sulawesi *National Seminar of Human Resources of Indonesian Geologtst* (Yogyakarta: Geological Engineering, Mineral Technology Faculty UPN "Veteran" Yogyakarta) pp 1–6
[7] Leeuwen T M Van and Pieters P E 2011 Mineral deposits of Sulawesi *The Sulawesi Mineral Recources 2011 Seminar MGEI-IAGI* ed N . Basuki (Manado: Masyarakat Geologi Ekonomi Indonesia) pp 1–130
[8] Marsden J . and House C . 2006 *The Chemistry of Gold Extraction* (Littlelon, Colorado: Society for Mining, Metallurgy, and Exploration, Inc. (SME))
[9] Hamilton W 1979 *Tectonics of the Indonesian Region* (U.S. Geological Survey)
[10] Effendi A C and Bawono S S 1997 Peta Geologi Lembar Manado, Sulawesi Utara 2nd Ed
[11] Hedenquist J W, Arribas R A and Gonzalez-Urien E 2000 Exploration of Epithermal Gold Deposits *SEG Rev.* **13** 245–77
[12] White N C and Hedenquist J W 1995 Epithermal Gold Deposits: Styles, Characteristics and Exploration *Publ. SEG Newsl.* **1** 9–13
[13] Simmons S F, White N C and John D A 2005 Geological Characteristics of Epithermal Precious and Base Metal Deposits *Econ. Geol.* **100th Anni** 485–522
[14] Evans A M 1993 *Ore Geology and Industrial Minerals* (Oxford: Blackwell Publishing)

Ore mineralogy of podiform-type chromite deposit in Tedubara area and its vicinity, Kabaena Island, Indonesia

S Septiana[1], A Idrus[1]*, F Zaccarini[2], G Garuti[2] and L D Setijadji[1]

[1]Departemen Teknik Geologi, Universitas Gadjah Mada, Jl. Grafika 2 Bulaksumur, 55281 Yogyakarta, Indonesia
[2]Department of Applied Geosciences and Geophysics, University of Leoben, Peter Tunner Str. 5, A-8700 Leoben, Austria
*E-mail: arifidrus@ugm.ac.id

Abstract: Kabaena Island is one of the areas in Southeast Sulawesi that has the potential for chromite deposits. This paper is aimed to provide information on the mineralogical aspects of chromitite from the research area. Back-scattered electron (BSE) imaging and chemical analysis of chromite were performed using an electron microprobe. Chromite deposits in the study area are of podiform-type. Chromite is massive and brecciated. Based on the content of Cr # = Cr / (Cr + Fe^{3+}) of the chromite, comprised between 0.61 and 0.74, the studied chromitite can be classified as Cr-rich. The only PGM (Platinum Group Minerals) found in several chromitite samples is laurite (RuS_2). PGM occurs enclosed in chromite crystals and in the contact with microfractures of chromite crystals.

1. Introduction

Chromite is one of the most important minerals as a source of metalliferous ore for industrial world such as stainless steel, gray cast iron, iron free high temperature alloys, and chromium plating to protect the iron surface [6]. The use of chromium in various industries has resulted in high demand for chromitites.

Chromitites can be found in several areas in Indonesia such as Sulawesi, Kalimantan, and Papua [4,10,11]. The chromite minerals found in this area are formed in ophiolite complexes with ultramafic rocks as host rocks. The chromite in ophiolite is usually enriched by Platinum Group Elements (PGE) [7]. Platinum Group minerals generally are formed in the chromite crystals, at the periphery of the crystal, or in the interstitial silicate gangue in chromite [12]. Information on the occurrence of PGM especially regarding the chromite deposits is still lacking. In Sulawesi Island, a research conducted by [11] indicated that PGM generally occur as inclusions in chromite or areas in contact with cracks. The most common PGM is laurite (RuS_2) along with minor irarsite (IrAsS) [11]. The PGM found included in chromite crystallize during the chromite precipitation at magmatic stage. In some cases, they can be modified during the post-magmatic at low temperature conditions. The PGM that contains the high melting and refractory platinum group elements (PGE) formed during the partial melting process in the mantle and then trapped as solid particles into chromite [12]. Chromite is more resistant to an alteration than silicates and sulfides, so the PGMs formed as inclusions in chromite are protected from serpentinization and alteration processes. Conversely, PGM that forms on the edge of chromite crystals or in the silicate gangue is easily altered and formed a secondary PGM [1]. This paper aims to provide information about ore mineralogy and chemical composition of chromite and to document the

IOP Conf. Series: Earth and Environmental Science **851** (2021) 012044 doi:10.1088/1755-1315/851/1/012044

presence of PGM in chromite deposits located in the Tedubara area of the Kabaena Island, which is part of the Ophiolite Belt of East Sulawesi.

2. Geological Setting of Kabaena Island

The East Sulawesi Ophiolite Belt has a wide distribution, from the upper end of the East Arm to the lower end of the Southeast Sulawesi Arm (figure 1). Ultramafic rocks are dominant in the Southeast Arm, while mafic rocks are more dominant in the north. Kabaena Island is part of the East Ophiolite Sulawesi Belt, and is mostly composed of ultramafic rocks, Cretaceous in age [9] (figure 2).

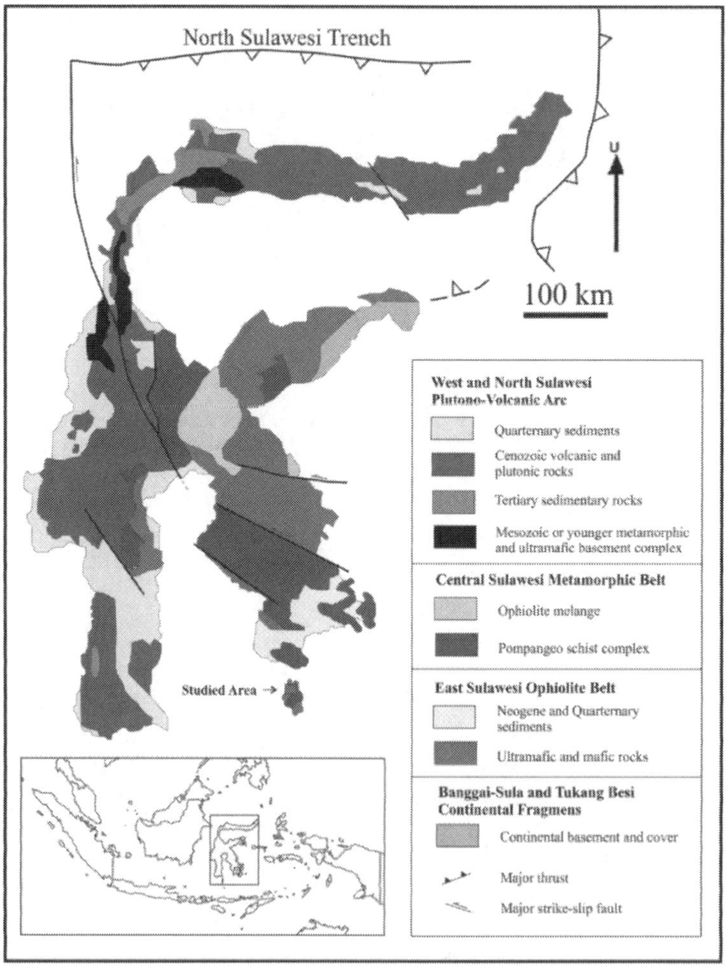

Figure 1. Map of Sulawesi Island lithotectonic units (modified from [5]), showing the distribution of the East Sulawesi Ophiolite (ESO) from the South Arm, through the East Arm to the Southeast Arm of Sulawesi Island.

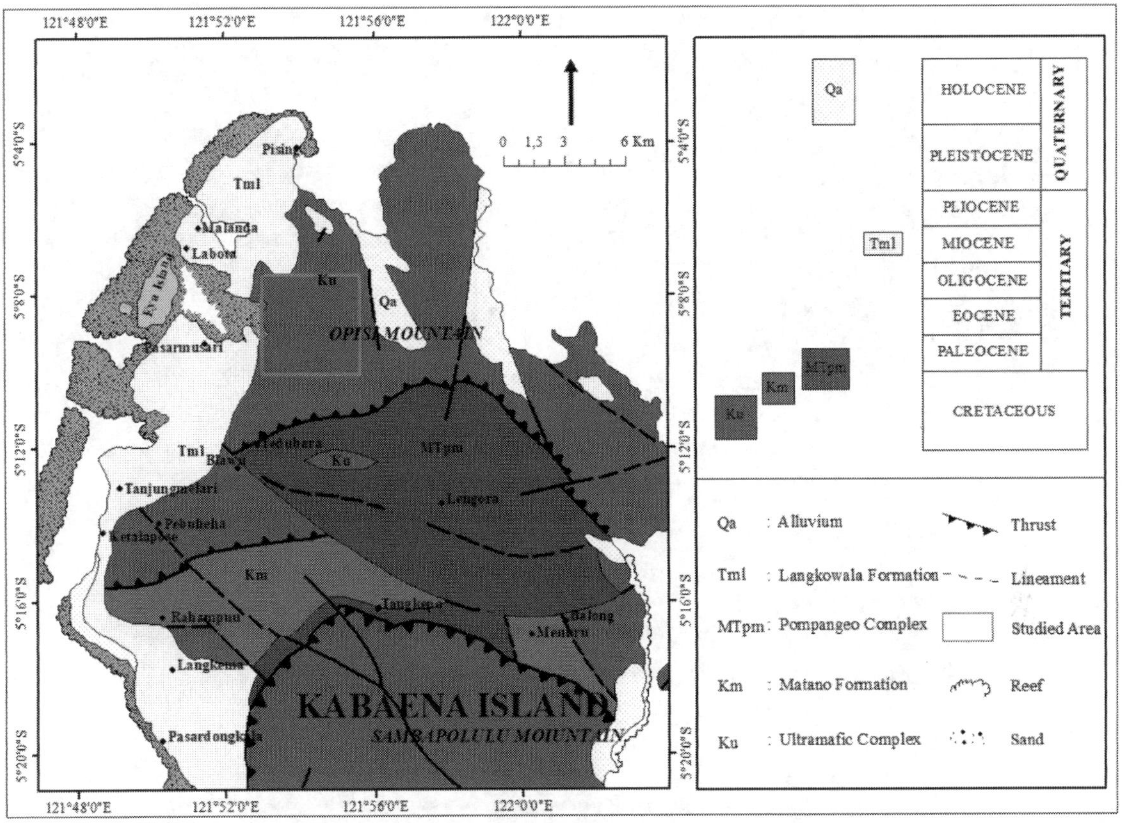

Figure 2. Regional geology of the study area (Modified from [8]).

Although data available on the tectonic setting of Kabaena Island are scarce, based on the reconstruction from [3], the age of Kabaena Island and the surrounding varies from the Cretaceous the Middle Miocene (± 15 Ma). Kabaena Island is located in the southern part of the forerunner to the Southeast Sulawesi Arm along with Muna Island which is separated by a subduction zone from the Banda Peninsula at that time. Kabaena Island is also being pushed closer to the Southeast Sulawesi Arm due to the division of the South Banda Sea that occurred during the Pleistocene (± 5 Ma). The ultramafic rocks exposed in Kabena Island and other parts of Sulawesi are estimated to have originated from spreading between the Indian and Australian plates. This is supported by research [5] on several ultramafic rock samples taken from the northern and southern parts of Kabaena Island which indicates a MOR affinity.

3. Sampling and Research Methods

The studied chromitites are mostly located in remote areas with difficult access because of the relatively dense vegetation and high hilly morphology as the dominant morphology. Chromite deposits have been exposed to the surface as a result of existing local community mining activities and now abandoned. Chromitite occur as lens-shaped deposits on the walls of outcrops rock as well as nodule-shaped minerals, associated with ultramafic rocks (figure 3). Based on field observation and shape of the mineralized body, the chromite deposits in the research area correspond to the podiform type [7].

Figure 3. Chromite deposits associated with ultramafic rock in the research area: (a) Outcrop showing the old mining activity; (b) Close-up view of ultramafic rock outcrop in the presence of a podiform chromite pod; (c) Massive and nodular chromitite.

Chromite samples were collected in 8 areas of Tedubara. They were previously studied by reflected-light microscope. Chemical analysis of minerals was carried out in all the collected samples by means of an electron microprobe using a JEOL JX A 8200 Superprobe (JEOL, Tokyo, Japan), installed in Eugen F. Stumpfl Laboratory, Leoben University in Austria. Chromite was quantitatively analysed using wave-light dispersive system (WDS) and PGM using energy dispersive system (EDS) The same instrument was used to obtain back-scattered electron (BSE) images of chromite and of PGM.

4. Results and Discussion

4.1. Chromite textures and composition

The studied samples represent massive chromitite, containing 60 to 90 volume % of chromite. They display the following textures: 1) massive and fractured, 2) lobate and 3) brecciated (figure 4). The silicates of the gangue mainly consist of serpentine, talc and chlorite. Few preserved crystals of olivine have been also recognized (figure 4c).

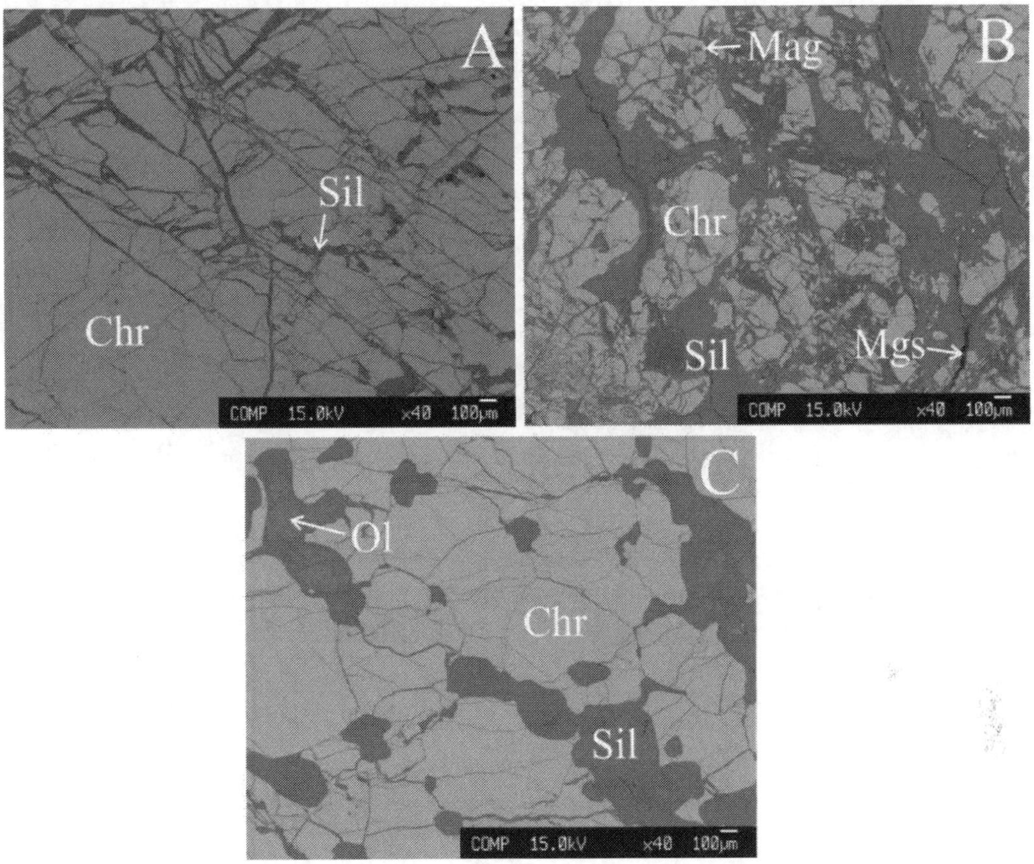

Figure 4. Back-scattered electron images showing the texture of the studied samples. (a) massive and fractured; (b) lobate; (c) brecciated. Abbreviation: Chr = chromite; Sil = silicate; Ol = olivine; Mgs = magnesite; Mag = magnetite.

Based on the electron microprobe analyses presented in Table 1, the Cr_2O_3 exhibits a spread in concentrations, varying from 48.65-56.04 wt % (average 51.59 wt %) The value variations (in wt %) of other major oxides are: Al_2O_3 (13.18-20.83), Fe_2O_3 (1.89-3.21), MgO (11.16-14.42), and FeO (12.8) -17.59). The amount of the trace elements that are comprised in the following range (wt%): TiO_2 (0.18-0.29), ZnO (0.01), V_2O_3 (0.07-0.19), and NiO (0.07-0.14). The SiO_2 content in all the analyzed samples is lower than 0.03 wt %.

The calculated atomic per formula unit (apfu) show that the values of Zn, Si, V, Ni and Mn are <0.002. The average values are the following: Cr (1.28), Al (0.65), Mg (0.62), Fe^{2+} (0.38), Fe^{3+} (0.06), and Ti (0.001). The Cr # = [Cr / (Cr + Al) in the analyzed chromitites are comprised between 0.61-0.74 and the Fe^{2+}# = [Fe^{2+} / (Fe^{2+} + Mg)] between 0.33-0.47. The high Cr # value (> 0.6) indicates that the chromitite in the research area can be classified as Cr-rich type [2].

Table 1. Average values (in wt %) of microprobe analysis data of chromite from Tedubara area, Kabaena Island.

Elements	Sample Codes							
	TBR01	TBR03	TBR04	TBR17	TBR30	TBR32	TBR33	TBR36
	(N = 15)	(N = 15)	(N = 15)	(N = 15)	(N = 15)	(N = 15)	(N = 15)	(N = 15)
SiO_2	0.03	0.03	0.04	0.02	0.03	0.04	0.03	0.03
TiO_2	0.29	0.22	0.23	0.18	0.19	0.24	0.24	0.22
Al_2O_3	20.08	14.72	14.43	20.83	13.18	18.74	17.96	20.65
FeO*	14.05	14.06	13.51	15.23	17.59	16.11	12.80	13.97
Fe_2O_3*	2.51	2.26	2.12	1.89	2.78	2.24	3.08	3.21
MgO	13.88	13.22	13.65	13.12	11.16	12.32	14.42	13.86
MnO	0.23	0.26	0.23	0.25	0.28	0.26	0.26	0.26
Cr_2O_3	49.09	55.23	56.04	48.65	54.80	49.95	51.35	47.62
NiO	0.09	0.10	0.09	0.07	0.07	0.07	0.09	0.07
ZnO	0.01	0.01	0.01	0.01	0.01	0.01	0.00	0.01
V_2O_3	0.08	0.07	0.09	0.10	0.14	0.11	0.10	0.10
Total	100.36	100.16	100.43	100.36	100.23	100.09	100.33	100.00
apfu								
Si	0.001	0.001	0.001	0.001	0.001	0.001	0.001	0.001
Ti	0.01	0.01	0.01	0.00	0.00	0.01	0.01	0.01
Al	0.73	0.55	0.54	0.76	0.50	0.69	0.66	0.75
Fe(II)	0.36	0.37	0.36	0.39	0.48	0.42	0.33	0.36
Fe(III)	0.06	0.05	0.05	0.04	0.07	0.05	0.07	0.07
Mg	0.64	0.62	0.64	0.60	0.54	0.58	0.67	0.64
Mn	0.006	0.007	0.006	0.007	0.008	0.007	0.007	0.007
Cr	1.20	1.38	1.40	1.19	1.40	1.24	1.26	1.16
Ni	0.002	0.002	0.002	0.002	0.002	0.002	0.002	0.002
Zn	0.0002	0.0003	0.0001	0.0001	0.0003	0.0003	0.0001	0.0002
V	0.001	0.001	0.001	0.001	0.002	0.001	0.001	0.001
Cr#	0.62	0.72	0.72	0.61	0.74	0.64	0.66	0.61
Fe^{2+}#	0.36	0.37	0.36	0.39	0.47	0.42	0.33	0.36

4.2. Platinum Group Minerals (PGM)

Few PGM have been found in the studied chromitites and, on the basis of preliminary EDS analyses, they are laurite ideally $(Ru,Os)S_2$. Laurite forms very polygonal and tiny grains (about 5 microns in size) and occurs in two different textural positions: 1) enclosed in chromite crystals (figure 5a-b) and 2) in the contact with microfractures of the host chromite (figure 5c-d).

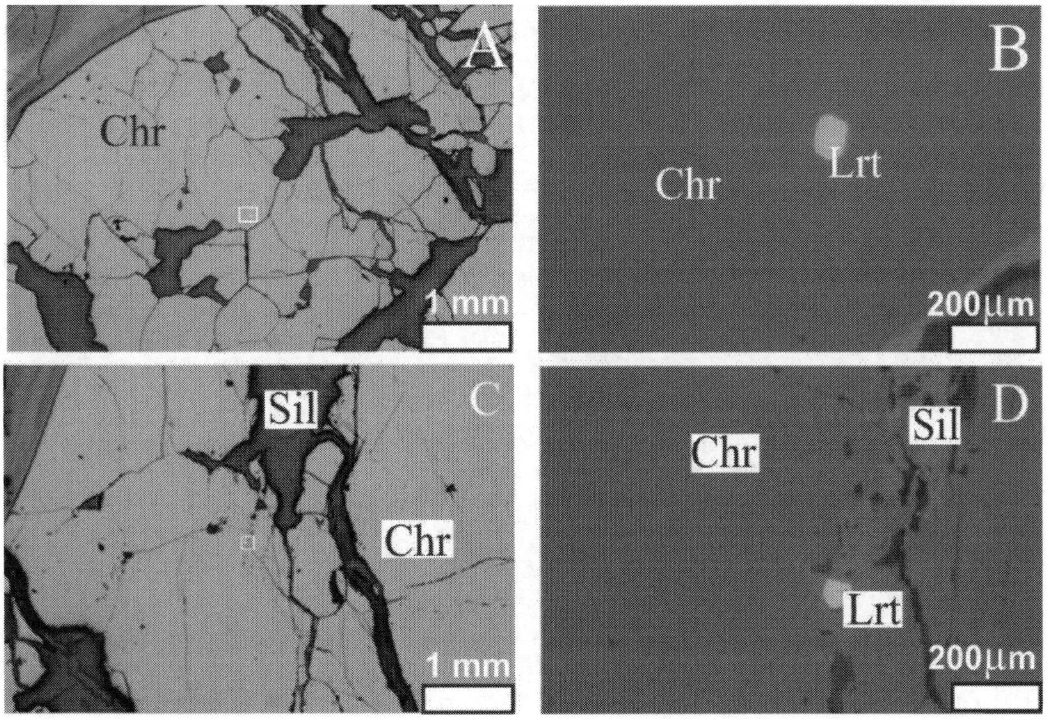

Figure 5. Reflected light microscope images showing the presence of PGM (laurite) associated with the chromitite from study area: (a) and (b) PGM as an inclusion in the chromite crystal; (c) and (d) PGM in the contact with microfractures in chromite crystals. Abbreviation: Chr = chromite; Sil = silicate; Lrt = laurite.

5. Summary and Conclusions

The investigated chromitites of Kabaena Island show a massive texture that during a post magmatic event had been fractured and brecciated. Most of the fractures are filled with secondary silicates which are interpreted to be the result of an alteration process. Few grains of olivine have been preserved in the silicate matrix. The electron microprobe analysis revealed that the chromitites can be classified as Cr-rich and the low TiO_2 content suggests that their composition is compatible with those of podiform chromitites. The only discovered PGM is laurite. Based on its morphology and composition, we can argue that it formed at high temperature, before or concomitantly with the host chromite.

Acknowledgments
The authors would like to thank the leader and residents of the Tedubara village residents who have been permitted to do this research in the study area. The UZAG is Leoben University in Austria. Those are highly appreciated.

References

[1] Garuti G., Zaccarini F., Proenza J A., Thalhammer O A. and Angeli N., 2012. Platinum-Group Minerals in Chromitites of The Niquelândia Layered Intrusion (Central Goias. Brazil): Their Magmatic Origin and Low-Temperature Reworking during Serpentinization and Lateritic Weathering. *Minerals* **2,** 365-84

[2] Gonzales-Jimenez J M., Griffin W L., Proenza J A., Gervilla F., O'Reilly S Y., Akbulut M. and Arai S., 2014. Chromitites in Ophiolites: How. Where. When. Why? Part II. The Crystallization of Chromitites. *Lithos* **189,** 140-158

[3] Hall R. and Sevastjanova I., 2012. Australian Crust in Indonesia. *Australian Journal of Earth Sciences* **59,** 827-44

[4] Ilhami A R., Nurhakim N. and Riswan R., 2019. Studi Keterdapatan Bijih Kromit pada Endapan Laterit di Kecamatan Karang Intan. Kabupaten Banjar. Provinsi Kalimantan Selatan. *Jurnal GEOSAPTA* **5,** 31-60

[5] Kadarusman A., Miyashita S., Maruyama S., Parkinson C D. and Ishikawa A., 2004. Petrology. Geochemistry and Paleogeographic Reconstruction of The East Sulawesi Ophiolite: Indonesia. *Tectonophysics* **392,** 55-83

[6] Lintjewas L., 2015. *Tipe Endapan Kromit di Daerah Konawe Utara Provinsi Sulawesi Tenggara.* LIPI, Bandung

[7] Mosier D L., Singer D A., Moring B C. and Galloway J P., 2012. *Podiform Chromite Deposits—Database and Grade and Tonnage Models.* US Geological Survey. Scientific Investigations Report 2012-5157, 45p

[8] Simandjuntak T O., Surono. and Sukido, 1993. *Peta Geologi Lembar Kolaka, Sulawesi.* Pusat Penelitian dan Pengembangan Geologi, Bandung

[9] Surono, 2013. *Geologi Lengan Tenggara Sulawesi.* Badan Geologi, Kementerian Energi dan Sumber Daya Mineral

[10] Wisi B N., 2017. Prospek Endapan Kromit pada Laterit Dosay. Pegunungan Cycloop. Sentani Barat. Jayapura. Papua. *Buletin Sumber Daya Geologi* **12**: 1-12

[11] Zaccarini F., Idrus A. and Garuti G., 2016. Chromite Composition and Accessory Minerals in Chromitites from Sulawesi. Indonesia: Their Genetic Significance. *Minerals* **6**: 1-23

[12] Zaccarini F., Garuti G., Pushcarev E. and Thalhammer O., 2018. Origin of platinum group minerals (PGM) inclusions in chromite deposits of the Urals. *Minerals* **8:** 379-390.

IOP Conf. Series: Earth and Environmental Science **851** (2021) 012045 doi:10.1088/1755-1315/851/1/012045

A preliminary study of geology and skarn of Cemorosewu Area, Bayat, Central Java, Indonesia

W N Muhammad[1], N I Setiawan[1*], S Husein[1], M Nukman[2]

[1] Geological Engineering Department, Engineering Faculty, Universitas Gadjah Mada, Yogyakarta, Indonesia
[2] Geophysics Department, Mathematics and Natural Sciences Faculty, Universitas Gadjah Mada, Yogyakarta, Indonesia
*E-mail: nugroho.setiawan@ugm.ac.id

Abstract. We present the first finding of skarn rocks in Cemorosewu area of East Jiwo Hill, Bayat, Central Java, Indonesia. The geological conditions of this area which related to the appearance of the skarn is also reported. The methods used in this research are DEM acquisition using drone to generate basemap for geological mapping and thin section petrographic analyses. The geological map shows that Cemorosewu area consists of metamorphic rocks (mica phyllite, graphite phyllite, with quartzite and marble lenses), sedimentary rocks (carbonate breccia and siltstone), and igneous rock (microdiorite). Based on the field observation and geological map, the regional metamorphic rocks are the oldest units in this area followed by sedimentary rocks and igneous rock which intruded both rocks. Skarn rocks were cropped out as a boulder along the Kluwihan creek with the maximum size of 8 m in length. The skarn consists of garnet, clinopyroxene, zoisite, actinolite, and minor quartz. Metasiltstone and skarn rock are suggested formed by contact metamorphism of microdiorite intrusion. The marble, which lenses within the phyllites, is suggested to be the protolith of the skarn formation in this area.

1. Introduction

Jiwo Hills in Bayat is well known for the occurrences of regional metamorphic rocks in Central Java, Indonesia, together with Luk-Ulo Complex and Ciletuh Complex in West Java [1]. The regional metamorphic rocks are considered to be the oldest rocks in this area, which are unconformably overlain by Paleogene and Neogene sedimentary rocks, which were intruded by Oligo-Miocene igneous rocks [2]. Metasomatic skarn was reported in the Pagerjurang area of West Jiwo as a product of contact metamorphism between microgabbro and marble [1; 3] even though similar geological condition with the West Jiwo, the skarn in East Jiwo have never been reported.

Skarn rocks crop out as a boulder along the Kluwihan creek of Cemorosewu area with the maximum size of 8 m in length (Figures 1 and 2). The purpose of this study is to investigate the geological conditions of the Cemorosewu area, East Jiwo related to the first finding of the skarn rock, by geological mapping and petrographic analyses.

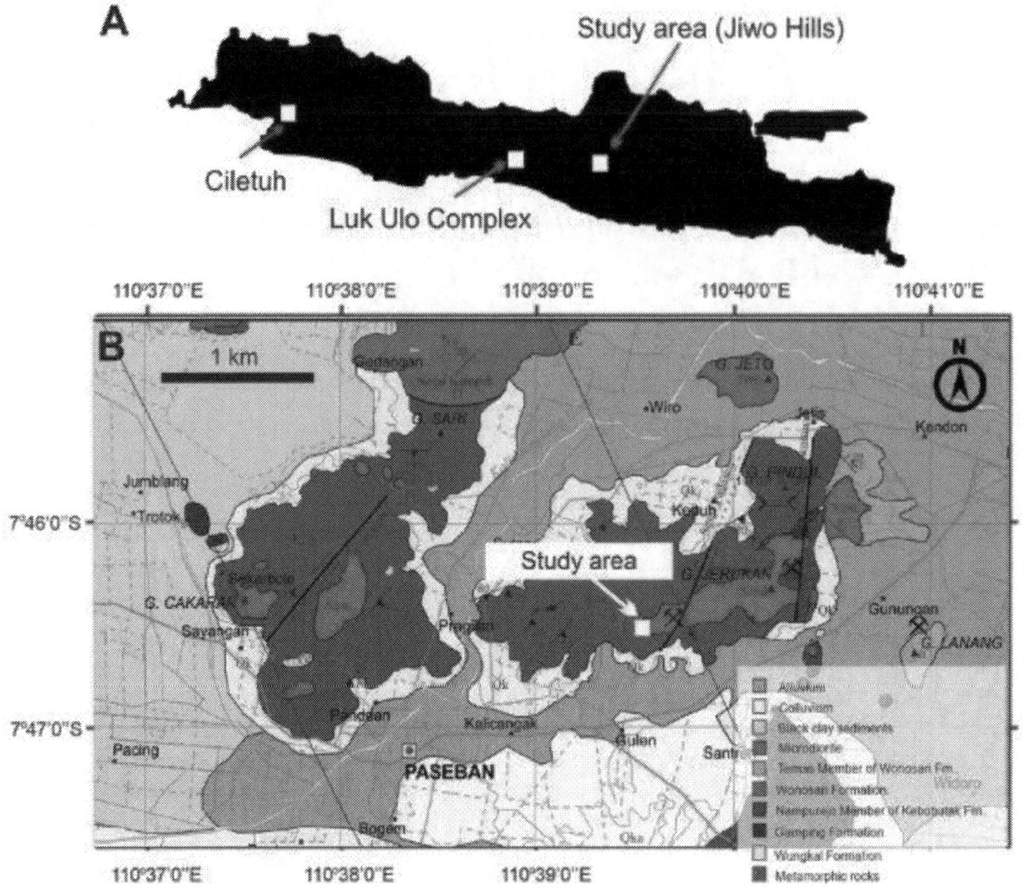

Figure 1. Regional geology map of Jiwo Hills and surrounding area [4]. The skarn rocks in East Jiwo located in the Cemorosewu area, East Jiwo.

2. Regional Geology

Jiwo Hills in Bayat is one of the areas that crop out the basement rocks in Central Java, Indonesia [1]. The Jiwo Hills is divided into West and East Jiwo by NE–SW trend of Dengkeng River (Figure 1). The metamorphic rocks (phyllite, marble, and quartzite) are considered the basement rocks in these areas with the age of Early Cretaceous [2; 4]. The crono-stratigraphic correlation suggested the regional metamorphic rocks unconformably overlain by Eocene sedimentary rocks of Wungkal and Gamping Formations (quartz sandstone and nummulitic limestone), Oligocene-Early Miocene volcaniclastic rocks of Kebo-Butak Formation, and intruded by Oligo-Miocene of dioritic to gabbroic igneous rocks [2; 4]. The carbonate rocks of the Wonosari Formation overlain the Kebo-Butak Formation in Middle-Late Miocene [2]. Furthermore, major faults found on the Jiwo Hills are showing four general orientations of NE–SW, N–S, NW–SE, and E–W [4].

3. Methodology

Detailed geological mapping was conducted in the Cemorosewu area of East Jiwo with the main target to understand the distribution of skarn rocks. The skarn rocks and the other lithology including igneous and sedimentary rocks, were collected and analyzed petrographically. Drone operation was conducted to the acquisition of elevation and imagery of the research area. The elevation data produced from drone operation is processed into a digital elevation model (DEM) to build the research base map of geological field mapping (Figure 2).

The surface geological mapping area is 284 x 272-m square area, which producing a geological map scale of 1: 1500 (Figures 2 and 3). This activity is carried out by directly observing the condition of geological objects of the outcrops in the research area. In total 21 field stations were used to record the geological conditions in this area (Figure 2). Twenty-one thin sections of rock samples were analyzed under the polarized-light microscope to determine the mineralogy and characteristics of rocks in the study area.

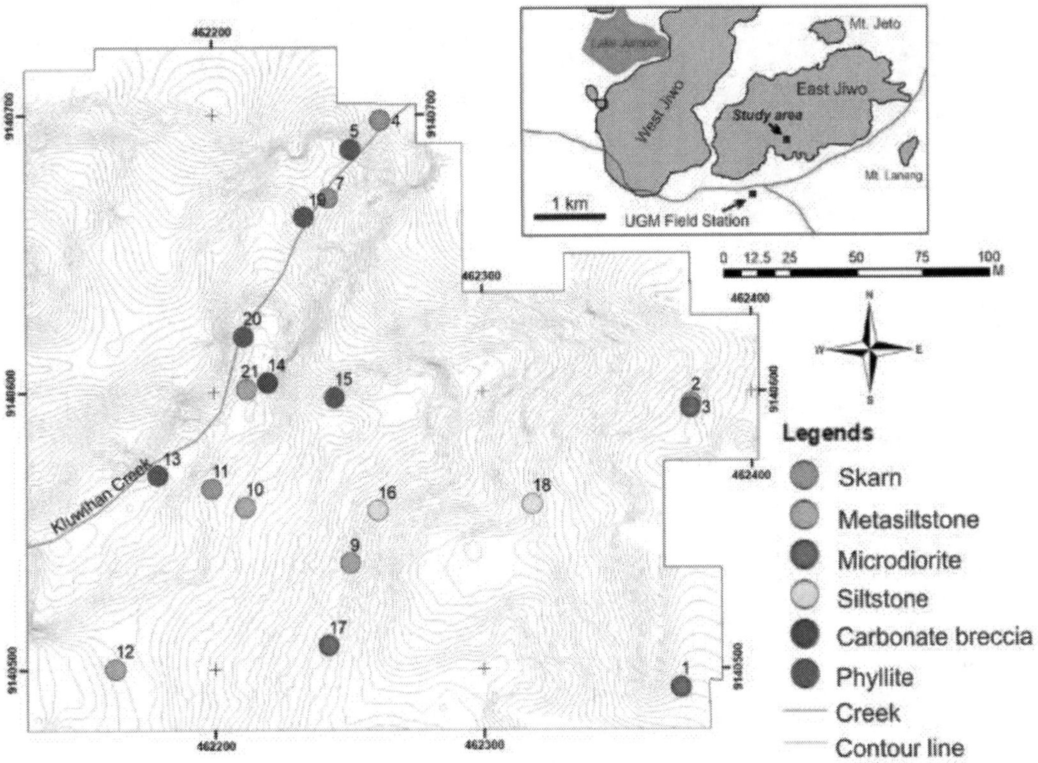

Figure 2. Observation stations of this study indicated by circles. The color of the circle denotes the lithology observed. Field photographs and thin section photos of some of the observation stations are shown on figures 4 and 5, respectively. The geological map is shown on figure 3. Station 6 is on the same location as station 5.

4. Geology of Cemorosewu area

The geological map, section and stratigraphic column resulted from this research is shown in figure 3. No major faults were recognized in this area, only 1 minor sinistral strike-slip fault recorded in the metasiltstone (Figures 3 and 4e). Based on the observed data from the field, there are 6 rock units observed in the field as follows from older to younger, respectively:

4.1. Phyllite with marble lenses unit

This unit covers about 50% of the study area. Based on the observation on station 15 (Figure 2) it has dark brown color with phyllitic structure, composed of platy minerals of mica. In several places, graphite phyllite rocks are found which dominated by graphites instead of micas. Both quartzite and marble lenses are common, marble lenses that are quite prominent are shown on the geological map. The field photograph and hand specimen are shown on figures 4a and 4h, respectively. The general foliation direction of this rock is N20°–60°E. Hydrothermal alteration processes are indicated by secondary quartz and hematite which found on a few places.

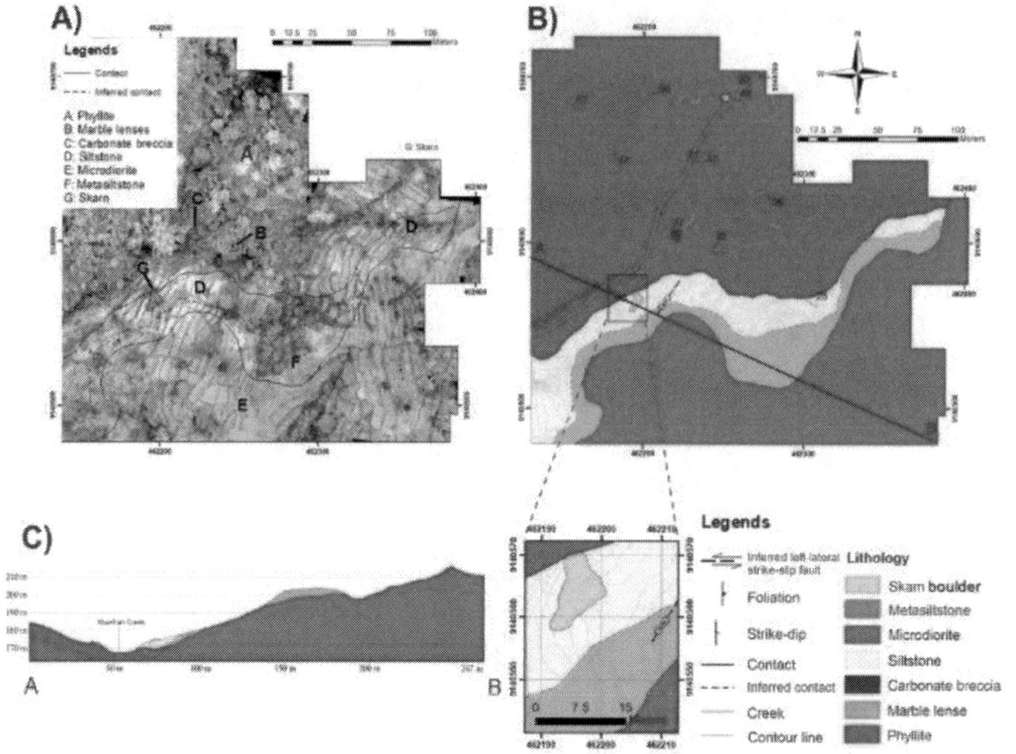

Figure 3. A) Drone imagery of the research area. B) geological map of the research area. C) geological section of A-B traverse line on the geological map.

4.2. Carbonate breccia unit

This unit is found as an outcrop with dimensions of 4 m long and 3 m wide, surrounded by phyllite rocks. The outcrop is located on station 14 in figure 2. The photograph of the outcrop can be seen on figure 4b. It is difficult to determine whether the outcrop is a float or in-situ rock. However, observations elsewhere in East Jiwo outside the study area indicate that the carbonate breccia lies unconformably over the phyllite. The lithological description of the breccia is gray-colored, gravel grain size, poorly sorted, angular grain shape, closed fabric, massive structure, consisting of metamorphic rock fragments and a matrix composed of carbonate material.

4.3. Siltstone unit

This unit covers about 10% of the study area. The description of the rock is based on field observation of station 18 on figure 2, which is brownish-gray color, good sorting, closed fabric, silt grain size, consisting of siliciclastic material with silt size, and a small amount of ash-sized volcanic materials. The photograph of the outcrop is shown on figure 4c. The rock has a layered sedimentary structure with an orientation of N118°E/28°. This unit lies over the marble-lensed phyllite unit and is intruded by the microdiorite unit, suggested forming metasiltstone.

4.4. Microdiorite unit

This unit covers about 30% of the study sites. The description of the rock based on field observation of station 17 of figure 2, is an intrusive igneous rock with a phaneritic texture of 1–2 mm crystal size, holocrystalline, subhedral crystal form, massive structure, major mineral compositions are plagioclase and pyroxene with few accessory minerals of hornblende. The photograph of the rock can be seen on figure 4d. This unit intrudes all rock units in the study area, which evidenced by the presence of contact metamorphism on the siltstone forming metasiltstone, and the occurrence of skarn.

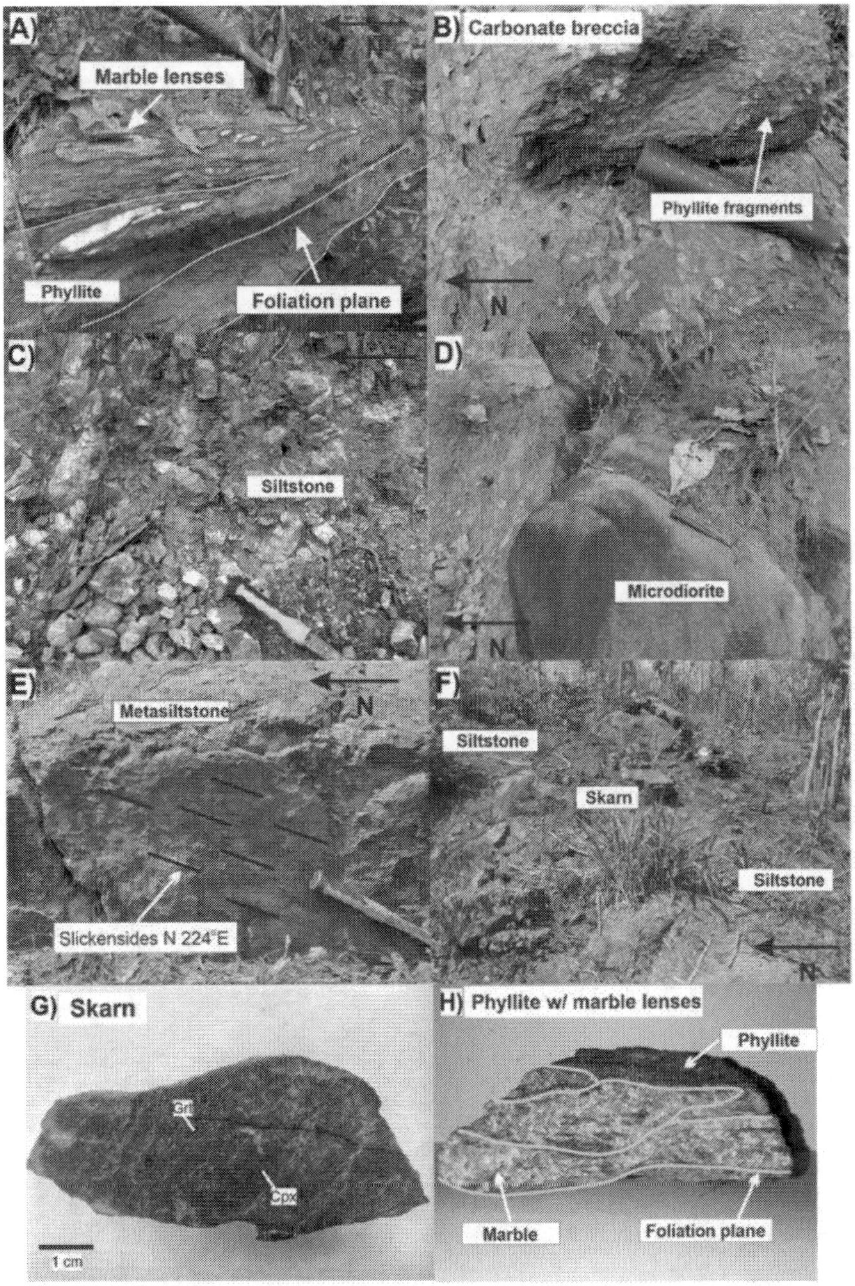

Figure 4. Field photograph of outcrops showing a variety of lithology found on the research area. Stations mentioned in this caption refers to figure 2. A) Phyllite with marble lenses denoted by red lines, approximately parallel to foliation plane on station 15. B) Carbonate breccia with phyllite fragments on station 14. C) Siltstone on station 18. D) Microdiorite on station 17. E) Metasiltstone showing striations that indicates a minor sinistral strike-slip fault on station 10. F) Skarn surrounded by metasiltstone on station 11. G) Photograph of hand sample of skarn taken from station 11. Grt: garnet; Cpx: clinopyroxene. H) Photograph of hand sample taken from station 15 showing marble lenses between phyllite.

4.5. Metasiltstone unit

This unit covers about 10% of the study sites. The description of the rocks based on field observation of station 10 of figure 2, is a light-gray brownish color with hornfelsic structure. This unit has resulted from the process of contact metamorphism between microdiorite and siltstone to form a baked margin. The field photograph is shown on figure 4e.

4.6 Skarn unit

This unit is only found in an outcrop of 8 m long and 6 m wide (Station 11 of figure 2), surrounded by metasiltstone. It is difficult to determine whether the outcrop is ex-situ or in-situ. Apart from being found in a fairly large outcrop, this unit is also found as a float along the Kluwihan creek with the trend of NW-SE (Figures 2 and 3). The lithological description of skarn rocks is a dark green rock with dark red spots, crystal size <1–3 mm, subhedral-euhedral crystal shape, consisting of dark red garnet minerals and greenish minerals clinopyroxene and epidote groups. The field photograph and the hand sample of the skarn rock is shown on figures 4f and 4g, respectively. It has foliated structure that formed by aggregates of red garnet, green clinopyroxene, and whitish clinozoisite. This structure is suggested from relict foliation of the Ca-rich regional metamorphic protolith (calc-silicate phyllite or marble).

Table 1. Summary of thin section analysis of the study area.

No	Lithology	Sample code	Mineral Composition															
			Qz	Ms	Chl	Cpx	Opx	Czo	Grt	Cal	Opq	Gr	Hem	Act	Pl	Zeo	Ser	Ep
1	Microdiorite	WNM 01	-	-	□	○	△	-	-	-	△	-	△	-	○	-	-	-
2	Metasiltstone	WNM 07A	□	-	△	○	-	-	-	-	△	-	△	-	○	-	-	-
3	Microdiorite	WNM 07B	-	-	□	○	△	-	-	-	△	-	□	-	○	-	-	-
4	Skarn	WNM 09	-	-	-	-	-	-	-	-	-	-	□	-	-	⊙	□	□
5	Phyllite	WNM 10A	○	□	○	-	-	-	-	□	△	△	△	-	△	-	-	-
6	Marble lense	WNM 10B	○	□	□	-	-	-	-	⊙	-	-	-	-	-	-	-	-
7	Skarn	WNM 11	-	□	□	□	-	○	△	-	△	-	△	△	-	-	-	-
9	Metasiltstone	WNM 20	○	-	-	□	-	-	-	-	△	-	□	-	□	-	-	-
10	Metasiltstone	WNM 22	○	-	-	□	-	-	-	-	△	-	□	-	□	-	-	-
11	Skarn	WNM 23	△	△	△	□	-	○	□	-	-	-	-	○	□	-	-	-
12	Metasiltstone	WNM 27	○	-	△	□	-	-	-	-	-	-	□	-	□	-	-	-
13	Phyllite	WNM 28	○	○	□	-	-	-	-	-	△	-	-	-	□	-	-	-
14	Phyllite breccia	WNM 29	□	□	□	-	-	-	-	□	△	-	□	-	-	-	-	-
15	Phyllite	WNM 30	○	○	□	-	-	-	-	-	△	□	-	-	□	-	-	-
16	Siltstone	WNM 33	○	-	□	□	-	-	-	-	-	-	△	-	□	-	-	-
18	Siltstone	WNM 36	○	-	-	△	-	-	-	-	-	-	△	-	○	-	-	-
19	Phyllite	WNM 41	○	△	○	-	-	-	-	-	△	-	□	-	△	-	-	-
20	Phyllite	WNM 44	○	△	○	-	-	-	-	-	□	-	□	-	△	-	-	-
21	Metasiltstone	WNM 45	○	-	△	□	-	-	-	-	-	-	△	-	□	-	-	-

- Absent △ Poor □ Moderate ○ Rich ⊙ Abundant

Qz = Quartz; Ms = muscovite; Chl = chlorite; Cpx = clinopyroxene; Opx = orthopiroxene; Czo = clinozoisite; Grt = garnet; Cal = calcite; Opq = opaque; Gr = graphite; Hem = hematite; Act = actinolite; Pl = plagioclase; Zeo = zeolite; Ser = sericite; Ep = epidote

5. Thin section petrographic analysis

The results of the petrographic analysis carried out on the samples taken from the study area are shown in table 1. Microdiorite from station 1 (Figure 5a) shows plagioclase, clinopyroxene, and chlorite-epidote minerals. Figure 5b from station 14 shows the thin section of carbonate breccia with phyllite fragments with matrix composed of carbonate mud. Figure 5c from station 10 shows a thin section of metasiltstone showing a very fine saccharoidal texture, consisted of quartz, plagioclase, and pyroxenes. Spotted texture of hematite is also observed. Figure 5d from station 18 is the thin section of siltstone, shows medium-sorted dominated by quartz, plagioclase, and pyroxene minerals, some of which have been oxidized to hematite. Figure 5e from station 15 shows a thin section of graphite-mica-chlorite phyllite characterized by a lepidoblastic texture appearance, intercalated by granoblastic minerals of quartz and albite. Figure 5f showing a contact between marble lenses and chlorite-quartz-muscovite phyllite, also found on station 15. The thin section of a skarn, that is actinolite-clinozoisite skarn from floats found on station 19, which shown in figure 5g, showing decussate texture composed of clinozoisite, actinolite, quartz, and garnet. Figure 5h from station 11 shows a thin section of skarn with decussate texture composed of clinozoisite, actinolite, quartz, and garnet. The actinolite content is more abundant.

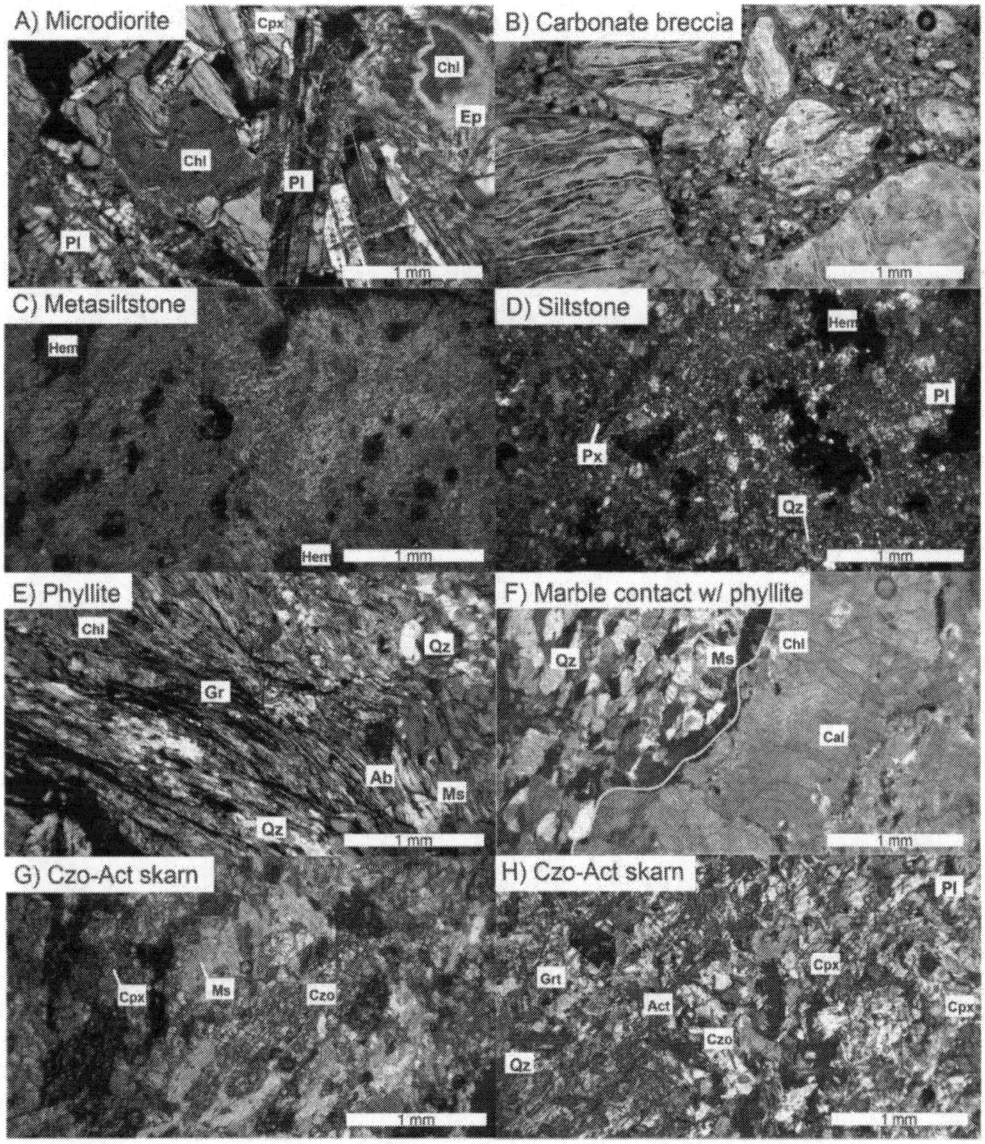

Figure 5. Thin section images of representative samples of the studied location. A) Microdiorite; B) Carbonate breccia; C) Metasiltstone; D) Siltstone; E) Graphite-chlorite-mica phyllite; F) Marble contact with phyllite; G) Clinozoisite-actinolite skarn; H) Clinozoisite-actinolite skarn. Explanation of these thin sections is on section 5 of this paper.

6. Discussion

The geological history of the research area is marked by the formation of regional metamorphic rocks in the form of mica phyllite with marble lenses. Mineralogical compositions and rock textures (Table 1) imply that the metamorphic rocks originated from pelitic intercalated with carbonate protoliths, thus forming marble lenses in the phyllite (Figure 4a). The boudinage shown by marble, as well as the phyllitic foliation structure, might indicate that the rocks experienced ductile deformation during the metamorphism.

The occurrence of basal carbonate breccias with phyllite fragments (Figures 4b and 5b) implies that the metamorphic rock was exposed followed by erosion, and lithified into breccia. Those are indicated by an unconformity between the metamorphic and the breccia rocks. Carbonate matrix and cement that form the breccia suggest that the rock was formed in the marine environment. The sedimentary rock of siltstone might indicate that the rock was deposited in the environment with low current velocity. After the siltstone has been deposited and lithified, the magmatic activity of intrusion occurred which interact with country rocks (metamorphic and sedimentary rocks), of which responsible for the contact metamorphism. The metasiltstone has suggested the product of the contact metamorphism between microdiorite and siltstone. The rock exhibit more compact with hornfelsic texture compared to the original rock of siltstone (Figures 4c and 4e). Common contact metamorphic minerals (andalusite, cordierite, etc.) are not identified in this rock, probably because the microdiorite formed as shallow intrusion with faster cooling processes. It is confirmed by the medium-grained of crystal size (1–2 mm) in the microdiorite (Figure 4a).

Skarn in the research area was found as floats with the maximum size of 8 m length along the Kluwihan creek (Figures 3b and 4f). Skarn is a calc-silicate rock made by metasomatic processes of igneous intrusion body with carbonate rocks [5]. There are 2 types of carbonate rocks as a candidate of the skarn protolith found in the research area, which are marble lenses in the phyllite and the carbonate breccia. However, carbonate breccia was only found as a single float in the research area (Station 14, figures 4b and 5b). Hence, highly possible of the protolith skarn rocks is marble lenses in the phyllite, which quite common to find within the metamorphic rocks elsewhere in Jiwo Hills [1; 4]. The major composition of highly calc-silicate rich minerals of pyroxene and epidote (Table 1) might be indicating as a calcic skarn type [6; 7; 8]. Furthermore, greenish color (Figure 4g) with high ratio of pyroxene and epidote to garnet might indicate that the skarn rocks in the Cemorosewu Area are on the zone of exoskarn with prograde assemblages [6; 7; 8]. The retrograde assemblages (clinozoisite, actinolite; Figure 5h) [6; 7; 8] were found in the skarn that located in the northeastern of study area (Stations 19 and 4; Figure 2). This might be implied that the heat source of the skarn formation more likely located in the southwestern of the research area, which is proved by the occurrence of microdiorite intrusion. However, this needs to be further confirmed by geochemical analyses. Other than skarn formation, indications of silicic alteration are also observed, mainly in the phyllite which shows brecciated phyllite with secondary quartz matrix, overprinted by quartz vein.

7. Conclusion

The lithology in the Cemorosewu area consists of metamorphic, sedimentary, and igneous rocks. The regional metamorphic rocks consist of mica phyllite, graphite phyllite, quartzite and marble lenses. Regional metamorphic rocks are the oldest units exposed in the study area. The contact metamorphic rocks exposed in the study area are metasiltstone and skarn. Sedimentary rocks found in the study area include siltstone and carbonate breccias with phyllite fragments. The sedimentary rock units unconformably overlay the regional metamorphic rock units. Igneous rock that appears in the study area is microdiorite, which intrudes the rocks above, manifested by metasiltstone and skarn. The major composition of skarn minerals are garnet, pyroxene, and epidote indicate that the skarn is a calcic skarn and part of the exoskarn zone. The prograde assemblage of skarn is located in the southwestern of the research area nearby the microdiorite intrusion. While retrograde assemblage of skarn is located in the northeastern of the research area. Further detailed investigation, including geochemical analyses, is needed to constrain this conclusion.

Acknowledgments
We would like to thank the Geological Engineering Department of UGM and RTA Program for providing the research grant to facilitate this research.

References
[1] Setiawan N I, Osanai Y and Prasetyadi C 2013 *Proc. Seminar Nasional Kebumian Ke-6 (Yogyakarta)* 12–24.
[2] Surono, Hartono U, and Permanadewi S 2006 *Geo-Res.* **15** 302–11.
[3] Dharmawan A G and Setijadji L D 2014 *Studi Ubahan Batuan Tipe Skarn di Desa Pagerjurang, Kecamatan Bayat, Kabupaten Klaten, Propinsi Jawa Tengah* [unpublished thesis] (Yogyakarta: Universitas Gadjah Mada) 164.
[4] Barianto D H, Margono U, Husein S, and Novian M I 2017 *Geological map of Wonosari Quadrangels, Jawa* (1408-31), Scale 1:50000 (Bandung: Geological Survery Centre)
[5] Harlov D E and Austrheim H 2013 *Metasomatism and the Chemical Transformation of Rock* (Berlin: Springer) p 806
[6] Meinert L D 1992 *Geosci. Canada* **19** 145–162.
[7] Fila-Wojcicka E 2000 *Acta Geol. Polonica* **50** 211–222.
[8] Kuscu I 2001 Turkish *Jour. of Earth Sci.* **10** 121–132.

IOP Conf. Series: Earth and Environmental Science **851** (2021) 012046 doi:10.1088/1755-1315/851/1/012046

Overpressure's top and generating mechanism in "APR" block, North Sumatra offshore area

A Pradana[1*], J Setyowiyoto[1], B Prasetyo[2] and W Haidar[2]

[1]Geological Engineering Department, Universitas Gadjah Mada
[2]Pertamina Hulu Energi North Sumatra Offshore
[*]E-mail: pradana.aryapradana.arya@gmail.com

Abstract. The offshore North Sumatra Basin is one of the hydrocarbon-prone basins in Indonesia. The Malacca Formation acts as the reservoir in this basin. However, there are some drilling problems caused by abnormal pore pressures. Overpressure analysis is carried out so that drilling activities are avoided from drilling problems such as kicks, blowouts, stuck pipes, loss circulation, and collapse. This study aims to analyze the overpressure in the "APR" block, offshore North Sumatra Basin, which includes determine the depth of the top overpressure and the mechanism of overpressure formation. In the research area, analysis of 3 exploration wells was carried out. Overpressure is analyzed based on well log data, and pressure test data using the Eaton method (1975) supported by other data such as LOT (Leak-Off Test) data, drilling reports, mudlogs. At Well A, top overpressure depth is ± 3235 ft TVDSS (true vertical depth sub sea). At Well P, top overpressure depth is ± 3065 ft TVDSS, and at Well R, top overpressure depth is ± 2901 ft TVDML (true vertical depth mud line). The overpressure generating mechanism in the study area is caused by the loading mechanism (disequilibrium compaction) and caused by the nonloading mechanism (hydrocarbon buoyancy).

1. Introduction

The North Sumatra Offshore area is a well-known productive basin in Indonesia. Several exploration wells at the "APR" Block had encounter drilling problems due to overpressure condition on the area. However, the characteristics of the overpressure, which includes top of overpressure on the area and generating mechanisms related to its geological environment, is still unknown.

This research uses three wells data at the "APR" Block, which will be analyzed to find the top of overpressure (Figure 1). It is highly important to know the characteristics because overpressure give heavy consequences in oil and gas industry, especially in drilling problems such as kick, blowout, and loss circulations. Therefore, overpressure analysis are important before drilling projects can be started at the research area.

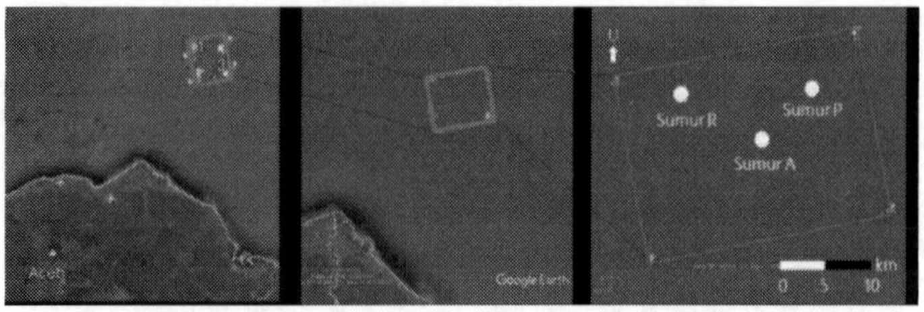

Figure 1. Research area shown within the red square.

2. Geological Setting

North Sumatra Offshore area belongs to the North Sumatra Basin, and located at the offshore southern side of Mergui Basin [3]. Physiographically speaking, the area is one of the back-arc basins in Indonesia and formed by the subduction of Indian Ocean Plate and Eurasian Plate [2]. Figure 2a shows Sumatra's regional geology map, and Figure 2b shows the vertical section of the North Sumatra Basin.

Structurally, none of any major fault or other kind of structures was found. Stratigraphic column [1] is shown on Figure 3 shows the target area of the overpressure analysis which includes Seurula Formation, Keutapang Formation, and Malacca Formation.

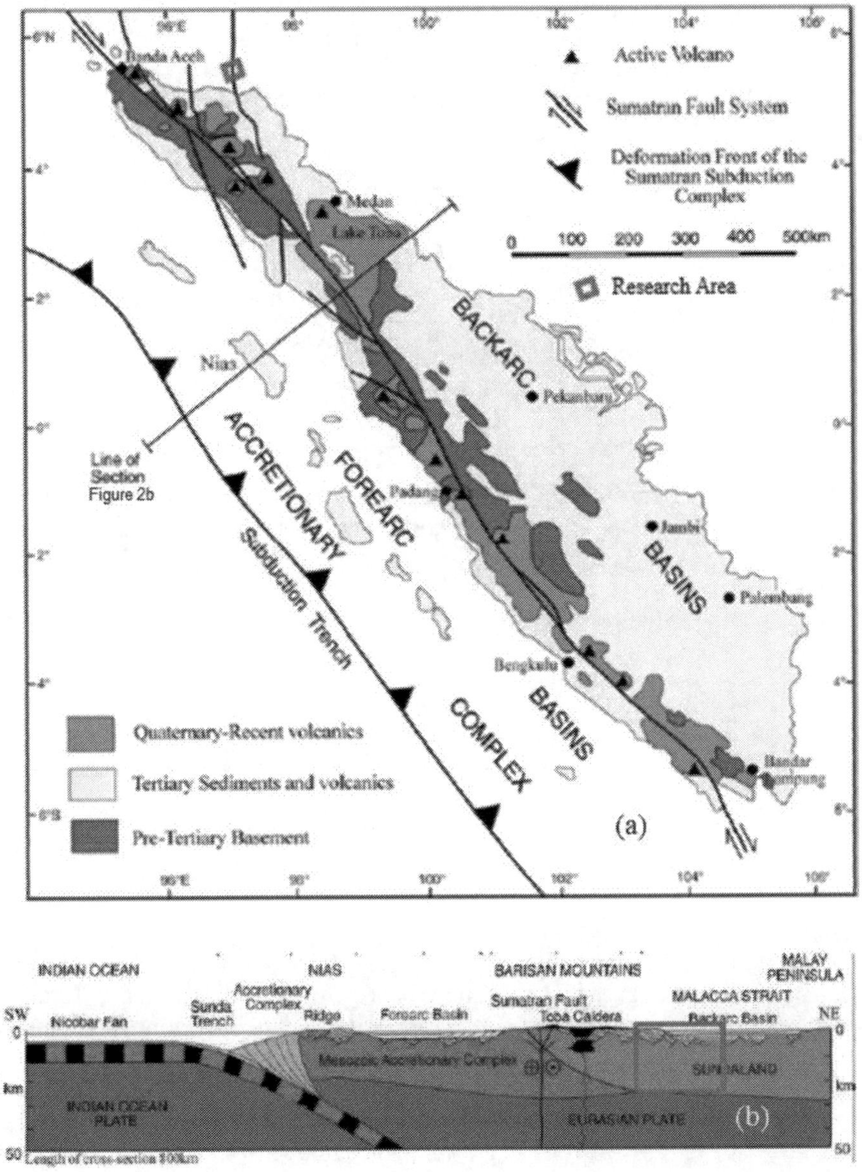

Figure 2. (a) Geologic Map of Sumatra [2], shows the distribution of stratigraphic and volcanism units. Research area shown by the red box. (b) Vertical section of the Sumatra's subduction system.

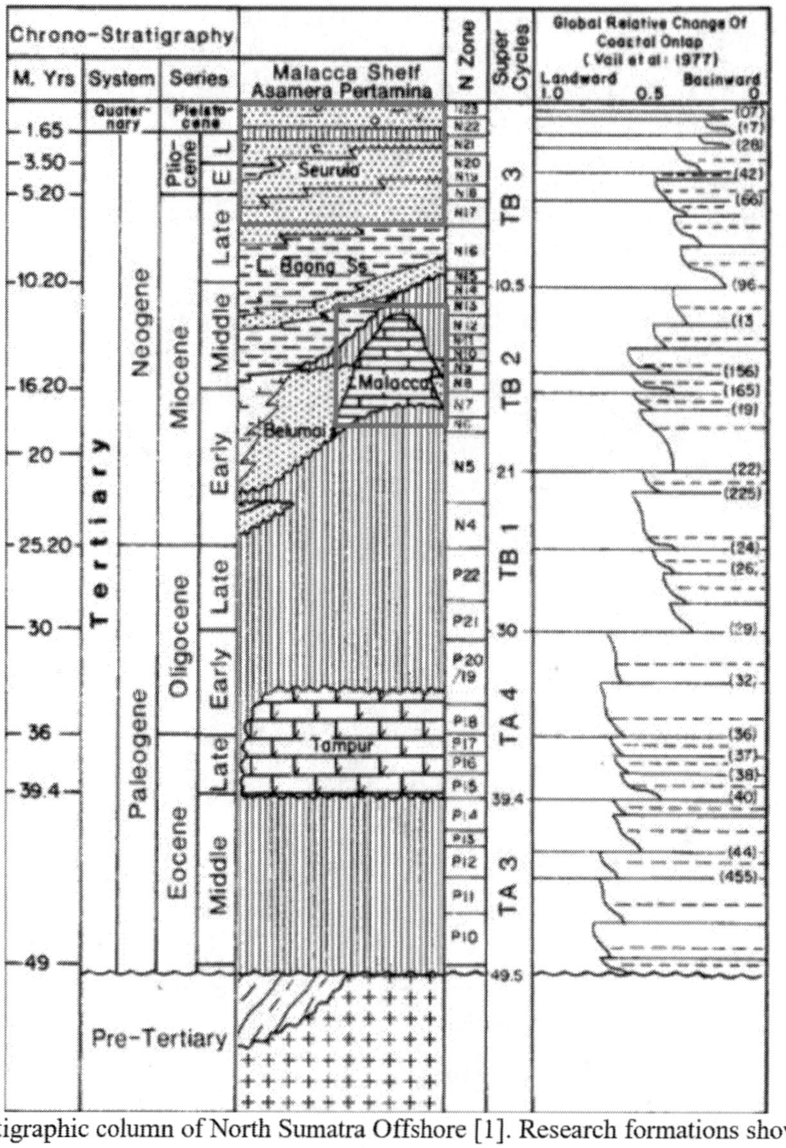

Figure 3. Stratigraphic column of North Sumatra Offshore [1]. Research formations shown by the red box.

3. Methods

This research uses wireline logs, direct pressure data, mudweights, drilling events, and leak-off test to identify the top of overpressure at each wells. Those data are presented as pressure-depth profile. Then, the profiles will be analized to determine top of overpressure and its generating mechanism.

Mudweight could be an indicator for the detection of overpressure, marked with the addition of mudweight. Drilling event i.e kick, loss, blow out, pull/drag, hole fill, connection gas, etc, may be caused by overpressure. Another data that we use to anlyze overpressure in this research is wireline log data.

The hydrostatic pressure is calculated using the default value of fresh water density gradient, and the lithostatic pressure is calculated using the wireline log data, which is the density log [5]. For the pore pressure calculation, the wireline log data processing will be conducted at first by discriminate between sand-rich and shalerich value from the shale volume value. After that, the sonic/vp log or the resistivity log on the shale intervals is used to create the normal compaction trend (NCT) and also to calculate the pore pressure value using Eaton's equation. Fracture pressure calculation is also done, with the leak-off test data and the pore pressure calculation itself [5].

The responses of wireline log data will show two basic types of overpressure generating mechanisms, which are loading and unloading [4]. Loading mechanism response will show constant trend of logs, caused by the dewatering rate is lower than the sedimentation rate, made the fluid couldn't escaped from the pore space (Figure 4a). Meanwhile, unloading mechanism will show reflecting pattern of log response from normal compaction trend, due to fluid expansion, made the pore space getting larger than before (Figure 4b). Aside from both loading and unloading mechanism, there are also another mechanism (nonloading mechanism) such as hydrocarbon buoyancy, aquathermal pressuring, hydraulic head, and osmosis. Those mechanism have their own characteristic of overpressuring from the pressure profile, for example the pore pressure profile from the hydrocarbon buoyancy's overpressure can be shown at Figure 4c.

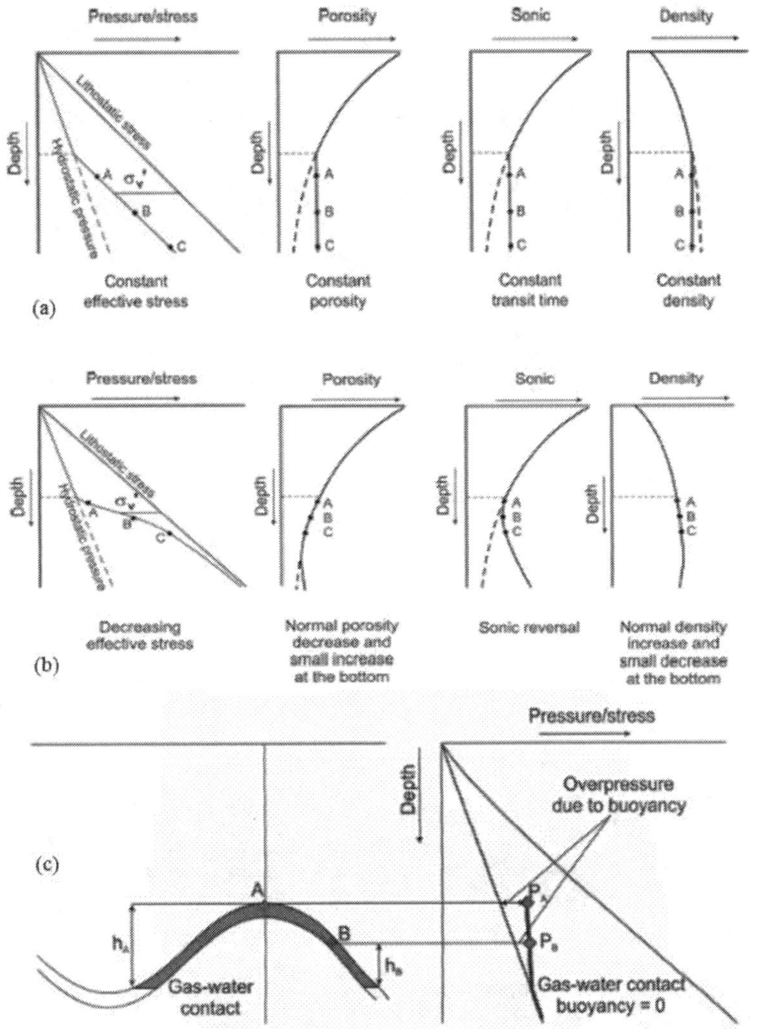

Figure 4. (a) Pressure-depth profile and wireline log response to loading mechanism. (b) Pressure-depth profile and wireline log response to unloading mechanism. (c) Example of pressure-depth profile to hydrocarbon buoyancy overpressuring.

4. Top of Overpressure Analysis

4.1. A-Well

At the A-well, the top of overpressure is at 3235 ft TVDSS, which is located in Keutapang Formation (Figure 5a). Top of overpressure is determined when the sonic log is deflecting from the normal compaction trend. From the gamma ray log, it shows decreasing value of shale volume, interpreted as decrease of the rock matrix percentage due to presence of the fluid. From the density and neutron logs, it also shows deflection from the NCT's (Figure 5b). There are also a drilling events which is well kick and increase at mud weight value. Increase of mud weight is happened due to increase of pore pressure value. The direct pressure measurement (which is Drill Stem Test) is also calculating the same value with the pore pressure prediction from vp log.

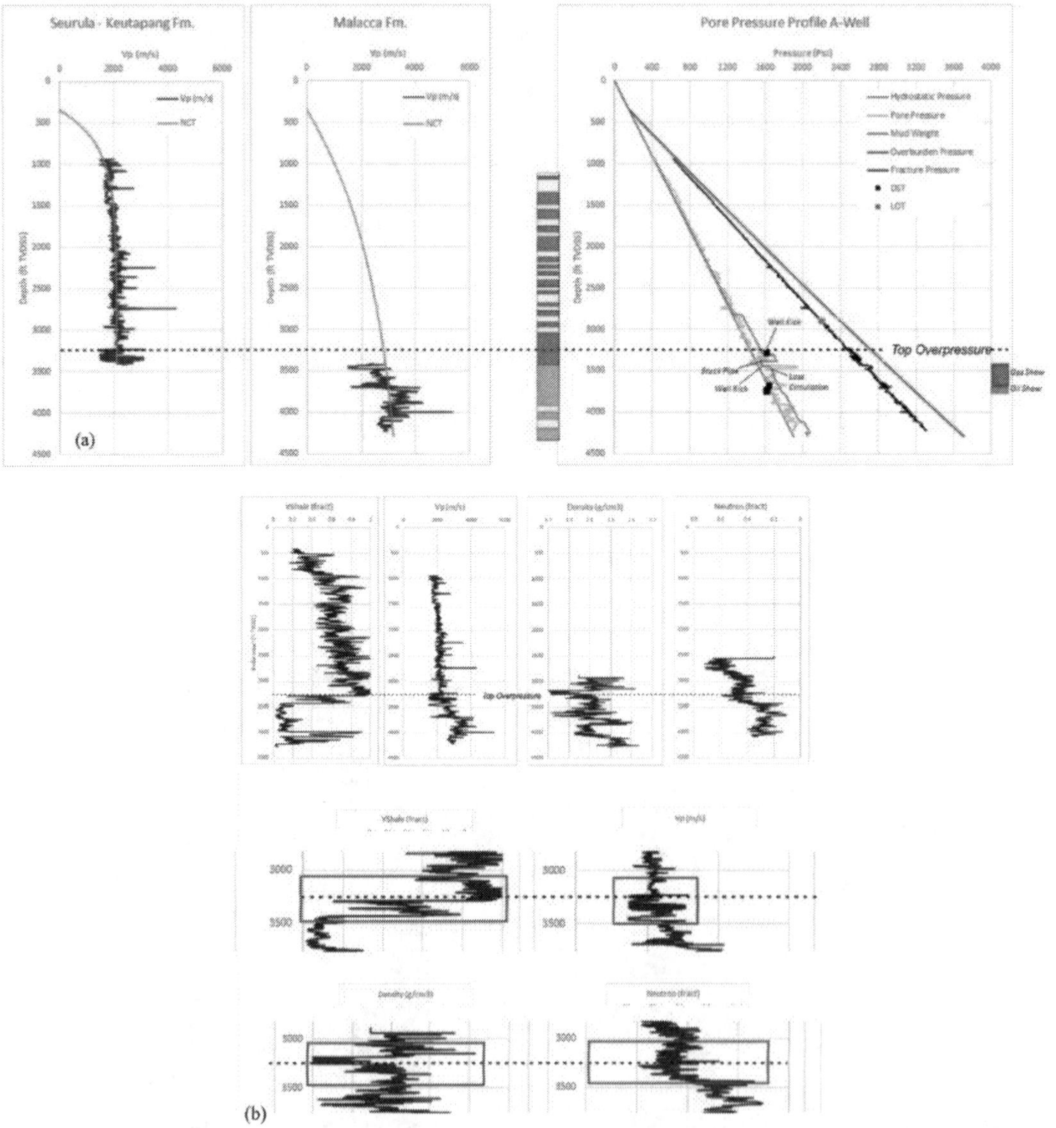

Figure 5. (a) Pore pressure profile of A-Well. (b) Wireline log responses of A-Well.

IOP Conf. Series: Earth and Environmental Science **851** (2021) 012046 doi:10.1088/1755-1315/851/1/012046

4.2. P-Well

At the P-well, the top of overpressure is at 3065 ft TVDSS, which is located in Malacca Formation (Figure 6a). Top of overpressure is determined when the sonic log is deflecting from the normal compaction trend. From the gamma ray log, it shows a significant decrease value of shale volume, interpreted as difference of lithology. From the density and neutron logs, it also shows deflection from the NCT's, even though there is no porosity log data at previous interval (Figure 6b). There are some lacks of wireline log data in this well at several intervals. These lacks made the overpressure analysis couldn't been done perfectly.

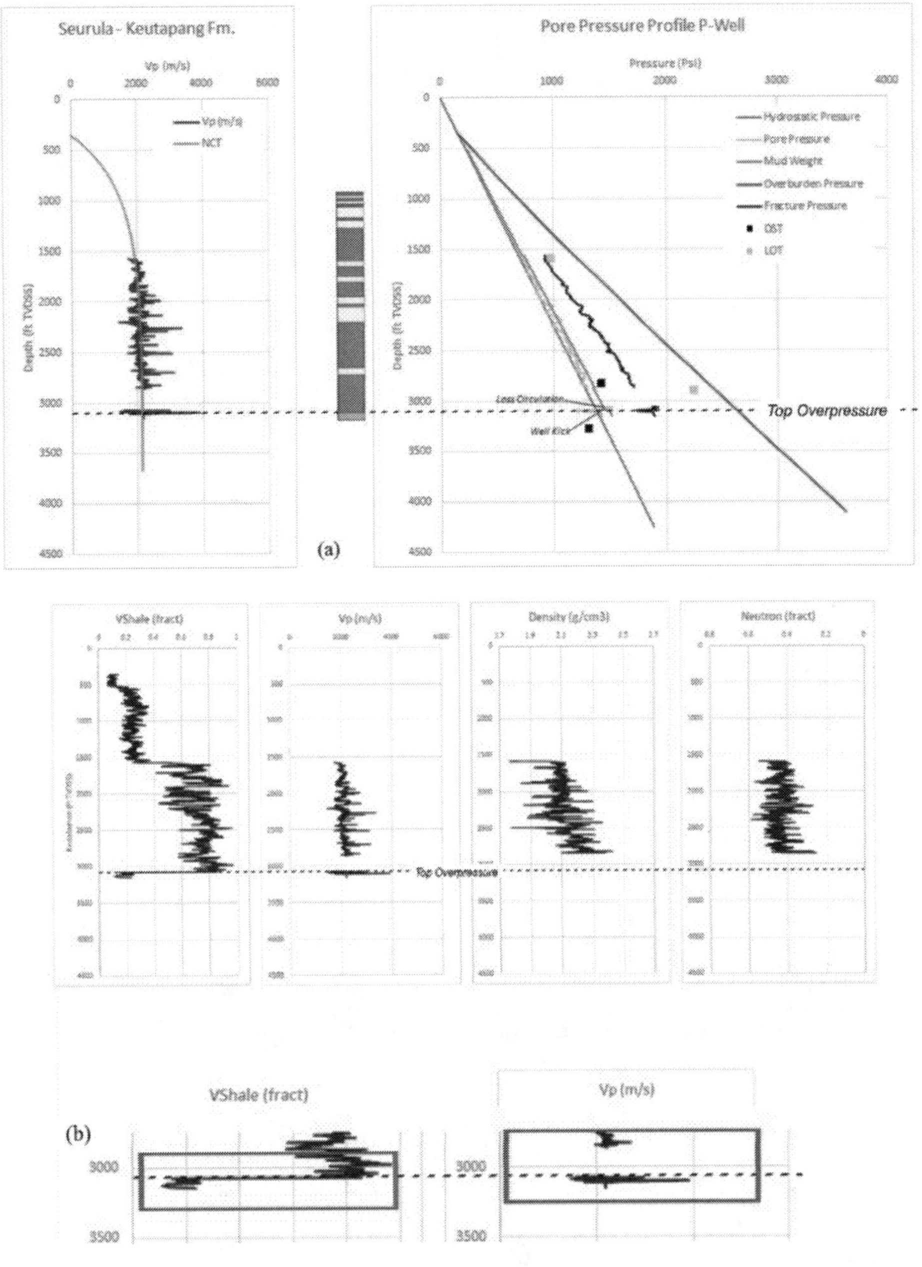

Figure 6. (a) Pore pressure profile of P-Well. (b) Wireline log responses of P-Well.

IOP Conf. Series: Earth and Environmental Science **851** (2021) 012046 doi:10.1088/1755-1315/851/1/012046

4.3. R-Well

At the P-well, the top of overpressure is at 2901 ft TVDML, which is approximately located in Keutapang Formation (Figure 7a). Top of overpressure is determined when the resistivity log is deflecting from the normal compaction trend. From the gamma ray log, it shows high value of shale volume, interpreted as high shale content (Figure 7b). None of the density and porosity logs are available in this depth. This lack made the overpressure analysis couldn't had been done perfectly. Also, there are no available information regarding to the mud weight and any direct pressure measurements, so the calculation couldn't be validated.

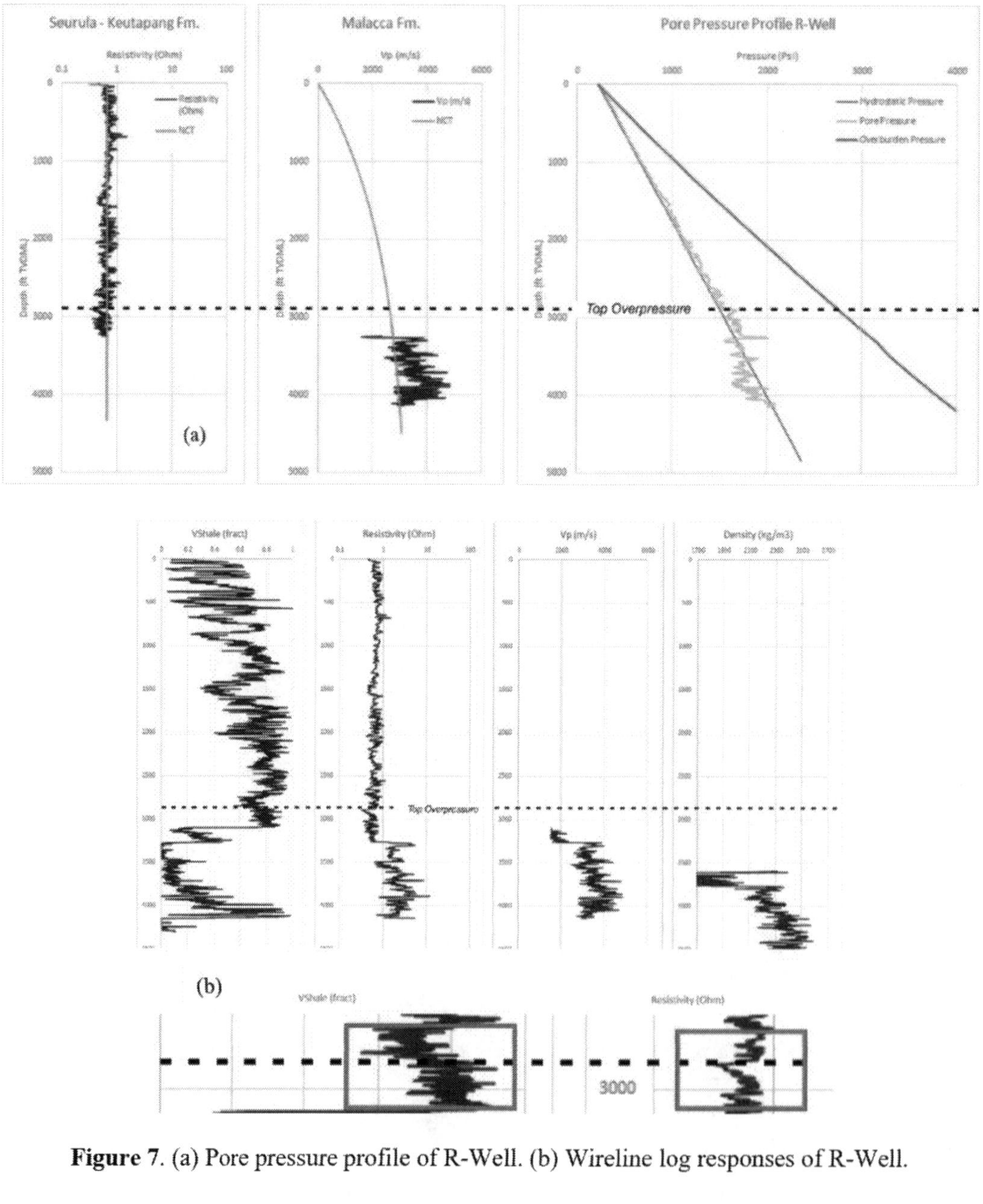

Figure 7. (a) Pore pressure profile of R-Well. (b) Wireline log responses of R-Well.

5. Overpressure Generating Mechanism Analysis

5.1. A-Well

Pore pressure profile shows a loading mechanism (disequilibrium compaction) response at 3235 ft TVDSS – 3435 ft TVDSS (Figure 8). This mechanism happened due to the undercompacted shale that decrease the dewatering rate, so the water would be trapped and increased the pore pressure (overpressure). Other intervals which are 3435 ft TVDSS – 3783 ft TVDSS, there is an increase of pore pressure in different lithology (carbonate). That interval is interpreted as the nonloading mechanism (hydrocarbon buoyancy). From the mudlog data, it is confirmed that the carbonate interval's act as hydrocarbon reservoir, which the overpressure is generated from the hydrocarbon column buoyancy.

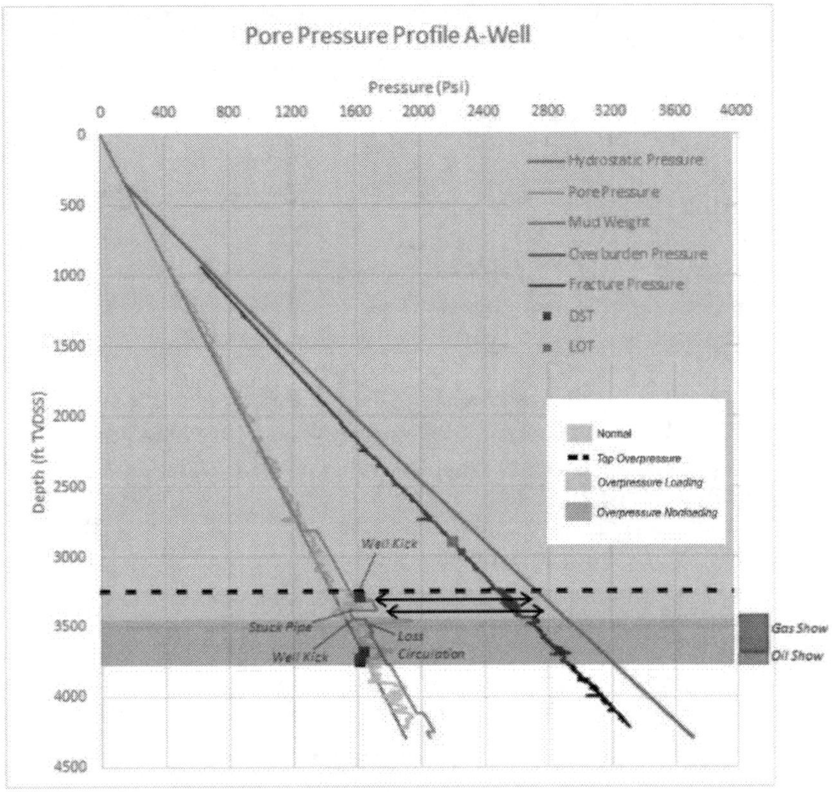

Figure 8. Pore pressure profile of A-Well and its overpressure generating mechanism.

5.2. P-Well

Pore pressure profile shows a nonloading mechanism (hydrocarbon buoyancy) response at 3065 ft TVDSS – final depth (Figure 9). This mechanism happened due to the presence of hydrocarbon, that has the buoyancy factor. Even though the mudlog did not have the cutting's data at the certain depth, but the resistivity log shows a high value, which can be one of the indicators of the presence from the hydrocarbon.

IOP Conf. Series: Earth and Environmental Science **851** (2021) 012046 doi:10.1088/1755-1315/851/1/012046

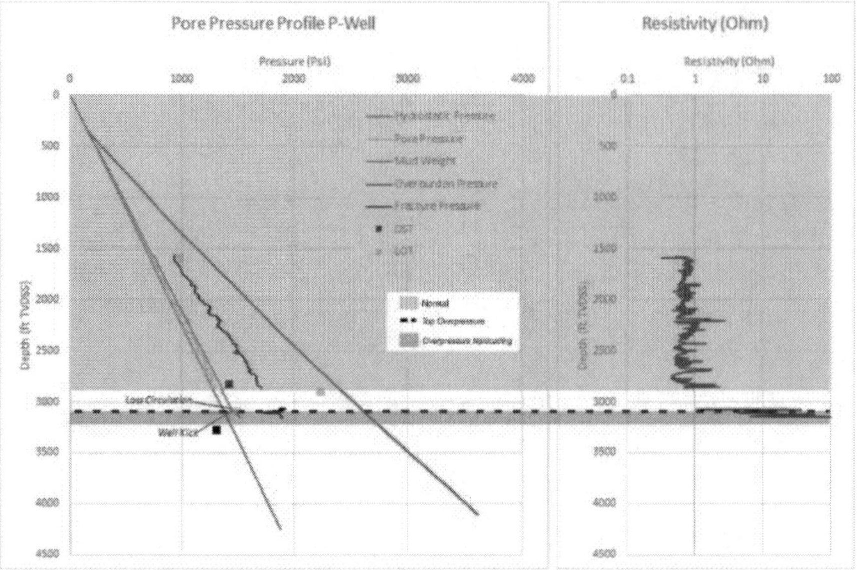

Figure 9. Pore pressure profile of P-Well and its overpressure generating mechanism. At nonloading mechanism zone, resistivity log shows higher value, indicating the hydrocarbon presence.

5.3. R-Well

Pore pressure profile shows a loading mechanism (disequilibrium compaction) response at 2901 ft TVDML – 3275 ft TVDML (Figure 10). Same as A-Well, this mechanism happened due to the undercompacted shale. Apart from the overpressure condition, there are also an underpressure condition which is marked with the pore pressure value that less than the hydrostatic. Regardless of the final data report unavailability, the underpressure condition was found in carbonate section (probably Malacca Formation), which shown by the low gamma ray value.

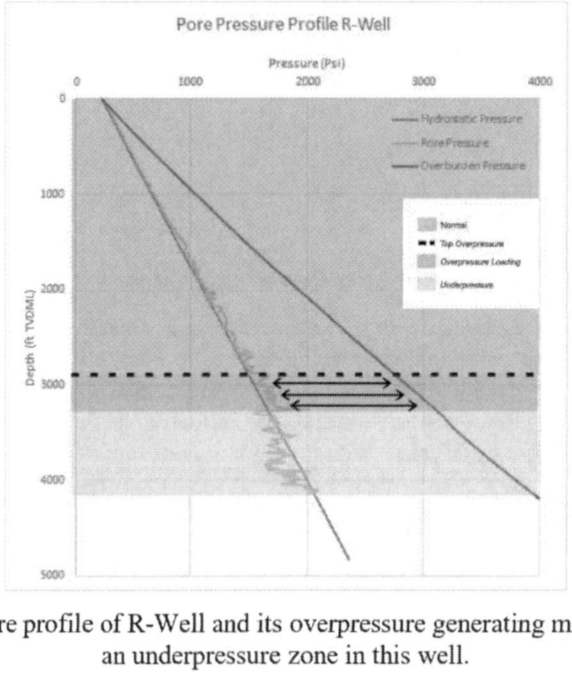

Figure 10. Pore pressure profile of R-Well and its overpressure generating mechanism. There is also an underpressure zone in this well.

IOP Conf. Series: Earth and Environmental Science **851** (2021) 012046 doi:10.1088/1755-1315/851/1/012046

6. Conclusion

At the "APR" Block, the overpressure occurrence is associated with shale lithology and presence of hydrocarbon. Tops of overpressure are detected by the wireline log data, at shale dominated sequence (A-Well and R-Well) or at the reservoir zone (P-Well). Overpressure generating mechanism analysis shows that the overpressure at the research area was generated by several mechanism, which are loading mechanism (disequilibrium compaction) and nonloading mechanism (hydrocarbon buoyancy).

Acknowledgments

Authors would like to thank Pertamina Hulu Energi North Sumatra Offshore (PHE NSO) for the opportunity to study overpressure in North Sumatra Offshore area.

References

[1] Anderson, B. L., Bon, J., dan Wahono, H. E. (1993). Reassessment of the Miocene stratigraphy, paleogeography and petroleum geochemistry of the Langsa Block in the offshore North Sumatra Basin. *Proceedings of 22nd Annual Convention & Exhibition, Indonesia Petroleum Association, 1*, 169-189.

[2] Barber, A. J., Crow, M. J., dan Milsom, J. S. (2005). *Sumatra: Geology, Resources, and Tectonic Evolution.* London: The Geological Society. 304p.

[3] Pertamina BPPKA. (1995). *Petroleum Geology of Indonesian Basins: Principles, Methods, and Application, Vol. 1 North Sumatra Basin.* Jakarta: Pertamina. 96p.

[4] Ramdhan, A. M. (2010). *Overpressure and Compaction in the Lower Kutai Basin, Indonesia.* Durham: Durham University [Dissertation, unpublished]. 329p.

[5] Zoback, M. D. (2007). *Reservoir Geomechanics.* New York: Cambridge University Press. 505p.

SEM-EDS imaging of multi-stage cleat minerals formation in the South Walker Creek Coals, Bowen Basin

A K Permana[1*] and C R Ward[2]
[1]Center for Geological Survey, Geological Agency, MEMR, Jl. Diponegoro No.57 Bandung, Jawa Barat 40122, Indonesia
[2]University of New South Wales, Kensington, Sydney, NSW - 2052, Australia
*E-mail: akpermana79@gmail.com

Abstract. The South Walker Creek coals are moderately to highly developed fracture systems. These fractures can be recognized from the top to the bottom of the seam, with both open and closed apertures. They are filled with several different minerals, occurs either as single-phase infillings (monomineralic), or as mixed mineral deposits (polymineralic). This paper presents scanning electron microscope (SEM) analysis of mineralogical association within coals fractures or cleats, as well as their origin and phase of development. SEM-EDS analysis indicates that the polymineralic cleats seem to be more common in the coal seam compared to monomineralic cleats. The polymineralic cleat infillings are composed of clay minerals, carbonates, and other minerals, such rutile and apatite, with minor occurrences of diaspore. This polymineralic association within the single cleat spaces represents a multi-phase minerals formation, as a result of different stages of epigenetic activity. The hydrothermal fluid circulation are responsible for this epigenetic process, they are re-opened previous minerals filled cleats and transformed pre-existing mineral within the cleats, and/or to remobilization earlier minerals, which then precipitated in the cleat as epigenetic minerals.

1. Introduction

Coal is an organic sedimentary rock that consists of organic matter and mineral matter [1,2,3]. The combustible organic particles (macerals) are composed of lithified plant debris, and occur as discrete layers of organic material each with its own physical and chemical properties [4]. The non-combustible mineral matter is mainly made up a variety of primary and secondary minerals, such as clays, carbonates, silicates, phosphates, sulphides and others [5]. Mineral matter in coal can be seen in various forms and identified by megascopic or microscope observations. Fine grained mineral matter, discrete bands, thin layers or partings, nodules and fissure infillings of minerals present in a coal deposit may often be examined by the naked eye or with a geological loupe. Smaller accumulations of mineral matter need to be studied under the microscope, for example thin bands, micro cleat or cell lumen fillings, or even replacements of cell structure [6,7,8].

Optical microscope studies have some limitations, it could identify some minerals, such clays and carbonates occur in association with particular macerals, but it is difficult to identify more specifically the type of carbonates and clay minerals involved using only optical techniques. Scanning electron microscopy coupled with energy-dispersive X-ray spectrometry (SEM-EDS) provides a basis for

IOP Conf. Series: Earth and Environmental Science **851** (2021) 012047 doi:10.1088/1755-1315/851/1/012047

determination of mineral composition to identify specific mineral types, and also to study the size, shape and their minerals association. SEM-EDS was used to characterize mineralogical origin, type and their association within the South Walker Creek Coals, Queensland, Bowen Basin. SEM-EDS provide better evaluation of the nature and formation processes of the minerals within the coal seams. SEM study indicates that the South Walker Creek coals contain three phases of mineralization, syngenetic, diagenetic and epigenetic. Clay mineral of kaolinite and Illite/Smectite occur as syngenetic phases, mainly derived from clastic sediment input during peat accumulation, however, some other clay minerals, chlorite and illite was probably combination of syngenetic phase with early diagenetic remobilisation and cell lumen or pore-filling precipitation. Carbonates such as calcite and ankerite, as well as dickite, and nacrite, along with additional kaolinite, illite, and chlorite were precipitated in cleats and fractures at a later stage as a result of epigenetic activity [9]. However, their minerals association and formation within the single cleat spaces were not clearly describe and poorly understood. This paper describes cleat mineral formation by using SEM-EDS analysis and provides insight multi-stage epigenetic process within the cleat spaces.

2. Regional Geology

The South Walker coals were deposited in the northern part of the Permo-Triassic Bowen Basin (figure 1). This basin is a part of the interconnected Sydney-Gunnedah-Bowen Basin System in eastern Australia. The Sydney-Gunnedah-Bowen Basin System extends along the western margin of the New England Fold Belt, and is confined to the west by the older Thomson and Lachlan Fold Belt. The complex tectonic and structural history of the Bowen Basin has been led to many different studies of its geological development. Hammond [9] considers that the basin formed during the Early Permian by crustal extension. Fielding [9] suggests that later in the Permian the Bowen Basin can be interpreted as a foreland sedimentary basin, formed by a compressive tectonic regime at a continent-ocean plate boundary. During the tectonic evolution of the Bowen Basin, complementary depositional processes were taking place in different areas. The stratigraphic sections have been correlated to sedimentation in the different phases of this tectonic evolution. They were deposited in terrestrial (fluvial) to intertidal and marine environments.

The coal bearing strata are mainly preserved on the protected tectonic platforms or synclinal margins. The Permian and Triassic sediments are mainly found in the Nebo Synclinorium, northern part of the Bowen Basin. Coal deposits in this area range from Early to Late Permian in age and occur in numerous stratigraphic units. There are mainly four economic coal bearing sequences in this area, the Collinsville Coal Measures, the Moranbah Coal Measures, the Fort Cooper Coal Measures and the Rangal Coal Measures. They were deposited in during regression and transgression phases of the Permo-Triassic sedimentation. The study area is located in the South Walker Creek within the Nebo Synclinorium on the eastern flank of the Carborough Syncline in the Northern Bowen Basin (figure 1a). It lies on the flat to gently undulating flood plain of Walker Creek, overlain by unconsolidated alluvium, comprising inter-bedded sand and clays with gravelly sand near the base. The Rangal Coal Measures is the most economic coal bearing in this area, which stratigraphically underlain comformably by the Fort Cooper Coal Measures and overlain by the Rewan Formation. The Rangal Coal Measures consists of fine to medium grained labile sandstone, grey siltstone, mudstone and coal seams, with 150 m in thickness (figure 1b). The most economic deposit of the Rangal Coal Measures at the South Walker Creek area is the Main seam, which splits to the north and the south to form the Main Tops and Main Bottoms coal seams.

The South Walker Creek coals are moderately affected by fault systems. Fractures and veins have been reported that have been mineralized by clays and carbonates. In addition, some dolerite dykes and sills occurring in some places in the Bowen Basin, and associated with the South Walker Coalas (figure 1b). These geological characteristics have influenced the coal properties and characteristics in the area. Permana [9] have been identified that the main seam in the South Walker Creek consists of mineralogical variation through the vertical sequence due to a combination of processes, including clastic sediment input to the peat swamp, thermal metamorphism due to burial sedimentation and hydrothermal influence. The distribution of the non-carbonate minerals assemblages cross cut the seam stratigraphy and appears to be primarily controlled by fault and other structural features; however, the illite–chlorite and diaspore-bearing assemblages resulted from hydrothermal fluid migration through the middle part of the coal seam. Tis indicate that some epigenetic minerals might be precipitated within the cleat spaces due to different epigenetic processes.

3. Methods

A series of representative coal polished sections from two boreholes (Borehole 11424 and 11852) of the South Walker Creek mine (Mulgrave and Walker Pits), Queensland, Australia, were analysed by using a Hitachi S3400-I scanning electron microscope (SEM) at the Mark Wainwright Analytical Centre, University of New South Wales. The SEM was equipped with backscattered electron (BSE) detectors and energy-dispersive X-ray spectra (EDS) microanalysis system, utilising NORAN System SIX software. The BSE image quality is crucial for SEM analysis, as the image was used and combined with EDS spectrum for characterization mineral matter in the coal samples. Application of SEM for coal analyses also requires specific of samples preparation. Coal samples are sensitive to the effect of the electron beam, which potentially damaged the surface of coal sampel. For this reason, the polished block of coal samples were coated with carbon to obtain a more conductive surface and to avoid charging effects on the specimens which could possibly destructed the samples.

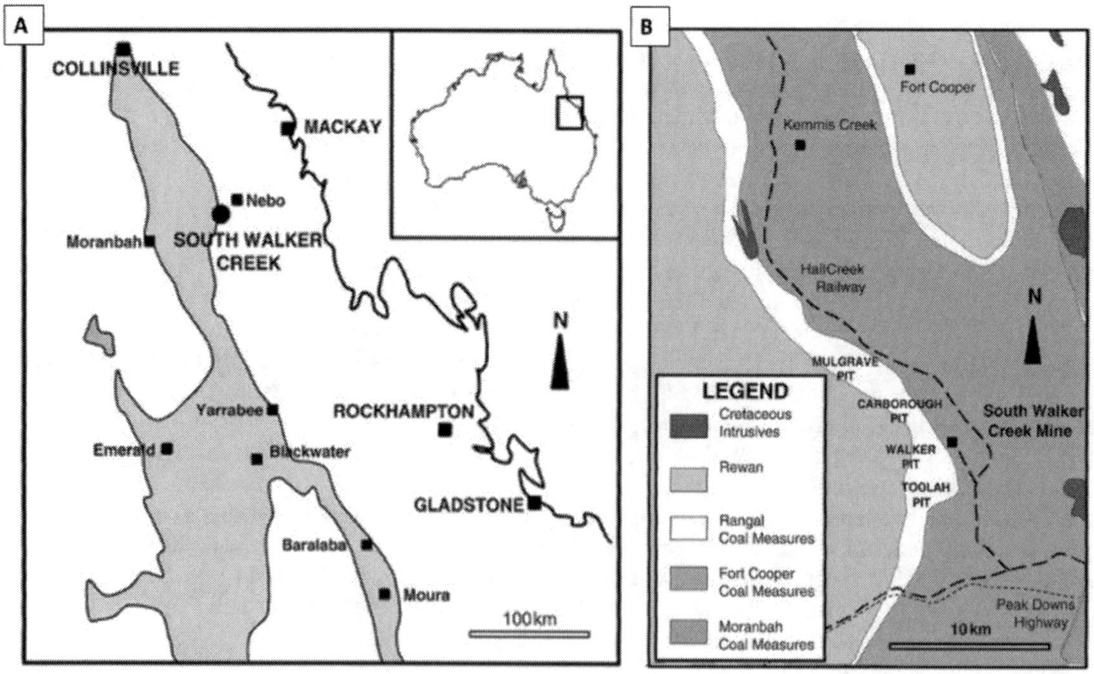

Figure 1. Location and regional geology of South Walker Creek Coal Mine; a) Distribution of the South Walker Creek coals deposit in the Bowen Basin; b) Simplified Geological map of the South Walker Creek area [9].

SEM basically has two modes of observation, SE (secondary electron) and BSE (backscattered electron) images. These observations give different information and it is desirable to have the BSE and SE images displayed simultaneously. The SE mode provides images similar to those from light microscopy in organic petrology and is used for determining morphology by contrasting grey levels. The BSE mode is used for distinguishing macerals from minerals, and further information on the minerals is easily obtained by EDS analysis. The characteristics of macerals from SE images have been summarized by [12]. They have also reported that different types of coal have different charging effects due to the interaction of electron beam with the coal surface. Bituminous and anthracitic coals have less charging effects, for example, than brown coals. A thinner carbon coating on the coal surface is the best way to get good BS detector results. Carbon is the best material for backscatter (BSE) work because it has a low atomic number. It is also suitable for SE work, although is not good as gold. The accelerating voltage is the most important variable during the SEM observation. This affects the image contrast and resolution. With the BSE mode, contrast increases with accelerating voltage but resolution of fine structure decreases due to the greater beam penetration and wider area of electron scattering. With the SE mode the images are better at lower accelerating voltage. Generally, the optimal voltage is the highest obtainable with minimum surface charging. This is commonly in the mid-range (15-20 kV) for coal studies [12]. X-ray (EDS) analysis is useful for identification of mineral particles and of minor elements in macerals. Huggins [13] has listed the minerals most commonly found in coal and how they can be recognized by energy-dispersive X-ray spectra under the SEM.

4. Results

4.1 Field and Core Observations

The coal seams of the South Walker Creek area are moderately structurally deformed due to compressional stress regime. This tectonic activity leads to the development of several major faults in the pits of South Walker Coal mine. The main fault structure is normal faults which structural trend is N-NE, whereas some thrust faults is also exposed in some places in the mine area.The marked structures and bedding shear deformation can be defined along the fault zone. The silickensides are a common marked structure found in the fault zone, with mostly coated by dickite minerals, and shearing also identified within the coal seams, mainly in the Toolah Pit of the South Walker Creek coal mine. The orthogonal set of cleats is visible at a megascopic scale, which is perpendicular or oblique to the coal seam bedding planes. This cleat set has distinct face and butt cleat features; it is well developed and continuous in the bright bands but less developed in the dull coal lithotypes.

Lithotype of the South Walker Creek coals show to be finely to moderately banded, dull and bright in lustre and highly cleated, with the cleats mostly filled by mineral matter. Core log descriptions indicate that the main seams is dominantly dull and bright (40-60% bright) coal lithotypes, and dull with numerous bright bands (10-40% bright), along with minor dull (< 1% bright) coal lithotypes (Australian Standards). The South Walker Creek coals show moderatelydeveloped face and butt cleat, with both open and closed apertures. The cleats are filled with carbonate and clay minerals. It has been observed that the cleat systems are commonly restricted to bright coal layers and terminate at the dull coal, and the brighter coal lithotypes mostly have a higher density of cleat spaces than the duller coal bands.

4.2 Optical Microscopy and SEM-EDS

The optical microscope with reflected light combined with SEM-EDS analyses was used for characterizing the mode and mineral occurrences in the coal seams [9]. These techniques were also applied to identify the cleat systems development and minerals formation within the cleat fractures, as well as the sequence of mineral formation. Based on lithotype descriptions of two cores from Borehole

IOP Conf. Series: Earth and Environmental Science **851** (2021) 012047 doi:10.1088/1755-1315/851/1/012047

11424 and 11852 (in the Mulgrave and Walker Pit areas) indicate that the orthogonal sets of cleats are mostly developed in the bright bands, and terminate in dull bands. This could suggest that the orthogonal cleat sets are of endo-microcleat origin, and were formed within the vitrinite macerals during diagenesis and coalification. Optical microscope and SEM-EDS observations show superimposed of polygonal and chaotic cleat sets. These types of cleat patterns may be products of post diagenesis processes, associated with tectonic stress, and related features such as such faults and shear zones. The polygonal set has a smaller aperture, often occurs as open cleats, and its likely cross-cut the chaotic set. The chaotic set typically has a larger aperture and is mostly filled with mineral matter. This may indicate that the (open) chaotic set post-dates the (mineralized) polygonal cleat set. Under the optical microscope and SEM-EDS, single cleats or micro-cleats are commonly restricted in the vitrinite macerals, but some cleats are terminated abruptly at the inertinite macerals (figure 2). Master cleats mostly pass through the inertinite macerals. These cleats are commonly infilled by minerals, and parallel inclined or perpendicular to the bedding planes. In particular cases, microcleats are composed of angular fragments of vitrinite to form a "mylonite", together with clay and/or carbonate minerals (figure 2a). Shear veins are also recognized in some samples, showing like "propagated veins", usually in the vitrinite macerals (figure 2b). SEM observation shows that the cleat widths apertures range from 1 to 100 μm, with lengths range from 20μm to more than 1 mm. The micro-cleats are dominated by straight line forms, although some curved or sub-curved and rhombic shapes are also present. Closed microcleats are typically more intensively developed than open microcleats. The cleat-filling minerals are mostly consisting of clay minerals, carbonates, diaspore, rutile, and apatite, and in many cases that multiple minerals precipitated together within a single cleat- filling minerals.

4.3 Cleat-Filled Minerals

On the basis of the SEM images, the cleats are filled with several different minerals, including clay minerals, carbonates, apatite, dickite and rutile. These minerals occur either as single-phase infillings (monominrelic) or as multiple-phase mineral deposits (polymineralic). The multiple-phase mineral deposit is mainly divided into three groups: (1) clay rich minerals, dominated by kaolinite, with minor occurrences of illite and chlorite, (2) carbonates, dominated by calcite and ankerite, with minor amounts of of siderite, and (3) other minerals, dominated by rutile and apatite, with limited occurrences of diaspore (figure 2).

Kaolinite is the most common clay mineral presence in the cleat fractures. Kaolinite rarely occurs as a single-phase cleat infillings, associated with other clay minerals, and non-clay minerals, such as rutile or anatase and quartz. At high magnification the kaolinite crystals occur as irregular flakes or booklet textures in the cleat aperture, in some cases, kaolinite cleat infillings are cross-cut by rutile and anatase. A combination of both rutile or anatase and chlorite may also be found within the kaolinite-filled cleat, as well as chlorite intergrowths within euhedral rutile or anatase patches. Chlorite always occurs in association with either kaolinite or illite. Chlorite usually forms intergrowths in the kaolinite cleat minerals, forming irregular flakes along cracks or cleavages of the kaolinite components. Quartz may also be found in association with the kaolinite cleat infillings, occurring as sub-rounded particles in the cleat space. Illite also mainly occurs as polymineralic associated with fluorapatite, rutile and calcite. Illite occurs at the end of cleat spaces within the inertinite macerals, associated with fluorapatite. In some cases, well crystallized illite cleat infillings are cross-cut by prismatic rutile and anatase. EDS pattern of rutile or anatase is indicated by a high peak of titanium (Ti), with the additional presence of zirconium (Zr) and silicon (Si) in the spectrum (figure 2c).

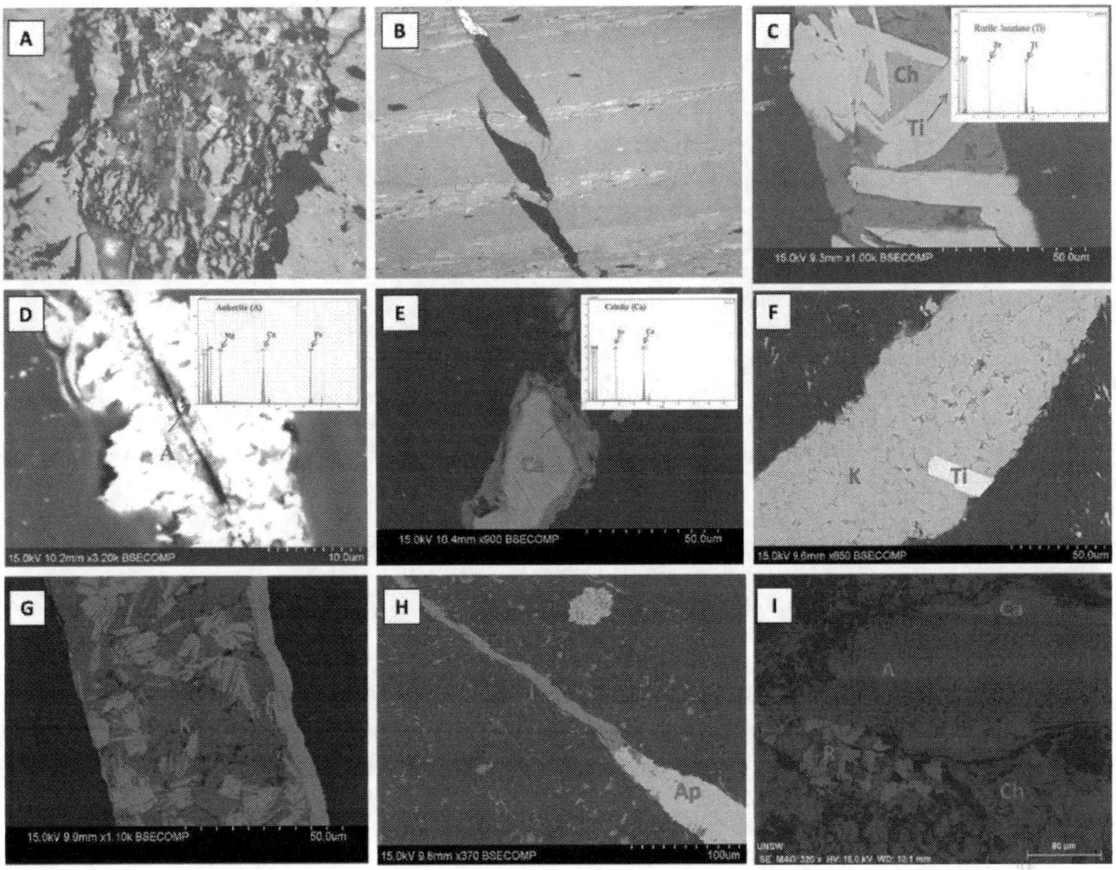

Figure 2. Optical Microscope and SEM Images shows cleat mineral formation, a) Mylonitic texture together with clay and/or carbonate minerals; b) Offset "striped shear veins", well developed in vitrinite; c) Ti (titanium) mineral associated with kaolinite (K) and chlorite (Ch), with high peak of titanium (Ti), with the additional presence of zirconium (Zr) and silicon (Si) in the spectrum; d) Monomineralic infilling of ankerite (A), with high peak of calcium (Ca), with lower of magnesium (Mg), and very low of iron (Fe) peaks; e) polymineralic infilling of kaolinite and calcite (ca), calcite shows a high single calcium (Ca) peak; f) Kaolinite (K) cleat infilling cross cut by Titanium mineral (Ti); g) Chlorite (Ch) intergrown with kaolinite (K), as well as lining in the cleat wall; h) Fluorapatite (Ap) cleat infillings, associated with illite (I) at the end of the cleat space; i) Interlayered calcite (Ca) and ankerite (An) in the cleat space, associated with titanium mineral (Ti) and chlorite (Ch).

Calcite and ankerite may occur as both single-phase and multiple-phase of cleat-filling mineral deposits. These minerals are commonly associated with kaolinite and illite, and in some cases with other clay minerals such as chlorite, diaspore and rutile or anatase. Monomineralic ankerite and calcite cleats are shown in figure 2d and 2e. The EDS pattern of these minerals indicates that the ankerite has a high peak of calcium (Ca), with lower of magnesium (Mg), and very low of iron (Fe) peaks (figure 2d), while calcite shows a high single calcium (Ca) peak (figure 2e). Calcite, as polymineralic cleat infillings is generally associated with kaolinite, illite and diaspore. Al rich minerals (diaspore) form linings on the cleat wall, with calcite in the middle of the cleat space. In most cases, however, calcite is found in association with illite and kaolinite. Illite and kaolinite occur as linings on the cleat wall, with calcite a later intergrowth in the middle of the cleat space. In some cases, the illite may also be cross cut by chalcopyrite. Although rarely occurring as the main mineral in the cleat, rutile (or anatase) and apatite are commonly found in association with clays and carbonate mineralcleat-fillings. Rutile/anatase commonly occurs cross cutting the kaolinite and illite in the cleat, usually as prismatic crystals. Rutile/anatase may also be found as a rim along the edge of kaolinite-filled cleats. In some cases, rutile or anatase may also found in association with calcite in the cleat space.

4.4 Multistage of Cleat Mineralization

The occurrence of intimate mixtures of minerals within single cleats makes it difficult to determine their mineral formation and origin, however SEM-EDS mapping can be obtained elemental composition of minerals within the cleat aperture. This allows identifying the minerals growth pattern from the rim to the middle of cleats spaces and interprets paragenetic sequence of cleat mineralization. Elemental mapping in carbonate-bearing cleat infills from the Carborough Pit of the South Walker Creek clearly shows that the aluminium (Al) and silicon (Si) are distributed in the outer part of the cleat space, well correlated with aluminosilcate rich minerals (kaolinite and chlorite). Other elements such as calcium (Ca), magnesium (Mg) and iron (Fe), however, are distributed in the middle of the cleat spaces associated with the occurrence of carbonates (e.g., calcite and ankerite). This may indicate that two or more generations of cleat mineralization, and that kaolinite and chlorite appear to have been formed earlier than the carbonates (figure 3).

Based on the minerals growth pattern and cross cutting relationship analysis under the SEM, the cleat mineralization in the South Walker Creek Coals may be divided into four stages: Silicate I (kaolinite and quartz), oxide (rutile or anatase), Silicate-II (illite-chlorite-diaspore-apatite), and carbonates (ankerite and calcite). Kaolinite and quartz were probably deposited in the early stages of cleat mineralization. Well crystallized kaolinite in association with sub-rounded quartz particles 1 -2 μm in diameter. This may indicate that the quartz was introduced to the cleat before the kaolinite.

Rutile or anatase commonly crosses cut perpendicular or oblique to kaolinite cleat infillings and occur as euhedral patches in the middle or on the edges of kaolinite cleat infills (figure 2f). This suggests that the rutile was deposited within the cleat at a relatively early stage, and was surrounded by later kaolinite deposition. At latter stages, kaolinite is developed at the edges of the cleat wall, with illite as irregular flakes attached to the kaolinite cleat infill. This may indicate that the kaolinite predates the illite in the cleat-filling material. Although the relationship is difficult to interpret, both the kaolinite and the illite then appear to have been invaded by diaspore-forming solutions in the last stage of mineral precipitation in this cleat fracture. Chlorite commonly occurs as inter-grown irregular flakes along cracks or cleavages of kaolinite cleat components (figure 2g). This may indicate that the chlorite mainly postdates the kaolinite. Chlorite may also occur in association with diaspore in the cleat space, as intergrowths in the diaspore cleat fills. This suggests that chlorite was deposited in early stage, and was surrounded by later diaspore formation due to the high concentration of Al in the hot solution within the cleat space.

Apatite occurs as an overgrowth component in the cleat associated with illite at the end of the cleat space (figure 2h). This indicates that apatite postdates illite formation. The final stage of cleat mineralization appears to have been deposition of the carbonate minerals, ankerite and calcite. These minerals are associated with kaolinite and illite within the cleat of the coal (figure 2i). Ankerite and calcite occur as elongate bodies, usually in the middle of illite and kaolinite cleat infills. This suggests that ankerite and calcite postdate the kaolinite and illite in the cleat fillings. Carbonate cleat mineralization commonly appears as layers, it seems that ankerite and calcite are inter-layered within the cleat space, with calcite being more likely have formed at a later stage than the ankerite.

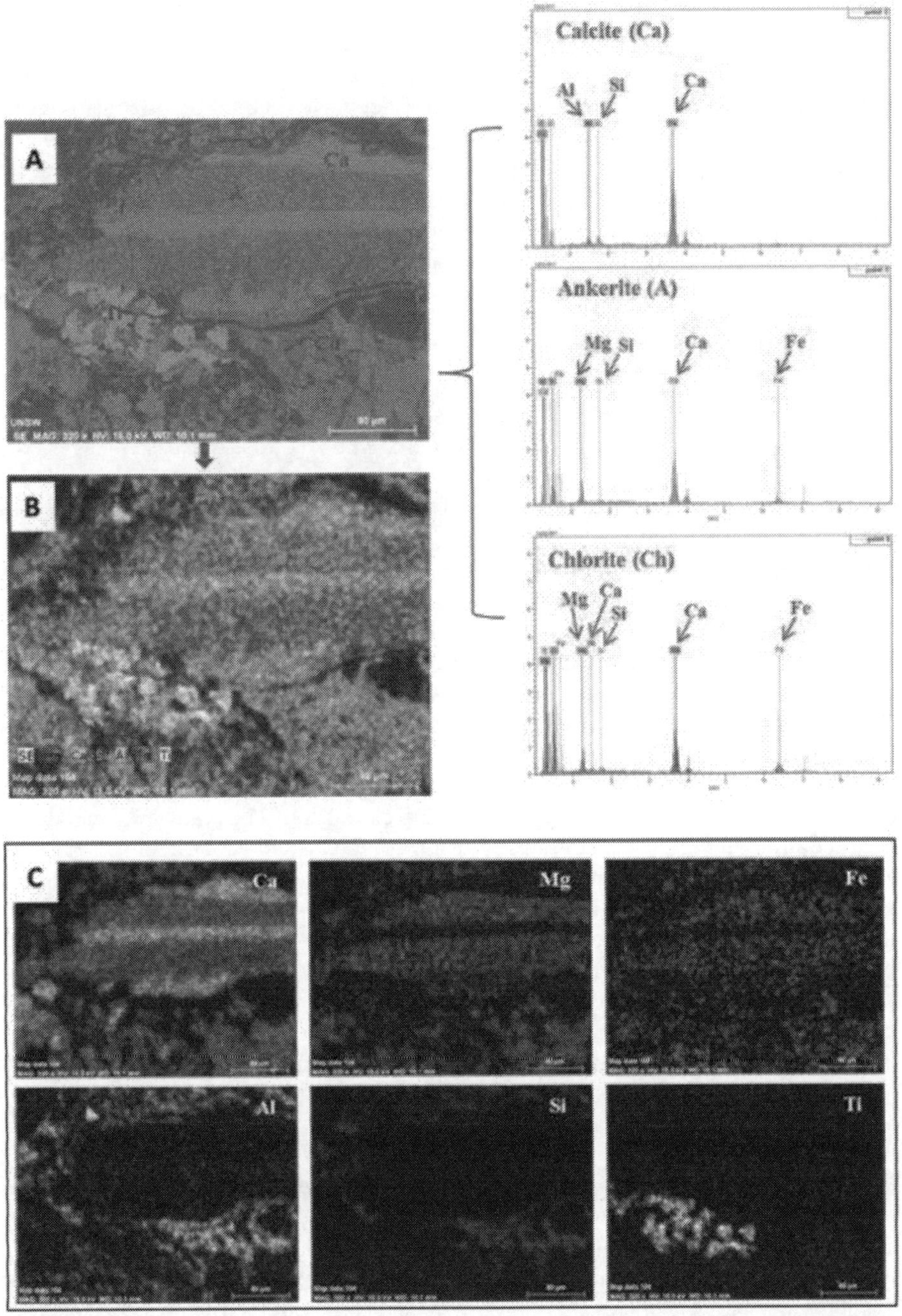

Figure 3. Photomicrograph of minerals growth pattern in the cleat spaces; a) Interlayered calcite (Ca) and ankerite (An) in the cleat space, associated with rutile or anatase (Ti) and chlorite (Ch); b and c: results of elemental mapping for cleat minerals.

5. Discussion

A complex geological history including factors such as igneous intrusion and tectonic events has been involved in the thermal maturation in the Bowen Basin. Mallett [14] have provided a comparison of thermal histories in different parts of the Bowen Basin. They have focused on three geological processes which may have impacted on the thermal maturation, burial by younger sediments, burial by tectonic processes such as folding and overthrusting, and periods of anomalous heat flow associated with intrusions. In particular in the northern Bowen Basin, they concluded that the Triassic events and Cretaceous heat flows were factors included in the thermal maturation of coals in the Nebo Synclinorium. Glikson [15] and Uysal [16] reveal that the paleogeothermal gradients in the northern part of the basin, including the South Walker Creek area, are relatively higher than the south area, due to hot fluid circulation along the fracture and fault zones. Fractures and cleats are favourable pathway for fluid movement in the coal seam and suggest that fluid circulation along the fault systems enhanced permeability zones in both mudrock and coal, as well as in the silisiclastic rocks of the northern Bowen Basin [15]. The South Walker Coals in the Northern Bowen Basin have been identified that mineralogical variation in the vertical sequence shows an unusual pattern, with kaolinite-rich assemblages are dominated in the top and bottom of the seam section and an illite-chlorite assemblages the middle, resemble a metamorphic association [9,17,18]. Carbonate cleat-filings, such calcite and ankerite are abundantin the middle part of the seam section, attributed to hot fluid circulation. Many of the cleat and fracture fillings are composed of intermixed minerals (polymineralic), with different phase of formation, which suggest multi-stage mineralisation process. The progressive illitization of kaolinite and illite/smectite, in coal seams may be due to either rank advance process at deeper burial and hot fluid flow circulation. The latter was probably responsible for the association of illite, chlorite and diaspore with kaolinite in the cleat mineralization process [9].

SEM-EDS observations show the evidence of multiphase process of cleat mineral formation in the South Walker Creek coals. Typical of layering carbonate minerals in the cleat space indicates that carbonates post-date the kaolinite, forming thin intercalated layers in the middle of otherwise kaolinite-filled cleat spaces, and with diaspore intergrowths in the carbonate-filled middle of the cleat space. The hydrothermal fluid brought high concentrations of Mg, Fe and Ca, and acting as a source of ankerite and calcite formation in the cleat. Al ions may also have occurred in the solutions, forming the diaspore in the middle of the cleat space (figure 4). This may suggest that the ankerite, calcite and diaspore are derived from precipitation of hydrothermal solution within the cleat space. Hatch [19] indicates that similar paragenetic sequences of multiphase mineral formation in coal cleat may have been formed by the variable composition of the fluid flow, which re-opened previous cleats and precipitated minerals episodically in the cleat space. This crack opening process is termed “crack-seal” deformation. Dietrich and Ramsay [20] demonstrated that in some fibrous vein infillings associated with low grade metamorphism “crack seal” developed by mutiple increments of microcrack opening followed by sealing of the microcrack by deposition material from solution.

The formation of epigenetic carbonates in the final stages of cleat mineralization in the present study area may indicate a similar crack-seal mechanism. The hot fluid flow in the permeable zones in particular cleat spaces re-opened the previous kaolinite-filled cleats, followed by sealing of the re-opened cleats by ankerite and calcite precipitated from the solution. Repeated crack seal increments involving calcite and ankerite in the cleat built up more complex fibrous veins. Emplacement of igneous intrusions in the study area, or possibly wider basin tectonics may have provided the hydrothermal system, injecting hot fluid solutions along the permeable zones of the coal seam. Dawson [21] reported that the isotopic signatures of the carbonate mineral indicate mixed or igneous sources as well as microbially derived carbonate for calcite within the Bowen Basin. Post-depositional formation of dawsonite ($NaAlCO_3(OH)_2$) in cleat space of the Permian coal from the Bowen-Gunnedah-Sydney (BGS) Basin is thought to be related to a basin-wide magmatic source, where the seepage of magmatic CO_2 associated with meteoric porewaters led to aluminosilicate breakdown and provided sodium (Na) and (Al) to from the dawsonite [22, 23]. A similar process may have introduced Al and Ca, as well as Mg and Fe, to the South Walker Creek coals, forming calcite, ankerite and diaspore in the permeable zones of the coal seam.

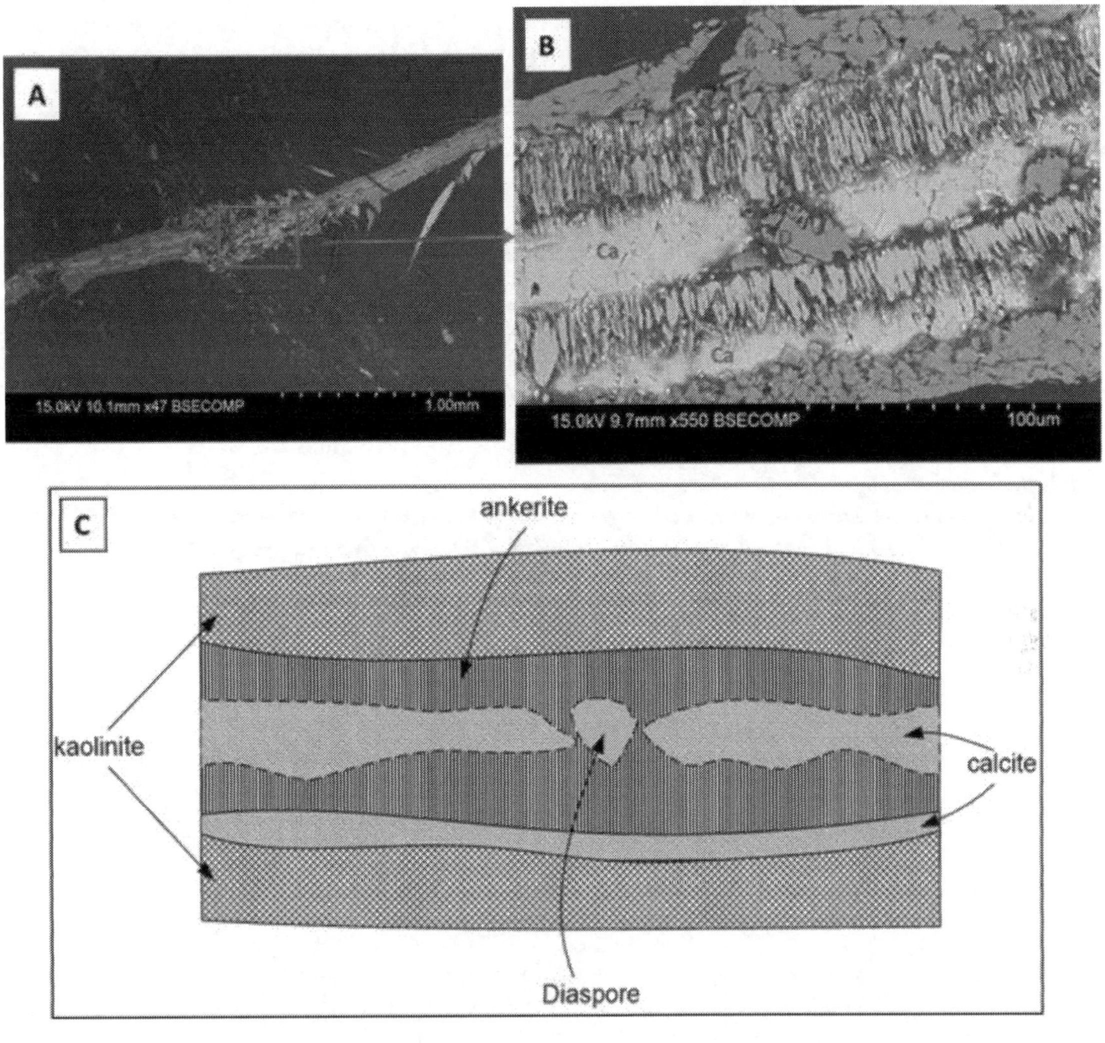

Figure 4. SEM Images illustrating possible crack-seal development; a) Fibrous cleat, composed of the inter-layered kaolinite (K) and carbonates; b) Enlarged view of image, representing multi stages of cleat mineral formation, ankerite (A) and calcite (Ca) occur as layering carbonate minerals in the cleat space, associated with kaolinite rims at the edges of the cleat wall, and diaspore intergrowths within the carbonates; C: Sketch of crack-seal development.

6. Conclusions

The main coal seam of the South Walker Creek is well developed cleat system. The cleats are both open and closed aperture, filled by minerals. The mineral formation in the cleats occurs either as single mineral (monomineralic) or inter-mixed mineral (polymineralic) and three types of cleat mineral formations. Optical microscope and SEM-EDS approach recognized the polymineralic association in the single cleat spaces represents a multi-phase minerals formation or crack sealing development, as a result of different stages of epigenetic processes. The paragenetic cleat mineral formation shows that first and second stages may be linked with the mineralogical composition of the coal and non- coal rock associated with the coal, formed due to remobilizing of earlier silicate rich material by ground water solutions, followed by precipitation within the cleat fractures. However, the two later stages may have been formed due to hot fluid movement in the permeable parts of the coal seam, followed by either transformation of pre-existing minerals within the seam, and/or remobilization of earlier mineral, which was then precipitated in cleat as epigenetic minerals.

Acknowledgments

This paper is dedicated to the memory of Prof. Colin R. Ward. This study was part of an MSc research programme supported by an Australian Development Scholarships in 2010 – 2011. Thanks are expressed to BHP Billiton Mitsubishi Alliance (BMA) Coal Pty Ltd for providing permission, coal samples and other data in this research study. In addition, I would like acknowledge to Joan Esterle of the University of Queensland, and to Stuart Davison of BMA, for providing samples and data of the South Walker Coals.

References

[1] Stach et al, 1982 *Textbook of Coal Petrology, 3rd ed.*: Berlin, Gebruder Borntraeger, pp. 535
[2] Ward C R 1984 *Coal Geology and Coal Technology*. Blackwell Oxford, pp. 345
[3] Thomas L 2002 *Coal geology: San Francisco, CA*, John Wiley & Sons, Ltd, pp. 384
[4] Taylor G H 1998 *Organic Petrology*. Gebruder Borntraeger, Berlin-Stuttgart, pp.704
[5] Ward C R 2002 Analysis and significance of mineral matter in coal seams. *International Journal of Coal Geology* **50**: 135-168
[7] Kemezys M and Taylor G H 1964 Occurrence and distribution of minerals in some Australian coals. *Journal of the Institute of Fuel* **37**:389-397
[8] Ward C R 1986 Review of mineral matter in coal. *Australian Coal Geology* **6**: 87-110
[9] Permana A K, Ward C R, Li Z and Gurba L W 2013 Distribution and origin of minerals in high-rank coals of the South Walker Creek area, Bowen Basin, Australia, *International Journal of Coal Geology* **116-117**, p 185-207
[10] Hammond R L 1987 The Bowen Basin, Queensland, Australia: an upper crustal extension model for its early history. In: Hower, J.C. and Gayer, R.A. (Authors), *International Journal of Coal Geology* **50**: 215–245
[11] Fielding C R, Falkner A J, Kassan J and Draper J 1990 Permian and Triassic depositional systems in the Bowen Basin. In: Beeston, J.W. (Ed.), *Proceedings Bowen Basin Symposium, Geological Society of Australia, Queensland Division*, Brisbane, p.21-25
[12] Hamilton L H and Salehi M R 1986 Use of scanning electron microscopy in coal petrology. *Journal of Coal Geology*, Geological Society of Australia, **8**: 77-85
[13] Huggins F E, Kosmack D A, Huffman G P and Lee R J 1980 Coal mineralogies by SEM automatic image analysis. *Scanning Electron Microscope* **1**: 531-540
[14] Mallett C W, Russel N and McLennan T 1990 In: Thermal history of the Bowen Basin. *Bowen Basin Symposium 1990 Proceedings*. Geological Society of Australia, p. 15-20

[15] Glikson M, Golding S D, Boreham C J and Saxby J D 2000 Mineralization in eastern Australia coals: a function of oil generation and primary migration. In: Glikson, M. and Mastalerz, M. (Eds.), *Organic matter and mineralization*. Kluwer Acaedmic Publisher, pp. 329-358

[16] Uysal I T, Glikson M, Golding S D and Audsley F 2000 The thermal history of the Bowen Basin, Queensland, Australia: vitrinite reflectance and clay mineralogy of late Permian Coal Measures. *Tectonophysics* **323** (1–2): 105-129

[17] Permana A K, Ward C R, Li Z, Gurba L W and Davison S 2010 Mineral matter in the high rank coals of the South Walker Creek area, Northern Bowen Basin. In Beeston, J.W. (Editor): *Bowen Basin Symposium 2010-Back in (the) Black*. Geological Society of Australia Inc. Coal Geology Group and the Bowen Basin Geologist Group, Mackay, October, 2010, pp. 27-34

[18] Permana A K 2011 Mineralogical variation and changes in the South Walker Creek coals, Bowen Basin, Queensland, Australia, University of New South Wales, Thesis, pp 343

[19] Hatch J R, Gluskoter H J and Lindahl P C 1976 Sphalerite in coals from the Illinois Basin. *Economic Geology* **71**: 613-624

[20] Ramsay J G 1980 The crack-seal mechanism of rock deformation. *Nature* **284**, 135-139

[21] Dawson G K W, Golding S D, Esterle J S and Massarotto P 2012 Occurrence of minerals within fractures and matrix of selected Bowen and Ruhr Basin coals, *International Journal Coal Geology* **94**, p 150-166

[22] Goldbery R and Loughnan F C 1970 Dawsonite and nordstrandite in the Permian Berry Formation of the Sydney Basin New South Wales. *American Mineralogist* **55**:477-490

[23] Baker J C, Bai G P, Hamilton P J, Golding S D and Keene J B 1995 Continental-scale magmatic carbon dioxide seepage record by dawsonite in the Bowen-Gunnedah-Sydney Basin system, eastern Australia. *Journal of Sedimentary Research* **3**: 522-530

IOP Conf. Series: Earth and Environmental Science **851** (2021) 012048 doi:10.1088/1755-1315/851/1/012048

An integrated approach using microtremor data for field development and reservoir monitoring: an example from Betung Field, Jambi

R. Pascarizativa[1], A Haris[2*] and M S A Alwi[3]

[1] Department of Physics, Faculty of Mathematics and Natural Sciences (FMIPA) Universitas Indonesia, Depok, 16424, Indonesia
[2] TAC PERTAMINA EP - PBMSJ, Jakarta, 12870, Indonesia.
[3] Geophysics Study Program, Faculty of Mathematics and Natural Sciences (FMIPA) Universitas Indonesia, Depok, 16424, Indonesia
*E-mail: aharis@sci.ui.ac.id

Abstract. Reservoir monitoring is one of the most important aspects in oil and gas field management, because it provides industries with the required data in hydrocarbon optimization. This is often carried out in Indonesia by the TAC Pertamina-Pusako Betung Muaro Senami Jambi (PBMSJ), an organization responsible for the management and coordination of oil well activities. In 2012, the organization conducted passive seismic data acquisition and series of drilling campaigns, which successfully increased oil production in Betung Block from 43 to 652 bopd. This campaign was successfully carried out through one and two wells in 2013 and 2014, respectively. Furthermore, the pre-2017 interpretation describes the Betung Anticline as a relatively symmetrical and low angle N-S trending field, which defies the South Sumatra Basin Northwest-Southeast regional trend. This anticline, bound by two high angle normal faults, is unable to define the LKO/OWC position, leading to the consideration of a more integral interpretation approach by the PBMSJ team in early 2017. The procedure was conducted through the use of a 2d seismic line bounded with new wells and SRTM imagery, to combine subsurface structural data with the surface fundamental analogue. In 2017, a passive seismic study was repeatedly conducted in the Betung field, with only 30 measuring points. The urgency in identifying the reservoir is essential within the petroleum industry, in order to produce sufficient hydrocarbons. It is also essential in confirming the potential uses of microseismic data in storage decision-making, through a dataset acquired in Betung, based on reservoir management.

1. Introduction

Betung Field is located on the western part of the Jambi area, approximately 11 km southwest and 10 km southeast of the Bajubang and Meruo Senami Provinces, respectively. This field covers an area of approximately 5 km2, which presently has 222 wells (Figure 1). The data obtained from the field is also very limited because, the entire old wells and records were abandoned and destroyed by NIAM, before the Japanese invasion during World War II. The Betung field signature product is light crude, as specific gravity ranges from 41-52 API.

The data consist of well formation tops based on logs, and two 2d seismic lines crossing the Betung field. Based on the use of a well-adjusted two-dimensional seismic interpretation, the structural map required further development of several areas, while also stating that the anticlinal structure in the southern area should be considered as an interesting prospect. However, there are insufficient data around the drilling location, due to the occurrence of several uncertainties. In addition, the anticline is a result obtained from the two-dimensional seismic interpretation, without any control of closely available surrounding well tops.

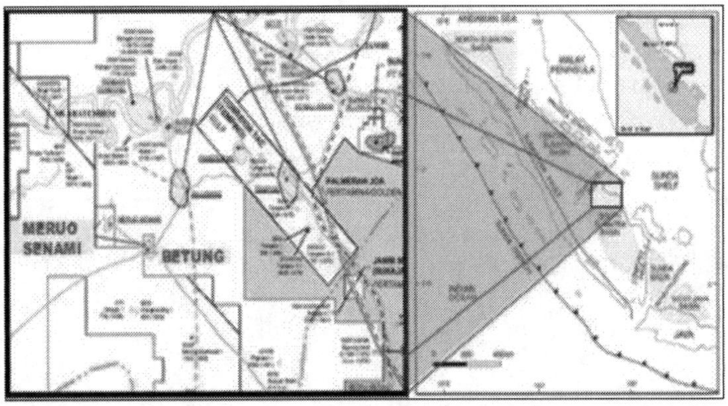

Figure 1. Location of Betung field.

Based on this stage, the creation of an additional approach to minimize uncertainties and drilling risk was considered. A low-frequency seismic analysis was also introduced to complement the previous exhausting geological interpretation.

The significant increase in oil production is further based on the evaluation and interpretation of passive seismic application, whose measurements emphasize very low magnitude character of natural tremors amplitude and frequencies ranging from 0.2–100 Hz. When this wave propagates impinge the surface of oil and gas-bearing in sand reservoirs, they resonance both in frequencies and amplitudes. Although in the stage of research, the frequency-related anomaly occurs within the range of 1–6 Hz, which are also confirmed in some areas of measurement.

The pre-2017 interpretation indicates that the Betung Anticline is asymmetrical and low angle N-S trending field bound by two high angle normal faults, as shown in Figure 2. This interpretation is unable to define the LKO/OWC position, and has failed to explain the level of passive seismic anomalies in the Betung Field (Figure 4a and 4b).

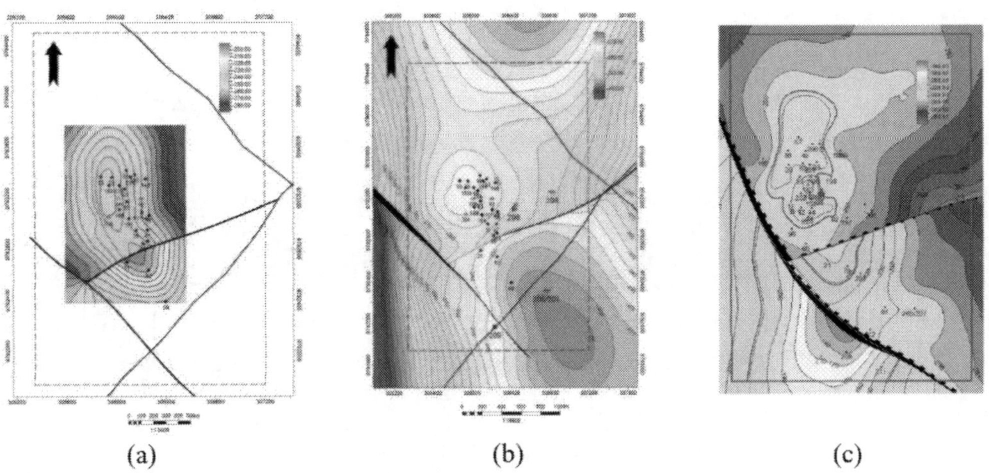

(a) (b) (c)

Figure 2. The evolution of Betung structural interpretation on a) POD original from Royal Dutch Shell, b) Inhouse POD 2011, and c) 2014 interpretation.

Based on the further development of the Betung Field, the PBMSJ team attempted to enhance the current interpretation in early 2017. They created a more integral approach of utilizing SRTM imagery and 2D seismic lines bound with new wells, in order to combine subsurface structural data with a surface fundamental analogy. Presently, the only parameters that PBMSJ utilizes are only three 2D seismic lines (line no.105, no. 173, and 09JS-001) with 13 sets of well log suites.

The result of a more careful seismic interpretation yields a more complex depth structure map, which consists of a new NW-SE anticline bounded with seven faults. This map delivers the alignment of the Betung anticline to NW-SE inversion, which is conformed with the transposition phase of the South Sumatra Basin, as shown in Figure 3.

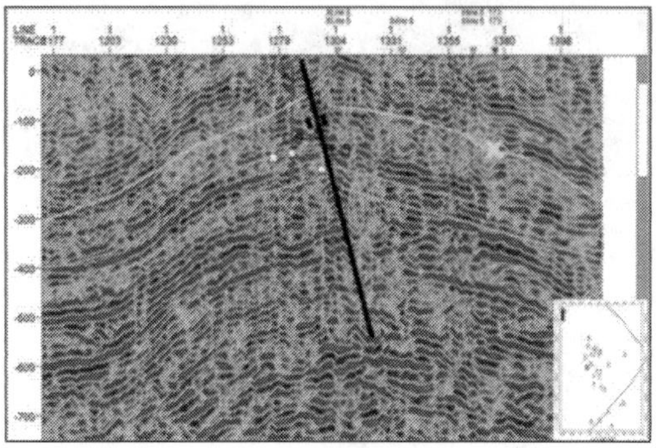
(a)

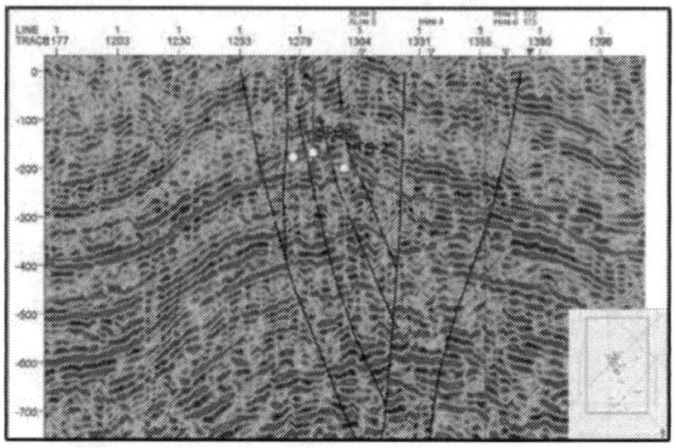
(b)

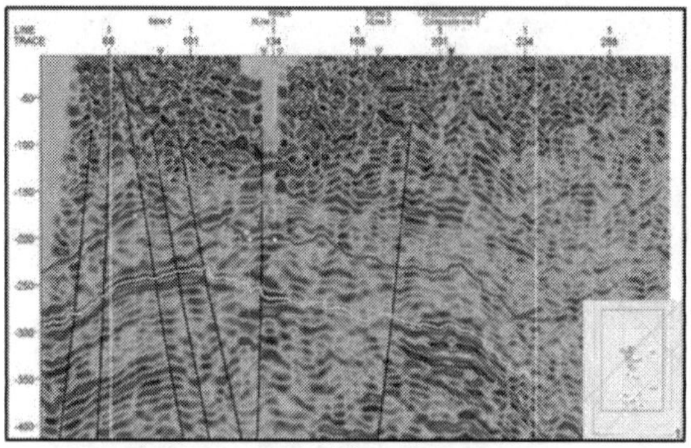
(c)

Figure 3. Seismic interpretation on a) Pre-2017 interpretation of 2d seismic line no.173, b) Post-2017 interpretation of 2d seismic line no.173, and c) 2017 horizon and fault interpretation in 09JS-001.

The previous interpretation stated that the Betung Anticline was a symmetrical and low angle N-S trending field, which also heightens the uncertainty of passive seismic data not related to the localization of anomalies (Figure 4a and 4b). However, the new interpretation is considered to have the ability in explaining the presence of a small anticlinorium complex that contains two NW-SE trending anticlines. These anticlines are symmetrical with low angle flanks, and bound by series of NW-SE faults. The trends of NW-SE normal and NE-SW cross faults further separate the west and south flanks from the main part of the field, respectively (Figure 4c). With more faults, orientation, and inter-fault relationship being interpreted, switched, and honored, the curiosity of a higher producing southern part of Betung Field was successfully explained. In addition, the unknown cause of passive anomalies in the southern part was previously explained by the presence of two anticlines that are symmetrical with low angle flank, and bound by series of NW-SE faults.

Furthermore, the passive seismic data acquisition was conducted with 60 and 30 points measurement station in 2012 and 2017, respectively. The data processing stage is based on mainly separating signal and noise, as the interpretation of processed information with vertical N-S and horizontal E-W components successfully increases the oil production in Betung Block. The evaluation result of amplitude anomalies from the ratio of vertical to horizontal magnitude within 1-6 Hz, indicates the location of trapped hydrocarbon irrespective of the depth.

The new interpretation also provides a clearer relationship between the structural and stratigraphical elements of southern Betung, as shown in Figure 4. Hydrocarbon occurrences within the southern part of the field further contributes to minimizing the drilling risk. Although the basic concept of passive seismic analysis and interpretation was still debatable, this technology was successfully applied in Betung. In addition, further methodology approaches of passive seismic and other supporting data are still needed to delineate the oil-bearing reservoirs.

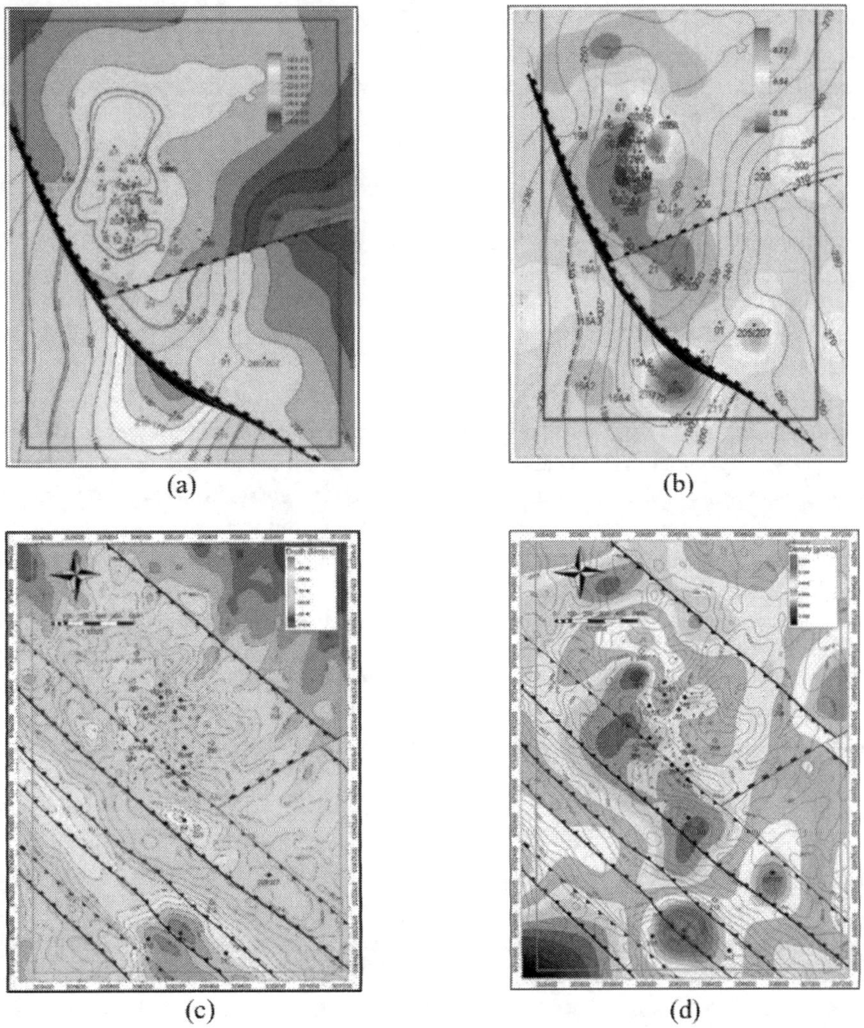

Figure 4. Comparison of Betung structural interpretation with passive seismic anomalies on a) pre-2017 structual interpretation, b) Pre-2017 structural interpretation overlaid with passive seismic anomalies, c) Post-2017 structural interpretation, and d) Post-2017 structural interpretation overlaid with passive seismic anomalies.

IOP Conf. Series: Earth and Environmental Science **851** (2021) 012048 doi:10.1088/1755-1315/851/1/012048

Reservoir monitoring is one of the most important elements in oil and gas field management, due to providing industry experts with the required data to maximize hydrocarbon recovery. A comprehensive technology of this process should be capable of recording changes in reservoir fluid saturations, measuring previous and present locations of displaced fluid fronts, as well as subsequently predicting methods by which these fluids are distributed in the future [1].

The identification of the capacity or distribution of the reservoir to produce hydrocarbon, especially oil, is important in the petroleum industry. This involves understanding the drainage radius or level (distance from the borehole) of oil extraction from the reservoir. Therefore, the role of the bubble map in describing the oil drain radius is very important. In addition, this new interpretation approach is also responsible for the reduction of drilling risk.

This study aims to enhance the understanding of reservoir monitoring, and to obtain the hydrocarbon prospect area in Betung, especially strengthening the probability of occurrences within the southern part of the field. An anomalous low frequency is utilized in order to understand the linearity between passive seismic data within 2012 and 2017, which is supported by geological data and a bubble map.

2. Problem Formulation

Based on the field development, reservoir monitoring, and utilization of passive seismic survey results (acquired in two different years), is this study eligible to provide an overview of storage distribution? Furthermore, the passive seismic surveys conducted twice in 2012 and 2017 serves as time-lapse monitoring, in order to describe the radius of hydrocarbon drainage or reservoir behaviour in the Betung structure.

3. Data and Method

Low frequencies were considered to be a noise in conventional seismic analysis, due to surface sources, including human activities, as well as atmospheric and water phenomena. However, the anomaly within low frequencies was hypothetically concluded as a hydrocarbon occurrence in passive seismic analysis. In contrast to the conventional active seismic, this method was carried out with continuous and passive (without any man-made sources) measures on signal or background noise above the earth surface. The measurements involving the use of passive seismic above the hydrocarbon reservoir also provided a response with the occurrence of spectrum attribute anomaly, at the frequency range of 1-6 Hz. Meanwhile, these anomalies were not shown in other locations outside the reservoir [2]. In addition, there were several theories and hypotheses related to the source of spectrum attribute anomaly in the low-frequency range. One of these principles was the interaction between the micro-wave and hydrocarbon within the reservoir pore, which emitted the new signal at the frequency range of 1-6 Hz [3].

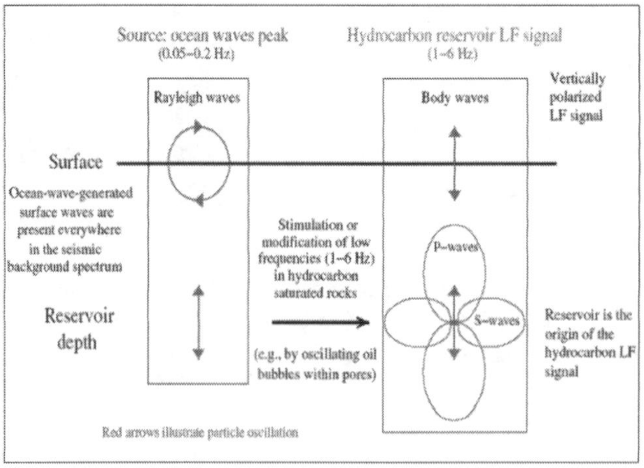

Figure 5. Natural seismic sources detected in passive seismic survey [2].

The anomaly of V/H (vertical to horizontal) and vertical spectrum ratios systematically correlated with the hydrocarbon occurrence, after the measurements on the points above the reservoir. The anomaly and amplitude peak in V/H was between the frequencies of 1-6 Hz [4] and 2-3 Hz [5], respectively. The equation used to obtain the ratio of V/H is [4],

$$V/H = \frac{V(f)}{H(f)} = \frac{V(f)}{\sqrt{X^2(f) + Y^2(f)}} \quad (1)$$

V/H Spectrum Ratio Attribute Equation. Where V(f) = Vertical component spectrum, H(f) = Horizontal component spectrum, X(f) = Horizontal component spectrum east-west, Y(f) = horizontal component spectrum north-south.

Furthermore, general workflow began with data acquisition, processing, and interpretation. Based on using three-component portable seismometer (X, Y, and Z), this acquisition process accomplished 60 and 30 stations in 2012 and 2017, respectively (Figure 6), near the wells and unexplored area. Each station also had measurement durations of approximately 70 mins.

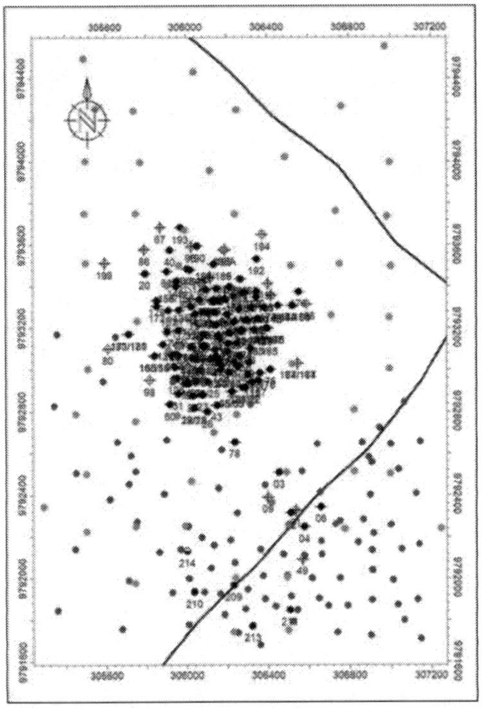

Figure 6. Basemap with two lines of 2d seismic (black lines). Purple dots are 2012 measurement stations and blue dots are the 2017 measurement stations.

The processing stage was performed to eliminate unwanted signals and noises, as well as change domain from time to frequency. After this, data objective was to improve the signal to noise ratio on the spectrum attribute. Based on the measurement results, quite a few noises were obtained, although they originated far away from human activity. However, there were several analytical stations located near the river or trees. To obtain a noise-free signal, backgrounds with strong disturbances at several intervals should be cut down.

The use of various spectrum attributes to identify the anomaly was further carried out, in order to interpret and analyze the low frequency passive seismic data. Furthermore, the vertical to horizontal (V/H) spectrum ratio was focused on, due to having better stability than other attributes, when measurement is conducted at different time intervals [7]. To obtain this exact range of frequency spectrum anomaly in 1-6 Hz, the spectrum attribute at the measurement area near the oil-producing well was used. This range (1-6 Hz) served as a variation with high amplitude anomaly, which often occurred in the hydrocarbon reservoir zone.

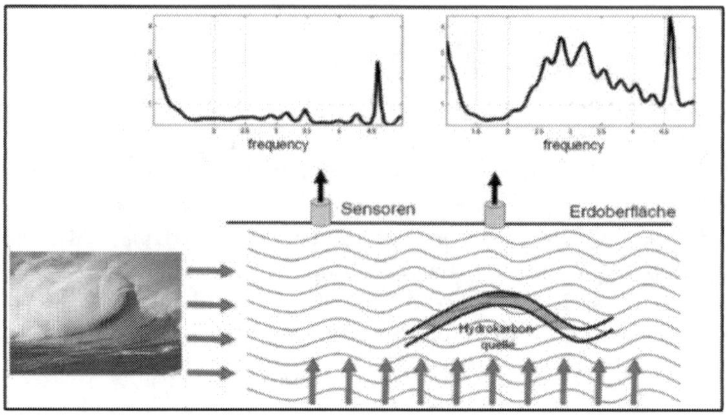

Figure 7. Signal response above reservoir and outside reservoir in frequency spectrum[6].

Based on the anomalies obtained, the passive seismic result was assumed to assist in monitoring the reservoir and plans for field development (location of the development well). Although the basic concept of passive seismic analysis and interpretation was still debatable, this technology was successfully applied in Betung. In maintaining continuity, it was necessary to monitor the production behaviour of reservoirs and wells, which were greatly influenced by the movement and presence of hydrocarbons below the surface.

Reservoir monitoring in the Betung field was also carried out by observing the results of passive seismic in two different years, with the distribution of the anomaly expected to be in line with oil drainage from each well, and illustrated through the bubble map. The results were combined in the form of a map, with that of the passive seismic anomaly distribution. These combinations further became an illustration of hydrocarbon distribution in the Betung area, based on providing additional data to determine the layout plan of the drilling well and production surveillance.

4. Results and Discussion

Based on the first passive seismic in 2012, the calibration of three wells was carried out, with each location having a different production history and vertical to horizontal (V/H) spectrum ratio. According to well 198 (oil and water producer) and 204 (suspension with 100% water production), the peak V/H values were 0.52 and 0.31 at 2.15 and 3.2 Hz, respectively. Meanwhile, the EL-BTH point that was further drilled in 2013 had a peak V/H value of 0.99 at 3.22 Hz, as well as the initial oil production of 126 BOPD and 0% water cut.

In 2017, a passive seismic study was repeatedly conducted in the Betung field, although only 30 measuring points were utilized. This was because production and development scenarios focused on the southern area of the Betung field (90% of the total production). In addition, the calibration was carried out at the same point as that of 2012. Based on Figure 8, the results of the amplitude spectrum were similar to the range of 1-4 Hz, at the CG-A and TT-BT points carried out in 2012 and 2017, respectively.

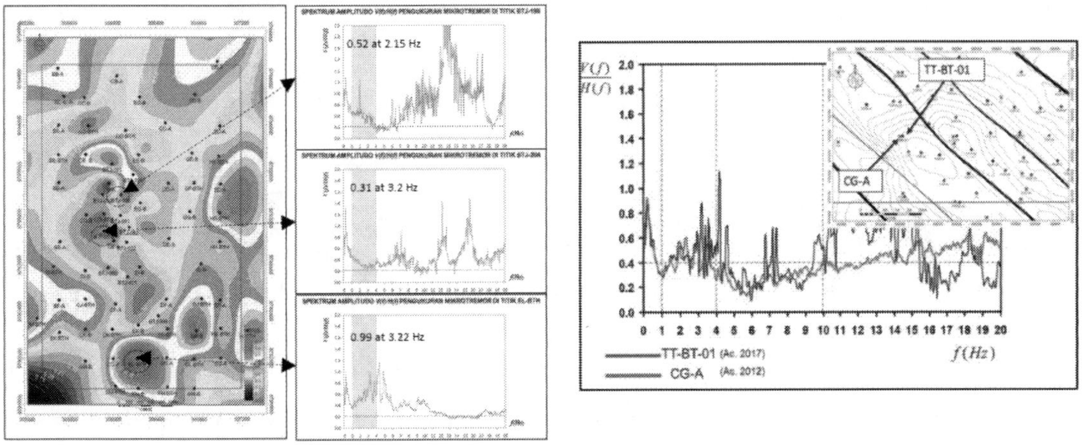

(a) (b)

Figure 8. Passive seismic calibration in 2012 (left) and passive seismic calibration in 2017 (right).

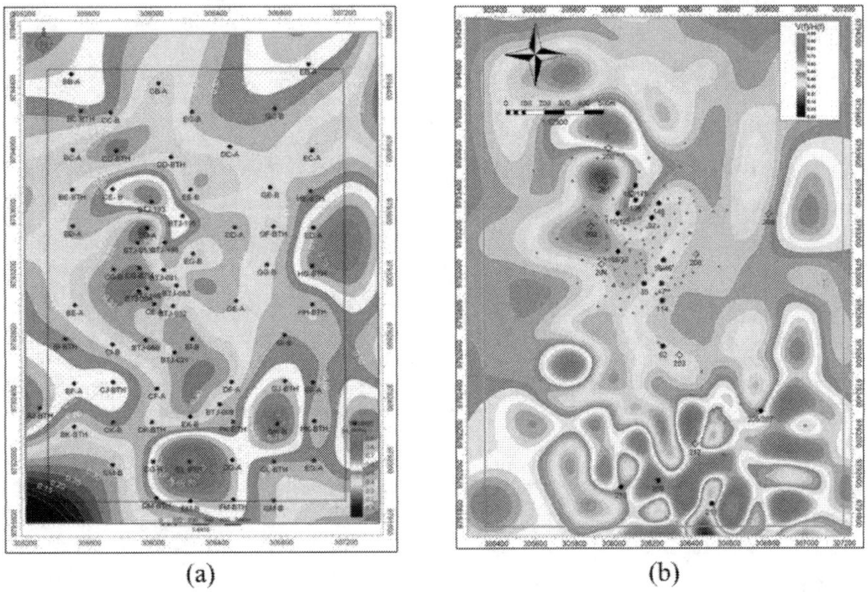

(a) (b)

Figure 9. Comparison of passive seismic anomalies map on a) 2012 and b) 2017 results.

Although the map results from the passive seismic did not directly show the position of the hydrocarbon or non-hydrocarbon fluid, they provided an overview of the V/H spectrum ratio distribution. Based on the map obtained in 2017, there were more widespread changes in the anomalies, when compared to the results of 2012. This was due to drilling and production activities in the southern area, where passive seismic described hydrocarbon anomalies or presence. To reservoir monitoring, the result was combined with the bubble map of the drainage radius or the cumulative oil production.

Based on limited data in the Betung field, the calculation of the oil drainage radius was carried out by using the volumetric method. This method considered the reservoir to be a tube and conducted calculation by using its volume with the cumulative production and storage properties. From this calculation, the reservoir was depicted in the form of a bubble map, which was expected to provide an overview of the oil amount in the area, although the shape of the drainage was not ideal as a sphere. This was due to the heterogeneity of the reservoir.

Table 1. Cumulative production and oil drainage radius.

Well	Cumulative Production (mbbl)	Volumetric Oil Drainage Radius (m)
BTJ-209	350	124
BTJ-210	429	159
BTJ-211	615	215
BTJ-213	254	130
BTJ-214	107	67

The wells with the bubble map of the drainage radius only produced oil, except the BTJ-214 that contained water. The well with the largest drainage radius in the Betung field was BTJ-211, which historically had a great result rate, since its initial production of 164 BOPD with zero water cut on July 2014.

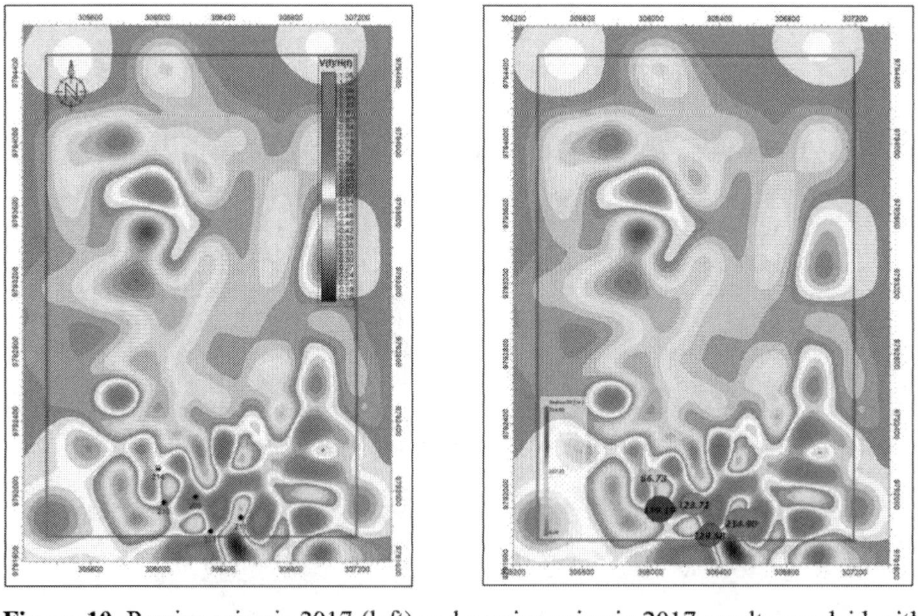

Figure 10. Passive seismic 2017 (left) and passive seismic 2017 results overlaid with oil drainage radius bubble map (right).

According to the integration data from 2012 and 2017, the structural reinterpretation, production monitoring, and anomaly distribution of passive seismic was linear with the oil drainage radius in the southern area wells. This was because the ability of the new method had the ability to estimate the presence of hydrocarbon fluid extraction, which was described as the cumulative production in each well. Furthermore, monitoring used a map of the oil drainage radius obtained from the calculation of cumulative production, reservoir properties, and recovery factor. The description of the bubble map also provided an approach in monitoring reservoir behaviour, which in this case was the oil drainage pattern.

Based on the combination of the passive seismic method and oil drain distribution, reservoir monitoring was influenced by several factors, including the utilized production technique, pump adjustment, and good storage management. These were carried out by regularly observing the fluid level, in order not to damage the reservoir.

5. Conclusion

The calibration stage was very important, in order to avoid the production of error or misinterpretation during data processing. This stage was carried out at three points in 2012, and was further repeated at the same location in 2017. The role of well data was also very important, as the interpretation and processing of G&G information was used to validate the results of passive seismic. This was carried out to determine the occurrence of trapped hydrocarbons with high vertical to horizontal (V/H) spectrum ratio.

The study indicated that there was a strong relationship between the distribution of anomalies and size of hydrocarbon drainage radius on the production field. Therefore, a time-lapse of oil drainage should be obtained in a certain area, with the ability to corroborate this data to field development, in terms of well replacement.

This study supported the microtremor or passive seismic signals theory, which provided a glimpse presence of hydrocarbon and its movement. Therefore, additional data from G&G and reservoir studies are needed, in order to strengthen the results of the passive seismic method.

Acknowledgments

The author is grateful to the Reservoir Geophysics, Department of Physics, Geophysics and Geology Study Program Universitas Indonesia for their support, and also to TAC PERTAMINA EP - PBMSJ for providing permission for this research data.

References

[1] Djuraev U, Jufar S R and Vasant P 2017 A review on conceptual and practical oil and gas reservoir monitoring methods. *Journal of Petroleum Science and Engineering* **152**:586-601

[2] Saenger E H, Torres A, Rentsch S, Lambert M, Schmalholz S M and Mendez-Hernandez E 2007 A hydrocarbon microtremor survey over a gas field: Identification of seismic attributes. In *SEG Technical Program Expanded Abstracts 2007* pp.1277-1281. Society of Exploration Geophysicists

[3] TAC Pertamina EP – PBMSJ – UGM 2012 Final Report: Survei Mikrotremor Lapangan Betung Meruo Senami Jambi, (Jakarta: Internal Report)

[4] Dangel S, Schaepman M E, Stoll E P, Carniel R, Barzandji O, Rode E D and Singer J M 2003 Phenomenology of tremor-like signals observed over hydrocarbon reservoirs. *Journal of Volcanology and Geothermal Research* **128**(1-3):135-58

[5] Ali M Y, Berteussen K A, Small J and Barkat B 2007 A low frequency, passive seismic experiment over a carbonate reservoir in Abu Dhabi. first break;25(11).

[6] Schmalholz S M, Podladchikov Y Y, Holzner R and Saenger E H 2006 Scientific strategy to explain observed spectral anomalies over hydrocarbon reservoirs generated by microtremors. In *First EAGE Passive Seismic Workshop-Exploration and Monitoring Applications* 2006 Dec 10 (pp. cp-21). European Association of Geoscientists & Engineers.

[7] Lambert M, Schmalholz S M, Saenger E H and Podladchikov Y Y 2007 Low-frequency anomalies in spectral ratios of single-station microtremor measurements: Observations across an oil and gas field in Austria. In *SEG Technical Program Expanded Abstracts* 2007 Jan 14 (pp. 1352-1356). Society of Exploration Geophysicists.

IOP Conf. Series: Earth and Environmental Science **851** (2021) 012049 doi:10.1088/1755-1315/851/1/012049

The testing of Qp and QsQp equation in Kujung I Formation, East Java, Indonesia

A Sadat[1], A Haris[1*], Haryono[2] and G A Dharmawan[2]
[1]Department of Physics, Faculty of Mathematics and Natural Sciences (FMIPA) Universitas Indonesia, Depok, 16424, Indonesia
[2]Saka Indonesia Pangkah Limited, Jl. Jend. Sudirman, Jakarta, 12910, Indonesia.
*E-mail: aharis@sci.ui.ac.id

Abstract. Hydrocarbon-bearing reservoir shows attenuation on seismic section, but the use of attenuation in hydrocarbon exploration is less popular than other attributes. This is due to the sophisticated and advanced technology required to measure attenuation. Meanwhile, some studies were designed to estimate attenuation by formulating rock quality factor equations from a wire-line log, although others run on sandstone samples. Therefore, this study aims to estimate rock quality factors in carbonate rock of Kujung Formation using three different methods, discriminate hydrocarbon-bearing reservoir, and identifying the impact of carbonate facies. The results showed the P-wave quality factor method is sensitive to discriminate high gas saturation on facies A. Meanwhile, the QsQp ratio is sensitive to discriminate shale and carbonate. The SQp and SQs methods are sensitive to discriminate hydrocarbon-bearing reservoirs and less affected by carbonate facies. The novelty of this is to estimate the rock quality factor from various methods and identify the impact of facies in hydrocarbon identification, especially for carbonate reservoirs in Northeast Java Basin. Therefore, a new understanding of facies impact on rock quality factor contributes to choosing the best methods in discriminating pore-filling fluid identification.

1. Introduction

The application of attenuation in hydrocarbon exploration is gradually known and is sensitive to identify fluid saturation, porosity, and fracture [1]. The dimensionless rock quality factor, Q, is used to measure the degree of seismic wave attenuation. Meanwhile, to measure the attenuation, full-waveform acoustic logging [2], vertical seismic profile with advanced processing technique [3], or core measurement [4] are required. These issues made seismic wave attenuation to become less popular than other attributes such as amplitude and frequency. An alternative option to estimate it is the use of a wire-line log. Also, some equations have been formulated to estimate rock quality factors in hydrocarbon exploration. Therefore, this study aims to estimate rock quality factors using three different methods, discriminate hydrocarbon-bearing and wet reservoir, as well as identify the impact of facies of Kujung I Formation.

2. Regional Geology

Kujung I formation is well known as the main producer in Northeast Java Basin, Indonesia, and it was deposited in shallow marine settings during the early Miocene. The tectonic quiescence took place during Kujung I time, and it consists of thick and massif carbonate succession. On the Northwestern edge of the basin, it is still underexplored, but the gas discovery of Lengo-1 has more exploration opportunity [5].

IOP Conf. Series: Earth and Environmental Science **851** (2021) 012049 doi:10.1088/1755-1315/851/1/012049

3. Material and method

This study was conducted in Northwestern Edge of Northeast Java Basin, Indonesia. Eight explorations wells were used to penetrate the Kujung I Formation. Only three were proven to contain hydrocarbon, while others were wet reservoirs. Furthermore, they were completed by wire-line logs, such as total gamma-ray (GR) and uranium-corrected gamma-ray (CGR), resistivity, density, neutron, and compressional slowness. Meanwhile, two wells were completed by a shear slowness log. The routine core data measurement used to calibrate the petrophysical analysis is analog to the nearest proven oil field. Also, there were no measured core data to validate the attenuation.

The facies identification was conducted using a type of gamma-ray and density log. Furthermore, the petrophysical analysis was conducted to determine the volume of shale, porosity, and water saturation. This was calculated by the deterministic approach. Subsequently, the compressional and shear slowness logs, as well as density logs play a significant role in calculating elastic rock properties.

The Q factor was calculated using three different methods. The first is the P-wave quality factor proposed by Klimentos and McCann [4], which was obtained from the rock sample measurement in the laboratory. It measured the magnitude of rock physic parameters, including attenuation under 40 MPa and 1 MHz. The result showed a linear relationship between porosity, shale content, and coefficient attenuation. Hence, the equation:

$$Q_P = (\pi f)/(0.0315PHIE + 0.241VSH - 0.132Vp) \tag{1}$$

where: Q_P = P-wave quality factor; PHIE = effective porosity; VSH = volume of shale; f = frequency, and Vp = P-wave velocity.

Secondly, the QsQp ratio formulated by Mavko et al [6] was obtained from Hudson's crack theory which proposed three theorems. The first theorem assumed that aligned crack orientation causes an anisotropy effect in reducing the propagated P-wave. The Equation is as follows:

$$\frac{Q_S}{Q_P} = \frac{1}{4}\frac{\left(3\left(\frac{V_P}{V_S}\right)^2 - 2\right)\left(\left(\frac{V_P}{V_S}\right)^2 - 2\right)^2}{\left(\left(\frac{V_P}{V_S}\right)^2 - 1\right)\left(\frac{V_P}{V_S}\right)^2} \tag{2}$$

The second theorem assumed that the crack is randomly distributed to introduce isotropy. The Equation is:

$$\frac{Q_S}{Q_P} = \frac{5}{4}\left[\frac{\left(\left(\frac{V_P}{V_S}\right)^2 - 2\right)^2}{\left(\left(\frac{V_P}{V_S}\right)^2 - 1\right)}\right]\left[\frac{2\left(\frac{V_P}{V_S}\right)^2}{3\left(\frac{V_P}{V_S}\right)^2 - 2} + \frac{\left(\frac{V_P}{V_S}\right)^2}{3\left(\left(\frac{V_P}{V_S}\right)^2 - 1\right)}\right]^{-1} \tag{3}$$

The third theorem assumed that the distribution of crack is random and the isotropy effect introduced causes a reduction of S-wave modulus. The Equation is as follows:

$$\frac{Q_S}{Q_P} = \frac{1}{\left(\frac{V_P}{V_S}\right)^2}\left[\frac{4}{3} + \frac{5}{4}\frac{\left(\left(\frac{V_P}{V_S}\right)^2 - \frac{2}{3}\right)\left(\left(\frac{V_P}{V_S}\right)^2 - \frac{4}{3}\right)^2}{\left(\left(\frac{V_P}{V_S}\right)^2 - \frac{8}{9}\right)}\right] \tag{4}$$

where: Q_P = P-wave quality factor; Q_S = S-wave quality factor; Vp = P-wave velocity; and Vs = S-wave velocity.

The last method is scaled P-wave (SQp) and scaled S-wave quality factor (SQs) proposed by Hermana et al [7]. This was adopted from Mavko et al first theorem which suggested that anisotropy is introduced by crack orientation. The equation was formulated by expanding the crack density factor into porosity and crack aspect ratio of sandstone. Hence, the equation is:

$$SQ_P^{-1} = \frac{5}{6}\frac{1}{\rho}\frac{((V_P/V_S)^2 - 2)^2}{((V_P/V_S)^2 - 1)} \tag{5}$$

$$SQ_S^{-1} = \frac{10}{3}\frac{1}{\rho}\frac{((V_P/V_S)^2)}{(3(V_P/V_S)^2 - 2)} \tag{6}$$

where: SQ_P = scaled P-wave quality factor; SQ_S = scaled S-wave quality factor; ρ = Bulk density; Vp = P-wave velocity; and Vs = S-wave velocity.

4. Results and discussion

Facies of Kujung I Formation can be distinguished into two types, namely A and B based on gamma-ray log type, shale content, and density log character. This is analog to the nearest well with conventional core data in figure 1a. Facies A is characterized by low and blocky gamma-ray log, massif, and thin reservoir with higher shale content and high-density log reading found interbedding with shale. Furthermore, the A was interpreted as backreef facies, meanwhile, B was characterized by low and blocky type gamma-ray log, massive and thick reservoir with lower shale content, and low-density log reading. Facies B was interpreted as reef flat. The facies of Kujung formation wells are seen in Figure 1b.

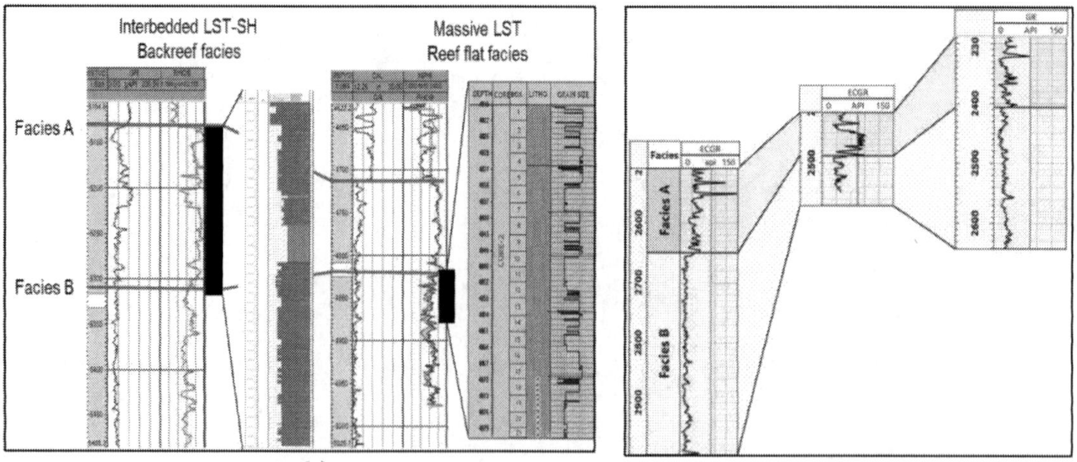

(a) (b)

Figure 1. a) Facies analog used to determine A and B. b) Well to well correlation showed the distribution of facies A and B each.

Shale volume was calculated by averaging gamma-ray and density-neutron logs calculation method. Subsequently, porosity calculation was conducted through the density-neutron log method. According to SAKA unpublished internal study (2020), tortuosity factor, cementation, and saturation exponent parameters used for water saturation are 1, 1.8-2, and 1.96-2 respectively. The water salinity was measured from sample analysis, which showed low saline water (20,000-30,000 ppm). Furthermore, the water saturation was calculated using the Archie equation model, and the result of the petrophysical

IOP Conf. Series: Earth and Environmental Science **851** (2021) 012049 doi:10.1088/1755-1315/851/1/012049

analysis is seen in Figure 2. Furthermore, the estimation of rock quality factor was conducted by firstly calculating dynamic elastic rock properties using a combination of compressional slowness, shear slowness, and density logs. This resulted in p-wave modulus (M), s-wave modulus (G), and VpVs ratio, and there is no measured core sample to validate the result.

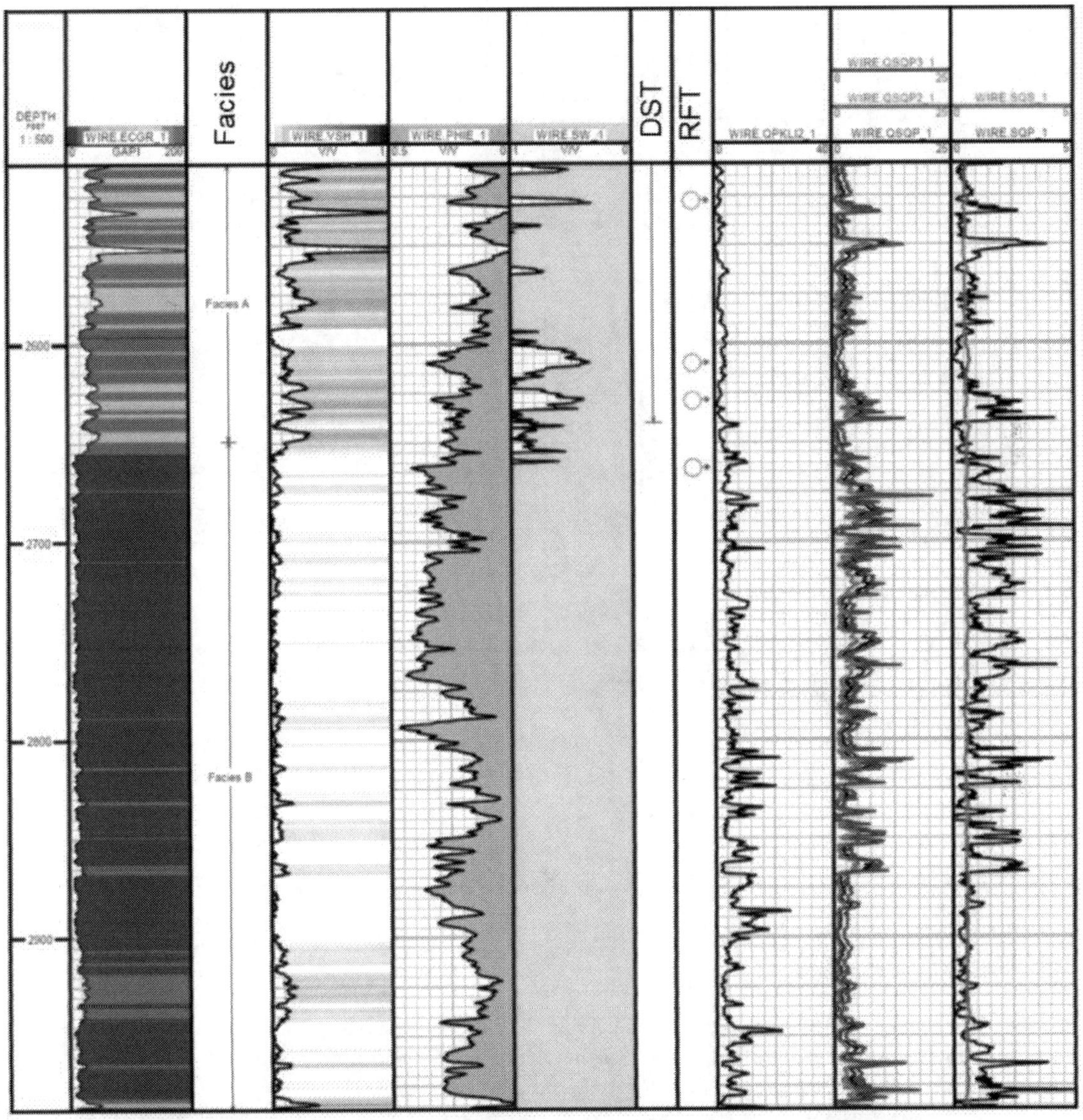

Figure 2. The petrophysical analysis result and its estimation of attenuation.

The result of the estimated rock quality factor can be seen in figure 3. The P-wave rock quality factor (Qp) by using Klimentos and McCann method showed discrimination of hydrocarbon-bearing and wet reservoirs on facies A in cross plot between Qp and VpVs (Figure 3b). Also, the reservoir with 20-50% water saturation was easily recognized from a cross plot in Figure 3b with Qp and VpVs ranging from 2-11 and 1.3-1.8 respectively. Although, in facies B, it was difficult to discriminate between hydrocarbon-bearing and wet reservoirs rocks as seen in Figure 3c. Furthermore, the VPVS ratio was more sensitive to discriminate the two reservoirs than Qp.

By using 3 theorems from Mavko et al method, the QsQp ratio can be estimated through 3 different equations. The result in Table 1 showed QsQp ratio was less sensitive to identify hydrocarbon reservoirs, as shown by a wide range value. However, facies A has a lower value and narrower range of QsQp ratio than B. Hence, the QsQp ratio is more sensitive to identify reservoir and non-reservoir lithology such as shale. The shale has a high QsQp ratio with 100% volume and water saturation.

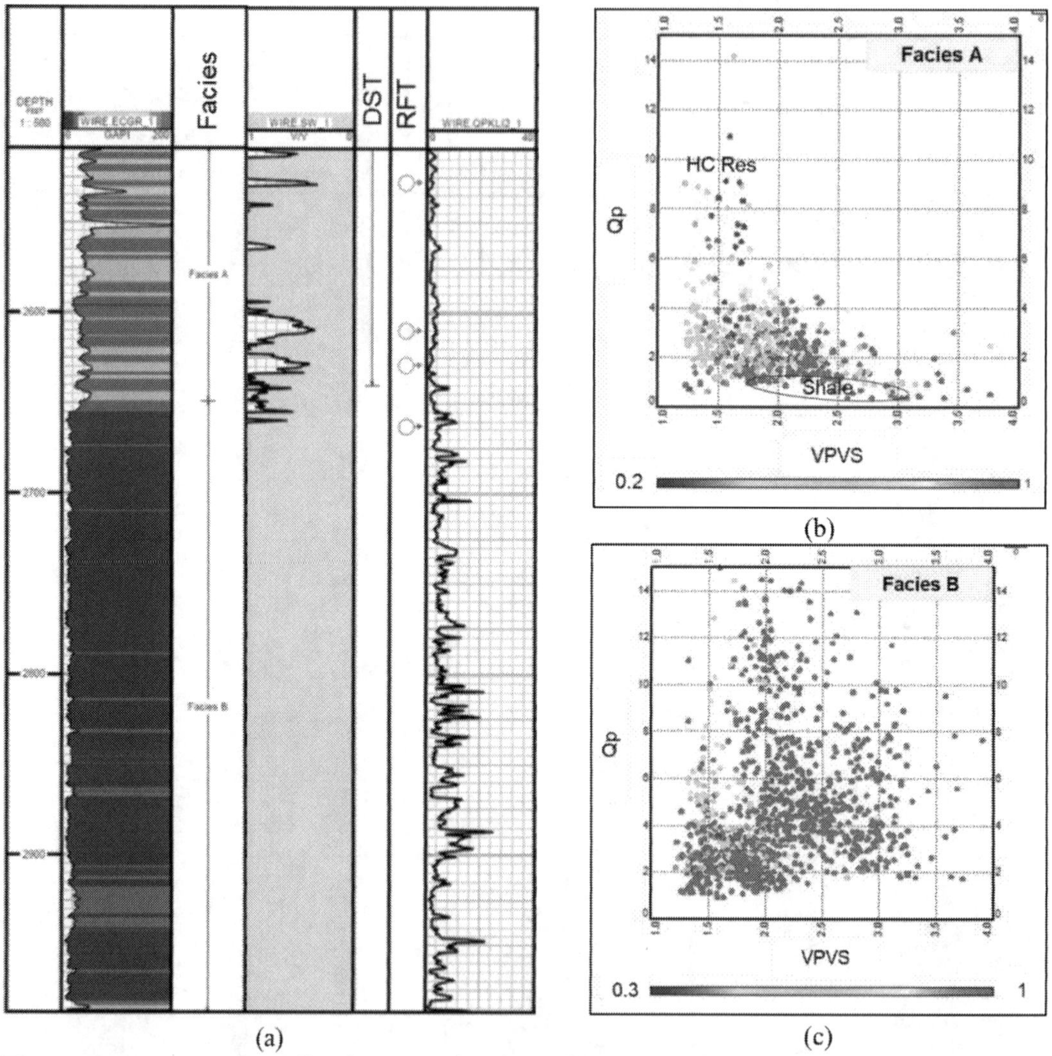

(a) (b) (c)

Figure 3. P-wave rock quality factor result using Klimentos and McCann Equation. a) well-based rock quality factor results and the cross plot between VpVs ratio and P-wave quality factor on b) facies A, and c) facies B which point color means water saturation.

Tabel 1. The summary of QsQp ratio of Mavko et al method.

	Facies A	Facies B	Hydrocarbon reservoir	Shale
QsQp	0-90	0-50	0-90	35-50
QsQp2	0-10	0-5	0-10	4-12
QsQp3	0.8-18	0.8-10	0.8-18	8-15

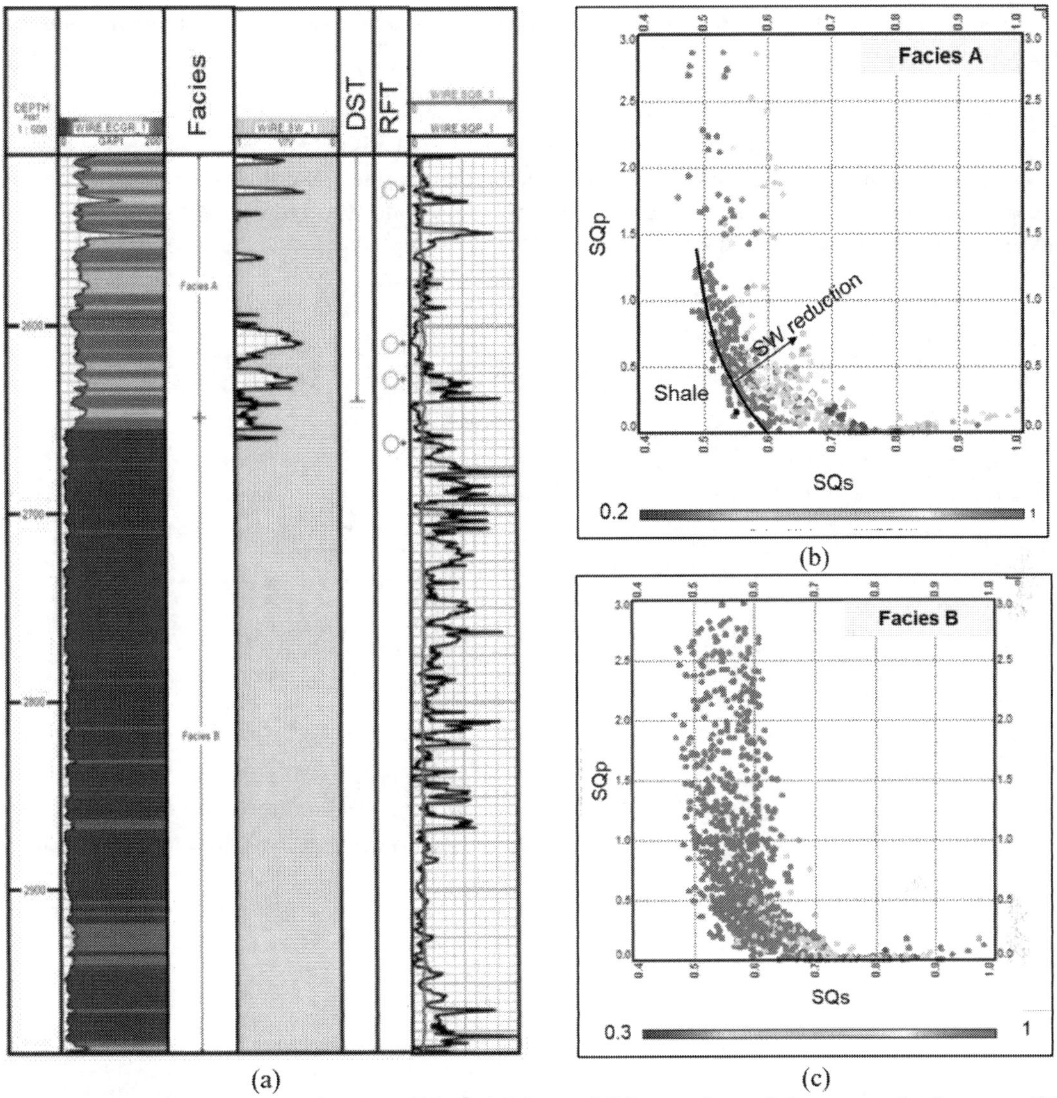

Figure 4. SQp and SQs result. a) well-based SQp and SQs results and the cross plot between SQp and SQs on b) facies A, and c) facies B which point color means water saturation.

The estimated inverse scaled-Q factor proposed by Hermana et al looks like the first equation result of the Mavko et al method in Figure 2. To discriminate between hydrocarbon-bearing and wet reservoirs, it is necessary to plot SQp and SQs values. In Figure 4b, SQp and SQs method gradually changed due to the reduction of water saturation on facies A. The hydrocarbon-bearing reservoir with 20-50% water saturation of SQp and SQs value ranges from 0.05-3 and 0.58-1 respectively. Meanwhile, the wet reservoir has SQp and SQs values ranging from 0.05-3 and 0.48-0.6. On facies B, a thin hydrocarbon-bearing reservoir was detected that SQp value ranges from 0.05–1 and SQs from 0.6–1. From all these facies, it is seen that SQs value was sensitive to detect water saturation and hydrocarbon indication. Therefore, with this method, facies give less impact to discriminate the hydrocarbon indication. In Figure 4b and 4c, both lithofacies shows gradual change as reduction of water saturation and increase in SQs value, even though facies B in (Figure 4c) looks more irregular than A (Figure 4b).

The facies affect the result of the estimated rock quality factor, and B will tend to produce a poor-quality result than A. Furthermore, low water saturation makes reservoir discrimination easier to identify, and the presence of high gas content affects wire-line log reading, especially for porosity and sonic logs. This directly affects the result of the estimated rock quality factor as seen in Qp, SQp, and SQs methods.

5. Conclusion

All the methods are applicable to estimate rock quality factors in carbonate reservoirs but only Qp, SQp, and SQs can discriminate pore-filling fluid. Furthermore, facies affect the estimated rock quality factor, and the QsQp ratio is sensitive to discriminate shale and carbonate. Therefore, a new understanding of facies impact on rock quality factor contributes to choosing the best methods to discriminate pore-filling fluid.

Acknowledgments

The authors are grateful to Saka Energi Indonesia for the permission to use the study data.

References

[1] Parra J O and Hackert C L 2006 *Journal of applied geophysics*, **58** 3 253-262.
[2] Cheng C H Toksöz M N and Willis M E 1982 *Journal of Geophysical Research: Solid Earth* **87** B7 5477-5484.
[3] Zhang Z and Stewart R R 2007 *CREWES Research Report* **19**.
[4] Klimentos T and McCann C 1990 *Geophysics* **55** 8 998-1014.
[5] Pireno G E and Roniwibowo A 2016 Proc. IPA Technical Symposium, Indonesia Exploration: Where From-Where To.
[6] Mavko G Dvorkin J and Walls J 2005 *Proc. SEG Technical Program Expanded Abstracts* 1469-72.
[7] Hermana M Ghosh D P and Sum C W 2016 *Proc. Offshore Technology Conference Asia.*
[8] Al-Sadi H N 1980 *Seismic Exploration: Technique and Processing* (Birkhäuser: SpringerLink)

IOP Conf. Series: Earth and Environmental Science **851** (2021) 012050 doi:10.1088/1755-1315/851/1/012050

Reservoirs resistivity correction factor in low resistivity pyritic sandstone reservoirs

S Winardi[1*], S S Surjono[1], D H Amijaya[1] and W Suryanto[2]

[1]Department of Geological Engineering, Universitas Gadjah Mada, Yogyakarta, Indonesia.
[2]Department of Physic, Universitas Gadjah Mada, Yogyakarta, Indonesia.
*E-mail: sarju_winardi@ugm.ac.id

Abstract The presence of pyrite in sandstone reservoirs will cause a problem known as low resistivity reservoirs case. The impact of pyrite volume in sandstone reservoirs is very important to be determined, especially to conduct its resistivity correction factor (Rcf). This research was done in the laboratory used nine sandstone pseudo-cores with various pyrite contents. Some data parameters such as voltage (V) and current (I) are directly measured and others (resistance R and reservoirs resistivity Rt) are calculated. The resistivity of pseudo-core data is calculated using the combined method between Ohm Law and Wenner. The result of this research shows that pyrite will reduce sandstone reservoirs resistivity exponentially. Pyrite will significantly reduce reservoirs resistivity if its presence > 4% and the Rt needs to be corrected. Plotting data between Rt and pyrite volume number for every Sw line can guide us to build a resistivity correction factor (Rcf). This Rcf will drive us to obtain the original resistivity reservoirs (Rto) from apparent well log resistivity reading (Ra) in case of low resistivity caused by pyrite mineral.

1. Introduction

The presence of pyrite in sandstone reservoirs will reduce the value of reservoir resistivity [1]. These reducing resistivity will cause some problems in log interpretation, mainly for water saturation (Sw) calculation. Reservoirs with resistivity below 5 ohm.m, which is well known as low resistivity reservoirs, will be indicated as water-bearing zones [2] & [3] if we analyze using conventional approaches. The relation between pyrite volume presence in reservoirs and its resistivity is very important to be determined. Then this laboratory-based research was conducted, aimed at two useful benefits i.e. (1) could set the threshold value of pyrite volume which significantly influenced the resistivity, and (2) could determine resistivity correction factor (Rcf). Rcf is needed in correcting apparent resistivity (Ra) which reads directly from resistivity log data and this Rcf can be used to obtain the original reservoir resistivity (Rto). Rto is reservoir resistivity in the assumption there is no pyrite presence (the reservoirs just consist of matrix resistive). It can be used to calculate correct Sw. This research has done in a laboratory using some pseudo-cores which have simulated contained various certain pyrite mineral volume.

2. Literature review

2.1. The presence of pyrite in sandstone

Iron sulfide (FeS_2) as the main component of pyrite will be formed in reductive environments, anaerobic, and rich in sulfur. Some examples of those environments are stagnant sea basins, tidal flat, and organic-rich lacustrine environments. Iron sulfide also can be found in sulfides hydrothermal veins at mid-oceanic ridges [4]. On other hand, pyrite minerals will widely be deposited in back swamp and bay fill deposits [5]. Pyrite is an opaque heavy mineral that can be found in sandstone as detritus, nodule, authigenic mineral, or replacement mineral. Pyrite is a conductive mineral and has a very low value of resistivity. Its resistivity can be compared to other conductive minerals (Table 1) that compiled from Parasnis [6], Milsom [7] and Telford *et al.* [8]. Some sandstone reservoirs in Indonesia (Table 2) are indicated contained by conductive minerals as well as pyrite [9].

Table 1. Resistivity value of pyrite compared to other conductive minerals.

Minerals	Resistivity (ohm.m)		
	Parasnis [6]	Milsom [7]	Teldford et al. [8]
Pyrite	10^{-4} - 10	$2x10^{-5} – 9x10^{-2}$	$3x10^{-7}$
Magnetite	10^{-2} - 10	10^{-2}	$5x10^{-5} – 5.7x10^{3}$
Hematite	$10^{-1} – 10^{2}$	$5x10^{-2}$	$3.5x10^{-3}$ - 10^{7}

Table 2. Some sandstone reservoirs in Indonesia with conductive minerals content including pyrite [9].

Area	Formations	Intervals	Conductive Minerals	Authors
Pagerungan	Ngimbang Clastics	Lower Sandstone Member	pyrite	Ebanks & Cook in Atkinson *et al.* [9].
	Pre-Ngimbang		pyrite, siderite	
Kurau	Lower Sihapas	Lithofacies-1	siderite , pyrite	Murphy in Atkinson *et al.* [9].
		Lithofacies-2	pyrite	
		Lithofacies-4	siderite, glauconite	
		Lithofacies-5	glauconite	
ONWJ	Upper Cibulakan	Facies 3-C,	siderite	Atkinson *et al.* [9].
		Facies 4-B, Facies 4-C	siderite	
		Facies 5	siderite	
Tuban	Ngrayong	Grigis Barat-1 core #1	siderite	Ardhana *et al.* in Atkinson *et al.* [9].
		Grigis Barat-1 core #2	glauconite, siderite, pyrite	
		Gondang-1 core#1 & 2	glauconite	
		Ngasin-1 core#1	glauconite, pyrite,	
OSES & NWJ	Talang Akar	TZ-1 #1	pyrite	Young & Atkinson in Atkinson *et al.* [9].
		TZ-1 #2	siderite	
NWJ	Jatibarang	OC-2 #3	pyrite, ferroan dolomite & ferroan calcite cement	Butterworth & Atkinson in Atkinson *et al.* [9].

2.2. Resistivity of Conductive Reservoirs

The reservoir model proposed by Archie [10] was the non-conductive model. Rock matrix in those models had been assumed resistive and conductivity of reservoirs came from saline formation water. Some authors then have developed different models. They are Patnode and Wyllie [11], Clavier *et al.* [12], and Givens [13]. They proposed another perspective that some components in the matrix could give a contribution to reservoirs conductivity.

IOP Conf. Series: Earth and Environmental Science **851** (2021) 012050 doi:10.1088/1755-1315/851/1/012050

Patnode and Wyllie [11] calculated the impact of the presence of conductive material into total resistivity. They used a slurry of silica and attapulgite (as a conductive fraction) and found some relation below.

$$\frac{1}{Rt} = \frac{Vs^{x}}{Rs}\left(1 - \frac{1}{F*}\right) + \frac{1}{F*Rf} \tag{1}$$

Rt = resistivity of *slurry*, ohm.m
Rs = resistivity of conductive solid material, ohm.m
Rf = resistivity of fluids, ohm.m
Vs = fraction of conductive material, %
x = heterogeneity factor
F* = apparent formation factor

Clavier *et al.* [12] observed the influence of pyrite on the conductivity of core samples (Figure 1). They found that samples with more abundant pyrite will give higher conductivity than smaller ones. They also illustrated the total conductivity model for saline formation water and conductive matrix or pyrite.

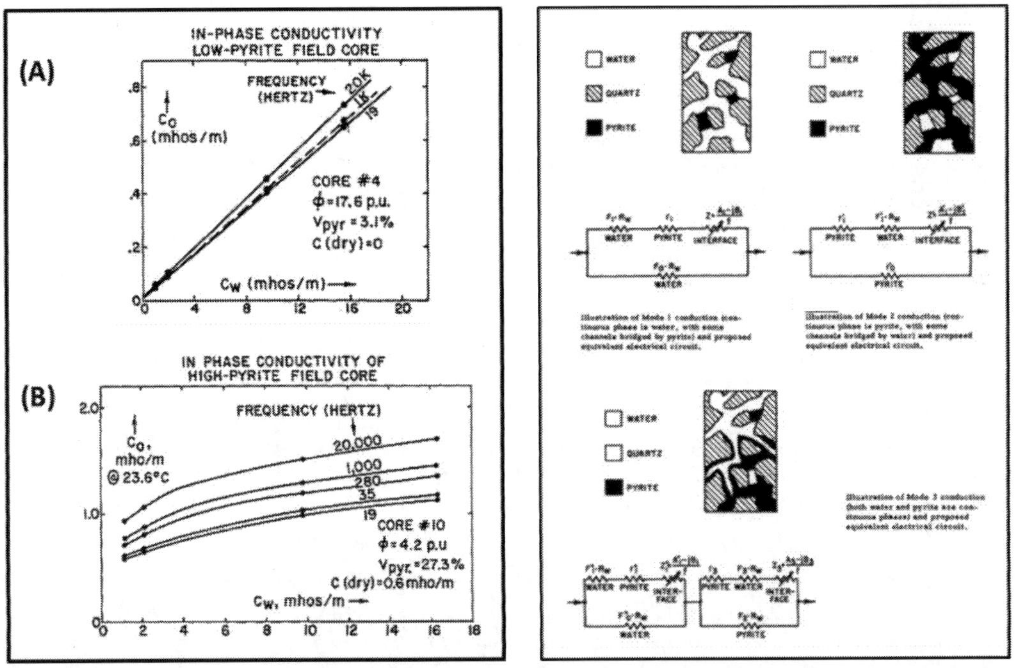

Figure 1. Analysis of the influence of pyrite on samples conductivity [12]. The left figure is a conductivity comparison between high content pyritic samples with low content of pyritic samples. The right figure is a combination model of total resistivity from a conductive matrix (pyrite) and saline formation water.

Givens [13] proposed a model called the conductive rock matrix model (CRMM). He found that conductive matrix could contribute to the total conductivity of rock. So total resistivity of saturated rock came from formation water resistivity and resistivity of the matrix. He arranged the relation as follows.

IOP Conf. Series: Earth and Environmental Science **851** (2021) 012050 doi:10.1088/1755-1315/851/1/012050

$$\frac{1}{Rt} = \frac{1}{Rpt} + \frac{1}{Rr} \qquad Sw = \left[\frac{Rw}{\emptyset_e^m}\left(\frac{1}{Rt} - \frac{1}{Rr}\right)\right]^{1/n} \qquad (2)$$

Rt = total resistivity of saturated rock
Rpt = resistivity of saturated pore
Rr = resistivity of matrix
Sw = water saturation
Rw = resistivity of formation water
Øe = effective porosity
m = cementation exponent
n = saturation exponent

Prayitno *et al.* [14] measured some pseudo-core that contain pyrite in the range of 0% to 34% (Figure 2). They did not mention the reservoir condition such as salinity, temperature, and porosity in their research. They found that pyrite can reduce the value of Rt, but the lowest of their measurement just reached around 10 ohm.m. Sw line of oil-bearing samples (Sw < 60%) still above 20 ohm.m. These results didn't yet represent low resistivity reservoir cases that usually the resistivity reach below 5 ohm.m. We tried to improve that research in such a way so the result would be meet the actual low resistivity cases.

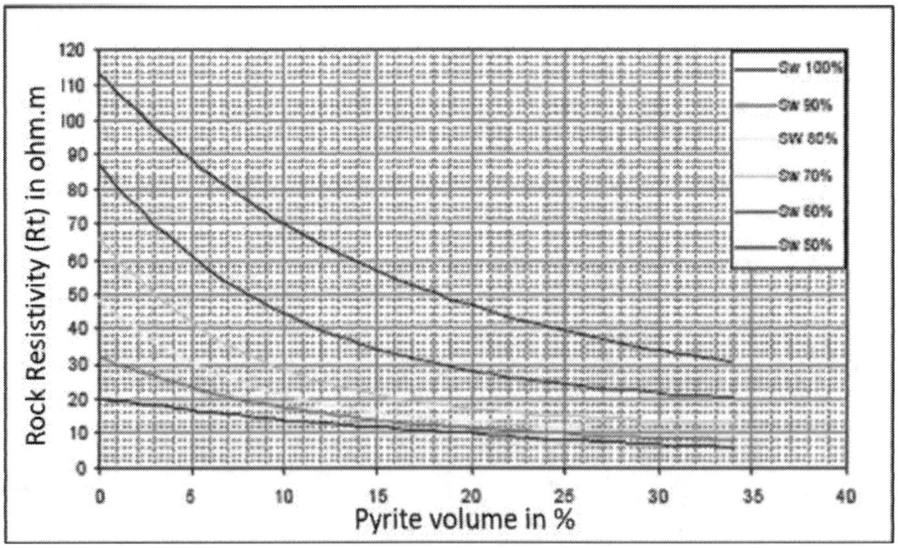

Figure 2. The relation of Rt and pyrite volume [14].

3. Research methodology

This research used sandstone pseudo-cores with various pyrite contents. Some data parameters are directly measured and others are calculated. The resistivity of pseudo-core data is determined using a combined method between Ohm Law and Wenner [15].

3.1. Data Parameter

The geometry of the pseudo-core has a total length of 0.20 m and its diameter is 0.038 m (left part of Figure 3). Pseudo-cores are made from medium sand size of quartz (around 0.3 mm), various pyrite volumes, and 5% cement. There are nine (9) pseudo-core samples with pyrite content from 0% to 25% and the porosity is set up for 40%. We used 20.000 ppm of formation water and data measured at surface temperature (about 25° C). Pseudo-cores are covered with PVC pipe, stainless steel at both of the end, and two electrodes between them (right part of Figure 3).

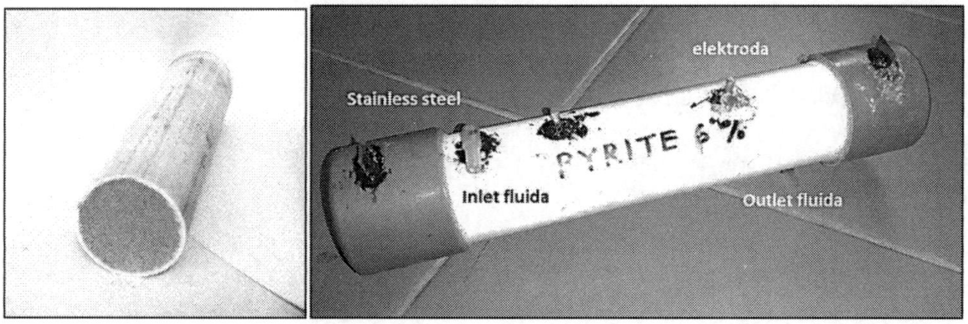

Figure 3. Pseudo-core sample geometry and the arrangement of electrodes at the sample. Some outlet and inlet fluid pipes are prepared in the fluids injecting process.

We measured data parameters during this research i.e. voltage (V in mV) and current (I in mA). Samples are saturated first with 100% formation water and injected by oil subsequently every 10% of volume. Then resistance (R) and resistivity (Rt) are calculated. So, we have parameters V, I, R, Rt, Sw, and volume conductive mineral (CM).

3.2. Data Measurement
The current was set up from a power supply of around 10 milliamperes and connected at the end of the sample. Then voltage was measured by ohm meter from electrodes set up in the body of the sample (Figure 4). Resistance is calculated from parameters I and V using Ohm Law and resistivity then calculated using a combination method of Ohm Law and Wenner Method [15] as expressed in equation 3.

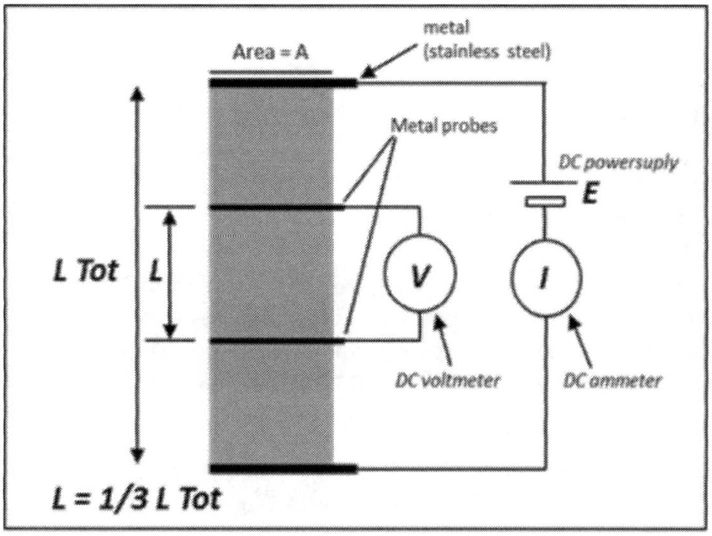

Figure 4. Measurement scheme of pseudo-core sample in the laboratory [15].

$$\rho = (V/I) * (A/L)*(0.04/L) \quad (3)$$

ρ = resistivity (ohm.m)
V = voltage (volt)
I = current (amper)
A = the acreage of sample section (m^2)
L = 1/3 total length of sample

4. Results

4.1. Characteristic of a pyritic sandstone sample

A thin section of pyritic sandstone from the pseudo-core sample shows the presence of quartz, pyrite, and some pores (left side of Figure 5). In a polarized microscope (Figure 5.A) pyrite (P) is observed as an angular shape opaque mineral between quartz (Q) and in the reflected microscope (Figure 5.B) it gives a special appearance of metal luster. XRD data analysis from pyrite grains that are used in the mixture of pseudo-core gives strong evidence of the pyrite peak curve (right side of Figure 5).

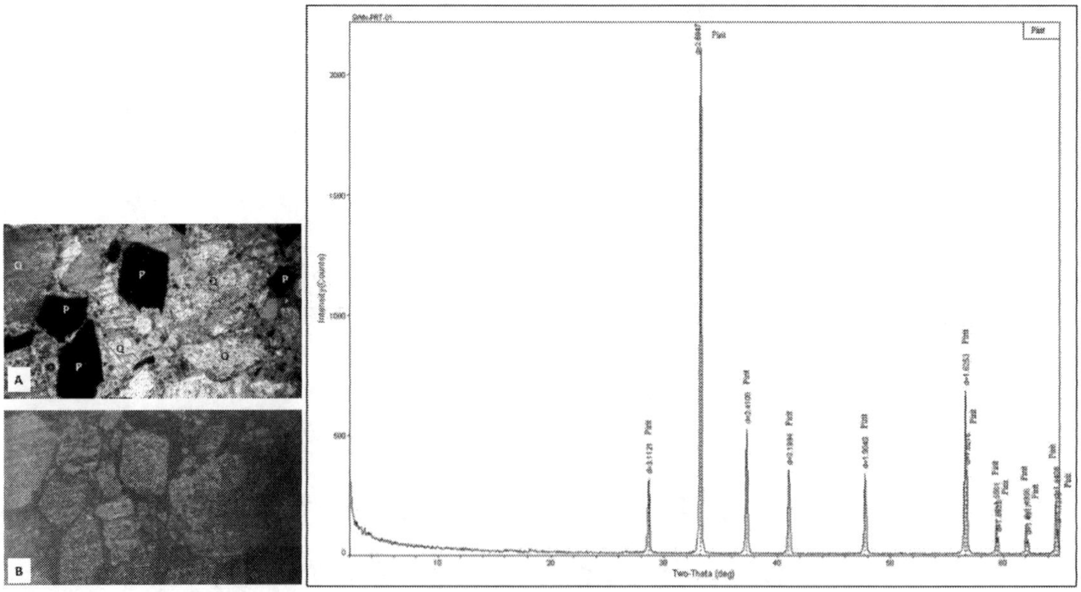

Figure 5. The appearance of pyrite minerals in sandstone. Figure 5.A show of thin section from polarization microscope and figure 5.B is a photograph of the same section using reflection microscope. The right side of figure 5 is XRD analysis from pyrite grains before used in pseudo-core forming grains.

4.2. Relationship Between Pyrite Volume and Rt

Measurement of voltage (in volt) of pyritic pseudo-core sandstone using 10 milliamperes of DC-current (0.01 A) gives range value from 1,6 – 8,3 volts (Table 3), whereas for non-pyritic pseudo-core sandstone gives values 3,25 – 13 volts. The values are controlled by the volume of pyrite (CM) in samples and its water saturation (Sw) value that injecting in samples. Resistance (R in ohm) can be calculated from those measurements (using Ohm Law) and gives values of R from 160 – 830 ohm (Table 4). Resistivity

(Rt) then derived from the value of resistance using the Winardi *et al.* approach [15] and gives values from 1.64 – 8.51 ohm.m (Table 5).

Plotting of each value of Rt vs CM for every Sw line gives the illustration of the relation between pyrite volume and decreasing value of Rt (figure 6). It is clear evidence that pyrite will reduce Rt value exponentially (Table 6). Compare to Prayitno *et al.* [14], this research shows the decreasing value of Rt can reach a value below 5 ohm.m and meet the common cases of low resistivity reservoirs. We calculate the percentage of decreasing value of Rt (Table 7) and it is observed when pyrite more than 4% Rt will reduce to around 50%. It means that samples with a pyrite content of more than 4% will give an apparent resistivity value (Rta) less than half of its original resistivity (Rto).

Table 3. The result of voltage (V in volt) measurement of pseudo-core.

	Sw	1,00	0,90	0,80	0,70	0,60	0,50	0,40	0,30	0,20
Pyrite Volume (%)	0	3,25	4,40	6,40	9,80	12,00	12,40	12,80	12,90	13,00
	2	2,90	3,03	4,27	6,00	7,22	7,43	7,50	7,90	8,30
	4	2,60	2,80	3,90	5,80	7,10	7,40	7,40	7,80	7,80
	6	2,30	2,40	3,80	5,00	6,60	6,80	7,00	7,20	7,40
	8	1,90	2,20	3,40	5,00	5,20	5,30	5,40	6,40	6,70
	10	1,90	2,20	3,20	4,20	4,90	5,50	5,70	6,00	6,40
	15	1,60	2,50	2,70	3,10	4,90	5,40	5,50	5,90	6,20
	20	1,80	2,40	2,55	3,10	3,60	3,90	4,20	4,50	5,20
	25	1,90	2,10	2,50	2,90	3,60	4,20	4,30	4,30	4,40

Table 4. The result of resistance calculation (R in ohm).

	Sw	1,00	0,90	0,80	0,70	0,60	0,50	0,40	0,30	0,20
Pyrite Volume (%)	0	325	440	640	980	1200	1240	1280	1290	1300
	2	290	303	427	600	722	743	750	790	830
	4	260	280	390	580	710	740	740	780	780
	6	230	240	380	500	660	680	700	720	740
	8	190	220	340	500	520	530	540	640	670
	10	190	220	320	420	490	550	570	600	640
	15	160	250	270	310	490	540	550	590	620
	20	180	240	255	310	360	390	420	450	520
	25	190	210	250	290	360	420	430	430	440

Table 5. The result of resistivity calculation (Rt in ohm.m).

	Sw	1,00	0,90	0,80	0,70	0,60	0,50	0,40	0,30	0,20
Pyrite Volume (%)	0,00	3,33	4,51	6,56	10,05	12,31	12,72	13,13	13,23	13,33
	2,00	2,97	3,11	4,38	6,15	7,40	7,62	7,69	8,10	8,51
	4,00	2,67	2,87	4,00	5,95	7,28	7,59	7,59	8,00	8,00
	6,00	2,36	2,46	3,90	5,13	6,77	6,97	7,18	7,38	7,59
	8,00	1,95	2,26	3,49	5,13	5,33	5,44	5,54	6,56	6,87
	10,00	1,95	2,26	3,28	4,31	5,03	5,64	5,85	6,15	6,56
	15,00	1,64	2,56	2,77	3,18	5,03	5,54	5,64	6,05	6,36
	20,00	1,85	2,46	2,62	3,18	3,69	4,00	4,31	4,62	5,33
	25,00	1,95	2,15	2,56	2,97	3,69	4,31	4,41	4,41	4,51

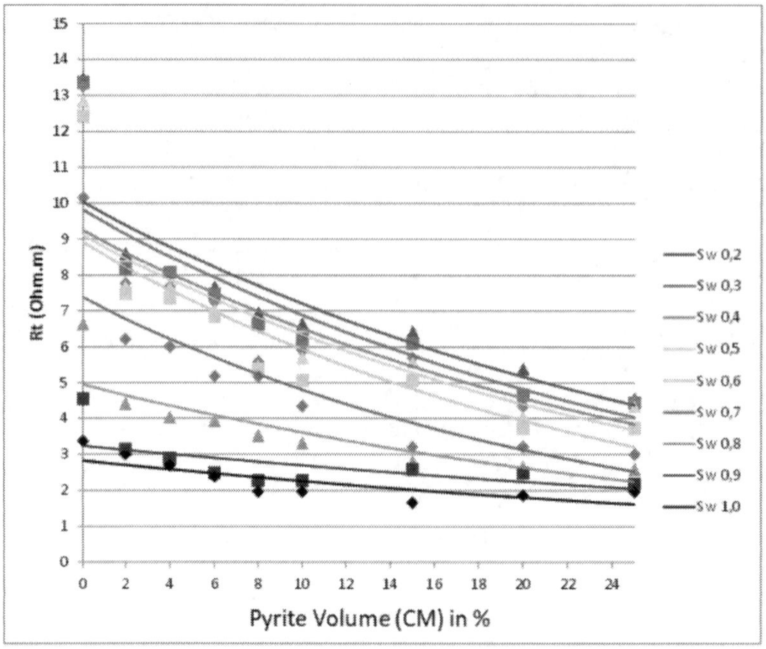

Figure 6. Data plotting between Rt, pyrite volume (CM) for every Sw line.

Table 6. The relation patterns of decreasing Rt for every Sw line.

Sw	Relation Rt vs CM	Regression
Sw 0,2	$y = 10,03e^{-0,03x}$	$R^2 = 0,83$
Sw 0,3	$y = 9,81e^{-0,04x}$	$R^2 = 0,83$
Sw 0,4	$y = 9,24e^{-0,04x}$	$R^2 = 0,75$
Sw 0,5	$y = 9,10e^{-0,04x}$	$R^2 = 0,77$
Sw 0,6	$y = 8,92e^{-0,04x}$	$R^2 = 0,82$
Sw 0,7	$y = 7,38e^{-0,04x}$	$R^2 = 0,84$
Sw 0,8	$y = 4,96e^{-0,03x}$	$R^2 = 0,80$
Sw 0,9	$y = 3.24e^{-0,02x}$	$R^2 = 0,46$
Sw 1,0	$y = 2,83e^{-0,02x}$	$R^2 = 0,62$

Table 7. Percentages of decreasing number of Rt caused by pyrite mineral.

	Sw	1,00	0,90	0,80	0,70	0,60	0,50	0,40	0,30	0,20
Pyrite Volume (%)	0,00	0,00	0,00	0,00	0,00	0,00	0,00	0,00	0,00	0,00
	2,00	0,11	0,31	0,33	0,39	0,40	0,40	0,41	0,39	0,36
	4,00	0,20	0,36	0,39	0,41	0,41	0,40	0,42	0,40	0,40
	6,00	0,29	0,45	0,41	0,49	0,45	0,45	0,45	0,44	0,43
	8,00	0,42	0,50	0,47	0,49	0,57	0,57	0,58	0,50	0,48
	10,00	0,42	0,50	0,50	0,57	0,59	0,56	0,55	0,53	0,51
	15,00	0,51	0,43	0,58	0,68	0,59	0,56	0,57	0,54	0,52
	20,00	0,45	0,45	0,60	0,68	0,70	0,69	0,67	0,65	0,60
	25,00	0,42	0,52	0,61	0,70	0,70	0,66	0,66	0,67	0,66

4.3. Threshold number

We obtained the volume of pyrite that will influence Rt significantly. We simulated the relation between Sw and Rt and put the number of Sw 0.55 as the threshold number of Sw cut-off. This number will be reached when Rt is around 6 ohm.m (Figure 7). Based on table 5, for Sw around 0.55 and the value of Rt around 6 ohm.m will be reached if the volume of pyrite more than 4%. It means that in oil zones (Sw<0.55) the presence of pyrite > 4% will give the value of Rt less than 6 ohm.m and cause low resistivity reservoir cases.

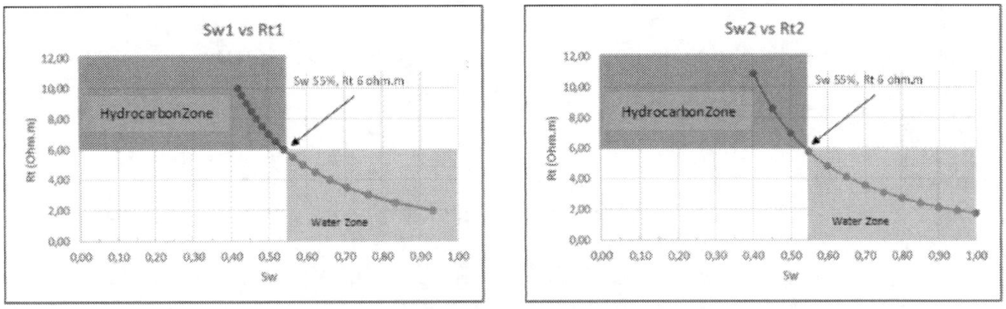

Figure 7. Simulation to obtain threshold number of Sw dan Ra. From the simulation hydrocarbon zone and water are distinguished using Sw 0.55 and Rt 6 ohm.m.

4.4. Resistivity correction factor

The resistivity correction factor will be calculated using this equation;

$$\boldsymbol{Rcf = \frac{Rt_o}{R_a}} \tag{4}$$

Rcf = resistivity correction factor
Rto = original resistivity reservoir
Ra = apparent resistivity reservoir

Therefore, every Sw line will have its Rcf. Table 8 shows the result of the Rcf calculation. This Rcf can be used to obtain corrected Rto.

Table 8. The result of calculation of Rcf for every Sw line and volume of pyrite contents.

SW	EQUATION	Rcf for every pyrite contents (%)															
		0	2	4	6	8	10	12	14	16	18	20	22	24	26	28	30
Sw 0,2	$y = 10{,}03e^{-0{,}03x}$	1,00	1,07	1,14	1,22	1,30	1,39	1,49	1,59	1,70	1,81	1,93	2,07	2,21	2,36	2,52	2,69
Sw 0,3	$y = 9{,}81e^{-0{,}04x}$	1,00	1,07	1,15	1,24	1,33	1,43	1,54	1,66	1,78	1,91	2,05	2,21	2,37	2,55	2,74	2,94
Sw 0,4	$y = 9{,}24e^{-0{,}04x}$	1,00	1,07	1,15	1,23	1,32	1,42	1,52	1,63	1,75	1,88	2,01	2,16	2,32	2,48	2,66	2,86
Sw 0,5	$y = 9{,}10e^{-0{,}04x}$	1,00	1,07	1,15	1,24	1,33	1,43	1,54	1,66	1,78	1,91	2,05	2,21	2,37	2,55	2,74	2,94
Sw 0,6	$y = 8{,}92e^{-0{,}04x}$	1,00	1,09	1,18	1,28	1,39	1,51	1,64	1,78	1,93	2,09	2,27	2,46	2,68	2,90	3,15	3,42
Sw 0,7	$y = 7{,}38e^{-0{,}04x}$	1,00	1,09	1,19	1,29	1,41	1,54	1,68	1,83	1,99	2,17	2,36	2,58	2,81	3,06	3,33	3,63
Sw 0,8	$y = 4{,}96e^{-0{,}03x}$	1,00	1,07	1,14	1,21	1,29	1,38	1,47	1,57	1,67	1,78	1,90	2,02	2,16	2,30	2,45	2,61
Sw 0,9	$y = 3.24e^{-0{,}02x}$	1,00	1,04	1,07	1,11	1,15	1,20	1,24	1,29	1,33	1,38	1,43	1,49	1,54	1,60	1,66	1,72
Sw 1,0	$y = 2{,}83e^{-0{,}02x}$	1,00	1,04	1,09	1,14	1,19	1,25	1,30	1,36	1,42	1,49	1,55	1,62	1,70	1,77	1,85	1,93

An example to use Table 8 in calculating **Rto**:

- From resistivity log data (Ra) we have 4 ohm.m. This interval has a pyrite content of 20% and Sw is predicted around 0.6.
- **Rto = Rcf x Ra.**
- From table 8, **Rcf** for **Sw** 0.6 and pyrite volume 20% is 2.27.
- So **Rto** = 2,27 x 4 ohm.m = 9,08 ohm.m.

5. Conclusions

This research gives strong evidence that pyrite will reduce sandstone reservoirs resistivity exponentially. The presence of pyrite in a sample for > 4% needs to be corrected, because it will reduce Rt > 50% and cause a low resistivity case (Rt < 6 ohm.m). Resistivity correction factor (Rcf) can be calculated from the graphic plot for every Sw line and the number of pyrite volume. This Rcf will drive us to obtain original resistivity reservoirs (Rto) from log resistivity reading when dealing with low resistivity cases caused by pyrite minerals.

Acknowledgments

Thank you very much to Sedimentology Laboratory of Geological Engineering Department and Geophysic Laboratory of Physic Department, Universitas Gadjah Mada, for deep collaboration and supporting this research.

References

[1] Hamada GM and Al-Awad MN 2002 Evaluation of low resistivity beds using nuclear magnetic resonance log *Eng Sci* **vol 14** no 1 47-61
[2] Boyd A, Darling H, Tobano J, Davis B, Lyon B, Flaum C, Klein J, Sneider, RJ, Sibbit A and Singer J 1995 The lowdown on low resistivity pay, *Oilfield Review* Autumn edition Schlumberger pp 4-18
[3] Gandhi A, Kubik P, Termina JJ, Rocque T and Volkmar M 2011, Petrophysics identifies low resistivity reservoirs *September 2011 Editor's Choice-Magazine* www.aogr.com/magazine/editors choice pp 1-12
[4] Bishop AC, Wooley AR *and* Hamilton WR 2005 *Philip's Guide To Minerals Rocks and Fossils* (London: Philip's) p 336
[5] Scholle PA 1979 *A Color Illustrated Guide To Constituents, Textures, Cements, And Porosities Of Sandstone And Associated Rocks* (Tulsa: AAPG) p 201
[6] Parasnis DS 1986 *Principles of Applied Geophysics* (New York: Chapman and Hall) p 402
[7] Milsom J 1989 *Field Geophysics* (New York: John Wiley and Sons Inc) p187
[8] Telfrod WM, Geldart LP and Sheriff RE 1990 *Applied Geophysics* 2nd ed (Cambridge: Cambridge University Press) p 744
[9] Atkinson CD, Scott J *and* Young R 1993 *Clastic Rocks and Reservoirs of Indonesia – A Core Workshop* (Jakarta: Indonesian Petroleum Association) p 229
[10] Archie GE 1942 The electrical resistivity log as an aid in determining some reservoirs characteristics *Petroleum Transactions of the AIME* **146** pp 54-67
[11] Patnode HW and Wyllie MRJ 1950 The presence of conductive solids in reservoirs rocks as a factor in electric log interpretation *Petroleum Transactions AIME* **Vol 189** pp 47-52
[12] Clavier C, Heim A and Scala C 1976 Effect of pyrite on resistivity and other logging measurements *SPWLA Seventeenth Annual Logging Symposium* p 34
[13] Givens WW 1987 A conductive rock matrix model (crmm) for the analysis of low-contrast resistivity formations *The Log Analyst* pp138-151
[14] Prayitno SH, Mardisewodjo M and Atmojo SM 2001 Pengaruh mineral pirit terhadap resistivitas batupasir dan aplikasinya pada kasus low resistivity *Proceeding Simposium Nasional IATMI* pp 1-7
[15] Winardi S, Suryono SS, Amijaya DH and Suryanto W 2019 Reservoirs resistivity measurement of pseudocore sample at laboratory scale based on Ohm's law and Wenner method *Proceeding of HAGI-IAGI-IAFMI-IATMI Joint Convention Yogyakarta* pp 1-8

IOP Conf. Series: Earth and Environmental Science **851** (2021) 012051 doi:10.1088/1755-1315/851/1/012051

Petrographic composition of coke product from Batu Ayau Formation Coking Coal, Kutai Basin, Central Kalimantan

M L Farhan[1], **D H Amijaya**[1*]
[1]Department of Geological Engineering, Faculty of Engineering, Universitas Gadjah Mada, Jalan Grafika 2 Yogyakarta 55281, Indonesia
*E-mail: hamijaya@ugm.ac.id

Abstract. The Batu Ayau Formation in the Kutai Basin has coking coal deposits that have the potential to be processed into coke. Analysis using organic petrography method is an important parameter to identify the composition of the coke product to determine the coke quality. The purpose of this study was to determine the composition of the coke products made from coking coal of Batu Ayau Formation in Kutai Basin, Central Kalimantan. Petrographic analysis on 12 samples obtained in Kal-Teng seam showed that the percentage of coke products varied. Carbon in the binder phase is 85.60 - 92 vol.%, then the form of carbon in the filler phase is 7.45- 13.64 vol.% and a small part of other than carbon material (miscellaneous categories) of 0-0.73% vol. The dominance of inert organic as a filler of the coke wall was identified. The studied coke which is originally made from coking coal with moderate volatile values shows that the coke is ideal in term of its petrographic composition.

1. Introduction

Coke is one type of coal products that has the ability to melt and make coherent residue when heated. That residue gets harden, which is called cake [1]. Coke is mostly known as metallurgical coke and used in factory processing especially for iron ore processing on blast furnace. In this process, the coke acts as a fuel as well as a reducing agent assists in the load-bearing process and provides a pathway for gas passage in the furnace later. The reducing agent function means that it separates oxygen from iron, while as an energy source, it is used to melt the ore. Coke should also have certain porosity because it is needed to pass the hot gas (aeration) in the furnace.

The need of coke for the metallurgical industry in Indonesia is still dependent on the supply of imported foundry coke. Domestic coke consumption is around 192.000 tons per year [2]. One of the uses of coke is to refine iron and nickel by using blast furnace [3]. This study will focus on coke products from Batu Ayau Formation coal in Kutai Basin, Central Kalimantan coal which has high-rank coal deposits. Petrographic analysis will be conducted to know the characteristics of the coke. Knowledge on the product of coke from Indonesian coking coal will be acquired from this study.

2. Geological Background

2.1. Regional Physiography and Tectonics

Physiographically this area is located in the Kutai Basin which is adjacent to Mangkalihat High and 2 fault zones, Bengalon and Sangkulirang Fault. Kutai Basin is bordered by Kuching orogenesis complex in the west and Makassar strait in the east. This basin is divided into 2 sub basins. They are Upper and

Lower Kutai Basin. The Upper Kutai Basin is located in the northwestern part where the research area is located, while the Kutai Bawah Basin is in the eastern part (see Figure 1).

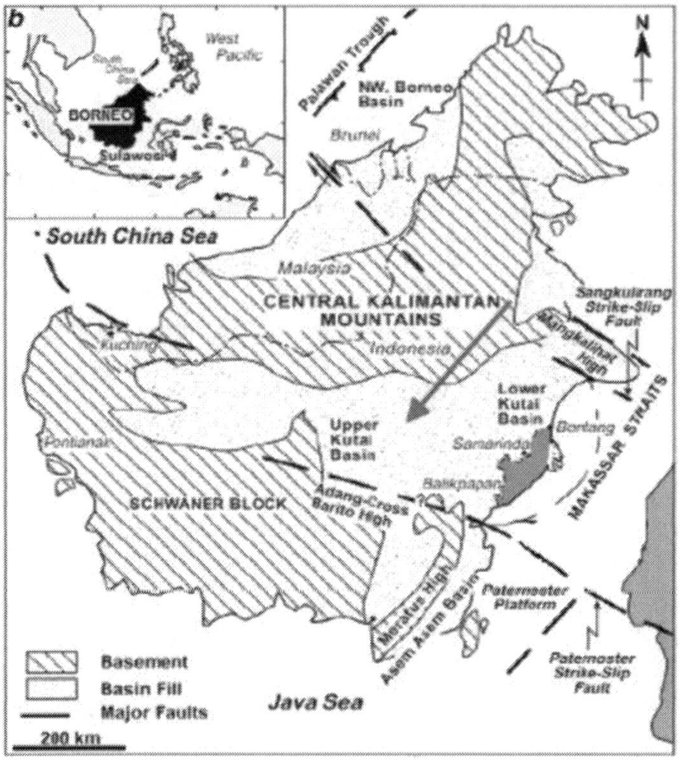

Figure 1. Physiography of Kutai Basin [4]. Studied area is shown by the (red arrow).

Kutai Basin is characterized by Tertiary structure, mainly the tectonic reversal or structure inversion. Inversion structure in Kutai Basin is formed by folding and simple fault propagation along planar fault [5].

2.2. Stratigraphy and Coal Characteristic

The stratigraphy in the study area is composed by Batu Ayau, Ujohbilang, and Karamuan Formation (see geological map in Figure 2). The Age of Batu Ayau Formation is Eocene, while Ujohbilang Formation was deposited in Oligocene. Karamuan Formation was formed in Miocene. There were also some intrusions occurred in that time, for example, Sintang Intrusion. Batu Ayau Formation in this area consists of sandstone, mudstone, carbonaceous siltstone with coal seam intercalation. The Batu Ayau Formation represents Middle-Late Eocene rift deposit which were deposited in fluvial-deltaic to outer shelf environment. Whereas Ujohbilang Formation is dominated by mudstone, some sandstone contains carbonates in some parts. This formation is the product of the sag phase in the Oligocene which caused deepening and changing to marine condition. Karamuan Formation consists of mudstone which contains carbonates in some parts, quartz sandstone, siltstone, and some fossiliferous carbonates and shale [6].

In the studied area, coal seams are found in the lower part of Batu Ayau Formation. Batu Ayau Formation coals in this area are considered as coking coal. It is characterized by volatile matter content of about 26.7 wt. % (adb). Petrographically the amount of vitrinite is more than 90 vol. % in average and Ro max of about 1 %. The carbon content is around 85 wt.%. CSN maximum is 9, the maximum fluidity value is about 1092 ddpm while HGI is 97 [7].

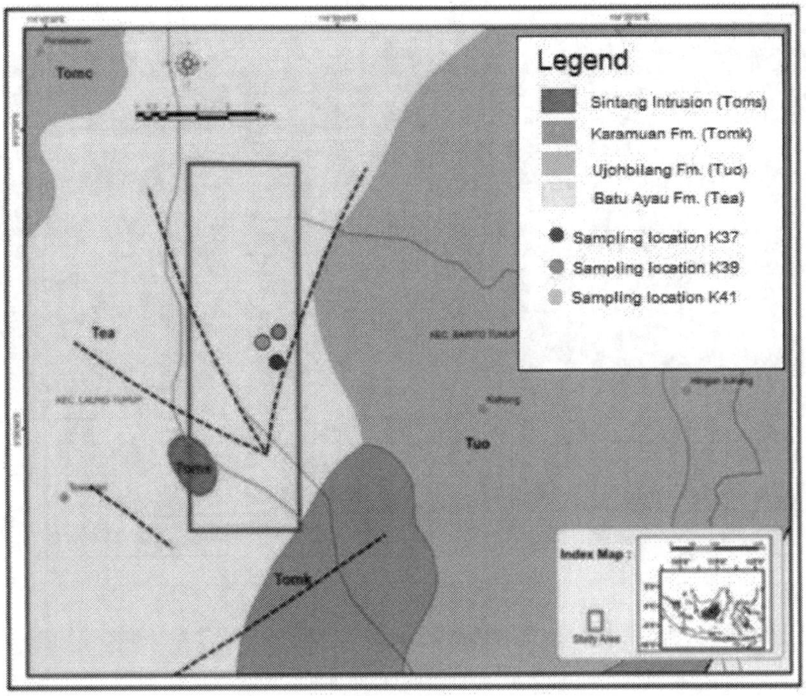

Figure 2. Geological map in the research area (red box) and surroundings [6. with modification].

3. Methodology

Field observation and coking coal sampling were conducted in the early phase of this research [see 7]. Coal lithotype identification was used to know the vertical succession of coal seam. The coal sampling was carried out based on coal lithotype by ply by ply. As many as 12 samples were taken from 3 seams which were named Kal-Teng (K) 37, 39, and 41 seams for this study. Samples were taken from Batu Ayau Formation in Murung Raya area, Central Kalimantan.

Laboratory work in form of organic (coke) petrography analysis was conducted at the Department of Geological Engineering, Universitas Gadjah Mada. The coke samples were produced by coking process at TekMIRA Laboratory in Bandung. Polished sections were used for organic petrography. Observation was conducted using a reflectance microscope with 40x magnification. The identification of coke products was based on coke microscopic methods [see 8,9].

4. Results

4.1. Field Data

The thickness of K. 37, K. 39 and K. 41 is 1.5 - 2.11 m (see Figure 3). The seams are characterized by the dominance of bright coal lithotype. Generally, the bright coal lithotype is in the form of glossy greyish black color, quite smooth surface and vitreous coal seam.

Figure 3. Appearance of the outcrop K. 37 (a), K. 39 (b), and K. 41 (c)

4.2 Petrography Data

The coke petrography which depicts the composition of coke products is shown in Table 1. The coke produced from all coking coals is dominated by binder phase carbon form as much as 85.6 – 92.00 vol.% and less dominant filler phase carbon form as much as 7.45 – 13.64 vol.% and a small portion of the miscellaneous material of about 0 – 0.73 vol. %. The average petrographic composition of the coke product for each seam is as follows, K-37 is on average composed by binder phase carbon of 88.09 vol.%, filler phase of 11.54 vol.%, and miscellaneous categories of 0.27 vol.%. K-39 is composed by binder phase carbon of 90.27 vol.%, filler phase of 9.5 vol.%, and miscellaneous categories of 0.22 vol.%. While K-41 is composed by binder phase carbon of 90.77 vol.%, filler phase of 9.13 vol.%, and miscellaneous categories of 0.09 vol.% (see graphic in Figure 4). Figures 5-7 show the microscopic appearances of coke in this study.

IOP Conf. Series: Earth and Environmental Science **851** (2021) 012051 doi:10.1088/1755-1315/851/1/012051

Table 1. Petrographic composition of studied coke.

Seam	Sample code	Binder Phase Carbon												Total (%)	Filler Phase Carbon							Total (%)	Miscellaneous Category								Total (%)
		Isotropic	*Incipient*	*Circular*			*Lenticular*			*Ribbon*					*Organic Inerts*		*Miscellaneous Inerts*			*Inorganic Inerts*			*Depositional Carbon*			*Additive Carbon*			*Miscellaneous Observations*		
				F	M	C	F	M	C	F	M	C			F	C	OC	BC	NCV	F	C		SOC	SC	PYC	CB	A	PC	GC	RC	
K. 37	K. 37B	0.91	2.73	2.18	24.91	36.91	1.82	2.00	4.18	1.45	2.73	5.82		**85.64**	4.18	4.18	0.36	0	0	1.82	3.09	**13.64**	0	0	0.73	0	0	0	0	0	**0.73**
	K. 37C	0.36	2.91	4.18	39.82	20.00	8.73	6.36	1.82	1.09	1.82	1.09		**87.82**	5.27	3.64	0.36	0	0	1.45	1.09	**11.82**	0	0	0	0	0	0	0	0	**0**
	K. 37M	0.55	2.55	15.27	26.36	20.73	9.82	7.27	2.55	1.45	1.82	0.73		**89.09**	5.45	2.91	1.09	0	0	0.73	0.73	**10.91**	0	0	0	0	0	0	0	0	**0**
	K. 37T	0.73	2.55	12.00	33.45	15.64	7.27	8.00	1.82	0.73	3.64	4.00		**89.82**	5.09	2.91	0	0	0	0.73	1.09	**9.82**	0	0	0.36	0	0	0	0	0	**0.36**
K. 39	K. 39B	0.73	0.91	32.91	20.18	4.36	20.55	8.18	1.64	0.36	1.27	0.91		**92.00**	4.36	2.00	0.55	0	0	0.73	0.36	**8.00**	0	0	0	0	0	0	0	0	**0**
	K. 39C	0.36	1.45	12.36	42.91	16.36	1.82	9.45	3.27	1.09	2.18	0.73		**92.00**	4.00	2.36	0	0	0	0.73	0.36	**7.45**	0	0	0.36	0.18	0	0	0	0	**0.55**
	K. 39M	1.09	1.09	4.36	40.36	28.00	2.18	4.73	1.82	2.18	2.55	1.09		**89.45**	5.45	2.55	0	0	0	1.09	1.45	**10.55**	0	0	0	0	0	0	0	0	**0**
	K. 39T	0.36	1.82	11.64	36.36	14.18	6.55	11.27	2.91	1.45	1.09	0		**87.64**	4	5.82	0	0	0	0.73	1.45	**12.00**	0.18	0	0	0.18	0	0	0	0	**0.36**
K. 41	K. 41B	0.36	1.82	26.18	38.36	3.27	7.82	10.36	0.73	1.82	0.36	0		**91.09**	6	1.82	0.18	0	0	0.56	0.36	**8.91**	0	0	0	0	0	0	0	0	**0**
	K. 41C	0.73	0.73	13.45	38.36	17.82	6.36	8.36	0.73	1.09	0.73	0.73		**89.09**	4.55	4	0.36	0	0	1.09	0.73	**10.73**	0	0	0.18	0	0	0	0	0	**0.18**
	K. 41M	0.36	1.82	3.64	43.27	19.27	5.82	7.27	1.09	3.27	5.45	0		**91.27**	2.36	6	0.36	0	0	0	0	**8.73**	0	0	0	0	0	0	0	0	**0**
	K. 41T	0.73	2.55	7.82	36.36	8.36	12.18	9.45	4	4.36	5.82	0		**91.64**	3.64	4.55	0	0	0	0	0	**8.18**	0	0	0.18	0	0	0	0	0	**0.18**

Note: F: fine; M: medium; C: coarse; OC: oxidized coal; BC: brecciated coal; NCV: non-coking vitrinite; SOC: sooty carbon; SC: spherulitic carbon ; PYC: pyrolytic carbon; CB: coke breeze; A: anthracite; AC: anthracite; PC: petroleum coke; GC: green coke; RC: reacted coke.

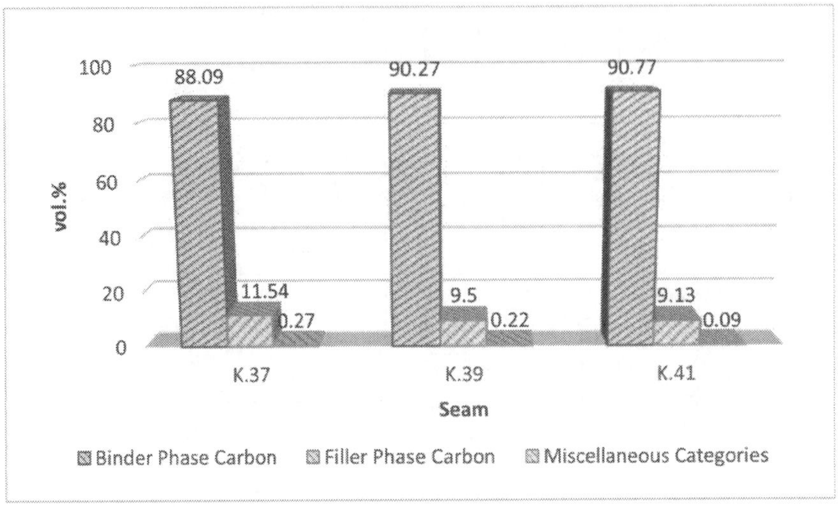

Figure 4. Average composition of coke products for each seam.

Binder phase carbon form

Binder phase product is the most dominant component in the studied coke. The binder phase is the transformation of reactive maceral in coking coal as the parent coal. The binder phases found were isotropic and anisotropic (incipient, circular, lenticular, and ribbon). The carbon isotropic form resembled a thin crack surface on a plain plate. Isotropic showed a relatively uniform shape and has no clear boundaries compared to the anisotropic form.

Anisotropic consists of various forms of carbon. Incipient is a form of carbon transition from isotropic to anisotropic. The shape of the incipient found in the sample was like an irregular slit. Circular, lenticular, and ribbon have a relatively compact shape compared to incipient. Circular was very dominant, which had a shape like a circle (oval and elongated). The form of circular carbon with medium and coarse size was more abundant than fine size. The lenticular had a flat, tapered shape that resembled a protruding lens. The ribbon had a wave-like shape.

Filler phase carbon form

The filler phase product quantity is less than the binder phase. The most common filler phases were organic inert and inorganic inert. Miscellaneous inert were found in small amounts in some samples.

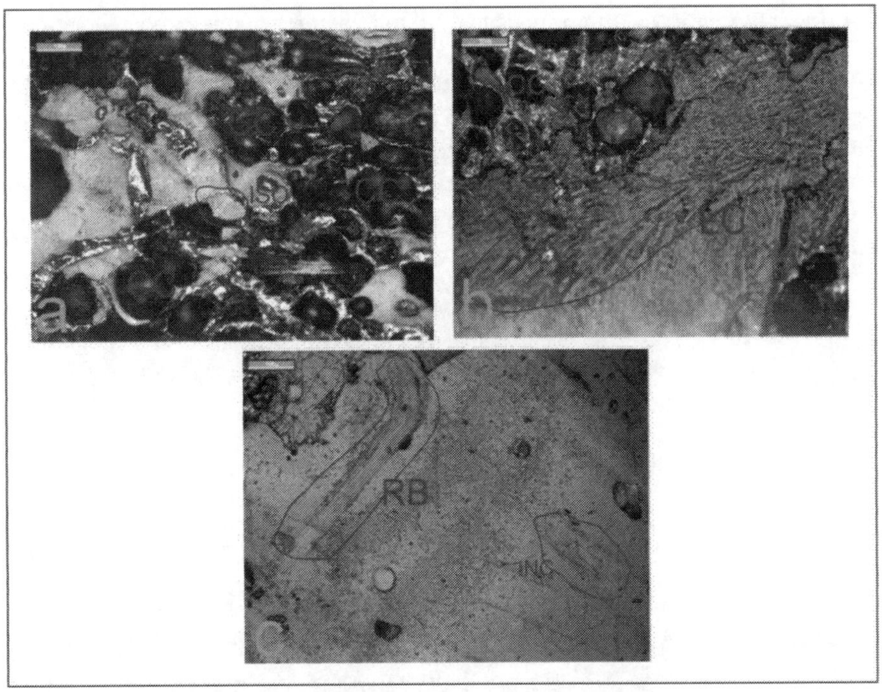

Figure 5. Microscopic appearance of coke product. (a) Circular carbon (CC) and isotropic (ISO) in K.41B sampel. (b) Lenticular carbon (LC) associated with circular carbon (CC) in K.39M sample. (c) Ribbon carbon (RB) and incipient (INC) in K.37T.

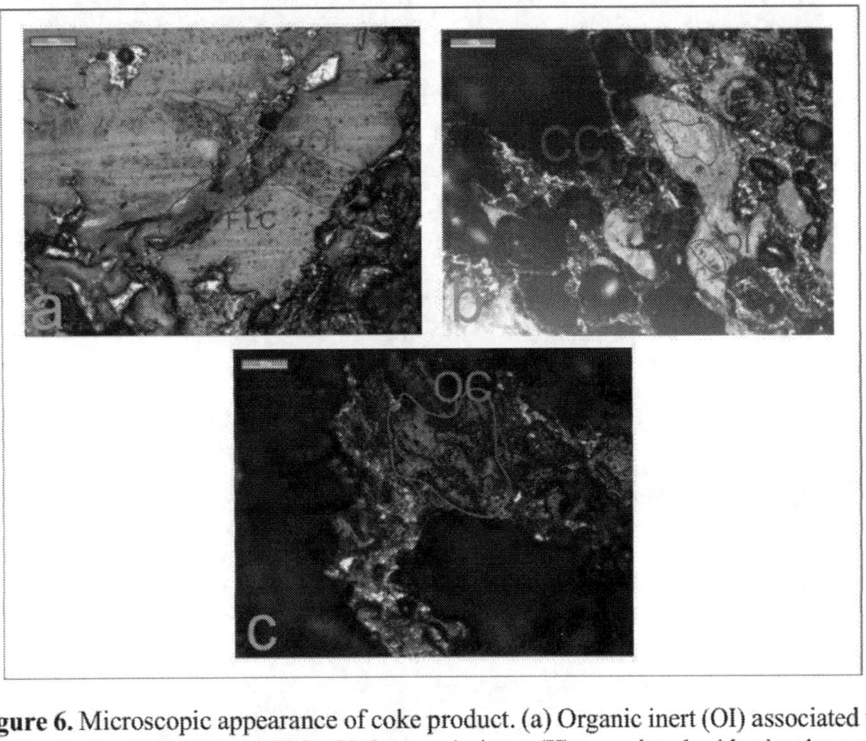

Figure 6. Microscopic appearance of coke product. (a) Organic inert (OI) associated with fine lenticular (F.LC) in K.37C. (b) Inorganic inert (II) associated with circular carbon (CC) in K.41B. (c) Oxidized carbon (OC) in K.39B.

Organic inert was like dark, smooth ribbons that fill the fractures in binder phase. They were divided into two sizes, fine and coarse. The two measures relatively had almost the same percentage in each sample. Inorganic inert was showing a form resemble black material that appears around binder phase product. Inorganic inert originated from minerals in coking coal. Oxidized coal had a rectangular shape that had cracks, low relief, and degasification pores because of the removal of dissolved gases from liquids.

Miscellaneous categories

There are miscellaneous products found in the samples, namely pyrolytic carbon (depositional carbon), coke breeze (additive carbons), and sooty carbon. Depositional carbons are a type of material that is deposited on the coke surface near the surface area in the coking process. Depositional carbons found in the study sample were pyrolytic carbon. Pyrolytic carbon had a smooth, brightly colored ribbon that filled the space between the binder phase and the filler phase.

Additive carbons are inert additives (anti fissurant) that are added during the coking process. The additive carbons found in the study sample were coke breeze. Coke breeze in the sample had a dense, fine-grained shape and bright in color. Sooty carbon is carbon particles produced by incomplete combustion of hydrocarbon. In the samples, it usually had a deformed irregular form and still showing the rest of original maceral.

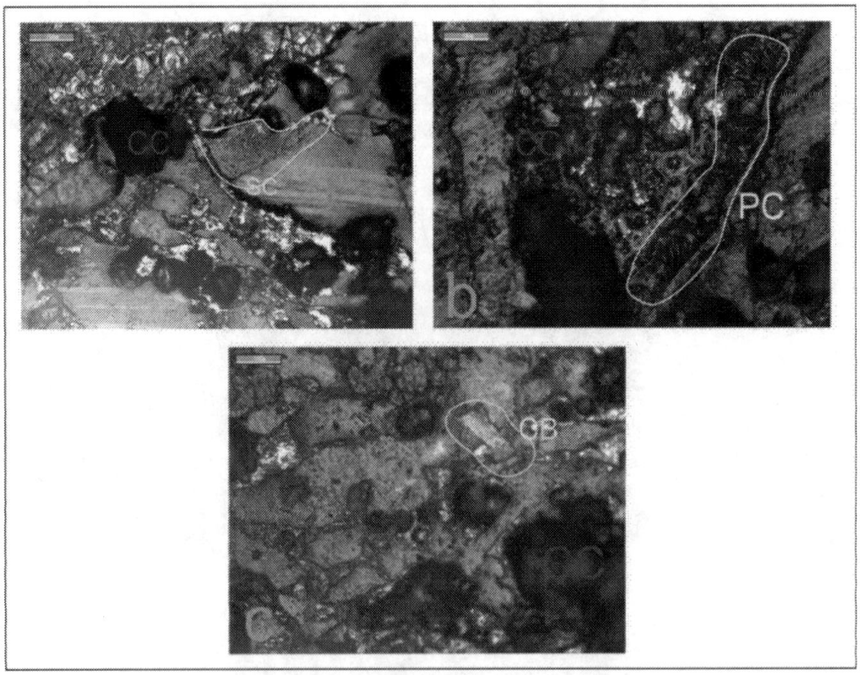

Figure 7. Microscopic appearance of coke product. (a) Sooty carbon (SC) associated with circular carbon (CC) in K.37 C. (b) Pyrolytic carbon (PC) associated with circular carbon (CC) on K.39C. (c) Coke breeze (CB) associated with circular carbon (CC) in K.39T.

5. Discussion

Coke is produced by heating coking coals in a special oven in a reducing atmosphere condition. Coking coals refer to bituminous class coals of high, medium, and low volatile rank that possess the ability to soften when heated, become plastic, and resolidify into a coherent mass. During the heating in the absence of air, it will melt vesiculate and harden into a spongelike mass of almost pure carbon. As the temperature of the coal increases, it becomes plastic, fusing together before resolidifying into coke particles. This phenomenon is known as the caking process. This process will produce particular characteristics in the petrographic composition of coke.

The composition of the coke product is one of important parameters for determining the microstructure of coke. Among others are pore size and coke walls, the form of carbon and comparison with the maceral composition of coking coal before processed into coke. In term of coking process, maceral can be divided into two part, namely reactive maceral and inert maceral. The reactive maceral in coking coal (vitrinite, liptinite, resinite, and reactive semifusinite) are softened during the carbonization process and will be transformed into binder carbon (binder phase carbon). These macerals tend to appear in a predominantly anisotropic (non-uniform) form of carbon. This is characterized with optical properties that vary when the polished sample is rotated and viewed by polarized light during microscopic observations. Meanwhile, inert maceral (inert semifusinite, fusinite, micrite, macrinite, and inertodetrinite) does not soften during carbonization, which will be transformed into filler carbon (filler phase carbon). These macerals tend to appear as isotropic (uniform) carbon with uniform optical properties. There is also miscellaneous category which is material entity other than carbon.

In general, the coke products from studied seams are dominated by binder phase products and less dominant filler phase products, as well as a small portion of the miscellaneous categories. The binder phase carbon product is divided into isotropic and anisotropic. Petrographic composition of the binder phase is shown in Table 2. The dominance of the binder phase carbon indicates that many macerals underwent a softening process, hardened again and bound inert maceral. The carbon binder phase can be used to determine the value of the degree of anisotropism and to see the level of volatility of the coal (see Gray, 1976).

Table 2. The percentage of dominant product of binder phase carbon in studied coke.

Coke Product	Percentage (Vol.%)
Isotropic	0,36 - 1,09%
Incipient	0,91 - 2,91%
Fine Circular	2,18 - 32,91%
Medium Circular	20,18 - 43,27%
Coarse Circular	4,36 - 20,73%
Fine Lenticular	1,82 - 20,55%
Medium Lenticular	2 - 11,27%
Coarse Lenticular	0,73 - 4,18%
Fine Ribbon	0,36 - 4,36%
Medium Ribbon	0,36 - 5,82%
Coarse Ribbon	0,36 - 5,82%

Anisotropic can be divided into incipient, circular, lenticular, and ribbon. Circular carbon dominates with a percentage of >50 vol.%. This confirms that the level of volatility in coal is relatively high and the degree of anisotropism is high. Fine size lenticular carbon is found to be the most abundant. This also shows that the sample has a relatively high volatile content. The size of the fine lenticular carbon material which dominates indicates that the strength of the coke is better. Lenticular with medium-coarse size and ribbon also indicates the presence of altered coal with a high degree of anisotropis but has a low volatile value.

The filler phase carbon products found can be categorized as organic inert, inorganic inert, and oxidized coal (composition is depicted in Table 3). The existence of the filler phase carbon indicates that there are some macerals that did not soften during the carbonization process and functioned as a filler in the binder phase carbon product. The filler phase carbon is used to determine the carbon associated with the inert maceral.

Table 3. The percentage of dominant product of filler phase carbon in studied coke.

Coke Product	Percentage (vol.%)
Organic Inert	6.36 – 9.82%
Inorganic Inert	0 – 4.91%
Oxidized Coal	0 – 1.09%

The organic inert carbon has a low level of anisotropism and still retains the original form of inert maceral. Organic inert is dominated the filler phase in the studied coke. Inert macerals such as some of the inertinite and liptinite maceral groups are thought to be the precursor of organic inert carbon. Organic inert carbon is useful for reducing the potential for cracks in the coke wall structure and increasing reactivity to CO_2. Inorganic inert carbon which is only found in small amounts indicates that the amounts of minerals in coal was also low. In contrast to organic inert carbon, this type of carbon is useful for reducing the reactivity of coke to inhibit gas diffusion. Oxidized coal is found in several samples in the study area due to weathering experienced during the coking process.

Table 4 shows the composition of miscellaneous categories found in organic petrographic observations. Those are pyrolytic carbon (depositional carbon) and coke breeze (additive carbons). Depositional carbon indicates an overheating process or heating process that happened too fast. The overheating process occurs on the wall or head of the furnace. This will increase the number of cracks and causes coking carbon to be stuck to the oven wall. However, there is also the possibility that depositional carbons were formed as a result of a condensation reaction or cracked hydrocarbon. Coke breeze is an additional inert carbon that is added during the carbonization process to protect the cathodic from corrosion. Sooty carbon is impure carbon particles produced by incomplete combustion of hydrocarbon. These miscellaneous carbons are found in very low quantities. This will not give a significant effect on the characteristic of the studied coke in general.

Table 4. The percentage of dominant product of miscellaneous categories in studied coke.

Coke Product	Percentage (vol.%)
Pyrolitic Carbon	0 – 0.73%
Coke Breeze	0 – 0.18%
Sooty Carbon	0 – 0.18%

Studied coke was produced from single coal (not blended coal). The composition is showing the characteristics of medium to high volatile coke. This indicates that the original coking coal can be used to produce coke as the primary material or blending material. The coke quality produced tends to have medium to high quality. This is also supported by the fact that the reactive maceral composition is relatively high but the coal has only a limited amount of inert maceral.

6. Conclusions

Based on this research, there are several characteristics found in the coke from Batu Ayau Formation coking coal from Central Kalimantan.

1. The coke is composed by coking products dominantly in form of carbon in the binder phase (85.6 - 92.00 vol.%), then in form of carbon in the filler phase (7.45 - 13.64 vol.%) and a small number

of miscellaneous material (0 - 0.73 vol.% vol). Coking product in miscellaneous categories are mainly pyrolytic carbon (depositional carbon) and coke breeze (additive carbons).
2. The coke quality produced tends to have medium to high quality.

Acknowledgements
Authors wish to thank Department of Geological Engineering, Faculty of Engineering, Universitas Gadjah Mada for supporting the study. Appreciation is given to PT. AKT for coal sampling permission. D. Tanggara was involved in the early phase of this study, to whom we also thank.

References
[1] Ward C R 1984 *Coal Geology and Coal Technology* (Singapore: Blackwell Scientific Publications) p **345**
[2] Supriyanto R H 2007 Penelitian Briket Kokas Pada Tungku Kupola Double Tuyer Meningkatkan Efisiensi Pengecoran Logam *Jurnal Sains dan Teknologi Indonesia* **9** 45
[3] Amarullah D 2009 Suatu Pemikiran untuk Memanfaatkan Potensi Batubara Formasi Tanjung di Daerah Lemo, Kalimantan Tengah sebagai Kokas *Buletin Sumber Daya Geologi (BSDG)* **4** 17.
[4] Witts D, Hall R, Nichols G and Morley R 2012 A New Depositional and Provenance Model For The Tanjung Formation, Barito Basin, SE Kalimantan, Indonesia. *Journal of Asia Earth Science* **56**, 77
[5] Satyana A H and Silitonga P D 1994 Tectonic Reversal in East Barito Basin, South Kalimantan : Consideration of the Types of Inversion Structures and Petroleum System Significance *Proc. Indonesian Petroleum Association (IPA) 23rd Annual Convention* (Jakarta) p **57**
[6] Supriatna S, Sudrajat A, and Abidin H Z 1995 Peta Geologi Lembar Muara Tewe Kalimantan, *Pusat Penelitian dan Pengembangan Geologi Indonesia.* (Bandung)
[7] Tanggara D, Amijaya D H, and Surjono S S 2018 Evaluation of Coking Properties Bituminous Medium Volatile Coal, Batu Ayau Formation, Kutai Basin, Central Kalimantan, *IOP Conference Series Earth and Environmental Science*, **212**, doi: 10.1088/1755-1315/212/1/012032
[8] Gray R J and Devanney K F 1986 Coke carbon forms: Microscopic Classification and Industrial Applications *IJCG* **6** 277
[9] Sharma R, Dash P S, Banerjee P K and Kumar D 2005 Effect of Coke Micro-Textural and Coal Petrographic Properties on Coke Strength Characteristics *ISIj International*, **45** 1820

IOP Conf. Series: Earth and Environmental Science **851** (2021) 012052 doi:10.1088/1755-1315/851/1/012052

Pedaro formation, equivalent of Plover sandstone at Savu Island, Outer Banda Arc

***L D Santy[1], A J Widiatama[2]**

[1]Centre for Geological Survey, Geological Agency, Diponegoro Street No. 57, B Build. 4th Fl., Bandung, West Java
[2]Institut Teknologi Sumatera, Room C 110, Terusan Ryacudu Street, Way Hui, Jati Agung Sub District, South Lampung
*E-mail: santylauti@gmail.com

Abstract. Plover Sandstone have been widely known as a good quality of oil and gas reservoir in NW Australia. As the continuity of NW Australia margin, outer Banda Arc become the distribution area of the equivalent of Plover Sandstone units. Therefore, a clear distinction and characterization of equivalent of Plover Sandstones distributed in this area is needed.
Thick unit of quartz rich sandstone is scrap out in south Savu Island. Refers to the location where the lithology is found widely distributed, the unit is suggested to be called as Pedaro Formation. The characteristic of the lithology is determined through detailed measured section from two trajectories and laboratory analysis, including fourteen samples of petrography analysis, eight samples of Scanning Electron Microscope (SEM) analysis, and three samples of X-Ray *Difraction* (XRD) analysis. The lower part of the unit is initiated by braided fluvial conglomerates which gradually become tidal sand flat association of shale layers, coal seams and sandstone insertion. While the upper part of the unit is consisting of thick bedded quartz sandstone with thin siltstone insertion, deposited in the shoreface environment. Those facies association developed at transgressive conditions in the estuarine environment. The characteristic of the sandstone unit of Pedaro Formation is thickly bedded quartz wacke to quartz arenite, white to light grey in color, moderately to very well sorted, mostly mature sand. Pedaro Formation is identified to be deposited at Early Jurassic in interior craton tectonic setting. The characterization of sandstone unit of Pedaro Formation is showing that the unit can be correlate to the equivalent of Plover Sandstone found in Timor. Sandstone unit of Pedaro Formation can play a role as reservoir candidate in petroleum system of Savu and surrounding area.

1. Introduction

Savu Island lies in the accretionary zone in the northern part of Timor Thrust. Stratigraphically, it is similar to the Timor Island [1,2], suggested the petroleum system of Savu region can be compared to Timor and other regions in the outer Banda Arc (Figures 1 and 2).

As the continuity of NW Australia margin, it has been widely known that the key elements of petroleum system in outer Banda Arc area occur within the Mesozoic succession, particularly in the Late Triassic and Jurassic section [3,4,5,6,7,8,9]. As a main reservoir target in NW Australia Margin is Plover Sandstone, a unit of quartz rich sandstone deposited inside Mesozoic Basin [7]. The unit consist of a series of widespread braided trunk river systems feeding a relatively narrow wave-dominated coastline and beyond, a wide marine shelf [7].

During the exploration of Banli-1 well in southern West Timor, quartz-rich sandstone was encountered, comformably overlied by Middle Jurassic Wailuli Formation [4]. The workers were determined it as Early Jurassic potential reservoir sequence of "Plover-equivalent". While, previous studies conducted in Savu Island have recognized the distribution of Late Triassic and Jurassic sequences in Savu Island [2,3,10,11]. However, the existence of equivalent of Plover Sandstone in Savu has not yet clearly described. Audley-Charles [3] and Rosidi *et.al.* [10], never mentioned the existence of the unit. While, Harris *et.al.* [2] mentioned about the Late Triassic unit of immature, poorly sorted, massive red sandstone that abundant with iron alteration and sandstone concrete, called as *'Pedaro Beds'*. The unit is used as one of the marker layers of Triassic sediment, found in the center part of Savu, because it is "considered" as a repetitive sandstone contain *Halobia sp.*

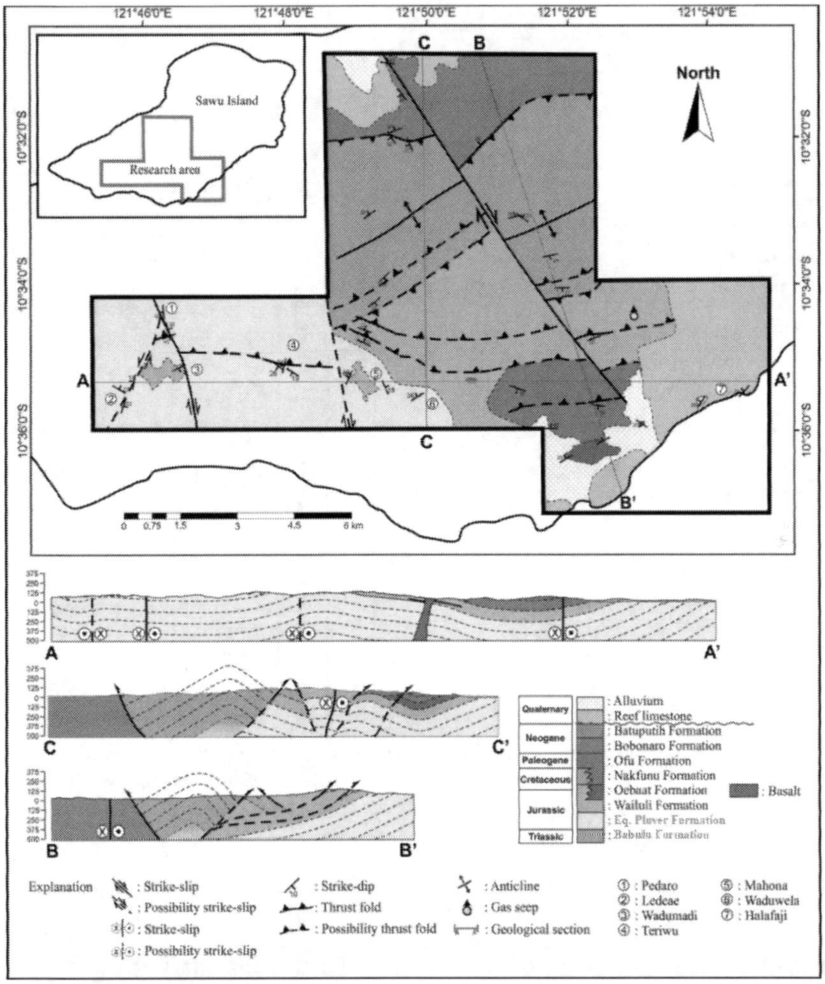

Figure 1. Geological map and profiles of Savu Island [12].

Fieldwork carried out by Centre for Geological Survey at 2014 has identified the unit of quartz rich sandstone distributed at south Savu area. The unit consist of thickly bedded quartz wacke to quartz arenite, white to light grey in color, moderately to very well sorted, mostly mature sand. The description of this unit is quite different to the one described by Harris *et.al.* [2]. Morphological features where the unit is distributed generally undulating hills, with gentle to slightly steep slopes. The unit is exposed in southern part of Savu Island, cropped out at Pedaro, Ledeae, Wadumadi,

Teriwu, Mahona, Waduwela, and Halafaji villages. In the geological map (Figure 1) this unit is marked by yellow colour. Geological reconstruction made by Widiatama [12] at Savu Island shows that the unit is Early Jurassic age, equivalent to the quartz-rich sandstone identified at well Banli-1 [4]. Therefore, this paper is trying to promote the unit as an equivalent of Plover Sandstone, and suggests that the facies succession where the unit is developed can be promote as a new stratigraphy formation called as Pedaro Formation, refers to the location where the unit is found.

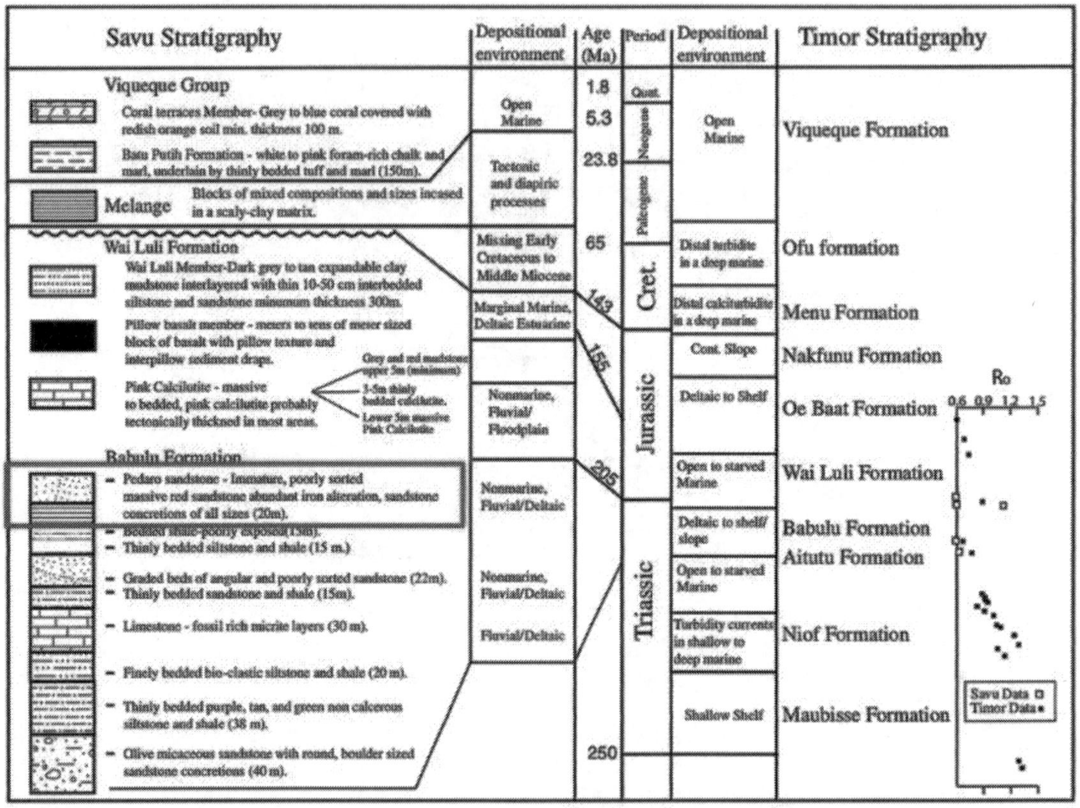

Figure 2. Stratigraphy of Savu compared to Timor Island [2]. Note that Pedaro sandstone is described as Triassic immature, poorly sorted massive red sandstone abundant iron alteration, and sandstone concretions of all sizes.

2. Methodology

Field data were obtained by conducting two geological trajectories, at East and West Savu. Field data was collected to obtain the description of lithologies, included sedimentary structures, measurement of geological structural elements, documentation, and rock sampling for laboratory analysis. All laboratory analysis are taken at GeolLabs of Centre for Geological Survey. Measurement of the stratigraphic section is carried out to determine the vertical and lateral distribution of each rock unit. Rock sampling was carried out systematically for thin section petrography and SEM-EDX analysis. Sixteen outcrop points has been observed.

Data processing includes sedimentological and stratigraphic analysis as well as laboratory analysis. Facies identification has been determined by observing the pattern of rock vertical section at the stratigraphic measurement column, correlation between lithological units, and correlation between West and East Savu trajectories.

Fourteen samples of sandstone unit of Pedaro Formation are taken for petrography analysis, represented lower, center and upper part of the formation. Eight samples have been taken for Scanning Electron Microscope (SEM) Analysis, using JEOL JSM-6360LA, equipped with the JEOL JED-2300 energy dispersive spectrometer (EDS) system. This tool can magnify objects up to 300,000 times and determine the levels of chemical elements semi-quantitatively. Three samples are taken for X-Ray Difraction (XRD) Analysis for knowing the composition of clay mineral intercalated between the sandstone layer. XRD Analysis is taken by equipment of X-ray Diffraction PANalytical X'Pert PRO PW3040/x0.

3. Result

3.1. Facies Succession of Pedaro Formation

Two measure sections have been carried out in Savu (Figure 3). In the East Savu trajectory, the 27 meters thick stratigraphy succession can be divided into three parts, the lower part which has a thickness of 9 meters consists of conglomerates which gradually become shale layers with coal seams at the bottom, and sandstone insertion. The sandstone layers tend to thickening and coarsening upwards. The middle part has a thickness of 9.25 meters consists of layers of quartz sandstone and siltstone. The sandstone layers tend to fining upward and thinning upward, while the silt layers tend to thicken upwards. The upper part has a thickness of 8.75 meters consists of layers of quartz sandstones, the sandstone layers tend to fining and thinning upwards.

On the West Savu trajectory, the measured stratigraphy of 39 meters can be divided into three parts, the lower part which has a thickness of 12 meters consists of shale with sandstone insertion. The sandstone layer tends to thickening and coarsening upwards turns into quartz sandstone with thin siltstone insertion. The middle part has a thickness of 19 meters consists of layers of quartz sandstone with an insertion of siltstone, the sandstone layer tends to thinning and fining upwards. The upper part has a thickness of 8 meters consists of quartz sandstone layers, the sandstone layer tends to thinning and the fining upwards.

The oldest outcrop of Pedaro succession consist of grain-supported conglomerates (Figure 4a). To the upper part can be found coal layer with a thickness of 30 cm (Figure 4b), which turns into an intersection of shale and sandstone. Shale with a fissile structure, light gray in color, layering of 3-10 cm alternating with fine-grained-very fine sandstones (Figure 4c).

In a younger direction, it will be found yellowish brown sandstones, fine sand to medium sand, 10-20 cm thick, parallel sedimentary structures, with abundant trace fossils. The uppermost part can be observed as light white to yellowish white sandstones, good to very well sorted, grain size medium to coarse sand, 20-60 cm thickness, parallel bed, tabular cross bed sedimentary structure, with trace fossils of Skolitos (Figure 4d).

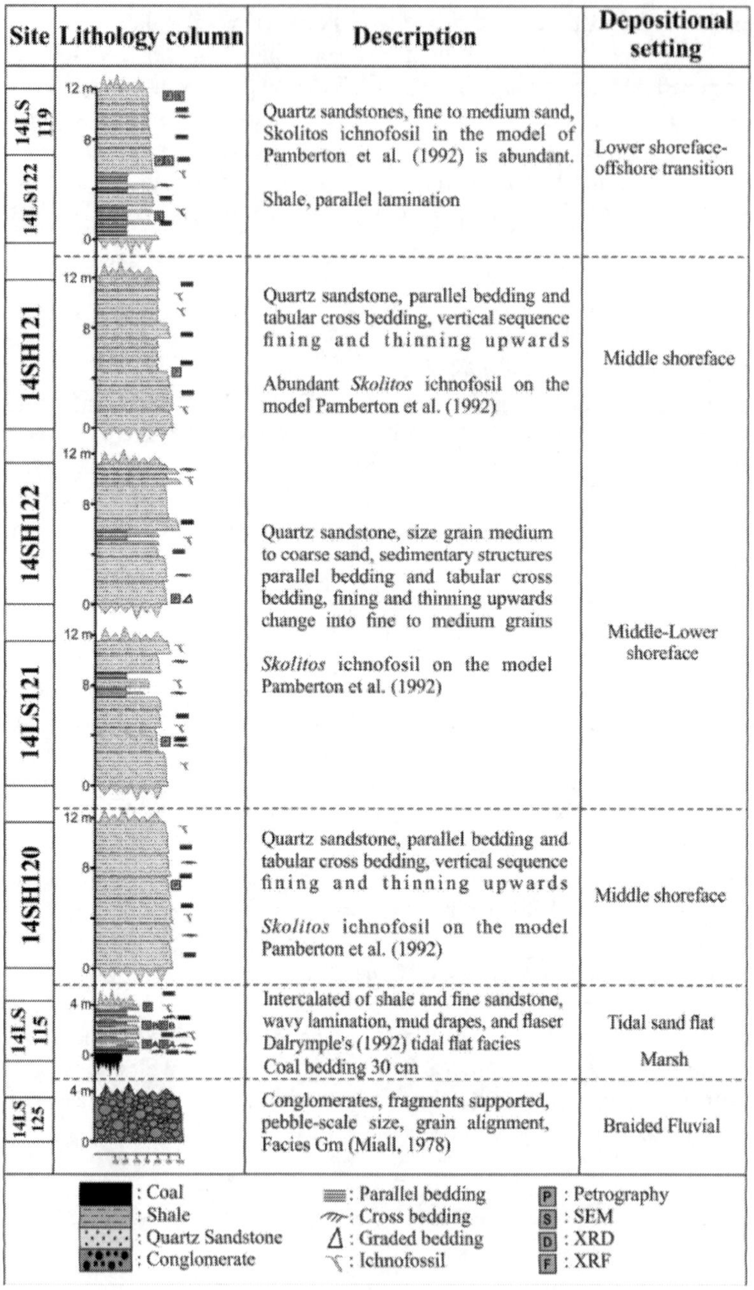

Figure 3. Measure section carried out in Savu Island to determine the succession of Pedaro Formation.

Figure 4. Outcrop of Pedaro Formation; (a) Conglomerate; (b) Coal layer; (c) Sand – shale intercalation; (d) quartz sandstone.

3.2. Origin of the Pedaro Sandstone

Fourteen petrographic analyses were carried out representing the lower, middle, and upper parts of the formation, summarized in Table 1. It can be seen that the composition of the sandstone is dominated by quartz arenite to quartz wacke (figure 5), while the shale is mudrock. Based on fourteen sandstone samples, provenance analysis of the Pedaro Sandstone shows that it is originated from an interior craton (Table 2).

Eight samples were carried out by SEM observations to determine more detailed microscopic characters, and three samples of XRD analysis which are summarized in Table 3. The Pedaro Formation shows illite as dominant of clay minerals, and in small amounts it can be identified smectite and illite-smectite that fill the pores (pore-filling) and cover the pores (pore-lining). The detritus quartz granules show secondary (autigenic) quartz growth and there are micro-fractures. XRD analysis showed that the illite clay minerals were of the nacrite type, while the smectite was of the nontronite, saponite, and montmorillonite types. The domination of illite and existing of nacrite can be interpreted as the result of metagenetic of the sediment in this area [13]. In addition, there are also faujasite sulfide minerals. The presence of iron sulfide (FeS_2) indicates a reducing environment.

IOP Conf. Series: Earth and Environmental Science **851** (2021) 012052 doi:10.1088/1755-1315/851/1/012052

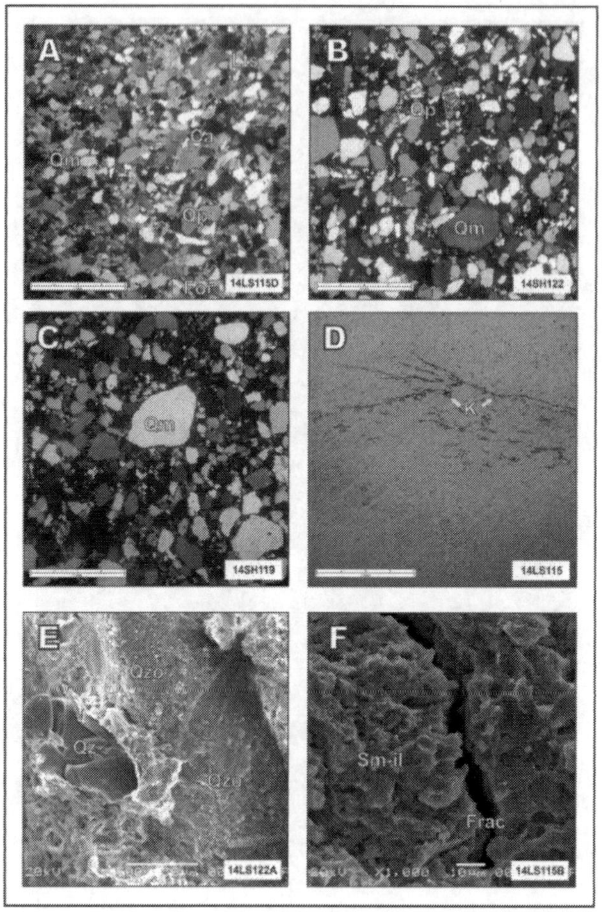

Figure 5. Fourteen thin section and eight SEM Analysis has been conducted for knowing the Pedaro characteristics; (a) thin section of quartz wacke with carbonate semen (Ca); (b) thin section of quartz arenite; (c) thin section of quartz arenite; (d) thin section of mudrock; (e) SEM of quartz mineral covered by growth quartz (Qzo); (f) Micro fracture formed at sandstone.

Table 1. Result of petrography analysis carried out at Pedaro Formation [12].

No	Sampels	Name	Q (%)		F (%)		Lithic (%)			Organic mat.	Matrix	Semen	Opaque minerals	Carbonate mud	Porosity	Other Grains
			Qm	Qp	KF	PF	Lv	Ls	Lm							
1	14LS115D	Quartz wacke	69.4	1.5	0.6	0	0	0	0	0	10.3	2	3.6	12.6	0	0
2	14LS116	Quartz arenite	69.9	12.1	2.7	0	0	1	0	0	7.3	2	5	0	0	0
3	14LS119	Quartz arenite	79.2	4.6	4.3	0	0	0	0	0	4.3	3	3.3	0	0	1.3
4	14SL121A	Quartz wacke	69.3	2.6	1.3	0	0	0	0	0	17.9	4	4.9	0	0	0
5	14LS122	Quartz wacke	66	2.6	1.3	0	0	0	0	0	18.5	3	7.6	0	0	1
6	14LS125	Quartz arenite	78.6	8.3	3.6	0	0	0	0	0	6.6	2	0.9	0	0	0
7	14SH119B	Quartz arenite	78.3	2.6	0	0	0	0	3	0	12.5	3	0.6	0	0	0
8	14SH122	Quartz arenite	80.3	4.3	0	0	0	0	5	0	7.3	2	1.1	0	0	0
9	14SH123	Quartz arenite	77.3	4.8	2.3	0	0	0	2	0	8.3	2	3.3	0	0	0
10	14SH124	Quartz arenite	75.3	4.3	0	0	0	0	1	0	11.8	2	4.6	0	0	1
11	14SH125	Quartz arenite	70.3	5.3	2	0	0	0	0	0	17.3	3	2.1	0	0	0
12	14SH126	Quartz wacke	75.8	1.8	2	0	0	0	0	0	16.3	2	1.1	0	0	1
13	14SH127C	Quartz arenite	76.7	1.6	1	0	0	0	0	0	9.9	2	1.8	4	0	3
14	14SH129	Quartz arenite	87.3	1.3	1	0	0	0	0	0	6.3	1.6	1.4	0	0	1.1

IOP Conf. Series: Earth and Environmental Science **851** (2021) 012052 doi:10.1088/1755-1315/851/1/012052

Table 2. Recapitulation of Provenance Analysis for Babulu and Pedaro Formation. It can be summarized that Pedaro Formation is originated from Craton interior [12].

For.	No	Sampels	Quartz (%)		Plagioclase (%)		Lithic (%)			Qt (%)	F (%)	L (%)	Q (%)	F (%)	Lt (%)	Lm (%)	Lv (%)	Ls (%)	Tectonic
			Qm	Qp	Kf	Pl	Lv	Ls	Lm										
Babulu Formation	1	14LS105A	41.6	3.0	13.0	0.3	0.0	8.1	0.0	67.6	20.2	12.3	63.0	20.2	16.8	27.0	0.0	73.0	*Recycled orogen*
	2	14LS105F	43.0	2.3	12.6	0.3	0.0	2.0	0.0	75.2	21.4	3.3	71.4	21.4	7.1	53.5	0.0	46.5	*Craton interior*
	3	14LS106B	11.6	10.0	5.0	0.0	0.0	26.4	0.0	40.8	9.4	49.8	21.9	9.4	68.7	27.5	0.0	72.5	*Recycled orogen*
	4	14LS106C	36.8	1.3	17.6	0.0	0.0	4.0	0.0	63.8	29.5	6.7	61.6	29.5	8.9	24.5	0.0	75.5	*Craton interior*
	5	14LS106D	20.6	3.0	6.6	0.0	0.0	15.9	0.0	51.2	14.3	34.5	44.7	14.3	41.0	15.9	0.0	84.1	*Recycled orogen*
	6	14LS107B	8.0	2.3	5.0	2.0	11.4	20.0	0.0	21.1	14.4	64.5	16.4	14.4	69.2	6.8	0.3	59.3	*Transitional arc*
	7	14LS123	50.7	5.3	13.1	0.0	0.0	1.3	0.0	79.5	18.6	1.8	72.0	18.6	9.4	80.3	0.0	19.7	*Craton interior*
	8	14SH101A	40.5	4.3	10.6	0.0	0.0	2.0	0.0	78.0	18.5	3.5	70.6	18.5	11.0	68.3	0.0	31.7	*Craton interior*
	9	14SH109A	38.9	3.5	20.0	0.0	0.0	1.0	0.0	66.9	31.5	1.6	61.4	31.5	7.1	77.8	0.0	22.2	*Craton interior*
Pedaro Formation	1	14LS115D	69.4	1.5	0.6	0.0	0.0	0.0	0.0	99.2	0.8	0.0	97.1	0.8	2.1	100.0	0.0	0.0	*Recycled orogen*
	2	14LS116	69.9	12.1	2.7	0.0	0.0	1.0	0.0	95.7	3.2	1.2	81.6	3.2	15.3	92.4	0.0	7.6	*Craton interior*
	3	14LS119	79.2	4.6	4.3	0.0	0.0	0.0	0.0	95.1	4.9	0.0	89.9	4.9	5.2	100.0	0.0	0.0	*Craton interior*
	4	14SL121A	69.3	2.6	1.3	0.0	0.0	0.0	0.0	98.2	1.8	0.0	94.7	1.8	3.6	100.0	0.0	0.0	*Craton interior*
	5	14LS122	66.0	2.6	1.3	0.0	0.0	0.0	0.0	98.1	1.9	0.0	94.4	1.9	3.7	100.0	0.0	0.0	*Craton interior*
	6	14LS125	78.6	8.3	3.6	0.0	0.0	0.0	0.0	96.0	4.0	0.0	86.9	4.0	9.2	100.0	0.0	0.0	*Craton interior*
	7	14SH119B	78.3	2.6	0.0	0.0	0.0	0.0	3.0	96.4	0.0	3.6	93.3	0.0	6.7	100.0	0.0	0.0	*Craton interior*
	8	14SH122	80.3	4.3	0.0	0.0	0.0	0.0	5.0	94.4	0.0	5.6	89.6	0.0	10.4	100.0	0.0	0.0	*Craton interior*
	9	14SH123	77.3	4.8	2.3	0.0	0.0	0.0	2.0	95.0	2.7	2.3	89.5	2.7	7.9	100.0	0.0	0.0	*Craton interior*
	10	14SH124	75.3	4.3	0.0	0.0	0.0	0.0	1.0	98.8	0.0	1.2	93.4	0.0	6.6	100.0	0.0	0.0	*Craton interior*
	11	14SH125	70.3	5.3	2.0	0.0	0.0	0.0	0.0	97.4	2.6	0.0	90.6	2.6	6.8	100.0	0.0	0.0	*Craton interior*
	12	14SH126	75.8	1.8	2.0	0.0	0.0	0.0	0.0	97.5	2.5	0.0	95.2	2.5	2.3	100.0	0.0	0.0	*Craton interior*
	13	14SH127C	76.7	1.6	1.0	0.0	0.0	0.0	0.0	98.7	1.3	0.0	96.7	1.3	2.0	100.0	0.0	0.0	*Craton interior*
	14	14SH129	87.3	1.3	1.0	0.0	0.0	0.0	0.0	98.9	1.1	0.0	97.4	1.1	1.5	100.0	0.0	0.0	*Craton interior*

Table 3. Recapitulation of SEM and XRD Analysis carried out at Pedaro Formation [12].

No.	Sample Code	Clay Mineral	Formation	Another Fitur	Behaviour	Clay Type (XRD)
1	14LS115	Illite, with minor clorite	*Mat; hexagonal plate*	–	*Pore-filling*	–
2	14LS115A	–	–	Pyrite	–	Nontronite; Faujasite
3	14LS115B	Smectite	*Crenulated to flaky*	Micro fractures	*Pore-filling*	Saponite and Monmorilonite
4	14LS119	Illite, with minor clorite	*Mat; hexagonal plate*	–	*Pore-filling and pore lining*	–
5	14LS121A	Illite	*Mat*	–	*Pore-filling and pore lining*	–
6	14LS122B	Smectite	*Crenulated to flaky*	Quartz growth and micro fracture	*Pore-filling and pore lining*	–
7	14LS125	Smectite with minor kaolinite	*Crenulated to flaky; with segmented plate*	Quartz growth	*Pore-filling and pore lining*	–
8	14SH126A	Illite - Smectite	*Crenulated-webby with filamentous*	Quartz growth from muscovite	*Pore-filling and pore lining*	*Nacrite*
9	14SH135B	Illite	*Mat*	Quartz growth	*Pore lining*	–

3.3. Age Determination

There were no fossils can be used to determine the age of Pedaro Formation. Based on the stratigraphic position, Pedaro Formation is comformably overlies by Wailuli Shales. While, the age of Wailuli has been determined using radiolarian fossils, showing Middle Jurassic age (Bajocian-Callovian) [12]. Therefore, it can be estimated that the age of Pedaro Formation is not younger than Middle Jurassic.

Contact between the Pedaro (Plover-equivalent) and the Babulu Formations can't be observed in the study area. Regionally, Charlton [6] describes that the Plover Formation is deposited in harmony above the Babulu Formation. Referring to Charlton [6] it is interpreted that in the research area, the Pedaro Formation is comformably overlain the Babulu Formation. Based on the reconstruction of the geological section it can be estimated that the thickness of this unit exposed in the study area at least 400 meters. This is suitable with the thickness of the Plover Formation equivalent in the Banli-1 well drilling, which shows a thickness of more than 100 meters.

3.4. Facies Association

Based on the measured stratigraphic section (Figure 3), the lower part of Pedaro Formation shows that it consists of conglomerates and shales. The conglomerate at the bottom has well-sorted grain-supported characteristics indicating it is deposited at a high energy environment. The rounded shape of the fragments shows intensive abrasion, the dominance of sedimentary rock fragments as fragments indicates that the sediment source comes from older sedimentary rocks that are exposed as a result of accumulation in the expansion tectonics or due to the exposing of land during lowering of sea level. The conglomerate corresponds to the Gm facies in Miall's classification [14], which is a feature of channel lag deposits. The association of channel lag deposited without material from the ocean is a characteristic of braided fluvial deposits [14]. Shale with coal seams overlapping conglomerates, showing parallel bed, parallel lamination, slurry cross-bedding, ripple, wavy layer, flaser layer and mud drapes indicates traction currents with different orientations controlled by tides, shedding of carbon material on shale. Shale is thought to be deposited in a tidal sand flat environment [15].

The middle section shows a sandstone deposition pattern with a thinning upward and fining upward layer, well sorted, and the matrix on the sediment indicates a little wave control in the deposition process. The sedimentary structure of parallel layers and laminations with coarse to fine sand-grain sized sediments is characteristic of the shoreface deposit [16]. The insertion of siltstone between the quartz sandstone layers contains *Skolithos* ichnofacies association indicates that the depositional environment is less influenced by waves, in the middle to lower shoreface environment.

The upper part shows repeated layers of quartz sandstone with a fining upward and thinning upward pattern. The sedimentary structure of parallel layers and laminations with coarse to fine sand-grain sized sediments is characteristic of the shoreface deposit [16]. The association of braided fluvial facies, tidal flat facies, and shoreface facies is characteristic of transgressive conditions in the estuarine environment [16]. The precipitation model for the quartz sandstone unit is depicted in Figure 6.

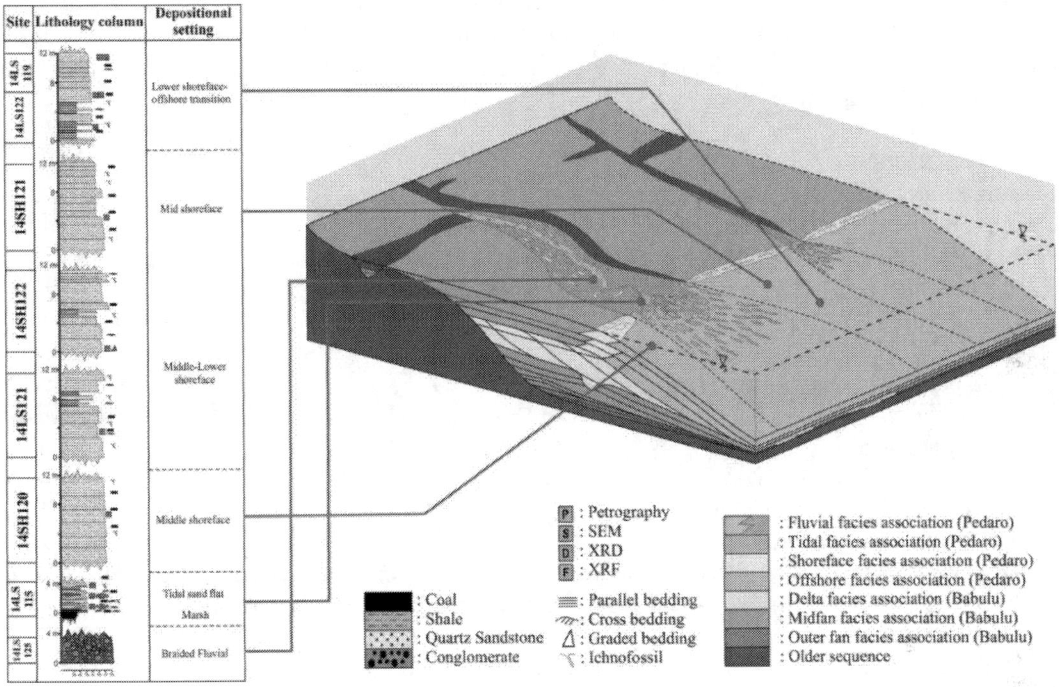

Figure 6. Depositional model of Facies Association of Pedaro Formation at braided fluvial, tidal flat, and shoreface environment.

4. Discussion

Early Jurassic sandstone has been described from the Banli-1 exploration well as the Plover equivalent Formation [4,6]. The Plover Formation is deposited in the passive margin of Gondwana in a delta environment with dominant sedimentation of quartz sand which is influenced by tidal currents [17]. On Tanimbar Island [18] there are Triassic to Early Jurassic-aged sandstones and Early Jurassic coal (outcrops in the form of boulders in poton) which are deposited in a fluviodic environment to tidal areas.

Physically and from age determination, thickly bedded quartz wacke to quartz arenite unit of Pedaro Formation can be compared to those of the Plover Formation in Australia and the Triassic-Jurassic Sandstone of Tanimbar Island with variations in depositional facies more towards the basin. Provenance analysis of quartz sandstones shows the origin of sediments from interior cratons [12].

The Pedaro Formation has some distinctive characteristic and facies succession, mapable in the south Savu area. It can be separated from the sandstone unit of Triassic Babulu Formation. The separation of Pedaro Formation from the rest of Babulu Formation sequence will help facilitate the mapping of hydrocarbon reservoir targets in Savu and its surrounding area.

5. Conclusion

Geological observation and laboratory analysis have been carried out to determine the characterization of quartz-rich sandstone unit cropped out at the south part of Savu Island as Plover equivalent. The facies unit consist of quartz sandstone, shale, conglomerate, and siltstone, deposited at fluvial and tidal environment in the lower part, gradationally changing to the upper part, becoming estuarine sediment deposite with the influence of marine wave, during transgressive event at Early Jurassic. The analysis of provenance suggested sediment origin from craton interior. Hereinafter, we suggested that the lithologic sucession carried out sandstone unit of "Plover equivalent" in Savu Island can be named as Pedaro Formation. Furthermore, we expected that the study of Pedaro as reservoir candidate and study of Savu petroleum system can be continued comprehensively in order to understand the petroleum play in the outer Banda Arc region.

Acknowledgement

The authors are grateful to all members of the 2014 Savu Basin Team who have helped in data collection, the GeoLab Laboratory of the Geological Survey Center, the Geological Agency Ministry of Energy and Mineral Resources Republic Indonesia that carried out G&G Survey at 2014 in Savu and Rote area, and also to Dr. Dardji Noeradi for his discussion and suggestions during draft writing.

References

[1] Tampubolon B T, Sameena Y 2009 Savu basin: a case of frontier basin area in eastern Indonesia *Proc. 33rd Indonesia Petroleum Association Annual Convention (Jakarta, Indonesia)*

[2] Harris R A, Vorkink M W, Prasetyadi C, Zobell E, Roosmawati N, and Apthorpe M 2009 Transition from subduction to arc-continent collision: Geologic and neotectonic evolution of Savu island, Indonesia. *Geosphere.* **5(3)** 152 – 71

[3] Audley-Charles M G 1987 Evolution of southern margin of tethys (north Australian region) from early Permian to late Cretaceous *(Tethys and Gondwana, Geol. Soc. Spec. Publ. 37)* ed M G Audley-Charles and A Hallam pp 79 – 100

[4] Sani K, Jacobson M I, and Sigit R 1995 The thin-skinned thrust structures of Timor Proc. *24th Indonesian Petroleum Association Annual Convention (Jakarta, Indonesia)* 277 – 93

[5] Milsom J 2000 Stratigraphic constraints on suture models for eastern Indonesia *J. of Asian Earth Sci.* **18** 761 – 79

[6] Charlton T R 2001 The petroleum potential of West Timor *Proc. 28th Indonesian Petroleum Association Annual Convention (Jakarta, Indonesia)* 301 – 17

[7] Barber P, Carter P, Fraser T, Bailie P, Myers K 2003 Paleozoic and Mesozoic petroleum systems in the Timor and Arafura seas, eastern Indonesia *Proc. 29th Indonesia Petroleum Association Annual Convention (Jakarta, Indonesia)*

[8] Matsui R, Shinbo E, Omokawa,M, Zushi T 2009 Quartz cementation and reservoir quality of the plover sandstone in the Abadi gas field *Proc.33rd . Indonesia Petroleum Association Annual Convention (Jakarta, Indonesia)*

[9] Zimmermann S, Hall R 2014 Provenance of Mesozoic sandstones in the Banda arc, Indonesia.*Proc. 33rd Indonesia Petroleum Association Annual Convention (Jakarta, Indonesia)*

[10] Rosidi H M D, Tjokrosapoetro S, and Gafoer S 1979 Peta geologi lembar Kupang, Timor *Pusat penelitian dan pengembangan Geologi (Bandung, Indonesia)*

[11] Roosmawati N and Harris R 2009 Surface uplift history of the incipient Banda arc-continent collision: geology and synorogenic foraminifera of Rote and Savu islands, Indonesia. *Tectonophysics* **479** 95 – 110

[12] Widiatama A J 2019 Sedimentasi batuan Mesozoikum pulau Sawu *Thesis Program Master Institut Teknologi Bandung (Unpublished)*

[13] De Segonzac, G D. 1970. The Transformation of clay minerals during diagenesis and low grade metamorphism: A review. *Sedimentology* **15**. Issue 3-4 pp 281 – 346
[14] Miall A D 1992 Alluvial deposites pp 119 – 142 *Facies model: Response to sea level change* **ed** Walker R G and James N p 409 (Geological Association of Canada, Ontario-Canada)
[15] Dalrymple R W 1992 Tidal depositional systems pp 195-218 *Facies model: Response to sea level change* **ed** Walker R G and James N p 409 (Geological Association of Canada, Ontario-Canada)
[16] Reinson G E 1992 Transgressive barrier island and estuarine systems pp 179-194 *Facies model: Response to sea level change* **ed** Walker R G and James N p 409 (Geological Association of Canada, Ontario-Canada)
[17] Tovaglieri F. and George A D 2014 Stratigraphic architecture of an early middle Jurassic tidally influenced deltaic system (Plover formation), Browse basin, Australian northwest shelf, *Marine and Petroleum Geology* **49** pp 59 – 83
[18] Charlton T R, de Smet M E M, Samodra H, and Kaye S J 1991 The stratigraphic and structural evolution of the Tanimbar islands, eastern Indonesia *J. of Southeast Asian Earth ScI* **6** pp 343 – 58

AUTHOR INDEX